The Biology and Conservation of Wild Canids

Canid diversity by Priscilla Barret.

(clockwise from top left: Arctic fox, bush dog, Ethiopian wolf, dhole, swift fox, African wild dog)

The Biology and Conservation of Wild Canids

Edited by DAVID W. MACDONALD
and CLAUDIO SILLERO-ZUBIRI

Wildlife Conservation Research Unit, University of Oxford

OXFORD
UNIVERSITY PRESS

Great Clarendon Street, Oxford OX2 6DP

Oxford University Press is a department of the University of Oxford.
It furthers the University's objective of excellence in research, scholarship,
and education by publishing worldwide in

Oxford New York

Auckland Bangkok Buenos Aires Cape Town Chennai
Dar es Salaam Delhi Hong Kong Istanbul Karachi Kolkata
Kuala Lumpur Madrid Melbourne Mexico City Mumbai Nairobi
São Paulo Shanghai Taipei Tokyo Toronto

Oxford is a registered trade mark of Oxford University Press
in the UK and in certain other countries

Published in the United States
by Oxford University Press Inc., New York

First published 2004

A catalogue record for this title is available from the British Library

Library of Congress Cataloging in Publication Data

(Data available)

ISBN 019 851555 3 (Hbk)

ISBN 019 851556 1 (Pbk)

10 9 8 7 6 5 4 3 2 1

Printed in Great Britain
on acid-free paper by
Antony Rowe Ltd, Chippenham, Witts

To our wives, Jenny and Jorgelina,
and children, Ewan, Fiona, Isobel, Max, and Pampita

Early steps from a hand-reared red fox cub © D. W. Macdonald.

Preface

Although it is positioned at the start, the Preface is—at least in this book—the last section to be written, and therefore this seems an appropriate place to take stock of the history, and the process, that has brought us to this point. On reflection, this book has had several different beginnings, of which the most immediate was the moment when we decided, in the late 1990s, that the year 2001 should be the one when we would draw together various major strands of our shared involvement with wild canids. We scheduled, for September of that year, the Canid Biology and Conservation Conference—held under the auspices of the IUCN/SSC Canid Specialist Group and hosted by Oxford University's Wildlife Conservation Research Unit (WildCRU). The intention was that this conference would serve three different goals—first, it would be a stimulating meeting of the world's specialists in wild canids, second, it would be a forum for the brain-storming that was to initiate the final stages of our work to edit the Second Canid Action Plan (the first had been published by Ginsberg & Macdonald in 1990), and third, it was to stimulate what we intended to become the most compendious work yet published on wild canids, their biology and conservation—this book is the result of that third goal. As it happens, the conference, with its 240 delegates from 38 countries was also a marvellous celebration—not only of the wondrous biology of the canids, but of the dedication and skill of the marvellous people who study them. It also happened to fall for one of us not only on the 20th anniversary of Chairmanship of the Canid Specialist Group, but also on a fiftieth birthday—so it was quite a party!

The concept of this book grew from the belief that to understand any group of animals—and certainly the family Canidae—one needs to slice the cake from two angles. The first slice takes a broad view of diversity and trends; for this reason, half the book is devoted to reviews. Each review is by three authors, each of them from a different institution and often a different continent, and each trio encompassing unsurpassable expertise in its field. The second essential slice takes a deep view of detail and intricacy; and that is why the other half of the book is devoted to case histories, some of which have continued over many years—and indeed thereby have outlived the miserable phase of biological fashion that deemed long-term field studies as no more than an excuse for 'more of the same'. Little did the advocates of that latter view appreciate that long-term data sets would emerge as amongst the richest veins sustaining modern ecology, as is amply demonstrated in the pages that follow.

In planning the book we obviously faced many considerations, and two deserve mention. First, after much discussion, we opted to omit chapters on domestic dogs. This decision was difficult, and might well have gone the other way; part of the reason was space—the wealth of material on wild canids already caused our text to grow replete until its covers (and the publishers) groaned, while to do justice to the fascination of domestic dogs could have expanded our scope by several chapters more. Anyway, our priority has been wild canids, and the science that underpins their conservation, which brings us to the second consideration, namely how this book relates to the Canid Action Plan, on which we have worked simultaneously (Sillero-Zubiri *et al.* in press). The answer is that while the two ventures have been separate, and have different motivations, there are synergies between them. First, in many cases the same experts have been involved. Second, while the Action Plan is intended to be a purely practical document, whereas this book is a scholarly one, our intention is that the science in this volume should underpin the practice advocated by the other. The two come together in our post script, Chapter 23.

Although the formal beginning of this trio of canid projects—the conference, the Action Plan and this book—was in the late 1990s, other roots run deeper. We are both members of the Wildlife Conservation Research Unit, and the origins of that Unit lay in the work of a maverick band of canid researchers known in the late 1970s as the Oxford Foxlot. It is a particular pleasure to us to note how many of our colleagues from this kennel have

David with hand-reared red fox cubs, Claudio with anaesthetised Ethiopian wolf © J. Macdonald, D. Gottelli.

emerged as substantial figures on the canid scene—indeed of fifty authors in this book, 15 of them trained in the WildCRU (and 11 secured doctorates there). In one sense, then, the roots of this book are buried in the origins and efforts of the WildCRU, and we offer an enormous vote of thanks to all the members of our unit, past and present, who have contributed to understanding canid biology. Our own infatuations with canids have even deeper roots—for one of us boyhood observations of Pampas foxes from horseback in the Argentine pampas, for the other, an early attempt to write a book on foxes when 11 years old was a first step in the exacting task of finding a way to make a living out of this particular inexplicable fascination (Macdonald 1987, p. 13). Looking through our photo libraries for pictures to illustrate this volume we unearthed youthful pictures below of ourselves with our first loves, and remembered just how much we owed to them.

Clearly, a book of this size and scope represents a cooperative effort. The conspicuous members of the pack are the authors—and we thank them for their hard work and good humour. Behind the scenes, however, are the many members of the Canid Specialist Group (www.canids.org), and such tireless volunteers as Michelle Nelson and Emma Harvey who researched the photos, Jim Scarff who mans the website and Mike Hoffmann who worked tirelessly on the distribution maps and much else besides. We thank them all, as we do also Ruairidh Campbell and Kerry Kilshaw who tracked down endless proof queries, and our colleagues at Oxford University Press, Ian Sherman and Anita Petrie.

A general lesson from these pages is how the burgeoning knowledge of wild canids over the last twenty five years has advanced in pulses—each spurt forward the result of a new innovation that allowed data to catch up, at least a little, with theory: night-vision scopes followed radio-collars, and were followed by molecular genetics. Now GPSs are upon us, miniaturisation advances apace and the price of molecular scatology tumbles—we are thus assured that new pulses of understanding will follow soon (at least, they will if minds full of technology retain a grip on natural history and field-craft). We hope that this book will provide a foundation for the next wave of canid discoveries and, in the meantime, kindle the fires of understanding and delight on which the futures of these glorious creatures will depend.

David W. Macdonald & Claudio Sillero-Zubiri
Wildlife Conservation Research Unit
and Lady Margaret Hall
Oxford

Contents

Contributors

Anders Angerbjörn, Department of Zoology, Stockholm University, Stockholm, S-106 91, Sweden

Marc Artois, ENVL, Unité Pathologie infectieuse, BP 83, 69280 Marcy l'Etoile, France

Cheryl S. Asa, Saint Louis Zoo, #1 Government Drive, St. Louis, MO 63110-1395, USA

Robert P. D. Atkinson, RSPCA, Wilberforce Way, Southwater, Horsham, West Sussex RH13 9RS, United Kingdom

Philip J. Baker, School of Biological Sciences, University of Bristol, Woodland Road, Bristol BS8 1UG, United Kingdom

Edward E. Bangs, US Fish and Wildlife Service, 100 North Park, Suite 320, Helena, Montana, 59601 USA

Luigi Boitani, Instituto di Zoologia 'Federico Rafaelle', Universita di Roma, Viale dell' Universita 32, Rome, 00185, Italy

Sarah Cleaveland, Centre for Tropical Veterinary Medicine, Royal School of Veterinary Studies, University of Edinburgh, Easter Bush, Roslin, Midlothian, EH25 9RG, United Kingdom

Orin Courtenay, Ecology and Epidemiology Group, Department of Biological Sciences, University of Warwick, Coventry CV4 7AL, United Kingdom

Scott Creel, Department of Ecology, Montana State University, Bozeman, MT 59717, USA

Brian L. Cypher, NPRC Endangered Species Program, Bakersfield, CA 93389, USA

Buddy B. Fazio, US Fish and Wildlife Service, PO Box 1969, Manteo, North Carolina 27954, USA

Martín C. Funes, Centro de Ecología Aplicada del Neuquén, Argentina

Eli Geffen, The Institute for Nature Conservation Research, Tel-Aviv University, Tel Aviv 69978, Israel

Eric M. Gese, National Wildlife Research Center, Department of Forest, Range, and Wildlife Sciences, Utah State University, Logan, UT 84322, USA

Dada Gottelli, Institute of Zoology, Zoological Society of London, Regents Park, London, NW1 4RY, United Kingdom

Stephen Harris, School of Biological Sciences, University of Bristol, Woodland Road, Bristol BS8 1UG, UK

Pall Hersteinsson, Institute of Biology, University of Iceland, Grensasvegur 12, Reykjavik, IS-108, Iceland

Jaime E. Jiménez, Laboratorio de Ecologia, Universidad de los Lagos, Osorno, Chile

A. J. T. Johnsingh, Wildlife Institute of India, Chandrabani PO, Chandrabani, Dehra Dun-358001, India

Kaarina Kauhala, Finnish Game & Fisheries Research Institute, Game Division, P.O. Box 6, Pukinmäenakio, Helsinki, Fin-00721, Finland

Brian T. Kelly, US Fish and Wildlife Service, PO Box 1306, Albuquerque, New Mexico 87103, USA

M. Karen Laurenson, Centre for Tropical Veterinary Medicine, University of Edinburgh, Easter Bush, Roslin, Midlothian, EH25 9RG, United Kingdom

Rurik List, Instituto de Ecología, Universidad Nacional Autónoma de México, 04510 DF México

Andrew J. Loveridge, WildCRU, Oxford University, Zoology Department, South Parks Rd., Oxford OX1 3PS, United Kingdom

Barbara Maas, Care for the Wild International, Tickfold Farm, Kingsfold, RH12 3SE, United Kingdom

David W. Macdonald, WildCRU, Oxford University, Zoology Department, South Parks Rd., Oxford OX1 3PS, United Kingdom

Jorgelina Marino, WildCRU, Oxford University, Zoology Department, South Parks Rd., Oxford OX1 3PS, United Kingdom

J. Weldon McNutt, Botswana Wild Dog Project, Box 13, Maun, Botswana

L. David Mech, US Geological Survey, The Raptor Center, 1920 Fitch Street, University of Minnesota, St Paul, Minnesota 55108, USA

Michael G. L. Mills, South African National Parks, Endangered Wildlife Trust and Mammal Research Institute, University of Pretoria

Axel Moehrenschlager, Centre for Conservation Research, Calgary Zoological Society, PO Box 3036, Station B Calgary, Alberta, T2M 4R8, Canada

Andrés J. Novaro, Centro de Ecologia Aplicada del Neuquen, Wildlife Conservation Society, CC 7, 8371 Junin de los Andes, Neuquen, Argentina

Rolf O. Peterson, School of Forest Resources and the Environment, Michigan Technological University, Michigan Technological University, Houghton, MI 49931, USA

Michael K. Phillips, Turner Endangered Species Fund, 1123 Research Drive, Bozeman, Montana 59718, USA

Katherine Ralls, Department of Zoological Research, National Zoological Park, Smithsonian Institution, Washington DC 20008, USA

Jonathan Reynolds, Predation Control Studies, The Game Conservancy Trust, Fordingbridge, Hampshire, SP6 1EF, United Kingdom

Gary W. Roemer, Department of Fishery and Wildlife Sciences, New Mexico State University, Las Cruces, New Mexico 88003, USA

Midori Saeki, National Institute for Land and Infrastructure Management, Ministry of Land, Infrastructure and Transport, Asahi 1, Tsukuba-city, Ibaraki 305-0804, Japan

Claudio Sillero-Zubiri, WildCRU, Oxford University, Zoology Department, South Parks Rd., Oxford OX1 3PS, United Kingdom

Marsha A. Sovada, Northern Prairie Wildlife Research Center, US Geological Survey, Jamestown, ND 58401, USA

Magnus Tannerfeldt, Department of Zoology, Stockholm University, Stockholm, S-106 91, Sweden

Richard H. Tedford, Department of Paleontology, American Museum of Natural History, Central Park West at 79th Street, New York, New York 10024-5192, USA

Blaire Van Valkenburgh, Department of Organismic Biology, Ecology, and Evolution, University of California, 621 Young Drive South, Los Angeles, California 90095-1606, USA

Arun B. Venkataraman, Asian Elephant Research and Conservation Centre, Centre for Ecological Sciences, Indian Institute of Science, Bangalore 560012, India

Carles Vilà, Department of Evolutionary Biology, Uppsala University, Norbyvägen 18D, S-752 36 Uppsala, Sweden

John A. Vucetich, School of Forest Resources and the Environment, Michigan Technological University, Houghton, MI 49931, USA

Xiaoming Wang, Department of Vertebrate Paleontology, Natural History Museum of Los Angeles County, 900 Exposition Blvd., Los Angeles, California 90007, USA

Robert K. Wayne, Department of Organismic Biology, Ecology and Evolution, University of California, Los Angeles, CA 90095, USA

Rosie Woodroffe, Department of Wildlife, Fish & Conservation Biology, University of California, One Shields Avenue, Davis, CA 95616, USA

PART I

Reviews

CHAPTER 1

Dramatis personae

Wild Canids—an introduction and *dramatis personae*

David W. Macdonald and Claudio Sillero-Zubiri

Male red wolf and pup © G. Koch.

To understand fully the modern view of the Canidae, and thereby to appreciate its excitement, one must know several things. Perhaps most revealing are the studies, often involving much of a lifetime's work, where biologists have burrowed deep into the intricate detail of the behaviour of a particular species or set of species in the wild. Fourteen of these Case Studies are presented in this book—encompassing between them sufficient species and themes to illustrate the revelations—from population processes to individual behaviour—that have repaid the ingenious application of a generation of innovative techniques, from radio-tracking to molecular analysis of paternity. These Case Studies also make conspicuous by their absence the species that have so far largely evaded in-depth investigation. However, to adapt the aphorism, understanding the trees is a prerequisite to seeing the wood, and so it is necessary to see each species and each behaviour as an element of the patterns of which they are a part. In this book that is achieved by six Reviews which reveal that, for example, one cannot understand the present without appreciating the past—of which it is merely a current snapshot, and emphasize how different truths are revealed by viewing the same creature in different ways and at the different scales of the molecular gel and the binocular, and that a grasp of yet more techniques and branches of knowledge is necessary to understand and solve the problems they face for the future. But to appreciate these patterns, an obvious essential is to be familiar with the actors on the stage, the *dramatis personae* of contemporary wild canids, so the purpose of this chapter is to introduce these, together with some of their features which are not covered elsewhere in the book.

Those familiar with taxonomy will not be surprised to know that even the question of how many species of canids there are does not have a simple answer, but we will say there are about 36 (Clutton-Brock *et al.* 1976; Wozencraft 1989; Ginsberg and Macdonald 1990 argue for various answers between 34 and 37, to which the recognition of *Canis lycaon* (Wilson *et al.* 2000) would add another). At first encounter, the most obvious thing about them (aside from their exceptional capacity to evoke strong feelings of charm, affinity, or loathing in human onlookers) is their diversity.

Interspecific variation

Canids range in size from Blanford's and fennec foxes (*Vulpes cana, V. zerda*) of which adult specimens can weigh less than 1 kg to the grey wolf (*Canis lupus*) exceeding 60 kg. Their distributions may be highly restricted—almost the entire Darwin's fox population (*Pseudalopex fulvipes*—Yankhe *et al.* 1996) occur only on one island and some unusual subspecies occur on one island each, such as island foxes (*Urocyon* littoralis) or Mednyi Arctic foxes (*Alopex lagopus semenovi*—Wayne *et al.* 1991b; Goltsman *et al.* 1996), whereas other species span several continents—about 70 million km^2 in the case of the red fox (*Vulpes vulpes*—Lloyd 1980). Their diets range from omnivory (with, at times, almost exclusive emphasis on frugivory or insectivory) to strict carnivory—and they glean these livings in habitats ranging from deserts to ice-fields, from mountain to swamp or grassland, and from rain forest to urban 'jungle' (reviews in Johnson *et al.* 1996; Macdonald 1992b). To do this they may travel home ranges as small as 0.5 km^2 (island fox—Roemer *et al.* 2001c) or as large and non-defencible, as 2000 km^2 in African wild dogs (*Lycaon pictus*—Frame *et al.* 1979).

Geographical variability in body size can be explained to some degree by differences in availability of food: small canids (e.g. fennec fox) are usually associated with arid and poor habitats in which only a small body mass can be supported year round, whereas large canids (e.g. Ethiopian wolf *Canis simensis* and African wild dog) are often associated with habitats in which prey is abundant. The maned wolf (*Chrysoscyon brachyurus* Fig. 1.1), unusual in its social organization for a large canid, lives in South American savannas and feeds largely on rodents and fruit (Dietz 1985). Geffen *et al.* (1996) suggest that low food availability probably constrains both the maned wolf's group and litter size (which is low at 2.2).

To accomplish such different lifestyles, the different canids have diverse adaptations; during the half-century of man-years for which we have—between us—researched the behaviour of canids, we have been fortunate to watch as red foxes used their special musculo-skeletal adaptations to launch into aerial strikes on mice (Henry 1996), as long-legged maned wolves bound over tall pampas grasses (Dietz 1984)

Figure 1.1 Maned wolf *Chrysocyon brachyurus* © F. C. Rodrigues.

and as short-legged bush dogs (*Speothos venaticus*) cartwheeled into handstands that projected their scent marks aloft (Macdonald 1996b). We have seen Ethiopian wolves hammer their reinforced snouts into rodent burrows (Sillero-Zubiri and Gottelli 1995a) and African wild dogs cram 3 days worth of food into their stomachs for transportation to their pups—stomachs that can hold at least 4.4 kg of meat (Reich 1981b), and possibly twice that amount (Creel and Creel 1995). The ability to regurgitate—a canid innovation—allows companions to feed a pregnant mother, whose ability to hunt is compromised in late pregnancy and early lactation—and in the case of African wild dogs to feed her pups for 2–3 months, and rear them even if their mother dies (Estes and Goddard 1967; McNutt 1996a). Intriguingly, despite watching plenty, we have never seen a vulpine fox of any ilk regurgitate—they are great carriers and cachers of food (Macdonald 1976, 1987)—from which we might deduce that regurgitation is a trait of the lupine—or wolf-like—canids, evolved only after their split 6 million years ago from the early vulpines—fox-like (Wayne *et al.* 1997). Insofar as regurgitation is an adaptation that facilitates aspects of the reproductive behaviour of modern lupine canids, this raises the questions of how the presumed ancestral absence of this adaptation constrained the societies of their ancestors.

Intraspecific variation

The impressiveness of the diversity between canid species was only increased when, early in our experience of them, the stunning extent of intraspecific variation in their biology began to emerge. Discovery of marked variations in fecundity and litter size in post mortem samples of red foxes throughout the length of Sweden (Englund 1970) anticipated the revelations of differences in behaviour between fox populations whose home ranges may span three orders of magnitude between habitats where their diets, spatial organizations, and society differ more than do those of many distantly related species (e.g. Macdonald 1981; Doncaster and Macdonald 1991; Macdonald *et al.* 1999; Baker and Harris, Chapter 12, this volume). Other examples rapidly accumulated—grey wolves live lives that vary between solitary and packs of 22 members (Messier 1985; Mech 2000a). Arctic foxes hold the record, at 19, amongst litter sizes (Ovsyanikov 1983), but between their populations mean litter sizes vary from 2.4 to 7.1 (Angerbjörn *et al.* in press), with lesser but nonetheless impressive regional variations in litter size between populations of grey wolves and red foxes (Voigt and Macdonald 1984; Mech and Boitani 2003). In order to cope with large litters Arctic foxes have twice as many teats as other canids of their size (Ewer 1973). Strikingly, some populations of Arctic foxes are essentially migratory over hundreds of kilometres (Eberhardt *et al.* 1983), while elsewhere they live in small territories occupied by close-knit matrilineal groups (Hersteinsson and Macdonald 1982; White 1992). In the case of golden jackals (*Canis aureus*), patterns of scent marking behaviour that had never been seen at low densities became conspicuous in a population living in large groups on small territories, and they also displayed such wolfish behaviours as mustering for territorial patrols (Macdonald 1979c).

Soon, reviews were accumulating that listed interpopulation variations that matched and exceeded interspecific ones (Macdonald and Moehlman 1982; Creel and Macdonald 1995; Geffen *et al.* 1996; Moehlman and Hofer 1997). The dimensions along which populations vary affect diverse aspects of canid lives. Thus on Round Island, Alaska, 71% of red foxes were polygynous when food was superabundant, whereas 100% were behaviourally monogamous when prey abundance declined, whereupon there was a concomitant decrease in litter size (Zabel and Taggart 1989). An even more subtle change was revealed amongst San Joaquin kit foxes (*Vulpes macrotis mutica*—Cypher *et al.* 2000): sex ratios at birth were male-biased during years of low food availability but female biased when fox abundance was low and the population increasing (Egoscue 1975).

With each species of canid seemingly sliding along a continuum of different behavioural possibilities, the question emerged: what determines the limits to intraspecific variation between the different canids and what determines interspecific differences in these limits—indeed, to what extent can the behaviour of one canid species be transformed into that more typical of another species simply by facing it with the correct combination of ecological circumstances (Macdonald 1983)? While this book is testament to the huge amount that has been discovered since this question first took shape, the answers remain incomplete.

Phylogenetic baggage

However, having dwelt on the burgeoning discoveries of diversity amongst the canids—a diversity that applies to contrasts at the levels of species, populations, and individuals—our lifetimes spent watching these creatures have simultaneously and paradoxically led to a realization of their sameness. This became particularly clear in a review of what characterized canids amongst the diversity of carnivore types (Macdonald 1992b). Clearly, species of canid differ: to watch members of a pack of bush dogs using each other's bodies to lever bones from a carcass, to see them slicing off hunks of meat from prey held in a companion's mouth, and to see them wriggle determinedly into the centre of a heap of the somnolent bodies of sleeping companions (Macdonald 1996b) is to realize that this is a different creature to the group of red foxes whose mood teeters jumpily on the divide between play and ferocity as they slam their flanks into each other with jaws agape (but no lupine snarl) in competition for a bloodied feather (Macdonald 1981). Notwithstanding these differences, the reality is that as we see Ethiopian wolves in the half-light milling around in a social hubbub prior to undertaking a border patrol, their actions and appearance replicate closely those of grey wolves we have watched in Minnesota or even golden jackals in Israel (Macdonald 1979c; Sillero-Zubiri and Macdonald 1998). And as we look at the meandering pair of crab-eating foxes (*Cerdocyon thous*) in the short grasslands of Brazil, it is hard to be sure they are not the same creature as the side-striped jackals (*Canis adustus*) we have watched in similar grasslands in Zimbabwe (Macdonald and Courtenay 1996; Loveridge and Macdonald 2002). Even the red foxes seen in the northern deserts of Saudi Arabia—less than half the weight of their Japanese conspecifics (Macdonald *et al.* 1999) seem scarcely distinguishable in demeanour and behaviour from the kit foxes watched in the arid lands of Mexico (List and Macdonald 2003). Certainly canids are intriguingly different but, equally certainly, all their differences are merely variations on a strikingly consistent theme.

The consistent themes of canid biology—their opportunism and versatility, their territoriality, their societies built from a foundation behavioural monogamy with its attendant dominance hierarchies, social suppression of reproduction and helpers—all shine through in the accounts that follow. But several canid commonalities have largely escaped attention elsewhere in this book, so we will briefly highlight them now.

Communicative canids

First, through postural, vocal, and olfactory signals, canids are highly communicative, and people are especially attuned to their signals because we are so frequently their recipient from our domestic dogs. Indeed, the body language of domestic dogs (described by Lorenz 1954; Scott and Fuller 1974) is scarcely a dialect of that documented in ethograms for grey wolves (Zimen 1981), or golden jackals (Golani and Keller 1975) and clearly part of the same 'linguistic family' as that of foxes (Tembrock 1962; see also Fox 1971)—although the sinuous lashing of the vulpine tail is a clearly different action to the wagging of a lupine tail (Macdonald 1987). Even though domestication has affected the domestic dog's repertoire (Coppinger 2002), the barks, growls, whines, and howls of wolves are heard daily in our backyards. Amongst wolf packs howling serves to maintain or increase distance, helping to establish and maintain exclusive territories and reduce the probability of encountering strange wolves or packs in areas of border overlap (Harrington and Mech 1979). As in other mammals, pitch and quality of voice are apparently characteristics used to express and assess an individual's fighting or resource-holding potential. Harrington (1987) suggests that lower pitched and harsher howls in wolves reflect greater hostility. Some canids produce less familiar sounds: the squeak of bush dogs (Kleiman 1972), whistling of dholes (*Cuon alpinus*—Fox 1984; Durbin 1998) are probably adaptations to keeping a hunting pack in coordinated contact in dense forest (Fig. 1.2), and the oddly un-doglike twittering of a social scrum of African wild dogs (van Lawick and van Lawick-Goodall 1970) may have its roots in a similar function. The first towering study of canid voices (and

Figure 1.2 Pack of dholes *Cuon alpinus* © K. Senani.

indeed it was part of a complete ethogram) was Tembrock's (1962) sonographic analysis of red fox voices. Interestingly, red foxes neither make growling vocalizations nor curl their lips in a fang-bearing snarl (a truth not always observed in lurid taxidermy specimens) but in the circumstances that might provoke a wolf to snarl they make variously staccato 'gekkering' noises with mouths agape (Macdonald 1987, 1992b). The growl and snarl, along with regurgitation, then, appear to fall on the lupine side of the subfamily divide. Nonetheless, both the wow-wow call and the shriek of the red fox are clearly recognizable as canid voices, and as they sound individually distinct to a human listener it seems likely that they are recognizable to vulpine ones. Indeed, this is demonstrably so between groups of Mednyi Island's Arctic foxes, amongst whom there was also evidence that the barks of family members were acoustically more similar to each other than to those of other foxes (Frommolt *et al.* 1997). Both male and female Arctic foxes bark, and do so particularly while they make territorial boundary patrols, and in response to barking by their neighbours (e.g. Naumov *et al.* 1981 cited in Frommolt *et al.* 2003).

A cocked leg is as emblematic a signal of canidness as there is—and is a visual as well as olfactory signal (Bekoff 1978a), and one that crosses the wolf–fox divide. So too do the presence of anal sacs, interdigital glands, and supracaudal (violet) glands. However, it is not clear how widely the use of lip and cheek glands by red foxes spreads throughout the family (Macdonald 1985). Studies of grey wolves (Mech and Peters 1977), Ethiopian wolves (Sillero-Zubiri and Macdonald 1998), coyotes (*Canis latrans*—Bekoff and Wells 1982), and red foxes (Macdonald 1979b), reveal that all douse their territories with token urinations at very high rates, and deposit their faeces at strategic (and often visually conspicuous) sites such as trail junctions, and sometimes with particular concentrations at borders (Macdonald 1980b).

Dispersal, disease, and body size

A second characteristic of canids is their propensity to long distance dispersal, and this has both theoretical and practical implications. Practically, the importance of dispersal in canid societies became clear when people began thinking about the epidemiology and control of wildlife rabies in terms of the ecology of red foxes—a line of thought first advocated by Macdonald (1977b) in Colin Kaplan's small but noteworthy book that itself caught the mood that was precursor to a new generation of joined-up thinking about rabies and other wildlife diseases. It soon became clear that fox dispersal distances were both long (e.g. Englund 1970; Storm *et al.* 1976; Lloyd 1980) and on average positively correlated with home range size and thus population density (Macdonald 1980c; Macdonald and Bacon 1982)—an observation that revealed the paradox that, all else being equal, rabies might be expected to occur at highest incidence in dense fox populations, but to spread fastest spatially in sparse populations. This in turn led to two questions that remain important a generation later, first, does the behaviour of healthy individuals provide a basis for modelling the behaviour of diseased ones—in the case of rabies the preliminary answer of yes came from a few highly influential foxes that developed rabies while being radio-tracked (David *et al.* 1982; Voigt *et al.* 1985). Second, in what ways might attempted control methods interact with the vector's behavioural ecology to cause nonlinear, and perhaps counterproductive outcomes—the suspicion that this was a problem for lethal control applies now not only to rabies (Macdonald 1995), but to wildlife management in general under the name of the perturbation hypothesis (Tuyttens and Macdonald 2000). Quarter of a century later, interest in canid diseases as a conservation issue has blossomed unrecognizably and is thus the subject of Chapter 6, although dispersal, despite some illuminating studies (Harris and Trewhella 1988) remains largely a black hole in conservation knowledge (Macdonald and Johnson 2001).

Canid dispersal is important to several aspects of evolutionary biology, including ideas on population genetics and fitness; an additional realm in which the theoretical importance of dispersal came into focus was as a primary factor whose costs were offset against possible benefits of group-living (Vehrencamp 1983; Macdonald and Carr 1995). Dispersal is generally assumed to be dangerous (perhaps 5–6 times more so than philopatry according to Waser *et al.* 1994). Where the costs of dispersal were high (and Ethiopian wolves appear to provide one such example, Sillero-Zubiri *et al.* 1996a), individuals may be more disposed to seek the benefits of joining (larger) groups, and all the more so if these costs are minimal (Macdonald 1983; Johnson *et al.* 2002). Ballard *et al.* (1987) illustrate the opposite case for Alaskan grey wolves in a hunted population (whereas Pletscher *et al.* (1997)) remind us yet again of the breadth of intraspecific variation by illustrating a case where grey wolves, too, face high dispersal costs). More than one set of costs and benefits may lead to group formation—for example, coyotes may form groups either where prey are large (Bowen 1981, 1982; Bekoff and Wells 1982, 1986) or where prey are small and abundant, but in both cases Barrette and Messier (1980) argue that a major factor in the formation of coyote groups is the high cost of dispersal imposed by habitat saturation. This topic is developed in Chapter 17, but beyond what is discussed in that chapter, there remain huge gaps in understanding of canid dispersal and its interaction with their mating and social systems. This is largely because a proper study of dispersal requires conditions that remain signally difficult to achieve: individuals and their circumstances (ecological and sociological) must be studied during all of three phases: pre-dispersal, dispersal, and settlement—despite some valiant efforts, this has never been achieved for any canid.

Distinctions in the sex ratio of dispersers reverberate through the major socioecological trends in the canidae. Although behavioural (but not necessarily genetic) monogamy is fundamental to canid societies (Kleiman 1977; Kleiman and Malcolm 1981), as case studies accumulated, Macdonald and Moehlman (1982) noted that canid social systems appeared to be size related (see also Creel and Macdonald 1995). In effect, canids can be categorized according to three size classes. These size/socioecology links are explored fully in Chapter 4, (Macdonald *et al.* Chapter 16, this volume), but they also demand mention here to set the scene. Small canids (<6 kg) are either largely monogamous (e.g. Blanford's, swift (*V. velox*) and kit foxes—Geffen and Macdonald 1992; Cypher *et al.* 2000; List and Macdonald 2003) or form small, loose knit groups with a female-biased sex ratio, from which young males tend to emigrate, and females stay in their natal range as helpers until a breeding opportunity arises (e.g. red and Arctic foxes, Macdonald 1979a; Hersteinsson and Macdonald 1982). Medium-sized canids (probably excluding the bush dog) (6–13 kg) have an equal adult sex ratio and emigration rate, and both sexes may be helpers and thus both sexes also disperse (golden, black-backed *Canis mesomelas* and side-striped jackals, coyotes, and crab-eating foxes, Bekoff and Wells 1982; Moehlman 1983; Macdonald and Courtenay 1996; Loveridge and Macdonald 2001). Larger canids (excluding the maned wolf—Dietz 1984, and perhaps the grey wolf—Packard *et al.* 1983) (>13 kg), in contrast, exhibit an adult sex ratio skewed towards males, female emigration and male helpers (e.g. Ethiopian wolves, dholes, African wild dogs—Kühne 1965; Johnsingh 1982; Sillero-Zubiri *et al.* 1996a—and perhaps the bush dog is an atypically diminutive member of this category, Macdonald 1996b). The first person to explore these trends analytically was Moehlman (1986, 1989) who analysed the comparative data then available to conclude that female body mass was positively related to gestation length, neonate mass, litter size, and litter mass. She developed the argument that canid mating systems and social organization arise as a result of the conflict between the effect of body size on reproductive traits and the constraints on females in obtaining resources for reproduction. The argument ran that females of large canids have large litters of relatively small, dependant pups. The period of dependency of these pups is, therefore, relatively long, and requires more male postpartum investment. In large canids, therefore, competition among females for males as helpers is likely to be more intense and is predicted to drive the system towards polyandry. Small canids in contrast produce small litters of more precocial cubs that require less parental investment. Because competition for male parental investment is reduced in this scenario, males can invest more time and resources in additional females, leading to polygyny. This influential idea was updated by Moehlman and Hofer (1997). They noted, *inter alia*, not only that larger canids tend to the largest litters and largest pups, and that there may be a tendency for them also to have the largest lifetime litter mass—all suggesting that they invest more in prepartum reproduction. They also noted that female reproductive suppression and the presence of helpers was most prevalent amongst larger canids, which also tend to eat larger prey, and to hunt in packs. They conclude that as energetic costs increase and the reproductive tactic is to produce more young per breeding attempt, there is a higher incidence of alloparental behaviour and reproductive suppression. This suggested to Moehlman and Hofer (1997) that increased reproductive output in canids may be an evolutionary consequence of selection that favoured reproductive suppression as a means of helper recruitment (Creel and Creel 1991 also found that litter mass, litter growth rate, and total investment were higher in communally breeding carnivores that had reproductive suppression).

The observation that larger canids kill larger prey also turns out to be an energetic necessity of prey availability (Carbone *et al.* 1999), and eating larger prey is associated, arguably as cause rather than effect, with a cascade of other ecological consequences that may facilitate group formation and much that follows from it (Kruuk and Macdonald 1985; Johnson *et al.* 2002) as discussed in Chapter 18. Indeed, that canid social systems would be influenced not only by the energetic costs of reproduction but also by ecological and demographic factors (as acknowledged by Moehlman and Hofer 1997) was the basis of an analysis by Geffen *et al.* (1996) that drew partly different conclusions to Moehlman's original interpretation. As elaborated in Chapter 4 (Macdonald *et al.* this volume; Geffen *et al.* 1992e) they suggest that much of the inter- and intraspecific variation in canid social structure can be explained by resource availability. Macdonald (1992b) noted that examples of female biased groups and male dispersal were thus far entirely confined to vulpine canids, and this and similar complications prompted Geffen *et al.* (1996) to control for phylogeny in an analysis that indicated that whereas neonate weight and litter weight are positively correlated with female weight, large canids do not have relatively smaller young. After controlling for phylogeny, neonate weight was independent of litter size, casting doubt on the notion of a general energetic linkage between these two variables. Geffen *et al.* (1996) suggest instead that changes in body size, litter size and social organization within the Canidae may be attributed to differences in food availability.

The interactions between dispersal, social system, and body size, and the network of related variables, remains full of puzzles, and is discussed for Carnivora as a whole by Macdonald (1992b, especially pp. 242–246). Comparative analyses of this topic are likely for some time to come to be distorted by two complications. First, the small number of field studies of each species and thus the impact on species' averages caused by only partial knowledge of their intraspecific variation. Second, the extent to which phylogeny should be considered a source of statistical dependence—there being no *prima facie* reason why the answer should be the same for all comparisons.

Sympatry and interspecific relations

Several chapters in this book mention competition, indeed hostility, between different species of canid. Since Hersteinsson and Macdonald (1982) first suggested that the red fox was a determinant of the geographical range of the Arctic fox, and Voigt and Earle (1983) reported that red foxes in Ontario existed only in the interstices between coyote territories (just as deer do between wolf territories, Mech 1977a), the idea of aggression between sympatric canids being an important force in their biology has been transformed within two decades from a smattering of anecdotes to a universal of canid community ecology. Tannerfeldt *et al.* (2002) showed that, in fact, Arctic foxes could scarcely breed within 8 km of a red fox den and when they tried, in most cases the red foxes killed their young. Indeed, intraguild aggression, as an expression of competition emerges as a commonplace of carnivore communities. Thus, among North American canids, grey wolves kill coyotes and red foxes (Berg and Chesness 1978; Carbyn 1982; Paquet 1992; Peterson 1995a), coyotes kill red foxes (Voigt and Earle 1983; Sargeant and Allen 1989), swift foxes (Scott-Brown *et al.* 1987), and kit foxes (Ralls and White 1995; White and Garrott 1997), and red foxes kill kit foxes (Ralls and White 1995) and Arctic foxes (Bailey 1992). A notable example is that within 10 years of the wolves' first arrival, coyotes had gone from Isle Royale (Allen 1979)—in this context, and if *C. lycaon* exists, it is interesting to wonder how hostilities flow between them and *C. lupus*, and from both wolf species to coyotes. In Europe, there is evidence of grey wolves killing red foxes (Macdonald *et al.* 1980) and red foxes killing Arctic foxes (Frafjord *et al.* 1989), although, so far, there are no such data on red foxes attacking the introduced raccoon dogs (*Nyctereutes procyonoides*—Kauhala *et al.* 1998b). In Africa, we have seen black-backed jackals tormenting cape foxes (*Vulpes chama*) in the Kalahari, and wild dogs kill bat-eared foxes (*Otocyon megalotis*—Rasmussen 1996a). Oddly, and against the generality that larger canids are hostile to smaller ones, black-backed jackals dominate the larger side-striped jackal (Loveridge and Macdonald 2002). Indeed, while in some canid communities there is evidence of character displacement (Dayan *et al.* 1989, 1992), in parts of Kenya where black-backed, golden, and side-striped jackals coexist they become more rather than less similar in size (Fuller *et al.* 1989; van Valkenburg and Wayne 1994), a result supported by findings in Zimbabwe (Chapter 16).

The demographic results of such hostilities (in addition to habitat preferences, and doubtless themselves influenced by landscape) probably explain why population densities of grey wolves and coyotes appear to be inversely related (Berg and Chesness 1978; Carbyn 1982) as are those of coyotes and red foxes (Linhart and Robinson 1972; Sargeant *et al.* 1987). Consequently, although wolves sometimes also kill them, red foxes are more numerous where wolves are found, benefiting from a corresponding decrease in coyotes. In contrast, the ranges of red and gray foxes (*Urocyon cinereoargenteus*) are reported to overlap commonly (e.g. Follman 1973; Sunquist *et al.* 1989). This leads to the obvious question of what determines, within a guild of canids, where the axes of animosity are strongest? For example, in parts of Africa the three species of jackal cohabit with both African wild dogs and bat-eared foxes—how does hostility vary between them? Two obvious predictions come to mind. First, the pairs of species most similar in size (and thus diet) might be in strongest competition and hence most likely to be hostile. Second, any given species might be most likely to harass only those species that are sufficiently smaller than itself to minimize the risk of injury (this idea assumes that the size difference is not a strong indicator of diminished competition, on the grounds that most canids have widely overlapping diets—itself rather a puzzle in terms of the minor differences in dental morphology reported by Dayan *et al.* 1992). There is some evidence in favour of both predictions. Grey wolves seem particularly hostile to coyotes, as do coyotes to red foxes—each perhaps harassing the next species down the ladder from itself. Similarly, C. Stuart and T. Stuart (personal communication) record two instances of black-backed jackal killing cape foxes. In contrast, there are no accounts of African wild dogs killing cape foxes; there are no records of wolves killing swift or kit foxes. On the other hand, coyotes do wreak havoc on swift and kit foxes, which are very much smaller than themselves. It is tempting to think that the mechanism behind intra-guild aggression between canids is rather undiscerning about particular size ratios—the sight of a red fox chasing an Arctic fox is so reminiscent of a dominant red fox chasing an inferior conspecific that it seems as if larger canid species may simply treat smaller canids as poor quality versions of themselves—a generality that comes to mind when observing the eagerness with which domestic dogs chase wild canids (and grey wolves readily kill domestic dogs, for example, Fritts and Paul 1989; Kojola and Kuittinen 2002).

This line of thought is further complicated by the fact that these intra-guild hostilities extend beyond the Canidae, and emerge as a general feature of interspecific relations within the Carnivora. Thus, while spotted hyaenas (*Crocuta crocuta*) and lions (*Panthera leo*) persecute African wild dogs (Creel and Creel 2002), red foxes kill pine martens (*Martes martes*—Lindström 1992). In terms of mechanisms and motivations, whether the tenor of the relationship between a red fox and a swift fox is the same as that between a red fox and a pine marten is unknown, but the outcome is much the same—the small predator ends up displaced or dead. Clearly, the consequences of intra-guild aggression face conservationists with difficult biological dilemmas, and even greater ethical ones.

Figure 1.3 Ancient hostilities: leopard *Panthera pardus* tormented by African wild dogs *Lycaon pictus* © M.G.L. Mills.

Canid geography

Canids are in flux. Sillero-Zubiri *et al.* (in press) estimate that over the last century the geographical ranges of seven species have increased, eight have decreased and nine have remained stable. The kaleidoscope of species diversity has changed: there are places where the grey wolf and the red fox have been replaced by what amounts to their 'ecological average', the coyote (once confined to mainly arid areas in western North America and now found in every state, province, and country north of Panama—Moore and Parker 1992; Reid 1997; Bekoff and Gese 2002; Gese and Bekoff in press). Contemporary canids are the most widely distributed family of the Carnivora, with members on every continent besides Antarctica (Ginsberg and Macdonald 1990). Africa, Asia, and South America support the greatest diversity with more than 10 canid species each (Johnson *et al.* 1996). Red foxes are sympatric with 14 other canids (from three geographical regions), golden jackals with 13 (from two regions), and grey wolves with 11 (from three regions). Within any one location, however, canid diversity is usually limited to 1–5 species. Five canids are endemic to just one country—the red wolf (*Canis rufus*), Ethiopian wolf, Darwin's fox, hoary fox (*Pseudalopex vetulus*), and island fox, with the Sechuran fox (*P. sechurae*) spanning two countries. Although the genera *Canis* and *Vulpes* are both found on North America, Europe, Africa, Asia, and were introduced by man to Australia, of the remaining eight genera six are restricted to one continent: *Chrysocyon, Otocyon, Pseudalopex, Speothos* (South America), *Cuon* (Asia), *Lycaon* (Africa); *Urocyon* is restricted to North and South America, whereas *Nyctereutes*, formerly restricted to Asia is now also introduced to Europe. At a species level, the numbers of species occurring on (and restricted to) each continent are Africa 13 (8), Asia 12 (2), Europe 7 (0), South America 11 (9), North America 10 (5), and Australia and Oceania 2 (0). Only three species are present in both the Old and New World: the Arctic and red foxes, and the grey wolf. Sillero-Zubiri *et al.* (in press) summarize the distributions of wild canid species by country: 79% of the world's 192 countries have wild canids (of which Sudan has the highest number, 10).

Dramatis personae

There are those—veterinarians notable amongst them—who wield definite articles in a way that shrinks all diversity to a handful of archetypes, with reference to 'The dog', 'The cat', or 'The rat'. A similarly injudicious habit amongst biologists is to talk about a tiny number of notable case studies as if they provided a balanced understanding of an entire taxon—in the case of canids—and in former days this led to overviews based precariously on observations of just a handful of species studied in a handful of places—with grey wolves and red foxes writ large amongst them. We are anxious that readers of this book—beguiled by the gratifyingly increased number of canids about which much is known—should not forget the still large number of others about which not much is known. To guard against this risk we introduce now a vignette account of all 36 canid taxa (including dingoes, here listed as a grey wolf subspecies *Canis lupus dingo*). Vignettes are presented in the same order as Wilson and Reeder (1993) whose systematics we follow closely; the only exceptions are *Vulpes macrotis* and *V. velox* treated here as separate species (Mercure *et al.* 1993), and *P. fulvipes* herewith given full specific status (Yahnke *et al.* 1996).

Arctic fox *Alopex lagopus* (Linnaeus, 1758)

The Arctic fox (Nasimovich and Isakov 1985; Audet *et al.* 2002) has a circumpolar distribution in all Arctic tundra habitats (Hersteinsson and Macdonald 1992). It breeds north of and above the tree line on the Arctic tundra in North America and Eurasia and Arctic islands, and on the alpine tundra in Fennoscandia, ranging from northern Greenland to

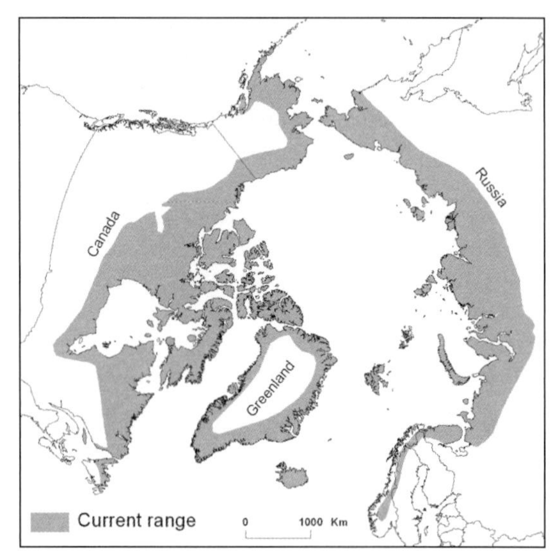

Artic fox © 2003 Canid Specialist Group and Global Mammal Assessment

the southern tip of Hudson Bay, Canada. The world population of Arctic foxes is in the order of several hundred thousand animals. Populations fluctuate widely between years in response to lemming numbers (Angerbjörn *et al.* 1995). Considering their dependence on cyclic lemmings (which occurred in 85% of faeces; Elmhagen *et al.* 2000), starvation is an important cause of Arctic fox mortality during some years (Garrott and Eberhardt 1982; Tannerfeldt and Angerbjörn 1998). The density of occupied natal dens varies from 1–3/100 km^2 (Boitzov 1937; Macpherson 1969) up to 8/100 km^2 (Hersteinsson *et al.* 2000). Combined group ranges contribute to territories from which occupants rarely stray (Hersteinsson and Macdonald 1982), and sizes vary with lemming abundance and habitat (10–125 km^2; Eberhardt *et al.* 1982; Frafjord and Prestrud 1992; Angerbjörn *et al.* 1997). The Arctic fox remains the single most important terrestrial game species in the Arctic, mainly because of their exceptional fur (Garrott and Eberhardt 1987; Hersteinsson *et al.* 1989), which has the best insulative properties among all mammals. In autumn, fox weight may increase by more than 50% as fat is deposited for insulation and reserved energy. They change between summer and winter pelage, thereby adjusting their insulating capabilities and enhancing their camouflaging potential. With the decline of the fur hunting industry, the threat of over-exploitation is lowered for most Arctic fox populations (but continued climatic warming may endanger some populations, such as in Fennoscandia).

	Male	Female
Weight	3.6 ± 0.4 kg, n = 478	3.1 ± 0.4 kg, n = 514
Head/body length	578 ± 31 mm, n = 89	548 ± 33 mm, n = 85

Ref: Angerbjörn *et al.* (in press)

Short-eared dog *Atelocynus microtis* (Sclater, 1882)

The short-eared dog (Berta 1986) is notoriously rare, and sightings are uncommon across its range. The species is poorly known by indigenous peoples of the Amazon basin and is not known to hold any special significance to them. The short-eared dog has been found in scattered sites from Colombia to Bolivia and Ecuador to Brazil, and is associated with undisturbed rainforest in the western Amazonian lowlands (Leite Pitman and Williams in press). They have been recorded in a wide variety of habitats, including terra firma forest, swamp forest, stands of bamboo, and primary succession along rivers (Peres 1991). Sightings of the species in rivers (and the presence of a partial interdigital membrane) suggest that the short-eared dog may be at least partly aquatic; fish form part of their diet (Defler and Santacruz 1994). No information on density is available, or on the continuity of the species' distribution within its extent of occurrence; the absence of any records from large areas suggests that the distribution may not be continuous within the extent of occurrence. Likely threats include disease and habitat loss.

Weight	9–10 kg
Head/body length	720–1000 mm

Ref: Nowak (1999)

Side-striped jackal *Canis adustus* Sundevall, 1847

The side-striped jackal (Atkinson 1997a) is well adapted anatomically and behaviourally for opportunism (Atkinson *et al.* 2002a; Loveridge and Macdonald 2003). Endemic to west, central, and southern Africa (excluding the southernmost part) (Kingdon 1977, 1997), side-striped jackals occupy a range of habitats, from game areas through farmland to towns within the broad-leaved savannah zones. The species is generally common, and apparently occurs in its highest densities in areas surrounding human settlement (e.g. around 1/km^2 in highveld commercial farmland in Zimbabwe, Rhodes *et al.* 1998). Side-striped

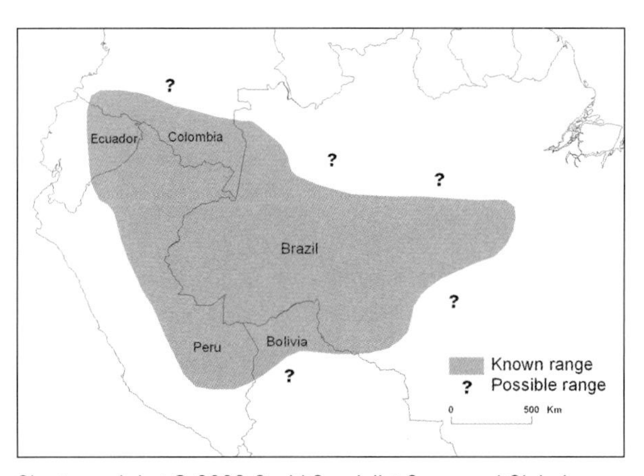

Short-eared dog © 2003 Canid Specialist Group and Global Mammal Assessment

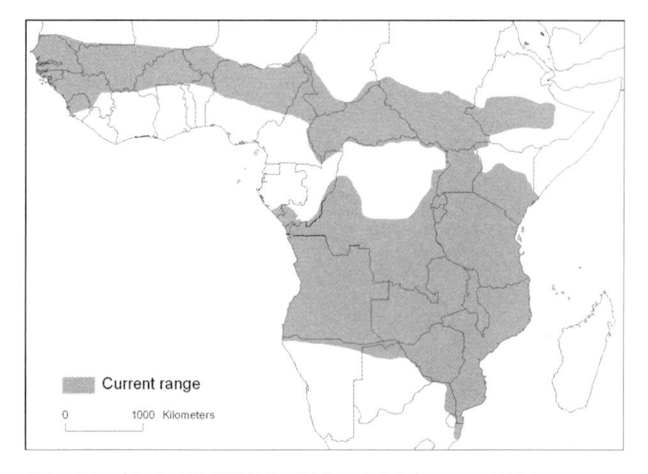

Side-striped jackal © 2003 Canid Specialist Group and Global Mammal Assessment

jackals occur solitarily, in pairs or family groups of up to seven individuals. In game areas of western Zimbabwe, home ranges varied seasonally from 0.2 to 1.2 km^2, whereas in highveld farmland, they were seasonally stable and >4.0 km^2 (Atkinson 1997b). Alloparental care of young occurs (Moehlman 1979, 1989). They are persecuted for their role in rabies transmission (Rhodes *et al.* 1998; Loveridge 1999; Loveridge and Macdonald 2001) and putative role as stock killers. It is unlikely that this persecution has an effect on the overall population, but indiscriminate culling through poisoning and snaring could affect local abundance. However, this species' dietary flexibility (Atkinson *et al.* 2002a) and ability to coexist with humans on the periphery of settlements and towns suggests that populations are only vulnerable in cases of extreme habitat modification, or intense disease epidemics.

	Male	Female
Weight	9.4 (7.3–12.4) kg, *n* = 50	8.3 (7.3–10.0) kg, *n* = 50
Head/body length	1082 (960-1165) mm, *n* = 50	1075 (1000-1170) mm, *n* = 50

Ref: Smithers (1983)

Golden jackal *Canis aureus* Linnaeus, 1758

The golden jackal is a typical representative of the genus *Canis*. The species is widespread in North and Northeast Africa, the Arabian Peninsula, western Europe, eastwards into the Middle East, Central Asia, the Indian subcontinent, and east and south to Sri Lanka and parts of Indo-China. The jackal features in mythological and cultural accounts of several civilizations: the ancient Egyptians worshipped the jackal-headed god Anubis, and the Greek gods Hermes and Cerberus probably derived their origins from the golden jackal. Due to their tolerance of dry habitats and their omnivorous diet (Fuller *et al.* 1989), the golden jackal can inhabit a wide variety of habitats, from the Sahel Desert to the evergreen forests of Myanmar and Thailand. The social organization of golden jackals is extremely flexible depending on the availability and distribution of food resources (Macdonald 1979c; Moehlman 1983, 1986, 1989; Fuller *et al.* 1989). Recorded home range sizes vary from 1.1 to 20.0 km^2 (van Lawick and van Lawick-Goodall 1970; Kingdon 1977) High densities are observed in areas with abundant food and cover. Nevertheless, over its entire range, except in protected areas, the jackal population is steadily declining. Traditional land-use practices are being steadily replaced by industrialization and intensive agriculture, while wilderness areas and rural landscapes are being rapidly urbanized. Jackal populations adapt to some extent to this change and may persist for a while, but eventually disappear from such areas. An estimated 80,000 jackals remain on the Indian subcontinent, but there are no estimates for Africa (Jhala and Moehlman in press).

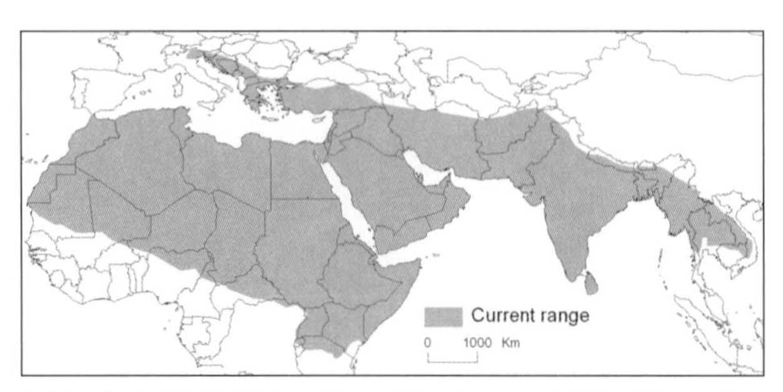

Golden jackal © 2003 Canid Specialist Group and Global Mammal Assessment

	Male	Female
Weight	8.8 (7.6-9.8) kg, *n* = 6	7.3 (6.5-7.8) kg, *n* = 4
Head/body length	793 (760-840) mm, *n* = 6	760 (740-800) mm, *n* = 3

Ref: Jhala and Moehlman (in press)

Coyote *Canis latrans* Say, 1823

The coyote (Young and Jackson 1951; Gier 1968; Bekoff and Gese 2002) is the most versatile of all canids, and their plasticity in behaviour, social ecology, and diet (Bekoff and Wells 1986; Gese *et al.* 1996a–c) allows them to not only exploit, but to thrive, in almost all environments modified by humans. Coyotes were believed to have been restricted to the southwest and plains regions of the United States and Canada, and northern and central

Mexico, prior to European settlement. With land conversion and removal of wolves after 1900, coyotes expanded into all of the Unites States and Mexico, southward into Central America, and north into Canada and Alaska. They continue to expand their distribution, occupying most areas between 8°N (Panama) and 70°N (northern Alaska), utilizing almost all available habitats including prairie, forest, desert, mountain, and tropical ecosystems. Coyote densities in different geographic areas and seasons vary from 0.01–0.09 coyotes/km^2 in the winter in the Yukon (O'Donoghue *et al.* 1997) to 0.9/km^2 in the fall, and 2.3/km^2 during the summer (post-whelping) in Texas (Knowlton 1972; Andelt 1985). Coyotes are a major predator of domestic sheep and lambs, and of game species (Andelt 1987; Knowlton *et al.* 1999; Lingle 2002; Sillero-Zubiri *et al.*, Chapter 5, this volume). In areas with predator control, losses to coyotes were 1.0–6.0% for lambs and 0.1–2.0% for ewes (USFWS 1978b). In areas with no predator control, losses to coyotes were 12–29% of lambs and 1–8% of ewes (McAdoo and Klebenow 1978; O'Gara *et al.* 1983). Notwithstanding, there appears to be no current threats to coyote populations and conservation measures have not been needed. Local reductions are temporary and coyotes remain abundant throughout their range.

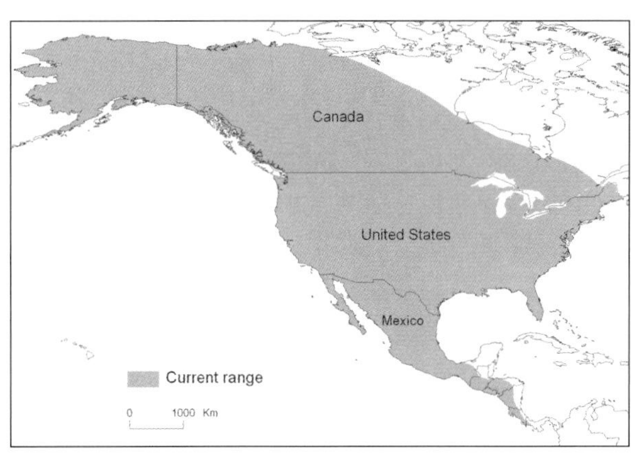

Coyote © 2003 Canid Specialist Group and Global Mammal Assessment

	Male	Female
Weight	11.6 (7.8-14.8) kg, $n = 86$	10.1 (7.7-14.5) kg, $n = 73$
Head/body length	842 (740-940) mm, $n = 38$	824 (730-940) mm, $n = 36$

Ref: Gese and Bekoff (in press)

Grey wolf *Canis lupus* Linnaeus, 1758

The largest wild canid (but with a huge geographical weight range from 12 to 62 kg—Mech and Boitani 2003), the grey wolf formerly was the world's most widely distributed mammal (a title now usurped by the versatile red fox), occurring throughout the northern hemisphere north of 15°N latitude in North America and 12°S in India and Arabian Peninsula (Harrington and Paquet 1982). Grey wolves may extend beyond the Sinai into Africa, where the controversial status of the little known *Canis aureus lupaster* awaits confirmation (Ferguson 1981). Poisoning and deliberate persecution due to depredation on livestock have reduced its original worldwide range by about one-third, and it has become extinct in much of Western Europe (Boitani 1995), Mexico and much of the United States (Mech 1970, 1974). Studies in the Unites States indicate that wolves are characterized by annual mortality rates of the order of 15–30%, with common causes of death including starvation, and being killed by humans or other wolves (Peterson *et al.* 1998). Since about 1970, legal protection, land-use changes, and rural human population shifts to cities have arrested wolf population declines, and fostered natural recolonization in parts of western Europe and the United States, and reintroduction in the western

Grey wolf © 2003 Canid Specialist Group and Global Mammal Assessment

United States (Carbyn *et al.* 1995; Mech and Boitani 2003). Remaining populations in northern habitats occur where there is suitable food, primarily wild and/or domestic ungulates such as white-tailed deer, moose, and reindeer. Wolf densities vary from about 0.08 to 0.008 km^2, being highest where prey biomass is highest (Mech and Boitani 2003). In summer, wolves hunt alone or in small groups, but in winter they hunt in packs, chasing prey for up to 5 km. Average daily food capture varies from 2.5 to 6.3 kg or more per individual per day, of which a proportion is lost to scavengers (mainly ravens, Vucetich *et al.* in press). Packs include up to 36 individuals, but smaller sizes (5–12) are more common. They occupy territories of 75–2500 km^2 depending on prey density, and these are maintained through howling, scent-marking, and direct killing (Mech 1970; Harrington and Mech 1983; Harrington 1987; Mech *et al.* 1998).

	Male	Female
Weight	40 (20–80) kg	37 (18–45) kg
Head/body length	1000–1600 mm	
Ref: Mech (1974)		

Dingo *Canis lupus dingo* (Meyer, 1793)

Primitive dingoes were associated with nomadic hunter–gatherer societies and later with sedentary agricultural population centres (Corbett 1995). Austronesian-speaking people transported dingoes from mainland Asia to Australia and Pacific islands 1000–5000 years ago (Corbett 1995). Europeans did not discover the dingo in Australia until the seventeenth century and taxonomists originally thought it was a feral domestic dog. In fact, cross-breeding with domestic dogs represents a significant threat to the long-term persistence of dingoes worldwide. In Australia, the proportion of pure dingoes (Thomson 1992a–c), based on skull morphometrics, has declined from about 49% in the 1960s (Newsome and Corbett 1985) to about 17% in the 1980s (Jones 1990). Today, pure dingoes occur only as remnant populations in central and northern Australia and throughout Thailand (Corbett 1995). Estimating dingo abundance is difficult because the external phenotypic characters of many hybrids are indistinguishable from pure dingoes. The density of wild dogs (dingoes and hybrids) in Australia varies between 0.03/km^2 and 0.3/km^2 according to habitat and prey availability (Fleming *et al.* 2001). Human control is a major cause of dingo mortality in Australia (Fleming *et al.* 2001). Dingoes eat a diverse range of prey types and over 170 species ranging from insects to water buffalo have been identified (Corbett 1995). The largest recorded home ranges (90–300 km^2) occur in the deserts of southwestern Australia (Thomson and Marsack 1992), compared to just 10–27 km^2 in forested mountains in eastern Australia (Harden 1985; McIlroy *et al.* 1986).

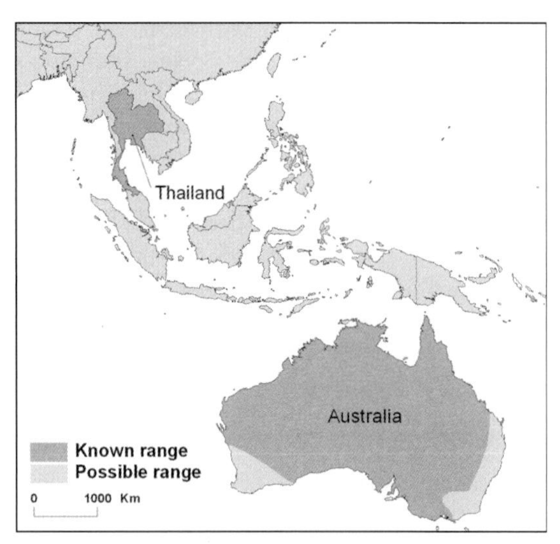

Dingo © 2003 Canid Specialist Group and Global Mammal Assessment

	Male	Female
Weight	15 (12–22) kg, *n* = 51	13 (11–17) kg, *n* = 38
Head/body length	914 (835–1110) mm, n = 50	883 (813–1010) mm, n = 38
Ref: Corbett (in press)		

Black-backed jackal *Canis mesomelas* Schreber, 1775

Somewhat fox-like in appearance, with a long pointed muzzle, and a longer premolar cutting blade than other jackal species (an indication of degree of

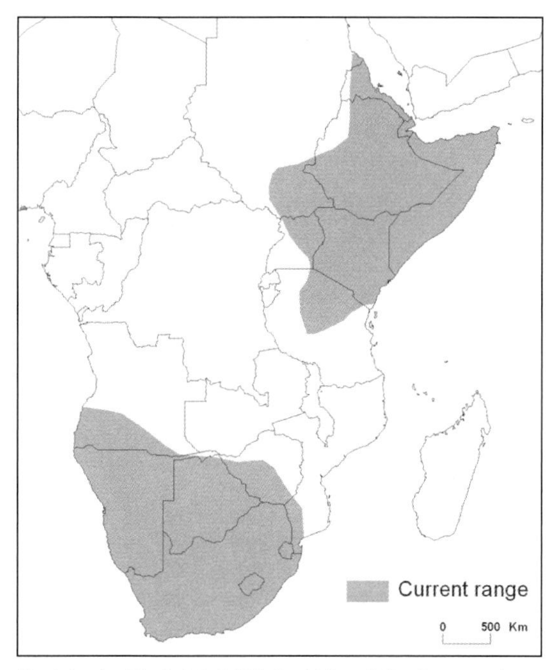

Black-backed jackal © 2003 Canid Specialist Group and Global Mammal Assessment

carnivory), the black-backed jackal has a disjunct distribution, occurring in two separate populations in East and southern Africa (Kingdon 1997), separated by as much as 1000 km (not unlike the bat-eared fox). Black-backed jackals are relatively unspecialized and well suited to an opportunistic lifestyle in a wide variety of habitats, including arid coastal desert, montane grassland, arid savannah and scrubland, open savannah, woodland savannah mosaics, and farmland (Loveridge and Macdonald 2002). In the Drakensberg Mountains of South Africa, Rowe-Rowe (1982) found densities of one jackal per 2.5–2.9 km^2. Diet typically includes small- to medium-sized mammals, reptiles, birds and birds' eggs, carrion, and human refuse (Loveridge and Nel in press). Alloparental care is well documented (Moehlman 1979, 1983). They appear well adapted to water deprivation which may explain their presence in the drier parts of the African continent. They occur in many livestock-producing areas, where they are considered vermin. Predation is usually localized and not extensive, but in certain areas losses up to 3.9% can result, or up to 18% on individual farms (Brand 1993). Where controlled herding is practiced losses amount to only 0.3–0.5% (Brown 1988). Jackals are also significant vectors of rabies in central southern Africa (Loveridge and Macdonald 2001). Nevertheless, population control efforts (e.g. use of dogs, poison, shooting, and gassing) appear largely ineffective and the species remains widespread in these areas today.

	Male	Female
Weight	8.1 (5.9-12.0) kg, $n = 59$	7.4 (6.2-9.9) kg, $n = 42$
Head/body length	785 (690-900) mm, $n = 65$	745 (650-850) mm, $n = 42$

Ref: Stuart (1981)

Red wolf *Canis rufus* Audubon & Bachman, 1851

The red wolf (Paradiso 1972) is intermediate in size between the coyote and grey wolf, and has been considered a fertile hybrid between the two species (Mech 1970; Wayne and Jenks 1991). Indeed, the taxonomic status of the red wolf has been widely debated (Nowak 1979, 2002; Phillips and Henry 1992): one line of recent genetic evidence suggests it is a unique taxon, while another proposes that red wolves and grey wolves in southern Ontario (*C. lupus lycaon*), are so genetically similar that they represent a separate species, *C. lycaon* (Wilson *et al.* 2000). The precise historical distribution of the red wolf is

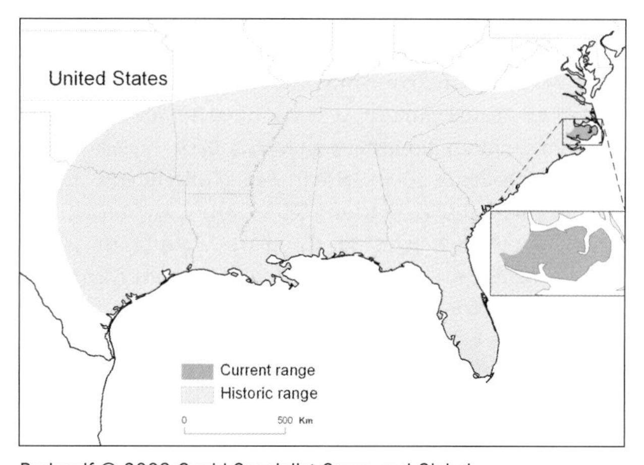

Red wolf © 2003 Canid Specialist Group and Global Mammal Assessment

equivocal, but the species was declared Extinct in the Wild by 1980, following years of human persecution (Riley and McBride 1972). In 1987 they were reintroduced into eastern North Carolina, and are now common within the roughly 6000 km^2 reintroduction area (USFWS 1989, 1992a; Phillips *et al.* 1995). In northeastern North Carolina, white-tailed deer, raccoon, and rabbits comprise 86% of prey species (Phillips *et al.* 2003). Home range size varies from 4 to 226 km^2, depending on habitat (Phillips *et al.* 2003). Although human persecution and other anthropogenic factors continue to impact numbers (e.g. road kills account for 25% of known red wolf deaths in the reintroduced population), hybridization with coyotes or red wolf *X* coyote hybrids is the primary threat to the species' persistence in the wild (Kelly *et al.* 1999). Projections are that the current red wolf population may be lost within 12–24 years if current levels of hybridization continue.

	Male	Female
Weight	28.5 (22.0-34.1) kg, *n* = 70	24.3 (20.1-29.7) kg, *n* = 61
Head/body length	1118 (1040-1250) mm, *n* = 58	1073 (990-1201) mm, *n* = 51

Ref: Kelly *et al.* (in press)

Ethiopian wolf *Canis simensis* Rüppell, 1835

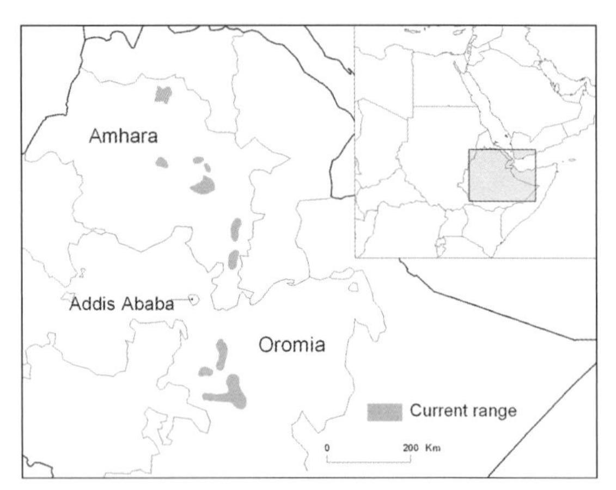

Ethiopian wolf © 2003 Canid Specialist Group and Global Mammal Assessment

The Ethiopian wolf (Sillero-Zubiri and Gottelli 1994) is a medium-sized canid with a reddish coat with distinctive white markings, long legs, and an elongated muzzle. It is confined to seven isolated mountain ranges in the Ethiopian highlands, at altitudes of 3000–4500 m a.s.l. (Gottelli and Sillero-Zubiri 1992; Marino 2003). More than half the population lives in the Bale Mountains, where highest wolf densities are found in short Afroalpine herbaceous communities (1.0–1.2 adults/km^2); lower densities are found in *Helichrysum* dwarf-scrub (0.2/km^2), and ericaceous heathlands and barren peaks (0.1/km^2). In Menz, wolf density was estimated at 0.2/km^2 using transect data (Ashenafi 2001). Ethiopian wolves live in packs of 3–13 adults (mean = 6), discrete and cohesive social units that share and communally defend an exclusive territory (6–13.4 km^2), but generally forage alone (Sillero-Zubiri 1994; Sillero-Zubiri and Gottelli 1995b; Sillero-Zubiri *et al.* 1996a), specializing on rodent prey (Sillero-Zubiri and Gottelli 1995a; Sillero-Zubiri *et al.* 1995a,b). Livestock predation has recently been reported as important in the heavily populated areas of Wollo and Simien (Marino 2003). Continuous loss of habitat due to high-altitude subsistence agriculture represents the major threat to Ethiopian wolves (Sillero-Zubiri and Macdonald 1997; Sillero-Zubiri *et al.* 2000). Sixty per cent of all land above 3200 m has been converted into farmland, and all populations below 3700 m are particularly vulnerable to further habitat loss. Rabies is the most dangerous and widespread disease to affect Ethiopian wolves, and is the main cause of mortality in Bale. Disease killed whole wolf packs in the early 1990s and accounted for a major population decline with losses of up to 75% (Sillero-Zubiri *et al.* 1996b; Laurenson *et al.* 1997, 1998). Hybridization with domestic dogs is also a problem (Gottelli *et al.* 1994).

	Male	Female
Weight	16.2 (14.2-19.2) kg, *n* = 18	12.8 (11.2-14.2) kg, *n* = 8
Head/body length	963 (928-1012) mm, *n* = 18	919 (841-960) mm, *n* = 8

Ref: Sillero-Zubiri and Gottelli (1994)

Crab-eating fox *Cerdocyon thous* (Linnaeus, 1766)

The crab-eating fox (Berta 1982) is a medium-sized (5–7 kg) canid, relatively common throughout its range from northern Colombia and Venezuela, south to Entre Ríos, Argentina (35°S), and from the Andean foothills in Bolivia and Argentina (67°W) to the Atlantic forests of east Brazil to the western coast of Colombia (1°N) (Berta 1987). The species occupies most habitats including marshland, savannah, woodland, and forests, and have been recorded up to 3000 m a.s.l. It readily adapts to deforestation, agricultural and horticultural development, and habitats in regeneration. Average densities range from 0.5 animals per km^2 in savannah/scrub mosaic in Brazil (Courtenay 1998) to 4/km^2 in the Venezuelan llanos (Eisenberg *et al.* 1979). Adults occupy stable territories of 0.48–10.4 km^2 (Sunquist *et al.* 1989; Macdonald and Courtenay 1996; Maffei and Taber in press). They commonly hunt as pairs accompanied by 1–3 adult-sized offspring (Montgomery and Lubin 1978). The dry season diet is predominantly small mammals, reptiles, and amphibians, with insect and fruit becoming more frequent in the wet season (Brady 1979; Motta-Junior *et al.* 1994). The population is generally considered stable and abundant, although there is the potential threat of local spill-over infection of diseases from dogs to wildlife (Courtenay *et al.* 1994, 2001). In addition, reports of poultry raiding by crab-eating foxes are widespread, which has led to their being shot, trapped, and poisoned indiscriminately. In Marajó, Brazil, 83% of 12 fox deaths between 1988 and 1991 were due to local hunters (Macdonald and Courtenay 1996). However there is no evidence that they represent a significant predator of lambs, or cause economic loss to farmers in wool-producing countries.

Crab-eating fox © 2003 Canid Specialist Group and Global Mammal Assessment

Weight	5.7 (4.5–8.5) kg, $n = 52$
Head/body length	658 (570–775) mm, $n = 61$
Ref: Courtenay and Maffei (in press)	

Maned wolf *Chrysocyon brachyurus* (Illiger, 1815)

The maned wolf is immediately distinguishable by its long, thin legs, long, reddish-orange fur, and large ears (Dietz 1985). It inhabits grasslands, *cerrado* forest, wet fields, and scrub forests of central South America from the mouth of the Parnaiba River in northeastern Brazil, south through the Chaco of Paraguay to 30°S in northern Argentina, and west to the Pampas del Heath in Peru (Dietz 1985; Rodden *et al.* in press). With their solitary habits (Dietz 1984; Silveira 1999; Bestelmeyer 2000) and relatively large home ranges, maned wolves live at low densities throughout their range. Dietz (1984) found that home ranges of pairs in Serra da Canastra National Park varied between 21.7 and 30.0 km^2, but elsewhere ranges are larger (up to 105 km^2—Silveira 1999; F. Rodrigues personal communication). Native folklore and superstitions contribute to the attitudes of local people to maned wolves (which range from tolerance to fear and dislike), yet, although it is one of the largest carnivores in the grasslands, the species is apparently not well known to a large segment of the population. About 50% of their diet is plant material, and they rarely prey on domestic animals (Dietz 1984; Motta-Júnior *et al.* 1996; Jácomo 1999). The most significant threat to maned wolf populations is the drastic reduction of habitat, especially due to conversion to agricultural and pastureland. The *cerrado* has been reduced to about 20% of its

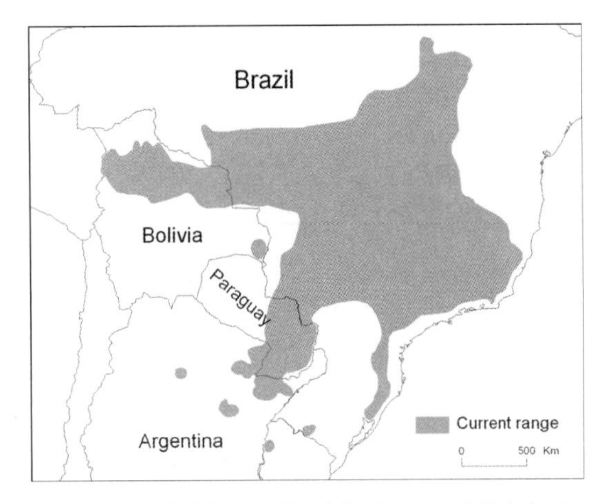

Maned wolf © 2003 Canid Specialist Group and Global Mammal Assessment

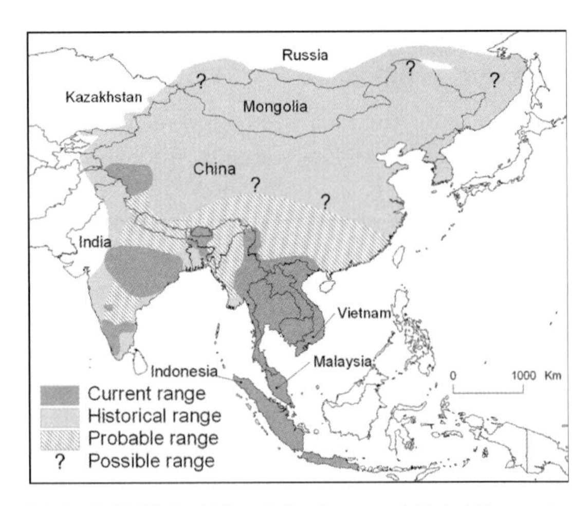

Dhole © 2003 Canid Specialist Group and Global Mammal Assessment

preserved original area (Myers *et al.* 2000), and only 1.5% of it is currently protected (Ratter *et al.* 1997). Road kills on highways are also responsible for mortality of approximately half of the annual production of pups in some reserves (Rodden *et al.* in press).

Weight	25.0 (20.5-30) kg, n = 16
Head/body length	1058 (950-1150) mm, n = 23
Ref: Rodden *et al.* (in press)	

Dhole *Cuon alpinus* (Pallas, 1811)

Dholes (Fox 1984) are large canids (typically 12–20 kg), usually having a reddish or brown coat and a darker, bushy tail. Their dentition is unique among Canidae, with one less lower molar. There are usually six or seven pairs of mammae, rather than the five pairs typical for *Canis* (Burton 1940). The species' known historical distribution covered much of East, South, and Southeast Asia. However, the dhole is presently extremelly rare in Russia, and there have been no recent reports from China (except Tibet) and Mongolia (Johnsingh 1985; Durbin *et al.* in press). Reported densities (all from a few protected areas in southern and central India) range from 0.095 dholes/km^2 to 0.3/km^2 (Durbin *et al.* in press). The dhole is found in a wide variety of forests, tropical grassland—scrub—forest mosaics and alpine steppe (up to 3000 m a.s.l.), but not desert regions. Dry deciduous and moist deciduous forest may represent optimal habitats. Dholes hunt mainly vertebrate prey, preferring medium to large ungulates like spotted deer (73% of biomass consumed in Bandipur, India; Johnsingh 1983) and sambar (17%). Dholes are communal hunters, occasionally forming male biased packs of over 30 animals but more often in hunting groups of <10, or even alone (Cohen 1977; Venkataraman *et al.* 1995; Venkataraman 1998). Sometimes they resort to killing livestock when their natural prey is diminished (Venkataraman *et al.* 1995). Throughout most of their geographical range dholes suffer from persecution for fear of stock predation (Durbin *et al.* in press). In India, bounties were paid for carcasses until the Wildlife Act of 1972, when dholes were given legal protection.

	Male	Female
Weight	15.8 (15.0-17.0) kg, n = 4	10-13 kg
Head/body length	970 (880-1050) mm, n = 3	
Ref: Durbin *et al.* (in press)		

African wild dog *Lycaon pictus* (Temminck, 1820)

A large, but lightly built, canid, the African wild dog (Creel and Creel 2002) was formerly distributed throughout sub-Saharan Africa, except for countries in West and Central Africa that were covered with rainforest. Occupying habitats including short grass plains, semi-desert, bushy savannahs and upland forest, wild dogs are rarely seen, and it appears that populations have always existed at very low densities; they reach their highest densities (e.g. 3.3 adults/100 km^2 in Hluhluwe-Umfolozi Game Reserve, South Africa) in thicker bush (Mills and Gorman 1997). Wild dogs are generalist pack predators (Fuller and Kat 1990; Creel and Creel 1995), hunting medium-sized antelope; whereas the dogs weigh 20–30 kg, their prey average around 50 kg, and may be as large as 200 kg. They are intensely social animals (Frame *et al.* 1979; Malcolm and Marten 1982; McNutt 1996a,b; Girman *et al.* 1997) and packs may number 30 adults and yearlings (Woodroffe *et al.* 1997). Packs are confined to relatively small areas (50–200 km^2) when they are young pups are at a den, but otherwise range widely (423–1318 km^2; Fuller *et al.* 1992b; Woodroffe *et al.* 1997). Wild dogs have very large litters for their body size, averaging 10–11 and occasionally as many as 21 (Fuller *et al.* 1992b). Competition with larger predators has a major impact on wild dogs' behaviour and population biology, with lions causing about 10% of mortality (Creel and Creel 1996). More than half of the mortality recorded among adults is caused directly by human activity, even in some of the largest and best-protected areas. They have disappeared from much of their former range—25 of the former 39 range states no longer support populations—and current population estimates suggest only 3000–5500 free-ranging wild dogs remain (Woodroffe *et al.* in press).

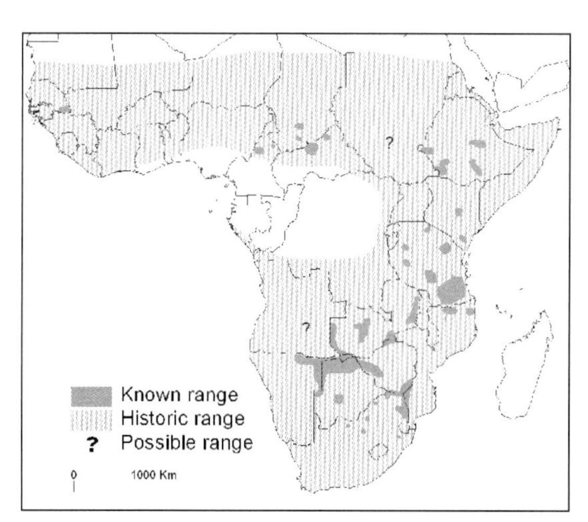

African wild dog © 2003 Canid Specialist Group and Global Mammal Assessment

	Male	Female
Weight	28.0 (25.5-34.5) kg, $n = 12$	24.0 (19.0-26.5) kg, $n = 12$
Head/body length	1229 (1060-1385) mm, $n = 16$	1265 (1090-1410) mm, $n = 15$

Ref: Woodroffe *et al.* (in press)

Raccoon dog *Nyctereutes procyonoides* (Gray, 1834)

The raccoon dog lineage diverged from other canids about 7–10 million years ago (Wayne 1993). As its name suggests, the raccoon dog (Judin 1977; Ikeda 1982; Kauhala *et al.* 1998c) is not unlike a raccoon in general appearance, and, uniquely amongst canids, hibernates in winter (especially in areas like southern Finland where winters are harsh). Adult raccoon dogs almost double their weight between June (4.5 kg on average) and October (8.5 kg) (Kauhala 1993). Originally restricted to the Far East, including the Japanese Archipelago, the raccoon dog, or tanuki, has often appeared in Japanese folklore. They have been raised for fur and were exported, mostly to the United States before the Second World War; their fur is still used in Japan. The Russians introduced raccoon dogs into the wild in the European part of the former Soviet Union to establish a valuable new fur animal in the wild. They are now widespread in northern and eastern Europe, thriving in moist forests with abundant undergrowth (Nasimovic and

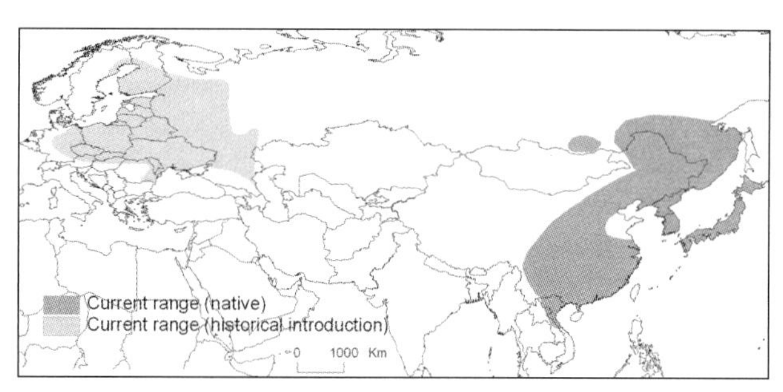

Raccoon dog © 2003 Canid Specialist Group and Global Mammal Assessment

Isakov 1985; Kauhala 1996a). In urban areas, raccoon dogs inhabit areas with as little as 5% forest cover. They are often found near water, and during autumn are more or less dependent on fruits and berries; small rodents are also important (Nasimovic and Isakov 1985; Kauhala *et al.* 1998b). In Japan, home range size varies from as little as 0.07 km^2 in an urban setting to 6.1 km^2 in a subalpine setting (Fukue 1991; Yamamoto *et al.* 1994). Elsewhere, ranges may be as large as 10 km^2 (Jedrzejewski and Jedrzejewska 1993; Goszczynski 1999; Kowalczyk *et al.* 2000). They face a number of significant threats (Kauhala and Saeki, in press), particularly road kills (110,000–370,000 per year in Japan), culling (4529 legal kills per annum in Japan), and parasitic infestations (especially scabies).

	Male	Female
Weight	4.5 (3.0-6.2) kg, n = 43	4.5 (3.0-5.8) kg, n = 29
Head/body length	556 (292-669) mm, n = 37	567 (505-654) mm, n = 24

Ref: Kauhala and Saeki (in press)

Bat-eared fox *Otocyon megalotis* (Desmarest, 1822)

The bat-eared fox is immediately recognizable by its conspicuously large ears. These small (3.9 kg) canids are unique amongst living, terrestrial eutherians in having four to five functional lower molars, and unique amongst modern canids in having three to four upper molars (Guilday 1962), yielding a dentition of 46–50 teeth, the largest number for any non-marsupial, heterodont, land mammal. Bat-eared foxes occur in two discrete populations (recognized subspecies), separated by about 1000 km, across the arid and semi-arid regions of eastern and southern Africa. The range of both subspecies overlaps almost completely with that of *Hodotermes* and *Microhodotermes*, termite genera prevailing in their predominantly insectivorous diet (Lamprecht 1979; Nel 1990). In the Serengeti, after leaving the den in the evening, groups of 2–3 individuals frequently patrol known *Hodotermes* patches in their 1–3 km^2 territories (Nel 1978; Nel *et al.* 1984; Malcolm 1986; Mackie 1988; Mackie and Nel 1989), calling each other to rich food patches with a

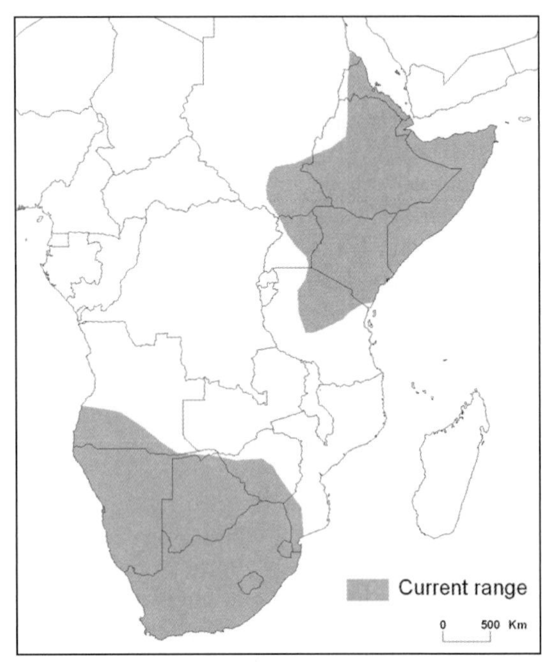

Bat-eared fox © 2003 Canid Specialist Group and Global Mammal Assessment

low whistle. Within a circumscribed habitat, numbers can fluctuate from abundant to rare depending on rainfall, food availability, breeding stage, or disease (Maas 1993b; Nel 1993); rabies and canine distemper can cause drastic population declines. Recorded densities range from 0.3 foxes/km^2 in South Africa (Mackie and Nel 1989) to 9.2 foxes/km^2 in Botswana (Berry 1978). Although common in protected areas across the range, they become increasingly uncommon in more arid areas and on farms in South Africa where they are occasionally persecuted because of the erroneous belief that they prey on young lambs. In East Africa, rabies and canine distemper are linked to reservoirs in domestic dogs, and in the Serengeti, disease caused 90.4% of mortality (Cleaveland and Dye 1995; Carpenter *et al.* 1998).

	Male	Female
Weight	4.0 (3.4-4.9) kg, n = 22	4.1 (3.2-5.4) kg, n = 29
Head/body length	529 (462-607) mm, n = 25	536 (467-607) mm, n = 29

Ref: Smithers (1971)

Culpeo *Pseudalopex culpaeus* (Molina, 1782)

Among South American canids, only the maned wolf is larger than the culpeo (Novaro 1997a). Males are on average 1.5 times heavier than females (Johnson and Franklin 1994a,b; Travaini *et al.* 2000a). The culpeo is distributed along the Andes and hilly regions of South America, from Nariño province of Colombia to Tierra del Fuego. Throughout its wide distribution, the culpeo uses many habitat types ranging from rugged and mountain terrain (up to 4800 m in the Andes), deep valleys and open deserts, scrubby pampas, sclerofilous matorral, to broad-leaved temperate southern beech forest in the south (Johnson 1992). The culpeo has the smallest molars of all South American foxes, and its relatively longer canines reflect its highly carnivorous diet. Up to 83% of the biomass of the culpeo diet in some areas is from exotic mammals (Crespo and De Carlo 1963; Miller and Rottmann 1976; Medel and Jaksic 1988; Novaro *et al.* 2000a). Culpeos are responsible for as much as 60% of the attacks by predators on small-sized livestock in Patagonia (Bellati and von Thüngen 1990), and, due to conflicts with humans (and because of their value as a furbearer), have been persecuted throughout their range. However, culpeos appear to withstand intense hunting levels, and still maintain viable regional populations (Novaro 1997b; Salvatori *et al.* 1999). When hunting pressure is reduced, culpeo populations usually can recover quickly. Density estimates (using a variety of methods) range from 0.2–1.3 individuals/km^2 for northwest Patagonia (Crespo and De Carlo 1963; Novaro *et al.* 2000b), to 0.3–2.6 individuals/km^2 in north central Chile (Jiménez 1993).

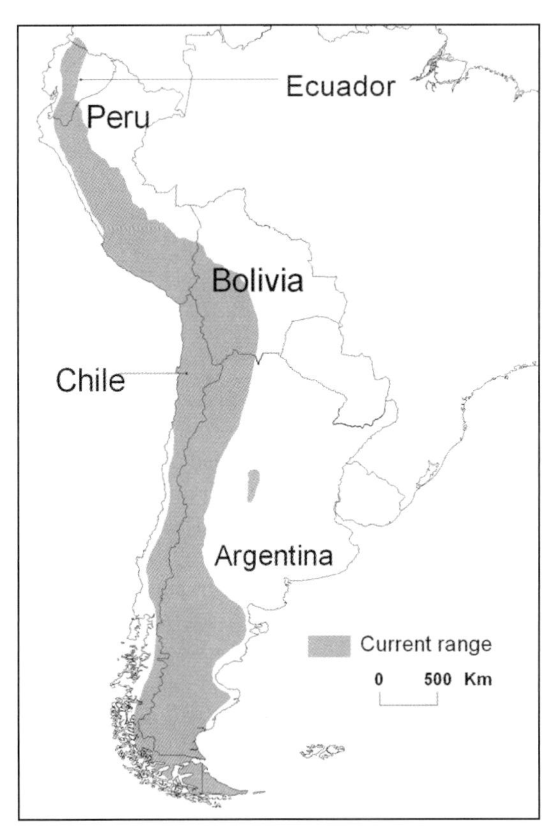

Culpeo © 2003 Canid Specialist Group and Global Mammal Assessment

	Male	Female
Weight	11.0 (8.5–12.3) kg, $n = 11$	8.5 (7.4–10.0) kg, $n = 15$
Head/body length	879 (810–925) mm, $n = 11$	832 (765–890) mm, $n = 15$

Ref: Jiménez and Novaro (in press)

Darwin's fox *Pseudalopex fulvipes* (Martin, 1837)

Until recently, Darwin's fox was known only from the 180 km by 60 km Island of Chiloé, off the coast of Chile where it was collected by Charles Darwin during his HMS *Beagle* voyage. More recently, this small (2–3 kg), stout fox of dark appearance was rediscovered 600 km away in the coastal mountains of the 68 km^2 Nahuelbuta National Park in mainland Chile (Medel *et al.* 1990). These two disjunct populations are thought to be relicts of a former wider distribution (Yahnke 1995; Yahnke *et al.* 1996). There are an estimated 500 foxes on Chiloé Island, and some 50–78 foxes on the mainland (Jiménez and McMahon, in press), the latter at an estimated density of 1.1 individuals/km^2 (E. McMahon, unpublished data). On Chiloé, overlapping home ranges are about 1.6 km^2 for males and 1.5 km^2 for females (Jiménez and McMahon, in press). Generally believed to be a forest-obligate species, in Chiloe, about 70% of their home ranges comprised old-growth forest (Jiménez 2000). Darwin's foxes have an omnivorous, highly opportunistic diet (Jaksic *et al.* 1990; Jiménez *et al.* 1990), and could be a key seed disperser for forest plants (49% of faeces contained seeds; Armesto *et al.* 1987). In Chiloé, they are well known for killing poultry and raiding garbage dumps, apparently with little fear of people and dogs, and even enter houses at night in search of food. While the island population remains relatively secure, the presence of dogs in Nahuelbuta National Park , may be the greatest conservation threat in the form of potential vectors of disease or direct attack. Recent habitat transformations and the rapid advance of the frontier of human impact could also have resulted in population declines. In Nahuelbuta, 74% of mortalities are due to natural causes while 26% are anthropogenic (McMahon 2002).

	Male	Female
Weight	3.3 (2.8–3.9) kg, $n = 7$	2.9 (2.5–3.7) kg, $n = 9$
Head/body length	540 (525–557) mm, $n = 6$	514 (480–550) mm, $n = 9$

Ref: Jiménez and McMahon (in press)

Chilla *Pseudalopex griseus* (Gray, 1837)

A small fox-like canid, lacking an interparietal crest, the chilla or South American grey fox is widespread in the plains and mountains on both sides of the Andes, from northern Chile (17°S) down to Tierra del Fuego (54°S) (where they were introduced in 1951 in an attempt to control introduced European rabbits; Jaksic and Yáñez 1983). They occupy steppes, pampas, and 'matorral' (scrubland) forests in southern Argentina and Chile (Olrog and Lucero 1981; Durán *et al.* 1985), but although they occur in a variety of habitats, they prefer shrubby open areas. Chillas generally inhabit plains and low mountains, but have been reported to occur as high as 4000 m a.s.l. They are tolerant to very different climatic regimes, from remarkably hot and dry areas, such as the Atacama coastal desert in northern Chile (<2 mm average annual rainfall, 22 °C mean annual temperature), to the humid regions of the temperate Valdivian forest (2000 mm, 12 °C) and the cold Tierra del Fuego. They are generally found in monogamous breeding pairs, and individual home range sizes ($n = 23$) varied between 2 and 3 km^2 (Johnson and Franklin 1994a–c). Although omnivorous (Jaksic *et al.* 1980; Medel and Jaksic 1988;

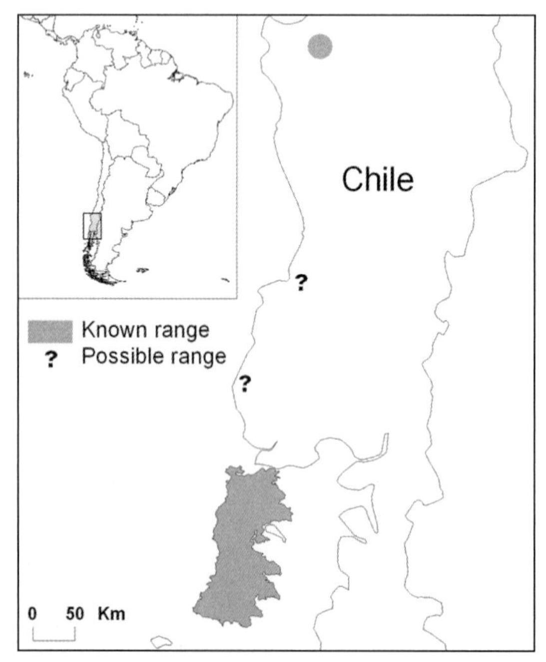

Darwin's fox © 2003 Canid Specialist Group and Global Mammal Assessment

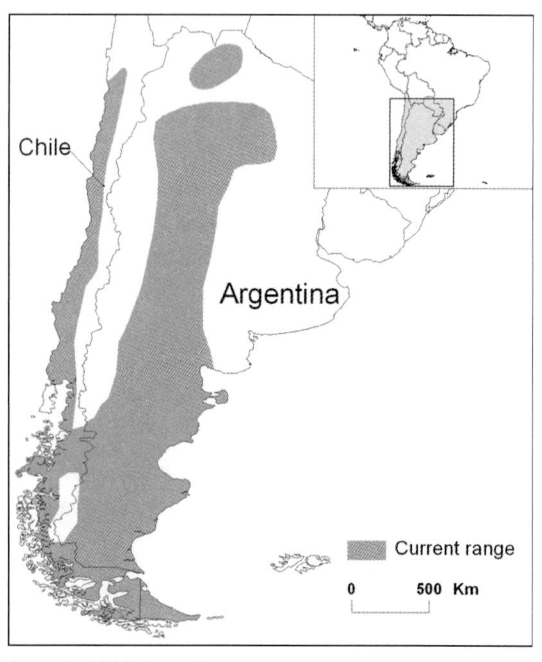

Chilla © 2003 Canid Specialist Group and Global Mammal Assessment

Rau *et al.* 1995; Campos and Ojeda 1996; González del Solar *et al.* 1997), a tendency to carnivory is apparent, and they have been considered a voracious predator of livestock, poultry, and game. This has led to their persecution, coupled with heavy hunting for pelts both in the past and present (Ojeda and Mares 1982; Iriarte and Jaksic 1986). Around 45% of the mortality documented by Johnson and Franklin (1994a) in Chile's Torres del Paine National Park resulted from either poaching or dog attacks.

	Male	Female
Weight	3.98 ± 0.09 (SE) kg, $n = 23$	3.34 ± 0.11 (SE) kg, $n = 21$
Head/body length	520 (501–540) mm, $n = 2$	566 (562–570) mm, $n = 2$

Ref: González del Solar and Rau (in press)

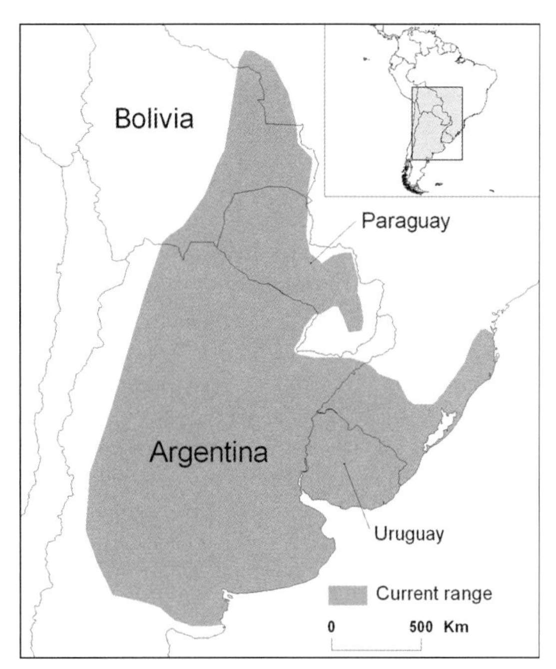

Pampas fox © 2003 Canid Specialist Group and Global Mammal Assessment

Pampas fox *Pseudalopex gymnocercus* (G. Fischer, 1814)

The Pampas fox is a medium-sized fox inhabiting the Southern Cone of South America, chiefly the Chaco, Argentine Monte, and Pampas ecoregions (Redford and Eisenberg 1992). It prefers open habitats and tall grass plains and sub-humid to dry habitats, but is also common in ridges, dry scrublands, and open woodlands. An adaptable carnivore, its diet shows great geographic variation and may include both wild and domestic vertebrates (Lucherini *et al.* in press). Pampas foxes are estimated to cause 2.9% of total lamb mortality in Uruguay (Cravino *et al.* 1997) and up to 6.9% in Argentina (Olachea *et al.* 1981). Predation on domestic stock has traditionally been one of the main reasons to justify their persecution by rural people, who have traditionally hunted foxes for their fur as an additional source of income. Hunting has been fuelled by State funded bounty systems, representing a real threat for the Pampas fox. Furthermore, much of the species' range has suffered massive habitat alteration (the Pampas grasslands have been largely obliterated by agriculture). Nevertheless, their adaptability has enabled them to remain common over most of their range, although there are little quantitative data on actual abundance. The highest reported density is in the Bolivian Chaco (1.8 individuals/km^2; Ayala and Noss 2000); in an Argentine Pampas area, Crespo (1971) found a density of 1.0 foxes/km^2, while Brooks (1992) estimated a density of 0.6 fox groups/km^2 for the Paraguayan Chaco, where fox abundance appeared to be correlated with annual rodent abundance. The taxonomic status of the Pampas fox and other related species is controversial (Massoia 1982; Zunino *et al.* 1995).

	Male	Female
Weight	4.6 kg, $n = 116$	4.2 kg, $n = 163$
Head/body length	648 (597–700) mm, $n = 10$	621 (535–683) mm, $n = 16$

Ref: Crespo (1971)

Sechuran fox *Pseudalopex sechurae* (Thomas, 1900)

At around 3.6 kg, the Sechuran desert fox is the smallest species of the genus *Pseudalopex*. Restricted to the coastal zones of northwestern Peru and southwestern Ecuador, between 3°S and 12°S, it occupies habitats including sandy deserts with low plant density, agricultural lands and dry forest (Cabrera 1931; Huey 1969; Langguth 1975). The small size and somewhat large ears may be an adaptation to desert life, as is their habit of nocturnal activity and denning during daylight hours. Their apparent ability to exist in areas with no standing water attests to their adaptation to arid habitats. A generalist, omnivorous species, Sechuran desert foxes often depend predominately on seeds or seed pods of species like *Prosopis juliflora* (algarrobo), *Capparis scabrida* (zapote), and *C. avicennifolia* (vichayo) (Huey 1969; Asa and Wallace 1990). The syrupy matrix surrounding the seeds may be the actual source of nourishment, and foxes may act as seed dispersers, improving the ability of seeds to germinate rapidly when sporadic rains occur (Asa and Cossíos *et al.* in press). Nevertheless, in Peru, rural inhabitants' attitudes towards the species are of persecution (68.3% of correspondents) or indifference (31.7%). Damage to domestic fowl and guinea pigs was cited by 65% of correspondents (D. Cossíos unpublished data). In Ecuador, habitat loss or reduction is considered the main threat (Tirira 2001).

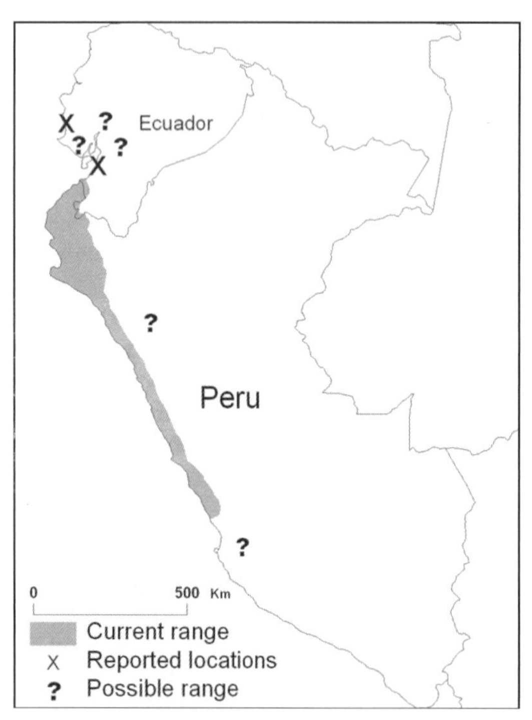

Sechuran fox © 2003 Canid Specialist Group and Global Mammal Assessment

	Male
Weight	3.6 (2.6–4.2) kg, $n = 4$
Head/body length	670 (500–780) mm, $n = 4$
Ref: Asa and Cossíos *et al.* (in press)	

Hoary fox *Pseudalopex vetulus* (Lund, 1842)

The hoary fox (Dalponte and Courtenay in press) is a slender, lightly built animal, weighing about 3.4 kg. The species is confined to Brazil, its core area of occurrence being the cerrado biome of the central Brazilian highlands (but see Costa and Courtenay, submitted) where it inhabits the grassland of open savannahs, but readily adapts to livestock pasture with rich insect sources. Although omnivorous, their diet appears predominantly insectivorous. Ground-dwelling harvester termites (*Synthermes spp.* and *Cornitermes spp.*), were recorded in 87% of faeces collected in six localities across its geographical range (Dalponte 1997; Silveira 1999; Juarez and Marinho-Filho 2002; Courtenay *et al.* submitted; J. Dalponte unpublished data). Nevertheless, they are killed indiscriminately as predators of domestic fowl, though they probably earn this reputation from crab-eating foxes which are formidable poultry raiders. Spot sightings in different habitats and localities revealed that groups were composed of single animals on 75% of occasions, followed by pairs (20%), and groups larger than two

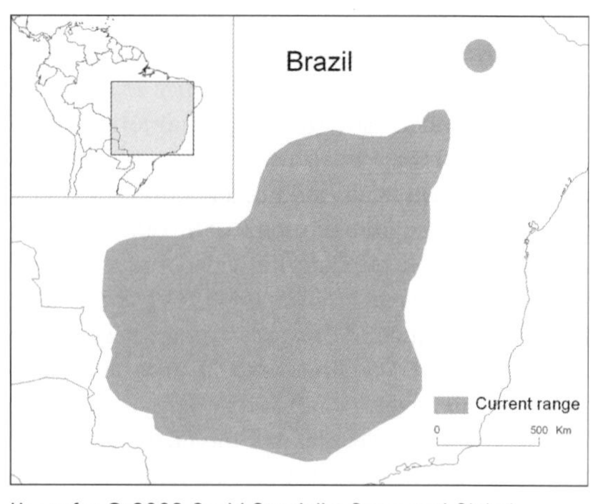

Hoary fox © 2003 Canid Specialist Group and Global Mammal Assessment

(4%) (J. Dalponte and E. Lima unpublished data). Both sexes care for the young (Courtney *et al.* submitted). Although data on abundance and population trends are lacking, their stronghold in the grasslands of central Brazil is threatened by habitat loss at a rate of 3% each year, largely in the interests of agriculture (Dalponte and Courtenay in press).

	Male	Female
Weight	3.3 (2.5–4.0) kg, n = 8	3.4 (3.0–3.6) kg, n = 3
Head/body length	587 (490–715) mm, n = 13	575 (510–660) mm, n = 6

Ref: Dalponte and Courtenay (in press)

Bush dog *Speothos venaticus* (Lund, 1842)

Considered by indigenous peoples to be one of the best hunters in the forest, the bush dog has a small, compact body (probably an adaptation to pursue burrowing prey and navigate through dense vegetation), short tail, and short legs with webbed feet, suggesting semi-aquatic habits (manifest by a penchant for diving in captivity). They are distinguished by several dental features, including a metaconule and hypocone on M_1, and a large, double-rooted M_2 (Berta 1987). Molecular analyses suggest bush dogs and maned wolves constitute a monophyletic group distinct from other South American canids (Wayne *et al.* 1997). Primarily carnivorous, bush dogs are most commonly observed hunting large rodents such as paca (*Agouti paca*) and agouti (Dasyprocta spp.) (53.1% and 28.1%, respectively, of reported sightings in central western Amazonia; Peres 1991). They are compulsively social (Kleiman 1972; Porton 1983; Macdonald 1996b), living and hunting in groups of 2–12 (Peres 1991). Bush dogs occur from extreme eastern Central America and northern South America to south through Paraguay and northeastern Argentina (Strahl *et al.* 1992; Aquino and Puertas 1997; Silveira *et al.* 1998). Isolated populations may also still occur in Ecuador and Colombia, west of the Andes. There is currently no information available regarding the species' density, and, despite its large distributional range and occurrence in a variety of habitats (i.e. cerrado and rainforest), has never been reported as abundant. Thus, it seems to be naturally rare throughout its range, independent of human disturbance. The only serious perceived threat is from habitat conversion and human encroachment.

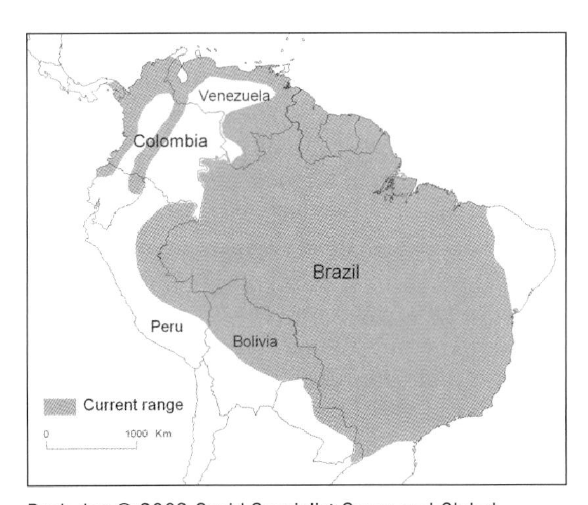

Bush dog © 2003 Canid Specialist Group and Global Mammal Assessment

Weight	(5–8) kg
Head/body length	630 (575–750) mm

Ref: Zuercher *et al.* (in press)

Gray fox *Urocyon cinereoargentatus* (Schreber, 1775)

A medium-sized fox, the gray fox is highly polytypic (up to 16 subspecies are recognized: Fritzell and Haroldson 1982), ranging from the southern edge of central and eastern Canada, and Oregon, Nevada, and Colorado in United States to northern Venezuela and Colombia, and from the Pacific coast of United States to the Atlantic and Caribbean oceans (Hall 1981; Fritzell and Haroldson 1982). It is widespread in forest, woodland, brushland, shrubland, and rocky habitats in temperate and tropical regions of North America, and in northernmost montane regions of South America (Harrison 1997). Although relatively common throughout their occupied range (reported densities range from 0.4/km^2 in California, Grinnell *et al.* 1937, to 1.5/km^2 in Florida, Lord 1961), gray foxes appear restricted to locally dense habitats where they are not excluded by sympatric coyotes and

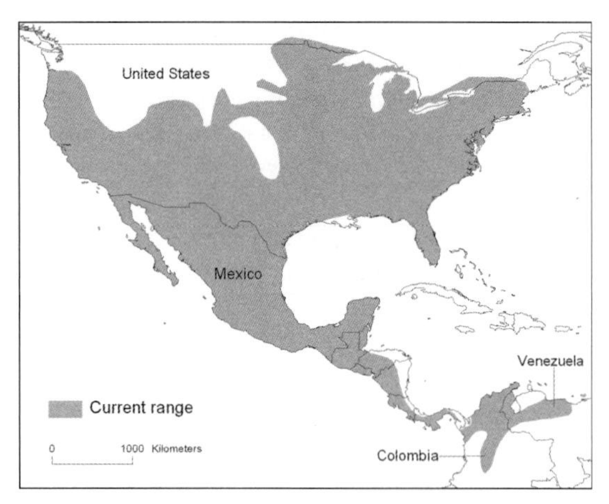

Gray fox © 2003 Canid Specialist Group and Global Mammal Assessment

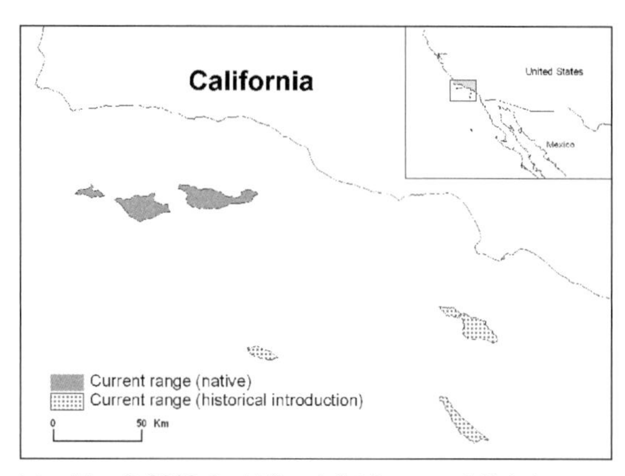

Island fox © 2003 Canid Specialist Group and Global Mammal Assessment

bobcats (*Lynx rufos*). Home range size, ranges from 0.8 (Yearsley and Samuel 1980) to 27.6 km^2 (Nicholson 1982). They are the most omnivorous of all North American fox species, and are notable tree climbers, able to climb branchless, vertical trunks to heights of 18 m, and to jump vertically from branch to branch (Feeney 1999). Monogamy with occasional polygyny is probably most typical in gray foxes (Trapp and Hallberg 1975), but few quantitative data are available. Trapping is legal throughout much of their range, and is likely to be the most important source of mortality where it occurs. In the United States, 90,604 skins were taken during the 1991/1992 season (Linscombe 1994). However, there is no evidence that regulated trapping has adversely affected gray fox population numbers, which appear stable throughout their range.

	Male	Female
Weight	4.0 (3.4-5.5) kg, $n = 18$	3.3 (2.0-3.9) kg, $n = 16$
Total length	981 (900-1100) mm, $n = 24$	924 (825-982) mm, $n = 20$

Ref: Grinnell *et al.* (1937)

Island fox *Urocyon littoralis* (Baird, 1858)

Island foxes are the smallest North American canid (1.8–2 kg, on average), representing a dwarf form of the mainland gray fox (this reduction in body size is likely a consequence of an insular existence). Island foxes are at least 30% smaller (Fritzell and Haroldson 1982), and typically have fewer caudal vertebrae (15–22; $n = 47$), than the gray fox (21–22; $n = 31$) (Moore and Collins 1995). Geographically restricted to the six largest of the eight California Channel Islands, each island population differs in genetic structure (and is considered a separate subspecies). They occur in all habitats on the islands including grassland, coastal sage scrub, maritime desert scrub, chaparral, oak-woodland, riparian, and dune, but exhibit substantial variability in abundance, both spatially and temporally. The home range size of the island fox is one of the smallest recorded for any canid. On Santa Cruz Island, fox home ranges varied between 0.15 and 0.87 km^2 (Crooks and Van Vuren 1996; Roemer *et al.* 2001c) depending on season and habitat type. Total island fox numbers have fallen within a decade from approximately 6000 individuals to less than 1500 in 2002 (Roemer 1999; Roemer *et al.* 2001a, b, 2002). Two populations in the southern Channel Islands have declined by an estimated 95% since 1994, and consist of 17 and <30 individuals (Chapter 9, this volume). Primary threats to the species include predation by golden eagles on the northern Channel Islands, and the possible introduction of canine diseases, especially canine distemper, to all populations. The small populations are especially vulnerable to any catastrophic mortality source.

	Male	Female
Weight	2.0 (1.4–2.5) kg, $n = 44$	1.8 (1.5–2.3) kg, $n = 50$
Head/body length	536 (470–585) mm, $n = 44$	528 (456–578) mm, $n = 50$

Ref: Roemer *et al.* (in press)

Indian fox *Vulpes bengalensis* (Shaw, 1800)

Morphologically, the Indian or Bengal fox (Johnsingh 1978b) is the most average vulpine fox. It is endemic to the Indian subcontinent and ranges from the foothills of the Himalayas in Nepal to the southern tip of the Indian peninsula. They avoid dense forests, steep terrain, tall grasslands, and true deserts, preferring semi-arid, flat to undulating terrain, scrub, and grassland habitats where it is easy to hunt and dig dens. Indian foxes are omnivorous, opportunistic feeders and generally consume any food that they can handle. The Indian fox features in several tales from the ancient Jataka texts and the Panchatantra where it is depicted as a clever and sometimes cunning creature. Despite these attributes and their widespread distribution, Indian foxes are nowhere abundant, and occur at low densities (0.04–0.06/km^2 to 1.62/km^2) throughout their range, with populations undergoing major fluctuations due to prey availability (Manakadan and Rahmani 2000). They are also quite sensitive to human modifications of habitat, and in some areas, such as Tamil Nadu, anthropogenic mortality is high with humans (often using dogs) killing foxes for their flesh, teeth, claws, and skin.

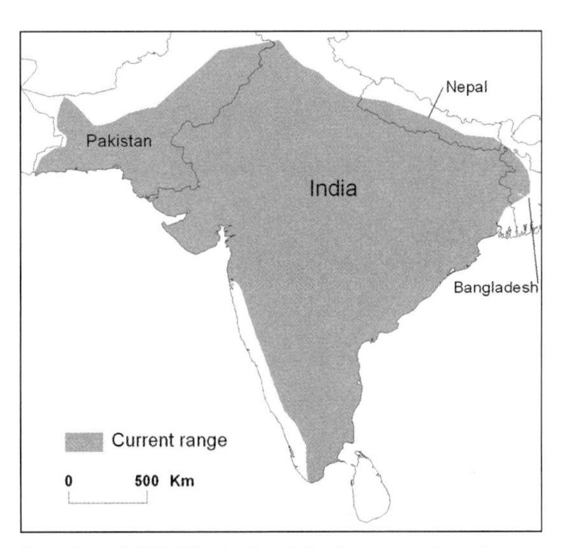

Bengal fox © 2003 Canid Specialist Group and Global Mammal Assessment

	Male	Female
Weight	2.7–3.2 kg	>1.8 kg
Head/body length	500 (390–575) mm, $n = 6$	472 (460–480) mm, $n = 3$

Ref: Johnsingh and Jhala (in press)

Blanford's fox *Vulpes cana* (Blanford, 1877)

At 1 kg, similar in body mass to the closely related fennec fox (Geffen *et al.* 1992a), Blanford's fox has an exceptionally long (323 mm), bushy tail, and curved, sharp, semi-retractile claws which enhance traction (Harrison and Bates 1991; Geffen *et al.* 1992b; Geffen 1994). They have been observed ascending vertical, crumbling cliffs by a series of jumps up the vertical sections (Mendelssohn *et al.* 1987). Present in arid mountainous regions of the Middle East eastwards to Afghanistan and recently recorded in Egypt (Harrison and Bates 1989;

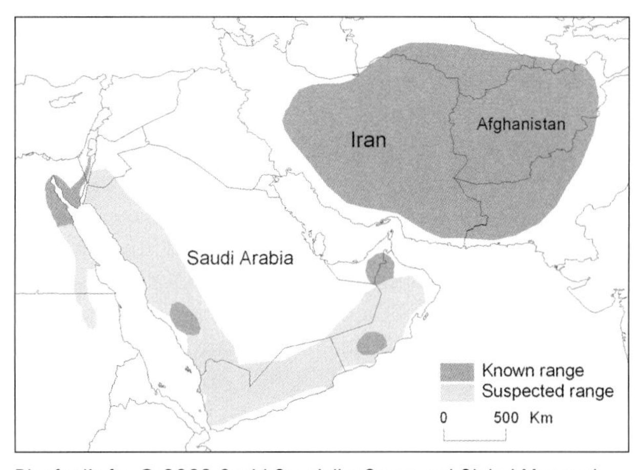

Blanford's fox © 2003 Canid Specialist Group and Global Mammal Assessment

Geffen *et al.* 1993; Peters and Rödel 1994). Although only discovered in Israel in 1981, they are fairly common, densities of 2.0/km^2 in Ein Gedi and 0.5/km^2 in Eilat have been recorded but abundance elsewhere is unknown. Strictly monogamous pairs use territories of *c.*1.6 km^2 that overlap minimally (Geffen *et al.* 1992c). At Ein Gedi, average distance travelled per night was 9.3 km, and nightly home range averaged 1.1 km^2 with little seasonal variation (Geffen and Macdonald 1992, 1993). Primarily insectivorous and frugivorous (Geffen *et al.* 1992b), they appear able to maintain water and energy balance on their diet alone. Daily energy expenditure near the Dead Sea was 0.63–0.65 kJ/g/day, with no significant seasonal difference (Geffen *et al.* 1992d). Although habitat loss is of limited concern for Israeli populations, human development in some areas, such as along the Dead Sea coasts, may threaten their survival.

	Male	Female
Weight	1.0 (0.8-1.3) kg, $n = 19$	1.0 (0.8-1.3) kg, $n = 5$
Head/body length	427 (385-470) mm, $n = 19$	411 (385-450) mm, $n = 17$

Ref: Geffen *et al.* (1992d)

Cape fox *Vulpes chama* (A. Smith, 1833)

At 3.6 kg, the Cape fox is the smallest canid and only true fox occurring in southern Africa. Widespread in the central and western regions of southern Africa, it is absent only in extreme southwestern Angola. It occupies mainly arid and semi-arid areas but in parts, such as the fynbos biome of South Africa, the species enters areas receiving higher precipitation and with denser vegetation. This fox has expanded its range over recent decades to the southwest where it reaches the Atlantic and Indian Ocean coastlines (Stuart 1981). Cape foxes mainly associate with open country, including grassland, grassland with scattered thickets, and lightly wooded areas. Small rodents are an important food, but hares, reptiles, birds, invertebrates, and some wild fruits are also taken (Bothma 1966; Lynch 1975; Stuart 1981).

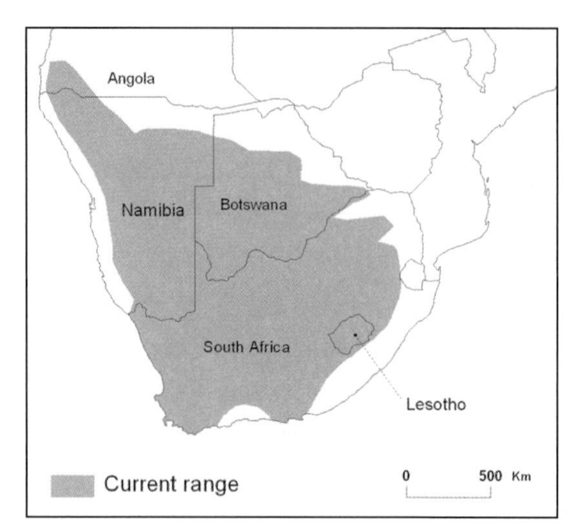

Cape fox © 2003 Canid Specialist Group and Global Mammal Assessment

Home ranges vary in size from 1.0 to 4.6 km^2 (Bester 1982). They are generally common to fairly abundant across most of their range, although problem animal control activities (aimed at black-backed jackals and caracals (*Felis caracal*)) have resulted in population reductions in some areas. Annual offtake resulting from problem animal control programmes averaged roughly 16% up to 1985, with no obvious declines in overall populations (Bester 1982). Populations are currently stable across their entire range, although the illegal, but widespread and indiscriminate, use of agricultural poisons on commercial farms poses a significant threat (Fig. 1.4).

	Male	Female
Weight	2.8 (2.0-4.2) kg, $n = 17$	2.5 (2.0-4.0) kg, $n = 11$
Head/body length	554 (540-610) mm, $n = 21$	553 (510-620) mm, $n = 15$

Ref: Stuart (1981)

Corsac fox *Vulpes corsac* (Linnaeus, 1768)

The corsac is a typical fox-like canid inhabiting the dry steppes, semi-deserts, and deserts from the lower

Figure 1.4 Cape fox *Vulpes chama* in the Kalahari © D.W. Macdonald.

Volga River and Iran to Mongolia, Manchuria, and Tibet (Ovsyanikov and Poyarkov in press). The species area consists of two parts—western and eastern—connected by a relatively narrow neck in Dgungar Gate and Zaysan Basin region. In recent years westward area expansion has been recorded, particularly into Voronezh region following active recovery of baibak (*Marmota bobac*) populations. Nevertheless, the modern distribution area is smaller than the historical range. One limiting factor is snow height in winter, as the species avoids areas where snow height exceeds 150 mm. Their presence appears to depend on distribution of ground squirrels and marmots whose dens they actively use as shelters (enlarging them) while hunting upon their owners. Home ranges vary from 1 to 40 km^2 and in some areas, as many as nine breeding dens per 15 km^2 have been recorded (reviewed by Ovsyanikov and Poyarkov in press). Corsac populations are highly variable (from <1 to 29 per 10 km^2, Blyznuk 1979 in Ovsyanikov and Poyarkov in press) and fluctuate significantly (Sidorov and Botvinkin 1987). In the twentieth century several catastrophic population declines were recorded, during which hunting on corsacs in the former Soviet Union was completely banned (Sidorov and Botvinkin 1987). In Turkmenistan, from 1924 to 1989, 103,500 corsac pelts were taken, which caused a significant decline during the same period (Ovsyanikov and Poyarkov in press). Current population status in many regions is not known, nor are major threats.

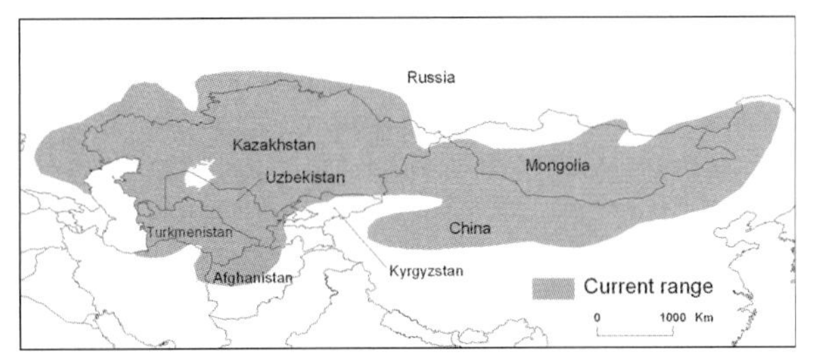

Corsac fox © 2003 Canid Specialist Group and Global Mammal Assessment

	Male	Female
Weight	2.7 (2.5–3.2) kg, $n = 22$	2.1 (1.9–2.4) kg, $n = 10$
Body length	500 (450–560) mm, $n = 22$	490 (450–500) mm, $n = 10$

Ref: Kyderbaev and Sludskyi (1981) in Ovsyanikov and Poyarkov (in press)

Tibetan fox *Vulpes ferrilata* (Hodgson, 1842)

The Tibetan or sand fox is small (3–4 kg) and seemingly compact with a soft, dense coat, a conspicuously narrow muzzle, and a bushy tail. They are widespread in the steppes and semi-deserts of the

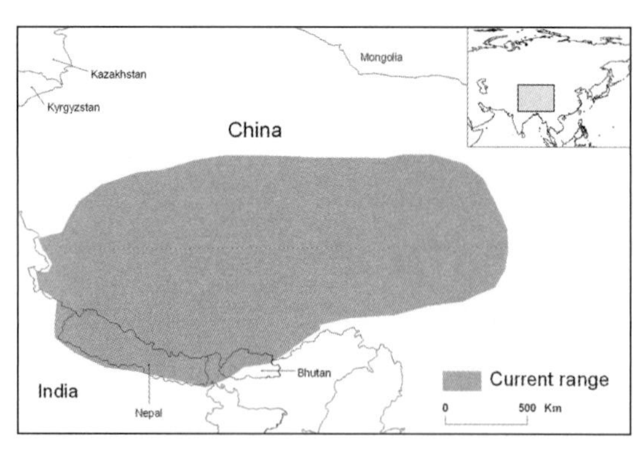

Tibetan fox © 2003 Canid Specialist Group and Global Mammal Assessment

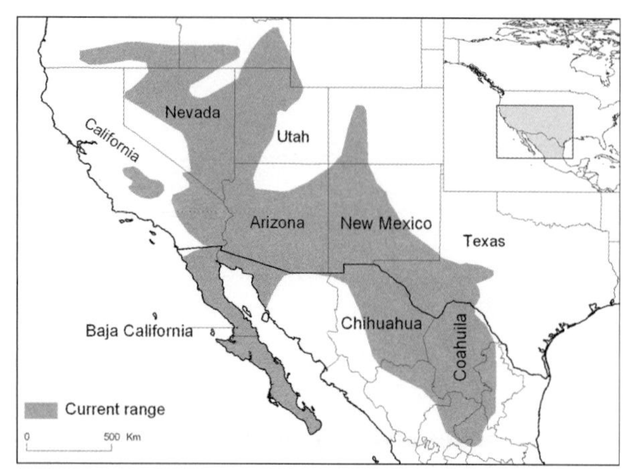

Kit fox © 2003 Canid Specialist Group and Global Mammal Assessment

Tibetan Plateau from the Ladakh area of India to east across China including parts of the Xinjiang, Gansu, Qinghai, and Sichuan provinces and all of the Tibet Autonomous Region. They are also present in Nepal, north of the Himalayas. The species is found in upland plains and hills from about 2500 to 5200 m in habitat consisting primarily of alpine meadow, alpine steppe, and desert steppe, where it feeds primarily on pikas—*Ochotona spp.*—and small rodents (Schaller 1998). The climate is harsh with temperatures reaching 30 °C in summer and dropping to −40 °C in winter. Their abundance depends partly on prey availability and partly on human hunting pressure. Nothing is known of overall status and trends (and very little on their biology), but the species is not considered threatened (Schaller and Ginsberg in press).

	Male	Female
Weight	4.1 (3.8–4.6) kg, $n = 7$	3.5 (3.0–4.1) kg, $n = 5$
Head/body length	587 (560–650) mm, $n = 7$	554 (490–610) mm, $n = 5$

Ref: Schaller and Ginsberg (in press)

Kit fox *Vulpes macrotis* (Merriam, 1888)

The kit fox is one of the smallest foxes on the American continent, and there are some eight recognized subspecies (McGrew 1979), inhabiting arid and semi-arid regions of western North America (encompassing desert scrub, chaparral, halophytic, and grassland communities, McGrew 1979; O'Farrell 1987). They will also use agricultural lands, particularly orchards, on a limited basis, and can inhabit urban environments. Kit foxes are well adapted to life in a warm, arid environment, and can obtain all necessary water from their food, although to do so must consume approximately 150% of daily energy requirements (Golightly and Ohmart 1984). Predation, mainly by coyotes, usually is the main source of mortality for kit foxes (commonly accounting for over 75% of deaths, Spiegel 1996; Cypher and Spencer 1998) although vehicles are a prime mortality factor in some areas. The main threat to the long-term survival of the kit fox is habitat conversion, mainly to agriculture but also to urban and industrial development. Considered common to rare, density fluctuates (e.g. 0.2–1.7/km^2 over 15 years on Californian study site Cypher *et al.* 2000) with annual environmental conditions, which are dependent upon precipitation. Overall, populations of the kit fox in Mexico are declining, while those in the United States are primarily stable with the exception of the San Joaquin kit fox, *V. m. mutica*, which is declining.

	Male	Female
Weight	2.3 (1.7-2.7) kg, $n = 8$	1.9 (1.6-2.2) kg, $n = 6$
Head/body length	537 (485-520) mm, $n = 7$	501 (455-535) mm, $n = 5$

Ref: List and Jimenez Guzmán (in press)

Pallid fox *Vulpes pallida* (Cretzschmar, 1827)

The pallid fox is one of the least-known canid species (Kingdon 1997). It is a small canid (Dorst and Dandelot 1970; Rosevear 1974; Happold 1987), distributed across the semi-arid Sahelian region of Africa bordering the Sahara, from Mauritania and Senegal through Nigeria and Cameroon to the Red Sea. They typically inhabit very dry sandy and stony sub-Saharan desert and semi-desert areas, but extend to some extent southwards into moister Guinean savannas. They therefore have a very extensive distribution within an unstable and fluctuating ecological band lying between true desert and the Guinean savannas. Although widespread, they are in most parts rare, and specific threats to their survival have not been established. They dig extensive burrows, 2–3 m deep and up to 15 m in length inhabited by several animals (Coetzee 1977).

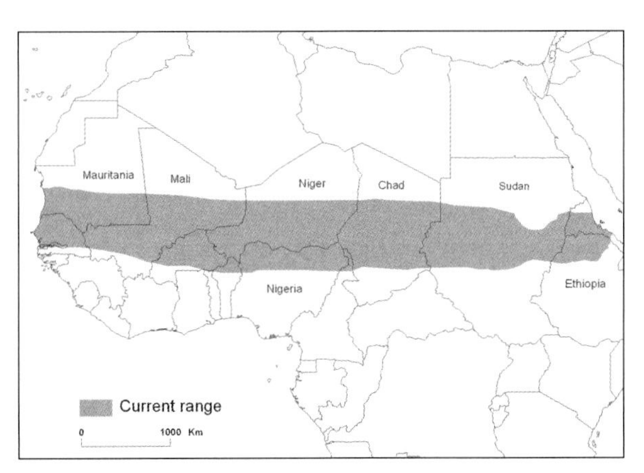

Pale fox © 2003 Canid Specialist Group and Global Mammal Assessment

Weight	2.0-3.6 kg
Head/body length	380-550 mm

Ref: Sillero-Zubiri (in press)

Rüppell's fox *Vulpes rueppellii* (Schinz, 1825)

As light as 1.5 kg, Rüppell's foxes are widespread in the arid biotopes of the desert and semi-desert regions of North Africa (north of 17°N) the northern limit of which is the northern fringes of the Sahara Desert. They are also present in arid regions across the Arabian Peninsula eastwards to Pakistan and northwest to Israel and Jordan. Their typical habitat includes sand and stone deserts, and they, like fennecs, are able to survive in areas without any available water. Generalist predators (Kowalski 1988) that hunt solitarily (Olfermann 1996), they are mainly crepuscular/nocturnal. They may be gregarious, having been sighted in groups of 3–15, and territories cover 10 km^2 (Lenain 2000) to 69 km^2, (Lindsay and Macdonald 1986). In a large, fenced, protected area of 2244 km^2 in Saudi Arabia, densities were 0.68/km^2 (Lenain 2000). There is no information on population size and trends, and density is usually low, but habitat loss, fragmentation and degradation, direct and indirect persecution by hunting, and indiscriminate use of poisons are the main threats to the species. In Israel, they are on the verge of extinction due to competitive exclusion by red foxes that are expanding their range following human settlement in the Negev Desert.

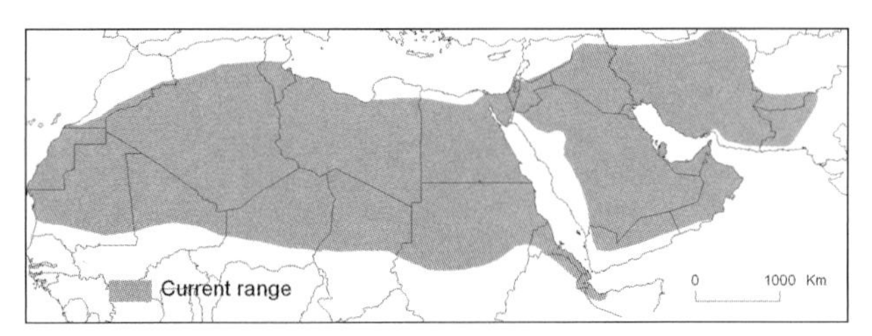

Rueppell's fox © 2003 Canid Specialist Group and Global Mammal Assessment

	Male	Female
Weight	1.6 (1.1-2.3) kg, $n = 179$	1.5 (1.1-1.8) kg, $n = 93$
Head/body length	462 (400-520) mm, $n = 35$	435 (345-487) mm, $n = 15$
Ref: Lenain (2000)		

Swift fox *Vulpes velox* (Say, 1823)

Swift foxes are native to short-grass and mixed-grass prairies of the Great Plains in North America (Egoscue 1979), though will den and forage in fallow cropland fields such as wheat (Jackson and Choate 2000; Sovada *et al.* 2003). They are distinguishable from other North American canids, except the kit fox, by black patches on each side of the muzzle, a black tail tip, and small body size (averaging just over 2 kg). The species is phenotypically and ecologically similar to the kit fox and interbreeding occurs between them in a small hybrid zone (<100 km) in New Mexico. The swift fox was common or abundant in much of its original range until the late 1800s to the early 1900s. Following extirpation from Canada by 1938, releases totalling 942 foxes between 1983 and 1997 have re-established a small population in Alberta, Saskatchewan, and Montana that now constitutes the northern extent of the species' range (Moehrenschlager and Moehrenschlager 2001). Current estimates for United States suggest that swift foxes are located in 39–42% of their historic range (Sovada and Scheick 1999). Average home range was 25.1 km^2 in western Kansas (Sovada *et al.* 2003) and 10.4 km^2 in Montana (Zimmerman *et al.* 2003). Swift foxes are opportunistic foragers, feeding on a variety of mammals, but also birds, insects, plants, and carrion (Kilgore 1969; Hines and Case 1991; Sovada *et al.* 2001). Predation by, and interspecific competition with, coyotes (Kitchen *et al.* 1999), and expansion of red fox populations, probably represent the two most serious limiting factors to swift fox recolonization of suitable habitat. Reported annual mortality rates range from 0.47 to 0.63 (Covell 1992; Sovada *et al.* 1998; Moehrenschlager 2000; Schauster *et al.* 2002; Andersen *et al.* 2003), and those of translocated foxes were similar to those of wild residents in Canada (Moehrenschlager and Macdonald 2003).

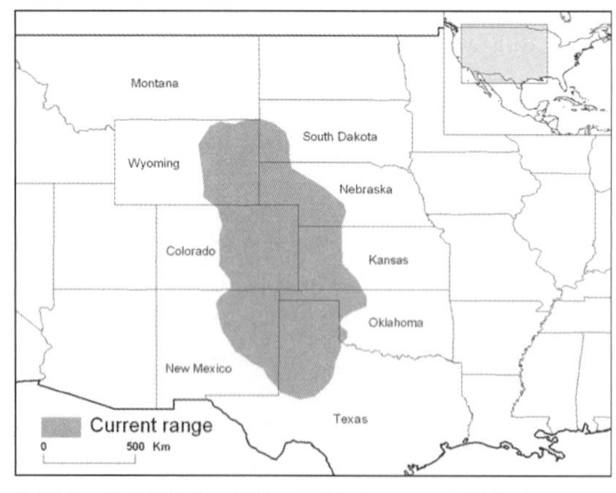

Swift fox © 2003 Canid Specialist Group and Global Mammal Assessment

	Male	Female
Weight	2.2 (2.0-2.5) kg, $n = 18$	2.0 (1.6-2.3) kg, $n = 9$
Head/body length	523 (500-545) mm, $n = 11$	503 (475-540) mm, $n = 10$
Ref: Harrison (in press)		

Red fox *Vulpes vulpes* (Linnaeus, 1758)

The red fox is the largest species in the genus *Vulpes* and has the widest geographical range of any member of the order Carnivora, nearly 70 million km^2 (Lloyd 1980; Macdonald 1987). Distributed across the entire Northern Hemisphere (its range in the United States having been extended through British imports) from the Arctic Circle to North Africa, Central America, and the Asiatic steppes (and introduced to Australia in 1800s), red foxes are recorded in habitats as diverse as tundra, desert, and forest, as well as in city centres (including London, Paris, and Stockholm). With this geographical variation comes immense variation in adult body size (head and body lengths range from 455 to 900 mm, and weights from 3 to 14 kg, Nowak 1999),

Figure 1.5 Fennec foxes *Vulpes zerda* © C. Dresner

home range (0.4 km^2 for urban foxes in Oxford, >30 km^2 in the Arctic—Voigt and Macdonald 1984), and density. Adaptable and opportunistic foragers (Macdonald 1976, 1977a), they have a long association with man and have been hunted at least since the fourth century BC. Alongside rabies (Voigt *et al.* 1985), people (roads, culling) are typically the major cause of fox mortality. In the United Kingdom, hunting on foot or horseback with packs of hounds probably kills 21,500–25,000 foxes annually, about 4% of total mortality (Macdonald *et al.* 2000c). In 1992–93, the red fox was the third most commercially important wild-caught furbearer in North America (Sheiff and Baker 1987). Nevertheless, populations are resilient and red foxes are generally common.

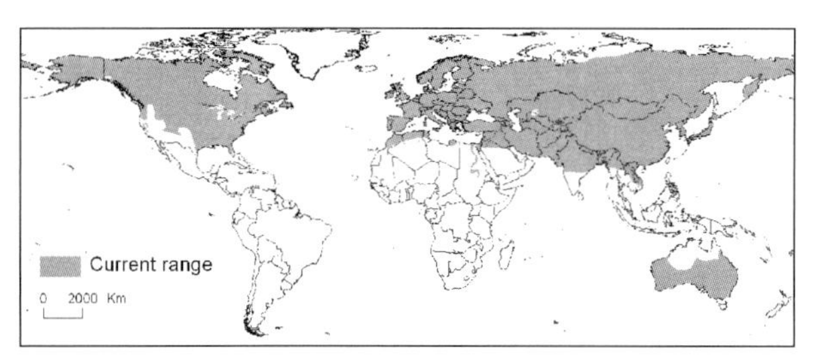

Red fox © 2003 Canid Specialist Group and Global Mammal Assessment

	Male	Female
Weight	6.3 (4.4–7.6) kg, $n = 20$	5.3 (3.6–6.5) kg, $n = 20$
Head/body length	660 (590–720) mm, $n = 11$	630 (550–680) mm, $n = 11$
Ref: Cavallini (1995)		

Fennec *Vulpes zerda* (Zimmermann, 1780)

With specimens as light as 0.8 kg, the fennec (sometimes referred to as the fennec fox) is the smallest canid species and is characterized by weak dentition, a rounded skull, and the largest ear-to-body ratio in the family—a possible adaptation to aid heat dissipation, and locating insects and small vertebrates (Fig. 1.5). Fennecs are widespread in the sandy deserts and semi-deserts of northern Africa to northern Sinai (Saleh and Basuony 1998); annual rainfall is <100 mm per year in the northern fringes of their distribution. They are nocturnal and this, coupled with their use of burrows during the day and the moisture content of their prey, probably contributes to their being the only carnivore of the Sahara living completely away from water sources (Noll-Banholzer 1979). Fennecs' physiology is adapted to high temperatures, and they only start to pant

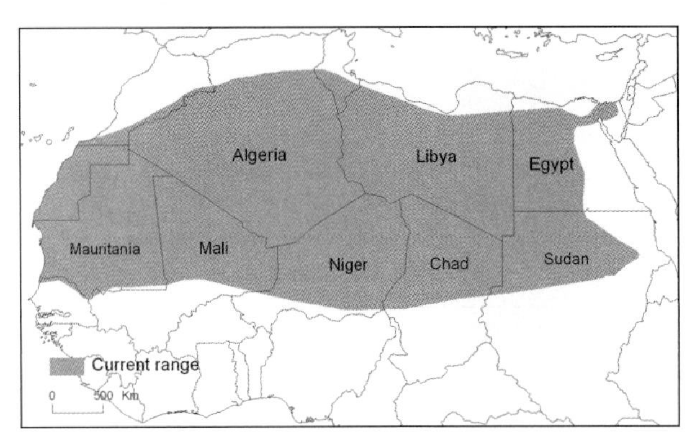

Fennec fox © 2003 Canid Specialist Group and Global Mammal Assessment

when the temperature exceeds 35 °C, but when they do they may reach as many as 690 breaths per minute (Macdonald 1992, p. 87). Following an exceptionally long copulatory tie (up to 2 h 45 min, Valdespino 2000; Valdespino *et al.* 2002), litters of 1–2 are born between March and April (Gauthier-Pilters 1967) or, in captivity, July (Petter 1957; Gauthier-Pilters 1962; Gangloff 1972; Bauman 2002). Dens may be huge, covering up to 120 m^2, with as many as 15 different entrances (Dragesco-Joffé 1993). Despite the primary threat from trapping for commercial use (furs, pet trade), they are thought to be relatively common throughout their range.

	Male	Female
Weight	1.5 (1.3–1.7) kg, $n = 2$	1.4 (1.0–1.9) kg, $n = 5$
Head/body length	392 (390–395) mm, $n = 2$	382 (345–395) mm, $n = 5$

Ref: Asa *et al.* (in press)

Studies and reconstructions of dire wolf (*Canis dirus*) and grey wolf (*Canis lupus*) from late Pleistocene Rancholebrea Tarpits, Los Angeles, California. Illustration by Pat Ortega.

CHAPTER 2

Ancestry

Evolutionary history, molecular systematics, and evolutionary ecology of Canidae

Xiaoming Wang, Richard H. Tedford, Blaire Van Valkenburgh, and Robert K. Wayne

The evolutionary history of canids (Family Canidae) is a history of successive radiations repeatedly occupying a broad spectrum of niches ranging from large, pursuit predators to small omnivores, or even to herbivory. Three such radiations were first recognized by Tedford (1978), each represented by a distinct subfamily (Fig. 2.1). Two archaic subfamilies, Hesperocyoninae and Borophaginae, thrived in the middle to late Cenozoic from about 40 to 2 million years ago (Ma) (Wang 1994; Wang *et al.* 1999). All living canids belong to the final radiation, Subfamily Caninae, which achieved their present diversity only in the last few million years (Tedford *et al.* 1995).

Canids originated more than 40 Ma in the late Eocene of North America from a group of archaic carnivorans, the Miacidae (Wang and Tedford 1994, 1996). They were confined to the North American continent during much of their early history, playing a wide range of predatory roles that encompass those of the living canids, procyonids, hyaenids, and possibly felids. By the latest Miocene (about 7–8 Ma), members of the Subfamily Caninae were finally able to cross the Bering Strait to reach Europe (Crusafont-Pairó 1950), commencing an explosive radiation and giving rise to the modern canids of the Old World. At the formation of the Isthmus of Panama, 3 Ma, canids arrived in South America and quickly established themselves as one of the most diverse groups of carnivorans on the continent (Berta 1987, 1988). With the aid of humans, *Canis lupus dingo* was transported to Australia late in the Holocene. Since that time, canids have become truly worldwide predators, unsurpassed in distribution by any other group of carnivorans.

Here, in the context of this volume on the modern canids, we place more emphasis on the subfamily Caninae, the latest of the three successive radiations of the Canidae. We do not attempt to cite all of the references in canid palaeontology and systematics, most of which have been summarized in the papers that we cite at the end of each section. Certain phylogenetic relationships of the Caninae are controversial, as reflected in the different conclusions reached on the basis of evidence from palaeontological/morphological or molecular research as presented below.

What is a canid?

Canids possess a pair of carnassial teeth (the upper fourth premolar and lower first molar) in the form of a shearing device, and thus belong to the Order Carnivora. Within the Carnivora, canids fall into the Suborder Caniformia, or dog-like forms. The Caniformia are divided into two major groups that have a sister relationship: Superfamily Cynoidea, which includes Canidae, and Superfamily Arctoidea, which include the Ursidae, Ailuridae, Procyonidae, and Mustelidae, as well as the aquatic Pinnipedia and the extinct Amphicyonidae.

As a cohesive group of carnivorans, living canids are easily distinguished from other carnivoran families. Morphologically there is little difficulty in recognizing living canids with their relatively uniform and unspecialized dentitions. However, the canids as

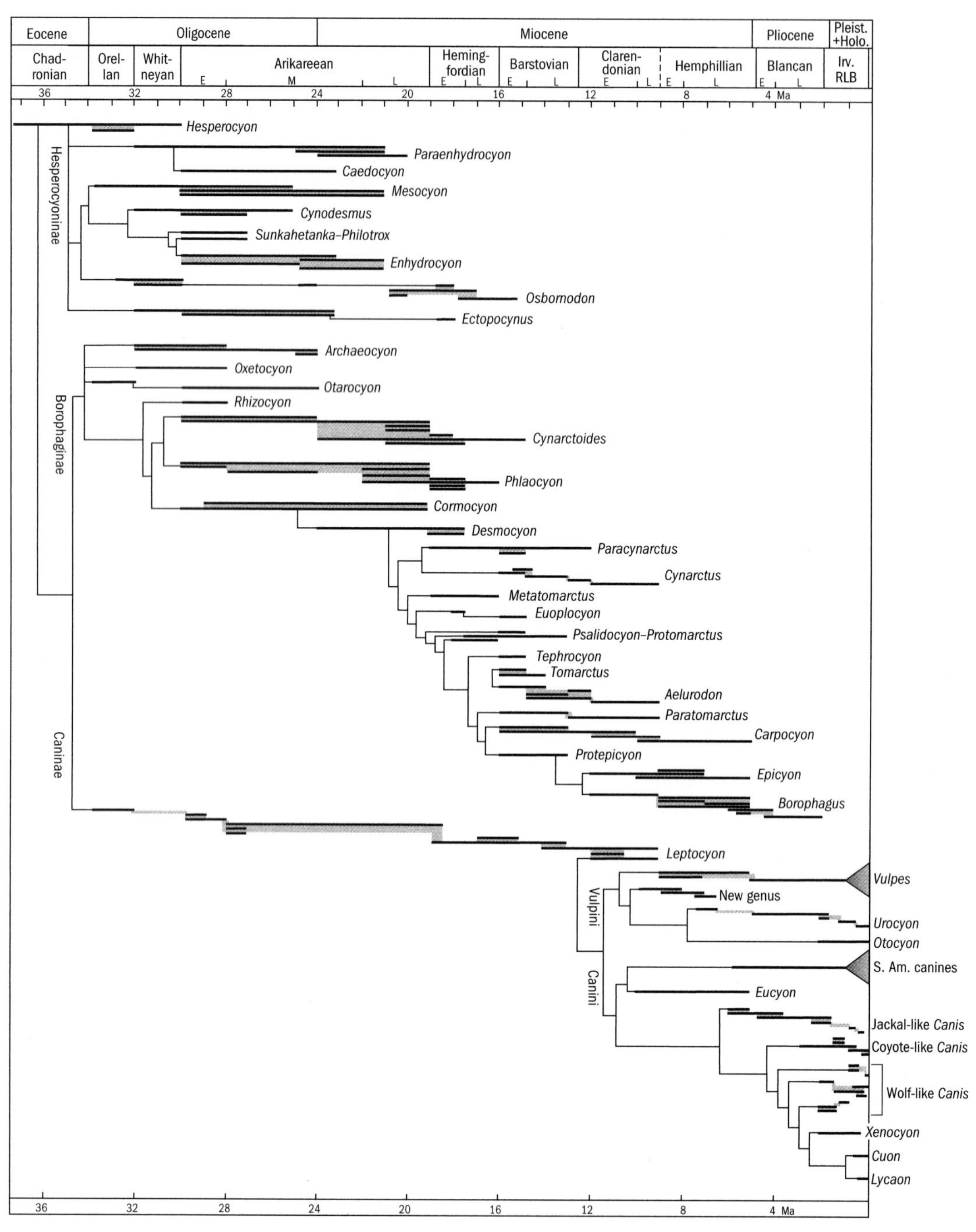

Figure 2.1 Simplified phylogenetic relationships of canids at the generic level. Species ranges are indicated by individual bars enclosed within grey rectangles, detailed relationships among species in a genus is not shown. Relationships for the Hesperocyoninae is modified from Wang (1994, fig. 65), that for the Borophaginae from Wang *et al.* (1999, fig. 141), and that for the Caninae from unpublished data by Tedford *et al.*

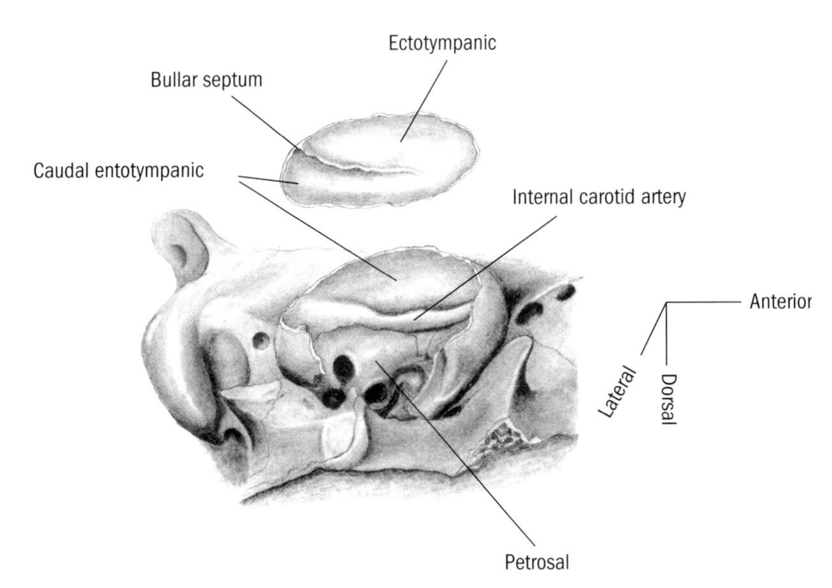

Figure 2.2 Ventrolateral view of basicranial morphology of a primitive canid, *Hesperocyon gregarius*, showing bullar composition and position of the internal carotid artery (see text for explanations). The ventral floor of the bulla is dissected away (isolated oval piece on top) to reveal the middle ear structures inside the bulla and the internal septum. Modified from Wang and Tedford (1994, fig. 1).

exemplified by the living forms are narrowly defined. Only a small fraction of a once diverse group has survived to the present day (Fig. 2.1). Canids in the past had departed from this conservative pattern sufficiently that paleontologists had misjudged some canids as procyonids. Similarly the extinct beardog Family Amphicyonidae, which belongs to the Arctoidea, is often placed within the Canidae, because of its unspecialized dentition.

How do we know a canid when we see one? A key region of the anatomy used to define canids is the middle ear region, an area that distinguishes most families of carnivorans (Hunt 1974), perhaps as a result of a widespread trend of ossifying bullar elements in independent lineages. Canids are characterized by an inflated entotympanic bulla that is divided by a partial septum along the entotympanic and ectotympanic suture (Fig. 2.2). Other features characteristic of canids are the loss of a stapedial artery and the medial position of the internal carotid artery that is situated between the entotympanic and petrosal for most of its course and contained within the rostral entotympanic anteriorly (Wang and Tedford 1994). These basicranial characteristics have remained more or less stable throughout the history of canids, allowing easy identification in the fossil record when these structures are preserved.

Evolutionary history

Among the living families within the Order Carnivora, the Canidae are the most ancient. The family arose in the late Eocene, when no other living families of carnivorans had yet emerged (two archaic families, Miacidae and Viverravidae, have a much older history but none survive to the present time). Furthermore, canids still maintain some features that are primitive among all carnivorans, to the extent that dog skulls are often used to illustrate a generalized mammal in zoological classrooms. Dentally, canids are closest to the ancestral morphotype of Carnivora. Canids have a relatively unreduced dental formula of 3142/3143 [numbers in sequence represent incisors, canines, premolars, and molars in the upper (left half before the oblique) and the lower (right half after the oblique) teeth] and relatively unmodified molars except for the morphology of the carnassials (P4, m1) typical of all carnivorans. In contrast, all other carnivoran families generally have a more reduced dental formula and highly modified cusp patterns.

From this mesocarnivorous (moderately carnivorous) conservative plan, canids generally evolve towards a hypercarnivorous (highly carnivorous) or hypocarnivorous (slightly carnivorous) dental pattern. In the hypercarnivorous pattern (Fig. 2.4(b,d))

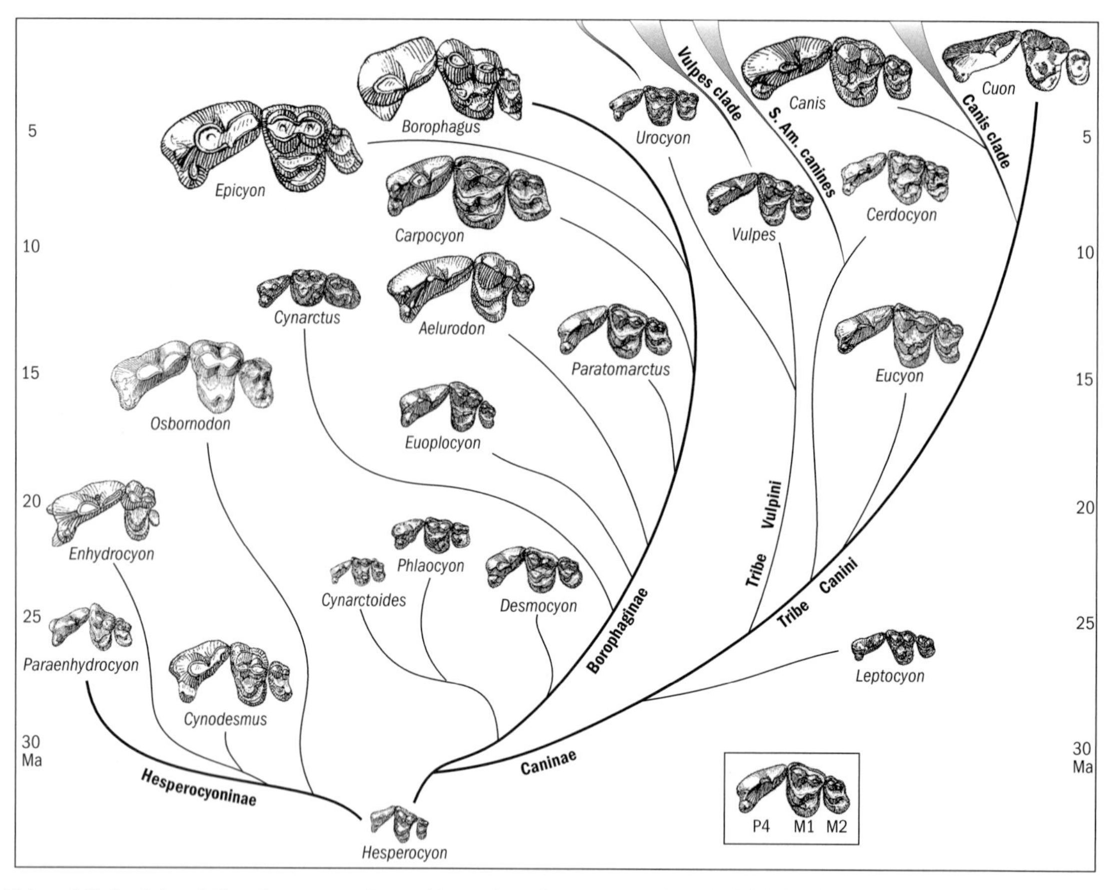

Figure 2.3 Dental evolution of representative canids as shown in upper cheek teeth (P4–M2). Generally the most derived species in each genus is chosen to enhance a sense of dental diversity. Species in the Hesperocyoninae are: *Hesperocyon gregarius*; *Paraenhydrocyon josephi*; *Cynodesmus martini*; *Enhydrocyon crassidens*; and *Osbornodon fricki*. Species in the Borophaginae are: *Cynarctoides acridens*; *Phlaocyon marslandensis*; *Desmocyon thomsoni*; *Cynarctus crucidens*; *Euoplocyon brachygnathus*; *Aelurodon stirtoni*; *Paratomarctus temerarius*; *Carpocyon webbi*; *Epicyon haydeni*; and *Borophagus diversidens*. Species in the Caninae are: *Leptocyon gregorii*; *Vulpes stenognathus*; *Urocyon minicephalus*; *Cerdocyon thous*; *Eucyon davisi*; *Canis dirus*; and *Cuon alpinus*. All teeth are scaled to be proportional to their sizes.

there is a general tendency to increase the size of the carnassial pair at the expense of the molars behind (see *Enhydrocyon, Aelurodon, Borophagus,* and *Cuon* in Fig. 2.3). This modification increases the efficiency of carnassial shear. A hypocarnivorous pattern (Fig. 2.4(a,c)) is the opposite, with development of the grinding part of the dentition (molars) at the expense of carnassial shear (see *Cynarctoides, Phlaocyon,* and *Cynarctus* in Fig. 2.3). This configuration was only possible in the sister-taxa Borophaginae and Caninae, which share a bicuspid m1 talonid (Fig. 2.4(c)). One of the major trends in canid evolution is the repeated development of hyper- and hypocarnivorous forms (see below).

Hesperocyoninae

The Subfamily Hesperocyoninae is the first major clade with a total of 28 species. Its earliest members are species of the small fox-like form, *Hesperocyon,* that first appears in the late Eocene (Duchesnean, 37–40 Ma) (Bryant 1992) and became abundant in the latest Eocene (Chadronian). By Oligocene time

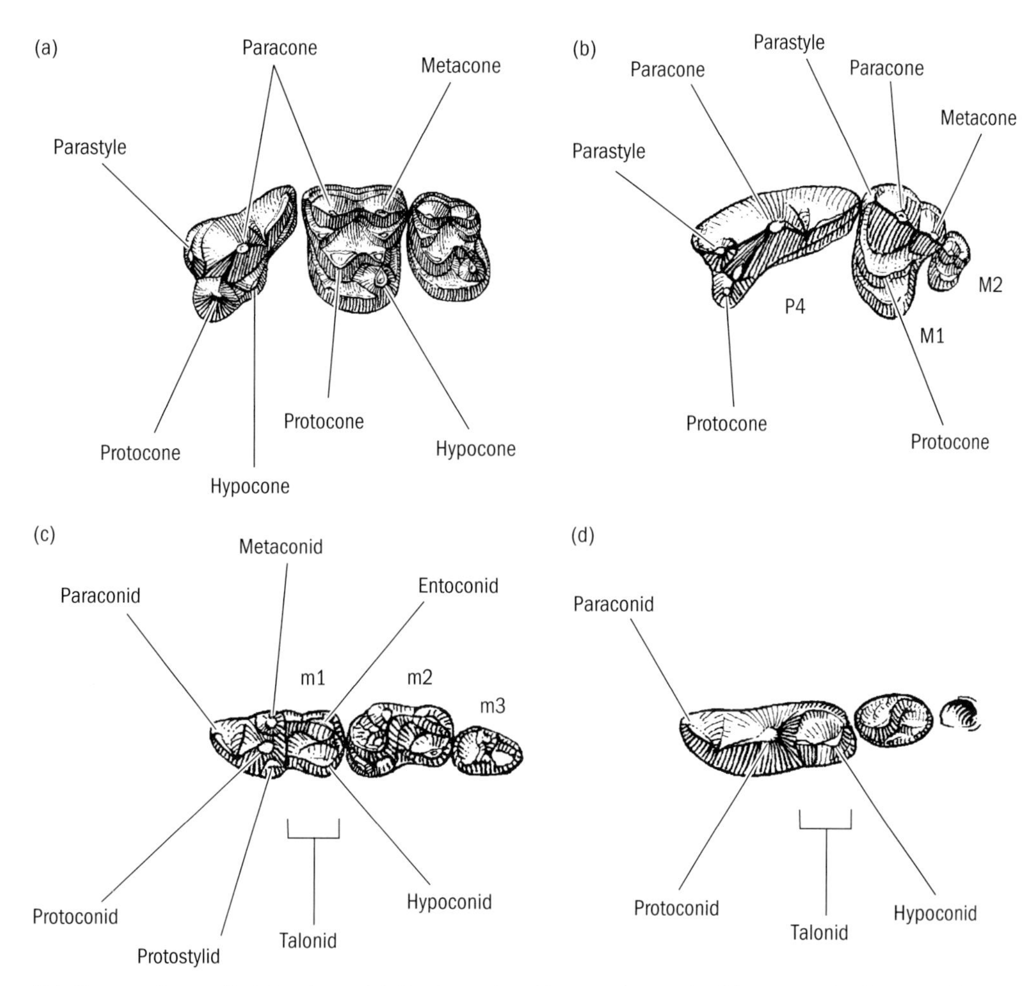

Figure 2.4 Hypercarnivorous (b, *Aelurodon* and d, *Euoplocyon*) and hypocarnivorous (a, *Phlaocyon* and c, *Cynarctus*) dentitions. In hypercarnivorous forms, the upper cheek teeth (b) tend to emphasize the shearing part of the dentition with an elongated and narrow P4, an enlarged parastyle on a transversely elongated M1, and a reduced M2. On the lower teeth (d), hypercarnivory is exemplified by a trenchant talonid due to the increased size and height of the hypoconid at the expense of the entoconid (reduced to a narrow and low ridge), accompanied by the enlargement of the protoconid at the expense of the metaconid (completely lost in *Euoplocyon*) and the elongation of the trigonid at the expense of the talonid. In hypocarnivorous forms, on the other hand, the upper teeth (a) emphasize the grinding part of the dentition with a shortened and broadened P4 (sometimes with a hypocone along the lingual border), a reduced parastyle on a quadrate M1 that has additional cusps (e.g. a conical hypocone along the internal cingulum) and cuspules, and an enlarged M2. The lower teeth (c) in hypocarnivorous forms possess a basined (bicuspid) talonid on m1 enclosed on either side by the hypoconid and entoconid that are approximately equal in size. Other signs of hypocarnivory on the lower teeth include widened lower molars, enlarged metaconids, and additional cuspules such as a protostylid.

(Orellan and Whitneyan, 34–30 Ma), early members of four small clades of the hesperocyonines had emerged: *Paraenhydrocyon*, *Enhydrocyon*, *Osbornodon*, and *Ectopocynus*. Hesperocyonines experienced their maximum diversity of 14 species during the late Oligocene (early Arikareean in 30–28 Ma), and reached their peak predatory adaptations (hypercarnivory) in the earliest Miocene (late Arikareean) with advanced species of *Enhydrocyon* and *Paraenhydrocyon*. The last species of the subfamily, *Osbornodon fricki*, became extinct in the early Barstovian (15 Ma), reaching the size of a small wolf.

With the exception of the *Osbornodon* clade, which acquired a bicuspid m1 talonid, hesperocyonines are primitively hypercarnivorous in dental adaptations with tendencies towards reduced last molars and trenchant (single cusped) talonid heels on the lower first molar. Although never reaching the extremes seen in the borophagines, hesperocyonines had modest development of bone cracking adaptations in their strong premolars. At least three lineages, in all species of *Enhydrocyon* and in terminal species of *Osbornodon* and *Ectopocynus*, have independently evolved their own unique array of bone cracking teeth. Hesperocyonines did not experiment with hypocarnivory.

Members of this subfamily were the topic of a monograph by Wang (1994). Two additional species of *Osbornodon* have since been added to the subfamily (Hayes 2000; Wang 2003). The evolutionary transition from plantigrade to digitigrade standing posture in early canids was explored by Wang (1993). A hereditary condition, osteochondroma, in the postcranials of early canids was documented by Wang and Rothschild (1992).

Borophaginae

From the primitive condition of a trenchant talonid heel on the lower first molar seen in the hesperocyonines, borophagines, and canines, on the other hand shared a basined (bicuspid) talonid acquired at the very beginning of their common ancestry (Fig. 2.4(c)). Along with a more quadrate upper first molar with its hypocone, the basined talonid establishes an ancestral state from which all subsequent forms were derived. Such a dental pattern proved to be very versatile and can readily be adapted towards either a hyper- or hypocarnivorous type of dentition, both of which were repeatedly employed by both borophagines and canines (Fig. 2.3).

The history of the borophagines also begins with a small fox-like form, *Archaeocyon*, in the late Oligocene. Contemporaneous with larger and more predatory hesperocyonines, these early borophagines in the late Oligocene and early Miocene (Arikareean) tended to be more omnivorous (hypocarnivorous) in their dental adaptations, such as *Oxetocyon*, *Otarocyon*, and *Phlaocyon*. One extreme case, *Cynarctoides* evolved selenodont-like molars as in modern artiodactyles, a rare occurrence of herbivory among carnivorans. These early borophagines are generally no larger than a raccoon, which is probably a good ecological model for some borophagines at a time when procyonids had yet to diversify.

After some transitional forms in the early Miocene (Hemingfordian), such as *Cormocyon* and *Desmocyon*, borophagines achieved their maximum ecological and numerical (i.e. species) diversity in the middle Miocene (Barstovian), with highly omnivorous forms, such as *Cynarctus*, that were almost ursid-like as well as highly predatory forms, such as *Aelurodon*, that were a larger version of the living African Wild Dog *Lycaon*. By then, borophagines had acquired their unique characteristics of a broad muzzle, a bony contact between premaxillary and frontal, multicuspid incisors, and an enlarged parastyle on the upper carnassials (modified from an enlargement of the anterior cingulum).

By the end of the Miocene, borophagines had evolved another lineage of omnivory, although only modestly in that direction, in the form of *Carpocyon*. Species of *Carpocyon* are mostly the size of jackals to small wolves. At the same time, the emergence of the genus *Epicyon* from a *Carpocyon*-like ancestor marked another major clade of hypercarnivorous borophagines. The terminal species of *Epicyon*, *E. haydeni*, reached the size of a large bear and holds record as the largest canid ever to have lived (Fig. 2.5). Closely related to *Epicyon* is *Borophagus*, the terminal genus of the Borophaginae. Both *Epicyon* and *Borophagus* are best known for their massive P4 and p4 in contrast to the diminutive premolars in front. This pair of enlarged premolars is designed for cracking bones, mirroring similar adaptations by hyaenids in the Old World. Advanced species of *Borophagus* survived the Pliocene but became extinct near the beginning of the Pleistocene.

The phylogeny and systematics of the Borophaginae were recently revised by Wang *et al.* (1999), which is the basis of above summary. Munthe (1979, 1989, 1998) analysed the functional morphology of borophagine limb bones and found a diverse array of postcranial adaptations, in contrast to the more stereotypical view that the hyaenoid dogs were noncursorial scavengers only. Werdelin (1989) compared

Figure 2.5 Reconstruction of *Epicyon saevus* (small individual, based on AMNH 8305) and *Epicyon haydeni* (large individual, composite figure, based on specimens from Jack Swayze Quarry). These two species co-occur extensively during the late Clarendonian and early Hemphillian of Western North America. Illustration by Mauricio Antón. (From Wang *et al.* 1999.)

the bone-cracking adaptations of borophagine canids and hyaenids in terms of evolutionary constraints within their prospective lineages.

Caninae

As in the hesperocyonines and borophagines, a small fox-sized species of *Leptocyon* is the earliest recognized member of the subfamily Caninae. Besides sharing a bicuspid talonid of m1 and a quadrate M1 with the borophagines, *Leptocyon* is also characterized by a slender rostrum and elongated lower jaw, and correspondingly narrow and slim premolars, features that are inherited in all subsequent canines. It first appeared in the early Oligocene (Orellan) and persisted through the late Miocene (Clarendonian). Throughout its long existence (no other canid genus had as long a duration), facing intense competition from the larger and diverse hesperocyonines and borophagines, *Leptocyon* generally remains small and inconspicuous, never having more than two or three species at a time.

By the latest Miocene (Hemphillian), fox-sized niches are widely available in North America, left open by extinctions of all small borophagines. The true fox clade, Tribe Vulpini, emerges at this time and undergoes a modest diversification to initiate primitive species of both *Vulpes* and *Urocyon* (and their extinct relatives). The North American Pliocene record of *Vulpes* is quite poor. Fragmentary materials from early Blancan indicate the presence of a swift fox-like form in the Great Plains. *Vulpes* species were widespread and diverse in Eurasia during the Pliocene (see Qiu and Tedford 1990), resulting from an immigration event independent from that of the *Canis* clade. Red fox (*Vulpes vulpes*) and Arctic fox (*Vulpes lagopus*) appeared in North America only in the late Pleistocene, evidently as a result of an immigration back to the New World.

Preferring more wooded areas, the gray fox *Urocyon* has remained in southern North America and Middle America. Records of the gray fox clade have a more or less continuous presence in North America throughout its existence, with intermediate forms leading to the living species *Urocyon cinereoargenteus*. Morphologically, the living African bat-eared fox *Otocyon* is closest to the *Urocyon* clade, although molecular evidence suggests that the bat-eared fox lies at the base of the fox clade or even lower (Geffen *et al.* 1992d; Wayne *et al.* 1997). If the morphological evidence has been correctly interpreted, then the Bat-eared fox must represent a Pliocene immigration event to the Old World independent of other foxes. A transitional form, *Protocyon*, occurs in southern Asia and Africa in the early Pleistocene.

Advanced members of the Caninae, Tribe Canini, first occur in the medial Miocene (Clarendonian, 9–12 Ma) in the form of a transitional taxon *Eucyon*. As a jackal-sized canid, *Eucyon* is mostly distinguished from the Vulpini in an expanded paroccipital process and enlarged mastoid process, and in the consistent presence of a frontal sinus. The latter character initiates a series of transformations in the Tribe Canini culminating in the elaborate development of the sinuses and a domed skull in *C. lupus*. By latest Miocene time, species of *Eucyon* have appeared in Europe (Rook 1992) and by the early Pliocene in Asia (Tedford and Qiu 1996). The North American records all predate the European ones, suggesting a westward dispersal of this form.

Arising from about the same phylogenetic level as *Eucyon* is the South American clade. Morphological and molecular evidence generally agrees that living South American canids, the most diverse group of canids on a single continent, belong to a natural group of their own. The South American canids are united by morphological characters such as a long palate, a large angular process of the jaw with a widened scar for attachment of the inferior branch of the medial pterygoid muscle, and a relatively long base of the coronoid process (Tedford *et al.* 1995). By the close of the Miocene, certain fragmentary materials from southern United States and Mexico indicate that taxa assignable to *Cerdocyon* (Torres and Ferrusquía-Villafranca 1981) and *Chrysocyon* occur in North America. The presence of these derived taxa in the North American late Miocene predicts that ancestral stocks of many of the South American canids may have been present in southern North America or Middle America. They appear in the South American fossil record shortly after the formation of the Isthmus of Panama in the Pliocene, around 3 Ma (Berta 1987). The earliest records are *Pseudalopex* and its close relative *Protocyon*, an extinct large hypercarnivore, from the Pliocene (Uquian, around 2.5–1.5 Ma) of Argentina. By the latest Pleistocene (Lujanian, 300,000–10,000 years ago), most living species or their close relatives have emerged, along with the extinct North American Dire Wolf, *Canis dirus*. By the end of the Pleistocene, all large, hypercarnivorous canids of South America (*Protocyon*, *Theriodictis*) as well as *C. dirus* had become extinct.

The *Canis* clade within the Tribe Canini, the most derived group in terms of large size and hypercarnivory, arises near the Miocene–Pliocene boundary between 5 and 6 Ma in North America. A series of jackal-sized ancestral species of *Canis* thrived in the early Pliocene (early Blancan), such as *Canis ferox*, *Canis lepophagus*, and other undescribed species. At about the same time, first records of canids begin to appear in the European late Neogene: '*Canis*' *cipio* in the late Miocene of Spain (Crusafont-Pairó 1950), *Eucyon monticinensis* in the latest Miocene of Italy (Rook 1992), the earliest raccoon-dog *Nyctereutes donnezani*, and the jackal-sized *Canis adoxus* in the early Pliocene of France (Martin 1973; Ginsburg 1999). The enigmatic '*Canis*' *cipio*, only represented by parts of the upper and lower dentition, may pertain to a form at the *Eucyon* level of differentiation rather than truly a species of *Canis*.

The next phase of *Canis* evolution is difficult to track. The newly arrived *Canis* in Eurasia underwent an extensive radiation and range expansion in the late Pliocene and Pleistocene, resulting in multiple, closely related species in Europe, Africa, and Asia. To compound this problem, the highly cursorial wolf-like *Canis* species apparently belong to a circum-arctic fauna that undergoes expansions and contractions with the fluctuating climate. Hypercarnivorous adaptations are common in the crown-group of *Canis* especially in the Eurasian middle latitudes and Africa. For the first time in canid history, phylogenetic studies cannot be satisfactorily performed on forms from any single continent because of their Holarctic distribution and faunal intermingling between the new and old worlds. Nevertheless some clades were localized in different parts of Holarctica. The vulpines' major centre of radiation was in the Old World. For the canines, North America remained a centre through the Pliocene producing the coyote as an endemic form. A larger radiation yielding the grey wolves (*Canis lupus*), dhole (*Cuon alpinus*), African wild dog (*Lycaon pictus*), and fossil relatives took place on the Eurasian and African continents. During the Pleistocene elements of the larger canid fauna invaded mid-latitude North America—the last invasion of which was the appearance of the grey wolf south of the glacial ice sheets in the latest Pleistocene (about 100 Ka).

A comprehensive systematic revision of North American fossil canines by Tedford *et al.* (in preparation) forms the basis of much of the foregoing summary. As part of that revision, the phylogenetic framework as derived from living genera was published by Tedford *et al.* (1995). Nowak (1979) published a monograph on the Quaternary *Canis* of North America. Berta (1981, 1987, 1988) undertook the most recent phylogenetic analysis of the South

American canids. Rook (1992, 1994) and Rook and Torre (1996a,b) partially summarized the Eurasian canids. The African canid records are relatively poorly understood but recent discoveries promise to advance our knowledge in that continent (Werdelin, personal communication). See also citations below for recent molecular systematic studies.

Molecular systematics

The ancient divergence of dogs from other carnivores is reaffirmed by molecular data. DNA–DNA hybridization of single copy DNA clearly shows them as the first divergence in the suborder Caniformia that includes seals, bears, weasel, and raccoon-like carnivores (Fig. 2.6). This basal placement is further supported by mitochondrial DNA sequence studies (Vrana *et al.* 1994; Slattery and Brien 1995; Flynn and Nedbal 1998), and recently studies of DNA sequences from nuclear genes (Murphy *et al.* 2001). Based on molecular clock calculations, the divergence time was estimated as 50 million years before present (Wayne *et al.* 1989). This value is consistent with the first appearance of the family in the Eocene, although it is somewhat more ancient than the date of 40 million years suggested by the fossil record (see above). Considering that first appearance dates generally post-date actual divergence dates because of the incompleteness of the record (e.g. Marshall 1977), the agreement between fossil and molecular dates is surprisingly good.

Evolutionary relationships within the family Canidae have been reconstructed using comparative karyology, allozyme electrophoresis, and mitochondrial DNA protein coding sequence data (Wayne and Brien 1987; Wayne *et al.* 1997, 1987a,b). Further, relationships at the genus level have been studied with mtDNA control region sequencing (a non-coding, hypervariable segment of about 1200 bp in the mitochondrial genome) and microsatellite loci (hypervariable single copy nuclear repeat loci) (Geffen *et al.* 1992; Bruford and Wayne 1993; Girman *et al.* 1993; Gottelli *et al.* 1994; Vilà *et al.* 1997, 1999). The protein-coding gene phylogeny, which is largely consistent with trees based on other genetic approaches, shows that the wolf genus *Canis* is a monophyletic group that also includes the dhole or Asian wild dog (*Cuon alpinus*). The grey wolf, coyote

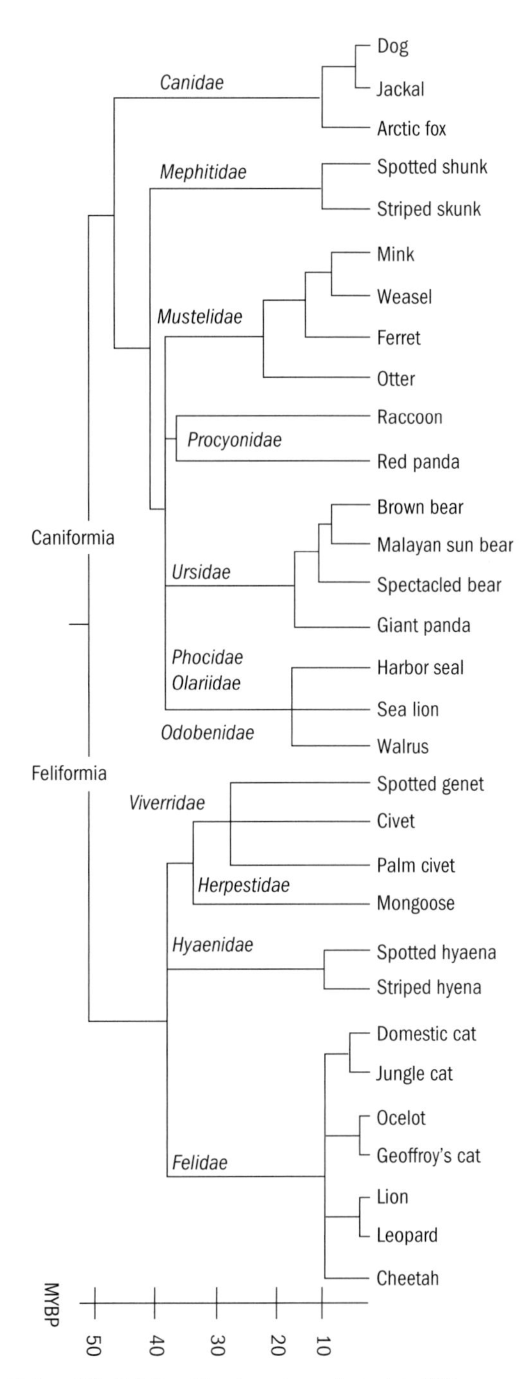

Figure 2.6 Relationship of carnivores based on DNA hybridization data (Wayne *et al.* 1989). Family and suborder groupings are indicated. Time scale in millions of year before present (MYBP) is based on comparisons of DNA sequence divergence to first appearance times in the fossil record.

(*Canis latrans*) and Ethiopian wolf (*Canis simensis*) form a monophyletic group, with the golden jackal (*C. aureus*) as the most likely sister taxon (Fig. 2.7). The black-backed and side-striped jackals (*C. mesomelas, C. adustus*) are sister taxa, but they do not form a monophyletic group with the golden jackal and Ethiopian wolf. Basal to *Canis* and *Cuon* are the African wild dog and a clade consisting of two South American canids, the bush dog (*Speothos venaticus*) and the maned wolf (*Chrysocyon brachyurus*). Consequently, although the African wild dog preys on large game as does the grey wolf and dhole, it is not closely

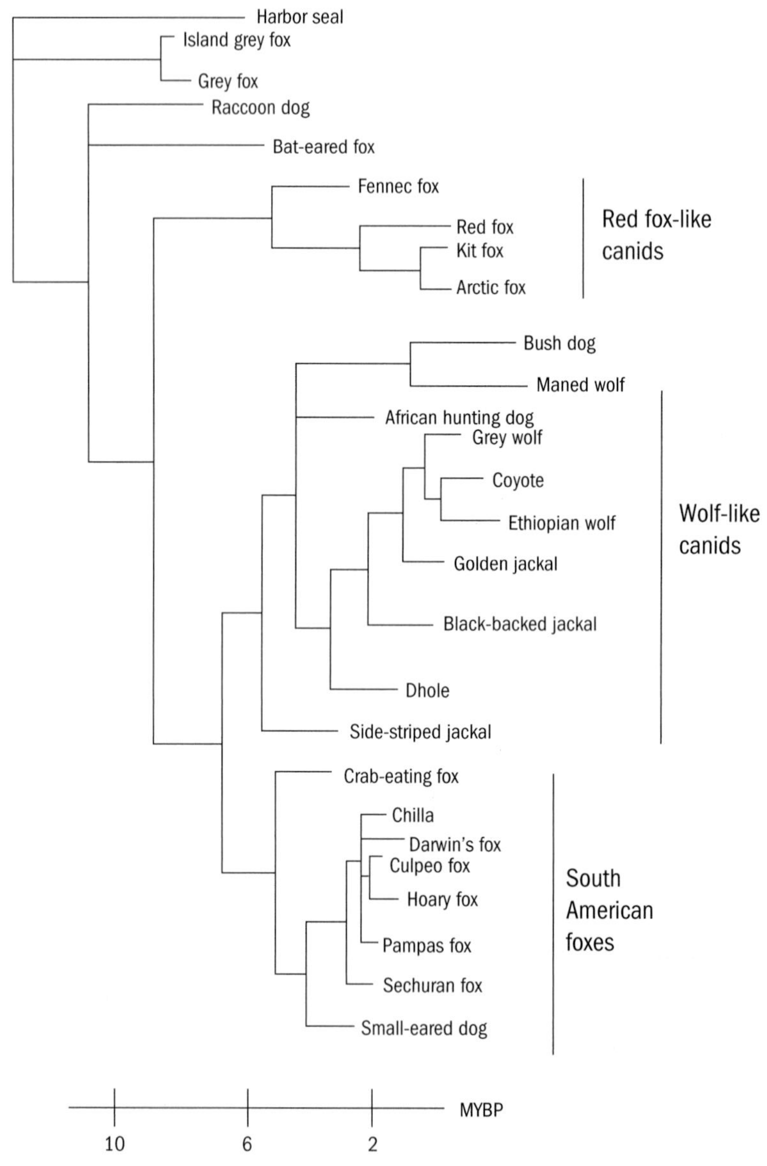

Figure 2.7 Consensus tree of 26 canid species based on analysis of 2001 bp of DNA sequence from mitochondrial protein coding genes (Wayne *et al.* 1997). See Geffen *et al.* (1992) for a more detailed analysis of the Red-fox like canids. Time scale in millions of year before present (MYBP) is based on comparisons of DNA sequence divergence to first appearance times in the fossil record.

related to either species but is sister to the clade containing these species. This phylogeny implies that the trenchant heeled carnassial now found only in *Speothos*, *Cuon*, and *Lycaon*, evolved at least twice or was primitive and lost in other wolf-like canids and the maned wolf.

The South American canids do not form a monophyletic group. *Speothos* and *Chrysocyon* are sister taxa that group with the wolf-like canids rather than the South American foxes. The large sequence divergence between the bush dog and maned wolf and between these taxa and the South American foxes suggests that they diverged from each other 6–7 Ma, well before the Panamanian land bridge formed about 2–3 Ma. Thus, three canid invasions of South America are required to explain the phylogenetic distribution of the extant species. These invasions are today survived by (1) the bush dog, (2) the maned wolf, and (3) the South American foxes. Further, within the South American foxes, divergence values between crab-eating fox (*Cerdocyon thous*), the short-eared fox *(Atelocynus microtis)*, and other South American foxes, suggest they may have diverged before the opening of the Panamanian land bridge as well (Wayne *et al.* 1997). The fossil record supports the hypothesis that the crab-eating fox had its origin outside of South America, as the genus has been described from late Miocene deposits of North America (3–6 Ma) (Berta 1984, 1987, see above). Consequently, only the foxes of the genus *Pseudalopex*, *Lycalopex*, and perhaps *Atelocynus*, might have a South American origin. Further, the generic distinction given to *Pseudalopex* and *Lycalopex* does not reflect much genetic differentiation, and in the absence of appreciable morphologic differences, the genetic data suggest these species should be assigned to a single genus.

A fourth grouping in the tree consists of other fox-like taxa, including *Vulpes*, *Alopex*, and *Fennecus* (Fig. 2.7) (Geffen *et al.* 1992; Mercure *et al.* 1993; Wayne *et al.* 1997). The Arctic fox, *Alopex*, is a close sister to the kit fox, *Vulpes macrotis* and both share the same unique karyotype (Wayne *et al.* 1987a). Basal to *Vulpes* is *Fennecus*, suggesting an early divergence of that lineage. Finally, *Otocyon*, *Nyctereutes*, and *Urocyon* appear basal to other canids in all molecular and karyological trees (Wayne *et al.* 1987a). The first two taxa are monospecific whereas the third includes the island Fox, *Urocyon littoralis* and the gray fox, *U. cinereoargenteus*. The three genera diverged early in the history of the family, approximately 8–12 Ma as suggested by molecular clock extrapolations.

In sum, the living Canidae is divided into five distinct groupings. These include the wolf-like canids, which consists of the coyote, grey wolf, jackals, dhole, and African wild dog. This clade is associated with a group containing bush dog and maned wolf in some trees and further, this larger grouping is associated with the South American foxes (Wayne *et al.* 1997). The red fox group is a fourth independent clade containing *Alopex*, *Vulpes*, and *Fennecus*. Finally, three lineages have long distinct evolutionary histories and are survived today by the raccoon dog, bat-eared fox, and grey fox. Assuming an approximate molecular clock, the origin of the modern Canidae begins about 10–12 Ma and is followed by the divergence of wolf and fox-like canids about 6 Ma. The South American canids are not a monophyletic group and likely owe their origin to three separate invasions. This group included the maned wolf, bush dog, crab-eating fox, and the other South American canids, which diverged from each other about 3–6 Ma.

Morphological and molecular phylogenies

Tedford *et al.* (1995) performed a cladistic analysis of living canids on morphological grounds. The result is a nearly fully resolved relationship based on an 18 taxa by 57 characters matrix at the generic level. This relationship recognizes three monophyletic clades in the canines: the fox group (tribe Vulpini), the South American canine group, and the wolf group containing hypercarnivorous forms (the latter two form the tribe Canini). Recent molecular studies (presented above), on the other hand, contradict some of these arrangements while maintaining other parts in the morphological tree (Fig. 2.8(a)).

Trees derived from 2001 bp of mitochondrial DNA (Wayne *et al.* 1997, and Fig. 2.7 of this chapter) tend to places the foxes near the basal part, the South American canines in the middle, and the wolves and wild dogs towards the terminal branches, a pattern that is consistent with the morphological tree.

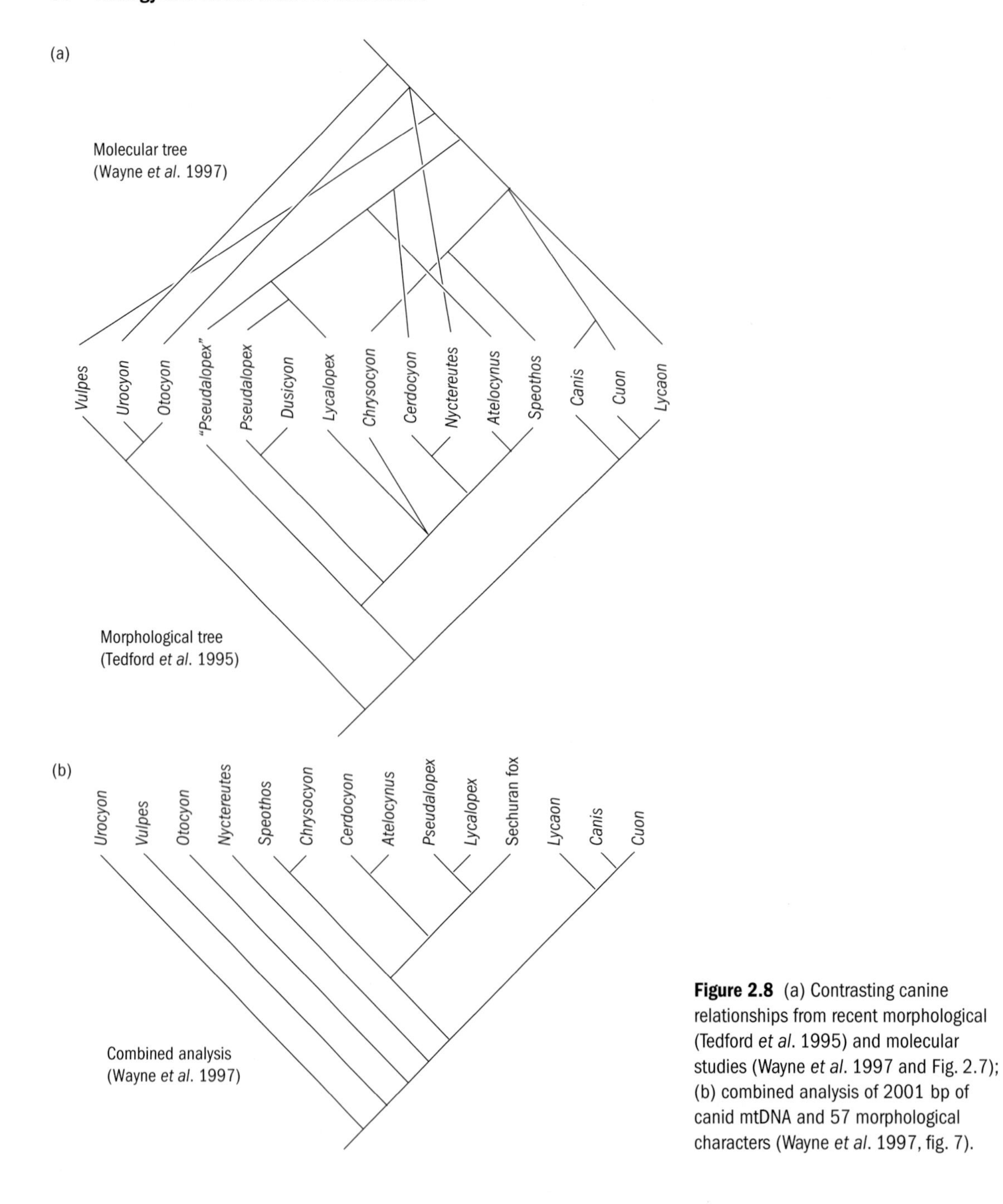

Figure 2.8 (a) Contrasting canine relationships from recent morphological (Tedford *et al*. 1995) and molecular studies (Wayne *et al*. 1997 and Fig. 2.7); (b) combined analysis of 2001 bp of canid mtDNA and 57 morphological characters (Wayne *et al*. 1997, fig. 7).

The detailed arrangements, however, differ in a number of ways. The foxes are generally in a paraphyletic arrangement in contrast to a monophyletic clade in the morphological tree. The gray fox and bat-eared fox are placed at the base despite their highly derived dental morphology compared with other foxes. Similarly, South American canines are no longer monophyletic under molecular analysis but form at least two paraphyletic branches. A glaring discrepancy is the Asiatic raccoon dog being allied to the foxes in the molecular analysis despite its numerous morphological characters shared with some South

American forms. Finally, molecular data suggest independent origins for the Asiatic and African hunting dogs in contrast to a sister relationship in the morphological tree supported by a large number of characters related to hypercarnivory.

Not surprisingly, there are increased agreements between the molecular and morphological results when the two data sets are combined in a total evidence analysis (Fig. 2.8(b)). Under such conditions, the South American canines (except *Nyctereutes*) become monophyletic, as does the clade including the wolf, dhole, and African wild dog.

Evolutionary ecology

Iterative evolution of hypercarnivory

One of the most remarkable features of canid history is their repeated tendency to evolve both hypocarnivorous and hypercarnivorous forms. As noted above, hypercarnivorous species evolved within each subfamily, and hypocarnivorous species evolved within two of the three (all but the Hesperocyoninae). Hypocarnivory was most fully expressed in the Borophaginae, where at least 15 species showed a tendency towards a dentition similar to that of living raccoons (Wang *et al.* 1999). Among the Caninae, the tendency has not been quite as strong, with only a single lineage, *Nyctereutes*, developing a markedly hypocarnivorous dentition. However, all three subfamilies include multiple species of apparent hypercarnivores with enhanced cutting blades on their carnassials, reduced grinding molars, and enlarged canines and lateral incisors. When and why did hypercarnivory evolve within each subfamily?

In two of the three subfamilies, Hesperocyoninae and Caninae, the evolution of hypercarnivory appears to have occurred at least partly in response to a reduced diversity of other hypercarnivorous taxa. The Hesperocyoninae evolved hypercarnivory early in their history (Figs 2.1 and 2.7) and the most advanced forms appear in the early Miocene (about 24–20 Ma) at a time when the two previously dominant carnivorous families had vanished. These two families were the Nimravidae, an extinct group of saber-tooth cat-like forms, and the Hyaenodontidae, a group of somewhat dog-like predators included in the extinct order Creodonta. The nimravids and hyaenodontids dominated the North American guild of large, predatory mammals in the late Eocene to mid-Oligocene (37–29 Ma), but faded rapidly in the late Oligocene, and were extinct in North America by about 25 Ma (Van Valkenburgh 1991, 1994). During most of their reign, hesperocyonines existed at low diversity and small (fox-size) body size, but as the hyaenodontids and nimravids declined in the late Oligocene, the early canids seem to have radiated to replace them. Most of these hypercarnivorous canids were jackal-size (less than 10 kg), with only the last surviving species, *Osbornodon fricki*, reaching the size of a small wolf (Wang 1994). In the early Miocene, large hypercarnivores immigrated from the Old World in the form of hemicyonine bears (Ursidae) and temnocyonine bear-dogs (Amphicyonidae). The subsequent decline to extinction of the hesperocyonines might have been a result of competition with these new predators (Van Valkenburgh 1991, 2001).

Hypercarnivory appears late in the history of the Caninae and represents at least several independent radiations in South America, North America, and the Old World (Figs 2.1 and 2.7). As was true of the hesperocyonine example, the South American radiation of large hypercarnivorous canids occurred at a time (2.5–0.01 Ma) when cat-like predators were rare or absent. It followed the elevation of the Panamanian land bridge around 2–3 Ma that allowed immigration between the previously separated continents. The canids that first entered South America found a depauperate predator community, consisting of one bear-like procyonid carnivoran, three species of carnivorous didelphid marsupials, one of which was the size of a coyote, and a gigantic, predaceous ground bird (Marshall 1977). With the possible exception of the rare ground bird, none of these species was a specialized hypercarnivore. Between 2.5 Ma and 10,000 years ago, 16 new species of canids appeared in South America, at least seven of which had trenchant heeled carnassials and clearly were adapted for hypercarnivory (Berta 1988; Van Valkenburgh 1991). They represent three different endemic genera, *Theriodictis*, *Protocyon*, and *Speothos*. In addition, there were three large wolf-like species of *Canis* in South America, *Canis gezi*, *Canis nehringi*, and *Canis dirus*, all of which were probably hypercarnivorous but retained a bicuspid heel on their carnassials. Of these only the dire wolf, *C. dirus*, evolved in North America. All but one of these ten hypercarnivorous canids of South America went extinct at the end of the Pleistocene (Van Valkenburgh 1991). The sole survivor, the bush dog (*Speothos*) is rarely sighted.

In the Old World, the evolution of hypercarnivorous canines occurred within the last 4 million years and did not coincide with an absence of cats. Large cats, both sabertooth and conical tooth forms, are present throughout the Plio-Pleistocene when the highly carnivorous species of *Canis*, *Cuon*, *Lycaon*, and *Xenocyon* appear (Turner and Antón 1996). However, their evolution might be a response to the decline of another group of hypercarnivores, wolf-like hyaenids. Hyaenids were the dominant dog-like predators of the Old World Miocene, reaching a diversity of 22 species between 9 and 5 Ma, but then declining dramatically to just five species by about 4 Ma (Werdelin and Turner 1996). Their decline may have opened up ecospace for the large canids and favored the evolution of hypercarnivory.

The remaining episode of hypercarnivory in canids occurred in the Borophaginae between 15 and 4 Ma (Van Valkenburgh *et al.* 2003). As was true of the Caninae, the hypercarnivorous species do not evolve early in the subfamily's history. Instead, they appear in the latter half of the subfamily's lifespan and only become prevalent in the last third (mid–late Miocene; Figs 2.1 and 2.7). In the late Miocene, borophagine canids were the dominant dog-like predators of North America, having replaced the amphicyonids and hemicyonine bears that had themselves replaced the hesperocyonines some ten million years earlier (Van Valkenburgh 1999). In the case of the Borophaginae, the evolution of hypercarnivory appears more gradual than in the other two subfamilies, and is not easily ascribed to opportunistic and rapid evolution into empty ecospace.

In all three subfamilies, there is a pattern of greater hypercarnivory and increasing body size with time (Fig. 2.9). Even in the Hesperocyoninae, where hypercarnivory evolves very early, large species with the most specialized meat-eating dentitions appear later (Wang 1994). This directional trend towards the evolution of large, hypercarnivorous forms is apparent in other groups of dog-like carnivores, such as the amphicyonids (Viranta 1996) and hyaenids (Werdelin and Solounias 1991; Werdelin and Turner 1996), and may be a fundamental feature of carnivore evolution. The likely cause is the prevalence of interspecific competition among large, sympatric predators. Interspecific competition tends to be more intense among large carnivores because prey are often difficult to capture and can represent a sizable quantity of food that is worthy of stealing and defending. Competition appears to be a motive for

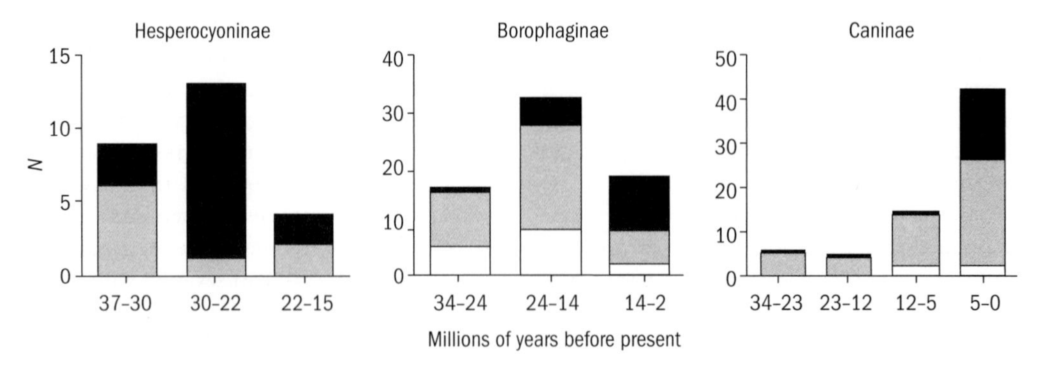

Figure 2.9 Iterative evolution of large hypercarnivores. Number (*N*) of hypocarnivorous (white), mesocarnivorous (grey), and large (>20 kg) hypercarnivorous (black) species over time in each of the three subfamilies. The few hesperocyonine species with trenchant-heeled carnassials estimated to have been less than 20 kg in mass were assigned to the mesocarnivorous category because they are assumed not have taken prey as large or larger than themselves. For the Hesperocyoninae and Borophaginae, their stratigraphic ranges were broken into thirds; for the Caninae, four time divisions were used because of the large number of species appearing in the past 5 million years. Species were assigned to dietary categories and body mass was estimated on the basis of dental morphology as described in Van Valkenburgh (1991) and Wang *et al.* (1999).

much intraguild predation because the victim often is not eaten (Johnson *et al.* 1996; Palomares and Caro 1999; Van Valkenburgh 2001). Larger carnivores tend to dominate smaller ones and so selection should favour the evolution of large body size. Large body size in turn selects for a highly carnivorous diet because of energetic considerations. As shown by Carbone *et al.* (1999), almost all extant carnivores that weigh more than 21 kg take prey as large or larger than themselves. Using an energetic model, they demonstrated that large body size brings with it constraints on foraging time and energetic return. Large carnivores cannot sustain themselves on relatively small prey because they would expend more energy in hunting than they would acquire. By taking prey as large or larger than themselves, they achieve a greater return for a given foraging bout. Killing and consuming large prey is best done with a hypercarnivorous dentition and so the evolution of large body size and hypercarnivory are linked. Of course, this does not preclude the evolution of hypercarnivory at sizes less than 21 kg, but it seems relatively rare. It has occurred in the Canidae as evidenced by the hesperocyonines and the extant Arctic fox, *Alopex lagopus*, and kit fox, *V. macrotis*. However, the two extant foxes do not have trenchant-heeled carnassials despite their tendency towards a highly carnivorous diet, and this may reflect regular, opportunistic consumption of fruits and invertebrates (Van Valkenburgh and Koepfli 1993).

Returning to the questions of when and why hypercarnivory evolves among canids, it seems that when and why are intertwined. That is, because of intraguild competition and predation, selection favours the evolution of larger size in canids and as a consequence, hypercarnivory. However, *when* this occurs it is largely a function of other members of the predator guild. In the case of the Hesperocyoninae, it occurred relatively early in their history because previously dominant large hypercarnivores were in decline or already extinct. In the case of the Borophaginae and Caninae, it did not occur until much later because other clades held the large hypercarnivorous roles for much of the Miocene. In all these examples, it appears as though the rise of large hypercarnivorous canids reflects opportunistic replacement rather than competitive displacement of formerly dominant taxa (Van Valkenburgh 1999).

The last one million years

All of the canids that are extant today evolved well prior to the late Pleistocene extinction event approximately 11,000 years ago. The same could be said of most, if not all, extant carnivores. In the New World, the end-Pleistocene event removed numerous large mammals, including both herbivores (e.g. camels, horses, proboscideans) and carnivores (e.g. sabertooth cat, dire wolf, short-faced bear). In the Old World, many of the ecological equivalents of these species disappeared earlier, around 500,000 years ago (Turner and Antón 1996). Consequently, all extant carnivore species evolved under very different ecological circumstances than exist at present. For example, the grey wolf today is considered the top predator in much of Holarctica, but it has only held this position for the last ten to eleven thousand years. For hundreds of thousands of years prior to that time, the wolf coexisted with 11 species of predator as large or larger than itself (Fig. 2.10). Now there are but three, the puma, black bear, and grizzly bear, and wolves are usually dominant over the first two species at least (Van Valkenburgh 2001). Thus, for most of its existence, the grey wolf was a mesopredator rather than a top predator, and so its morphology and behaviour should be viewed from that perspective. Given the greater diversity and probable greater abundance of predators in the past, interspecific competition was likely more intense than at present. Higher tooth fracture frequencies in late Pleistocene North American predators provide indirect evidence of heavy carcass utilization and strong food competition at that time (Van Valkenburgh and Hertel 1993). Intense food competition would favour group defence of kills and higher levels of interspecific aggression. Perhaps the sociality of the wolf and the tendency of some carnivores to kill but not eat smaller predators are remnant behaviours from a more turbulent past.

The only canid to go extinct in the North American end Pleistocene was the dire wolf, *C. dirus*. the grey wolf, coyote, and several foxes survived. In addition to the dire wolf, two bears and three cats went extinct, all of which were very large (Fig. 2.10). Can we learn something about the causes of current predator declines by examining the winners and losers in the late Pleistocene? Examination of the

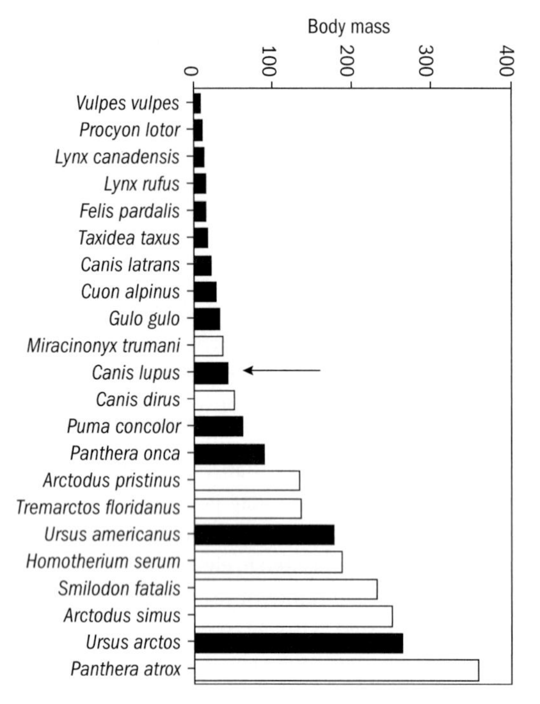

Figure 2.10 North American Pleistocene carnivorans arranged by body mass. Black bars represent extant species, and white bars represent extinct species. Arrow indicates the grey wolf (*Canis lupus*). Data from Van Valkenburgh and Hertel (1998).

loser species reveals that they tended to be the more specialized members of their clades; they were larger (Fig. 2.10) and tended to be more dentally specialized for hypercarnivory (Van Valkenburgh and Hertel 1998). Remarkably, two of the species that went extinct, the dire wolf and sabertooth cat (*Smilodon fatalis*), are five times more common in the Rancho La Brea tar pit deposits than the next most common carnivore, the coyote (*C. latrans*). This suggests that the dire wolf and sabertooth cat were dominant predators at this time, comparable to the numerically dominant African lion and spotted hyaena of extant African ecosystems. The extinction of the apparently successful dire wolf and sabertooth cat implies there was a major perturbation to the ecosystem in the late Pleistocene. Their demise and that of the other large hypercarnivores suggest that large prey biomass dropped to extremely low levels. Supporting this are the parallel extinctions of 10 of the 27 species of raptors and vultures (Van Valkenburgh and Hertel 1998).

In the late Pleistocene, the largest meat-eaters, both avian and mammalian, were the most vulnerable. Is this the case today for canids? Of the three large hypercarnivorous canids, the dhole, grey wolf, and African wild dog, only the wild dog is highly endangered. Among living canids in general, species that appear to be most at risk tend to be insular (Darwin's fox, island fox) or restricted to limited habitats (Ethiopian wolf), or just very poorly known species (e.g. short-eared dog, bush dog). Indeed, it is a bit difficult to answer the question of which of the living species are most endangered because we have so little information on many of the smaller taxa. Nevertheless, it does seem that the end Pleistocene extinction is not a good analogue for what is happening at present, at least in terms of which is most vulnerable. Then, it was the largest, most abundant, and most carnivorous. Now it seems more often to be smaller mesocarnivores that are at risk due to small population size exacerbated by habitat loss. In both the end-Pleistocene and at present, the hand of humanity looms large as a cause of predator declines. Initially, the damage was largely due to overhunting of both prey and predator, and to this we have added significant habitat loss. Survivors of the current crisis are likely to be both dietary and habitat generalists, such as the coyote.

Acknowledgements

We thank David Macdonald and Claudio Sillero for the invitation to write this contribution and David Macdonald for his critical review. This research is funded in part by grants from the National Science Foundation (DEB 9420004; 9707555).

CHAPTER 3

Population genetics

Population and conservation genetics of canids

Robert K. Wayne, Eli Geffen, and Carles Vilà

The island fox (*Urocyon littoralis*) survives in small isolated populations depleted of genetic diversity © G. Roemer.

Introduction

Canids are highly mobile carnivores. Grey wolves (*Canis lupus*) commonly disperse over 50 km before reproducing and small canids often disperse a dozen kilometers or more. Over generations, this dispersal results in the movement of genes across landscapes and may counteract genetic differentiation due to drift between even widely separated populations (e.g. Vilà *et al.* 1999). Additionally, topographic and environmental barriers to migration that would cause geographic discontinuities in the distribution of less mobile species may not deter dispersal in canids. Many canids, such as grey wolves, coyotes (*C. latrans*) and red and culpeo foxes (*Vulpes vulpes, Pseudalopex culpaeus*) are habitat generalists and can live in and disperse across a diversity of natural and human altered habitats. In these canids, the more relevant factors that influence dispersal may be the movements of prey (Carmichael *et al.* 2001), density effects (Roemer *et al.* 2001a), or kinship (Lehman *et al.* 1992; Girman *et al.* 2001). Also, in many species, a history of dynamic changes in abundance and distribution are superimposed on current demographic conditions. For example, genetic analysis of grey wolves and coyotes has suggested range contractions during glacial maxima followed by reinvasion across several continents (Frati *et al.* 1998; Vilà *et al.* 1999). This confounding influence of history is well exemplified in brown bears (*Ursus arctus*) where current genetic structure differs dramatically from genetic patterns in the late Pleistocene (Leonard *et al.* 2000; Barnes *et al.* 2002).

Mobility influences the degree to which interspecific hybridization affects the genetic composition of hybridizing species (Lehman *et al.* 1991; Jenks and Wayne 1992; Mercure *et al.* 1993; Wilson *et al.* 2000; Wayne and Brown 2001). The width of a hybrid zone is a function of the distance travelled from birth to place of first reproduction and the degree of natural selection against hybrids (Barton and Hewitt 1989). If selection is weak, hybrid zones may span a considerable distant in highly mobile species and interspecific gene flow may strongly affect the genetic heritage of hybridizing forms (Jenks and Wayne 1992; Wilson *et al.* 2000; Wayne and Brown 2001). Clear examples of this phenomena exist in Canidae, and are perhaps more dramatic than in any other carnivore species. Hybridization has contributed to the genetic extinction of red wolves (*Canis rufus*) in the wild, has greatly compromised the genetic composition of the Great Lakes wolf, and has flooded New England with wolf/coyote hybrids (Nowak 1979; Lehman *et al.* 1991; Jenks and Wayne 1992; Roy *et al.* 1994a,b, 1996; Wilson *et al.* 2000, in review). Further, hybridization between domestic dogs and Ethiopian wolves (*C. simensis*) (Gottelli *et al.* 1994) and perhaps between grey wolves and domestic dogs in certain areas may have implications for conservation (Vila and Wayne 1999; Andersone *et al.* 2002; Randi and Lucchini 2002; Vilà *et al.* 2003a).

In this chapter, we discuss patterns of genetic variation and subdivision in a wide variety of canids. In general, the degree of subdivision increases with decreasing body size as might be expected considering that small canids have smaller dispersal distances. We discuss how levels of variation are influenced by demographic history and ecological and topographic barriers and the effect interspecific hybridization has on the genetic composition of canid populations. For each case study, we end with a discussion of conservation implications.

Molecular genetic approaches

The first studies of population genetic variability in canids examined variation in allozymes (Table 3.1) which are forms of an enzyme that have similar activity but differ in amino acid sequence and represent different alleles for the same locus. These alleles are separated by charge and size in a gel matrix under an electric field (electrophoresis). In general, canids have only low to moderate levels of allozyme polymorphism and consequently, population level studies have low resolution (Ferrell *et al.* 1978; Wayne and O'Brien 1987; Wayne *et al.* 1991a,b; Kennedy *et al.* 1991; Randi 1993; Lorenzini and Fico 1995; Frati *et al.* 1998). More recent studies analysed sequence variation in mitochondrial DNA (mtDNA) assessed through indirect or direct sequencing techniques (Table 3.1). The mitochondria are cytoplasmic organelles containing multiple copies of a small circular DNA molecule, about 16–18,000 base pairs (bp) in length in mammals that codes for proteins which function in electron transport (Fig. 3.1). Hundreds of mitochondria may occur within the cell and thus mitochondrial genes are several fold more abundant than their nuclear counterparts. The relative abundance of mitochondrial versus nuclear genomes facilitates sequencing, especially of samples having low concentrations of DNA or damaged DNA such as in museum or ancient samples or in trace samples of organisms such as faeces, hair, or saliva (Kohn *et al.* 1999; Wayne *et al.* 1999; Hofreiter *et al.* 2001). Additionally, in mammals, mtDNA sequences have a mutation rate that is 5–10 times greater than that of nuclear genes. Consequently, closely related species and populations may have accumulated diagnostic mtDNA mutations in the absence of similar changes in the nuclear genome. Finally, with only a few

Table 3.1 Genetic studies of canid populations by marker type

	Species	Site	Populations	Haplotypes or alleles	*V*	Source
Allozymes	Black-backed jackal	Africa	6	1.6/8	Ns	Wayne *et al.* (1990b)
	Island fox	USA	7	2.3/7	0.019[a]	Wayne *et al.* (1991a)
	Golden jackal	Africa	1	1.8/8	Ns	Wayne *et al.* (1990b)
	Grey wolf	East NA	4	2.0/5	0.059[a]	Wayne *et al.* (1991b)
	Grey wolf	Italy	1	2.3/4	0.028[a]	Randi *et al.* (1993)
	Grey wolf	Italy	1	2.1/7	0.036[a]	Lorenzini and Fico (1995)
	Grey wolf	NW Canada	9	2.2/5	0.030[a]	Kennedy *et al.* (1991)
	Mexican wolf	Mexico	2	1.7/3	0.158[a]	Shields *et al.* (1987)
	Red fox	Europe	10	1.1/9	0.153[a]	Frati *et al.* (1998)
	Side-striped jackal	Africa	4	1.8/8	Ns	Wayne *et al.* (1990b)
Microsatellites	African wild dog	Africa	7	4.0/11	0.643[a]	Girman *et al.* (2001)
	Coyote	N. America	6	5.9/10	0.583[a]	Roy *et al.* (1994)
	Ethiopian wolf	Ethiopia	2	2.4/9	0.241[a]	Gottelli *et al.* (1994)
	Golden jackal	Kenya	1	4.8/10	0.412[a]	Roy *et al.* (1994)
	Grey wolf	Italy	1	4.3/18	0.440[a]	Randi and Lucchini (2002)
	Grey wolf	N. Europe	2	5.8/19	0.640[a]	Vilà *et al.* (2003)
	Grey wolf	NA	7	5.0/10	0.528[a]	Roy *et al.* (1994)
	Grey wolf	NW USA	2	4.7/10	0.605[a]	Forbes and Boyd (1996)
	Grey wolf	NW USA	6	5.4/10	0.587[a]	Forbes and Boyd (1997)
	Grey wolf	Scandinavia	5	5.0/4	0.588[b]	Sundqvist *et al.* (2001)
	Grey wolf	Scandinavia	1	2.9/29	0.510[a]	Ellegren (1999)
	Island fox	Sta. Cruz I.	2	2.5/10	0.450[a]	Roemer *et al.* (2001a)
	Mexican wolf	Mexico	3	1.9/10	0.281[a]	Garcia-Moreno *et al.* (1996)
	Red fox	Australia	4	3.2/7	0.467[a]	Lade *et al.* (1996)
	Red fox	Switzerland	5	6.6/11	0.658[a]	Wandeler *et al.* (2003)
	Red wolf	N. America	1	5.3/10	0.507[a]	Roy *et al.* (1994)
	Red wolf	N. America	2	5.5/10	0.576[a]	Roy *et al.* (1996)
Minisatellites	Grey wolf	Isle-Royale	5	18.1	40.1[c]	Wayne *et al.* (1991b)
	Mexican wolf	Mexico	3	14.9	18.3[c]	Fain *et al.* (1995)
	Island fox	USA	6	21.8	12.0[c]	Gilbert *et al.* (1990)
	Island fox	USA	7	21.9	17.9[c]	Wayne *et al.* (1991a)
MHC	Coyote	USA	1	4.5/3	0.036[d]	Seddon and Ellegren (2002)
	Grey wolf	NA and N. Europe	2	12.3/3	0.055[d]	Seddon and Ellegren (2002)
	Red wolf	USA	1	4	0.833[a]	Hedrick *et al.* (2002)
mtDNA	African wild dog	Africa	5	6	0.009[e]	Girman *et al.* (1993)
	African wild dog	Africa	7	8	0.014[e]	Girman *et al.* (2001)
	Arctic fox	Scandinavia	3	10	0.659[d]	Dalen *et al.* (2002)
	Black-backed jackal	Africa	6	4	0.045[e]	Wayne *et al.* (1990a)
	Island fox	USA	7	12	0.046[e]	Wayne *et al.* (1991a)
	Coyote	NA	17	32	0.291[a]	Lehman and Wayne (1991)

Table 3.1 (Continued)

Species	Site	Populations	Haplotypes or alleles	*V*	Source
Coyote	NA	13	24	0.020[e]	Lehman *et al.* (1991)
Golden jackal	Africa	1	2	0.001[e]	Wayne *et al.* (1990a)
Grey wolf	Europe	9	20	0.500[d]	Randi *et al.* (2000)
Grey wolf	Isle-Royale	5	9	0.018[e]	Wayne *et al.* (1991b)
Grey wolf	Italy	1	1	0.000[e]	Randi *et al.* (1995)
Grey wolf	NA	9	9	0.006[e]	Lehman *et al.* (1991)
Grey wolf	Scandinavia	2	4	0.021[e]	Ellegren *et al.* (1996)
Grey wolf	Worldwide	30	24	0.744[d]	Vilà *et al.* (1999)
Grey wolf	Worldwide	26	18	0.790[a]	Wayne *et al.* (1992)
Kit fox	N. America	10	21	0.012[e]	Maldonado *et al.* (1997)
Kit fox	USA	10	24	0.700[a]	Mercure *et al.* (1993)
Mexican wolf	Mexico	2	1	0.000[e]	Shields *et al.* (1987)
Red fox	Europe	10	18	0.484[d]	Frati *et al.* (1998)
Red wolf	N. America	2	6	0.025[e]	Roy *et al.* (1996)
Red wolf	N. America	5	5	0.025[e]	Wayne and Jenks (1991)
Side-striped jackal	Africa	4	2	0.002[e]	Wayne *et al.* (1990a)

Note: Ns = not specified. Number of haplotypes are indicated for MHC and mtDNA analyses. Mean number of alleles and loci scanned are indicated for allozymes and microsatellites, and mean number of fragments is indicated for minisatellites. *V* is a measure of genetic variation.

[a] Observed heterozygosity (Ht).

[b] Fst or Nst.

[c] Average proportion difference (APD).

[d] Nucleotide diversity (π).

[e] Sequence divergence.

exceptions, mtDNA is maternally rather than biparentally inherited, and there is no recombination. Therefore, phylogenetic analysis of mtDNA sequences within species provides a history of maternal lineages that can be represented as a simple branching phylogenetic tree (Avise 1994, 2000).

The first studies of variation in mtDNA sequences were indirect and utilized a panel of restriction enzymes that cut DNA at sites in the mitochondrial genome where specific 4 or 6 bp sequence patterns are located. For each enzyme, the cut fragments are separated by size electrophoretically. Differences in the fragment pattern between individuals indicate nucleotide changes at the restriction site, and allow the number of sequence changes overall to be estimated between different mitochondrial genomes (haplotypes; e.g. Wayne *et al.* 1989a, 1991a,b; Lehman *et al.* 1991). Such restriction fragment length polymorphisms (RFLP, Table 3.1) provided the first estimates of sequence variation within populations and allowed the relationships of populations to be reconstructed (see Avise 1994). Beginning in the late 1980s, the advent of the polymerase chain reaction (PCR) in combination with new DNA sequencing techniques, made population level sequencing studies feasible. DNA sequence studies used nucleotide difference to develop a more precise reconstruction of historical demographic events such as colonization and gene flow (Fig. 3.1; Avise 1994, 2000). Both restriction fragment analysis (e.g. Lehman *et al.* 1991; Wayne *et al.* 1992; Randi and Lucchini 1995; Pilgrim *et al.* 1998) and, more recently, mtDNA sequencing

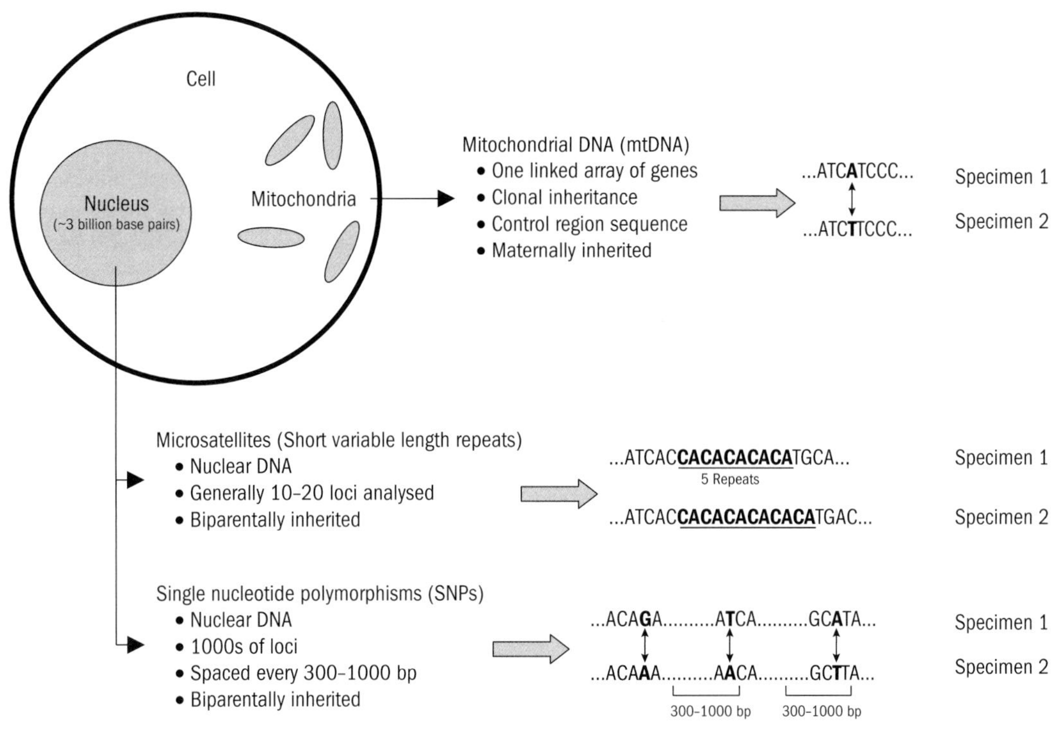

Figure 3.1 Comparison of modern molecular genetic techniques. MtDNA typing requires the sequencing of one or more DNA segments. For population comparisons, commonly, part of the control region is sequenced because it is highly variable. Comparison of sequences allows different haplotypes to be identified (two haplotypes differing by an A to T transition are shown). Such sequence differences can be used to reconstruct the evolutionary relationship of haplotypes (e.g. Figs 3.3 and 3.5). Microsatellite alleles differ in the number of simple repeat units they contain and consequently can be identified by differences in size and do not have to be sequenced. These loci are highly polymorphic and abundant in the nucleus of plants and animals. Single nucleotide polymorphisms are a new type of marker useful for quantifying sequence variation in nuclear genes and non-transcribed regions. The data are analogous to sequence polymorphism in the mitochondrial genome but are much less variable and biallelic, that is, each SNP generally has just two alleles.

by PCR (Vilà *et al.* 1999; Randi *et al.* 2000; Wilson *et al.* 2000) have been applied to canids (Table 3.1). Hypervariable regions of the mitochondrial genome, such as the control region, have been the focus of recent sequencing efforts. The estimates of diversity based on RFLP or sequencing analysis include the number of haplotypes, measures of the sequence difference among these haplotypes (e.g. nucleotide diversity) and the spatial and phylogenetic relationship of haplotypes from different populations (Avise 2000; Emerson *et al.* 2001; Posada and Crandall 2001).

However, phylogenetic trees based on mtDNA record the history of only a single linked set of genes. Additionally, because of the smaller effective population size, levels of mtDNA variability are more severely affected by changes in population size than are nuclear loci (Avise 1994). Recently, nuclear loci with high mutation rates that can easily be surveyed in large population samples have been identified. These include multi-locus DNA fingerprints (Burke *et al.* 1996) in which complex DNA banding patterns simultaneously represent approximately 10–20 unspecified minisatellite loci. However, a generally more desirable approach involves microsatellite loci which are composed of tandem repeats of short sequences 2–6 bp in length (Fig. 3.1; Bruford and

Wayne 1993; Hancock 1999). Microsatellite loci are often preferable because simple sequence repeats can be amplified by the PCR from minute or highly degraded samples of DNA such as bones, hair, and faeces (Roy *et al.* 1994a, 1996; Taberlet *et al.* 1996a; Foran *et al.* 1997; Kohn and Wayne 1997; Kohn *et al.* 1999; Fedriani and Kohn, 2001; Smith *et al.* 2001; Lucchini *et al.* 2002). Additionally, each locus is scored separately either through autoradiography or staining of acrylamide gels or by an automated sequencer (Bruford *et al.* 1996). These methods allow for the identification of the two alleles inherited from both parents at each locus (both alleles are detected, as in co-dominant markers). Since half of the alleles are shared between parent and offspring, these methods also led to a robust determination of family relationships. Consequently, microsatellite data can be analysed by traditional population genetic approaches developed for co-dominant loci (Goldstein and Pollock 1997; Rousset and Raymond 1997; Smith *et al.* 1997a; Bossart and Prowell 1998; Girman *et al.* 2001; Roemer *et al.* 2001a).

In contrast, minisatellite, randomly amplified polymorphic DNA (RAPD), and amplified fragment length polymorphisms (AFLP) approaches (Avise 1994, 2000; Smith and Wayne 1996) are generally multi-locus (many loci are simultaneously analysed at the same time), and a heterozygote may not be distinguishable from a homozygote genotype (i.e. are not co-dominant). Thus, these procedures require additional assumptions for statistical analyses. A panel of 10 or fewer microsatellite loci may be sufficient to quantify components of variation accurately within and among populations and to study individual relatedness within social groups (Bruford and Wayne 1993; Queller *et al.* 1993; Roy *et al.* 1994b; Ellegren *et al.* 1996; Forbes and Boyd 1996; Smith *et al.* 1997a; Bossart and Prowell 1998; Girman *et al.* 2001; Roemer *et al.* 2001a).

Recently, researchers studying humans have identified single nucleotide polymorphisms (Fig. 3.1) and microsatellites on the Y chromosome (Cooper *et al.* 1996; White *et al.* 1999; Jobling and Tyler-Smith 2000). These markers allowed a new paternal view on evolutionary patterns that complemented studies on maternally inherited mtDNA and biparentally inherited nuclear genes (Jorde *et al.* 2000). Consequently, Y chromosome studies represent an independent test for hypotheses based on mitochondrial sequences or microsatellites (Pritchard *et al.* 1999; Seielstad *et al.* 1999; Thomson *et al.* 2000) and permit estimation of sex-biased migration and dispersal (Seielstad *et al.* 1998). Y chromosome studies are still uncommon in other mammal species (Boissinot and Boursot 1997; Hanotte *et al.* 2000), but with the development of canine specific markers (Olivier and Lust 1998; Olivier *et al.* 1999; Sundqvist *et al.* 2001), this new kind of population genetic and phylogenetic studies is possible in canids (Sundqvist *et al.* 2001; Vilà *et al.* 2003a).

Finally, in grey wolves and other canids, genetic variation at the major histocompatability complex (MHC) has now been studied (Hedrick *et al.* 2000, 2002, Aguilar *et al.* in press). The MHC is a 4 million bp segment of DNA in vertebrates containing over 80 genes arranged in three functional classes (Hedrick 1994; Edwards and Hedrick 1998). Unlike other genetic markers, overdominance, frequency dependence, and geographically varying directional selection influence variation at the MHC and may maintain high levels of polymorphism (Potts and Wakeland 1993; Hedrick 1994; Edwards and Hedrick 1998). The specific factors hypothesized to affect polymorphism at the MHC are parasite-mediated selection (Hughes *et al.* 1994; Black and Hedrick 1997), sexual selection (Potts *et al.* 1991), and in some instances, maternal–fetal interactions (Haig 1997). MHC class II molecules are responsible for the presentation to the immune system of foreign (exogenous) antigens on cytotoxic and helper T lymphocytes. Thus, a widely accepted hypothesis to explain high heterozygosity at MHC class II genes is that a wider array of pathogens will initiate an immune response if individuals are heterozygous (i.e. balancing selection; Hughes and Nei 1992; Black and Hedrick 1997). Therefore, assessing patterns of genetic variation in the MHC may be critical to documenting the selective forces that influence gene frequencies and the genes which affect fitness (Aguilar *et al.* in press).

Review of case studies

The Ethiopian wolf

The Ethiopian wolf, is the most endangered living canid (Gottelli and Sillero-Zubiri 1992; Sillero-Zubiri and Macdonald 1997; Marino 2003a,b). The total

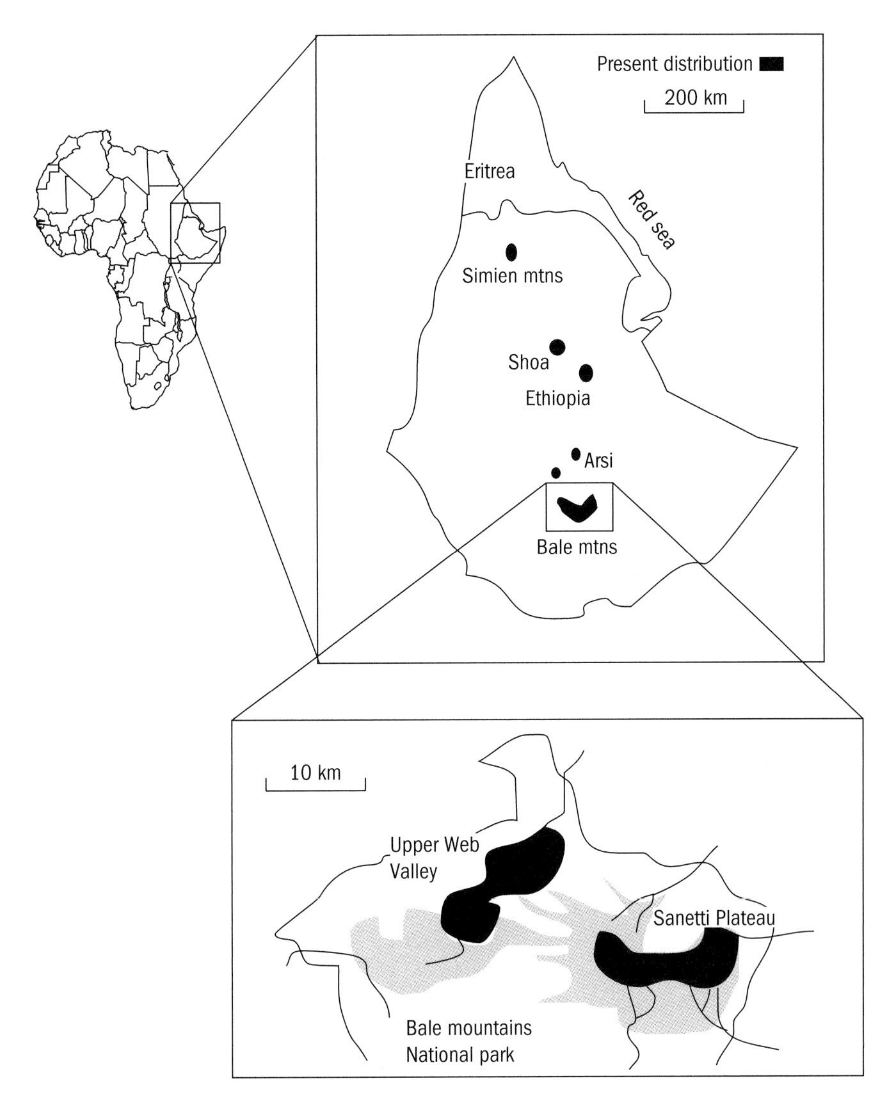

Figure 3.2 A map of sampling localities indicating extant populations of Ethiopian wolves and the area sampled in the genetic study of Gottelli *et al.* (1994).

population in 1999 was less than 500 individuals. The species is dispersed across the Ethiopian highlands above 3000 m in small, highly isolated populations (Fig. 3.2; Gottelli and Sillero-Zubiri 1992; Marino 2003b). Ethiopian wolves are more than twice the size of African jackals, are wolf-like in morphology, and have an organized grey wolf-like pack structure, although the rate of extra-pair copulation is likely to be much higher than that in grey wolves (Mech 1987; Sillero-Zubiri and Gottelli 1995a; Sillero-Zubiri *et al.* 1996a; Smith *et al.* 1997a). Phylogenetic analysis of mtDNA sequences showed that the closest living relatives of Ethiopian wolves are probably grey wolves and coyotes (Fig. 3.3; Gottelli *et al.* 1994; Wayne and Gottelli 1997; Vilà *et al.* 1999). An evolutionary hypothesis consistent with these results is that Ethiopian wolves are a relict form remaining from a Pleistocene invasion of a wolf-like progenitor into East Africa. The current extent of Ethiopian high altitude moorland habitats is only 5% of the area existing after the last Ice

Age (Yalden 1983; Kingdon 1990; Gottelli *et al.* 1994). Consequently, the geographic range and numerical abundance of Ethiopian wolves likely has decreased during the Holocene. More recently, habitat loss and fragmentation due to human population growth and agriculture have accelerated the decline of Ethiopian wolves.

Genetic variation and population differentiation

The RFLP and sequence analyses of mtDNA showed that the two populations in the Bale Mountains had very low variability, as all wolves had the same mtDNA haplotype (Table 3.1; Fig. 3.3). Variability of microsatellite loci was also low in these populations. The average value of heterozygosity for 10 microsatellite loci was only 46% and mean allelic diversity 38% of that commonly found in other wolf-like canids (Table 3.1). Such low levels of heterozygosity are consistent with an equilibrium effective population size of only a few hundred individuals (Gottelli *et al.* 1994). However, a preliminary analysis of 134 bp of mtDNA of control region sequence from

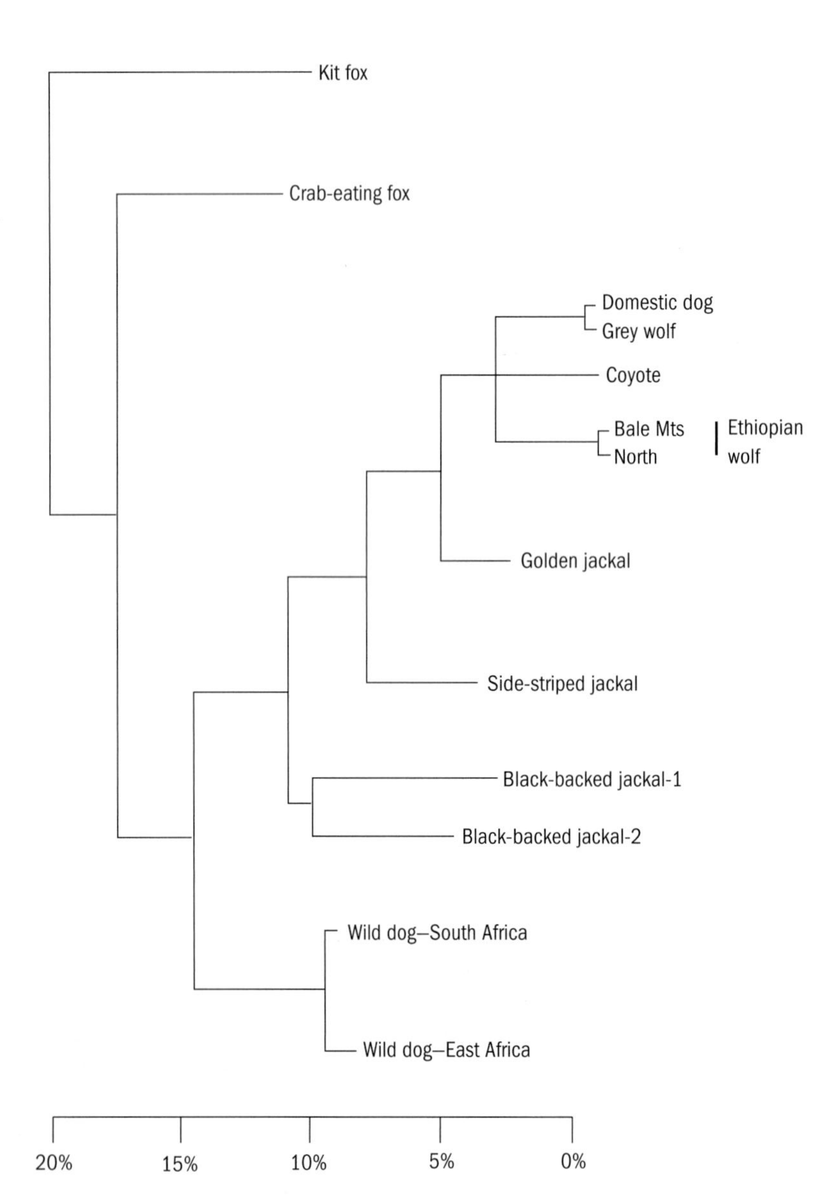

Figure 3.3 A strict consensus tree of the two most parsimonious trees obtained from phylogenetic analysis of 2001 bp of mtDNA sequence (Gottelli *et al.* 1994). Two individuals were sequenced for each species. The two East African black-backed jackal sequences (1 and 2) are representatives of the two divergence mtDNA haplotypes found there (Wayne *et al.* 1990a). The kit fox sequence was used to root the tree.

two museum skins from an unknown Northern Ethiopian wolf population revealed two unique substitutions (Fig. 3.3; Roy *et al.* 1994a). Recent analysis of other populations show further evidence of genetic differences (Gottelli *et al.* in press).

Hybridization with domestic dogs

Although loss of variation and inbreeding in isolated populations are concerns for endangered species, an additional problem for Ethiopian wolves is hybridization with domestic dogs. Genetic analysis showed that suspected hybrid individuals in a population in the Sanetti Valley of the Bale Mountains preserve had microsatellite alleles not otherwise found in Ethiopian wolves, but were present in domestic dogs. In contrast, these individuals had mtDNA haplotypes identical to those in 'pure' Ethiopian wolves (Gottelli *et al.* 1994), a result consistent with field reports that interspecific matings only involved male domestic dogs and female Ethiopian wolves (Sillero-Zubiri and Gottelli 1991). Additionally, parentage analysis found that a single litter had both a wolf and dog as fathers, showing that multiple paternity occurs in wolves and can involve both species. Dogs not only hybridize with Ethiopian wolves and compete with them for food, but also are reservoirs of canine diseases (Sillero-Zubiri *et al.* 1966b).

Conservation implications

The sharply lower levels of variation in the Ethiopian wolf reflect a long history of population declines compounded by recent habitat fragmentation (Gottelli *et al.* 1994). However, perhaps a greater concern than the reduced levels of genetic variation is the vulnerability of the few remaining populations to diseases such as rabies, which is already thought to have eliminated about one-half of the Bale Mountain population (Sillero-Zubiri *et al.* 1996b, Sillero-Zubiri and Macdonald 1997) and to other stochastic demographic effects (Lande 1988). Inbreeding depression may occur in canids (Laikre and Ryman 1991; Laikre *et al.* 1993; Fredrickson and Hedrick 2002; but see Kalinowski *et al.* 1999) and may conceivably influence the persistence of the population (e.g. Lacy 1997; Seal and Lacy 1998). Loss of genetic variation in small populations may also influence the ability of individuals to adapt to changing conditions (Frankham *et al.* 2002). However, Ethiopian wolves actively avoid inbreeding (Sillero *et al.* 1996), thus decreasing the rate at which genetic variation is lost and mitigating the effect of inbreeding. In one population, the loss of unique Ethiopian wolf characteristics may result from interbreeding with dogs; however, this threat may be restricted to that locality (Wayne and Gottelli 1997). Ethiopian wolves are not being bred in captivity (Ginsberg and Macdonald 1990) and the genetic results suggest that a reservoir of pure wolves should be protected and bred in a captive setting as a source for reintroduction should efforts to sustain the wild population fail. Finally, other populations of Ethiopian wolves should be surveyed genetically so that a balanced program of captive and *in situ* management can be constructed that maintains historic levels of variation within and gene flow between populations (Wayne and Gottelli 1997; Crandall *et al.* 2000).

African wild dog

The African wild dog, *Lycaon pictus*, once ranged over most of Africa south of the Sahara, inhabiting areas of dry woodland and savannah (Fig. 3.4; Ginsberg and Macdonald 1990). However, due to habitat loss, hunting, and disease, many populations have vanished or are severely reduced in number. The extant populations are highly fragmented and total no more than several thousand individuals (Ginsberg and Macdonald 1990; Fanshawe *et al.* 1997; Ginsberg and Woodroffe 1997). Importantly, the western and Kenyan populations are nearing extinction, yet these populations are not represented in zoos as only South African wild dogs are kept in captivity. Populations in South Africa currently are stabilized in protected areas (Fanshawe *et al.* 1997).

Genetic variation and population differentiation

The genetic diversity within and among seven populations of African wild dogs was determined based on mtDNA control region sequences and eleven microsatellite loci (Table 3.2; Girman *et al.* 2001). Blood samples and museum bone samples were collected from 280 individuals (Fig. 3.4). Analysis of mtDNA nucleotide diversity suggested that wild dog

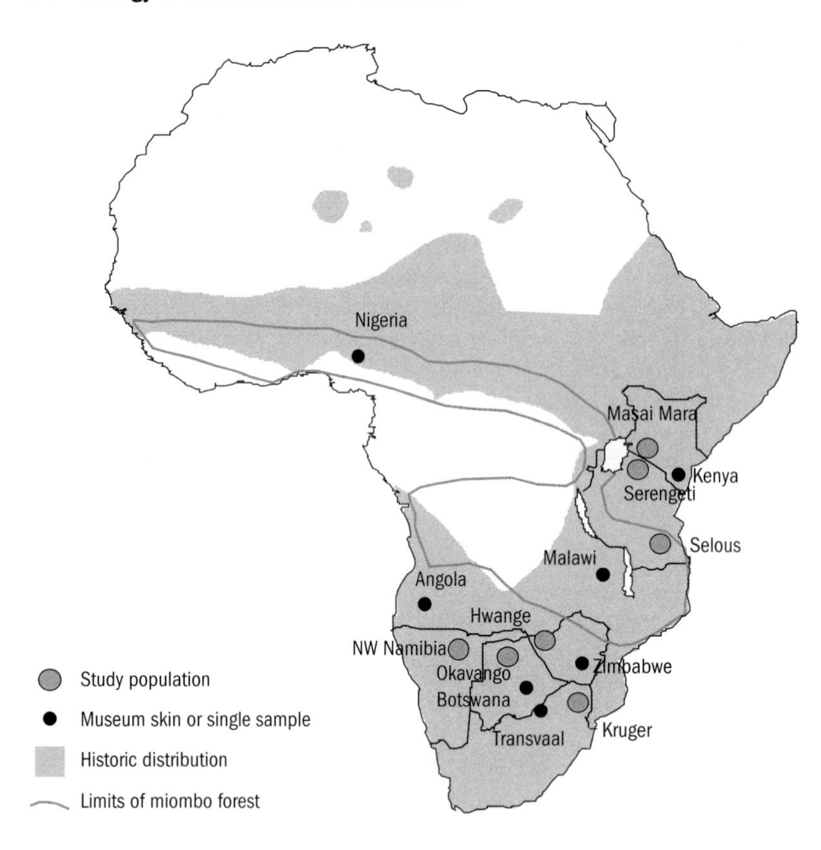

Figure 3.4 Locations of study populations and museum samples from Girman *et al.* (2001). Historic distribution of wild dogs according to Creel and Creel (1998). Limits of the distribution of 'miombo' (*Brachystegia-Julbernardia*) forests modified from Matthee and Robinson (1997).

Table 3.2 Genetic diversity in populations of the African wild dog

	mtDNA			Microsatelites				
Population	*n*	Hp	$\hat{Q}_T$	*n*	Loci	*N*/locus	Al	He
Hwange	28	5	0.0077	22	11	21.0	4.0	0.653
Kruger	94	2	0.0013	94	11	93.8	3.9	0.555
Mara/ Serengeti	27	2	0.0048	28	11	26.6	4.2	0.622
NW Namibia	6	1	0.0000	6	11	5.80	3.4	0.618
Okavango	42	5	0.0120	31	11	29.7	4.0	0.605
Selous	31	2	0.0114	22	11	17.8	4.4	0.665
Captive	37	1	0.0000	30	11	28.4	2.6	0.431

Note: The number of mtDNA haplotypes (Hp), nucleotide diversity ($\hat{Q}_T$), mean number of microsatellite alleles per locus (Al), and mean expected heterozygosity (He; unbiased estimate; Nei 1987) are indicated. Modified from Girman *et al.* (2001).

populations historically have been small relative to other large carnivores. However, recent population declines due to human-induced habitat loss have not caused a dramatic reduction in genetic diversity. Levels of diversity in microsatellite loci do not show strong evidence of recent or historic population decline relative to other carnivores (Tables 3.1 and 3.2). Further, the levels of genetic polymorphism estimated from the microsatellite data were relatively similar in all seven populations (Table 3.2). Although the average sample size for each population varied greatly (5.8 in the NW Namibia population to 93.8 in the Kruger population), the mean number of alleles per locus ranged only between 3.4 and 4.4. Heterozygosity values were also similar, ranging from 0.56 for the Kruger population to 0.67 for the Selous population. The heterozygosity of a captive South African population was lower (0.50) and the mean number of alleles per locus was only 3.3.

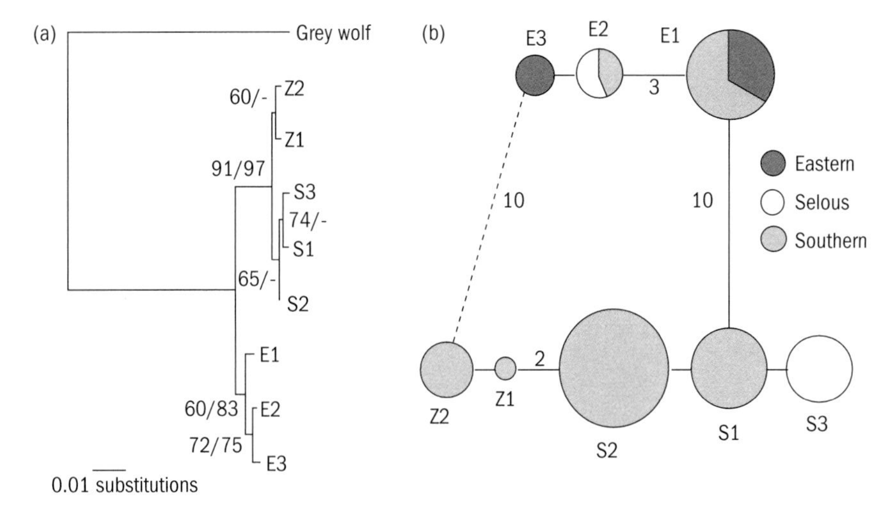

Figure 3.5 Phylogenetic relationships of wild dog control region haplotypes. (a) Neighbour-joining tree based on Tamura and Nei (1993) distance (gamma correction, $\gamma = 0.5$). The ratio by each node is the bootstrap percentage of support when higher than 50% in 1000 neighbour-joining (numerator) and maximum parsimony (denominator) trees. (b) Minimum spanning network. The size of the nodes indicates the frequency of the haplotype in the entire sample. The frequency of each haplotype in Eastern (Masai Mara and Serengeti), Selous, and Southern (all others) populations are indicated by shading. The number of substitutions differentiating haplotypes is shown on each branch when different from one. An alternative link between northern and southern haplotypes is indicated with a dashed line.

Mitochondrial and microsatellite loci showed significant differentiation between populations. Eastern and Southern populations may have been historically isolated. One historic and eight recent mtDNA haplotypes were found that defined two highly divergent clades (Fig. 3.5). In contrast to a previous more limited mtDNA analysis (Girman *et al.* 1993), sequences from these clades were not geographically restricted to eastern or southern African populations. Rather, a large admixture zone was found spanning populations from Botswana, Zimbabwe, and southeastern Tanzania. Genetic differentiation between populations was significant for both microsatellite and mitochondrial markers and unique mtDNA haplotypes and alleles characterized the populations (Φ_{ST}, 0.158–0.935 for mtDNA; θ, 0.041–0.140 for microsatellites). However, gene flow estimates (Nm) based on microsatellite data were moderate to high (range 1.53–5.88), greater than one migrant per generation. In contrast, gene flow estimates based on the mtDNA control region were lower than expected (range 0.04–2.67). Given the differences in the mode of inheritance of mitochondrial and nuclear markers, the results suggest a male bias in long distance dispersal. However, dispersal distance has been found to be similar for males and females in a Botswana population (McNutt 1996b), so the genetic results could indicate a higher frequency of male dispersal. Finally, mitochondrial and microsatellite population trees differed with regard to the association of east and southern populations (Fig. 3.6). Past and present distribution of the 'miombo' (*Brachystegia-Julbernardia*) forest, and grassland as well as the barrier imposed by the rift valley are biogeographic factors that may explain the current distribution of genetic variability (Fig. 3.4; Girman *et al.* 2001). However, West African populations, represented by a single sample from a museum specimen define a distinct branch suggesting a history of genetic isolation.

Conservation implications

Our previous results (Girman *et al.* 1993) suggested that eastern and southern African wild dog

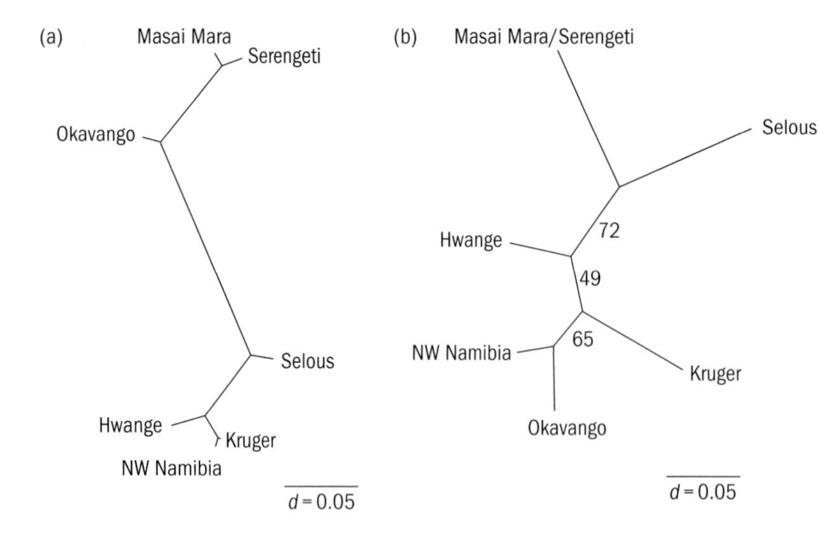

Figure 3.6 Neighbour-joining trees of wild dog populations. (a) Control region topology based on the average sequence divergence between sequences in each population. (b) Microsatellite topology based on Nei's (1978) unbiased genetic distance between populations. Bootstrap support in 1000 replicates is indicated by the branches.

populations were morphologically and genetically differentiated and formed reciprocally monophyletic units which supported their classification as evolutionarily significant units warranting separate conservation (Moritz 1994). However, the more recent and extensive genetic survey finds that these two haplotype clades co-occur over much of the current geographic range, which likely reflects natural mixing of previously isolated populations (Fig. 3.5; Girman *et al.* 2001). Consequently, genetic management should aim at mimicking observed levels of gene flow between contiguous populations within this admixture zone (Crandall *et al.* 2000; Wayne and Brown 2001). Individual-based models of wild dog population dynamics also suggest that even low rates of migration between populations can demographically stabilize populations otherwise at risk of extinction (Vucetich and Creel 1999). However, in African wild dogs, genetic differentiation of microsatellite loci increases with distance, and eastern and southern African populations may be morphologically distinct. Consequently, translocations between geographically distant southern and eastern populations are not advised because adaptive differences may exist (Crandall *et al.* 2000). For the Masai Mara and Serengeti, where African wild dogs are endangered (Fanshawe *et al.* 1997), the Selous region would be an appropriate source of individuals for reintroduction at the level of a few migrants per generation. Additionally, because the genetic results suggest more frequent dispersal and/or longer dispersal distances in males than in females, the population management strategy should focus on the more frequent translocation of males to replicate natural processes. West African population should be a high priority for research and conservation given evidence of genetic distinction and their perilous population status.

Finally, our genetic analyses have shown that despite recent population declines, the genetic diversity of several populations was still high. However, the populations that were studied were among the largest in Africa (Fanshawe *et al.* 1997) and thus may not be indicative of genetic diversity in smaller, more isolated populations. Given increasing habitat loss and fragmentation, a future decline in genetic variation appears likely (Fanshawe *et al.* 1997; Ginsberg and Woodroffe 1997; Girman and Wayne, 1997). To ameliorate this decline, population sizes should be kept as large as possible given the remaining habitat area. Additionally, gene flow should be facilitated by maintaining corridors that link populations, and when this is not possible, through translocation at historic levels as indicated by genetic data. The maintenance of genetic variation, especially the component that influences fitness, is critical to population persistence and the future evolutionary response of hunting dogs to changing environmental conditions (Crandall *et al.* 2000).

Grey wolf

The grey wolf, is an interesting species from a population and conservation genetic perspective. The wolf had the largest geographic range of any canid, and exists in a wide range of habitats from cold tundra to the warm deserts of the Old and New World. Because grey wolves are the most mobile canid, genetic differentiation between populations connected by appropriate habitat is expected to be low. However, wolves vary geographically in body size and pelage suggesting selection as the cause for differentiation despite high levels of gene flow (e.g. Smith *et al.* 1997b). For example, selection for differences according to habitat type (e.g. Tundra versus boreal forest) or prey (migratory versus resident, large versus small; Kolenosky and Standfield 1975; Peterson *et al.* 1998; Carmichael *et al.* 2001) could presumably cause differentiation despite gene flow. Understanding the dynamics between gene flow and selection is essential for testing alternative hypotheses about differentiation (Koskinen *et al.* 2002). Further, wolf populations need to be connected by effective corridors. Despite the high potential mobility of wolves, habitat fragmentation and habitat loss can dramatically affect the demography and genetic variability of wolf populations. Specifically, we might expect populations isolated by habitat fragment and in decline due to habitat loss to have less genetic variation within and high levels of differentiation between populations. For example, western European populations have reduced mtDNA variation within populations but often have unique mtDNA haplotypes (Wayne *et al.* 1992, see below). Similarly, by reducing the effective population size of isolated populations, predator control programs may cause declines in genetic variation, an increase in levels of inbreeding and a disruption of social hierarchies (Ellegren *et al.* 1996; Ellegren 1999; Vilà *et al.* 2003b; Wayne and Vilà 2003). Alternatively, populations that are controlled may also become population sinks if immigration is common which may enhance genetic variation (Frati *et al.* 1998; Wang and Ryman 2001). A critical goal of population genetic analysis is to test alternative predictions about population variability and differentiation given current and historical population changes.

Genetic variation

MtDNA and microsatellite variability within large interconnected wolf populations is generally high (Table 3.3). Large populations in the Old and New World have several mtDNA control region or mtDNA RFLP haplotypes (Wayne *et al.* 1992; Vilà *et al.* 1999; Randi *et al.* 2000) and have high values of nucleotide diversity. Similarly, the average number of microsatellite alleles and average heterozygosity of North American wolf populations, except for Mexican wolves, is 5.0% and 54%, respectively (Roy *et al.* 1994b). These values are similar to those in other vertebrate populations (Avise 1994, 2000). However, mtDNA variation is greater in coyotes than in grey wolves (Lehman and Wayne 1991; Vilà *et al.* 1999) as might be predicted from the higher abundance of the former species (Voigt and Berg 1987; Ginsberg and Macdonald 1990). Although only a small number of coyotes were sampled and their distribution is restricted to North America, the average sequence divergence observed among coyotes is 4.2% as compared to 2.9% among grey wolves sampled from throughout the world (Vilà *et al.* 1999).

Nucleotide and genealogical measures of diversity can be used to reconstruct the historic and recent

Table 3.3 Genetic diversity of populations of the grey wolf

	mtDNA RFLP			Microsatellites			
Population	*n*	Hp	*H*	*n*	Loci	A	He
Vancouver	15	1	0.000	20	10	3.4	0.421
Alaska	50	2	0.350	19	10	4.1	0.536
Alberta	9	3	0.556	20	10	4.5	0.605
Minnesota	35	3	0.259	20	10	6.3	0.532
Central Ontario/ south Quebec	12	2	0.485	24	10	6.4	0.593
West Ontario/ north Quebec	8	1	0.000	14	10	4.1	0.533
NW Territories	71	3	0.463	30	10	6.4	0.547

Note: The number of mtDNA haplotypes (Hp), Gene diversity (*H*), mean number of microsatellite alleles per locus (A), and mean expected heterozygosity (He; unbiased estimate; Nei 1987) are indicated. Modified from Roy *et al.* (1994) and Lehman *et al.* (1992).

demographic history of wolves and coyotes (Vilà *et al.* 1999). Genealogical measures of nucleotide diversity suggest that grey wolves were more abundant than coyotes in the past and that both species declined throughout the late Pleistocene. In general, nucleotide diversity data imply a decline in grey wolves from over 5 million breeding females (about 33 million wolves) worldwide in the late Pleistocene to about 173,000 breeding females (1.2 million wolves) in the recent past. Today less than 300,000 wolves exist worldwide (Ginsberg and Macdonald 1990). This contrasts with a decline followed by a very recent increase in coyotes. As suggested by nucleotide diversity values, coyote numbers decreased from about 3.7 million breeding females (about 18 million coyotes) in the late Pleistocene to 460,000 breeding females (2.2 million coyotes) in the recent past. This drop was followed by an increase to about 7 million coyotes today (Vilà *et al.* 1999 and references therein). These differences between abundance estimates for the recent past and today may reflect habitat loss and direct persecution that reduced wolf populations but increased coyote numbers and distribution (see discussion in Lehman *et al.* 1991 and Vilà *et al.* 1999).

Population bottlenecks

Dramatic demographic declines or population bottlenecks have been historically documented for some wolf populations and genetic studies have found them to contain less genetic variation. For example, the Italian wolf population declined dramatically in the eighteenth and nineteenth centuries due to habitat loss and predator-control programmes (Randi 1993; Randi and Lucchini 1995; Randi *et al.* 2000; Scandura *et al.* 2001). By the 1970s, only about 100 wolves were left in Italy, mostly in the central and southern Apennine Mountains (Zimen and Boitani 1975). Extensive mtDNA RFLP and mtDNA control region sequencing studies showed these wolves to have a single mitochondrial haplotype, which represents lower diversity than that in other Old World populations (Wayne *et al.* 1992; Vilà *et al.* 1999; Randi and Lucchini 1995; Randi *et al.* 2000; Scandura *et al.* 2001). Moreover, the Italian wolf haplotype is unique, and is otherwise found only in French wolves, a population recently founded by wolves from Italy (Taberlet *et al.* 1996b; Lucchini *et al.* 2002; Valière *et al.* 2003). However, levels of microsatellite variation approach that in large wolf populations (Randi *et al.* 2000; Scandura *et al.* 2001; Table 3.3).

Scandinavian wolves have likewise declined over the past few hundred years to the point of near extinction in the 1970s. However, a new group of wolves was discovered in southern Sweden in the early 1980s, and this is thought to be the founding stock of the current Scandinavian population, estimated to be about 100 individuals in 2000. MtDNA and microsatellite studies suggested that the current population is reduced in genetic variation and that variability was being lost over time (Ellegren *et al.* 1996; Vilà *et al.* 2003b). The Scandinavian population has 71% of the variation in the large neighbouring population of Finland and Russia and is fixed for a single mtDNA haplotype. The level of inbreeding observed in the Scandinavian wolves is similar to that of the Swedish captive population (Ellegren 1999) in which inbreeding depression was detected (Laikre and Ryman 1991; Laikre *et al.* 1993). The Scandinavian population has a single control region haplotype and unique microsatellite alleles that were not found in the captive population, excluding the possibility that it had been founded by individuals released from captivity. A recent study of Y chromosome microsatellites (Sundqvist *et al.* 2001) has supported these results as two Y chromosome haplotypes were found in the extant Scandinavian wolf population, and they were different from the only one found in the captive wolves. The reconstruction of the genotype of the founder pair for 19 microsatellite markers, mtDNA and microsatellites in the sex chromosomes (X and Y) suggests that the Scandinavian population was founded by two individuals that successfully migrated from the Finnish–Russian population and established a breeding pack in 1983 (Vilà *et al.* 2003b). During one decade, the population remained small and highly inbred, and the arrival of a new male migrant, reproducing for the first time in 1991, allowed the temporary population recovery of the population and avoided extreme inbreeding (Vilà *et al.* 2003b).

The Mexican wolf (*C. lupus baileyi*) has declined to extinction in the wild due to habitat loss and an extensive extermination programme in the first half of the twentieth century. Two of the three captive Mexican wolf populations had fewer microsatellite alleles and reduced heterozygosity (García-Moreno *et al.* 1996; Hedrick *et al.* 1997). Moreover, only two mtDNA haplotypes were found in the three captive populations (Hedrick *et al.* 1997). The total founding population numbered about seven. In the past, only the certified lineage, founded from three individuals of known Mexican wolf ancestry, was used in the captive breeding programme. However, genetic analysis established a close relationship among the three captive populations and found no evidence of dog, coyote, or Northern grey wolf ancestry (García-Moreno *et al.* 1996; Hedrick *et al.* 1997). Consequently, to preserve the maximum genetic diversity of the Mexican wolf, plans to interbreed the three populations were developed. Like captive Swedish wolves, Mexican wolves showed signs of inbreeding depression (Fredrickson and Hedrick 2002; however, see also Kalinowski *et al.* 1999).

Founding events also can cause population bottlenecks. Inbreeding over time in such populations may contribute to the population decline (Frankham *et al.* 2002). Isle Royale, an island in northern Lake Superior, was founded by a single pair of grey wolves that crossed an ice bridge from the Canadian mainland about 1949 (Mech 1966). Thereafter, this isolated and well-monitored population increased to over 50 wolves by 1980, but then dramatically declined to a dozen or fewer individuals by 1990 (Peterson and Page 1988; Peterson unpublished data). For several years no new litters were born. Disease and changes in food abundance were first suggested as causes for the decline, but both became increasingly improbable explanations when no evidence of disease was found in serological surveys and when wolf numbers did not increase as expected when their main prey, the moose, increased (Peterson and Page 1988; Peterson *et al.* 1998; Peterson 1999). Molecular genetic analysis showed that the Isle Royale wolf population possessed a single mtDNA haplotype, and only one-half the level of allozyme heterozygosity observed in an adjacent mainland population. Furthermore, results of a multi-locus DNA fingerprint survey suggested that the Isle Royale wolves were related about as closely as full siblings or parent–offspring pairs in captivity (Wayne *et al.* 1991a). Inbreeding depression was suggested as the explanation for the population decline as occasionally observed in captive wolf populations (Laikre and Ryman 1991; Laikre *et al.* 1993; Fredrickson and Hedrick 2002; however, see Kalinowski *et al.* 1999). Alternatively, after the population crash to a single breeding pair, only wolves that recognized each other as siblings or parent–offspring were available for pair bonding and they may have avoided breeding (Wayne *et al.* 1991a). Thus, behavioural avoidance of incest may have prevented the formation of additional breeding pairs until individuals from litters with no temporal overlap were produced. A similar explanation has been suggested to explain the lack of growth for the Scandinavian wolf population for one decade, until a migrant arrived and allowed non-incestuous matings (Vilà *et al.* 2003b).

Genetic differentiation

Grey wolves show evidence of genetic differentiation on regional and continental scales. Wolves in the Old and New World do not commonly share mtDNA haplotypes as defined by RFLP (Wayne *et al.* 1992) or by control region sequencing (Vilà *et al.* 1999). However, a network analysis of these data indicates that the New World was invaded multiple times by wolves representing distinct haplotype clades (e.g. haplotypes 28, 29, 30, 31, 32, and 33 in Fig. 3.7). Alternatively, a single invasion followed by geographic lineage sorting is a possibility although work in progress on DNA from Arctic permafrost specimens accumulated over the last 50,000 years supports multiple invasions (Leonard *et al.* in preparation).

The degree of genetic subdivision among populations differs in wolves of the Old and New World (Wayne *et al.* 1992; Randi 1993; Roy *et al.* 1994b; Randi and Lucchini 1995; Ellegren *et al.* 1996; Forbes and Boyd 1996, 1997; Ellegren 1999; Vilà *et al.* 1999; Randi *et al.* 2000; Scandura *et al.* 2001). In the Old World, mtDNA data suggest that most populations are genetically differentiated with the exception of neighbouring populations such as those in Spain and Portugal or recently invaded areas such as France, where Italian wolves have migrated (Taberlet *et al.* 1996c; Vilà *et al.* 1999; Randi *et al.* 2000). In Western Europe, genetic subdivision may reflect recent habitat fragmentation that occurred over the past few 100 years with the loss of forests and, more importantly, a dramatic decrease in the size of all wolf populations due to human persecution (Wayne *et al.* 1992; Vilà *et al.* 1999). Finally, in Asia, new mitochondrial DNA evidence supports two subspecies

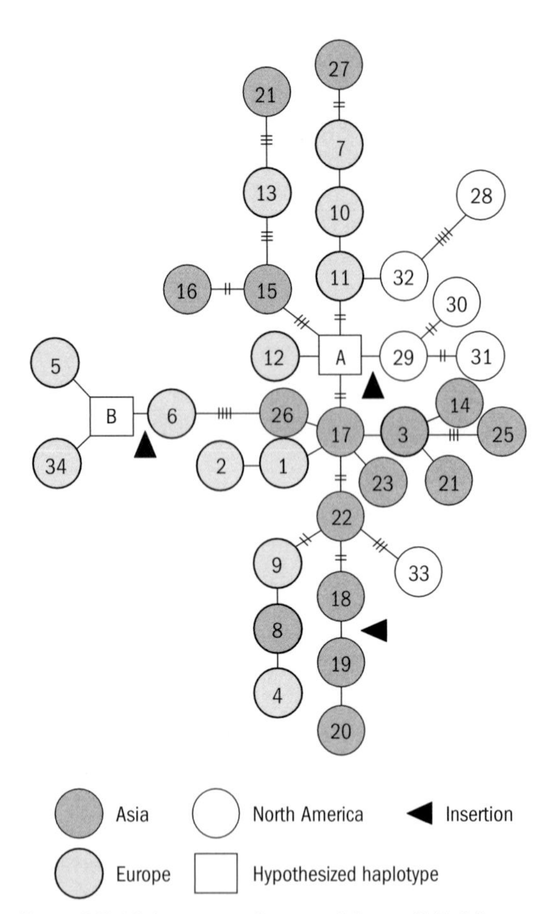

Figure 3.7 Minimum spanning tree of Grey wolf (1-34) haplotypes based on analysis of control region sequences (Vilà *et al.* 1999). Shading of circles indicates continent where each haplotype was found.

of highly distinct Himalayan and lowland Indian wolves (*C. lupus pallipes* and *C. l. chanco*, respectively). (Aggarwal *et al.* 2003; Sharma *et al.* 2003)

However, genetic divergence between populations does not show a consistent relationship with geography. For example, mtDNA control region sequences from China (haplotypes 17, 22, and 23 in Fig. 3.7) and Spain (haplotypes 1, 2, and 4) are more similar than are those from Spain and Italy (haplotype 5). Moreover, some localities, such as Greece, contain highly divergent control region sequences (haplotypes 3, 6, 10, and 11). This pattern suggests the effect of superimposed invasions following the many glaciations of the Pleistocene as well as recent gene flow. With each glacial retreat and the advance of forest into formerly glaciated areas, new waves of immigrating wolves may have added genetic diversity to refugial populations (see Hewitt 2000), resulting in very poorly defined patterns of mtDNA sequence differentiation across populations (Vilà *et al.* 1999). Consequently, the degree of genetic similarity between populations may depend more on the population specific history of immigration and demography rather than on the geographic distance separating populations (e.g. Leonard *et al.* 2000; Barnes *et al.* 2002).

The presence of genetic subdivision in Europe contrasts with the patterns in North America where clinal variation in microsatellite alleles may exist over short distances (Forbes and Boyd 1996, 1997) although it is less apparent at a continental scale (Roy *et al.* 1994b). Similarly, mtDNA haplotypes are shared across large distances (Wayne *et al.* 1992; Vilà *et al.* 1999) but some geographic patterns also are evident. For example, mitochondrial RFLP haplotype W3 was common in Alaska and Northwest Territories but absent from populations in eastern Canada (Wayne *et al.* 1992). Conversely, RFLP haplotype W1 was absent in Alaskan wolves but common in eastern Canada. A similar pattern was observed for mitochondrial control region sequences (Vilà *et al.* 1999). Conceivably, these weak clinal patterns reflect prior Pleistocene isolation in a southern and Alaskan refugia followed by expansion and intergradation during interglacials (Hewitt 2000). Water barriers and differences in prey may also result in differentiation. For example, a recent study found that wolves specializing on different caribou herds in the Canadian Northwest as well as populations on Banks and Victoria Islands were differentiated (Carmichael *et al.* 2001). Finally, another level of complexity is suggested by the recent finding that the Great Lakes wolf population may have been a distinct red wolf-like canid, *Canis lycaon*, which is now interbreeding with grey wolves that have migrated into eastern Canada after the last glaciation and coyotes which have entered the region in the past 100 years (see below; Wilson *et al.* 2000). Regardless, North American grey wolves proved not to be as dramatically structured and reduced in variation as their Old World counterparts as evidenced by the observation that population variability was high and levels of differentiation were low (Wayne *et al.* 1992; Roy *et al.* 1994b, 1996; Vilà *et al.* 1999).

The most highly differentiated North American grey wolf population is the Mexican wolf. Except for a reintroduced experimental population, this subspecies was thought to be extinct in the wild and exists only in three captive populations, each initiated by a small number of founders (García-Moreno *et al.* 1996; Hedrick *et al.* 1997). Two of the captive Mexican wolf populations displayed a single divergent mtDNA haplotype found nowhere else (33 in Fig. 3.7) that is more closely related to a subset of Old World haplotypes than to any New World haplotype. This suggested that Mexican wolves shared a more recent ancestry with wolves from the Old World. Likewise, the most similar control region haplotypes in the New World are five substitutions (2.2%) and one insertion–deletion event different from the Mexican wolf, whereas the most similar haplotypes found in Eurasian populations are three substitutions (1.3%) different from that of the Mexican wolf (Fig. 3.7, Vilà *et al.* 1999). Further, the basal position of the Mexican wolf sequences in phylogenetic trees and analysis of historic museum specimens suggests that the Mexican wolf is a relict form stemming from an early invasion of grey wolves from Asia (Wayne *et al.* 1992; Vilà *et al.* 1999; Leonard *et al.* in preparation).

Wolf–coyote hybridization

Interbreeding between highly mobile species, such as wolves and coyotes, may result in the development of large hybrid zones. The grey wolf once ranged throughout most of North America and parts of Mexico, but over the past few hundred years, wolves have been eliminated from the United States and Mexico. Similarly, the red wolf (*Canis rufus*) was exterminated by about 1975 from throughout its historic distribution which included much of the southeastern United States, although it has since been reintroduced to a refuge in North Carolina (Parker 1987). Coyotes interbred extensively with red wolves as they approached extinction (Nowak 1979) and consequently, mtDNA haplotypes and microsatellite alleles otherwise unique to coyotes are found in red wolves (Wayne and Jenks 1991; Roy *et al.* 1994b). However, an extensive genetic analysis characterizing microsatellite and mtDNA variation in coyotes, grey wolves, and historic and recent red wolves found no markers unique to red wolves. Instead, only haplotypes and microsatellite alleles identical or very similar to those in grey wolves and coyotes were found (Roy *et al.* 1994a,b, 1996). Consequently, an origin of the red wolf through hybridization of grey wolves and coyotes in historic times or earlier was postulated (Wayne and Jenks 1991; Roy *et al.* 1994b, 1996; Reich *et al.* 1999).

Evidence of hybridization between grey wolves and coyotes from Minnesota and eastern Canada was revealed by analysis of mtDNA and microsatellite loci (Lehman *et al.* 1991; Roy *et al.* 1994b). Coyotes invaded Minnesota about 100 years ago and then moved into eastern Canada and New England within the last 50 years (Nowak 1979; Moore and Parker 1992). MtDNA analysis of recent wolves from the Great Lakes region found a high proportion of haplotypes similar to those in coyotes. The frequency of coyote haplotypes in grey wolves increased to the east from 50% in Minnesota to 100% in Southern Quebec. Hybridization between coyotes and wolves in disturbed areas of eastern Canada where wolves had become rare through predator control efforts and habitat loss, but coyotes had become common was hypothesized as the explanation of these results. No sampled coyote had wolf-like haplotypes, suggesting that the predominant cross was between male wolves and female coyotes followed by a backcross to wolves (mtDNA is maternally inherited). This conclusion that wolves and coyotes had hybridized was also supported by microsatellite analysis showing that wolves from this area were genetically more similar to coyotes than to wolves elsewhere and indicated 2–3 successful hybridization events per generation (Roy *et al.* 1994b). Thus, the genetic data imply both that significant hybridization has occurred between the two species and that introgression of coyote genes into the wolf population has occurred over a broad geographic region.

New genetic results question these conclusions (Wilson *et al.* 2000, 2003). Detailed genetic analysis of eastern Canadian wolf-like canids and coyotes has found divergent mtDNA control region haplotypes with a distribution centred at Algonquin Provincial Park, Ontario. These divergent haplotypes appear to be phylogenetically similar to those of red wolves, which in turn are grouped with haplotypes of coyotes. These results may indicate that the smallish Grey wolf that formerly inhabited the Great Lakes areas, *C. lupus lycaon*, and the red wolf, are the same species, designated as *C. lycaon* (Wilson *et al.* 2000). These authors suggest that the Algonquin wolf is a native New World wolf-like form that evolved independently, from North American coyote-like ancestors (see Nowak 2002 for alternative view). This interpretation of new genetic data presents a novel paradigm that should be tested with additional genetic and morphologic data from populations of coyotes and grey wolves (e.g. Hedrick *et al.* 2002).

Finally, a recent analysis of coyotes in the southeastern United States has shown that one dog haplotype appeared in multiple individuals across a large area (Adams *et al.* 2003). This suggests an ancient coyote–domestic dog hybridization event when the first coyotes were expanding into eastern habitats formerly occupied by red wolves.

Wolf–dog hybridization

In the wild, hybridization between grey wolves and dogs is likely to be most frequent near human settlements where wolf density is low and habitats are fragmented, and where feral and domestic dogs are common (Boitani 1983; Bibikov 1988). The genetic integrity of wild wolf populations has been a concern among some conservationists (Boitani 1984; Blanco *et al.* 1992; Butler 1994). The majority of wolf populations show no evidence of hybridization (Vilà and Wayne 1999). However, genetic studies have detected limited wolf–dog hybridization in Bulgaria, Italy, Latvia, Scandinavia, and Spain (Dolf *et al.* 2000; Randi *et al.* 2000; Andersone *et al.* 2002; Randi and Lucchini 2002; Vilà *et al.* 2003a, in preparation).

Pack structure and mating systems

Generally, a wolf pack consists of a mated pair, their immediate offspring, and adult or subadult helper offspring from previous years (Mech 1999b). In areas that have sufficient resources to support many wolves, packs develop well-defined territories, and interpack aggression can be intense (Mech 1999). Thus, within each pack most members are closely related, and they are less related to neighbouring packs. Microsatellite analyses of wolf packs in Denali National Park, Alaska and in Northern Minnesota confirmed these relationships (Lehman *et al.* 1992; Smith *et al.* 1997a). Additionally, genetic fingerprinting showed that wolf packs are not inbred and offspring disperse into neighbouring packs, despite high levels of interpack aggression, or form new packs nearby. In Minnesota, for example, multilocus genetic fingerprinting showed that seven interpack genetic similarity values were as large as those between known siblings or parent–offspring pairs (Lehman *et al.* 1992) and in Denali National Park, six such interpack connections were discovered. However, no interpack genetic similarity was observed among wolf packs in the Inuvik region of Canada's Northwest Territories. The difference in the number of relatedness connections among the three populations may indicate higher genetic turnover in populations such as Inuvik, where wolves are heavily controlled.

Conservation implications

Several conservation implications are suggested by the genetic results. First, because the endangered Mexican grey wolf is genetically and physically distinct, and historically isolated from other grey wolves (Nowak 1979), the breeding of pure Mexican wolves in captivity for reintroduction into the wild is advised. Second, because most wolf populations in North America are not strongly differentiated genetically and gene flow is high among populations, reintroduction need not include only the nearest extant populations as source material. Although the reintroduced Yellowstone wolves are slightly different from naturally recolonizing wolves in Montana, the minor difference is not a conservation concern (Forbes and Boyd 1997). However, reintroducing wolves from populations where hybridization with coyotes has occurred is perhaps not advisable (see below). Finally, genetic analysis of recolonized populations in Montana and France has found that high levels of genetic variation can be preserved (Forbes and Boyd 1997; Scandura *et al.* 2001).

The grey wolf has been divided into as many as 32 subspecies worldwide (Hall and Kelson 1959). Nowak (1995) suggested that the 24 North American subspecies should be reduced to 5. However, rates of gene flow among North American wolf populations are high, and differentiation by distance characterizes the genetic variation of wolves at some geographic scales. In this sense, typological species concepts may be inappropriate because geographic variation in the wolf is distributed along a continuum rather than being partitioned into discrete geographic areas delineated by fixed boundaries. A focus on locality-specific adaptations to prey size or climate (e.g. Thurber and Peterson 1991; Carmicheal *et al.* 2001) or size variation with latitude may be a more appropriate guide to conservation rather than arbitrary boundaries of a continuously distributed and highly mobile species (Crandall *et al.* 2000). Finally, although contemporary wolf populations in Europe appear more genetically subdivided than their North American counterparts (Wayne *et al.* 1992; Vilà *et al.* 1999; Randi *et al.* 2000), the North American pattern might well reflect the ancestral condition in Western Europe prior to habitat fragmentation and population decimation. Therefore, efforts to increase gene flow among European wolf populations to levels similar to that in North America could be defended.

The possible presence of a hybrid zone between a native northeastern wolf species, *C. lycaon*, and coyotes and grey wolves (see above) complicates taxonomic and conservation recommendations. If *C. lycaon* is a distinct species, conspecific with the red wolf, then captive breeding and conservation efforts *in situ* may be urgently needed. If *C. lycaon* is a hybrid between grey wolves and coyotes that is due to human-induced habitat changes and predator control efforts, then further conservation efforts may not be warranted (Jenks and Wayne 1992; Wayne and Brown 2001). For the hybridization process to be of conservation concern, even hybridization between a unique North American wolf and other canids, it should be caused by human activities rather than natural processes, such as glacial-induced range expansions. Additional genetic data involving multiple mitochondrial, nuclear, and Y chromosome markers are needed to better test alternative hypotheses for the origin of the red wolf and the Algonquin wolf. Finally, wolf–dog hybridization is a non-natural occurrence that fortunately may be of concern only in a few European populations (see above).

Perhaps of greater concern is the loss of genetic variation in isolated wolf populations in the Old World (see above). Inbreeding depression has been documented in captivity (Laikre and Ryman 1991; Laikre *et al.* 1993; Federoff and Nowak 1998; Fredrickson and Hedrick 2002; but cf. Kalinowski *et al.* 1999). Italian, Scandinavian, and Isle Royale wolves have levels of average relatedness approaching inbred captive populations (see above), and could conceivably suffer a decrease in fitness that would eventually affect population persistence (Mace *et al.* 1996; Hedrick and Kalinowski 2000). High levels of gene flow likely characterized Old World populations in the past, so there is reason to restore past levels of gene flow in parts of Europe, either through habitat restoration and protection along dispersal corridors or through translocation. Future research should be aimed at monitoring and predicting genetic changes that will occur in wolf populations and trying to determine any possible population effects.

One outstanding question is the genetic effect of wolf harvesting. Preliminary fingerprint data suggested that one heavily controlled population had fewer kinship ties and more genetic turnover than two protected ones (see above, Lehman *et al.* 1992; Williams *et al.* 2003). If interpack kinship affects social stability and pack persistence (Wayne 1996), then control plans that minimize the effects on genetic population structure may need to be considered. Another need is to assess the extent to which population genetic variation is relevant to survival and reproduction and hence to the persistence of populations. The role of the MHC in immunity to infectious disease is an important candidate in this regard (Hedrick and Kim 1999; Hedrick and Kalinowski 2000; Hedrick *et al.* 2000, 2002; Seddon and Ellegren 2002; Aguilar *et al.* in press). Finally, new genetic techniques such as microsatellite analysis, Y chromosome haplotyping (Sundqvist *et al.* 2001; Vilà *et al.* 2003a) and faecal DNA typing promise a new understanding of wolf mating systems and of the role of kinship on behaviour. Faecal DNA analysis is a non-invasive approach that can be used to census populations, document patterns of mating and kinship, and assess sex ratios and population differentiation (Kohn and Wayne 1997; Kohn *et al.* 1999; Luchinni *et al.* 2002).

The African jackals

The golden (*Canis aureus*), black-backed (*C. mesomelas*), and side-striped jackals (*C. adustus*) are medium-sized and morphologically similar African canids. All have been studied with molecular genetic techniques (Wayne *et al.* 1989a, 1990a,b). The three species are found in savanna and acacia woodland habitats in sub-Saharan Africa. The side-striped and black-backed jackals have nearly parallel distributions from Kenya to South Africa whereas the golden jackal has a distribution restricted to East Africa but also includes much of the Eurasian subcontinent. The golden jackal has probably entered Africa only in the last half million years (Van Valkenburgh and Wayne 1994). All three species can be found in the same habitats although the side-striped jackal is clearly more common in closed woodland, whereas the other jackals frequent plains and open acacia woodlands (Fuller *et al.* 1989). Additionally golden jackals may be more diurnal than the black-backed which has a crepuscular activity cycle (Fuller *et al.* 1989). All three species take vertebrate prey although the side-striped jackal include more non-vertebrate prey and fruits in their diet (Van Valkenburgh and Wayne 1994 and references therein). The overlap in distribution and prey suggests interspecific competition, which is further supported by morphological studies showing divergence in tooth size and shape (Van Valkenburgh and Wayne 1994). The presumed divergence in activity and diet in sympatry support evolutionary hypotheses that can be further tested with genetic data. First, what are the evolutionary reasons for the general size and phenotypic convergence among the three jackal species? Presumably, competition could be occurring but if sympatry is recent there could be a time lag in response (Wayne *et al.* 1989a). Alternatively, competition could cause divergence not in size, as is common in carnivores (Van Valkenburgh 1994), but along other more subtle traits. Second, is ecological and behavioural divergence sufficient to maintain reproductive isolation?

Population variation and differentiation

The three jackal species have substantial levels of mtDNA RFLP and protein variation. In a sample of 26 golden, 29 black-backed, and 8 side-striped jackals from East Africa, 10, 12, and 3 haplotypes were found (Table 3.1). Levels of protein heterozygosity approached 10%, a value that is moderate to high for canids and other vertebrates (Table 3.1). The three species differed in the amount of sequence variation between haplotypes (Fig. 3.8). The amount of sequence divergence with the golden and side-striped jackals was low, but with East African black-backed jackal two divergent sequence clades were found that differed by 8–12% sequence divergence, which is as much as that commonly found between canid genera (Wayne *et al.* 1997). Moreover, the rate of molecular evolution appeared to differ by 50% in the two lineages implying a dramatic acceleration of sequence evolution in one lineage. However, this apparent difference was dependent on the outgroup used because when a wide range of outgroups was included in the phylogeny, the rate difference was not as great (Wayne *et al.* 1997). Finally, the amount of differentiation among East African populations was in general low; however, the presence of the two divergent lineages only in the Rift Valley and not elsewhere suggested it was an obstacle to dispersal. Further, new research (Wayne *et al.* in preparation) suggests that South African black-backed jackals have haplotypes from only one of the two divergent clades in Fig. 3.8, but within this clade there is high sequence diversity. This result implies that the clade originated in South Africa (see Wayne *et al.* 1989a) and suggests a genetic division between East and South Africa similar to that in wild dogs (Girman *et al.* 2001).

The molecular data clearly suggests that a time lag does not explain the lack of size divergence among the three species of African jackals. Relative to two sympatric South American fox species that have diverged by 50% in size, the three jackal species have an order of magnitude more sequence divergence. Thus, ecological divergence can occur rapidly, in a few 100,000 years or less (Wayne *et al.* 1989a). In African jackals, interspecific competition appears to have primarily caused divergence in diet, habitat use, and time of activity (Fuller *et al.* 1989; Van Valkenburgh and Wayne 1994). Further, the recent insinuation of the golden jackal into the East African jackal guild appears to have caused a subtle response in tooth dimensions of the East African black-backed jackals that is absent from its South African conspecifics. Instead of size divergence, the three species show a size convergence that may be caused by competitive interactions with a diverse array of carnivorous species below and above the size category inhabited by the three jackals species (Van Valkenburgh and Wayne 1994). Finally, ecological or behavioural divergence in the three species seems sufficient to maintain reproductive isolation as no haplotypes were shared among them.

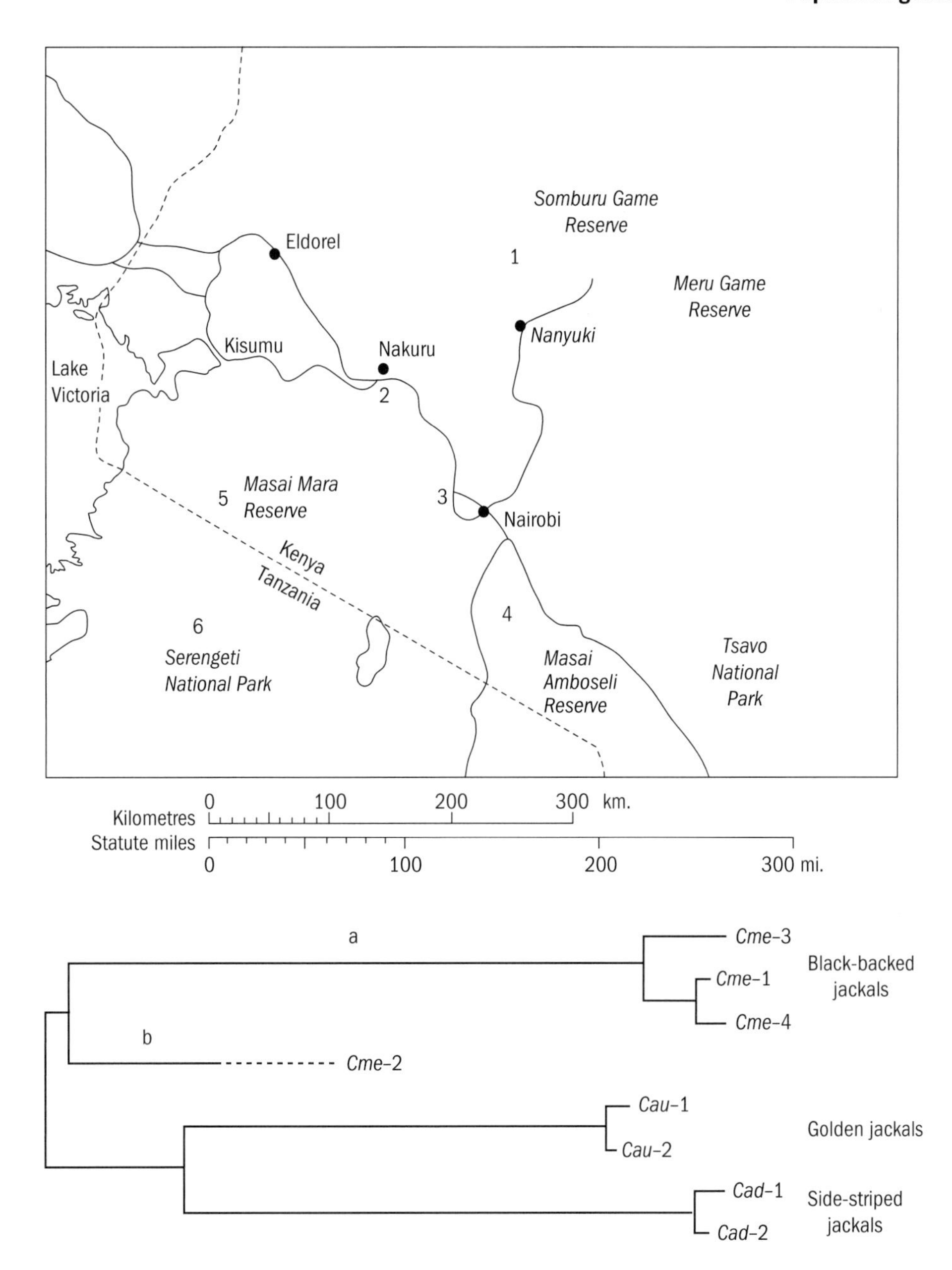

Figure 3.8 Sampling localities and phylogeny of mtDNA haplotypes of African jackals (Wayne *et al*. 1989a, 1990a).

Conservation implications

Jackals have a genetically diverse ancestry (Fig. 3.8) and appear to have distinct ecological roles in their community. The preliminary genetic survey of African jackals shows that they do not interbreed in the wild. However, as in wolf-like canids, interbreeding may occur under circumstances where jackals are rare and other canids abundant or may occur in human altered habitats. There have been heresay reports of jackals interbreeding with dogs in urban settings and populations under a wide variety of ecological conditions should be surveyed to assess the possibility of hybridization. The genetic data indicate that some regional subdivision may exist among black-backed populations across the Rift Valley and at larger geographic scales between East and South African populations. These genetic differences among populations imply that topographic barriers and distance are important considerations in choosing genetic units for conservation. Within populations, high levels of variation suggest that numbers are sufficient to maintain population-level adaptation.

Kit fox and swift fox

Small canids such as foxes may have limited dispersal ability and be less able to traverse topographic barriers. Moreover, due to shorter dispersal distances, small canids may show a more pronounced pattern of genetic differentiation with distance and population subdivision. The small arid land foxes of North America are habitat specialists and relatively poor dispersers. In California, for example, the kit fox of the San Joaquin Valley, whose range is circumscribed by the coastal mountain range to the west and the Sierra Nevada mountain range to the east, is considered a distinct subspecies (*Vulpes macrotis mutica)* and is protected by the US Endangered Species Act (Hall 1981; O'Farrell 1987). Populations to the east of the Rocky Mountains are collectively referred to as swift foxes (*Vulpes velox*), and those to the west as kit foxes (*V. macrotis*) (Fig. 3.9). However, the two forms hybridize in north-central Texas and are recognized as conspecific by some authors (Packard and Bowers 1970; Rohwer and Kilgore 1973; Nowak and Paradiso 1983; O'Farrell 1987; Dragoo *et al.* 1990).

Population variation and differentiation

Results of mtDNA RFLP and cytochrome b sequence analyses suggest that genetic divergence is related to the distance between populations and the severity of the topographic barriers separating them (Mercure *et al.* 1993). A survey of 75 foxes from the north and south San Joaquin Valley identified three RFLP haplotypes that together defined a significant monophyletic group (Figs. 3.9, 3.10 and 3.11). A major genetic subdivision within the kit–swift fox complex distinguished populations from the east and west side of the Rocky Mountains, consistent with the taxonomic distinction between *V. macrotis* and *V. velox*. The divergence between these taxa was nearly as great as that between them and the Arctic fox (*Alopex lagopus*) (Fig. 3.10), often classified in a separate genus (Fig. 3.9). Furthermore, within each of the two major kit–swift fox mtDNA clades, genetic distances among populations tended to increase with geographic distance (Mercure *et al.* 1993). The distinct phylogeographic pattern in the kit–swift foxes contrasts with the lack of pattern observed in coyotes and grey wolves (Lehman and Wayne 1991) and suggests that the kit and swift fox may be distinct species. However, the two forms hybridize in a contact zone in New Mexico and microsatellite evidence indicates hybridization occurs freely within the hybrid zone (Dragoo and

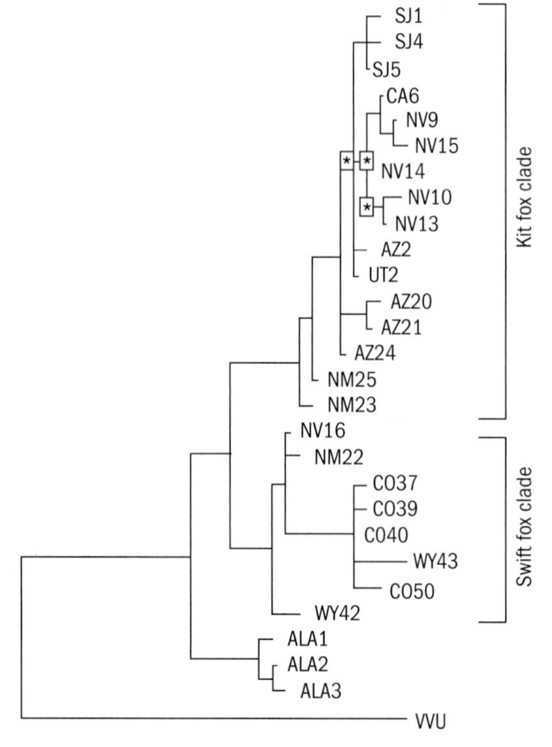

Figure 3.9 Phylogenetic tree of kit and swift fox haplotypes based on analysis of mtDNA restriction site data (Mecure *et al.* 1993). Haplotype codes indicate individual ID numbers and localities (see Mecure *et al.* 1993). Localities are coded as follows: SJ, San Joaquin Valley; NV, Nevada; CA, Mohave, California; UT, Utah; AZ, central and southeast Arizona; NM, New Mexico; CO, Colorado; WY, Wyoming. The New Mexico (NM) locality is in the kit–swift fox hybrid zone. Outgroup species are: ALA, *Alopex lagopus* (Arctic fox) and VVU, *Vulpes vulpes* (red fox).

Figure 3.10 Arctic fox cubs *Alopex lagopus* © L. Dalén.

Wayne 2003; Dragoo in preparation; Figs. 3.10 and locality 4 in 3.11).

The Mexican kit fox (*V. macrotis zinseri*) is an endangered subspecies that is restricted to the arid plains of the Sierra Madre Mountains of Mexico (Fig. 3.11). Because of its intermediate geographic position, the Mexican kit fox could be assigned to either kit or swift fox clades. Eight hundred base pairs of cytochrome *b* sequence and three hundred and sixty base pairs of control region sequence were analysed to determine if Mexican kit foxes were genetically distinct and to assess their relationship to kit and swift foxes of the United States (Maldonado *et al.* 1997). Two haplotypes were found in foxes from Nuevo León and Coahuila, Mexico that differed by a single base pair substitution (Figs. 3.11). These haplotypes were phylogenetically grouped with those from kit foxes in Arizona, Utah and New Mexico but differed from them by 1–4 substitutions (0.3–1.2%) in control region sequence. Consequently, these results suggest that the Mexican kit foxes are most closely related to the kit fox and populations west of the Rockies.

Kinship and group structure

Kit fox society appear to be structured in large part by kinship bonds. A genetic and observational study of 35 San Joaquin kit foxes found that kit foxes can be solitary and paired and that paired foxes may have an additional adult joining the group (Ralls *et al.* 2001). This additional adult was related to the mated pair in two cases suggesting it was an offspring of a previous year. Moreover, neighbouring females tended to share high relatedness and may share dens suggesting lifelong relationships with offspring. However, unrelated individuals may sequentially use the same den and be found in social groups. In the former instance, use of the same den by unrelated individuals may be unsuccessful pairing events. Overall, the population deviated from random mating, reflecting the non-random distribution of relatives. These results suggest that reintroduction efforts need to consider kinship ties in any genetic management or restoration plan.

Conservation implications

The San Joaquin and Mexican kit fox are genetically distinct populations that are related to kit foxes west of the Rocky Mountains. This degree of distinction suggests a limited history of isolation and provides some support for special preservation efforts. Topographic barriers such as the Colorado River or habitat barriers appear to influence geographic differentiation, but the predominant pattern within clades is one of geographic differentiation with distance. The scale of differentiation with distance is much finer in kit/swift foxes than in large canids reflecting differences in dispersal abilities and suggesting that a larger

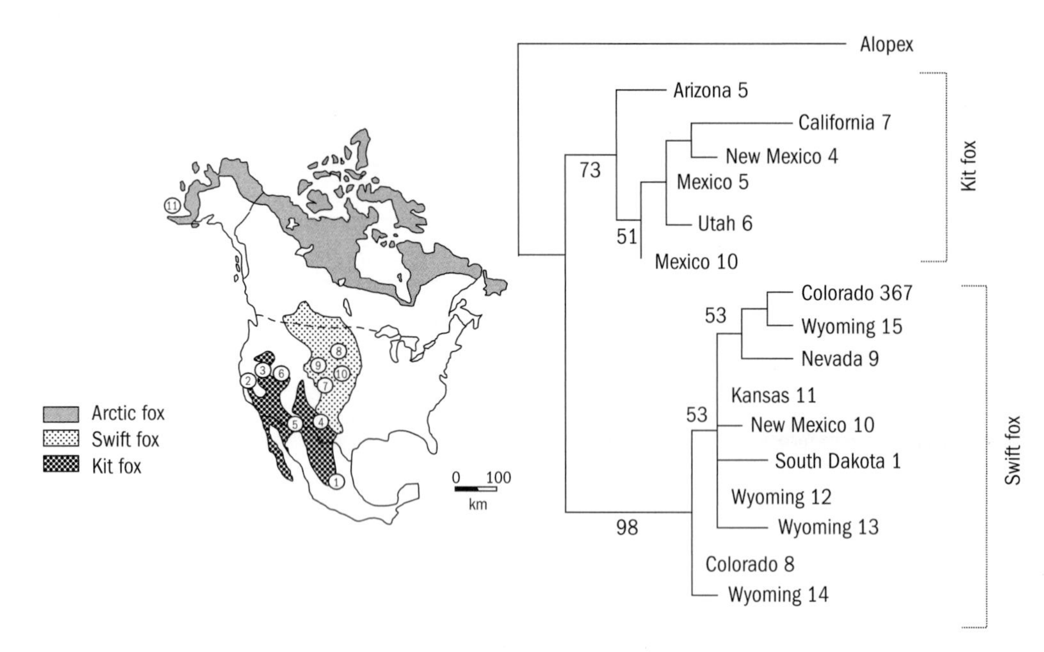

Figure 3.11 Present geographic range and sampling localities of kit foxes, swift foxes, and arctic foxes from Maldonado *et al.* (1997). Phylogeny of haplotypes with numerical id is shown.

number of genetic units of conservation concern can be defined in small canids. With respect to the design of reintroduction programmes, source stocks for small canids should in general be drawn from smaller geographic areas than the large canids. For example, given the mtDNA findings (Figs 3.9 and 3.11), the recent use of foxes from Colorado and South Dakota rather than New Mexico or Texas as a source for a reintroduction into Canadian Province, Saskatchewan, appears to have been appropriate (Scott-Brown *et al.* 1987). However, the Mexican kit fox has unique haplotypes and merits conservation concern as they are distinct and isolated from their conspecifics in the United States. Levels of variation do not seem critically low, however, and haplotype diversity was similar to that in other canid populations (Table 3.1).

Island fox

The island fox, *Urocyon littoralis*, is an endangered species found only on the six Channel Islands off the coast of southern California (Gilbert *et al.* 1990; Wayne *et al.* 1991b). The island fox is an insular dwarf, about two-thirds the size of its mainland ancestor, the gray fox, *Urocyon cinereoargenteus* (Collins 1991a; Wayne *et al.* 1991b). As suggested by the fossil and geological record, about 16,000 years ago, the three northern islands, which at that time were connected to one another, were colonized by foxes from the mainland. As sea level rose, 9500–11,500 years ago, the northern islands were separated. About 4000 years ago, foxes first arrived on the southern Channel Islands and were probably brought there by Native Americans. This succession of events, combined with estimates of population size and island area, allows predictions about relative levels of genetic variation in the island populations (Table 3.4). For example, the smallest and last founded populations should have the least genetic variability (San Miguel and San Nicolas Islands, Table 3.4). The largest and first founded populations should have the highest levels of variation (Santa Cruz and Santa Rosa Islands). Past population sizes vary from several hundred to several thousand individuals (Table 3.4, Wayne *et al.* 1991b; Roemer *et al.* 2002).

The ecology of island and continental populations may differ because the finite area of small islands

Table 3.4 Genetic diversity of each island population of the island fox

Island	Ne	DI	He(a)	mtDNA	He(f)	V_r
San Miguel	163	9500	0.008	0.031	0.13	0.62
Santa Rosa	955	9500	0.055	0.031	0.34	0.97
Santa Cruz	984	11000	0.041	0.026	0.19	1.12
San Nicolas	247	2200	0.000	0.000	0.00	0.00
Santa Catalina	979	2300	0.000	0.075	0.45	2.06
San Clemente	551	3800	0.013	0.000	0.25	2.15

Note: The effective population size (Ne), duration of isolation (DI in years), allozyme heterozygosity (He(a)), average number of substitutions per nucleotide for RFLP mtDNA (mtDNA), minisatellite fingerprint DNA heterozygosity (He(f)), and variance in repeat scores of 19 microsatellite loci (V_r). Modified from Goldstein *et al.* (1999) and Wayne *et al.* (1991).

constrains dispersal and gene flow. Moreover, the simple structure of island communities diminishes the intensity of biological interactions, such as interspecific competition and predation (Stamps and Buechner 1985; Adler and Levins 1994). Island populations typically have higher and more stable population densities, increased survivorship, reduced fecundity, and decreased dispersal distances (Adler and Levins 1994). Additionally, changes in social ecology including reduced aggression towards conspecifics, reduced territory size, increased territory overlap with neighbours, and an abandonment of territoriality may occur (Stamps and Buechner 1985). These changes have been termed the 'Island Syndrome' (Adler and Levins 1994). In island foxes these ecological, behavioural, and genetic predictions of the island syndrome can be tested. Specifically, because dispersal opportunities on small islands are more limited than on the mainland, dispersal distances should be smaller than for mainland canids of similar body size. Additionally, the consequences of shorter dispersal distances may include higher local levels of kinship among island foxes, a greater degree of population substructure, and possibly a higher frequency of incestuous matings (Stamps and Buechner 1985; Adler and Levins 1994). Second, due to reduced metabolic requirements, it has been suggested that home range size should be smaller (Gittleman and Harvey 1982). Moreover, home range overlap should be greater and territoriality either reduced or absent (Stamps and Buechner 1985).

Population variation and differentiation

In general, predictions about genetic variation and island area and colonization time were supported by molecular genetic analyses (Table 3.4; Gilbert *et al.* 1990; Wayne *et al.* 1991b; Goldstein *et al.* 1999). The small, late colonized, San Nicolas population was invariant in all genetic markers surveyed including multi-locus DNA fingerprints and 19 microsatellite loci (Gilbert *et al.* 1990; Goldstein *et al.* 1999; Roemer *et al.* 2001a, 2002; Table 3.4). Only inbred mice strains show a similar lack of variation but no other wild population except the inbred eusocial naked mole rats approaches this level of monomorphism (Jeffreys *et al.* 1985; Reeve *et al.* 1990). Similarly, foxes from the smallest island, San Miguel, had low levels of variation. In contrast, those from the large islands, Santa Catalina, Santa Rosa, and Santa Cruz had higher levels of variation. However, the Santa Cruz island population, although founded early had lower levels of mini satellite variation (He(f)) than expected and Santa Catalina, although founded last, had the highest levels of variation (Table 3.4). Analysis of mtDNA RFLP polymorphisms suggested that Santa Catalina island may have been colonized multiple times from Southern and Northern islands. Finally, although protein loci also showed the predicted pattern, Santa Rosa and Santa Cruz islands had higher levels of heterozygosity than implied by levels in other genetic markers perhaps suggesting the action of selection on these protein loci (Table 3.4). Similarly, high variability in the MHC of San Nicolas island foxes indicates the action of strong balancing selection (Aguilar *et al.* in press).

All populations were well differentiated (Fig. 3.12; Wayne *et al.* 1991b; Goldstein *et al.* 1999). The island foxes did not share RFLP haplotypes with the mainland grey fox and some populations had unique haplotypes. For example, within the southern group of islands, the small population on San Nicolas possessed a unique mtDNA haplotype. Similarly, island populations had unique multi-locus fingerprint bands and microsatellite alleles and differed in allele frequencies. Consequently, foxes could be correctly classified to island of origin, and haplotype trees

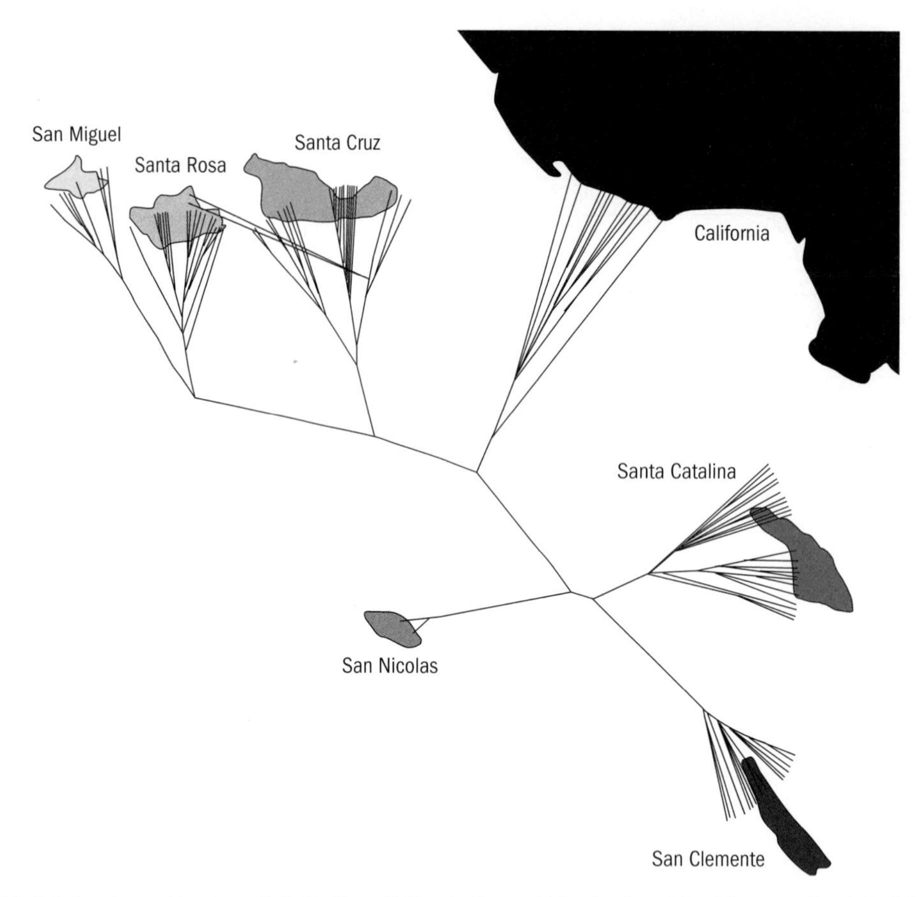

Figure 3.12 Relative size and location of six Southern California Channel Islands where island foxes are found. Individual lines represent multi-locus genotypes and their relationship based on 14 microsatellite loci (Goldstein *et al.* 1999). Two foxes have genotypes grouped with those of Santa Cruz foxes suggesting they were translocated to the island. All others are grouped with island of origin.

resolved an evolutionary history of colonization consistent with the archaeological record (Fig. 3.12; Wayne *et al.* 1991b; Goldstein *et al.* 1999).

The island syndrome

Island syndrome predictions were tested by simultaneous analysis of spatial distribution, relatedness, and paternity in Santa Cruz island foxes from Frazer point (Roemer *et al.* 2001a). Home range size was assessed by radio-telemetry and mark-recapture data. Patterns of relatedness among foxes within a single area and genetic differences between foxes from two areas were assessed using 10 microsatellite loci. The results showed that Santa Cruz island foxes have diverged from other mainland foxes in aspects of their demography, social ecology, and genetic structure. Dispersal distances in island foxes were very low (mean = 1.39 km) relative to other canids of similar size (Roemer *et al.* 2001a). Home range size was the smallest (mean = 0.55 km^2) and density is nearly the highest recorded of any canid species (2.4–15.9 foxes/km^2). As inferred from genetic and observational data, the island fox social system consists of mated pairs that maintain discrete territories. Overlap among mated pairs was high whereas overlap among neighbours, regardless of sex, was low. However, island foxes are not strictly monogamous, 4 of 16 offspring resulted from extra-pair fertilizations. Mated pairs were unrelated, however, suggesting inbreeding avoidance. Genetic subdivision was apparent between populations separated by only 13 km (F_{ST} = 0.11). These observations are consistent with the island syndrome and suggest that limited dispersal opportunities imposed by small body size and the limited area of the island are the primary influences on the demography and social structure of the island fox.

Conservation implications

Fox populations on five of six islands have decreased dramatically over the past 10 years (Roemer *et al.* 2002). On the three Northern islands, the decline was due to predation by eagles (Roemer *et al.* 2001b, 2002). On Santa Catalina island, the decline was due to a distemper epidemic and on San Clemente, a more gradual decline likely reflected predator control efforts of the shrike reintroduction programme (Roemer and Wayne 2003). In each case, genetic management of the remaining population is needed. On the Northern islands, captive breeding is necessary to restore the wild populations, and preliminary studies have suggested that the captive population has sampled a limited subset of variation in the wild implying additional founders would be a beneficial addition to the captive breeding programme (Aguliar and Wayne unpub. data; Gray *et al.* in preparation). On San Miguel island, the wild population may be extinct, but on the other two Northern islands, a dynamic exchange of wild and captive born foxes to enrich genetic variability is conceivable. On Santa Catalina island, foxes have disappeared from about 90% of the island and a captive breeding programme of survivors has been established to assist in replenishing the lost populations. However, a wild reservoir of over 150 foxes exists on the far western end of the island and genetic data indicated they provide a more genetically variable source for reintroduction (Aguliar and Wayne unpub. data). On San Clemente island, several hundred individuals remain in the wild and significant genetic loss is unlikely to have occurred. However, there should be immediate efforts to stabilize the population and prevent further decline (Roemer and Wayne, 2003).

The genetic results suggest that each island population should be treated as a separate conservation unit. Further, low levels of genetic variation in each island population relative to mainland grey foxes imply that they may be more vulnerable to environmental changes (Frankham *et al.* 2002). In previous conservation plans, the species has been treated as a single taxonomic unit with a combined population of about 8000 individuals (California Code of Regulations 1992). However, as is now clear, by virtue of their isolation and small size, the islands are more vulnerable than an equivalently sized mainland population. Each island should be designated an independent unit with regard to conservation and at least five populations should be considered in immediate danger of extinct (Roemer and Wayne 2003). Study of captive populations combined with careful genetic management may allow successful reintroduction and more informative management of wild populations in the future.

Darwin's fox

On Chiloé Island off the west coast of Chile, Charles Darwin observed and was the first to describe a small endemic fox, *Pseudalopex fulvipes* (Fig. 3.13). Darwin's fox has the smallest geographic range of any living canid (Osgood 1943; Cabrera 1958), and the unique island temperate rainforest it inhabits is not duplicated elsewhere. There are perhaps less than 500 foxes currently in existence and none in zoos. Darwin's fox is distinctive in having a small body size, short legs, and abbreviated muzzle (Osgood 1943). Related foxes, widespread on mainland Chile, and from which Darwin's fox presumably arose, are the South American chilla, *P. griseus*, and the culpeo fox, *P. culpaeus*. The former is about 50% larger than Darwin's fox, and often assumed to be conspecific with it (Honacki *et al.* 1982; Wozencraft 1993). The culpeo is generally the largest of the three, but size variation within both the chilla and the culpeo is so extreme that these species are difficult to distinguish in some areas (Fuentes and Jaksic 1979).

Darwin's fox was thought to be recently isolated from mainland foxes given that the channel separating Chiloé from the continent is only about 5 km wide, and the island was likely connected to South America when sea levels were lower during the last glaciation (*c.*13000 years BP; Yahnke 1994). However, recent reports of Darwin's fox on the mainland in

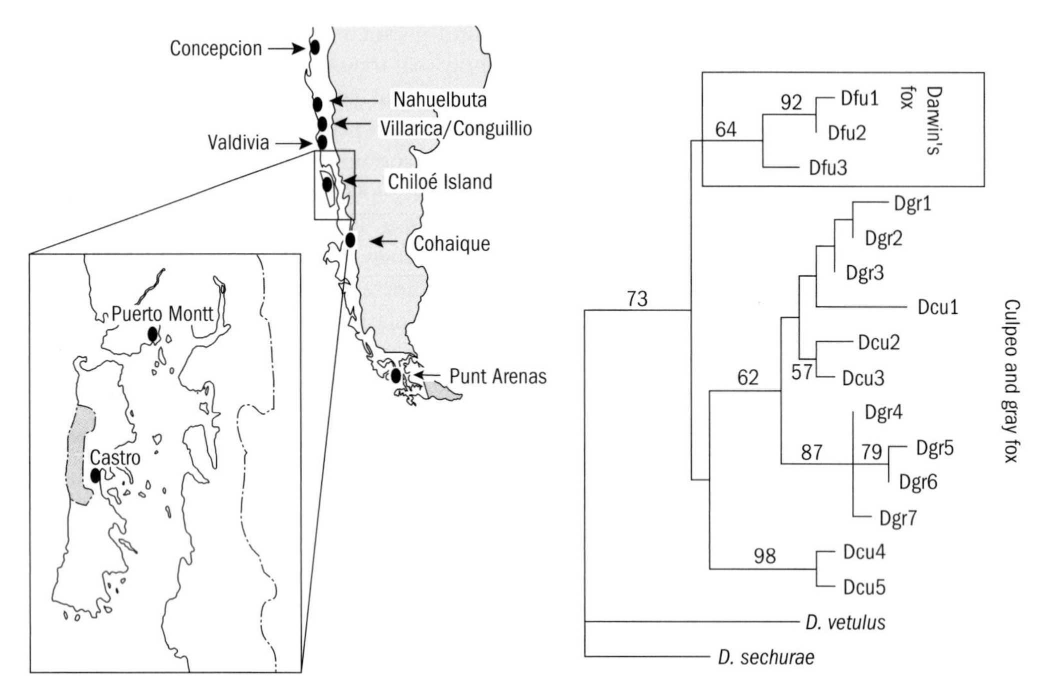

Figure 3.13 Sampling localities and phylogenetic tree of Chilean foxes based on analysis of mtDNA control region sequence (Yahnke *et al.* 1996). A Sechuran fox (*Pseudalopex sechurae*) from Peru and a hoary fox (*Lycalopex vetulus*) from Brazil are used as outgroup species. Bootstrap values are given at the base of nodes and indicate relative support for various clades.

central Chile, in Nahuelbuta National Park 350 km from the coast (Fig. 3.13; Medel *et al.* 1990) where they are sympatric with the mainland chilla, suggested that Darwin's fox may be a distinct species.

Genetic variation and differentiation

Control region sequences were obtained from two Darwin's foxes on the mainland and one from Chiloé Island as well as from chilla and culpeos from the mainland (Yahnke *et al.* 1996). Genetic variation was high in the mainland foxes. Phylogenetic analysis of sequences confirmed the close genetic relationship of mainland and island population of Darwin's fox, and suggested that they may be distinct populations within a single species (Fig. 3.13). Darwin's fox appears to have diverged early in the radiation of Chilean foxes, and is at least as divergent from the chilla and culpeo as the latter two are from one another. These results indicate that Darwin's fox is a relict form, having evolved from the first immigrant foxes to Chile after the land bridge formed between North and South America about 2–3 million years ago (Marshall *et al.* 1979; Webb 1985; Yanke *et al.* 1996).

Conservation implications

Darwin's fox is genetically distinct and appears to be the progenitor of mainland fox species. The genetic results suggest it had a previous distribution on the mainland, rather than having been introduced there by humans. Darwin's fox also has a morphology unlike mainland foxes, and occupies a restricted and unique temperate rainforest habitat. Recent surveys have revealed that small and isolated mainland populations may still exist, or may have existed until very recently (Vilà *et al.* in press). Darwin's fox needs to be considered a distinct species of urgent conservation importance. The island population needs greater protection and plans for reintroduction to the mainland need to be considered. No captive populations are currently established. Captive breeding and observation of Darwin's fox might provide understanding of the species that will assist in its conservation.

Arctic fox

The Arctic fox has a circumpolar distribution and with the grey wolf, shares the snowy environments of the tundra. Although wolves have been surveyed genetically across their geographic range, only a few populations of Arctic fox have been surveyed in Fennoscandia and Siberia (Dalen *et al.* 2002). Limited variation was found differentiating Fennoscandian and Siberian foxes as might be expected in a species of limited mobility. In contrast, high levels of variation were found within populations. These data suggest that moderate population numbers are sustained over considerable time periods. A more extensive survey of Arctic foxes worldwide confirms these moderate levels of within and between population variations (Waidyaratne *et al.* in preparation)—a result which contrasts with that of grey wolves in which low levels of within and between population variation were found.

Red fox

The red fox (*Vulpes vulpes*) is a widely distributed generalized, medium-sized canid that is found throughout Eurasia and North America. It adapts well to urban and pastoral settings and has been introduced to Australia and other islands. The genetics of invading species and genetic structure that results in natural and artificial habitats is of interest as is their evolutionary history in Europe and elsewhere.

Population variation and differentiation

Population variability of red foxes appears to be moderate. A study of 120 European foxes based on 9 polymorphic allozyme loci and analysis of cytochrome *b* variation in a subset of 41 foxes found levels of heterozygosity ranging from 0% to 4.4%. A total of 18 cytochrome *b* haplotypes was found. East and Western European foxes appeared to be genetically differentiated perhaps reflecting different ice age refugia or colonization waves after glaciation events (Frati *et al.* 1998). Studies of Australian foxes have also found evidence for differentiation. In one microsatellite study, foxes that colonized Phillip Island in about 1912 were differentiated from their mainland counterparts suggesting limited gene flow, founder effect, or natural selection (Lade *et al.* 1996). Microsatellite differentiation was also found between urban populations in Melbourne, Australia that was larger than that found in rural fox populations or between Phillip Island and mainland foxes (Robinson *et al.* 2001). This suggests that foxes cannot easily disperse across urban landscapes. Another study comparing urban and rural foxes in Switzerland (Wandeler *et al.* 2003) suggests that urban population are founded by a small number of individuals, producing strong differentiation by random genetic drift. However, once the urban population grows, the genetic differences may be eroded. Finally, an allozyme and cytochrome *b* study of hunted and non-hunted fox populations in Central Europe and the Mediterranean area showed that foxes in protected areas have lower levels of variation (Frati *et al.* 2000). Although this result needs to be confirmed with other genetic markers and additional populations, it suggests that hunted populations are genetic sinks where immigrant foxes more readily enrich the gene pool.

Conservation implications

Foxes have moderate level of variation, but this level differs among populations. The variability may have to do with habitat loss and isolation of small populations and it may be larger in hunted than non-hunted populations. Foxes appear more genetically differentiated than larger canids and are more differentiated when isolated on islands and when separated by urban barriers. This suggests that population continuity can readily be disrupted by natural and artificial barriers and genetic variation lost. Consequently, genetic management needs to be considered if fox populations are to be sustained with appreciable levels of variation and to prevent them from disappearing in the absence of migration.

Conclusions and perspectives

In general, the smaller fox-like canids show higher levels of variation between and within populations. These differences reflect higher densities and lower levels of mobility in small canids. Insular canids, such as the island fox, Darwin's fox and the Isle Royale wolf, have the lowest levels of genetic variation but high levels of differentiation from mainland populations. A similar pattern is evident when habitats have been subdivided and populations isolated by human activities. For example, Scandinavian and Italian wolves have low levels of variation within populations but high levels of differentiation reflecting a recent history of isolation and population bottlenecks (Randi *et al.* 2000; Vilà *et al.* 2003b). The most endangered canid, the Ethiopian wolf, has the lowest levels of variation of any studied canid. In contrast, African wild dog genetic patterns appear dominated by ancient vicarious events such as Pleistocene isolation of south and eastern populations followed by intermixing. However, populations in Kruger National Park and Kenya may have recently lost genetic variation due to population bottlenecks. Finally, interspecific hybridization may occur in disturbed populations, especially if one species is rare and the other abundant and the rate of encounters is high due to the presence of concentrated resources such as refuse dumps. Hybridization with domestic dogs threatens to obscure the unique genetic characteristics of one of the largest remaining populations of the endangered Ethiopian wolf but does not appear as a consequential threat to grey wolves.

Molecular genetic analysis supports species distinction for Darwin's fox, kit and swift foxes, and the island fox. Analysis of populations within species have uncovered important genetic and phenotypic units including each of the island populations of island fox, the San Joaquin and Mexican kit fox, the Mexican wolf and Algonquin wolf, West and South African wild dogs, and New and Old World wolves. These distinct conservation units warrant separate breeding and *in situ* management. Interbreeding should be avoided in the absence of evidence for inbreeding depression (e.g. Hedrick and Kalinowski 2000). The next phase in genetic research on canids should focus on the study of both neutral and fitness related genes so that both history and population adaptation can be assessed. This information will be valuable to conservation programs (Crandall *et al.* 2000).

Acknowledgements

We would like to thank K. Koepfli and B. Van Valkenburgh for helpful comments. Work was supported by the US Fish and Wildlife Service, the National Science Foundation, Nature Conservancy, WolfHaven, the Smithsonian Institution, National Geographic Society, El Padron, and the Genetic Resources and Conservation Program and the Academic Senate of the University of California. A. Moehrenschlager would like to thank Husky Energy Inc. and the Wildlife Preservation Trust Canada for its funding, leadership and logistical support.

CHAPTER 4

Society

Canid Society

David W. Macdonald, Scott Creel, and Michael G. L. Mills

African wild dog *Lycaon pictus* © D. W. Macdonald.

Canids in context

Canids are wonderfully diverse, and a review of their behavioural ecology could take many directions. We have chosen to dwell on why some canids live in groups, while others do not, and to ask what shapes their societies? (Fig. 4.1) As a preamble we explore two comparative analyses for the Carnivores as a whole. These ask how body size, and thus metabolic needs, link with prey size and home range size?

Larger carnivores, even correcting for phylogeny, tend to eat larger prey, and this applies to canids too (Fig. 4.2) (Carbone *et al.* 1999). Intriguingly, the order divides into two groupings, smaller carnivores eating principally small prey, and big ones eating larger prey, with a jump between them at about 20 kg (over 90% of the 139 species fall clearly into one or other of these categories); there are some canids in each category, and for them this prey mass discontinuity is very marked. Since there is no scaling reason why small canids should not take larger prey, this discontinuity requires explanation. Small prey, tend to be abundant and easy to catch, but they are small and, in the case of invertebrates, their availability can be heavily weather-dependent. Do tiny prey place a limit on the energy available to carnivores, and hence to their body size? To estimate a maximum carnivore mass that could be sustained on invertebrates, Carbone *et al.* (1999) used Gorman *et al.*'s (1998) net rate model, which assumes that carnivores divide their time between resting (Tr) and hunting (Th). While they are resting, species spend energy at a rate (Er) determined largely by their body size, and while hunting they spend it faster (again at

Figure 4.1 African wild dogs *Lycaon pictus* (a) setting out on a hunt (b) juveniles playing (c) complex social interactions (d) amicable greeting (e) communal sleep-heap (f) thirsty, but showing restraint and not drinking while watching crocodiles © D. W. Macdonald.

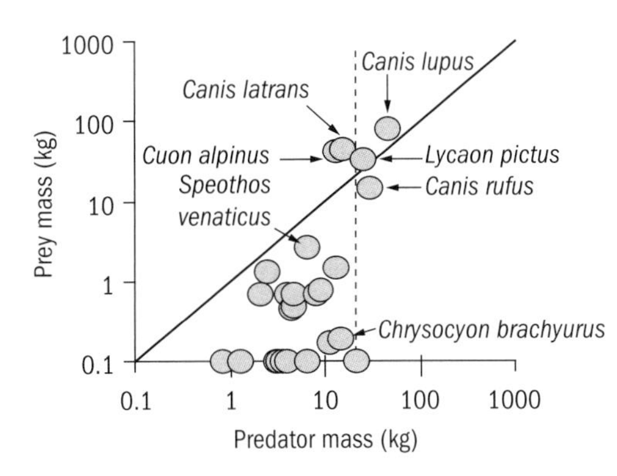

Figure 4.2 Large canids eat much larger prey. From Carbone *et al.* (2000).

a rate determined by their size). To balance its energetic books, an individual needs a certain minimum hunting time (Th) over the 24 h of the day. Solving this equation reveals that there are not enough hours in the day to gather sufficient small prey to sustain carnivores bigger than 21.5 kg—a value close to the discontinuity observed in nature.

Canids for whom their average prey size is <2 kg (often very much less) can, and therefore do, capitalize on the easy living provided by small prey (Fig. 4.2). Getting bigger requires a quantum change in lifestyle, tackling prey much closer to their own body size, and in this particular data set, the five species that have crossed this divide are the grey (*Canis lupus*) and red wolves (*C. rufus*), the African wild dog (*Lycaon pictus*), the dhole (*Cuon alpinus*), and the coyote (*C. latrans*). Intriguingly, the maned wolf and African wild dog both hover on the body weight divide but have opposite lifestyles.

By using doubly labelled water to measure the energy consumption of Kruger's wild dogs, Gorman *et al.* (1998) solved the net rate equation directly and thus demonstrated that, according to their body weight, while they are resting—which they do for about 21.5 h daily—wild dogs are expected to consume 0.11 MJ/h. The field data revealed a Daily Energy Expenditure (DEE) of 15.3 MJ—equivalent to roughly 3.5 kg of ungulate meat daily. For the equation to balance, the dogs must be using 3.14 MJ/h during the 3.5 h of hunting, and to pay these fuel bills they must secure 4.43 MJ/h of hunting. In short, hunting is so expensive that a deterioration in the measured hunting success would lead to an exponential increase in the time needed to break even energetically. If that deterioration were, for example, a loss of 25% of their kills to spotted hyenas (*Crowta crowta*), this would push up the daily hunting time from 3.5 to 12 h per day. Perhaps that is why wild dogs are scarce where hyaenas thrive, especially in open habitats where detection of wild dogs feeding is high, (Creel and Creel 1996; Mills and Gorman 1997) and in arid areas where the food supply is ephemeral.

Thus, as a background to this review: (1) energetic constraints may shape canid behaviour, (2) intraguild competition may be a major force in canid communities, and (3) conservation of one species may raise awkward ethical questions regarding the management of another (Macdonald and Sillero-Zubiri Chapter 1, this volume).

But what of home range size, at least partly the product of body size and feeding ecology? Where resources are randomly distributed, animals are expected to partition space into approximately exclusive home ranges. Thus, all else being equal, an increase in group metabolic needs demands *pro rata* an increase in range size to maintain per capita intake. Mammalian home ranges scale allometrically with body mass (or with the combined metabolic needs of the group, at approximately $M^{0.75}$) (Harestad and Bunnell 1979; Gittleman and Harvey 1982). However, in reality, resources are often not randomly distributed, and therefore there are circumstances under which home range size is not predictable from the scaling of metabolic needs alone. Johnson *et al.* (submitted a) explore whether departures from the predicted allometry reveal relationships between spatial organization and resource availability. They argue that points, like those for wolves, wild dogs, or raccoon dogs, that deviate from the familiar allometry do so because of non-random resource dispersion. With one exception, the canids revealed on Fig. 4.2 to eat unusually large prey all have positive residuals, whereas several other canids, and many other carnivores have negative residuals. Furthermore, as group size increases, there is a tendency for home range sizes to become smaller relative to those predicted by metabolic needs, an effect statistically attributable entirely to species with smaller than expected ranges (Fig. 4.3). From this starting point, we will now review canid behavioural ecology before attempting a unifying explanation of why amongst canids the ranges of some social species are smaller than expected

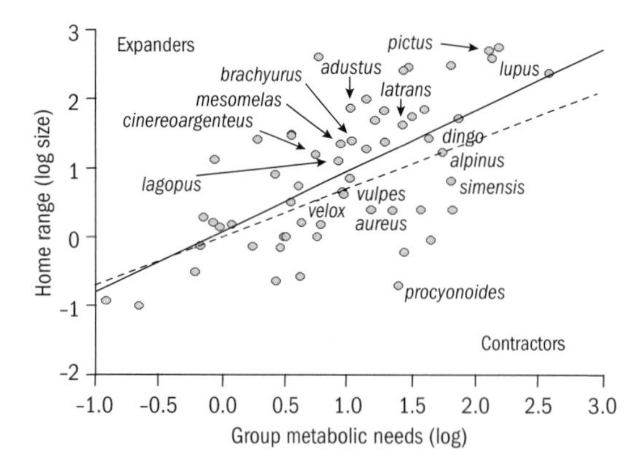

Figure 4.3 Departures from the allometry between home range size and group metabolic needs (dotted line corrected for phylogeny). (From Johnson *et al.* submitted).

for their body size, whereas others—notably those eating large prey—have larger ranges than expected.

Why do some canids live in groups?

Creel and Macdonald (1995) describe five general mechanisms or conditions that may select for sociality in carnivores. Two conditions reduce the costs of tolerating conspecifics: (1) abundant prey (or other resources), rich or variable prey patches, or rapid prey renewal, lead to low costs of tolerating conspecifics, in terms of foraging success; (2) constraints on dispersal opportunities such as lack of suitable habitat, low availability of mates, or intraspecific competition may favour the retention of young past the age of maturity. Three general mechanisms identify the benefits of tolerating conspecifics: (3) groups may use strength of numbers in the acquisition and retention of resources, for example, hunting in groups may increase foraging success, where prey are large or difficult to kill, groups may also fare better in territorial defence and intra- and interspecific competition for food, especially at large kills; (4) groups may be less vulnerable to intra- and interspecific predation; and (5) groups may help to meet the costs of reproduction through alloparental care.

Two general predictions are that: first, group size is likely to be less where dispersal opportunities are greater and, second, cub survival to weaning is likely to increase in the presence of alloparents. Amongst carnivores, and canids in particular, groups generally form by the delayed dispersal of kin of either sex. With this in mind, the first proposition is based on the assumption that an individual's decision of whether to remain in its natal group reflects the likely costs and benefits of biding versus dispersing; if the opportunities for establishing new territories increase, the risk of dispersal is lowered and therefore the incentive for philopatry diminished, leading to a reduction in group size (Macdonald and Carr 1989). The second proposition is based on the assumption that individuals are likely to behave in ways that benefit their kin (Hamilton 1967), and the intuition that behaviour as seemingly helpful as provisioning food is likely to be beneficial (but see below).

Behavioural selective pressures for group living

Group formation and group size are governed by a set of costs and benefits that have received copious attention by behavioural ecologists (Bertram 1978; Krebs and Davies 1993; Wilson 2000). Each potential benefit carries potential costs: for example, animals living together may catch each other's parasites (Hoogland 1979), and may control them by mutual grooming (as in the case of badgers; Macdonald *et al.* 2000, Stewart *et al.* 2003). Those hunting together have to share their spoils (analysed in detail for African wild dogs by Creel and Creel 2002), and those sharing a mate may also face sharing his support (as for male red foxes provisioning the litters of two vixens (Zabel and Taggart 1989)).

Strength of numbers

Benefits of hunting in groups

The notion that canids hunt together in order more effectively to overwhelm prey too challenging to be hunted alone is so intuitively plausible that, as an explanation for pack living, it became dogma long before researchers discovered just how difficult it would be to support empirically. An influential, but unsubstantiated, article by Wyman in 1967 stated that black-backed jackals (*Canis mesomelas*) in the

Serengeti were more successful in killing gazelle fawns (*Gazella spp.*) when they cooperated in deflecting the mother's defences. Lamprecht's (1978a) attempt to verify this claim, yielded scarcely more than a hint that cooperation pays for golden jackals (*C. aureus*), and no evidence at all that it benefits black-backs (taking only cases where 1–2 adult gazelles defended the fawn, Table 4.1). In Zimbabwe, the improvement in the success of pairs, in comparison to singletons, of black-backs hunting for springhares was just significant (Macdonald *et al.* Chapter 16, this volume) and the correlation between the occurrence of springhares in the diet and periods when mated pairs of jackals operated together could be evidence that cooperation facilitated capture of these nimble rodents. Conclusions are similarly hazy regarding coyotes; there is a correspondence between places where they prey on elk (*Cervus elaphus*) and mule deer (*Odocoileus hemionus*) and those where coyotes occur in large groups (Bekoff and Wells 1980, 1982), highlighting the need to separate cause and effect because of the interesting possibility that only where prey are large can packs be large (see below); furthermore, although Bowen (1981) wrote that 'Cooperation is generally necessary' for capture of large prey, Camenzind (1978a,b) concluded that 'the apparent major advantage of the coyote pack . . . lies in its ability to maintain ownership of the carrion food source and *not* in actual depredation of that animal'. Similar uncertainty surrounds the advantages of cooperative hunting by dholes (Chapter 21, this volume), whereas the picture for grey wolves almost invariably emphasizes the downside of cooperative hunting: Thurber and Peterson (1993) graph the declining intake of food per wolf per day with larger pack sizes on Isle Royale (Fig. 4.4), and Schmidt and Mech (1997) repeat this result from a synthesis of many studies.

Table 4.1 Numbers of hunts by jackals hunting alone or in groups, and the numbers of kills made (Lamprecht 1978a)

	Golden jackal		Black-backed jackal	
Group size	Hunts	Kills	Hunts	Kills
1	4	0	3	2
2/3	4	4	5	4

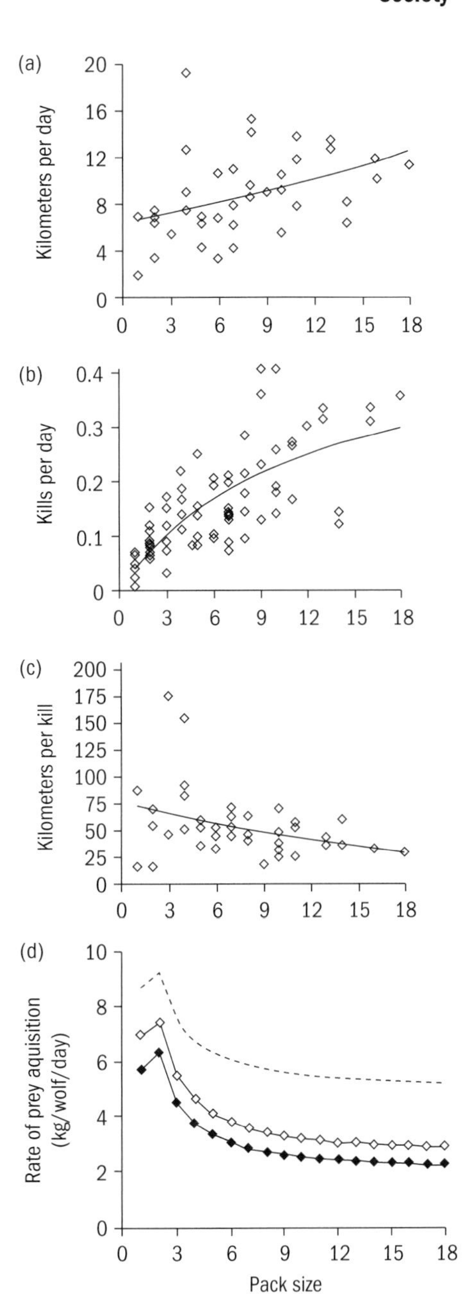

Figure 4.4 Relationship between wolf pack size, on Isle Royale, and (a) distance travelled per day, (b) number of kills per day, (c) distance travelled between kills, and (d) relationship between pack size and mean daily per capita rate of energy acquisition (dashed line is the net rate of the combinations of values explored in a sensitivity analysis; open circles are the combination with the greatest values and steepest slope; the closed symbols had the lowest values and the shallowest slope). (From Vucetich *et al.* 2003.)

African wild dogs tend to select less fit prey (Pole *et al.* 2003) and although even a single wild dog can kill a large prey, Fanshawe and Fitzgibbon (1993) (Fig. 4.5) provided the first hard evidence that the hunting success of African wild dogs increased with pack size, at least up to four adults (NB. all the pairs which killed wildebeest (*Connochaetes taurinus*) actually killed calves). The hint of a decline in success with larger packs is, interestingly, more marked when they hunted gazelle. Again the tendency of larger packs to hunt larger prey (Fig. 4.5c) raises interesting questions about cause and effect (see also Fuller and Kat 1990).

Creel (1997) shifted the focus from the bulk of prey eaten to the profit and loss account of catching them. Specifically, he demonstrated that while the slightly U-shaped relationship of pack size against kJ killed/dog/day (or per hunt) might suggest that at least initially cooperative hunting is disadvantageous, this conclusion is changed (Fig. 4.6) when taking account of the facts that larger packs travel shorter distances—and have higher success—per attempted hunt. Clearly, it is necessary to take account not merely of the benefits of hunting (or doing anything else) in a pack of given size, but also of the costs. Various currencies (summarised in Creel and Creel 2002), each give different answers for one population of dogs. Measuring kg killed/dog/hunt gives a slightly U-shaped relationship with group size, kg killed/dog/km travelled is essentially linear, whereas kg killed/dog/km of fast chase gives an inverted U-shaped (suggesting an intermediate hunting pack size is optimum). By considering net kj/obtained/dog/day Creel (1997) produced a measure of net benefit rather than of efficiency, and net benefit increased significantly with group size (see

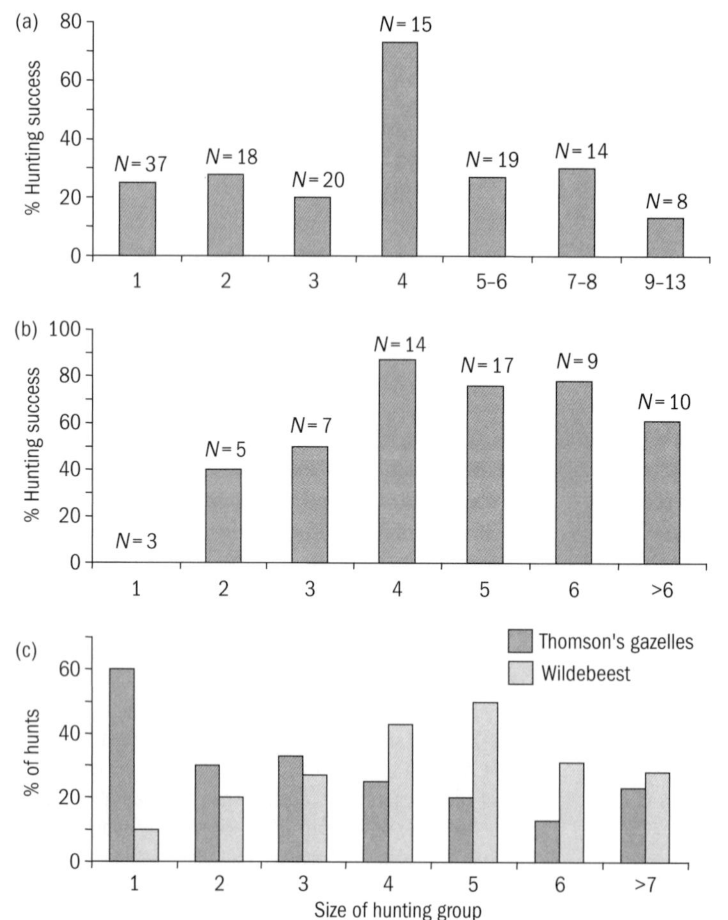

Figure 4.5 Hunting success of wild dogs chasing (a) gazelles, (b) wildebeest, and (c) larger packs hunting larger prey. (From Fanshawe and Fitzgibbon 1993.)

also Courchamp *et al.* 2002, Fig. 4.6a). Energetically, packs should be as large as possible, and certainly bigger than the modal pack size of 10 observed. However, very large hunting parties need more kills to meet the requirements of each hunter, so hunting success per dog may decrease above a certain group size (Creel and Creel 2002). Thus, hunting in large groups (but not too large) will provide the optimal ratio of benefit to cost (Fanshawe and Fitzgibbon 1993; Creel and Creel 1995; Creel 1997). Vucetich *et al.* (2003) report a similarly elegant approach to wolf pack hunting success (see Fig. 4.4), which reveals that in the absence of scavengers, members of larger packs secure reduced foraging returns and would do best by hunting in pairs.

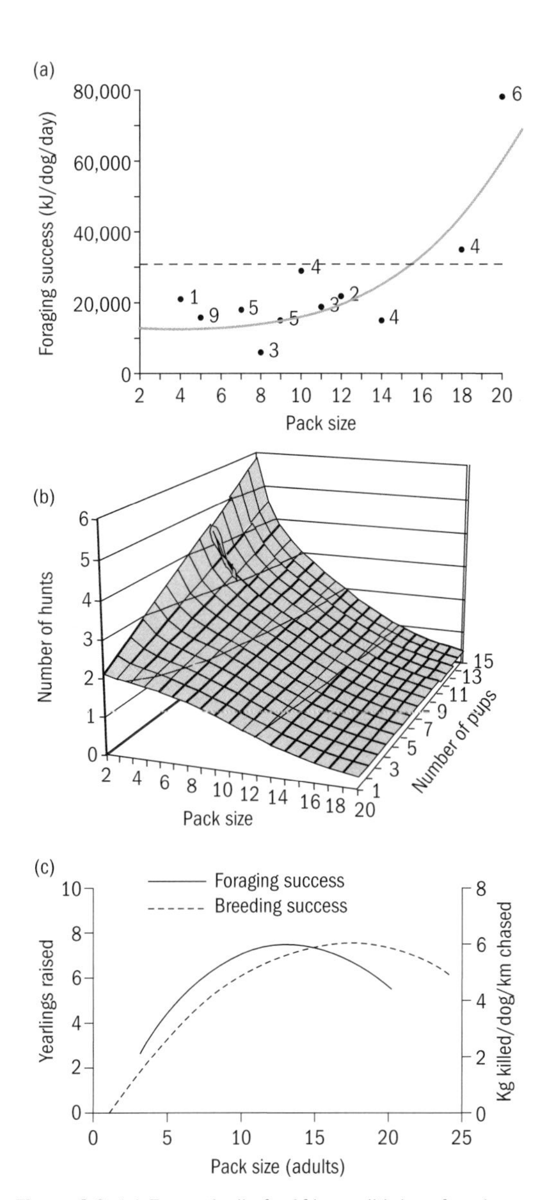

Figure 4.6 (a) Energetically, for African wild dogs foraging success increases with pack size (Courchamp *et al.* 2002), (b) there are trades-off between the pack size, frequency of hunts, and number of pups raised, and (c) the optimal pack size may differ between measures of foraging and breeding success (Creel and Creel 1995).

Hunting is only one of the, often mutually exclusive, demands on a canid's time. In the context of pup rearing, Courchamp *et al.* (2002) describe how, regardless of the use of a pup guard, smaller packs will theoretically have to increase the number of hunts (and the associated costs and risks) to raise the same number of pups as do larger packs (Fig. 4.6b). A pair would have to undertake an additional five hunts a day if they attempt to guard the babies also.

Group protection of kills

Intraspecific

Interference competition at kills can be considerable and the outcome of such competition generally appears to be affected by group size (Creel and Macdonald 1995) and habitat type (carcass theft being a greater risk in open habitat). Irrespective of the contribution of collaboration to making a kill, Bekoff and Wells (1980, 1982) describe how larger groups of coyotes are more successful at defending, or indeed stealing, carcasses from their neighbours. Similarly, in Israel, a large pack of golden jackals habitually stole food from a smaller pack (Macdonald 1979c), and Moehlman (1989) wrote 'cooperative groups of jackals are more successful in defending and feeding on carcasses'.

Interspecific

Interspecific interference competition is a major component of canid ecology. In the Serengeti, black-backed jackals and African wild dogs, respectively, lose up to 30% and 86% of Thomson's gazelle (*Gazella thompsonii*) kills to spotted hyaenas (Lamprecht 1978; Fanshawe and Fitzgibbon 1993). However, exchanges of ownership are not straightforward to interpret as carcasses often change hands after they have been abandoned (Mills 1989a,b); furthermore, species differ in their abilities to consume a carcass: a jackal may leave a bone untouched, which will be split by a wild dog for its marrow and completely digested by a spotted hyaena (Creel and Macdonald 1995). It is not clear what function of aggregate body weight and numbers of sets of teeth determines the algebra of victory, but it is clear that strength of numbers is important: Schaller (1967) reports that packs of dholes are capable of driving a tiger (*Panthera tigris*), 10 times their size, from kills. African wild dogs can defend a carcass from spotted hyaenas provided they outnumber the challenging hyaenas although each dog is about a third to a quarter of the body weight of a hyaena (Estes and Goddard 1967; Malcolm 1979; Reich 1981; Fanshawe and Fitzgibbon 1993). Larger packs of wild dogs are able to repel marauding hyaenas for longer, capitalizing meanwhile on the greater speed at which they can dismember the carcass and bolt it down (Malcolm and Marten, 1982; Fuller and Kat, 1990; McNutt 1996a); in contrast, up to 19 defending dogs failed to prevent the 7 instances of theft by between 1 and 8 lions (*P. leo*) witnessed by McNutt (1996a). Interspecific clashes over kills is one facet of a broader hostility within guilds of sympatric canids (Macdonald and Sillero-Zubiri, Chapter 1 this volume).

Scavenging can be a vital part of canid ecology, either as the perpetrators (e.g. Roth (2003) reveals that arctic fox productivity can be determined by access to seals killed by polar bears) or as its victims. In the latter case, Peterson and Ciucci (2003) report 10 red foxes simultaneously scavenging from one wolf kill, and another instance when 135 ravens did so. They present data by C. Promberger to show that a lone wolf might lose two-thirds of a moose kill to scavengers, a pair of wolves, half of their kill, but a pack of 10 would lose only 10%. Modelling these results, Vucetich *et al.* (2003) show that the greater food-sharing costs in larger packs of wolves are more than offset by the smaller losses to scavengers and increased rates of prey acquisition.

Group size and territorial defence

Strength of numbers may be important in territorial defence. Bekoff and Wells (1982) (Table 4.2) found that their study pack of coyotes responded pragmatically to intruders, invariably attacking small groups but generally avoiding confrontation with larger ones.

Table 4.2 Relationship between the numbers of intruders and the percentage of occasions on which resident coyotes retaliated. (Bekoff and Wells 1982.)

No. of intruders	Frequency of retaliation (%)	*N*
1–3	100	17
4	67	6
5	59	17
9	29	7

Members of a larger group of red foxes (*Vulpes vulpes*) observed in Israel generally chased off members of a smaller group (Macdonald 1984). On Mednyi Island in the Bering Sea, the cubs of Arctic foxes (*Alopex lagopus*) may be attended at the den by various number of adults and the more of these guards present at a den, the more likely one of them is to launch an attack on a passing intruder (mobbing occurred in 42.6% of the 54 cases where only one animal was present at the den, but in 87.9% of the 66 cases where two or more animals were present (Frommolt *et al.* 2003; Krutchenkova *et al.* submitted). In Ethiopian wolves (*Canis simensis*) and bat-eared foxes (*Otocyon megalotis*), territorial clashes almost always result in retreat of the smaller pack (see Sillero-Zubiri *et al.* Chapter 20 and Maas and Macdonald Chapter 14, this volume), and Creel and Creel (1998) report 10 clashes between packs of wild dogs and in every case the larger routed the smaller.

Territorial clashes amongst Minnesotan wolves, accounted for almost all natural adult mortality (Mech 1977b). Similarly, in Selous, 7 wild dogs (4 adults and 3 pups) were killed during inter-pack fights (16% of 45 known-cause deaths; a further 53% of deaths were infanticide—also caused intraspecifically, but within the packs through infanticide) (Creel and Creel, 1998). Intriguingly, despite large samples, observations of Ethiopian wolves have not revealed such ferocity of encounters (Sillero-Zubiri *et al.* Chapter 20, this volume).

Linked to territorial defence is communication in canids, as reviewed briefly by Macdonald and Sillero-Zubiri (Chapter 1, this volume). Macdonald (1980b, 1985) proposed that larger groups could collectively generate more faeces, and thus might have open to them forms of territorial signalling (e.g. border

Figure 4.7 A family of eight bat-eared foxes *Otocyon megalotis* harrass a spotted hyaena © B. Maas.

latrines) less available to singletons, in the same way that many voices add impact to a wolf pack's howl (Harrington and Asa 2003).

Anti-predator behaviour

The advantages of grouping in the context of predation include the selfish herd, dilution, vigilance, and group defence.

Predation, particularly as an expression of intraguild competition, is an enormous force in canid ecology. Amongst swift foxes (*Vulpes velox*) raptors are an important threat, as are coyotes (Moehrenschlager *et al.* 2003). Although we are unaware of data from canids to support the intuition that more pairs of eyes, ears, or nostrils increase the likelihood of detecting predatory danger, wild dogs will aggressively mob potential predators if these threaten the pups (e.g. Kühme 1965; Estes and Goddard 1967), and bat-eared foxes form larger groups to mob hyaenas (Fig 4.7) than they do when mobbing jackals (Maas and Macdonald Chapter 14, this volume)—they will even mob leopards (*Panthera pardus*) (Mills personal observation). The ferocity and organization of wild dog pack action can drive off spotted hyaenas, jackals, leopards, or even lions (Malcolm and Marten 1982; Creel and Creel 1995; McNutt 1996a), and is more effective for larger packs. Wild dogs tend to avoid lions (Creel and Creel 2002), which are responsible for 43% of their natural mortality in Kruger (Mills and Gorman 1997). Heavy predation by lions (Mills and Gorman 1997; Woodroffe and Ginsberg 1997a; Vucetich and Creel 1999) and kleptoparasitism by hyaenas (Carbone *et al.* 1997; Gorman *et al.* 1998) are important factors to which large packs of wild dogs seem likely to be adaptated, but this awaits quantitative test. Carbone *et al.* (1997) concluded that hunting pack sizes of four or five wild dogs maximize prey profitability in the presence of kleptoparasitic hyaenas. As Macdonald and Sillero-Zubiri (Chapter 1, this volume) stress, intra-guild hostility between canids is a major force in their distributions and demography, and can affect their sociology too. For example, Arjo and Pletscher (1999) report that in northwestern Montana, before wolf re-colonisation in 1980, most coyotes travelled alone (62%) or in pairs (29%). However, by 1997, and after wolf re-colonisation, they more commonly travelled in pairs (48%) or larger groups (33%). This may reflect the need for vigilance and defence. In a similar vein, coyote group sizes have changed in Yellowstone National Park following the reintroduction of wolves (Phillips *et al.* Chapter 19, this volume).

Cooperative breeding

Helpers

Canid pups have a prolonged period of dependency on adults and are commonly tended by both parents (Kleiman and Eisenberg 1973). Canids (although not vulpine ones) are the only carnivoran family to regurgitate (Macdonald and Sillero-Zubiri, Chapter 1, this volume), and males and females also provision pregnant and lactating mates or fellow group members.

Since the 1970s, an accumulation of studies has suggested that collaborative care of young may be a more fundamental (and certainly more ubiquitous) feature of canid society than the historically much vaunted cooperative hunting. For example, non-breeding female red foxes guard and play with cubs, may even split them between several dens, and sometimes feed them at least as diligently as does their mother (Macdonald 1979). The original list of species for which non-breeding 'helpers' fed and tended the young—African wild dogs, jackals, red and Arctic foxes, grey and Ethiopian wolves (Kühne 1965; Macdonald 1979; Moehlman 1979; Fentress and Ryon 1982; Hersteinsson and Macdonald 1982; Malcolm and Marten 1982; Sillero-Zubiri *et al.* Chapter 20, this volume)—has expanded almost in direct proportion to the number of species studied, revealing alloparental care by non-breeding adults as a widespread trait of the family. Bat-eared foxes are a partial exception in that all adult females in a group may produce young and suckle communally (Maas and Macdonald, Chapter 14, this volume). Provisioning the mother during pregnancy and lactation allows her to guard her young more continuously while also directing more energy into gestation and lactation (Oftedal and Gittleman 1989) (this is a contribution which bat-eared fox fathers are uniquely unable to make since their exclusively insect prey is not portable, Maas and Macdonald, Chapter 14, this volume). Moehlman (1983) found that food provisioning increased with one and two non-breeder helpers in black-backed jackals, but not in golden jackals. Alloparental care appears self-evidently helpful, and the classic demonstration that this is so is Moehlman's (1979) that pup survival to weaning increased with numbers of helpers amongst black-backed jackals. Similar, if less conclusive, evidence exists for red foxes, including a case where a helper reared the cubs of an ailing mother (Macdonald 1979, 1978a,b), and for coyotes where Bekoff and Wells (1982) found a non-significant relationship between the number of adults attending the den and the number of the pups surviving to 5–6 months (but a significant relationship with the percentage of pups surviving), and African wild dogs (Malcolm and Marten, 1982). However, sometimes the demonstration that helping, however assiduous it may appear, translates into improved reproductive success has proven difficult.

Indeed, proof that alloparental behaviour benefits the recipients amongst mammalian societies remains rare (Jennions and Macdonald 1994) and that proof has remained elusive even for some of the most conspicuously helpful carnivores (e.g. Doolan and Macdonald 1999). Thus, although the relationship between pup survival and number of helpers was significant for black-backed jackals by the time of Moehlman's 1979 publication (and by the time Moehlman (1983) linked to an increased rate of provisioning with one and two non-breeding helpers), comparable data gathered simultaneously on allopatric groups of golden jackals did not yield a detectable benefit from helpers until a further 20 years of research had passed (Moehlman and Hofer 1997). Amongst red foxes, Zabel and Taggart (1989) found cub survival was the same for polygynous versus monogamous females on Round Island. Total litter size was also not affected in Arctic foxes in the north Pacific Commander Islands (either at emergence or on weaning) by either group size or the presence of non-breeding females (whether they were seen to be involved as helpers or not) (Table 4.3) (Krutchenkova *et al.* submitted). Amongst grey wolves, helpers appear to increase the survival of the young only when food is abundant (Waser *et al.* 1996). When food is scarce, offspring in large wolf groups actually survive less well (Harrington *et al.* 1983; see also Harrington and Mech 1982; Peterson *et al.* 1984). The Ethiopian wolf is a typical cooperative breeder; groups of up to 13 adults collaborate in scent marking and defending their territory (Sillero-Zubiri and Gottelli 1995a; Sillero-Zubiri and Macdonald, 1998), and in feeding, grooming, playing with and defending the young. The young of only one female are reared in each pack annually. Young male wolves remain in their natal territories beyond physiological maturity and never disperse (Sillero-Zubiri *et al.* 1996a). Most pack members brought solid food to the pups, but breeders contributed significantly more food than did non-breeders, and females more than males. Although the breeding females and putative father spent more time at the den on average than did other wolves, the proportion of time for which pups were left unattended was inversely correlated to the number of non-breeding helpers in the pack. Non-breeding females were particularly attentive to the pups. Pack size may, thus, improve anti-predator behaviour, since babysitters were active deterring and chasing potential avian and mammalian predators, such as eagles and spotted hyaenas. Guarding may also protect the pups from visiting wolves from other packs. While increased numbers of helpers amongst Ethiopian wolves led to a decline in the work rate of mothers provisioning pups, there was no detectable increase in pup survival (Sillero-Zubiri *et al.* Chapter 20, this volume).

The contribution of non-breeding pack members amongst African wild dogs can take the form of babysitting or provisioning. Thanks to babysitters, mothers—often, by their status, experienced hunters—can return to the chase 3 weeks after parturition, but 8 weeks before pups can follow the pack. The babysitter chases predators away, ensures that the pups do not stray, and warns them to go down the den if there is threat of danger (Kühme 1965; Malcolm and Marten 1982). Pups are entirely dependent upon adults for providing them with meat until about 12 months old (McNutt 1996a), so the survival of litters depends on helping (Fuller *et al.* 1992b,c). Although the generality seems indeed to be that pup survival is correlated with pack size (Vucetich and Creel 1999), successive analyses of the impact of pack size on pup survival amongst wild dogs in the Serengeti illustrate some complexities. Using data up to 1980, Malcolm and Martin (1982) found a positive, although not significant, correlation between pup survival to 1 year and pack-size *excluding* yearlings in the Serengeti; when Burrows (1995) analysed similar data for subsequent years (1985–91), this relationship was significant only if yearlings were *included*. Although yearlings are not yet experienced hunters, their presence contributes to foraging efficiency. It is generally the alpha male or the alpha female which leads the hunt, but yearlings nonetheless contribute to several aspects of cooperative hunting, including the diminution of kleptoparasitism, by decreasing the carcass access time (increased cleaning efficiency) and increasing the ratio of dogs/hyaenas (see Fanshawe and Fitzgibbon 1993; Carbone *et al.* 1997), and regurgitating meat to the begging pups back at the den (Estes and Goddard 1967; Malcom and Marten 1982) (i.e. transport more food back to the den: Kühme 1965). For these reasons, the pack may chose to use an adult pup guard when the cost of doing so is offset by the presence of yearlings to increase the hunting party. In the Selous, the Creels' data suggest that you can have too much of a good thing: while the number of yearlings successfully raised increases with pack size, the relationship is dome shaped (Fig. 4.6).

Tackling the question of whether small packs can afford babysitters, Courchamp *et al.* (2002) re-analysed Malcolm and Marten's (1982) data from the Serengeti to reveal a nonlinear relationship between the number of pup guarding occurrences and the ratio of adults to pups. There seemed to be a threshold set at two pups per adult below which pup guarding becomes much less likely. If one takes 10.31 as the average litter size (calculated over 165 litters from Fuller *et al.* 1992; Burrows *et al.* 1994; Maddock and Mills 1994), then this threshold is at 5.16 adults. This suggests that there is a cost to pup guarding when fewer adults are in charge of more pups (Fig. 4.6). To test this empirically, Courchamp *et al.* (2002) analysed 246 hunts in Zimbabwe; 33% were undertaken by the whole pack, but in all of the remaining 167 hunts, one pack member remained with the pups as a guard. Hunts that took place during the night (when the risk of pup predation is high) invariably involved a guard remaining with the pups, whereas during 39% of diurnal hunts pups were left alone. Furthermore, pup guarding was significantly more likely in larger packs (the eight packs of less than five individuals left a guard with the pups in 34.7 ± 0.1% of their hunts, while the five packs of more than five left a guard in 88.5 ± 0.1% of their hunts).

Related phenomena may be invalid care and adoption. Macdonald (1987) records adult foxes feeding an injured adult group member, and amongst wild dogs incapacitated and older members of the pack are tolerated at kills (Estes and Goddard 1967), and may be fed by regurgitation (Rasmussen unpublished data). A possibly related phenomenon may be adoption of unrelated pups, of which all four cases documented by McNutt (1996a) involved 'smaller-than-average groups'. Adoption has also been recorded in red foxes (Macdonald 1979; von Schantz 1984a) and African wild dogs (Estes and Goddard 1967).

Allosuckling

Perhaps the most extreme form of alloparental behaviour is allosuckling, which has been seen in red, Bengal (*Vulpes bengalensis*), bat-eared, and Arctic foxes, as well as grey and Ethiopian wolves, chillas (*Pseudolopex griseus*), coyotes, and African wild dogs. Various mechanisms may lead an allosuckling female to be lactating. For example, as an aspect of reproductive suppression (see below), a subordinate female may lose her pups through the dominant's infanticide (e.g. wild dogs: van Lawick 1974; Frame *et al.* 1979; Malcolm and Marten 1982; dingos: Corbett 1988). For example, a subordinate female Ethiopian wolf that attempted to split the pack and breed independently recruited the assistance of two subadults she had previously helped to rear, but failed after the dominant female's intervention and death of the pups. The subordinate and her helpers then returned to assist at the dominant's den (Sillero-Zubiri *et al.* Chapter 20, this volume).

Alternatively, a female may lose her own cubs due to incompetence attributable either to inexperience or induced by social repression. Macdonald (1980, 1987) describes how a formerly successful breeding female red fox, having lost her dominance, so over-anxiously tended a subsequent litter that it consequently succumbed, whereupon she calmly and conscientiously nursed the cubs of the new dominant. Similarly, the death of a subordinate's litter amongst bush dogs (*Speothos venaticus*) (Macdonald 1996b), appeared to be due to over-anxious mothering. Per capita litter sizes of communally nursing bat-eared foxes were lower than those of neighbouring females without allo-nurses (Maas and Macdonald, Chapter 14, this volume). These examples raise the possibility that in addition to infanticide by dominant females, a second mechanism leading to litter reduction in communally breeding canids is mis-mothering akin to a 'tug-of-love'over the cubs.

A third possible route to allosuckling is that two females may both rear litters within a group and may nurse them communally (e.g. red foxes: Macdonald 1979, 1984; Zabel and Taggart 1989; wild dogs: Malcolm and Marten 1982; Mills unpublished data). The only canid for which this is generally the case is the bat-eared fox (Maas and Macdonald, Chapter 14, this volume). A fourth and more remarkable route is when females appear to lactate spontaneously (e.g. Ethiopian wolves, Sillero-Zubiri *et al.* Chapter 20, this volume). This may follow an aborted pregnancy or pseudo-pregnancy, the latter being widely reported in domestic dogs, and Macdonald (1980) suggested that this might be an adaptive function of pseudo-pregnancy (see also Creel *et al.* 1991). Although the mechanism remains uncertain (Packer *et al.* 1992), five of eight allosuckling Ethiopian wolves had shown no visible signs of pregnancy (Sillero-Zubiri *et al.* Chapter 20, this volume).

Measuring the benefit, or otherwise, of having two suckling females in a group of canids has been confounded by the fact that benefits of the extra milk supply may be offset by behavioural tension between the mothers, insofar as such instances may generally be 'failures' of reproductive suppression. Nonetheless, Malcolm and Martin (1982) suggest that amongst wild dogs, communal suckling allows alternate rest periods to the two mothers (which may thus participate in hunts sooner) and might increase survival of the two litters by allowing the transfer of more maternal antibodies to both the litters (Roulin and Heeb 1999). However, the evidence that allo-suckling is helpful is far from conclusive. Amongst Ethiopian wolf packs, Sillero-Zubiri *et al.* (Chapter 20, this volume) found that the survival to weaning of pups nursed by two females was significantly worse than those nursed by only one (although there may have been longer term reward). Similarly, amongst Mednyi island Arctic foxes, the number of cubs weaned per lactating female was significantly higher in groups with one, as opposed to two, lactating females (Krutchenkova *et al.* submitted, Table 4.3). Finally, the number of cubs reared by each communally nursing bat-eared fox was significantly less that reared by one female (Maas and Macdonald, Chapter 14, this volume).

What evolutionary explanation lies behind helping in those instances where it appears to be unhelpful? One possibility is that the consequences of helpers may become apparent only in the long term, and may be conditional upon circumstances (Emlen 1991).

Table 4.3 Numbers of cubs weaned per lactating female by pairs of Mednyi Island Arctic foxes without helpers, with non-breeding helpers and in groups with more than one communally lactating female (Krutchenkova *et al.* submitted)

Pairs with	No helper	Helpers	>1 lactating
Mean $\pm$ SD cubs weaned	4.1 $\pm$ 1.7	4.2 $\pm$ 1.5	2.09 $\pm$ 0.9
N pairs	42	10	28

Another is that the survival of a given litter may be the wrong measure—rather by lightening the load of pup care, helpers may increase the life-time reproductive success of mothers rather than the survival of offspring. A different, and yet more speculative, possibility is that there are circumstances—perhaps those of communally nursing Mednyi Arctic foxes or Ethiopian wolves—where the trait is maladaptive; a genetic blueprint for 'helping' may have been fixed early in canid evolution and occasionally manifest under circumstances where it is not advantageous.

Reproductive suppression

Canid society is typified by reproductive suppression exerted by dominant females on their subordinates (reviews in Macdonald 1978; Macdonald and Moehlman 1982; Moehlman 1986, 1989; Creel and Macdonald 1995). This has been reported for grey wolves (Packard *et al.* 1983, 1985), African wild dogs (Malcolm and Marten 1982; van Heerden and Kuhn 1985; Creel *et al.* 1997), Ethiopian wolves (Sillero-Zubiri *et al.* 1998, Chapter 20, this volume), and most small canids. Indeed, reproductive suppression was recorded in at least 44% of 25 species for which there was information (Moehlman and Hofer 1997), and is thus conspicuous by its absence in bat-eared foxes (Maas and Macdonald, Chapter 14, this volume). For example, female wolves can breed as yearlings but rarely do so in the wild before the age of 3. Subordinate female wolves rarely have offspring (Van Ballenberghe 1985; Peterson *et al.* 1984; Ballard *et al.* 1987) and generally lose those they do have (Peterson *et al.* 1984).

The degree of suppression, however, varies both among and within species (Creel and Waser 1991, 1994), and subordinates of wolves, African wild dogs and red foxes do reproduce, albeit at lower rates than do dominants (Malcolm 1979; Reich 1981; Packard *et al.* 1983; Macdonald 1987; Fuller and Kat 1990; Creel *et al.* 1997). At 40% of 25 dens in the Kruger National Park, more than one female produced pups, yet only about 9% of all pups genetically examined were offspring of the subordinates (Girman *et al.* 1997). In the Masai Mara Reserve, several females bred in 38% of packs (Fuller *et al.* 1992). In high density red fox populations where interactions with the dominant vixen are high, subordinate females do not usually breed, although they may breed more successfully in other populations (Macdonald 1980, 1987).

It is important to distinguish between suppression of endocrine cycles and suppression of reproductive behaviour. In grey wolves, age at first ovulation varies between 10 and 22 months and can be delayed by both social and environmental factors. However, once a female has cycled, anoestrus is rare, and reproductive failure is generally attributable to lack of copulation. Endocrine suppression amongst female canids usually affects the pre-ovulatory stages (wolves: Packard *et al.* 1983, 1985; African wild dogs: van Heerden and Kuhn 1985, Creel *et al.* 1997).

Behavioural suppression appears to be more common in males: dominant males may directly prevent subordinates from mating (wolves: Packard *et al.* 1983; African wild dogs: Malcolm 1979; Reich 1981; red foxes: Macdonald 1979, 1980). Nonetheless, multiple paternity has been reported in Ethiopian wolves (Gottelli *et al.* 1994) and wild dogs (Girman *et al.* 1997). Creel and Macdonald (1995) suggest that such complications in mating systems may be widespread: territorial male red foxes make frequent excursions beyond their territories during the mating season, during which itinerant males also make incursions into territories (Macdonald 1987). In contrast, the habit of travelling as a pair may make philandering much lessening for bat-eared foxes (Wright 2004, Maas and Macdonald, Chapter 14, this volume). Although the mating system of Ethiopian wolves is similar to that found in grey wolves (dominant pair and its offspring), female Ethiopian wolves have been observed to copulate with males from neighbouring packs more often than with the alpha male in their own pack (Sillero-Zubiri and Gottelli 1991; Gottelli *et al.* 1994). Furthermore, while dominant female Ethiopian wolves appear to mate only with the dominant male within their own group, they will mate with males of any status from neighbouring groups (Sillero-Zubiri *et al.* Chapter 20, this volume).

Female versus male helpers

The benefits provided by the helper may vary according to the sex of the helper. For instance, only females have the potential to allosuckle whereas one early study suggested that males provide more solid food to pups per capita than do females in African wild dogs (Malcolm and Marten 1982) as they do in grey wolves (Fentress and Ryon 1982).

Behavioural factors summary

Membership of larger groups may bring canids advantages due to combinations of improved foraging efficiency, breeding success, and survivorship. In some cases, and it has been argued for African wild dogs, this may be reflected as an Allee effect: a positive feedback loop of poor reproduction and low survival culminating in failure of the whole pack (Courchamp and Macdonald 2001; Courchamp *et al.* 2002). This would lead to a dome-shaped distribution of pack size, with populations subject to inverse density dependence at low density and direct density dependence at high density, exactly as observed by Creel (1997) for African wild dog packs.

Ecological selective pressures for group formation

While behavioural benefits may be amongst the selective pressures favouring sociality in wild canids, ecological factors create the framework within which these pressures operate, and dictate the balance of costs and benefits between group membership and dispersal. Alexander (1974) was the first to suggest that groups of genetic relatives, and helping within such groups, are a secondary consequence of group living that is initially favoured by some other ecological reason, of which a crucial component is resource dispersion (Macdonald 1983).

Resource dispersion

Irrespective of the current functional advantages to a species of group living, ecological factors such as the dispersion of resources, including food, water, or shelter, affect the costs of grouping and may render them negligible. The idea that certain patterns in resource availability might facilitate group formation by making coexistence feasible (even in the absence of any sociological benefit to grouping) without competition, grew especially out of observations on badgers (Kruuk 1978), and was formalized as the Resource Dispersion Hypothesis (RDH) by Macdonald (1983) and Carr and Macdonald (1986). The RDH is that groups may develop where resources are dispersed such that the smallest economically defensible territory for a pair (or whatever is the minimum social unit) can also sustain additional animals (Macdonald 1983).

The idea has been widely reviewed (e.g. Kruuk and Macdonald 1985; Macdonald and Carr 1989; Woodroffe and Macdonald 1993; Johnson *et al.* 2002), refined (Bacon *et al.* 1991; Bacon and Blackwell 1993), and criticized (von Schantz 1984; Revilla 2003, but see Johnson *et al.* 2003 and Johnson and Macdonald 2003). It is easily visualized in a landscape where resources are dispersed in discrete patches and where pairs (or whatever is the minimum breeding unit) establish the smallest territories necessary to ensure sufficient resources (usually food) even in seasonal lows. If this territorial collection of food patches is shareable (perhaps because individual patches are rich, or because of the dispersed nature of smaller patches), then this can allow a group of animals to use the same territory at no extra cost to the primary occupants. If the availability of resources is patchy in space and time, then a large number of potential food patches must be included in the territory to guarantee some 'critical' probability (Cp_α) of encountering enough usable food patches in a feeding period. Assuming a certain frequency distribution of availability across all patches, one can calculate the proportion of feeding periods on which the total amount of resources available will exceed Cp_α (Carr and Macdonald 1986). It is this excess that permits secondary individuals (willing to live with marginally lower food security) into the territory at little or no cost to the primary pair (Fig. 4.8). Thus, the spatial distribution of resources in time and space will determine territory configuration and size, which will generally be independent of group size. The overall degree of variability of resources in the environment, however, will determine group size (Fig. 4.8). This principle applies equally to the more realistic situation of contours of richness, rather than patches.

The RDH thus offers an explanation of variance in group size regardless of whether individuals gain from each other's presence or not. Not only may it

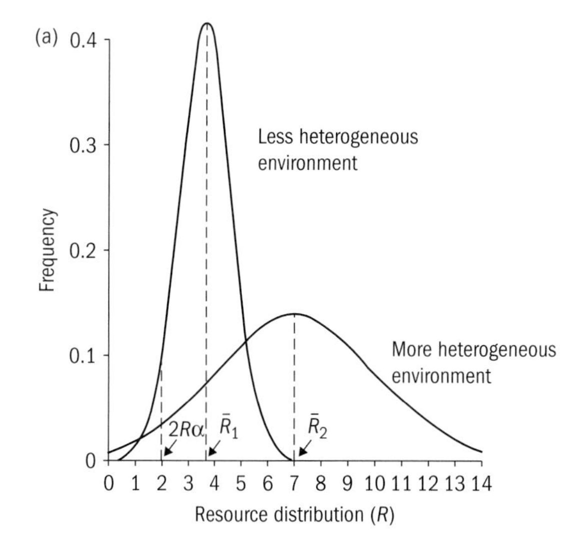

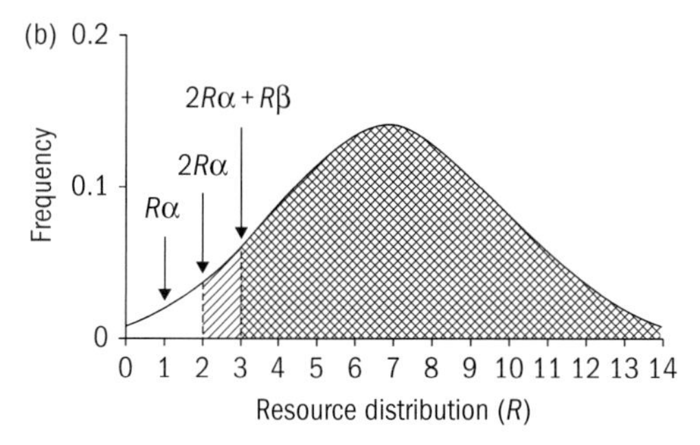

Figure 4.8 The RDH (from Carr and Macdonald 1986). If resource patches have a certain probability of availability, then several must be simultaneously defended to guarantee some probability of finding enough food for a primary pair ($2R\alpha$) on a given night. A frequency distribution of availability across all patches (here, arbitrarily, $N = 1$–14) indicates the proportion of nights on which the total amount of resources available will exceed $2R\alpha$. A secondary can join the territory when their own resource needs are met on top of those of the primaries ($2R\alpha + R\beta$). The integrals of the distribution illustrate critical probabilities, the proportion of times that such conditions occur. Obviously, changing the shape of the distribution will not alter $R\alpha$ and $R\beta$, but it will alter the critical probabilities associated with them, leading to a different prediction for group size.

apply to current societies, whether or not they display cooperation, but it may describe the conditions that favoured the evolution of sociality. It provides an explanation as to why (1) group size may generally be independent of territory size. Instead, it predicts (2) territory size is determined by the dispersion of resource patches, while (3) group size is independently determined by (a) the heterogeneity and (b) the richness of those resource patches. Where heterogeneity in the environment is greater, group sizes are larger, because it extends the distribution such that Cp_α is a smaller percentage of the total resources available over time (Fig. 4.8).

While RDH can only be tested by manipulative experiment (Von Schantz 1984; Kruuk and Macdonald 1985), several field studies have revealed, generally *a posteriori*, that its correlative predictions accord with observations (mostly assuming that habitat types are synonymous with resource patches). Although the RDH has been applied to taxa varying from birds (Davies *et al.* 1995) to rodents (Herrera and Macdonald 1987), it has been most influential in studies of Carnivora. While the resources in question may be water (e.g. coatis, Valenzuala and Macdonald 2002) or dens (e.g. badgers, Doncaster and Woodroffe 1993, Macdonald *et al.* in press), they are generally food. A food patch may include any large prey (e.g. a large carcass for a brown hyaena (Mills 1982), or groups of small prey (e.g. a school of fish for an otter (Kruuk and Moorhouse 1991), a tree of fruit for a kinkajou (Kays and Gittleman 1995), a mound of termites for an aardwolf (Richardson 1987), or cluster of worms for a badger (Kruuk and Parish 1982)). Canid examples that have invoked the hypothesis are red foxes (Macdonald 1981), Arctic foxes (Hersteinsson and Macdonald 1982), Blanford's foxes (Geffen *et al.* 1992c), crab-eating zorros (*Cerdocyon thous*) (Macdonald and Courtenay 1996), and bat-eared foxes (Maas and Macdonald, Chapter 14, this volume). The most interesting evidence against the hypothesis comes from Baker and Harris's (Chapter 12, this volume) study of urban red foxes.

The same principle applies to temporal variation even in homogenously dispersed resources; for example, red foxes depending on a cyclic vole population might configure their territories to sustain them through trough years, and accommodate extra group members in peak years (Lindstrom 1993; von Schantz 1984), and Moehlman (1989) invoked a similar argument to explain variation in group size within constant territory sizes in golden jackals, and a similar temporal emphasis of RDH might apply to kit foxes or bat-eared foxes adapting their territories to periods of drought (Egoscue 1975; White *et al.* 1996; Maas and Macdonald, Chapter 14, this volume) or crab-eating foxes adapting theirs to periods of flooding (Macdonald and Courtenay 1996) (Fig. 4.9).

The Arctic foxes on Mednyi Island might be interpreted as an extreme case of RDH based on a single, indivisible but variously rich patches (Goltsman *et al.* submitted). Their groups comprise 2–6 adults, feed largely around seabird colonies, which are highly clumped around the coastline. Classifying the coastline into 500 m stretches, some are poor, with almost no vertebrate prey, and few fox groups hold territories on these; others contain up to 3000 breeding seabirds (generally in a clump extending over 0.5–2.5 km), and are the focus of fox breeding territories. The number of adult foxes, lactating females, and helpers were all significantly correlated with the richness of the single, rich, shareable, but seemingly indivisible food patch in each territory (Fig. 4.10).

Figure 4.9 Crab-eating fox *Cerdocyon thous* © A. Gambarini.

None of these examples touch on pack-hunting canids, such as bush dogs (Fig. 4.11), dholes, African wild dogs, and wolves, but we will offer below a synthesis to explain how the RDH concept provides a framework for explaining their social organizations too.

Dispersal

Dispersal is poorly understood in canids, yet it is a crucial topic both to conservation (Macdonald and Johnson 2001) and to understanding life-history processes (Waser 1996) and, specifically, group living. Under RDH, the dispersion of resources may facilitate group formation, but the benefits of group living and the costs of dispersal will determine whether individuals opt for philopatry or dispersal. As Macdonald and Carr (1989) model, the balance of advantage will shift with circumstances; thus the fact that female wolves in Alaska are least likely to emigrate in hunted populations may be because this higher mortality improves their chances of attaining breeding status in their natal group (Ballard *et al.* 1987). Amongst coyotes, while larger groups may enjoy fitness benefits in the efficiency of securing and retaining prey (Bowen 1978, 1981), Messier and Barrette (1982) conclude that the major selective force for larger social groups is delayed dispersal in saturated habitats. One extreme of this continuum of dispersal costs is illustrated by Ethiopian wolves in the Bale Mountains, which are effectively restricted to 'islands' of afro-alpine meadow, making dispersal an unpromising

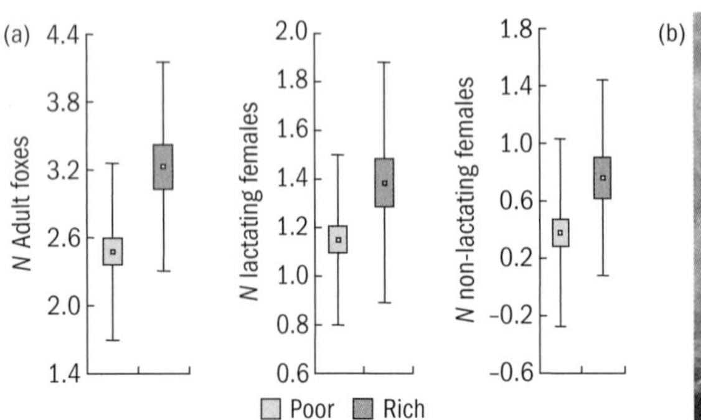

Figure 4.10 (a) Group sizes of resident Arctic foxes in rich versus poor territories on Mednyi Island. Numbers of adult foxes, lactating foxes, and helpers versus habitat richness (Krutchenkova *et al.* submitted). (b) Arctic fox *Alopex lagopus* © D. W. Macdonald.

Figure 4.11 Bush dog *Speothos venaticus* © C. and T. Stuart.

option. However, philopatry brings with it the risk of inbreeding, which may be why breeding females surreptitiously seek liaisons with neighbouring males. Within groups, females effectively accept mating with only the dominant male, but during cross-border liaisons, they mate indiscriminately (Sillero-Zubiri *et al.* 1996a). This must blur the genetic discontinuities between neighbours, as does the behaviour of crab-eating foxes reported by Macdonald and Courtenay (1996). These crab-eaters lived in groups of 2–5 individuals. Some dispersers settled in territories at the borders of the natal range, and (seemingly not accompanied by their new mate) returned intermittently to their original territory in amicable company with their parents (one male tending the next generation of his siblings during his return visits). Of four dispersing males, two subsequently returned to their natal group following the deaths of their mates at least 3–13 months after their initial dispersal, in one case after breeding elsewhere. In short, a superficially straightforward territorial system actually involves neighbourhood settlement, intermittent returns home, and, occasionally, the ultimate return of a disperser. These family ties affected the tenor of social encounters: those between unrelated neighbours tended to be at borders and to be hostile, whereas those between related neighbours tended to occur during incursions and to be amiable.

Examples of both phenomena, neighbourhood relatedness and return from dispersal, are mounting amongst canids and may be widespread (e.g. wolves: Lehman *et al.* 1992; wild dogs: Girman *et al.* 1997; bat-eared foxes: Maas and Macdonald, Chapter 14, this volume). Inbreeding may also occur locally between related pack founders derived from neighbouring packs (e.g. Mech 1987; Wayne *et al.* 1991b), circumstantial evidence suggests it is the norm amongst bat-eared foxes in the Serengeti where, through natal philopatry, 7 out of 54 females were mounted by their father, and one by their brother (Maas and Macdonald, Chapter 14, this volume).

The ethology of dispersal is poorly understood. Even the trigger for dispersal remains obscure. Although a general assumption was that the subordinate youngsters of a generation were the most likely to disperse (and there is some evidence for this in red foxes, Macdonald 1987), Bekoff (1978b) argued for the interesting possibility that amongst coyotes it was the most robust individuals that dispersed. The first outcome of food shortage amongst both wolves and African wild dogs seems to be not that young helpers disperse but that pups starve (Malcolm and Marten 1982; Harrington *et al.* 1983)—explicable

Figure 4.12 Sociality amongst adult red foxes (*Vulpes vulpes*). (a) adult sister group mates mutually groom (b) subordinate vixen greets her mother (c) a squabble between two female group members and (d) play between adult females.
© D. W. Macdonald.

perhaps in terms of parents investing where they are likely to secure the greatest returns. Furthermore, despite the general assumption that parents aggressively drive the young away, there is rather little evidence of this and Harris and White (1992) emphasize instead a decline in affiliative behaviour by the breeding pair in red foxes. In black-backed jackals, agonistic behaviours are only observed between same-sex helpers (Moehlman 1983).

Some sense of the generally unknown sociological mysteries of canid dispersal is given by the tantalizing case histories detailed by Mech (1970) for grey wolves. For them the time between emigrating and settling varies between 1 week and 12 months, but averages less than a month for females and more than 4 months for males; furthermore, when older individuals disperse, they may generally cross fewer territories than do younger dispersers before settling (Gese and Mech 1991). Mech (1987) documents the behaviour of 300 radio-tagged wolves from 25 contiguous packs in Minnesota, including one female who had travelled some 4117 km^2 before settling, and whose descendents illustrate every variant of dispersal: one female dispersed, paired, lost her litter, and returned to her natal territory permanently, another did the same but dispersed again a few weeks later; some had a period of days separated from the pack but within the natal territory before departing, one female lived this way for months; one male returned home intermittently for a year while courting a neighbouring female; some moved to adjoining territories, some moved far. From studies such as this, some generalizations emerge. Most canids, like most carnivores, disperse from their natal home range at sexual maturity (reviewed by Waser 1996); for example, wolves generally disperse at 2 years of age (Fritts and Mech 1981; Fuller 1989), whereas most foxes do so in their first winter. Storm *et al.*'s (1976) monumental tagging study was the first of several to reveal that amongst yearling red foxes, a greater proportion of males dispersed, and dispersed further, than did females (Englund 1970; Lloyd 1980). However, Macdonald and Bacon (1982) noticed that the beeline distances of dispersing red foxes, while varying greatly between populations, tended to cross a rather constant number of territory diameters. In contrast to male bias in red foxes, tagging studies found no sex bias in dispersal distance or tendency amongst grey wolves (Mech 1987; Gese and Mech 1991), although genetic evidence suggests that either males engage in more long-range dispersal or males suffer greater mortality en route, than do females (Lehman *et al.* 1992). At the other end of the spectrum lie African wild dogs and Ethiopian wolves, amongst which dispersal predominates amongst females (Frame *et al.* 1979; Fuller *et al.* 1992; Sillero-Zubiri *et al.* 1996a; Creel and Creel 2002)—although, mindful as ever of intraspecific variation, note that McNutt (1996b) describes a population of African wild dogs in which males dispersed later, in larger groups and further than did females, and Girmen *et al.* (1997) found no genetic evidence for sex-biased dispersal. (Creel *et al.* Chapter 22, this volume).

Dispersal is generally thought to be expensive, but what if the property market is full of vacancies? Formerly, on Mednyi Island, Arctic foxes occupied effectively contiguous coastal territories. However, in the aftermath of a mange outbreak in the 1970s, 90% of former territories—including several rich ones—were unoccupied (Kruchenkova *et al.* 1996). Considering that non-breeding helpers appeared to gain no indirect fitness benefit and communal breeders end up, per capita, with smaller litters than females breeding alone (see Table 4.3), Goltsman *et al.* (submitted b) predicted that secondary members would opt for what appeared to be cost-free dispersal, and that male-biased dispersal would explain the female-biased group composition. On the contrary, despite the predicted sex bias in dispersal (*c.*90% of males versus <40% females dispersed), females rarely emigrated from large groups despite the availability of vacant rich ranges. There was no difference in the sex ratio of dispersers from poor or rich natal territories, but there was a significant preponderance (2.25 : 1) of female cubs at weaning on rich territories. In short, groups of this long-isolated subspecies appear to get larger on richer territories not because of changes in the proportion of daughters dispersing, but rather because of changes in the proportion of daughters in emerging litters. Why super-numery female group members do not disperse to vacant territories is a matter of speculation (but the phenomenon is broadly in accord with predictions made by Julliard (2000)). An intriguing parallel is that amongst bat-eared foxes, the sex ratio of cubs at emergence swung significantly from 52% female during a year when the breeding population was entirely comprised of pairs, to 67% during the following years when most territories were occupied by groups (Maas and Macdonald, Chapter 14, this volume).

Comparative trends

Allometries

Moehlman (1986, 1989) reported that female body mass was positively correlated with gestation, neonate mass, litter size, and litter mass, and that from these corollaries of size flow generalizations about interspecific differences in adult sex ratio, dispersal, mating, and neonate rearing systems. As introduced in Chapter 1 (Macdonald and Sillero-Zubiri, this volume), Moehlman (1986, 1989) argued that female body weight was the driver of a cascade of effects: females of large canids have large litters of relatively small, dependant neonates, with prolonged dependency therefore requiring more male postpartum investment, competition among females for males as helpers therefore drives the system towards polyandry. Small canids, she argued, produce smaller litters of more precocial neonates that require less parental investment; the lesser demand for paternal investment enables males to invest in additional females (polygyny) (developments of this idea by Moehlman and Hofer (1997) are mentioned on p. 9 Chapter 1, this volume).

However, a review by Geffen *et al.* (1996; see also Geffen *et al.* 1992) did not support the body mass/allometry hypotheses and suggested, instead, that much of the inter- and intraspecific variation in canid social structure can be better explained by environmental variation in resource availability. Their phylogenetic analysis of canid life histories confirmed that neonate weight and litter weight are positively correlated with female weight, but suggested that large canids do not have relatively smaller young. Although the correlation between maternal weight and litter size was significant, it explained only 26% of the observed variance, and the regression slope was less than one. After controlling for phylogeny (and controlling for female body weight), neonate weight was independent of litter size (implying that there is no general energetic linkage between these two variables).

In short, Geffen *et al.*'s analysis casts doubt on the idea that body size imposes different energetic constraints on reproduction in small and large canids, and thereby is the ultimate influence on their social organization. Rather, they proposed that the high correlation and isometric relationship between neonatal weight and female body weight implied that the size of the young is constrained either by female body size directly or by some allometric correlate of female body size, for example, pelvic width (suggested to limit neonate size in primates, Leutenegger and Cheverud 1982). Litter size, in contrast, is only weakly and non-isometrically correlated with female body weight, suggesting that litter size may be adjusted in response to the availability of resources. Geffen *et al.* (1996) suggested, therefore, that variance in female pre-birth investment can be adjusted only by varying litter size—red and arctic foxes, and wolves are amongst canid species exhibiting decreases in litter size with decreases in prey abundance (Macpherson 1969; Harrington *et al.* 1983; Lindström 1989; Angerbjörn *et al.* 1991; Hersteinsson and Macdonald 1992). On this view, changes in body size, litter size, and social organization within the Canidae may be attributed primarily to differences in food availability. Thus, small canids (e.g. fennec fox, *Vulpes zerda*) are usually associated with arid and poor habitats in which only a small body mass can be supported year round, whereas large canids are often associated with habitats in which prey are at least very abundant (e.g. Ethiopian wolves) and more generally, abundant and large (e.g. African wild dog, grey wolf)—and the special impact of abundant, herding, ungulate prey on canid society will be explained below. The maned wolf (*Chrsocyon brachyurus*) unusual in its social organization for a large canid, lives in South American savannas and feeds largely on rodents and fruit (Dietz 1985); perhaps low food availability constrains both group and litter size (which is low at 2.2). Similarly, Cypher *et al.* (2000); see also Moehrenschlager *et al.* Chapter 10, this volume), show for the San Joaquin kit fox (*V. macrotis mutica*) that demographic parameters, such as fox density and growth rates, are dynamic and fluctuate widely under variable environmental conditions, tracking primary productivity determined by rainfall a year earlier. Pregnancy rates are similar between years, but neonatal survival is reduced in years of low precipitation (Spiegel and Tom 1996). Reproductive success, and litter size, tend to be low during periods of low food availability, when sex ratios at birth are male-biased, but is female biased when fox abundance is low and the population is increasing (Egoscue 1975; Spencer *et al.* 1992). Similarly, coyote group size may depend on relative prey size: in habitats where mule, deer, and elk are important prey items, coyotes have delayed pup dispersal and form larger groups (Bowen 1978, 1981; Bekoff and Wells 1982, 1986). Where small rodents are the main prey, group size tends to be smaller and dispersal occurs earlier (Bekoff and Wells 1982), but there are populations that feed on high density rodents in habitats in which dispersal is difficult and form larger groups (Andelt 1982).

A further example is provided by a study of red foxes on Round Island, Alaska (Zabel and Taggart 1989) that demonstrated a shift from 71% polygyny when food was superabundant to 100% monogamy when prey abundance declined, with a concomitant decrease in litter size. Geffen *et al.* (1996) suggest that if an increase in food availability permits an increase in litter size, then males could afford to invest in more than one female only when prey is especially abundant.

Synthesis

Finally, to a synthesis. Can these disparate facets of socio-ecology be brought together? First, Kruuk and Macdonald (1985) noted that starting from a minimum defensible territory there are two possibilities for group formation: one alternative is contractionism, that is building up a group in so far as extra members can be squeezed into the minimum territory that will support a pair, but no further. Instances of decoupling of group and territory sizes provides circumstantial evidence that contractionist groups exist and RDH not only suggests a mechanism to explain how they arise but also predicts that their territories will be small relative to that predicted from their group metabolic needs, because their prey occurs in rich, heterogeneously available patches (see Fig. 4.3). Often, these prey are invertebrates and fruit, hunted alone by canids whose sociality is characterized by spatial groupings (Fig. 4.13c). Examples may include some populations of red, Arctic and crab-eating foxes, and raccoon dogs, and we suspect occasional cases among most other small foxes too. Bat-eared foxes provide an unusual case where there appears to be no consistent relationship between group size and territory size, but where group members often forage as a party (Maas and Macdonald, Chapter 14, this volume).

The alternative to contractionism is that the benefits of sociality may be so great, that it pays the group to expand to a bigger territory. Ethiopian wolves seem to illustrate such expansionism, and have home ranges of a size predicted by their group metabolic weight (Sillero-Zubiri, Chapter 20, this volume). However, some canids with unexpectedly large ranges are at least sometimes also expansionists (wolves and coyotes, see graphs originally published in Macdonald 1983). Why? Again RDH provides a possible mechanism and testable predictions. These canids specialize on large ungulate prey, which have several relevant characteristics: they run away and must be chased over large distances; they form clumped, mobile herds; they are big, itself a form of clumping, and they attract kleptoparasitism. All these features may explain not only why canids chasing such prey have larger home ranges than expected for their body size *irrespective* of sociality, but why they also create conditions where the smallest territory necessary for a pair of hunters is likely to contain sufficient prey to support additional group members. And because their prey is big, these canids can not only eat together, they can benefit by hunting together, and become mob operators (Fig. 4.13a). Kelptoparasitism certainly makes it advantageous to eat together. These benefits of society are amongst the forces that we would expect to favour expansionism in some ecosystems for mob operators such as the wolves, wild dogs, dholes, coyotes, and bush dogs. How might this model accommodate different group-biomass to range size ratios even between canid species? For example, African wild dogs have much larger ranges than do the others. The answer is that the combined nature and mobility of their prey, their hunting tactics, and their need to evade interguild competitors simply mean that the smallest range that is sustainable for the wild dog's particular

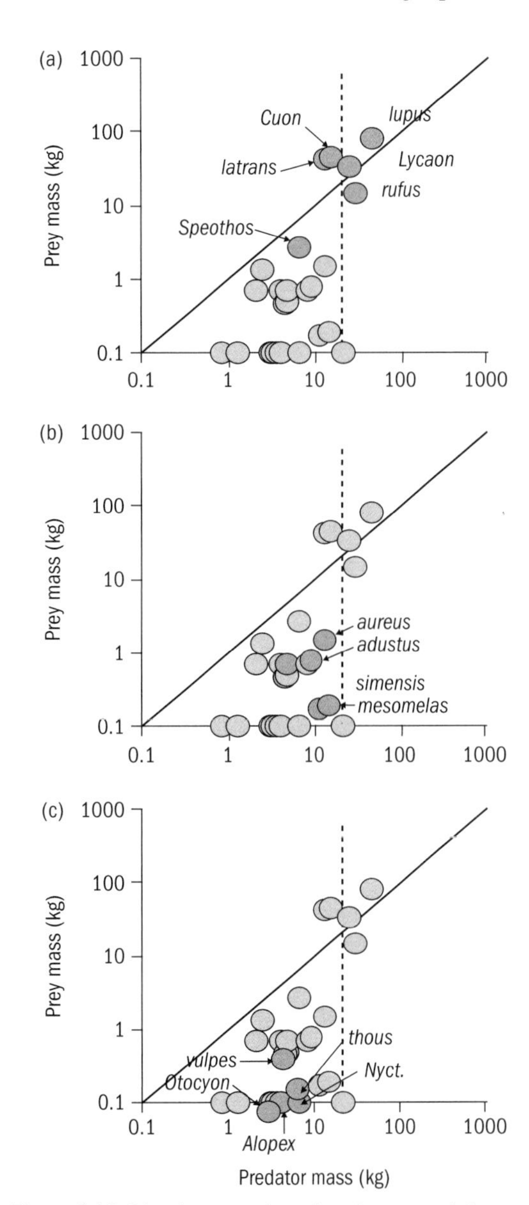

Figure 4.13 Diet shapes society—there is an association between the relationship between predator and prey masses and three broad categories of social grouping (a) The mob operators, (b) The middle way groupers, and (c) The spatial groupers. In each case, the species fitting a given category are indicated by darker shading.

lifestyle is much larger than the ranges required by the lifestyles of most populations of, say, wolves (or indeed, by the spotted hyaenas and lions from which the wild dogs are fugitive).

Between these extremes are medium-sized canids, occupying intermediate ecological niches and having intermediately cohesive societies (Fig. 4.13b). Revisiting Carbone *et al.*'s (1989) original plot, in conclusion, variants of RDH might explain why canids, whether small or large, originally evolved towards sociality, whether as spatial groupers or mob operators. RDH may also explain *both* negative and positive deviations by social canids from the home range size predicted by their metabolic needs (Fig. 4.3 Johnson *et al.* submitted). One type of resource clumping, characterized by patches of rapidly renewing or heterogeneously available small prey enables more secondary group members to be squeezed into the space required by primaries, and in nature these prey tend to occur at net higher density than the average, so groups of these canids occur in smaller territories than expected for their collective body weight. Another type of resource, characterized by large, fleet-footed mammalian prey, also allows sharing because of their mobility and body size, but in ranges that are larger than expected because clumps of these prey are characteristically very widely dispersed. In both cases, the point is that ecological circumstances create conditions that diminish or obliterate the costs of group formation. However, it will be a matter of local, autecological circumstances whether a particular species or population opts to form groups up to the size potentially accommodated in such shareable enclaves, or even to form larger groups requiring expansionist territories, or not to share at all.

Acknowledgements

We thank Lauren Harrington for the skilful compilation of diverse reference material, and Dominic Johnson for stimulating discussions of RDH. M. G. L. Mills thanks SAN Parks, the Endangered Wildlife Trust and the Tony and Lisette Lewis Foundation for support.

CHAPTER 5

Management

Management and control of wild canids alongside people

Claudio Sillero-Zubiri, Jonathan Reynolds, and Andrés J. Novaro

Coyote *Canis latrans* © E.M. Gese.

Introduction

Canids command attention in a way that is disproportionate to their number of species or abundance, chiefly because they so frequently and successfully contest human interests. Often they compete with man as predators upon unwillingly shared resources, targeting domestic animals and game. Some of the larger canids may occasionally even maul or kill people. A further reason for canid–human conflict, as explored by Woodroffe *et al.* (Chapter 6, this volume), is that canids are involved with diseases that can be harmful to people and their domestic animals.

As a result of such conflicts, many canid species have a long history of persecution by man, often well coordinated, at national scale, and state-funded (e.g. African wild dogs *Lycaon pictus*, Woodroffe 2001b). Although some canid species have gone extinct as a result (e.g. Falklands wolf *Dusicyon australis*, see below), many have been notoriously resilient to widespread and sustained persecution. For instance, coyotes (*Canis latrans*), jackals (*Canis adustus; C. aureus; C. mesomelas*), red foxes (*Vulpes vulpes*), and three *Pseudalopex* fox species in the southern cone of South America are all thriving despite tremendous hunting pressure for the pelt

Figure 5.1 Kit fox *Vulpes macrotis* © C. Van Horn Job.

trade or as targets of eradication campaigns. Other canids currently have an improving conservation status. Of these, several are medium-sized opportunists that have extended their distributions recently, sometimes aided by the removal of larger carnivores (Macdonald and Sillero-Zubiri, Chapter 1, this volume), sometimes due to flourishing in new, man-made environments (there are urban populations of red foxes, coyotes, and even kit foxes *Vulpes macrotis,* Fig. 5.1). Thanks to changing public opinion, legal protection, and habitat recovery, the grey wolf (*Canis lupus*), is returning to areas in Europe and North America where it was long ago hunted to extinction (Wydeven *et al.* 1995; Breitenmoser 1998). In contrast, other canid species are threatened and restricted in distribution (e.g. Ethiopian wolf *Canis simensis*, red wolf *C. rufus*, Darwin's fox *Pseudalopex fulvipes*, island fox *Urocyon littoralis*) (see Macdonald and Sillero-Zubiri, Chapters 1 and 23, this volume).

In this chapter we explore why canids frequently find themselves in conflict with humans, and how managers and conservationists have tackled these conflicts. We distinguish approaches based on prevention, deterrence, or removal of individual problem animals and those directed at populations.

The nature of the problem: why canids antagonize humans

Outlining the biological basis for conflict

Canids have traditionally been viewed as adversaries to be avoided or killed (Kruuk 2002). In European cultures at any rate (Reynolds and Tapper 1996), wolves, coyotes, wild dogs, jackals, and foxes have been persecuted for hundreds of years, and in some cases extirpated in the wake of expanding human populations.

Conflicts between canids and humans persist because some canid populations have the ability to recover quickly after population reductions (Gittleman 1989; Harris and Saunders 1993). Factors that contribute to this resilience are: high productivity under favourable conditions, which can be due to large litter sizes and high proportions of females breeding (Harris and Smith 1987; Clark and Fritzell 1992; Lindström 1992); dispersal ability, facilitating recolonization, and population recovery (Zarnoch *et al.* 1977; Gese *et al.* 1989; Clark and Fritzell 1992); and dietary eclecticism. However, even canids can be outgunned by high levels of culling. For instance the Malvinas zorro or Falklands wolf disappeared in 1876, due to the activity of fur traders, and poisoning by settlers to control sheep predation, worsened by the wolves' unwary behaviour and exposed, treeless habitat (Nowak and Paradiso 1993). Intensive culling has also had an impact on grey wolves in North America (Mech and Boitani 2003, in press), but less so on coyotes (Knowlton *et al.* 1999), though the historical near-absence of coyotes in the prairie pothole region has been attributed to culling (Sargeant *et al.* 1994). A detailed example of the complex regional impact of culling is that of the red fox in parts of England and Wales (Heydon and Reynolds 2000a,b; Heydon *et al.* 2000).

On the other hand, several canid species have been a resource for humans, exploited for subsistence, commercial profit, or hunting for sport. Their commercial exploitation has fluctuated depending on supply and demand (Johnson *et al.* 2001, Johnson in press). Until the advent of fur-farming in the twentieth century, the fur trade concentrated on wild populations, notably coyote, red fox, Arctic fox (*Alopex lagopus*), and three of South America's Southern Cone foxes, often having an impact on the density and demographic structure of these populations (Ginsberg and Macdonald 1990; Hersteinsson and Macdonald 1992; Novaro 1997b). Much of the pioneering history of northern latitudes was built on trade in fur, much of it originally worn by foxes. In some areas, trapping may largely be the prerogative of ranch- or farm-hands motivated primarily to reduce predation on livestock or poultry. However, even for these people the sale of fur can be economically important (e.g. up to 26% of their annual income in Patagonia, Novaro *et al.* Chapter 15, this volume).

Either way (pest or resource), people have historically tended to look at canids over the barrel of a gun. Increasingly, more tolerant attitudes are emerging, but these may be least evident amongst those people living closest to wild canids (Sillero-Zubiri and Laurenson 2001). Furthermore, the versatility of many canids enables them to flourish in anthropogenic landscapes (everywhere on farmland, but even in cities, e.g. Macdonald and Newdick 1982; Baker and Timm 1998; Baker and Harris, Chapter 12, this volume), bringing them into conflict with humans far beyond the borders of 'protected areas' (Woodroffe and Ginsberg 1998).

The ability of large and medium-sized canids, such as grey wolves, African wild dogs, dholes (*Cuon alpinus*), and coyotes to kill large prey means they can cause substantial economic damage through depredation on livestock. This, together with occasional attacks on people, explains their intense history of persecution. Perceptions, though, may exaggerate reality. Grey wolves vividly illustrate this public relations problem (Kellert *et al.* 1996). Although they seldom attack people in North America and documented livestock losses are very low, wolves are often blamed for livestock attacks and are still widely feared. Folklore undoubtedly demonizes the wolf (e.g. Gipson *et al.* 1998), and the biological studies that underpin the new view of wolves have been done only in the last half-century.

Persecution appealed as a reaction to conflict with canids probably because it involved revenge and action rather than frustration and inaction. For many large carnivore species it was undeniably effective in population terms. Arguably it has also been a major selective force, encouraging fearfulness of people (see also Howard 1990). Thus, coyote were substantially less diurnally active after a decade of intensive control activity (Kitchen *et al.* 2000), Ethiopian wolves revert to more nocturnal habits when persecuted (Sillero-Zubiri and Macdonald 1997), grey wolves have adopted more secretive and nocturnal activity patterns in parts of Europe where they coexist with people (Vila *et al.* 1995; Ciucci *et al.* 1997), and red foxes are more diurnal where undisturbed, and dramatically more approachable beyond those parts of Europe and North America where they have been persecuted (Macdonald 1987). Whereas worldwide most direct human–wolf encounters are actually trouble-free, confident or aggressive behaviours are found in more remote wolf populations, sometimes leading to attacks on humans (McNay 2000; Linnell *et al.* 2002). Some of these differences may be genetically determined. This is suggested by fur-farm observation of the flightiness of different red fox colour morphs (Keeler 1975); by the deliberate breeding of 'domesticated' red foxes (Belyaev and Trut 1975; Trut 1999); and by differences in trappability between red foxes in much of Europe (shy) compared with those in the middle East or oceanic islands (bold) (D. W. Macdonald, personal communication).

Attacks on humans and disease transmission

Despite prejudice to the contrary, predation by wild canids on humans is rare. Grey wolves have caused no human deaths in North America for a century (Mech 1970), unlike pumas, *Puma concolor* (Seidensticker and Lumpkin 1992) or bears (Herrero 1985). In contrast, there are many documented cases of wolves killing people in the Old World (Kruuk 2002; Linnell *et al.* 2002), but few were fatal during the twentieth century (with the exception of India, see below). African wild dogs are feared across Africa as 'ruthless killers' (e.g. Bere 1955), but their attacks on humans are rare or non-existent (Creel and Creel 2002). No attack by dholes has been documented.

Most recent recorded attacks by wild canids on humans involve rabid animals. Individuals in the 'furious' phase may bite several people—often children—in a single attack (Linnell *et al.* 2002; see Woodroffe *et al.* Chapter 6, this volume). With the eradication or reduction of rabies, the incidence of wolf attacks has dropped disproportionately (Linnell *et al.* 2002), but cases are still reported from Asia and the Middle East. The first diagnoses of rabies in South American wild canids were reported only recently, with cases in crab-eating foxes (*Cerdocyon thous*) and culpeos (*Pseudalopex culpaeus*) that attacked people in Argentina (Delpietro *et al.* 1997; M. C. Funes personal communication).

Exceptionally, in India there is good evidence that wolves select children as prey (Kruuk 2002). In the last 20 years, 273 children have been reported killed by wolves in the Indian states of Andhra Pradesh, Bihar, and Uttar Pradesh—areas where wolf ecology brings them into contact with shepherd children (Jhala and Sharma 1997). Attacks are also associated with habituation and provocation. Coyote attacks on people are also known, although rarely fatal (Baker and Timm 1998). In the United Kingdom, a few cases of red foxes in urban areas biting or scratching infants have been reported in the newspapers in recent years.

Predation on livestock

The greatest source of human–canid conflict is competition for resources, whether domestic animals, or for wild or captive bred 'game' species conserved and managed for hunting. The history of conflicts over domesticated stock traces back to the development and spread of herding societies (Reynolds and Tapper 1996).

Livestock depredation produces important economic losses throughout the world, and canids are often the main culprits. In United States, 79% of sheep and 83% of cattle depredations are due to canids, with 15% and 18% of these losses (respectively) due to domestic dogs and the rest due mostly to coyotes (Data for 1999 and 2000 from website of National Agricultural Statistics Service, USA Department of Agriculture, www.usda.gov/nass). Estimates of sheep losses to wild canids in United States were $19–38 million in 1977, $75–150 million in 1980, $83 million in 1987, and $16 million in 1999 (US Fish and Wildlife Service 1978; Wade 1982; Terrell 1988; all cited in Knowlton *et al.* 1999). Cattle and calf losses to predators in the United States represented US$52 million in 2000. These economic losses promote high investments in control efforts. Farmers and ranchers in the United States invest $9 and $185 million a year on non-lethal methods alone to prevent predator loss of sheep and cattle, respectively (National Agricultural Statistics Service).

Domestication selects against 'wild' behaviours, making stock tractable but also susceptible to predators (Hemmer 1989). Predation on domesticated animals, from chickens to cattle, is the root of a deeply ingrained antipathy towards wild canids throughout the world. Although larger carnivores are more conspicuous and attract particular wrath, the collective damage of smaller species such as jackals, coyotes, and foxes may be greater (Macdonald and Sillero-Zubiri 2002). Conflict has been exacerbated by changes in husbandry over the last century. In particular, the economics of modern farming generally precludes once-traditional livestock-guarding practices. Where stock-guarding traditions have vanished during the historical absence of wolves, re-colonization by wolves now provokes furious public complaint and requests for compensation (Treves *et al.* 2002).

Livestock depredation by canids is highly variable in space and time, making generalizations misleading (Knowlton *et al.* 1999). Fifty per cent of sheep producers in the United States, for example, reported annual losses to coyotes that were less than 5% of their stocks, but nearly a quarter of producers reported losses greater than 15% (Balser 1974; Gee *et al.* 1977; cited in Knowlton *et al.* 1999). Factors that may affect coyote predation rates are sheep breed (which embodies size, group cohesiveness, mothering ability, etc.), sheep and predator management practices (e.g. confinement, shed lambing, etc.), aspects of coyote biology, and environmental conditions (Knowlton *et al.* 1999). Coyotes also differ individually, with most confirmed kills of sheep being attributed to breeding, territorial coyotes (Shivik *et al.* 1996; Conner *et al.* 1998; Sacks *et al.* 1999b). Finally, the availability of wild prey has opposite short- and long-term effects on canid depredations. In the short term, abundant wild prey can reduce canid predation on livestock due to prey switching (Meriggi and Lovari 1996; Sacks and Neale 2002; reviewed in Knowlton *et al.* 1999; Novaro *et al.* Chapter 15, this volume). In the longer term, abundant wild prey can increase canid abundance, leading to increased predation on livestock (Wagner 1988; Stoddart, unpublished data cited in Knowlton *et al.* 1999).

Similarly, in the United Kingdom, lamb losses to red foxes are typically estimated or reported to be 1–2% of lambs born (Macdonald *et al.* 2000c), though reported loss for individual flocks can reach 15% (Heydon and Reynolds 2000b). The low average levels must be viewed against the historical background of culling (resulting in currently suppressed fox density in some regions—Heydon and Reynolds 2000a) and existing husbandry practices. For example, the risk posed by Arctic foxes to lambs in Iceland increases with distance from the farmstead (Hersteinsson and Macdonald 1996). Furthermore, Macdonald *et al.* (2000c) suggest that only a minority of red foxes in England—even where sheep husbandry is widespread—try to, or have the opportunity to, kill lambs. There is no hard evidence to indicate that lamb-killing behaviour is related to age, sex, breeding status, or genetics of the fox. Factors likely to restrict opportunity include protective behaviour by ewes, good husbandry, and predator swamping.

Management of canids that are important sheep predators over large regions (e.g. coyotes, dingoes *Canis lupus dingo*, culpeos, red foxes) has concentrated in the past on reducing population sizes at regional scales (Wagner 1988; Corbett 1995; Novaro 1995; Knowlton *et al.* 1999; Macdonald *et al.* 2000c). However, contracting agricultural economies change this relationship. In Argentina, for example, the Patagonian steppe was devoted to sheep ranching during most of the twentieth century, and the densities of sheep predators were reduced by hunting throughout the region—the annual tally in culpeo skins averaged 15,000–20,000 during the 1970s and 1980s (Rabinovich *et al.* 1987; Novaro 1995). Culpeos were reported to kill 7–15% of lambs annually and were second only to starvation among causes of lamb mortality (Howard 1969; Bellati and von Thungen 1990). During the last two decades 90% of sheep ranches in southern Neuquén province have switched to cattle production, and many ranches in the least productive areas of Patagonia have been abandoned, resulting in a 51% decline of the total sheep stock (INDEC 2002). These changes, combined with lower fur prices, have lead to reduced hunting pressure on culpeos (as evidenced by a 70% decline in annual fox numbers killed) and increased fox densities (from $0.5 \pm 0.1/km^2$ in 1989–92 to $1.0 \pm 0.2/km^2$ in 2000–02 in southern Neuquén; Novaro 1995; see also Novaro *et al.* Chapter 15, this volume). Increased culpeo densities, in turn, have resulted in severe sheep depredation problems (24–40% of lamb stock killed per year, Novaro *et al.* Chapter 15, this volume) that are spatially concentrated on the remaining sheep ranches and native Indian reservations. A similar trend in land use and in the spatial scale of predation problems has occurred in the United States. There the number of sheep declined by 75% in the 50 years prior to the 1980s (Wagner 1988) and was reduced by another 25% between 1991 and 1996 (US Department of Agriculture 1997 cited in Knowlton *et al.* 1999). Simultaneously, depredation problems have concentrated in the remaining, isolated ranches that still raise sheep (Hacket 1990 cited in Sacks *et al.* 1999a), requiring a management scale change from regional population reduction to local control (Wagner 1988; Sacks *et al.* 1999a).

Historically, livestock husbandry displaced wild ungulate prey (Yalden 1996), reducing the availability of wild prey for carnivores and favouring altered patterns of predation. The hope that restored populations of natural prey, alongside appropriate husbandry practices, would solve the issue of predation on livestock (e.g. reintroduction of red deer *Cervus elaphus* to parts of Italy to ameliorate impact of recovering of wolf numbers—Boitani 1992) has proved to be largely unfounded. A few studies (e.g. Meriggi and Lovari 1996; Sacks and Neale 2002; Pía *et al.* 2003) have quantitatively assessed prey selection by canids where both livestock and wild prey are available. Sacks and Neale (2002) found that coyotes kill sheep in proportion to their availability. Consumption of large wild prey (deer) was negatively correlated with sheep predation rate during the lambing period, while there was no correlation between deer consumption and sheep predation at other times (Sacks and Neale 2002). They concluded that in a prey-rich area the coyotes minimized time spent acquiring food rather than maximizing net energy gain, and therefore recommended that sheep should be held in as small an area as possible, particularly at times when they are most vulnerable, such as during lambing.

Livestock losses often lead to an increased farmers' antagonism to wild carnivores and to any associated conservation project, with the overall negative impact on conservation activities in emotional and even political terms often exceeding the actual financial cost of predation (Mech 1970). This impact on 'tolerance' is explored by Macdonald and Sillero-Zubiri (Chapter 23, this volume). But this is not a black and white issue: the economic balance sheets for different canid control strategies and different agricultural objectives can be extremely complex. Macdonald *et al.* (2003) explore the bio-economics of red fox population control in the United Kingdom as an example. For an arable farmer with no livestock or game interests, the balance is in favour of no control because foxes reduce rabbit grazing on crops, leading to increased yields. This predation saves £1.63–9.25/ha in year 1, rising to £3.12–38.58 by year 3 (the range representing the outcome of different model choices). At the other extreme, a rural community dependent on an extensive low-husbandry sheep farming system obtains a net break-even by maintaining historical suppression of fox numbers on a regional basis, at a cost of £0.28/ewe (for comparison, this is one tenth of routine veterinary costs), despite the fact that average lamb losses to foxes are 1–2%.

Predation on game species

Wildlife managers and gamekeepers also may perceive wild canids as a threat. Throughout history, predators have been seen as competitors for man's wild prey (chiefly deer and game birds) and as a result have been killed. For example, in royal hunting preserves of Europe, wolves and other large carnivores were killed to protect deer populations. In Britain, the practice of predator control to benefit game species emerged with the development of large, privately owned sporting estates in the nineteenth century. Currently, an estimated annual tally of 70–80,000 red foxes is killed by a force of *c.*3500 professional gamekeepers, chiefly motivated to preserve game birds for sport shooting (Tapper 1992; Macdonald *et al.* 2000c). Recorded fox culls on a subset of *c.*500 shooting estates have been increasing steadily since the early 1960s (Tapper 1992). Paradoxically, these higher culls probably reflect less effective control at a regional scale, because the number of professional gamekeepers fell by about 90% between 1911 and 1981 (Tapper 1992). The consequent decrease in culling intensity probably allowed the previously suppressed red fox population to increase in many regions. As a result, gamekeepers now encounter and kill many more foxes than they would have done 100 years ago. Additionally, there has been a seasonal shift in emphasis from spring and summer to autumn and winter culling, again resulting in higher numbers encountered and killed. The effectiveness of so many independent control efforts at a local scale is difficult to judge (Heydon and Reynolds 2000a,b). However, Tapper *et al.* (1996) demonstrated by controlled experiment in a case study that local and seasonally targeted fox culling, combined with similarly focused culling of other common predator species, allowed grey partridge (*Perdix perdix*) productivity to increase 3.5-fold over 3 years.

In North America there was an ethos of reducing canid numbers in order to increase the populations of game, even inside national parks (Dunlap 1988; Clark *et al.* 2001a; Grandy *et al.* 2003). From the 1930s, park management strategy changed to include canids as part of the ecosystem. Nevertheless, many hunters believe that competition with canids reduces hunting opportunities. Deer hunting is a popular sport on mainland Europe and North America and today deer hunters are among the most vocal opponents to wolf reintroduction, although predictive models suggest some fears may be misplaced (e.g. Singer and Mack 1999).

A counter-argument is that canids target sick and infirm animals that would otherwise perish of natural causes and are actually not sought after by hunters. In Alaska, grey wolves and bears were blamed for low moose (*Alces alces*), and caribou (*Rangifer tarandus*) densities, and thus for reduced hunting quotas. Gasaway *et al.* (1992) confirmed that predation by bears and wolves was the main factor limiting moose at low densities in one 9700 km^2 study area in Alaska/Yukon. However, wolf control resulted in prey increases only when wolves were seriously reduced over a large area for at least 4 years, and there is no evidence of ungulate increases persisting appreciably after predator control ceased (National Research Council (U.S.) Committee on Management of Wolf and Bear Populations in Alaska 1997).

Conflict with threatened wildlife species

The introduction of alien carnivores by man can have a catastrophic impact on resident faunas, either through predation or competition, and often results in extinctions, extirpation, or range contractions (Macdonald and Thom 2001). The dingo, introduced to the Australian continent as long as 11,000 years ago, may have displaced by competition both the tylacine (*Thylacinus cynoecephalus*) and the Tasmanian devil (*Sacophilus harrisi*) (Lever 1994). Comparison of kangaroo and emu (*Dromaius novaehollandiae*) populations inside and outside the 'dingo fence' in South Australia suggests that dingoes limit and probably regulate these prey species (Newsome *et al.* 1989; Pople *et al.* 2000). Similarly, red fox control by culling in the wheat-belt of Western Australia allowed two rock wallaby (*Petrogale lateralis*) populations to increase by 138% and 223%, compared with 14% and 85% declines at nearby sites without fox control (Kinnear *et al.* 1998).

Many ground-nesting seabirds are dependent on predator-free islands for nesting, and introduced foxes can decimate them (Bailey 1993; Reynolds and Tapper 1996). The Arctic fox, for example, had a large impact on several Arctic seabird colonies, either where introduced by man (Bailey 1992) or where they have naturally invaded islands (Birkhead and Nettleship 1995). In California, for instance, introduced red foxes are threatening rare clapper rails (*Rallus longirostris*) and salt marsh harvest mouse (*Reithrodontomys rariventris*) (Reynolds and Tapper 1996), and they also kill endangered San Joaquin kit foxes (*Vulpes macrotis mutica*) (Ralls and White 1995). In the United Kingdom, there is a debate over whether control of indigenous red foxes is necessary to ensure the persistence of endangered indigenous bird populations such as capercaillie (*Tetrao urogallus*), stone curlew (*Burhinus oedicnemus*), and sandwich terns (*Sterna sandvicensis*). Part of the issue is that staple food resources which sustain foxes at observed levels are alien species (e.g. rabbit *Oryctolagus cuniculus*, brown hare *Lepus europaeus*, pheasant *Phasianus colchicus*), themselves supported by human agriculture (rabbits, hares) or artificially supplemented (pheasant). These abundant resources allow foxes to persist at densities where they can exert a commanding influence in the population dynamics of much less common prey species. Thus in one study (summarized by Reynolds and Tapper 1996), 85% of fox diet consisted of rabbit and brown hare. Foxes were key determinants of grey partridge density, yet all the grey partridges killed by predators could not have made more than 2% of annual fox food requirements.

To illustrate the complexity of managing predators it is necessary to look at the cascade of effects not only on their prey but also within their guild (Tuyttens and Macdonald 2000). Due to their broad range of sizes, canid species may be top or meso-predators, and thus can have dramatically different ecological roles. Killing them, therefore, may also have different impacts on their communities (Henke and Bryant 1999; Berger *et al.* 2001). In addition, because larger canids may interfere with smaller ones, changes in the abundance of larger canids affect not only other predators and their prey, but also smaller canids (Johnson *et al.* 1996; Palomares and Caro 1999; Linnell and Strand 2000; Tannerfeldt *et al.* 2002). The removal of wolves, as well as bears, from the Yellowstone ecosystem 150 years ago, for example, triggered a cascade effect by allowing increased density of moose, whose increased foraging pressure on riparian vegetation negatively affected the diversity of Neotropical migrant birds (Berger *et al.* 2001).

Coyotes play a keystone role—their removal can lead to increased abundances of mesopredators, including gray foxes (*Urocyon cinereoargenteus*), and in turn reductions in the abundances and diversity of prey, such as small rodents and jackrabbits (*Lepus californicus*) (Soulé *et al.* 1988; Vickery *et al.* 1992; Sovada *et al.* 1995; Henke and Bryant 1999). These effects have to be weighed in the balance against attempts to reduce coyotes with a view to protecting livestock and game (Ransom *et al.* 1987; Canon 1995; Nunley 1995).

It appears that red fox populations increased in California when coyotes declined following urbanization. Red foxes are excluded as coyotes kill them (so where red foxes are inimical to conservation aims, managers may tolerate coyotes; see Moehrenschlager *et al.* Chapter 10, this volume). Similarly, introduced Arctic foxes in the Pribiloff islands were eliminated by the introduction of sterile red foxes (Bailey 1992). Arctic and red fox breeding-site distribution suggests that red foxes exclude endangered Arctic foxes from high quality, low elevation breeding habitat (Hersteinsson and Macdonald 1982, Tannerfeldt *et al.* 2002). The generality of inter-guild competition in canid ecology is summarized by Macdonald and Sillero-Zubiri (Chapter 1, this volume), and is vividly illustrated by the plight of island foxes (Roemer Chapter 9, this volume). Such conflicts raise awkward ethical issues (see Macdonald 2001; Sillero-Zubiri and Laurenson 2001), a further illustration being the case of protected grey wolves in India which prey upon, and may limit the numbers of, the endangered blackbuck antelope (*Antelope cervicapra*) in Velavadar National Park (Jhala 1994).

Approaches to solving canid–people conflict

Approaches to resolve human–canid conflicts include those that improve tolerance through education and cost-sharing; and those methods that attempt, either by lethal or non-lethal means, to reduce or remove the problem. The latter may tackle the problem by:

(1) changes in the protection given to prey;
(2) changes in the behaviour of the animals involved, either prey or predator;
(3) removing the offending individual(s); or
(4) undertaking to control canid population density.

Approaches aimed to increase people's tolerance

The negative impact of canids tends to affect well-defined communities, be it small-scale shepherds in Africa (Kruuk 1980) or Spain (Blanco *et al.* 1992), British gamekeepers (Macdonald *et al.* 2000c), or commercial cattle farmers in western United States (Kellert 1985). Often, the lack of recognition of a community's problem worsens conflict and the mere act of listening can help. Recognition that residents have legitimate concerns can reduce the resentment that is directed at individual animals as 'surrogates' for distant government officials and elitist environmentalists (e.g. Ethiopian wolf persecution following the 1991 government overthrow in Addis Ababa—Gottelli and Sillero-Zubiri 1992). The 'human dimension' of carnivore conservation is increasingly recognized (Clark *et al.* 2001b).

Conserving canids amounts to more than just saving the animals; it includes also the decision process by which human communities identify and solve the problems of sharing the land with carnivores (Clark *et al.* 2001a). Understanding of the problem may involve researching complicated biological questions. The human dimension is illustrated by the contradictory views of proposals to reintroduce grey wolves in the United States: some see them as a threat to livestock, others as a boost to ecotourism (in Macdonald *et al.* 2002a,c). Enck and Brown (2002) pointed to the hazards of unrealistic expectations: some of the communities in northern New York State that hope to reap the benefits of ecotourism were socially and economically ill equipped to do so. Sillero-Zubiri and Laurenson (2001) stressed that a prerequisite to reintroducing carnivores was an assessment of not only the habitat but also the local human community to accommodate them. A vivid case study is the success delivered by the unprecedented public outreach to all stakeholders before wolves were released in Yellowstone (Fritts *et al.* 1997; see Phillips *et al.* Chapter 19, this volume).

Participation of land owners/local communities in management

Co-management of habitat and wildlife with local communities within development programmes is increasingly seen as essential for conservation outside protected areas. African wild dogs have benefited in southern Africa from community-based initiatives and the establishment of large private nature reserves and conservancies, where income from cattle farming (susceptible to drought) is being replaced by tourism, trophy hunting, and game-ranching.

In the United States, Defenders of Wildlife has created the Proactive Carnivore Conservation Fund. They share with ranchers the costs of actions to prevent livestock depredation, buying livestock guarding dogs, erecting electric fencing, hiring 'wolf guardians' to monitor wolves in sheep range by radio-telemetry and chasing them away when they get close to livestock (Nina Fascione, personal communication). They have paid more than US$270,000 in compensation to ranchers for losses due to wolf attacks since 1995.

Improve economic benefits

Demonstrating the economic benefits of canid conservation can require ingenuity. Traditionally, canid skins have represented a perk to farmhands. In Patagonia, ranch-hands have annual incomes of US$500–1200, and fox pelts represent 2–14% of their annual income (Funes and Novaro 1999), and up to 26% in some years (Novaro *et al.* Chapter 15, this volume). When the value of red fox skins soared in the early 1980s British farmhands began to harvest them seriously (Macdonald and Carr 1981). However, for the most part the pelt value merely causes people who would anyway kill these canids to take the time to skin them.

Job creation in research, conservation, and management activities

Canid conservation projects can create local employment. For instance the Ethiopian Wolf Conservation Programme (EWCP) employs 25 local people full-time, at a cost of US$47,000 a year in wages); EWCP staff have gained in status and respect in the community and the programme continues to receive constant requests for employment (S. Williams personal communication; www.ethiopianwolf.info). Similarly, the Painted Dog Conservation Project in Zimbabwe has employed former poachers for conservation jobs (G. Rasmussen, personal communication). At the other end of the spectrum common canid species generate jobs; for example, in the United Kingdom red foxes support the employment of *c.*3500 game-keepers and an estimated 6000 people associated with traditional fox-hunting with hounds (Macdonald *et al.* 2000c), and coyotes and dingoes support similar control industries.

Non-consumptive recreational use

There is no doubt that large carnivores are a major attraction for tourists. High-profile and visible canid species such as African wild dogs, grey wolves, Ethiopian wolves, dholes, and maned wolves may be capable of supporting, at least partially, a sustainable tourist trade, but many canids are frustratingly secretive. Although the big cats were the main draw to a Zimbabwe national park when all nationalities were pooled, wild dogs ranked top for Zimbabweans, and second among South Africans (Davies 1998). Visitors to Yellowstone take pleasure in the expectation of seeing or hearing wolves; more than 20,000 visitors to Yellowstone have observed wolves since 1995 (www.nps.gov). Ethiopian wolves are the chief attraction to visitors to the Bale Mountains National Park, and maned wolves are a highlight of tourism to the protected grasslands of Argentina and Brazil. Nonetheless, expectations of revenue should not be exaggerated. For example, in Ethiopia's Bale Mountains, where income from tourism is often given as a justification to the local community for the presence of a park, the number of tourists visiting each year is numbered only in the hundreds, many use their vehicles rather than local guides or horses, and the sums reaching the local community are not great. Furthermore, the susceptibility of tourism to fluctuations in the global economy, and to political instability, makes it unwise to base conservation entirely on economic values.

Consumptive recreational use

Hunting of large carnivores can increase their value, offset livestock losses, and provide a way to dispose of known 'problem' animals. Regulating these activities within a sustainable framework can be difficult.

The results of public harvest of grey wolves (defined to exclude aircraft- and snowmobile-assisted hunting) intended to increase deer may not have been clear-cut in terms of population biology (e.g. Kenai Peninsula—Peterson *et al.* 1984b, north of Anchorage—Gasaway *et al.* 1992), but produced a very positive response from sport hunters who felt empowered by their cost-effective participation in a programme necessitating minimal government involvement (Boertje *et al.* 1995).

Sharing the cost of conservation across a wider society

Primm (1996) notes that the costs of tolerating carnivores may be very unevenly spread, and that society should share the burden with the afflicted individuals. This raises complicated questions about who owns, or takes responsibility for, wildlife. Compensation schemes seek to share the burdens of tolerating predators, and to be effective require strong institutional support and clear guidelines. They also require quick and accurate verification of damage, prompt and fair payment, sufficient and sustainable funds, and measures of success (Nyhus *et al.* 2003).

In Italy, for example, the local government compensates 100% of the value of livestock killed by wolves, bears, and even feral dogs (Cozza *et al.* 1996). In France, recolonization by wolves has been closely monitored, with the provision of community guards and official damage assessors. In 2001, 372 attacks by wolves on livestock involved 1830 dead sheep, and cost over €300,000 in compensation (Dahier 2002). After wolves returned to Montana in 1987, Defenders of Wildlife (see above) have paid out more than US$270,000 to more than 225 ranchers to compensate for 327 cows, 678 sheep, and 34 other animals killed by wolves in Idaho, Montana, and Wyoming (Nyhus *et al.* 2003; N. Fascione personal communication). However, a pilot study suggests that for every calf killed by wolves and found by the cattle producer, as many as 5.7 additional wolf kills may have gone undetected (Nina Fascione, personal communication).

The opposite of compensation is a bounty scheme, such as that involving approximately 1000 Patagonian sheep producers, the majority of whom get 5–10 bounties every year. As an example, US$10–25 were paid for *c.*50,000 bounties between 1996 and 2001 in two Patagonian provinces alone (Wildlife Agencies of Río Negro and Chubut provinces unpublished data, Novaro *et al.* Chapter 15, this volume). Of course, compensation schemes do nothing to alleviate the problem, rarely deal with full costs, are open to corruption and can involve an expensive bureaucracy. Some of these difficulties may be circumvented by community-based insurance schemes (Nyhus *et al.* 2003). A novel way of sharing the cost of living with carnivores is to add a premium price to goods produced by 'predator-friendly' farms (L. Marker personal communication).

Education: improving aesthetic and moral benefits to community

Perceptions of predator problems often exaggerate the reality, and education programmes can target this by delivering accurate information and increase people's tolerance and appreciation for wildlife (Conover 2002). Some canid species can act as flagships for projects in order to gain public support for habitat conservation—grey and Ethiopian wolves fit this mould, which could be extended to African wild dogs, dholes and perhaps even Darwin's or island foxes.

Approaches that seek to reduce the scale of the problem

Efforts to deal with predation directly range from preventing contact between predators and their potential prey—either by modifying husbandry practices, exclusion fences or guarding—to dealing with problem animals, to dealing wholesale with problem populations (Fig. 5.2). There are both scientific and ethical problems with lethal control of canids, fostering an interest in non-lethal alternatives (Treves and Woodroffe in press) and greater selectivity in lethal control (Treves 2002).

Preventing contact between canids and potential prey

Changes in livestock husbandry

Extensification of husbandry systems, the lack of supervision of livestock (and the increase of some

Figure 5.2 Poster of 'co-existing with Coyotes' Program.Vancouver British Columbia © Stanley Park Ecological Society.

hitherto threatened species) has led to an increase in livestock predation. Risk tends to increase with herd size, distance from people and buildings, proximity to thick cover, and carcasses left in the open (e.g. Mech *et al.* 2000; Kruuk 2002; Ogada *et al.* 2003). Conscientious husbandry is essential, and can be straightforward, such as improved vigilance, preventing livestock from straying, and returning herds to enclosures at night (e.g. Kruuk 1980, 2002). Specific husbandry practices, however, must be developed for the particular situation of each producer group. Some recommended practices only delay predation or have undesirable side effects; penning animals at night, for example, is costly and frequently deteriorates pastures locally (Knowlton *et al.* 1999). In southern Chile and Argentina, where puma and culpeo populations have increased, a shift from sheep to cattle husbandry, and further diversification have been proposed (Johnson *et al.* 2001), a move encouraged by falling wool prices and increased predation.

Setting barriers that exclude predators

People have built barriers to protect stock from predators since time immemorial. In northern Kenya a simple thorn-bush *boma* (corral) can make a big difference with 90% of stock killed outside the enclosures that would have protected them (Kruuk 1980), and appropriate fencing can deter predation by coyotes on sheep (e.g. Linhart *et al.* 1982; but see Thompson 1978; Nass and Theade 1988). Of course, fences can have undesirable environmental costs, for instance cutting thorn fences contributes to habitat loss (Kruuk 2002), fencing may curtail wildlife movements, and wire stolen from electric fences may be used by poachers to make snares (G. Rasmussen personal communication). Nonetheless, mobile electric fences can prevent predator attacks in Romania Carpathian mountains, where wolves and bears annually kill *c.*1.5% of the sheep stock (Mertens *et al.* 2002). Sheep losses in camps with electric fences (averaging US$6.70 per camp) were only 2.6% of the losses in other camps without electric fences. Thus an electric fence (that costs approximately US$250), would be paid for by the reduction of livestock losses in 1 year alone. However, installation costs tend to be prohibitive at a larger scale, to the extent that they render fencing impractical as a means of preventing coyote or culpeo predation on sheep production systems in the western United States and Argentine Patagonia (Knowlton *et al.* 1999). As a cheaper alternative to wire fences Musiani and Visalberghi (2001) propose fladry, a line of red flags hanging from ropes traditionally used to hunt wolves in eastern Europe. Tests of fladry on captive wolves showed some potential; wolves tended to avoid red flags set <50 cm apart and never crossed flag lines intersecting their usual stereotyped routes, even when the daily food ration was placed on the other side.

Fences have been used to protect ground-nesting birds (Greenwood *et al.* 1995; reviewed by Reynolds and Tapper 1996). For example, wire-mesh fences successfully excluded Arctic foxes from the nests of Alaskan pectoral sandpipers *Calidris melanotos* (Estelle *et al.* 1996; see also Minsky 1980; Beauchamp *et al.* 1996). Alternatively, some ground nesting birds can be persuaded to nest on elevated or safe platforms out of reach of canids (Conover 2002).

Use of livestock guarding animals

Sometimes canids can be deterred from livestock by using guarding animals (Andelt 1999, 2001; Rigg 2001; Coppinger and Coppinger 2002). Dogs have been used to guard stock for millennia (Rigg 2001) and more recently donkeys or llamas (*Llama glama*) have been recruited for this purpose (Meadows and Knowlton 2000; Andelt 2001). A good livestock guarding dogs needs to be independent, intelligent, stubborn, trustworthy, show high aggressiveness to predators, and be attentive to sheep (Andelt 1999; Knowlton *et al.* 1999; Rigg 2001). Several breeds have been selected for this purpose; Kuvasz and Caucassian guarding dogs are used in Slovakia and elsewhere in eastern Europe (Rigg 2001), while Italian Maremmano-Abruzzese, French Great Pyrenees, Turkish Akbash, and Anatolian shepherds, Hungarian Komondor, Yugoslavian Sarplaninacs, and Spanish Mastiffs have now been introduced far and wide to serve as herd protectors. One study in the United States rated Akbash as more effective than Great Pyrenees and Komondors (Andelt 1999). Effectiveness is limited by large flocks, rough terrain, thick cover, and limited training and supervision (Knowlton *et al.* 1999). Unfortunately, guarding dogs may be too expensive for all but the owners of large flocks, and even large flocks are often seasonally fragmented. Evaluations of effectiveness of guard dogs to protect sheep from coyotes have had mixed results (e.g. Linhart *et al.* 1979; Andelt 1992; Knowlton *et al.* 1999), and recent studies indicate llamas can be more useful in preventing sheep losses to coyote predation (Meadows and Knowlton 2000).

The advantages of llamas are that they can be kept in fenced pastures, require no special feed, and have a working life three times that of dogs (Knowlton *et al.* 1999; Meadows and Knowlton 2000). Leadership, alertness, and body weight of llamas correlated with aggression towards dogs (Cavalcanti and Knowlton 1998) with single large males scoring highest. Furthermore, it is not necessary to rear llamas with sheep or train them, and one gelded male llama may protect 250–300 sheep (Andelt 2001). However, while pilot trials in western United States indicate that llamas deter coyote predation on sheep they appear less effective than guard dogs (reviewed by Rigg 2001). Donkeys (preferably a single female with foal) are naturally aggressive towards canids, and are very adaptable (Andelt 2001). Donkeys have been tested to repel wolves in Switzerland (Landy 2000), and in Texas a tenth of 11,000 sheep and goat growers used donkeys in 1989 and 59% rated them as good or fair at deterring predators (Andelt 1999; Rigg 2001).

Dealing with problem animals

Discriminating the guilty from the innocent

Traditional canid control has not targeted the individual culprit. Gipson (1975) showed that trapping could kill 60–70% of coyotes that were innocent bystanders. The goal should be to minimize human–canid conflict, while minimizing impact on innocent individuals (Treves 2002; Treves and Karanth 2003), indeed it may be more efficient to remove culprit individuals from a canid population than to attempt population control (Conner *et al.* 1998; Blejwas *et al.* 2002).

What is the evidence that problem behaviour is confined to a proportion of the canid population? Linnell *et al.* (1999) questioned the existence of 'problem individuals' and proposed that most individual large carnivores will at least occasionally kill accessible livestock that they encounter. However, the axis of shy–bold behaviours (Wilson *et al.* 1994) does have a genetic basis in canids (see also Belyaev and Trut 1975; Keeler 1975). Second, losses would be much greater if all canids were responsible. For instance, lamb predation by red foxes typically accounts for only 1–2% of lambs born in UK conditions, and is patchy in occurrence (Heydon and Reynolds 2000a). In Wisconsin, USA, more than two-thirds of grey wolf packs did not attack nearby livestock (Wydeven *et al.* 2003). Amongst coyotes, territorial pairs may be disproportionately involved in stock-killing (Sacks *et al.* 1999b). Selective removal of problem animals may not only have a disproportionate reward in terms of public approval, but in principle also selects against any heritable or learned traits that inclines individuals to stock-killing (Treves 2002).

Disruptive and aversive stimuli

Strobes producing light and sounds that may startle or frighten have been tested on canids by Shivik *et al.* (2003). Candidate deterrent sound effects have included the noises of helicopters, gun-fire, people yelling, and breaking glass, but predators tend to habituate swiftly (Bomford and O'Brien 1990). Habituation may be reduced by targeting the alarm only to the moment when the culprit is, for example, killing or eating from a carcass (Shivik *et al.* 2003), but that is difficult in practice. Such so-called 'contingent disruptive stimuli' can be achieved by devices activated by particular triggers on collars worn by target predators (Shivik *et al.* 2002, 2003). However, this hi-tech approach is at present beyond most everyday applications.

Canids may be trained to associate predation on livestock or valuable game with a negative experience (Shivik *et al.* 2003). Any stimulus that causes discomfort, pain or other negative experience to the predator may potentially be conditioned as an aversive stimulus. Using electric shock, Badridze *et al.* (1992) successfully conditioned a pack of captive-bred wolves to move away from sheep prior to their release in Georgia. Similarly Andelt *et al.* (1999) used an electronic dog-training collar to deter captive coyotes from killing domestic lambs. Shivik *et al.* (2002) failed to train wild grey wolves that were previously involved on livestock damage not to attack livestock during a pen trial using dog-training collars.

Breck *et al.* (2002) developed the radio-activated guard (RAG); strobe lights and sirens were activated when a predator wearing a collar-mounted device approached the livestock pasture. RAG boxes deterred wolves from attacking cattle: although the wolves activated the RAG boxes some 15 times during 2 months, not a single calf was killed, whereas 16 calves were killed in unprotected pastures. A similar method involves a sound activated aversive conditioning (SAAC) collar, responding to inexpensive bells worn by vulnerable free-ranging livestock. When livestock is disturbed by an approaching predator fitted with a SAAC collar the bells ring and cause the intruder to receive a shock from the collar resulting in aversive conditioning (Shivik and Martin 2001). Approaches such as these are likely to be affordable only to affluent stakeholders.

Conditioned taste aversion

Conditioned taste aversion (CTA) refers to a specific behavioural pattern whereby mildly poisonous substances cause a deep-seated and lasting aversion to associated tastes. Gustavson *et al.* (1976) first suggested that the use of an emetic such as lithium chloride could be a useful management tool for problem predators. CTA differs from other learned aversions in that the aversive stimulus is not immediately discernable by the animal and as such is more useful for stopping an animal from eating certain foods rather than for limiting killing behaviour (Conover and Kessler 1994). Attempts have been made since the 1970s to exploit this behaviour in wildlife—and especially canid—management, but because of poor experimental design, results have been equivocal and therefore controversial (Reynolds 1999). Satisfactorily designed experiments are extremely difficult to carry out with wild predators. A recent attempt with red foxes and grey partridge (Reynolds *et al.* 2000, unpublished data) identified several obstacles that ultimately prevent deployment in UK conditions. These included the difficulty of dosing an adequate proportion of foxes using referent baits; the involvement of protected non-target species (Eurasian badgers *Meles meles*); and the lack of any already-registered product suitable for widespread environmental use.

An encouraging trial of the principle is provided by Macdonald and Baker (2004), who showed that a family of captive red foxes was successfully conditioned to avoid untreated milk after drinking foul-tasting milk (laced with Bitrex™, a bitter substance that they were unable to detect except by taste). In a separate experiment the capacity of ziram (a food-based repellent) to create learned aversions to untreated foodstuffs was tested. Red foxes showed direct aversions to treated baits and learned aversions to untreated baits, made shorter visits to the trial station after trying baits (S. Baker *et al.* unpublished data). Generalized aversion has an advantage over CTA in that because the experience of foul taste on sampling is immediate it involves no ambiguity as to which prey is associated with the negative experience, and it may therefore effectively confer protection upon untreated prey (Macdonald and Baker 2004). Despite these tantalizingly successful trials, the reality is that practical application of such techniques faces many hurdles.

Translocation of problem animals

Translocation of individual problem animals is routine for North American pumas and bears guilty of trespassing in urban areas or killing livestock, and has occasionally been applied to grey wolves and African wild dogs with mixed success. Of 107 wolves translocated in northern Minnesota following depredation or harassment of livestock 17% were shot, or recaptured at least once, for re-offending (Fritts *et al.* 1985). Overall, the mortality of translocated wolves was not higher than that of wolves already resident in the area, but pack mates failed to stay together and travelled long distances with some animals returning home; 9 of 32 wolves tracked (28%) returned to within 10 km of their capture site (Fritts *et al.* 1984). Animals translocated more than 64 km did not return to their capture sites. Five wild dogs that were translocated to South Africa's Kalahari Gemsbok National Park after their pack-mates had been shot by farmers split into two groups and disappeared within a few months (Frame and Fanshawe 1990 in Woodroffe *et al.* 1997). More recently, translocation attempts in Zimbabwe involving the wholesale relocation of wild dogs pack away from farmland have showed some promise (G. Rasmussen personal communication).

Overall, it appears that translocation is too expensive, time consuming and technically complex to be suitable for common canids, but for endangered species it may be cost-effective (see also Linnell *et al.* 1997; Treves and Karanth 2003). From both a conservation and an animal welfare perspective, the risk that translocated animals may be doomed must be thoroughly explored: release into habitat already occupied by conspecifics can lead to intraspecific aggression, social disruption of the residents, and mortality.

Dealing with problem populations

Even blanket control of over-abundant canids is likely to involve inadvertent selection, for example, by concentrating on individuals at problem sites. The objection that an unfocused cull is inefficient loses impact if there is a sport interest in the cull itself. Thus in contrasting this blanket culling approach with alternatives, the questions to be answered are whether it is effective to reduce conflict, whether it is cost-efficient, and whether alternatives exist that are demonstrably less wasteful of human effort and/or canids. A crucial consideration is whether the cull reduces the target population sufficiently to diminish the measured nuisance for a useful period. A potential risk is that compensatory movements (a vacuum effect) or breeding (density dependent) rapidly negates any benefit measured in terms of reduced damage, or causes other counter-productive perturbation (Sacks *et al.* 1999a; Tuyttens and Macdonald 2000; Blejwas *et al.* 2002). Obviously, such complications can be planned for, as illustrated by the 'dingo fence' in Queensland, Australia (Allen and Sparkes 2001), or by a 'cordon sanitaire' (Jensen 1966).

Intense local culling may create a geographical patchwork of source and sink populations

(Pulliam 1988). This type of dynamic has been observed in coyotes (Pyrah 1984), red foxes (Allen and Sargeant 1993), and culpeos (Novaro 1997b). It may be typical of some bounty schemes where the 'additive' component of culling remains insufficient to make mortality exceed productivity, and local culling mortality is easily 'made good' through reproduction and dispersal. The result is that the bounty-provider simply pays for an increased regional harvest (e.g. bounties proved ineffective for dingo control—Allen and Sparkes 2001, citing Harden and Robertshaw 1987; Smith 1990). At the other extreme, where culling is geographically widespread and its impact on the population high, replacement may be very low. Obviously this is the situation that especially concerns canid conservationists.

Sustainable harvest rather than control also results when culling becomes an activity in its own right, as for instance fox-hunting in the United Kingdom, winter fox shooting in western continental Europe, or trapping in central North America and southern South America. Here the aims of the operator may involve a higher canid density than do the aims of the livestock producer or game manager. He may be willing to (and logically should!) pay for this, squaring the differences in aims between livestock producers, hunters, and conservationists. We would not suggest that exploitation of canids be promoted where none exists—but the scenario illustrates the successful sharing of costs and risks through a wider society, as in the CAMPFIRE schemes. Also people who are interested in a sustainable harvest of canids can be allies of conservationists, because they want to have high canid densities, whereas livestock producers would rather have low (or zero!) canid densities. This is the case in Argentine Patagonia, where the fur traders association opposes the state bounty system to reduce canid densities.

Among culling techniques, poisoning deserves special mention here, because it starkly exemplifies a trade-off between utility (cost-efficiency) on the one hand, and on the other hand conservation (target-specificity) and humaneness. Even intensive poison campaigns suffer the same limitations on effectiveness as described above, wherein geographical coverage, effort, and accurate targeting all contribute to achieving aims. Because of this, the trade-off of cost against other qualities is a non-trivial choice.

Chemo-sterilants and immuno-contraception have been proposed on numerous occasions for canid population control (Asa 1992; Boyle 1994; Tyndale-Biscoe 1994; Newsome 1995; Tuyttens and Macdonald 1998). Immuno-contraception is the preferred approach in Australia to eliminate red foxes in the interest of marsupial conservation, and considerable funds have been allocated to its development. In individual-based simulation modelling involving realistic parameters for population density, growth, and dispersal, a very high 'hit rate' was found to be necessary to achieve population control in the red fox (Macdonald *et al.* 2000c).

A rarely considered aspect of population control is the degree to which it sets the background for other approaches. Where a canid population is regionally suppressed (e.g. Heydon and Reynolds 2000a) the level of conflict may be lower, and the effectiveness of local control measures higher, than would otherwise be the case. To pursue the example cited, poultry are raised out-of-doors in Norfolk, England with far lower levels of physical security than would be required elsewhere in the country, because fox density is low in the entire region, making local fox control a viable option for poultry farmers.

Although non-selective culling tackles the canid population as a whole, its aim is not necessarily either long term or far-reaching. For instance, in controlling red fox predation in the context of wild game-bird management, temporary local suppression of numbers during the breeding season is sufficient to allow enhanced game-bird productivity and thus to allow a harvest (Tapper *et al.* 1996). The effect on fox density may be impossible to measure directly, although its impact on game-birds is demonstrable experimentally.

For operator and biologist alike, it is extremely difficult to distinguish those situations in which culling is effective and moderate, from those in which it is ineffectual and wasteful (or damaging). Even in developed countries, it can be impossible to quantify culling intensity with reasonable accuracy (Heydon and Reynolds 2000b), and assessment of its impact in population terms may need to be indirect (Heydon and Reynolds 2000a). Where there is real conservation concern for a targeted species, regulation of both culling and recording is highly desirable, but may be impractical in countries unaccustomed to such bureaucracy.

Conclusions

Despite the unpromising circumstances of increasing human population and habitat loss, only a quarter of all canid species are considered threatened (Sillero-Zubiri *et al.* in press, Macdonald and Sillero-Zubiri, Chapter 23, this volume); most canid species are holding out, and some thrive. Their future, as for so many other creatures, lies in conservation initiatives that recognize the dual importance of their need for large, linked areas of suitable habitat and of the development of the human communities alongside which canids must live. In many cases, sensitive education must challenge deeply engrained cultural prejudices about wild canids, whereas the sources of genuine conflict must be identified, understood, and whittled away. Where conflict remains it will often be appropriate for wider society to lift the burden, or risk, off individual producers in the interest of preserving species. Clearly, such topics range far from biology, through such fields as economics, development, politics, and beyond. In so doing they demand a breadth of vision and knowledge that is seldom accounted for in the training or career structure of today's conservationists (Macdonald 2001). A now outmoded view characterized people, often already disadvantaged rural people, as the problem; for the future it seems essential, on the contrary, that they become part of the solution (Sillero-Zubiri and Laurenson 2001).

We have shown that conflict between wild canids and people exists, and necessitates management, for both imperilled and abundant species. The problems faced by these two categories clearly differ in detail, but both merit the attention of conservationists, and both may be susceptible to similar approaches using the same tools. For example, management of both rare and abundant canids is likely to involve a mix of strategies drawn from a wealth of disciplines, to involve changes in animal husbandry, exploration of selective, non-lethal methods, and complicated evaluations of costs and benefits (measured in such incommensurable currencies as biodiversity, money, and ethics). In affluent countries, the public, welfare organizations, and wildlife managers increasingly find lethal control unpalatable; indeed, it seems likely that the science of animal welfare will become an important implement in the conservationists' toolkit, along with expertise in the human dimension of environmental management. The hi-tech revolution surely has much to offer, but so too do traditional livestock husbandry methods. Ultimately, while the practicalities must surely be underpinned by innovative science, the solution to canid conflict rests on value—the value that people (often very poor people) place on canids. For centuries, and around the world, few other animals have so regularly been burdened with the epithet 'the only good one is a dead one' as have canids. Changing that view is a matter of changing values, and a capacity to achieve that change will be the measure of good management.

Acknowledgements

We would like to thank Jorgelina Marino and David Switzer, who commented helpfully on earlier versions of the manuscript. Claudio Sillero-Zubiri is supported by the Born Free Foundation, through WildCRU's People & Wildlife Initiative, Jonathan Reynolds is supported by the Game Conservancy Trust, and Andrés Novaro by the Argentine Research Council (CONICET) and the Wildlife Conservation Society.

CHAPTER 6

Infectious disease

Infectious disease in the management and conservation of wild canids

Rosie Woodroffe, Sarah Cleaveland, Orin Courtenay, M. Karen Laurenson, and Marc Artois

Red foxes *Vulpes vulpes* © J. M. Macdonald.

Introduction

Why worry about canid diseases?

Infectious disease is increasingly recognized as an important factor influencing the dynamics of wildlife populations (e.g. Hudson *et al.* 2002). Canid diseases in particular cause concern for two reasons (Macdonald 1993). First, widespread species such as red foxes (*Vulpes vulpes*) and coyotes (*Canis latrans*) may carry infections such as rabies, leishmaniasis, and hydatid disease that can be transmitted to people and livestock, sometimes leading to a need for management of wild canid populations. Second, populations of threatened canids such as Ethiopian wolves (*Canis simensis*), African wild dogs (*Lycaon pictus*), and island foxes (*Urocyon littoralis*) may be at risk of extinction through the effects of virulent infections such as rabies and canine distemper, sometimes needing management to protect them from infection. Both issues may be important to the conservation of wild canids, because widespread lethal control has been an important component of past attempts to protect people and domestic animals from some canid-borne zoonoses.

Are canids especially susceptible to infectious disease?

There are only 36 species of foxes, wolves, jackals, and dog, yet canids appear frequently in reviews of the impacts of infectious disease on wildlife populations (e.g. Young 1994; Funk *et al.* 2001; Cleaveland *et al.* 2002), suggesting that they may be particularly susceptible to disease outbreaks. There are several possible reasons for this. First, canids' ecology may expose them to infection. Their trophic position exposes them to infections carried by prey as well as by conspecifics (e.g. grey wolves, *Canis lupus*, apparently contract brucellosis from consuming infected caribou, *Rangifer tarandus*, Brand *et al.* 1995; African wild dogs may contract anthrax from consuming infected prey, Creel *et al.* 1995). In addition, intolerance of larger carnivores for smaller competitors (which emerges as a general rule, see Chapter 1) may increase contact among wild canid species, presenting another possible route of infection (e.g. African wild dogs reintroduced to Etosha National Park, Namibia, died of rabies after killing and consuming an infected jackal, *Canis mesomelas*; Scheepers and Venzke, 1995). Close contact among social group members (including frequent licking and grooming), as well as scent communication through potentially infectious faeces and urine (e.g. Macdonald 1985) may also increase canids' exposure to infections carried by conspecifics.

A further—and undoubtedly important—reason for canids' apparent susceptibility to infection derives from the fact that the domestic dog (*Canis familiaris*) is a canid. As members of the *Canis* genus, domestic dogs are closely related to some wild canids and share receptivity to numerous pathogens. Most dog populations are not regulated by diseases, but by humans (Wandeler *et al.* 1993). Large, high-density human populations in rural communities are often associated with large populations of dogs (The dog:human ratio most commonly lies between 1:10 and 1:6 but recent investigations often found this ratio to be considerably underestimated; WHO 1988; Wandeler *et al.* 1993; Cleaveland and Dye 1995; Kitala *et al.* 2000). These large populations of dogs may be able to sustain multiple infections and act as a persistent source of infection for other species. Where dogs are permitted to move around freely, contact with wild canids is likely to increase (Laurenson *et al.* 1997; Rhodes *et al.* 1998). Small canids' tolerance of human encroachment may bring them into close contact with domestic dogs—whether it be red foxes susceptible to rabies, mange, or hydatids in towns (e.g. Macdonald and Newdick 1982; Chapter 12) or crab-eating foxes *Cerdocyon thous* susceptible to zoonotic visceral leishmaniasis (ZVL) in Amazonian villages (Courtenay *et al.* 2001; Macdonald and Courtenay 1993). Such contact may be exacerbated by wild canids' ability to increase in density in response to food resources associated with human refuse. Larger, more wide-ranging canids may encounter domestic dogs when they move beyond the boundaries of protected areas—hence disease transmission from domestic dogs may constitute an anthropogenic 'edge effect' influencing nominally protected populations (Woodroffe and Ginsberg 1998).

The ecology of wildlife disease

Understanding the need—or otherwise—for managing canid diseases, and evaluating alternative management strategies, demands an understanding of disease dynamics in host populations. Pathogens can persist in host populations only if each infected host, on average, infects one or more susceptible hosts. If the average number of new hosts infected per case (termed R_0) drops below one, then the pathogen population will die out (Anderson and May 1991). Persistence, then, requires a supply of susceptible hosts. If a virulent pathogen is introduced to a naïve population of susceptibles, infection may initially spread rapidly, generating an epidemic. However, if hosts that have been infected are no longer available to the pathogen (either because they are immune for life, or because the infection kills them), then pathogens are likely to die out (and, hence, the epidemic will fade out) as they run out of susceptible hosts. In large or rapidly breeding populations, births may provide a sufficient supply of new susceptible hosts to allow the pathogen to persist following the epidemic; however, in smaller populations, this supply will often be insufficient to allow persistence. If immunity is short lived, then hosts can be reinfected and persistence is more likely.

These basic epidemiological concepts have several important consequences. Disease management frequently aims to eradicate pathogens from host populations. Vaccination is one way to achieve this. By inducing immunity in hosts that have not been exposed to natural infection, vaccination reduces the supply of susceptible hosts available to the pathogen. If vaccination coverage is high enough, R_0 can be forced below one, and the pathogen will die out. For microparasites, the proportion to be vaccinated (pc) is calculated as $pc = 1 - (1/R_0)$ (Anderson and May 1991). Hence, the higher the value of R_0, the greater the proportion of hosts that must be vaccinated to achieve eradication. For example, typical values of R_0 for endemic urban dog rabies range between 1 and 3 indicating that, on average, about 70% of dogs must be vaccinated to achieve rabies eradication—this is confirmed by empirical studies (Coleman and Dye 1996; Kitala *et al.* 2001). Culling can also be used to eradicate infection from wildlife populations, because it, too, can reduce the density of susceptible hosts below the threshold needed for pathogen persistence (e.g. Barlow 1996).

Situations in which canid diseases cause concern, however, are more complex than these simple examples, because they almost invariably involve transmission between host species. Governments institute measures to control rabies in wild canids not (usually) because of concerns for the well-being of wild foxes, jackals, or wolves, but because they wish to avoid transmission of this extremely unpleasant infection to people and domestic animals. Rabies does not persist in human populations and human-to-human transmission has been documented on only extremely rare occasions (Fekadu *et al.* 1996). People are thus a 'spillover' host, contributing little or nothing to the persistence of the pathogen in its primary host species. This situation almost exactly parallels that of threatened canids. Populations of species such as Ethiopian wolves, island foxes, and African wild dogs are too small and isolated for highly pathogenic infections to persist inside them. Like humans, they are 'spillover' hosts that become infected through contact with more abundant 'reservoir' host species. These multi-host systems involving generalist pathogens may lead to complex management decisions, and involve political and social issues that are sometimes in conflict with conservation.

Wild canids as reservoirs of zoonotic disease

As discussed above, human societies may become concerned about diseases of wild canids that can be transmitted to people or their domestic animals. We choose to present case studies of two infections, rabies and ZVL, which illustrate the scale and complexity of the wild canid issue, and the importance of understanding infection dynamics for reservoir incrimination.

Rabies

Rabies is a serious infection that has a major impact on human lives, particularly in developing countries. For example, 30,000 people are estimated to die annually from rabies in India alone (WHO 1995). To put wild canids' contribution to this problem into perspective, of 534 human deaths investigated worldwide in 1995, 88.8% were attributed to domestic dog bites, with only 46 (8.6%) attributed to wildlife (including 30 (5.6%) cases attributed to bats; WHO 1995). In Western Europe, where domestic dogs' movements are well controlled, dogs are well vaccinated, and red foxes are a confirmed rabies reservoir (Artois *et al.* 2001), fewer than 3% of people given treatment for presumed rabies exposure had been directly exposed to foxes (Table 6.1). However, many of the dog rabies cases may have originated in the fox reservoir, and people's far closer contact with domestic dogs (relative to foxes) predisposes them to infection via this route.

Wild canids may be self-sustaining rabies reservoirs in some areas (e.g. red foxes in Western Europe; Blancou *et al.* 1991). However, in other regions, field studies and modelling have indicated that wild populations occur at densities too low for rabies to persist without cross-species infection from other hosts. For example, Rhodes *et al.* (1998) predicted that rabies could not persist in wild populations of side-striped jackals (*Canis adustus*) alone, and presented data suggesting that domestic dogs acted as an alternative host providing constant re-infection. Cleaveland and Dye (1995) reached similar conclusions, using cross-correlation analyses to show that jackal rabies cases followed those in dogs, rather

Table 6.1 The proportion of people treated for presumed rabies exposure in Western Europe in 1995 that had had direct contact with red foxes.

Country	Total people treated	Cases attributed to Domestic dogs (%)	Cases attributed to Foxes (%)
Belgium	334	46 (14%)	56 (17%)
France	6005	4087 (68%)	126 (2.1%)
Luxembourg	56	23 (41%)	5 (8.9%)
Netherlands	19	0 (0%)	8 (42%)
Portugal	33	25 (76%)	1 (3%)
Spain	458	330 (72%)	7 (15%)
Total	6905	4511 (65%)	203 (2.9%)

Source: Data from WHO (1995).

than vice versa. These authors explored the possibility that a small number of infectious 'carrier' animals might promote persistence within domestic dog populations, but an alternative explanation is that occasional reinfection from other host species could promote persistence of the pathogen. Interactions between domestic dogs and multiple wild species could generate even more complex dynamics (e.g. Loveridge and Macdonald 2001).

Culling

Culling has been widely employed in attempts to control rabies in wild canid populations, and may have had regional conservation implications. For example, during a campaign to control rabies in Alberta, Canada, 50,000 red foxes, 35,000 coyotes, and 4200 grey wolves were killed in an 18-month period (along with 7500 lynx, *Lynx canadensis*, 1850 bears, *Ursus* spp., 500 striped skunks, *Mephitis mephitis*, and 164 cougars, *Puma concolor*; Ballantyne and O'Donoghue, 1954). Trapping, shooting, gassing, and poisoning of foxes, coyotes, and wolves has been widely practised in Eurasia and North America in attempts to control rabies (Macdonald 1980).

Culling has successfully eradicated rabies in a handful of cases (generally in restricted areas; Macdonald 1980). However, this approach is now considered an ineffective means of protecting people and domestic animals from wildlife rabies (Blancou *et al.* 1991; Funk *et al.* 2001). Host species such as red foxes, coyotes, and jackals are socially and physiologically flexible animals, forming populations with a remarkable capacity to recover from control, and to react to it with perverse consequences (Tuyttens and Macdonald 2000; Frank and Woodroffe 2001). Immigration, increased litter sizes, and improved survival have all been recorded in canid populations subject to control (Frank and Woodroffe 2001). Far from eradicating infection, inducing these sorts of population responses has the potential to *increase* the supply of susceptible hosts needed to fuel endemic infection. Indeed, recognition of the way that culling affects the social behaviour of red foxes and, hence, the dynamics of rabies infection (e.g. Macdonald and Bacon 1982; Macdonald 1995), contributed to the search for an alternative approach to the control of rabies in Western Europe. This ecological approach to wildlife disease—previously the preserve of veterinarians and health officials—saw early fruits in the still-relevant chapters of Bacon (1985).

Vaccination

Vaccination has been used with great success to control rabies in Western Europe and North America. Modelling approach very early has suggested that reservoir immunization can be efficient (Anderson *et al.* 1981; Bacon 1985) namely in the context of fox behavioural ecology (*idem.*). It has the advantage of not disrupting the natural density-dependent population dynamics and spatial organization of the host species; it is also more humane (Macdonald 1980). Initial field trials of oral rabies vaccine, concealed inside baits made from chicken heads, were carried out in Switzerland in 1978 (Steck *et al.* 1982). Following these initial promising results, the programme was spread across much of Western Europe, with up to 60% of foxes immunized (Pastoret and Brochier 1999). Dramatic decreases in rabies incidence have been recorded where well-coordinated oral vaccination programmes have been implemented, while infection has persisted in untreated areas (Artois *et al.* 2001). In Europe, rabies control in wild canids appears to be fully achievable where appropriate and well-coordinated baiting programmes are carried out, although at considerable cost (Artois *et al.* 2001). As Macdonald (1995) emphasized, the crucial lesson from attempts to control rabies in red foxes is that while approximately the same proportion of foxes ate vaccine baits as had in earlier campaigns eaten poisoned baits, the former controlled the disease where the latter had failed to do so. The only plausible difference between the two approaches is their impact on the contact rate (R_0) amongst the survivors—the perturbation effect—vindicating the view (radical only 25 years ago (Macdonald 1980)) that understanding canid social behaviour is relevant to controlling their epizootics. Successful rabies control programmes have also been carried out in North America, involving oral vaccination of coyotes and racoons (*Procyon lotor*; Fearneyhough *et al.* 1998; Roscoe *et al.* 1998; Mackowiak *et al.* 1999). Oral rabies vaccines have also been tested for potential use on jackal rabies in Southern Africa, although no field trial has yet been carried out (Bingham *et al.* 1999). There are some safety concerns surrounding the use of orally delivered live rabies vaccines in areas of high biodiversity, where it is difficult to carry out safety trials on all species that might consume the baits. For example, earlier trials showed that a live vaccine strain different from that tested in Bingham *et al.* (1999) was protective for jackals but induced clinical rabies in baboons (*Papio* sp; Bingham *et al.* 1995). Destruction of the vaccine when baits were placed in direct sunlight also created a need for hand-placing baits in shaded areas (rather than distributing them by aircraft), potentially increasing the cost of vaccination operations. However, it is important to note that these concerns originate primarily from the use of live rabies vaccines. The use of a recombinant rabies vaccine, where only part of the rabies virus (the surface glycoprotein gene) is incorporated into a *Vaccinia* virus carrier (which was used to eradicate smallpox in humans), avoids such issues as this vaccine cannot cause rabies (Brochier *et al.* 1996). This type of vaccine is also much more thermostable.

Zoonotic visceral leishmaniasis

ZVL is a vector-borne disease of humans and domestic dogs resulting from infection by the protozoan *Leishmania infantum* (also known as *L. chagasi*), transmitted by phlebotomine sandflies. It occurs in 70 countries in Latin America, Africa, Europe, and Asia, where it is considered a major public health and/or veterinary problem. Throughout the world, the domestic dog acts as the principal reservoir ('source host'), whereas humans are considered incidental and non-infectious hosts. Typically, 50% of infected dogs present clinical signs of canine ZVL (Bettini and Gradoni 1986) including alopecia, dermatitis, chancres, conjunctivitis, onychogryphosis (excessive nail growth), lymphadenopathy (enlarged lymph nodes), and emaciation, usually followed by death.

Wild canids also acquire infection, and the available data for three species (red foxes, crab-eating foxes, and golden jackals *Canis aureus*; Table 6.2) indicate, in some cases, infection prevalences as high as in endemic domestic dog populations. This suggests that wild canids may be important additional reservoirs for human infection. However, the epidemiological significance of wild canids to peridomestic transmission will depend not only on infection rates, but on (1) their ability to transmit infection to sandflies, (2) their relative contribution to transmission in the presence of infectious dogs, and (3) the likelihood that they can (re)-introduce the pathogen into uninfected dog populations (e.g. following ZVL control in dogs) (Fig. 6.1).

Crab-eating foxes

The only wild canid species for which these questions have been fully addressed is the crab-eating fox during extensive behavioural, ecological, and epidemiological studies of free-ranging populations in Marajó, Pará state, Amazon Brazil, where their role has been compared to that of sympatric domestic dogs (Courtenay *et al.* 1994, 2002a,b; Quinnell *et al.* 1997).

The initial evidence incriminating South American foxes came from parasite isolations from the skin, viscera, or blood of 4/33 animals caught in an endemic foci in northeast Brazil (Deane and Deane 1954a; Deane 1956), followed soon after by isolations from a further 7/173 animals caught in the same region (Alencar 1959, 1961). These specimens were originally identified as hoary foxes *Pseudalopex vetulus* (= *Dusicyon vetulus*) though most certainly in error, since there is recent evidence that they were

Figure 6.1 Crab-eating fox *Cerdocyon thous*, host of *Leishmania infantum* in Brazil © O. Courtenay.

crab-eating foxes (Courtenay *et al.* 1996). In Amazon Brazil, 14 of 49 (28.6%) crab-eating foxes were similarly found to be infected, including 11/26 animals from Marajó (Lainson *et al.* 1969, 1987; Lainson and Shaw 1971; Silveira *et al.* 1982; Lainson *et al.* 1990). The ability of this host to infect the sandfly vector *Lutzomyia longipalpis* was shown for a single naturally infected animal in advanced stages of the disease, which infected 10/10 *L. longipalpis* that fed on it (Deane and Deane 1954b). The infectious nature of the fox was later confirmed by Lainson *et al.* (1990), who observed an asymptomatic animal to infect 4/54 *L. longipalpis* 15 weeks after its experimental inoculation with a local *Leishmania infantum* fox strain. Deane and Deane (1955) and later Lainson *et al.* (1969) reasonably concluded that foxes were an important ZVL wildlife reservoir. This acquired general acceptance in the scientific literature.

These studies and anecdotal accounts that foxes predated domestic fowl from peridomestic animal huts known to harbour large numbers of *L. longipalpis* (Lainson 1988)—and thus a potential source for human infection—prompted studies of these foxes in nature in Marajó, with the aim to clarify their role in peridomestic transmission between domestic dogs, humans, and sandflies, and in a putative wildlife transmission cycle between foxes and sylvatic populations of sandflies independent of domestic dogs.

Behavioural ecology, infection, and disease

Behavioural observations to reveal the spatio-temporal distribution of foxes relative to sandfly populations were monitored during >3500 h by radio-telemetry and direct observation (Macdonald and Courtenay 1996). These data showed (1) adult foxes to maintain stable territories of 532 ha (range: 48–1042, $n = 21$), (2) foxes did not use dens, instead they slept above ground in thick vegetation often on the edge of pineapple and manioc plantations; (3) fox territories comprised a monogamous breeding pair and between 1 and 3 adult-sized offspring which dispersed when 18–24 months old; (4) foxes had a general habitat preference for wooded savannah (34%), and scrub (31%) depending on territory location, season (wet versus dry), and social status. For example, elevated habitats were favoured in the wet season when low lying savannah was inundated, and widespread flooding forced social groups to share use of higher ground; (5) dispersed offspring maintained amicable relations with parents and often returned to their natal range, in some cases following the death of their mate.

Our entomological studies (Lainson *et al.* 1990; Macdonald and Courtenay 1993) revealed that *L. longipalpis* did not occur in large densities in sylvatic habitats that foxes frequented: no *L. longipalpis* were caught in fox sleeping sites or surrounding areas of savannah, though small numbers did occur in residual gallery forest where few foxes had access and rarely hunted. In contrast, high densities of this sandfly are regularly captured in peridomestic animal pens in rural settings (Macdonald and Courtenay 1993; Kelly *et al.* 1997), where chickens act as an important blood source. Though parasites do not biologically develop in chickens (hence fowl are usually considered 'dead-end' hosts), chickens and livestock are nevertheless a principal lure for blood-seeking sandflies to the domestic environment.

Of 24 radio-monitored foxes, 92% entered 1–3 villages per night where they spent approximately 40 min (range: 0–242 min) each night in potential contact with peridomestic sandfly populations (Macdonald and Courtenay 1993; Courtenay *et al.* 2001). Seroprevalence (IFAT) varied between fox social groups, and number of villages visited, however, we found no statistical association between the probability of a fox or fox group being seropositive and the habitat type it visited or time spent in villages. Furthermore, the possibility that sick foxes behaved differentially and/or avoided detection was not supported by our longitudinal observations of seropositive (potentially sick) foxes: none of the foxes presented overt clinical signs of ZVL nor were there significant variations in spatial behaviour between ecologically matched seropositive *versus*. seronegative animals (Courtenay *et al.* 1994; Macdonald and Courtenay, 1996). Neither did we detect infection to increase the natural mortality rate, which was 0.325 per year (95% CL: 0.180–0.587). This high mortality rate we showed to be predominantly due to human persecution (Courtenay *et al.* 1994; Macdonald and Courtenay 1996) similar high mortality rates were recorded amongst Marajo dogs, but were largely the result of disease (Courtenay *et al.* 1991, 2002a).

Progressive longitudinal studies of this wildlife population's infection rates using a broader array of diagnostic techniques revealed cumulative prevalences of 78% (29/37) by serology (enzyme linked immunosorbent assay, ELISA), 23% (8/35) by polymerase chain reaction (PCR), and 38% (8/21) by parasite culture, with point prevalences of 74% (serology), 15% (PCR), and 26% (culture); the incidence of patent infection was 0.10–0.12 per month, which were similar or higher than equivalent estimates in sympatric domestic dogs (Courtenay *et al.* 2002a,b; Quinnell *et al.* 1997). Fox infection rises rapidly in the young fox population, followed by a decline in the proportion parasite positive in the older age classes, suggesting that older foxes successfully clear parasites and/or become less susceptible to infection/re-infection. This is not accompanied by loss of seropositivity or decline in antibody titre with time (Fig. 6.2) (Courtenay *et al.* 1994, 2002b). A similar pattern of infection is seen in domestic dogs (Quinnell *et al.* 1997), however unlike foxes, seropositive dogs often present progressive ZVL disease. The asymptomatic nature of *L. infantum* infection in wild canids generally is illustrated in Table 6.2.

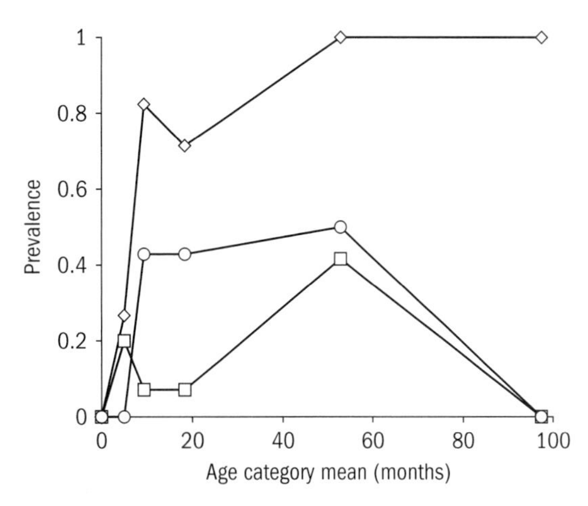

Figure 6.2 The age-prevalence of *L. infantum* infection in crab-eating foxes. Prevalence was assessed by serology (◇), PCR (□) or *in vitro*/*in vivo* culture (○). Values are shown for mean ages of age-class: 0–6, 7–12, 13–24, 25–84, and 85–114 months. (Redrawn from Courtenay *et al.* 2002b.)

Infectiousness and contribution to transmission

A crucial variable in epidemiological models is the probability that the disease will be transmitted from host to vector, in this case from the fox to the sandfly. This can be measured by xenodiagnosis, a technique that involves exposing potential reservoirs to colony-reared (i.e. parasite free) vectors, which are then screened for the presence of the parasite. In this case, 26 free-ranging Marajo foxes were caught and brought into captivity where female sandflies were given the chance to feed on them on 1–3 independent occasions. ($n = 44$ feeding trials,

Table 6.2 Natural infections of *L. infantum* in selected canids in endemic regions

Number of parasite positive/N (seropositive/N)	Number of symptomatics/N positives	Region/country	Source[a]
Crab-eating fox *C. thous*			
11/26 (13/25)	0/11 (0/13)	Marajó, Pará, Brazil	Silveira *et al.* (1982); Lainson *et al.* (1990) Courtenay *et al.* (1994)
8/37[b] (29/37)	0/8 (0/29)	Marajó, Pará, Brazil	Courtenay *et al.* (2002b)
3/23	0/3	Belém, Pará, Brazil	Lainson *et al.* (1969, 1987) Lainson and Shaw (1971)
4/33	3/4	Sobral, Ceará, Brazil	Deane and Deane (1954a, 1955) Deane (1956)
1/11	0/1	Corumba, Mato Grosso, Brazil	Mello *et al.* (1988)
7/173	0/7	Ceará, Brazil	Alençar (1959, 1961)
Red fox *V. vulpes*			
4/5 (4/5)	0/4 (0/4)	Setubal region, Portugal	Santos *et al.* (1996)
9/50 (18)	0/9	Imperia, Liguria region, Italy	Mancianti *et al.* (1994)
4/71 (14/61)	0/4 (0/14)	Setubal region, Portugal	Abranches *et al.* (1982, 1983, 1984)
3/64	0/3	Alhama, Spain	Marin Iniesta *et al.* (1982)
2/99	1/2	Cevennes, France	Rioux *et al.* (1968)
2/150	1/2	Cevennes, France	Lanotte (1975)
≥1/ 68	0/1	Grosetto, Tuscany, Italy	Bettini *et al.* (1980); Pozio *et al.* (1981b)
0/169 (0/22)	–	Alcacer do Sal, Portugal	Abranches *et al.* (1982, 1983, 1984)
0/24 (0/7)	–	Lisbon region, Portugal	Abranches *et al.* (1982, 1983)
–(11/16)	–(0/11)	Priorat, Tarragona, Spain	Saladrigas (1992)
–(3/5)	–	Alto-Douro, Arrabida, Spain	Semiao *et al.* (1996)
50/67[c]	–	Guadalajara, Spain	Criado-Fornelio *et al.* (2000)
2/19	0/2	C. Asia	Maruashvili & Bardzhadze (1966)
1/36	0/1	C. Asia	Maruashvili & Bardzhadze (1966)
1/10	0/1	Iran	Nadim *et al.* (1978)
1/10 (2/10)	0/1 (0/2)	Iran	Edrissian *et al.* (1993)
–(1/20)	–	Israel	Baneth *et al.* (1998)
Golden jackal *C. aureus*			
–(4/53)	–	Israel	Baneth *et al.* (1998)
1/20	1/1	Iran	Nadim *et al.* (1978)
4/161 (6/48)	1/4 (1/6)	Iran	Hamidi *et al.* (1982)
2/30 (5/30)	0/2 (0/5)	Iran	Edrissian *et al.* (1993)
≥5/nd	–	C. Asia	Latyshev *et al.* (1961); Lubova (1973) Dursunova *et al.* (1965)
–(3/46)	–(0/3)	Israel	Shamir *et al.* (2001)

[a] Sources before 1998 are fully cited in Courtenay *et al.* (1998).
[b] Molecular diagnosis by PCR and/or culture.
[c] Molecular diagnosis by PCR.

$n = 1469$ female *L. longipalpis*—the vector). Despite the fact that 81% of the foxes were currently or recently infected, and flies fed on foxes to full engorgement, these trials proved that crab-eating foxes were rarely infectious: a conservative estimate of the proportion of sandflies infected were <1% for seropositive, PCR, or culture-positive foxes. In striking contrast, longitudinal xenodiagnosis of the sympatric infected domestic dog population during the same time period showed 43% of them to be infectious to a median 11% of sandflies. Applying these estimates in a deterministic multi-host model of ZVL, Courtenay *et al.* (2002b) calculated that the maximum contribution of infected foxes to overall transmission in nature was only 9% compared to 91% by domestic dogs. Moreover, the basic reproduction number R_0, for foxes was lower ($R_0 < 1$) than the expected threshold for parasite persistence. Thus, despite high infection rates, the incapacity of foxes to infect a large fraction of biting vectors suggests that the species represents a parasite 'sink', rather than 'source', host. Their poor 'reservoirial capacity' to transmit *Leishmania infantum* is probably attributed to the asymptomatic nature of fox infection since asymptomatic infected dogs similarly contribute <1% to all transmission events compared to >99% by symptomatic dogs (Courtenay *et al.* 2002a).

Control: (re)-introduction of infection into Leishmania-free populations

The epidemiological significance of foxes to ZVL transmission will depend on their ability to maintain the parasite in the population independent of infectious dogs (i.e. in a self-sustaining sylvatic cycle), such that successful elimination of the parasite in the dog population will not be short lived due to the (re)-introduction of infection into clean dog and sandfly populations by infectious foxes. Indeed, the evidence above suggests that there is no such sylvatic cycle, supported by additional observations showing that;

1. *Leishmania* isolates from crab-eating foxes are indistinguishable from isolates obtained from local dogs and *Lutzomyia longipalpis* (Mauricio *et al.* 1999, 2001). All isolates are considered to be *L. infantum* MON-1, which is responsible for the human and canine disease throughout the geographical distribution of ZVL.

2. Vector infection prevalences are typically <1%; thus successful transmission requires a large number of contacts to receive an infectious bite: sylvatic vector populations in Marajo fox territories appear to be rare (Lainson *et al.* 1990), and the foxes live at low densities (0.55 animals per km^2, SE: 0.071, range: 0.273–0.769, $n = 7$ territorial groups (Macdonald and Courtenay 1996; Courtenay 1998), suggesting that fox contact with sylvatic flies is low, particularly in habitats that foxes frequent most (Macdonald and Courtenay 1993, 1996).

3. In contrast, foxes have substantial contact with peridomestic habitats where *Lu. longipalpis* populations are large. These collective data thus strongly suggest that ZVL 'spills over' into the fox populations from dog-infected sandfly populations in the peridomestic setting. In the event ZVL is eliminated in dog populations, it is therefore unlikely that foxes could (re)-introduce the parasite.

The geographical range of the crab-eating fox extends from Venezuela to Argentina (Courtenay and Maffei, in press), and while it is possible that there is geographical variation in ZVL susceptibility, few wild canids have been examined outside Amazon Brazil. Proposed wildlife hosts of ZVL in Mediterranean Europe, and middle Asia countries include the red fox and golden jackal, though nothing is known of their capacities to infect sandfly vectors. The general absence of the disease in wild canids complies with the idea of a non-pathogenic parasite–host relationship in an indigenous wildlife reservoir, as proposed for the crab-eating fox in South America (Lainson *et al.* 1987). The alternative view is that the greater diversity of *L. infantum* strains in the Old World compared to the New World suggests that *L. infantum* was initially introduced into the Americas from the Old World (Momen *et al.* 1993), possibly via an infected domestic dog in post-Colombian times (Killick-Kendrick 1985), or earlier via a wild canid host presumably during the radiations of the Canidae in the late Miocene to early Pleistocene (Berta 1987; Martin 1989).

Implications for ZVL control

Strategies to prevent (as opposed to treat) human ZVL disease in many endemic countries include vector control (house spraying with residual insecticides) and/or reservoir control (culling infected dogs). There is no effective anti-*Leishmania* human or canine vaccine, and 70% of dog infections relapse following canine chemotherapy. Although depopulation of wild canids has not been part of any ZVL public/veterinary health policy, current information, together with lessons learnt from the unsuccessful culling programme of domestic dogs in Brazil (Dye 1996; Courtenay *et al.* 2002a) suggest that culling wild canids would not be an efficacious ZVL control option, irrespective of whether infection is controlled in domestic dogs. By contrast, we anticipate that successful control of infection (and infectiousness) in dog populations e.g. by protecting dogs against sandfly bites, will result in *L. infantum* elimination in foxes.

Disease as an extinction threat to wild canids

Infectious disease is an important and intractable extinction risk to many canid species. Disease has caused dramatic die-offs and local extinctions of several canid populations (Table 6.3). Both theoretical predictions and empirical observations indicate that highly pathogenic infections cannot persist in small, isolated populations (Lyles and Dobson 1993). As described above, epidemics can spread only when each infected host transmits infection to one or more susceptible hosts; in small populations epidemics tend to 'burn out' as the majority of hosts either die or become immune. It is not surprising, therefore, that most disease outbreaks in small, isolated populations of threatened hosts tend to involve generalist pathogens that 'spill over' from other, more common, host species (Table 6.3). For example, a rabies outbreak in a newly reintroduced population of African wild dogs in South Africa was linked to a simultaneous outbreak in sympatric jackals (*C. mesomelas*; Hofmeyr *et al.* 2000).

Infectious disease may also suppress the viability of canid populations in less dramatic ways. In particular, infections that increase pup mortality may cause slower declines, or reduce populations' ability to recover from perturbations. For example, cub deaths from otodectic mange depressed the only population of the Mednyi Arctic fox (*Alopex lagopus*

Table 6.3 Local extinctions and crashes of canid populations known to have been caused by infectious disease

	Population size				
Host species	Before outbreak	After outbreak	Pathogen	Source of infection	Refs
African wild dog	50–70	0	Rabies[a]	Domestic dogs	Gascoyne *et al.* (1993), Kat *et al.* (1995)
	4	0	Rabies	Jackal	Scheepers and Venzke (1995)
	12	3[b]	Rabies	Jackal	Hofmeyr *et al.* (2000)
	10 packs[c]	5 packs[c]	Rabies	Jackal	J.W. McNutt (personal communication)
	12	0[c]	Distemper	Domestic dogs?	Alexander *et al.* (1996)
Ethiopian wolf	53	12	Rabies	Domestic dogs	Sillero Zubiri *et al.* (1996)
	23	11	Rabies[a]	Domestic dogs	Sillero Zubiri *et al.* (1996)
Island fox	1340[d]	150[d]	Distemper	Domestic dogs	Timm *et al.* (2000)
Blanford's fox	4[c]	1	Rabies	Red fox	Macdonald (1993)
Red fox	46[c]	0	Mange	Unknown	Harris and Baker (2001)

Notes

[a] Cause of deaths inferred from confirmed diagnoses in contiguous population.

[b] Survived when captured, isolated and vaccinated.

[c] Study population within a larger contiguous population.

[d] Approximate figures. Most of the remaining population protected from exposure by a physical barrier to disease spread.

semenovi; Goltsman *et al.* 1996). Likewise, the annual seroprevalence of parvovirus—a virus that mainly kills pups—was negatively correlated with the recruitment of wolf (*C. lupus*) pups in Minnesota (Mech and Goyal 1995).

Population viability analyses indicate that disease represents an important extinction risk for some canid populations (Ginsberg and Woodroffe 1997; Vucetich and Creel 1999; Roemer *et al.* 2000b; Haydon *et al.* 2002). In general, highly pathogenic infection such as rabies threatens the persistence of small- and medium-sized populations, with less pathogenic infections threatening only smaller populations (such as relict populations and those in the process of recovery through reintroduction; Woodroffe 1999; Haydon *et al.* 2002). This creates a need for management tools to protect threatened canid populations—especially those that are small and isolated—from the potentially deleterious effects of pathogenic infections. Since the diseases that represent the most serious threats to wild canid populations (primarily rabies and canine distemper) tend to be maintained in other host species, conservation interventions may include direct vaccination or treatment of threatened hosts themselves, management of reservoir hosts in attempts to

Table 6.4 Management options for disease control for wild canids (modified from Laurenson *et al.* 1997)

Option	Advantages	Disadvantages	Likely benefits/chance of success
Do nothing	Cheap, easy, evades controversy	Population viability not guaranteed	Depends on local situation
Reduce disease in **reservoir** species		No guarantee of protection in target	
Vaccination	No intervention with target	Expensive logistics cost, welfare, cultural attitudes	High
Culling	Effective vaccines available	Limited effectiveness	Not sustainable
Limit reproduction	Can be very effective	Effective methods not yet available over large areas	High in theory, but may not be practicable
Treatment	Therapy availability depends on pathogen	Limited effectiveness	Weak
Reduce disease in **target** species			
Vaccination	Direct protection	Effectiveness vaccines not always available	High over the short term (emergency plan), but may be good strategy in some species/situations if feasible and cost/effective (e.g. small pops wolves
Treatment		Often unfeasible	Last chance in emergency situation
Prevent **contact** between target and reservoir			
Fencing/physical barrier	No intervention	Often unfeasible	High
Restraining domestic animal reservoir		Cultural constraints/conflict with dog function	Medium on continental situation
		Long term	High on islands
Translocation of reservoir (e.g. limit human activities in protected areas)		Feasibility	Medium

eradicate infections from entire ecosystems (or at least to reduce regional disease incidence), and limitation of contact between threatened and reservoir hosts (Laurenson *et al.* 1997; Table 6.4). It is worth mentioning, however, that in theoretical terms, the simplest and perhaps most sustainable way to reduce disease risks to small populations is to strive to transform them into larger populations. This will be impossible where all or most suitable habitat is occupied, but it may be an option is some regions, and should not be dismissed.

Direct vaccination as a conservation tool

Direct vaccination of threatened hosts has been used as a conservation tool on a number of occasions (reviewed in Hall and Harwood 1990; Woodroffe 1999). In no case has it been demonstrated that vaccination leads to higher survival—however, this is because most cases have been 'crisis' interventions dealing with acute disease risks, when no animals have been left as unvaccinated controls. In one case, administering a live vaccine without safety trials led to disastrous consequences (Carpenter *et al.* 1976), but if vaccines are safe and effective, then their use has the potential to improve the viability of canid populations severely threatened by infectious disease. Population viability modelling suggests that vaccinating 20–40% of an Ethiopian wolf population against rabies could markedly reduce extinction risks, although in very small populations (25 animals) higher coverage is required to remove extinction risk (Haydon *et al.* 2002). This level of coverage is quite low in comparison with the vaccination coverages needed to eradicate rabies in reservoir hosts; the reason it appears effective in conservation terms is that rabies is not eradicated from the system. Instead, vaccination is used to protect a core of target 'spillover' hosts, which permits a population to persist—despite some mortality of unvaccinated animals—when rabies outbreaks occur. This theoretical finding that even comparatively low vaccination coverages can promote the viability of target host populations offers a promising approach to conserving highly endangered populations with technically feasible levels of vaccination.

The value of direct vaccination assumes, of course, that safe, effective vaccination protocols are available.

African wild dogs *Lycaon pictus* interacting © J. Ginsberg.

This is not yet the case for Ethiopian wolves, and research has been instigated to examine the feasibility of this approach. Vaccines have, however, been tested and used in wild populations of two other endangered canid species, the African wild dog and the island fox; these are discussed in greater detail below.

African wild dogs

Rabies and, to a lesser extent, canine distemper represent acute threats to the persistence of small wild dog populations. However, attempts to protect wild dogs from these threats by direct vaccination have met with mixed success.

Rabies

Free-ranging wild dogs have received inactivated rabies vaccines in four different projects (Table 6.5). In all cases, a proportion of dogs subsequently died, with rabies confirmed as a cause of death in three of the four (Table 6.5). Inactivated rabies vaccines cannot in themselves cause rabies—their viral components are killed and cannot revert to virulence. It is worth mentioning that all of 68 captive wild dogs survived 12 months after being given inactivated rabies vaccine for the first time (Woodroffe in preparation). The most likely explanation for the deaths—from rabies—of rabies-vaccinated dogs in the wild is that the single dose of inactivated rabies vaccine recommended for domestic dogs does not confer adequate protection. Although four wild dogs given a single dose of Madivak (Hoescht) in Frankfurt Zoo all seroconverted (Gascoyne *et al.* 1993b), none of 25 wild dog pups given a single dose of Dohyrab (Solvay Duphar) in captivity in Tanzania showed evidence

Table 6.5 Fates of all free-ranging wild dogs known to have been vaccinated against rabies

Location	Animals	Date last vaccinated	Vaccine used	Last sighting	Fate	Rabies confirmed? (no. carcasses tested)	Time to death[a]	References
Masai Mara, Kenya	3 yearlings (Aitong Pack)	Jun–Jul 1989	Imrab	Aug–Sep 1989	2 died, 1 disappeared	Yes (1)	2 months	Kat *et al.* (1995), P. Kat and L. Munson (personal communication)
Masai Mara, Kenya	3 adults and 4 pups (Intrepids pack)	Jan 1990	?	Dec 1990	Probably died	No (0)	–	Alexander and Appel (1994), P. Kat (personal communication)
Masai Mara Kenya	1 adult (Ole Sere pack)	Dec 1989	?	Jan 1991	Died	Yes (1)[b]	13 months	Alexander and Appel (1994), P. Kat (personal communication)
Etosha, Namibia	4 adults[c]	Dec 1989[d]	Rabisin	Aug 1990	Died	Yes (?)	9 months	Scheepers and Venzke (1995), L. Scheepers (personal communication)
Serengeti, Tanzania	13 adults (Ndoha pack)	Sep 1990	Madivak	Jan–May 1991	Probably died	No (0)[e]	–	Gascoyne *et al.* (1993a,b)
Serengeti, Tanzania	16 adults and 5 pups (Salei pack)	Sep 1990	Madivak	May–Jun 1991	Probably died	No (0)	–	Gascoyne *et al.* (1993a,b)
Madikwe, South Africa	3 adult males[c]	Dec 1994	Rabisin	Sep 1997	Died	No (0)	33 months	Hofmeyr *et al.* (2000), J. Bingham (personal communication)
Madikwe, South Africa	3 adult females[c]	Probably Feb 1995	?	Oct 1997	Died	Yes (3)	32 months?	Hofmeyr *et al.* (2000), J. Bingham (personal communication)

[a] Time between rabies vaccination and death from rabies.
[b] Rabies virus detected by only one of two labs that examined samples.
[c] Involved in reintroduction programmes.
[d] Vaccinated annually before that date.
[e] Rabies in Serengeti wild dogs was confirmed from the carcass of a member of the unvaccinated Mountain pack in August 1990 (Gascoyne *et al.* 1993).

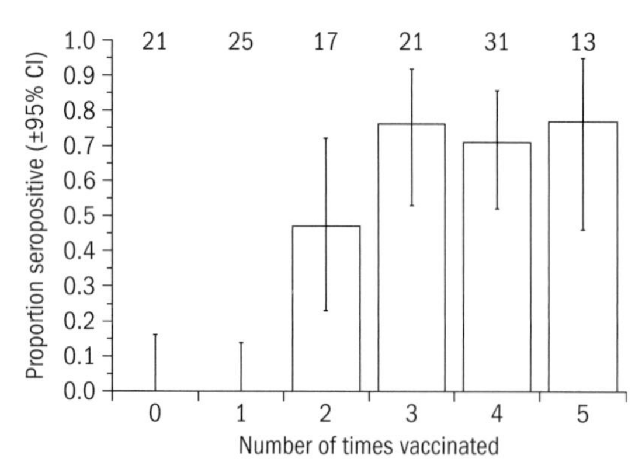

Fig. 6.3 The proportion of captive African wild dogs found to be seropositive for rabies (>0.5 international units rabies serum neutralizing antibodies) when first tested following administration of inactivated rabies vaccine (Dohyrab; Solvay Duphar). Figures indicate the sample size for each treatment; note that the same individuals were tested repeatedly in this study, so some dogs appear in all six columns. (Data from Visee *et al.* 2001.)

of seroconversion when tested 10 weeks later (Visee 1996). Repeated administration of the vaccine led to a higher proportion of dogs becoming seropositive, although at no point were all of the animals treated found to be seropositive (Fig. 6.3). It is not known to what extent detectability of rabies serum neutralizing antibodies reflects immunity to rabies in African wild dogs. Nevertheless, these results suggest that wild dogs may mount a poor immune response to a single dose of inactivated rabies vaccines, which may explain the vaccines' failure—as administered in the past—to prevent wild dogs from contracting rabies when exposed to it in the wild.

Rabies vaccination of wild dogs in the Serengeti ecosystem (Serengeti National Park, Tanzania and the environs of Masai Mara National Reserve, Kenya) attracted considerable attention when this intervention was blamed for causing the local extinction of wild dogs (Burrows 1992). It was suggested that vaccination, perhaps in combination with stress imposed by capture and some form of social stress, compromised wild dogs' immune systems leading to the reactivation of quiescent rabies infections (Burrows 1992; Burrows *et al.* 1994). Burrows and his co-authors argued that such reactivation would be followed by transmission of the virus to pack members that had not been handled, leading to rapid death of entire packs. This hypothesis has been discussed at length in the scientific and popular press, and has influenced government agencies' attitudes to handling and radio-collaring of wild dogs (reviewed in Woodroffe 2001). However, this hypothesis is based on the supposition of a number of occurrences—that wild dogs had contracted non-fatal rabies infection, that handling (including vaccination) induced chronic stress, and that stress could reactivate a quiescent rabies infection—none of which is supported by the available data (reviewed in Macdonald *et al.* 1992; Creel *et al.* 1996; Woodroffe 2001a). This indicates a vanishingly small probability that vaccination—or any other form of handling—caused the extinction of the Serengeti wild dog population (Woodroffe 2001). It is more likely that rabies vaccination simply failed to prevent local extinction, either because of the limited response to the vaccine described above, or because another disease, such as canine distemper, contributed to the die-off (Alexander *et al.* 1994; Cleaveland *et al.* 2000).

African wild dogs are very difficult to capture, especially when they are not fitted with radio-collars (new projects typically take around 6 months to capture their first animals; Woodroffe 2001). This difficulty limits the usefulness of inactivated rabies vaccines, which need to be administered several times if they are to provide protection. New data suggest that live rabies vaccines may be more immunogenic than inactivated ones (Knobel *et al.* 2002). These vaccines can be administered orally. Although wild dogs very rarely scavenge, in preliminary trials both captive and free-ranging wild dogs took baits and good coverage was achieved (Knobel *et al.* 2002). It is difficult to control which species consume baits, raising issues of vaccine safety in non-target species (Bingham *et al.* 1995), although the use of new recombinant rabies vaccine, and very controlled administration of baits, may circumvent this concern.

Distemper

We are not aware of distemper vaccines ever having been administered to free-ranging wild dogs. A principal reason for this is that the early modified live-distemper vaccines used widely on domestic dogs may induce clinical distemper and death when administered to wild dog pups (McCormick 1983; van

Heerden *et al.* 1989; Durchfeld *et al.* 1990). Inactivated distemper vaccines are presumably safer in that they cannot revert to virulence; unfortunately they are far less immunogenic in wild dogs and other wild canids (Montali *et al.* 1983; Visee 1996). For example, only 5 of 19 captive wild dogs given three doses of a inactivated distemper vaccine showed any evidence of protection. Repeated administration of this vaccine led to higher levels of seroconversion; however, an outbreak of wild-type distemper subsequently led to the deaths of 49 of the 52 dogs in this colony (Visee *et al.* 2001; van de Bildt 2002) indicating that seropositivity was not evidence of protection in this case.

Given this failure of inactivated distemper vaccine to provide protection, it is worth considering the risks associated with administering more immunogenic modified live vaccines. Seven wild dogs given new-generation modified live vaccine showed antibody responses with no ill effects (Spencer and Burroughs 1992). In a more recent survey, a questionnaire sent to 35 zoos holding wild dogs indicated that 74% had administered distemper vaccines and, of 13 that gave details of vaccine brands, 12 had used modified live vaccines (Woodroffe in preparation). Among 115 wild dog pups traced for 3 months after being given modified live distemper vaccines for the first time, 2 died (both before 1988) with symptoms that might have indicated vaccine-induced distemper (although this diagnosis was not confirmed in either case). Assuming (conservatively) that both pups had in fact died of vaccine-induced distemper, the risk of causing death through administering modified live distemper vaccines can be estimated as 2/115 = 1.7%, with a 95% chance that the true risk falls below 6.1% (Woodroffe in preparation). While it is difficult to calculate an acceptable level of risk, these figures may at least help to inform decisions about the use of modified live distemper vaccines in small wild dog populations facing acute extinction risks, given the fact that distemper has caused deaths of whole wild dog packs (Alexander *et al.* 1996) and 94% mortality across age classes in a captive population (van de Bildt 2002).

Neither inactivated nor modified live distemper vaccines appear perfect for use in African wild dogs. This places a high priority on testing of new recombinant vaccines (see below) on this species.

Island fox

Being isolated on six small islands, the island fox has little history of exposure to common canine pathogens; in particular, in 1992, there was no evidence of past exposure to canine distemper virus, suggesting that mortality could be high if distemper were ever introduced (Garcelon *et al.* 1992). The very high mortality associated with a distemper outbreak on Santa Catalina Island in 1999 demonstrated the need for a tool to protect recovering populations from distemper on this and the three other islands that had experienced major declines (Timm *et al.* 2000) (Fig. 6.4).

As for African wild dogs, there was legitimate concern about using either inactivated or modified live distemper vaccines on a critically endangered species: inactivated because they are not highly immunogenic for wild canids (Montali *et al.* 1983), and modified live because they have induced clinical distemper and death in several species, including the closely related gray fox (*Urocyon cineroargenteus*; Henke 1997). For this reason, the decision was taken to trial a new recombinant distemper vaccine, vectored by canary pox virus (Timm *et al.* 2000). This vaccine cannot replicate or shed CDV or CPV in mammals. The vaccine was trialled initially on six wild-caught foxes held in captivity; all six seroconverted and no ill-effects were detected (Timm *et al.* 2000). Testing was therefore expanded to the wild fox population remaining on the western part of Santa Catalina

Figure 6.4 The subspecies of island fox *Urocyon littoralis* endemic to Santa Catalina Island was decimated by canine distemper virus in 1999 © G. Roemer.

Island not reached by the epidemic (S. Timm personal communication).

Island foxes' tractability as a study animal (they are easily recaptured for sequential vaccination and serum sampling) permitted this field testing of vaccination protocols. Whether vaccination (which was carried out some time after fade-out of the epidemic) has contributed to the ongoing recovery of the Santa Catalina island fox population is unknown. Moreover, in the absence of challenge experiments, it is impossible to be certain that vaccination confers protection from infection. However, with all six island fox subspecies at risk of similar outbreaks, and three critically depressed by other factors (Coonan 2002), the existence of a distemper vaccination protocol known to be safe and likely to be effective for use in free-ranging island foxes is a valuable addition to the toolkit for conservation of this critically endangered species.

Epidemiological and evolutionary concerns about direct vaccination

Direct vaccination of threatened hosts is unlikely to eradicate infection from an entire system, since infection usually 'spills over' from more abundant (often domestic) reservoir hosts. Threatened hosts will therefore remain in contact with the pathogen (although spillover may be a rare event). Since susceptible animals are constantly born into the population, protection will wane if vaccination cover is halted. This creates a need to maintain vaccination in perpetuity unless disease threats can be alleviated by other means. Disease control programmes for threatened species may therefore require long-term funding.

Concern has been expressed that, by protecting hosts from natural infection, vaccination may impede natural selection for heritable resistance to disease (discussed in Woodroffe 1999). However, if spillover occurs rarely, selection for such resistance may be weak (Laurenson *et al.* 1997). The evolutionary costs of vaccination are likely to be extremely low for rabies (against which little natural immunity appears to exist in wild canids; Baer and Wandeler 1987) but higher for less virulent infections. Because the net benefit of vaccination will depend on the virulence of the pathogen, commercial preparations containing vaccines against several canine pathogens, of varying virulence, may sometimes be unsuitable for use in wild canids.

Management of infection in reservoir hosts

Threatened canid species may also be protected from infectious disease through management of reservoir hosts. Simulations of Ethiopian wolf populations suggest that reducing disease incidence in sympatric domestic dogs can substantially reduce the risks of local wolf extinction (Haydon *et al.* 2002). This approach to disease control directly parallels the control of zoonotic diseases such as rabies and ZVL in domestic dogs and abundant wild canids; the successes and failures of past culling and vaccination efforts therefore provide important lessons for the conservation of rare canids threatened by infectious disease.

In evaluating management options, it is essential to determine which species are involved in maintaining infections that are likely to impact the threatened population. Domestic dogs are an important reservoir host, but wild canids have also been implicated in a number of cases (Table 3.3). This is an important distinction, because management options are quite different for wild and domestic canids.

Limiting host density by culling or fertility control

Domestic dog populations are usually limited by human decisions (Perry 1993); thus it ought to be technically (if often not practically) possible to limit domestic dog densities to levels at which dangerous infections such as rabies cannot persist (e.g. Cleaveland and Dye 1995). Cultural attitudes towards dogs vary widely, but in most developing countries their usefulness is acknowledged, and even if individual dogs are perceived as problem animals, the attitude of the general public is not favourable to mass culling. Food and shelter are provided to dogs from a range of levels of dependence on people from true pets up to almost entirely feral individuals. The limitation of dog densities to below the threshold density capable of supporting infections such as rabies or distemper is theoretically possible, but as far as we are aware, it has never been achieved at a sustainable level. Given humans' role in limiting dog populations, dog numbers may be most effectively controlled through changing social attitudes to dogs. However, this will be difficult if dogs fulfil important tasks (e.g. as guards). Moreover, where human densities are high, even comparatively low dog : human ratios may generate populations large enough to represent a disease risk to local wildlife.

As discussed above, past attempts to limit rabies spread to humans by culling wild canids seem to have mainly met with failure due to inefficient removal and rapid recovery of the populations subject to control. There is no reason to suppose that this approach would be any more successful if implemented to protect threatened species from disease. Moreover, it would not be appropriate to use poisoning—an important technique used to cull wild canids in the past—in areas where threatened species might also consume baits and succumb.

Canid population densities might also be managed through various forms of fertility control. Modelling suggests that this approach has the potential to control infection, although it is expected to reduce disease incidence much more slowly than either culling or vaccination, and would depend on the turnover rate of the population (Macdonald and Bacon 1982; Barlow 1996). Female sterilization by surgical means has been used as an adjunct to reduce conflict between Ethiopian wolves and dogs within wolf range in the Bale Mountains in Ethiopia. However, this approach is costly, as well as being culturally and logistically difficult. Moreover, since human populations move seasonally in and out of the area, the dog population is not closed and new pups can easily be obtained. The impact of this programme is currently being assessed (EWCP unpublished data).

Fertility control would be even more difficult to achieve among wild canids, although initial investigations of immunocontraceptive vaccines, which target reproductive proteins, have shown encouraging results for red foxes in France and Australia Bradley 1994 was working on immunocontraception etc and modified viruses). Social suppression of reproduction—a common feature of wild canid societies—can interfere with fertility control attempts and reduce (or even reverse) its effectiveness (Caughley *et al.* 1992). Fertility control was rejected as a means of controlling tuberculosis in badgers (*Meles meles*) for this reason (Krebs *et al.* 1996). Oral contraceptives are available for use in wildlife (reviewed in Tuyttens and Macdonald 1998) but, like poisons, their use in areas occupied by threatened populations would probably be inappropriate. Despite these concerns, immunocontraception—especially if it could be combined with vaccination—may hold some promise for the future management of disease reservoirs.

Vaccination

Experience from the control of rabies risks to humans and livestock (see above) suggests that vaccination of both domestic dogs and wild canids may be powerful tools for the protection of threatened species from acute disease threats.

Ongoing studies in Ethiopia and Tanzania are evaluating the potential of domestic dog vaccination as a means of protecting Ethiopian wolves and other large carnivores from rabies and distemper. In rural Tanzania, results demonstrate that a simple central-point vaccination strategy, resulting in vaccination of 60–65% of dogs adjacent to Serengeti National Park, has significantly reduced the incidence of rabies in dogs and risk of exposure to people, with opportunities for transmission to wildlife also decreasing (Cleaveland *et al.* 2003). In Ethiopia, no case of rabies or CDV has been reported within wolf range within the Bale Mountains National Park since widespread dog vaccination began in 1998, and rabies cases in dogs and other species have occurred primarily at the edge of vaccination zones, with the overall incidence in dogs and humans very much reduced.*

Concern has been expressed that vaccination of disease reservoirs (Suppu *et al.* 2000)—especially domestic dogs—could remove an agent of population limitation and lead to increased host density (Moutou 1997). This could be potentially damaging, especially if vaccine cover were to be halted (Woodroffe 1999). However, preliminary studies

* In late 2003, an outbreak of rabies killed over 50 Ethiopian Wolves in the Bale Mountains National Park. The Ethiopian authorities granted permission for inactivated rabies vaccine to be administered to free-ranging wolves on an experimental basis, a move strongly supported by a joint statement of the IUCN Canid and Veterinary Specialist Groups. At the time of writing, 60 wolves have been vaccinated without ill-effects, although it is too early to evaluate the effectiveness of the intervention.

indicate that, while dog vaccination in northern Tanzania has led to a significant decline in mortality rates, population growth rates have not increased (Cleaveland *et al.* 2002). It appears as though the reduced demand for puppies has lowered recruitment rates. Studies to examine this issue are also underway in Ethiopia.

Rabies control programmes intended to protect people in North America and Europe have achieved eradication through vaccination campaigns carried out across extremely large areas (tens or hundreds of thousands of km^2). Few, if any, conservation programmes would be able to fund operations of this scale. However, since rabies control provides widespread benefits to a region, an integrated approach involving public health, livestock development, and wildlife conservation agencies may provide a feasible and cost-effective approach to larger scale programmes. The size of area to be covered depends upon the population density and ranging behaviour of both reservoir and threatened host species; for example, the area to be covered to protect Ethiopian wolves (home range 6–11 km^2 per pack; Sillero-Zubiri and Macdonald 1997) would be smaller than that needed to protect African wild dogs (home range 400–1200 km^2 per pack, Woodroffe *et al.* 1997). Haydon *et al.* (2002) suggest that domestic dogs would need to be vaccinated in a *cordon sanitaire* up to 15 km wide surrounding Ethiopian wolf habitat (about 2500 km^2 to protect the 1080 km^2 of wolf habitat in the Bale Mountains National Park, with the total area to be vaccinated depending on the shape of habitat patches; Laurenson *et al.* in press); both the area of habitat and the width of the *cordon sanitaire* would likely be much greater for wild dogs. As both Ethiopian wolf and wild dog populations remain surrounded by human-altered landscapes inhabited by domestic dogs, regional eradication is near-impossible, and vaccination cover would have to be maintained in perpetuity.

Large size Canid populations can recover after a contagious disease outbreak such as fox mange in Scandinavia (Linstrom and Hornfetldt 1994) and fox rabies in western Europe (Chautan *et al.* 2000). There is, nevertheless, no simple explanation for the observed trend of increasing fox populations. In Scandinavia, where environmental conditions are sub optimal for foxes, vole availability seems to remain the regulating factor. In western Europe, both environmental conditions and human influence could have been conjugated to allow a steady-increase in fox populations over the second half of the 19th century: eradication of rabies could have released the potential of fox populations to continue a long-term increase than began before the merge of the epizootic. Nonetheless, the fox population increase has been associated with complaints from various categories of citizens, including hunters for whom this abundance has the potential to cause problems, namely, the increase of echinococcosis infections (Giraudoux *et al.* 2001). In such a context of pest control, the temptation to use infectious disease as a means to limit problem species can be attractive (Dosbon 1988). It has been suggested that a species-specific virus has the potential to limit feral cats on small islands (Courchamo and Sugihara 1999). However, following the controversial results of the introduction of myxomatosis in European rabbit populations, this strategy has never really been considered in the "Old World" (Artois 1997). Recent studies developed in Australia aim at delivering baits containing genetically modified Canine herpes virus that harbour genes coding for contraceptive proteins: using bait will limit the probability that the virus can spread among the fox population (Reubel and Lin 2001).

Managing interactions between host species

In theory, the most effective means of limiting disease transmission to threatened hosts is to prevent contact with reservoir hosts. This approach has been adopted to protect bighorn sheep (*Ovis canadensis*) from pneumonia and scabies carried by domestic sheep (*Ovis aries*); domestic sheep are simply barred from buffer zones surrounding bighorn populations (Jessup *et al.* 1991). It is comparatively easy to control the distribution of livestock in this way, but controlling the behaviour of free-ranging domestic dogs is more difficult, and limiting contact between wildlife species is likely to be near-impossible.

In South Africa, fencing, used to prevent wildlife leaving many private and public reserves and domestic animals entering, may reduce contact between wild canids inside and wild or domestic canids outside. This may help to explain the unusual absence of evidence of exposure to CDV and canine parvovirus among wild dogs inside Kruger National Park (Van Heerden *et al.* 1995). However, making these fences impermeable to small carnivores is very difficult: indeed, wild dogs in Madikwe died of rabies, which was thought to have been brought into the park by jackals (Hofmeyer *et al.* 2000).

Attempts to limit Ethiopian wolf contact with domestic dogs have involved community education programmes encouraging people to tie up their dogs and to keep them at home. Collars and chains were supplied, but the people used them for other purposes. Tying up dogs may conflict with their function as guards and cleaners. It is probably impossible to accustom older dogs to accept being tied up, thus training would have to start with the next generation of pups. Cultural resistance from dog owners, however, is likely to hinder such efforts.

In gazetted national parks there is rarely any need to tolerate domestic dogs; unaccompanied dogs should 'be captured and humanly euthanised if not reclaimed, or shot on sight if no other appropriate way can be used.' Indeed, such management is often practiced—formally or informally—on public and private lands. If domestic dogs cannot be excluded entirely, owners should be required to provide evidence that their dogs have up-to-date vaccinations against rabies (and, where appropriate, other infections such as distemper and parvovirus; Woodroffe *et al.* 1997). Similar measures have been proposed to protect island foxes from disease on Santa Catalina Island, California (Timm *et al.* 2000).

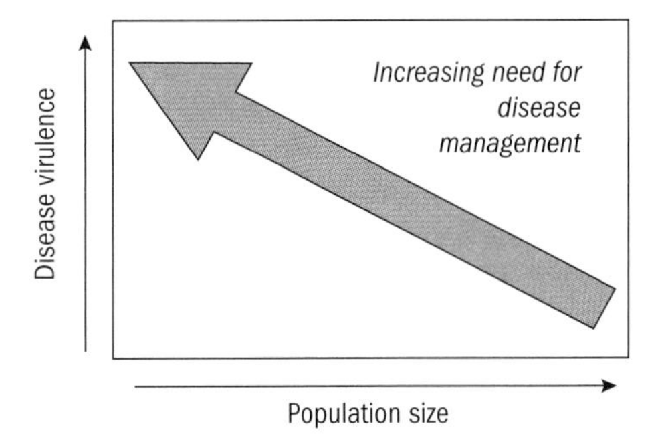

Fig. 6.5 Schematic representation of how the need for intervention may vary according to host population size and the virulence of the infection concerned.

Which approach is best?

Infectious disease is a threat to wild canids that conservationists are ill equipped to manage. Lack of information hinders management of this newly recognized threat—there are no established models to follow, and some early and unsurprising failures have attracted damaging controversy (Woodroffe 1999). This makes it difficult to assess which approach is most likely to meet with success. However, it is important to recognize that the decision not to intervene is in itself a decision (and may be a viable management option in some cases). Table 6.4 provides a comparison of the costs and benefits of various options. In general, intervention will be most warranted in small, isolated populations and for highly pathogenic infections (Fig. 6.5). It may be more appropriate, given ecological and evolutionary considerations, not to intervene in larger populations, and for less virulent infections. Where intervention is warranted, direct vaccination of threatened hosts may be an effective approach where safe, effective, and practicable vaccination protocols are available. Vaccination of domestic dogs may be a promising approach where these are the disease reservoir; vaccination of wildlife reservoirs will be more problematic.

It is important to bear in mind that in many circumstances, intervention will not be appropriate. Disease is an important component of natural ecosystems, and periodic die-offs—though they may cause suffering for the animals concerned—represent an ecological and evolutionary process worthy of conserving. Intervention is warranted only when there is a realistic concern that such die-offs will eradicate populations of conservation concern. This threat will be greatest—and intervention most appropriate—where the expansion of human populations has fragmented natural habitat, making wild populations both smaller (hence less able to persist in the face of disease outbreaks) and in greater contact with domestic dogs (hence increasing risks of disease exposure).

Conclusions

Infectious disease has important impacts on wild canid populations, both directly (through the effects of mortality and morbidity on population dynamics and persistence) and indirectly (through the impacts of human attempts to control zoonotic diseases). Indeed, even the abundance, dispersion, and community composition of canids has been much affected by people, so there is a sense in which people have affected the pre-disposition of their populations to be susceptible to catastrophic disease in the first place. However you look at it, wildlife disease has emerged as highly relevant to conservationists (Macdonald 1996). Close parallels exist between the control of canid-borne zoonoses and the protection of threatened canid hosts: in both cases, a generalist pathogen (often rabies) persists in large populations of a reservoir host, and 'spills over' into a secondary host where it cannot persist. Unfortunately (and somewhat surprisingly) such multi-host systems have been neglected by both theoretical and empirical epidemiologists, and there are currently few control models to follow. The management of disease outbreaks in threatened canids, in particular, has been hampered by lack of information on which management tools will be most effective, and how they should be implemented. Basic data are urgently needed on the ecology and epidemiology of infections such as rabies and distemper in natural ecosystems. Disease surveillance will help to identify (or eliminate) potential disease risks to wild populations, and testing of vaccination protocols on captive populations may accelerate responses to future outbreaks in the wild.

Acknowledgements

We are very grateful to the Canid Specialist Group members who contributed comments and suggestions to an earlier draft of this paper. Mme Duchene at AFSSA Nancy was, as in so many circumstances, particularly helpful in providing appropriate literature references to Marc Artois.

CHAPTER 7

Tools

Tools for canid conservation

Luigi Boitani, Cheryl S. Asa, and Axel Moehrenschlager

Pall Hersteinsson releasing a live-trapped Arctic fox *Alopex lagopus* in Iceland

Introduction

Canids, in their diversity and abundance, have arguably generated more conflict with humans than any other family of mammals. This has resulted in intense persecution of many canids, especially the larger species whose populations have been drastically reduced in number and distribution. Of the 35 extant canid species, at least 11 are endangered or vulnerable and all of the large species are threatened in parts of their range. Thus, conservation is increasingly the primary purpose for many studies on canids.

As a family, canids are faced with numerous problems and the result of these problems on individual species depends on their biology, their distribution, and their ability to adapt to change. While some have adapted well, generalist canids have flourished as human populations have grown around the planet, many have not. Attempts by humans to limit canids are as old as human history, mostly due to livestock losses and competition with hunters for prey. In addition to direct persecution, humans often create problems for canids in more subtle ways. For example, carnivore introductions and invasions, sometimes even of canids themselves, have caused disease transmission, hybridization, or interference competition that negatively impact endemic canid populations (Boitani 2001; Macdonald and Thom 2001; Wayne and Brown 2001). The reduction of suitable habitat through, for example, urban spread, agriculture, or resource-extraction is one of the greatest threats facing canids throughout the world. Even protected areas are often inadequate in maintaining genetically viable populations of canids that have large home range requirements (Woodroffe and Ginsberg 1998). If suitable habitat is available, the persistence of canids is also dependent upon ecosystem factors such as prey availability (Fuller and Sievert 2001) and interspecific competition (Creel *et al.* 2001), which may or may not be influenced by human actions.

Conservation is rarely, if ever, accomplished by a single action on an animal population rather, it involves multi-stage and multidisciplinary process that develops through at least four different stages: (1) research to understand the reasons for the problem and identify possible solutions; (2) public education and information; (3) stake-holder involvement; and (4) implementation of conservation actions (Macdonald 2001). As such, conservation requires an extended range of skills that belong to a variety of disciplines, from sociology and psychology to ecology, genetics, and all fields of biology, and in order for it to be effective, calls for the application of several different tools in a coordinated and integrated approach.

Many good books have been written on research and conservation techniques (e.g. Bookhout 1994; Krebs 1999; Boitani and Fuller 2000; Sutherland 2000; Elzinga *et al.* 2001) where a variety of tools are described in detail and the many potential sources of methodological errors are discussed. Tools relevant to canid conservation are a subset of those relevant to any mammalian conservation issue and can be found in these generalist textbooks, but we wish to present here a selection of those we believe are most frequently used in canid conservation. Conservation tools for canids are as variable as the diversity of problems they face and the multifaceted approaches involved in the development of appropriate solutions. In this chapter, we focus primarily on rapidly evolving conservation tools that use recent advances in several fields of conservation biology and take into account sociological aspects. We consider the use of conservation tools in (1) assessing the status of wild canid populations; (2) limiting canids that become too numerous; (3) restoring canids that are threatened; (4) protecting canids in an ecosystem context; and (5) influencing socio-political change for canid conservation. As the outlook for many canid species is increasingly desperate, we urge conservationists to reach into this conservation toolbox, apply as many components as possible, and spur on the development of additional tools so that canids around the world can co-exist with humans for generations to come.

Assessing the status of wild canid populations

Genetic tools for canid conservation

The expanding knowledge, technologies, and applications of genetic techniques represent one of the most rapidly evolving and exciting tools for canid conservation. Conceptually, the scope varies from the formation of phylogenetic trees across taxa to the identification of a single individual. A description of the large variety of molecular techniques that are utilized for these purposes is covered in depth in Chapter 3, so here we summarize only briefly a sample of applications. Conservation genetics can be utilized to determine relationships between canid taxa on genus, species, or subspecies levels. For example, close relationships have been demonstrated between swift *Vulpes velox* and Arctic foxes (*Alopex lagopus*) (Geffen *et al.* 1992e; Mercure *et al.* 1993), as well as between fennec and Blanford's foxes (*Vulpes zerda, V. cana*), but significant distances have been identified between grey and bat-eared foxes (*Urocyon cinereoargenteus, Otocyon megalotis*) (Geffen *et al.* 1992). Numerous and sometimes conflicting methods are used for conservation genetics to discern species boundaries (Goldstein *et al.* 2000), and their results have profound implications for canid conservation. For example, Yahnke *et al.* (1996) determined that Darwin's fox in Chile (*Pseudalopex fulvipes*), previously thought to be a subspecies of *P. griseus*, is actually a distinct species. Consequently, Darwin's foxes are now considered one of the world's most critically endangered canids, which will receive increasing research and conservation attention. The same scenario could unfold for wolves in eastern Canada, which some argue should be classified as a distinct species, namely *Canis lycaon* (Wilson *et al.* 2000).

Taxonomic differentiation on a subspecies level is also critical for the conservation of imperilled canid populations. For example, mitochondrial DNA sequence data support the classification of Mexican kit foxes as a distinct subspecies (*Vulpes macrotis zinseri*) (Maldonado *et al.* 1997), thus warranting particular protection. Conservation genetics has further uses in determining hybridization occurrences. On the one hand, Vila and Wayne (1999) determined that, while wolf-dog hybrids had been observed, significant introgression of dog markers into the European wolf genome had not occurred. On the other hand, coyote (*Canis latrans*) genes have had significant introgression into the wolf genome in regions of North America (Lehman *et al.* 1991; Wayne *et al.* 1992). While conservation genetics can be used to identify the boundaries between canid taxa, numerous applications also exist to better understand population demographics. For example, Randi *et al.* (2000) found that Italian wolves had a unique haplotype, whereas 26 wolves from outbred populations in Bulgaria had seven haplotypes. These findings confirmed the lack of mitochondrial DNA variation among Italian wolves, and the combination of low effective population size with low genetic variability now threatens population viability to such an extent that a controlled demographic increase is recommended. Inbreeding can have physiological results and seems to affect the body size of Mexican wolves (*Canis lupus baileyi*; Fredrickson and Hedrick 2002).

The potential use of dogs to find faeces for genetic analyses (Smith *et al.* 2001), and the extraction of faecal DNA yields numerous applications on a population level (Paxinos *et al.* 1997; Mills *et al.* 2000). Such methods have, for example, been used to estimate population size and sex ratios of coyotes and the potential also exists to identify individual wolves (Lucchini *et al.* 2002) and their home range sizes. An examination of microsatellite variation within and among populations of island foxes (*Urocyon littoralis*) on California's Channel Islands, suggests that the pattern of variation across unlinked microsatellite loci can also be used to test whether populations have been growing or remained constant (Goldstein *et al.* 1999).

Finally, within populations, genetic techniques can be utilized to determine the relationships among individuals and gene flow. Microsatellites have been used to determine kinship (Queller 1992), and Girman *et al.* (1997) were able to establish that African wild dogs (*Lycaon pictus*) in Kruger National Park had the expected pack composition of unrelated pair mates, subdominant close relatives, and related offspring of the breeding pair. Contrary to expectation, kit foxes that shared dens in the San Joaquin Valley of California (*V. macrotis mutica*) were often unmated neighbours, that may have been together during unsuccessful attempts at pair formation (Ralls *et al.* 2001). Moreover, assessments of relatedness throughout populations are increasingly utilized to also estimate dispersal distances (Girman *et al.* 1997; Taylor *et al.* 2000).

Monitoring population status

The status of carnivore populations can be monitored with an array of traditional techniques, which may differ as one attempts to determine the presence, distribution, abundance, or demography of populations (see Gese 2001). The presence of canids such as African wild dogs, grey wolves, and red foxes (*Vulpes vulpes*), has been determined using questionnaires, interviews, and sighting reports (Harris 1981; Fuller *et al.* 1992a; Fanshaew *et al.* 1997). Hunting returns of red foxes (Erickson 1982; Heydon and Reynolds 2000a) and road mortalities can also be used to determine the presence and, with caution, the relative abundance of canids over time.

Aerial surveys have been used to locate dens of kit and red foxes (Trautman *et al.* 1974; O'Farrell 1987) and, with transect methods, to determine the abundance of coyotes (Todd *et al.* 1981). This method is only effective in relatively open habitats where the individuals and dens can be seen relatively easily. Alternatively, ground-surveys can be conducted. Breeding den counts have been used as an index of canid abundance. For instance Insley (1977) surveyed red fox dens, which can be identified by fresh soil above ground in winter, the smell of fox urine, faeces, or uneaten prey remains. Trained dogs can indicate whether a den is, or has recently been, occupied (Reynolds and Tapper 1994). Vocalization response surveys have also been utilized to determine the abundance of coyotes and wolves (Okoniewski and Chambers 1984; Fuller and Sampson 1988).

Scent-station surveys, where a canid is attracted with a scented tablet or food-item and subsequent tracks are recorded on 1-m radius sifted dirt or smoked track plates, can be used to determine relative canid numbers over time (Sargeant *et al.* 1998). Scent-posts have been utilized to determine the relative abundance of gray foxes *Urocyon cinereoargenteus* (Conner *et al.* 1983) and coyotes (Linhart and Knowlton 1975) for example. The reluctance of animals to step on the substrate, misidentification of tracks, seasonal changes in animal movements, and loss of scent-stations due to weather can make this method challenging to use (Gese 2001a). Remote cameras can be used in conjunction with scent-stations and track plates to identify the species more accurately and, in the case of visibly tagged animals, to estimate population abundance using mark-recapture estimators.

Spotlight surveys are conducted by slowly driving a truck, shining two spotlights of at least 500,000 candlelight power on both sides of the road, and identifying observed canids through binoculars. The subsequent sightings per kilometre index can be used with program DISTANCE (Buckland *et al.* 1993) to estimate canid densities. This method has been used to compare the relative abundance of coyotes and kit foxes between prairie dog town and grassland areas in northern Mexico over time (Moehrenschlager and List 1996). The accuracy of spotlight surveys is dependent upon the sightability of canids in different habitats and the number of survey replicates (Ralls and Eberhardt 1997).

Transects for scat or tracks can be sampled opportunistically to determine canid presence. Reynolds and Tapper (1994) pointed out that red fox tracks can be counted on dirt roads and firebreaks while their faeces can be located on prominent sites such as molehills, stones, tussocks of grass, and trail junctions. Systematic transects for scat or tracks can also be used as indices of canid distribution and abundance over time, but the accuracy and precision of such estimates is highly dependent on sampling effort, transect length, and weather (Gese 2001).

Radio-tracking not only allows for the identification of individuals, but it also provides demographic parameters such as survival, fecundity, dispersal, and recruitment (Fig. 7.1). Moreover, density estimates can be established by estimating seasonal home range movements, range overlaps, and habitat use. Radio-tracking can elucidate the comparative ecology of

Figure 7.1 Radio-collared kit fox *Vulpes macrotis* © C. Wingert.

geographically separate species such as Arctic, kit, and swift foxes (Tannerfeldt *et al.* 2003), or the interspecific interactions of sympatric canids. Radio-tracking has shown, for example, that red foxes can escape coyotes through habitat partitioning (Major and Sherburne 1987; Sargeant *et al.* 1987; Theberge and Wedeles 1989). As the technology of GPS collars continues to improve in terms of decreasing collar weight, increasing battery-life, and improved data retrieval systems, such applications will be increasingly attractive, affordable, and accurate in the future.

Genetic techniques can also be used to assess and monitor the presence, abundance and genetic structure of populations (Gese 2001) opening up new frontiers in non-invasive sampling that were unthinkable only a few years ago. Laboratory techniques for obtaining genetic data from free-ranging animals using non-invasive samples such as faeces and hairs are rapidly being refined. Although several limitations still apply (Taberlet *et al.* 1999, 2001; Waits *et al.* 2001), these techniques provide new possibilities for examining social structure, dispersal, and pack interactions while providing insight into aspects of behavioural ecology that were, until recently, only possible through direct observation of the animals (see below). Endocrinological studies, which are an important component of conservation efforts, continue to explore the use of non-invasive sampling techniques. For example, the nutritional condition of free-ranging wolves has been assessed from urine collected in the snow by measuring the urinary urea nitrogen/÷creatinine (UN/÷C) ratio as an indicator of nutritional restriction (Del Giudice *et al.* 1992; Moen and Del Giudice 1997).

Faecal steroid hormone analysis can aid in distinguishing males from females in some canid species (maned wolf (*Chrysocyon brachyurus*) Wasser *et al.* 1995; Mexican wolf: J.E. Bauman, unpublished data) but likely not in others (e.g. island fox: J.E. Bauman, unpublished data). Differentiating pregnant from non-pregnant females is more challenging, since in canids an obligate pseudopregnancy, with similar levels of estradiol and progesterone, follows ovulation even in females that are not with males (Asa 1998; Asa and Valespino 1998). The maned wolf may be an exception, in that Wasser *et al.* (1995) and Velloso *et al.* (1998) found a difference between pregnant and pseudopregnant females, but that difference may not be great enough to be used for diagnostic purposes in individual females. Although there is a commercially available assay for the hormone relaxin (Synbiotics Corp.) that can diagnose pregnancy in dogs (Steinetz *et al.* 1987), Mexican wolves (J.E. Bauman, unpublished data), and possibly other canid species, it requires a blood sample.

When canids become pests—limiting the many

Conflict management

The major threat to large canid conservation is conflict over livestock losses, thus the reduction of this problem is of paramount importance for canid conservation. Here we use the term 'conservation' in its broad sense of sustainable use of a resource without compromising the interests of future generations (World Commission on Environment and Development 1987). This definition is needed if we wish conservation to be acceptable to the widest possible range of stakeholders, because not all species and/-or populations need to be totally protected, some may be exploited and others may be managed or controlled (Johnson 2001; Macdonald 2001). Canids provide an extended range of examples for a great variety of conflict management schemes. These range from the full protection needed for the African wild dog (Woodroffe *et al.* 1997) to the flexible system of wolf control in many Canadian provinces (Cluff and Murray 1995) as well as the widespread system of compensation for livestock losses to wolves adopted by many European countries (Boitani 2000).

Several tools are available for conflict reduction, but they can be conveniently grouped under three major headings: prevention, mitigation, and control. Prevention tools include all means of precluding canids from attacking domestic animals and include fencing, repellents aversive conditioning, husbandry patterns guard dogs (Sillero-Zubiri and Laurenson 2001; Chapter 5, this volume). Mitigation tools consist mainly of compensation for losses. The purpose of compensation schemes is never to reduce the number of losses, but only to make them more acceptable to people (Ciucci and Boitani 1998); however, in many European countries compensation schemes are the only conservation tool used to mediate wolf/livestock conflicts. Finally, control tools include all means of reducing canid populations, such as population control and the removal or translocation of problem animals (Harris and Saunders 1993; Sillero-Zubiri and Laurenson 2001, and Chapter 5, this volume). However, Linnell *et al.* (1999) critically reviewed the concept of problem animals and concluded that there is little evidence that some individuals or groups in a carnivore population kill a disproportionate share of livestock. They suggested that most individuals of large carnivore species will occasionally kill the livestock they encounter and if this is true, individual control would need to remove most individuals that might encounter livestock (Linnell *et al.* 1997).

However, according to the 'breeding pair' hypothesis (Till and Knowlton 1983), coyote depredation is significantly greater in pairs with young, since they switch to larger prey such as lambs to meet the energetic demands of the pups. The hypothesis was confirmed in a recent study of vasectomized male and tubally ligated female coyotes that maintained territories but preyed less on sheep than control packs (Bromley and Gese 2001). These results suggest that permanent sterilization or reversible contraception may be reasonable strategies for reducing predation, at least in some canid species.

Each of the above tools is potentially useful for managing conflicts and they have been extensively discussed elsewhere, but a cautionary note is needed. No tool seems capable of achieving a long-term solution unless it is firmly put in the context of a comprehensive management strategy that provides for prevention, mitigation and, when needed and feasible, some population control.

Contraception

Contraception has been used in zoos for genetic management and to limit reproduction of surplus animals for more than 25 years, but attempts to control free-ranging animal populations have been less successful. Contraception in canids presents problems beyond those encountered with most other mammals. Commonly used steroid hormone-based contraceptives are effective in canids, as they are in most other mammals, but in canids, as well as other carnivores, they can be associated with potentially deadly side-effects (Jabara 1962; Frank *et al.* 1979; Concannon *et al.* 1980; Asa and Porton 1991).

Porcine zona pellucida vaccines that prevent sperm from attaching to the zona pellucida, the coating of the unfertilized oocyte, can be delivered by dart, but booster injections must be given annually. Although they do not affect general health, they can cause local damage to the ovary, which may not be reversible (Mahi-Brown *et al.* 1988).

Alternative methods are being evaluated and are hoped to be commercially available soon. The most promising is a gonadotropin releasing-hormone (GnRH) agonist that suppresses production of gonadal steroids in both males (testosterone) and females (estradiol and progesterone), preventing the production of mature sperm and eggs, respectively (Bertschinger *et al.* 2001). There are currently no known or suspected deleterious physiological effects associated with this treatment.

When selecting a contraceptive method, potential effects on behaviour and social interactions should also be considered. In this regard, GnRH agonists produce effects similar to castration or ovariectomy. Castration in particular can cause profound changes in male-typical behaviour, but this very change makes it a poor choice for population control, since males without testosterone are unlikely to seek or remain with a female. Not only does this disrupt family groups, but it also leaves the female available for other intact males to court and fertilize. Likewise, with the removal of ovarian steroids in females, the absence of oestrous behaviour is likely to result in the eventual dissolution of previously formed pair bonds (see Asa 1996a,b).

The preferred method of fertility control in canids is either surgical or chemical vasectomy. This method spares gonadal hormones and merely prevents the passage of sperm during copulation. The only observable change is the absence of offspring. Although there has been concern that the absence of young might result in the dissolution of pairs bonds, disrupting basic social structure, vasectomies of five male wolves in four packs in Minnesota demonstrated that sterile males would continue to hold territories and remain with their mates, at least during the 1–7 years these packs were monitored (Mech *et al.* 1996). Sterilization of females, although more difficult to accomplish, was evaluated in red foxes and found not to effect territory size, dispersal, or survival (Saunders *et al.* 2002).

Chemical (Pineda 1977) rather than surgical vasectomy can make field application more practical, but still requires capture and handling. Because the efficacy of contraception and sterilization for population control has not been widely evaluated, many questions remain (Tuyttens and Macdonald 1998). In a computer simulation comparing vasectomy versus removal for wolf control, Haight and Mech (1997) found that variables such as rates of immigration could profoundly affect success.

When canids become threatened—restoring the few

Captive breeding

The role of captive breeding in conservation continues to meet with controversy. At the extreme, some have argued that it is better for a species to go extinct than exist solely in captivity. While it is true that captive breeding will not be necessary or appropriate for many species, it has been essential to the recovery of three critically endangered canids. The red wolf and Mexican grey wolf, considered extinct in the wild, were successfully bred in zoos, in cooperation with the US Fish and Wildlife Service, and reintroduced back into the wild. Following population crashes of the island foxes on two of the islands, all remaining foxes on those islands were brought into captivity to protect them until the cause of the decline could be identified and controlled. Captive colonies have been established on other islands as well, after population declines were detected. These captive breeding programs were predicated on the anticipation of eventual reintroduction. The question remains whether a species extinct in the wild, with no prospect of reintroduction, should be maintained indefinitely in captivity. Fortunately, at least in the case of canids, we have not been faced with this decision.

Even zoos, that used to see their primary role in species conservation to be serving as 'Noah's Arks', now realize that they have a greater responsibility, including involvement in and support of in situ conservation projects (Hutchins and Conway 1995). Zoos contribute to conservation by bringing public attention to the plight of animals, by inspiring an appreciation for individual species, and by providing extensive conservation education programs. In addition, research on captive animals can produce basic biological data needed for population modelling, for reintroduction programs, and management of wildlife reserves. Most new techniques for application in the field are also best developed and validated under the controlled conditions of captivity. Although it is critical to maintain balance in any program, this extensive list of contributions made by zoos addresses another objection to reliance on captive programs: the concern that they divert attention and resources from the real reasons for species endangerment, for example habitat loss, hunting, etc. (Balmford *et al.* 1995).

Although most captive breeding programs rely primarily on zoos, private and governmental facilities are sometimes also involved. The American Association of Zoos and Aquariums (AZA) coordinates the world's most extensive captive breeding program, but there are roughly parallel programs in other regions, notably Europe and Australia. The work of the World Association of Zoos and Aquariums (WAZA) and the Conservation Breeding Specialist Group (CBSG) of the IUCN helps coordinate these programmes on a more global level.

The structure and operation of AZA programs serves as a good model for captive breeding programs. The basic unit of the captive management program is the Studbook, which contains the pedigree of each species. The committees that oversee captive management and endangered species recovery include the AZA Species Survival Plans (SSP), Population Management Plans (PMP), Taxon Advisory Groups (TAG), Scientific Advisory Groups (SAG), and Conservation Action Partnerships (CAP).

Species survival plans are cooperative programs that monitor and manage genetic diversity of all captive individuals of a species as a single population. In the AZA there are currently SSPs for the red wolf, Mexican wolf, maned wolf, and African wild dog. PMPs provide genetic and demographic management recommendations for species with studbooks

Figure 7.2 Reintroduction of captive-bred red wolf *Canis rufus* © US Fish and Wildlife Service.

that are not covered by SSPs. In the AZA there is currently a PMP for the fennec fox, and one is recommended for the island fox. TAGs coordinate, facilitate, and review cooperative management and conservation programs for a group of related species. Canids are covered under the AZA Canid and Hyenid TAG. A primary responsibility of TAGs is to create regional collection plans that allocate available captive space among species of the taxon in an attempt to maximize management and conservation goals. The Canid and Hyenid TAG Regional Collection Plan evaluated each species and its appropriateness for a captive programme by using a decision-tree approach. Key elements included whether there was a viable captive population, the ability to link with field programmes, value to conservation education or research, and the willingness of individuals to volunteer to lead each programme. The number of species that can be included is limited by current and anticipated availability of space.

SAGs serve as advisors to the other conservation committees and programmes. They currently include Behavior/Husbandry, Biomaterials, Contraception, Data Management, Nutrition, Reintroduction, Reproduction, Small Population Management, Systematics, and Veterinary Medicine. CAPs coordinate projects and programmes by region rather than by taxon. CAPs with canid programmes include Paraguay (bush dogs (*Speothos venaticus*)), Brazil (maned wolves), and Mesoamerica/Caribbean (Mexican wolf).

Research with captive animals

Although captivity seldom approximates the conditions an animal faces in the wild, much can be learned from captive animals, particularly because many environmental variables can be controlled. Pedigree and social history are usually known. In addition, animals are easily accessible and already habituated to human presence. These factors facilitate collection of basic life history data, such as growth rates, age at puberty, litter size, inter-birth intervals, and reproductive life span. Such data are important for planning conservation programmes for each species, since they are the basis for calculations of reproductive output and for constructing population models such as Population and Habitat Viability Analysis (PHVA). Although reproductive data can vary depending, for example, on nutrition and social competition, information from captivity can provide a basis for comparison with observations from the wild. This is especially valuable for more cryptic species.

Captive animals also provide the opportunity to test or validate techniques or equipment to be used in the field. For example, hormone assays are best validated in the controlled conditions of captivity with samples from known individuals. The development of faecal hormone assays has made it possible to assess not only reproductive condition (e.g. Monfort *et al.* 1997) but to study the relationship between 'stress' hormones and social status or degree of disturbance (Creel *et al.* 1997a,b), which can have important implications for conservation action.

New equipment designs, such as radio-transmitters or collars, can be much more easily tested with captive individuals than free-ranging ones. It can be useful to try lures, marking techniques or traps on captive animals first. It can also be useful, before embarking on a field study, to become familiar with a species in a captive environment, where they can be easily viewed. Ethograms can be constructed, and, although they will likely need some modification in the field, valuable time can be saved and preparations made by first observing captive animals (wolf: Zimen 1982; golden jackal: Golani and Keller 1975; red fox: Macdonald 1980a).

Gene banking

Captive animals can serve as a genetic reservoir themselves but can also be a source for gene banking. Because of the success of sperm and embryo cryo-preservation for humans and domestic animals, there is an increasing expectation that gene banking should be used for endangered species. Unfortunately, these expectations cannot be easily met. The fundamental problem is the differences among species in the response of sperm and embryos to such manipulation. Seldom has it been possible to transfer a technique from a closely related domestic species to a non-domestic one without modification. Compounding the difficulty for canids is that developing these techniques for the dog has proven more challenging than for other domestic species. Semen collection by electroejaculation and sperm cryopreservation has been successful in grey wolves, Mexican wolves, and red wolves, but post-thaw sperm survival is poor compared with most other mammalian species (cf. Koehler *et al.* 1998 for red wolf). However, there are active research programmes for Mexican and red wolves that are building on results from the recent increase in work on assisted reproduction in domestic dogs.

Assisted reproduction

Other applications of assisted reproduction techniques to genetic management include artificial insemination (AI), in vitro fertilization (IVF), and embryo transfer. These manipulations have been considered primarily useful within captive populations, for example, to accomplish pairings of distant or behaviourally incompatible individuals. However, transferring genes between geographically separated small populations to effect outbreeding can be equally important for free-ranging animals (for reviews see Hewitt and England 2001).

Unfortunately, as with gene banking, assisted reproduction techniques have proven difficult in dogs. Further compounding the problems inherent in developing these methods in canids is that the approaches that are proving successful with dogs are not practical with wild canids because of the regular, intensive handling required. For example, AI in dogs requires daily blood samples to detect an increase in progesterone as a marker of ovulation. The female is usually inseminated twice on two separate days. Following each insemination she may be held in a head-down position for 10–20 min to facilitate sperm transport into the uterus. In wild canids, handling for daily blood is not likely to be possible and although faecal hormone assays can measure progesterone, their levels tend to be more variable from day to day than are blood levels, making detection of a meaningful change more challenging. In regard to the insemination itself, repeated anaesthetisations may themselves reduce the chances of fertilization, meaning that only one insemination may be practical. In order to increase the likelihood of success, the AI should be intra-uterine, not vaginal, which requires identifying an insemination catheter appropriate for the species. Thus, even if the technique can be made to work, if it is considered too invasive or disruptive, it will not be broadly applied, restricting its usefulness.

Although *in vitro* fertilization requires even more extensive and invasive handling, it does not accomplish more for genetic management than does AI, reducing the justification for directing limited resources towards its development. IVF becomes useful only for individual males with sperm not vigorous enough to traverse the female reproductive tract, since with IVF sperm are placed in a dish with direct access to ova. However, harvesting and maturing those ova require that the female receive hormone stimulation followed by either surgical or ultrasound-guided retrieval of ova, techniques still not successful in dogs.

Because ova do not usually survive freezing, the only way to preserve female genes at this time is in embryos. These can be created via IVF or more simply flushed from the uterus following natural or artificial insemination. Apart from the desire to preserve female genes, production or flushing of embryos is needed for programmes hoping to transfer genes between populations in different countries. Although sperm are significantly easier to collect and handle, sperm viability is compromised by the cleaning needed to insure against transfer of disease. Embryos, in contrast, are more likely to survive such procedure and so are more likely to be approved for international shipment. However, efforts are underway to develop sperm washing protocols that satisfy importation requirements but that do not substantially affect viability.

Reintroductions and translocations

Another tool for enhancing threatened or endangered populations is reintroduction, which is essentially: 'An attempt to establish a species in an area which was once part of its historical range, but from which it has been extirpated' (Kleiman and Beck 1994). Reintroductions can be a powerful conservation tool for canids. For canid reintroductions to be effective, several questions must be answered regarding the need, feasibility, potential pitfalls, and measures of success before they are initiated (Macdonald *et al.* 2002).

1. *When should reintroductions be used for canids?* Successful reintroductions require that a number of species-specific, environmental, and bio-political criteria are met (Kleiman and Beck 1994). There should be a need to augment the wild population, sufficient founder stock should be available, and extant wild populations should not be jeopardized by the reintroduction (Kleiman and Beck 1994; Woodford and Rossiter 1994). The species' biology should be well understood, appropriate reintroduction techniques should be known, and sufficient resources should be available for the programme. The original causes for the species' extirpation should be removed and sufficient unsaturated, protected habitat should be available. Reintroductions should conform to legal requirements, be supported by both government and non-government agencies, and have minimal negative impacts on local people (Kleiman and Beck 1994). Reintroductions of many IUCN red data book species are not attempted or are unsuccessful (Wilson and Stanley Price 1994), perhaps because the necessary requirements for success are so numerous and difficult to satisfy. Many canid species are ill-suited for reintroduction when: (1) their large home range requirements can only be satisfied in extensive protected areas which might not be available (Woodroffe and Ginsberg 1998), (2) local people oppose the reintroduction of species that prey on domestic livestock or threaten humans (Phillips 1995; Woodroffe and Ginsberg 1999) or; (3) the extensive planning and implementation required for reintroductions (Fritts *et al.* 1997) is prohibitively expensive.

2. *When can reintroductions be considered successful?* Although proposed measures of success for reintroductions vary widely, some examples exist, such as: (1) breeding by the first wild-born generation; (2) a 3 year breeding population with recruitment exceeding adult death rate; (3) an unsupported wild population of at least 500 individuals; and (4) establishment of a self-sustaining population (Seddon 1999). Debates ensue about the minimum effective population size that is required for population sustainability. A minimum effective population size of 50 individuals has been proposed to avoid short-term deleterious effects of inbreeding depression (Soule 1980; Franklin 1980), but some argue that this number is too small when populations are exposed to environmental stress (Reed and Bryant 2000). A minimum population size of 500 is thought to maintain sufficient genetic variability in quantitative characters (Franklin 1980; Reed and Bryant 2000). However, this number has also been debated extensively. While Franklin and Frankham (1998)

believe an effective population size of 500–1000 is generally appropriate, Lynch and Lande (1998) maintain that 1000–5000 individuals should be considered a minimum. If supportive breeding is used to supplement wild populations, the effective wild population size to prevent inbreeding depression will decrease but the variance effective size, which represents a minimal loss of heterozygosity, can potentially increase (Ryman *et al.* 1995).

Threatened populations, and species undergoing reintroductions in particular, face deterministic and stochastic factors that, depending on their individual and combined impacts, may drive populations to extinction (Lande 1993). Deterministic factors include: (1) *Allee effects*, which may set minimum population sizes as investigated by Courchamp and Macdonald (2001); (2) *Edge effects*, which may make patch sizes too small for population persistence or too isolated to allow for successful dispersal; and (3) *Local extinctions and colonization*, which, if habitat patches for a metapopulation are degraded or eliminated, may cause the extinction of the entire population. Stochastic factors are: (1) *Demographic stochasticity*, where the intrinsic variation in reproduction and survival may cause the extinction of small populations; (2) *Environmental stochasticity*, which involves continuous small or moderate perturbations that affect the survival or reproduction of all individuals and; (3) *Random catastrophes*, which are large environmental perturbations that produce sudden major reductions in population size at random times (Lande 1993).

Depending on assessments of minimum effective population size and species-specific life histories, reintroduced canid populations may need to be composed of several thousands of individuals to be genetically viable. Minimum population sizes might increase further as the likelihood and magnitude of potential demographic or stochastic factors on reintroduced populations increase.

3. *Canid reintroductions using captive-breeding or translocation* Many of the world's most endangered canids are not captive-bred and few have been reintroduced (Ginsberg 1994). However, reintroductions using captive breeding and/or translocations have been attempted with African wild dogs, grey wolves, red wolves, Mexican wolves, San Joaquin kit foxes, and swift foxes. Reintroductions of endangered African wild dogs (Woodroffe and Ginsberg 1999a) using captive-bred or wild caught individuals have been attempted in South Africa, Namibia, Zimbabwe, and Kenya since 1975 (Woodroffe *et al.* 1997; Woodroffe and Ginsberg 1999b). However, nine of the ten attempts have failed to produce wild offspring (Woodroffe and Ginsberg 1999b; Moehrenschlager and Somers, in press). Grey wolf reintroductions of five and four animals in Alaska and Michigan respectively, failed (Henshaw *et al.* 1979; Weise *et al.* 1979) but a translocation of 107 wolves in Minnesota to minimize livestock depredation was successful (Fritts *et al.* 1984, 1985). The recent reintroduction of Canadian grey wolves to Yellowstone Park in the United States has been effective, as released wolves have survived, reproduced, and the population has increased (Smith *et al.* 2003a). In an attempt to restore extirpated red wolves to the eastern United States, a captive-breeding programme was initiated and 63 animals were released by 1995. By that time 42 animals survived in the wild, of which 36 were wild born (Phillips 1995), but the fact that current red wolves are wolf-coyote hybrids (Wayne 1995) and released wolves hybridize with coyotes (Nowak 1995), has compromised the scope of the programme. A reintroduction of San Joaquin kit foxes in California failed with annual fox mortality rates of 97%, primarily because of coyote predation (Scrivner *et al.* 1993). Fourteen swift foxes that were experimentally released into low-density swift fox areas in South Dakota dispersed over large distances and 50% survived, but the programme was discontinued (Sharps and Whitcher 1984). Although some reintroduction attempts have been successful and Yellowstone wolves are thriving, no canid reintroduction has arguably produced a self-sustaining wild population to date (Fig. 7.3).

Translocations: the case study of swift foxes for reintroduction in Canada

With the release of 942 individuals from 1983 to 1997, the reintroduction of swift foxes to Canada represents the most extensive reintroduction programme for canids to date. In many ways, the Canadian swift fox reintroduction has met Kleiman and Beck's (1994) reintroduction criteria. Extirpation

Figure 7.3 The grey wolf *Canis lupus* re-introduction to Yellowstone is one of the few successful re-introductions of canids © W. Campbell.

of the species from the northern periphery of its range in the 1930s substantiated a need for subsequent restoration. Populations in the central United States were apparently healthy and therefore sufficient founder stock was available without jeopardizing wild populations. Support for the reintroduction came from government and non-government agencies and swift foxes did not have negative impacts on cattle ranchers in the release areas. One primary shortcoming was that the species biology of swift foxes was poorly understood and specific causes of the extinction in Canada were largely anecdotal. Consequently, it was not certain whether the original factors causing the decline of the species had been removed when releases began and the reintroduction was acknowledged as an experiment in reintroduction biology.

Although captive breeding was primarily utilized during the early phases of the swift fox reintroduction, the question remained whether wild swift foxes from the United States could be successfully translocated to the Canadian prairie. Dispersal and survival rates of 56 Canadian resident foxes were compared to 29 Wyoming swift foxes, which were translocated to Canada between 1994 and 1996 and tracked for up to 850 days, respectively, after release (Moehrenschlager and Macdonald 2003). While only 36.4% of resident juvenile foxes had dispersed from natal home ranges at 9.5 months of age, 93% of translocated foxes dispersed, normally within the first week after release. Translocated adults moved further than juveniles and daily movement rates were significantly greater than those of established resident foxes for the first 50 post-release days. After only 4 days, translocated foxes had dispersed 17.3 km on average, already exceeding final mean dispersal distances of resident male and female foxes. Survival rates of translocated foxes were significantly greater among males than females and, regardless of sex, smaller as dispersal distances increased. Since foxes that dispersed over large distances frequently died, successful foxes had smaller mean and total dispersal distances than non-breeders.

As survival rates of translocated foxes were similar to those of resident foxes and greater than captive-bred foxes in previous trials, It was concluded that translocation could serve as a highly effective re-introduction tool. Several recommendations emerged. First, since survival rates were lower for females than males, female-biased release cohorts may provide a more balanced sex ratio for founding breeders within the population. Second, translocating juveniles could be effective because, having lower dispersal distances than adults, they are more likely to establish localized populations within small, protected prairie patches. Furthermore, removing juveniles during the dispersal period may be less damaging to source populations than the removal of experienced adult breeders with established home range boundaries. The final recommendation was the use of soft-release pens to acclimate animals to the release sites before they are set free, since survival and reproductive success were highest for foxes with small post-release dispersal distances (Moehrenschlager and Macdonald 2003).

Canid protection in an ecosystem context

Protected areas

Protected areas are one of the fundamental tools of nature conservation although the significance of their role depends on their size, location within the landscape context, and management efficiency. Most protected areas do not seem adequate to ensure the long-term persistence of any large canid species. Woodroffe and Ginsberg (1998) compared the persistence of ten species of carnivores, including wolves, dhole, and African wild dogs, with reserve sizes within the species' historical ranges. They found that species had persisted in reserves above a critical size, which they identified as varying from 723 km^2 for dhole (*Cuon alpinus*) to 3606 km^2 for the African wild dogs. They also found that female home range size was a good predictor of critical reserve size, while population density had no significant effect. Although these results may have been affected by several hidden environmental and historical factors, they point to the varying order of magnitude of reserve size, which has allowed for the longer persistence of some canid populations.

However, a far more important and difficult question is how to define the size of a protected area that can maintain a viable population. The difficulties arise first with the definition of a viable population and the reliability of current tools for assessing it (Beissinger and Westphal 1998; Fieberg and Ellner 2000; White 2000; Coulson *et al.* 2001b), and secondly with identification of the area needed to host that population. Past estimates of minimum reserve size to protect a large canid population have never been more precise than the 'several thousand km^2' suggested by Fritts and Carbyn (1995) for wolves and the 10,000 km^2 suggested by Woodroffe *et al.* (1997) for African wild dogs. The shape of a protected area can affect its efficiency in protecting animal populations and Woodroffe and Ginsberg (2000) have suggested that the magnitude of the edge effect is related to a species' ranging behaviour: when a species is wide ranging, a relatively larger proportion of the population is exposed to the edge effect, compared to a species with smaller home ranges. However, in spite of a few interesting theoretical insights on the optimal design of protected areas and the availability of several priority setting exercises (Ginsberg 2001), we are still far from having a robust set of principles to guide the practical implementation of a reserve network. In short, we have not progressed much beyond the suggestion that reserves be as large as possible and with the shortest possible boundaries.

If protected areas are to be an effective conservation tool for small canids (although we cannot predict the effects of population fragmentation), they cannot be expected to ensure large canid survival unless they encompass enormous areas—as large as entire regions. This approach has been proposed for North America (Soule and Terborgh 1999), where its feasibility remains to be seen, but it seems unrealistic for most of the rest of the world. Generally, conservation, especially of species living at low density and on large home ranges, can be achieved only through the integrated management of a reserve network and coexistence with humans on unprotected land. However, defining a minimum set of reserves is extremely difficult with current tools for setting priorities (Ginsberg 2001) and the paucity of data on distribution and ecological requirements of many canid species. Protected areas will continue to be a primary tool for canid conservation only if they are an integral component of more comprehensive conservation strategies that extend over the population's entire range and include all aspects of the animal/human interface (Boitani 2000; Woodroffe 2001).

Range distribution modelling

Accurate information on population status and distribution are essential for any conservation strategy but they are rarely available in the quality and quantity needed for useful conservation action. Fieldwork to gather distribution data is not feasible, especially for many small size canids that are nocturnal and elusive, and a range of modelling techniques has been proposed to obtain species distribution from species-habitat relationships (Busby 1991; Morrison *et al.* 1992; Corsi *et al.* 2000). Habitat suitability models have long been used by government agencies and conservation organizations, and have grown in accuracy and complexity as satellite data and geographic information systems (GIS) became available.

The use of GIS has enabled not only simple identification of the area, which represents a species' total range, but also analysis of the quality of patches within this area. The classic presentation of the distribution ranges of animal species drawn on maps encloses the area in which an observer has a chance of finding individuals. This is generally termed the species' Extent of Occurrence (EO). Traditional EO is of little use today, as it generally lacks an explicit indication of the probability threshold used to draw the boundaries and fails to convey important information concerning the distribution pattern within the species' range in relation to environmental factors. The Area of Occupancy (AO) represents the areas really occupied by the species within the EO (Gaston 1991), and is indeed the most crucial piece of information needed for implementing an effective conservation plan. In practice, the concept of AO represents a better approximation of the EO, and the robustness of its representation depends on the species' biology and the quality and scale of the data used.

Various strategies have been adopted to develop GIS-based distribution models. Habitat Suitability Indices have been used to draw species distribution maps (Donovan *et al.* 1987), whereas Gap Analysis uses the distribution of vegetation types and some 'umbrella' species to evaluate biodiversity patterns (Scott *et al.* 1993; Hollander *et al.* 1994). Different correlation models between species distribution and environmental variables (e.g. vegetation, land use, altitude, etc.) have been proposed (Miller 1994) and used to analyse the 'internal anatomy' of distribution ranges. Corsi *et al.* (2000) described the two basic techniques in modelling distribution ranges, the deductive and the inductive approach. The deductive approach describes the species' environmental preferences, as derived from available literature, in terms of environmental variables. Based on this description, one or more experts provide a ranking of suitability for each different combination of environmental layers observed within the area under analysis. The Area of Occupancy appears as a patchwork of more or less suitable areas. The inductive approach uses the environmental information obtained at the geographical location of species' presence (points and/or areas) to build a function that is capable of ranking the entire study area according to a continuous suitability index. Even though the results of the second approach are more objective, data on species location are not always of the quality needed to ensure reliable modelling outputs.

Using logistic regression based on two variables, Mladenoff *et al.* (1995), Mladenoff and Sickley (1998) evaluated the wolf habitat in the northern Great Lakes region and the potential areas for wolf recovery in the northeastern United States; on the other hand, Corsi *et al.* (1999) used the Mahalanobis statistics to assess the area of occupancy of the wolf in Italy. Both these exercises validated the models through independent data sets. The Areas of Occupancy of all African canid species have been assessed by Boitani *et al.* (1999) as part of a larger project on the distribution of all medium and large mammals of Africa. In this project, the EO was used to discriminate between expected and possible presence of the species and both the deductive and inductive approaches were used to cover the entire African continent. Validation of these models was carried out through a direct ground-truthing effort in four African countries (Boitani *et al.* 1999). In summary, GIS distribution models, even using data often limited in quality and quantity, provide a powerful tool for wide scale analyses that are urgently needed if conservation is to be effectively moved from a single species to a community and ecosystem approach (Linnell and Strand 2000).

Influencing change

Legislation and planning context

A critical set of conservation tools is concerned with providing the legal and planning context for management; though these tools are necessary, they remain ineffective unless substantiated by real action on animal and human populations. International treaties such as the Convention on International Trade of Endangered Species (CITES) or the Convention for the Conservation of European Endangered Species and their Habitat (Bern Convention) provide the necessary mechanism for species conservation across international boundaries. At a national level, specific pieces of legislation are needed to justify government action, fund availability, and project implementation. For example in the United States, the Endangered Species Act has been a fundamental tool for the conservation of many species in the last three decades (Clark 1993), including the San Joaquin kit fox, and for reintroducing wolves into Yellowstone National Park, while in Europe the Habitat Directive has been the primary legal tool for obtaining funds for wolf conservation.

Conservation planning at a species and/or national levels is the purpose of the IUCN Species Action Plans. At an international level, these documents rarely have any legal power and their credibility relies on the prestige of the issuing organization (e.g. IUCN—World Conservation Union). Three notable examples are the Action Plans for Canids (Ginsberg and Macdonald 1990), the African wild dog (Woodroffe *et al.* 1997), and the Ethiopian wolf *Canis simensis* (Sillero-Zubiri and Macdonald 1997). On a national and continental level, Action Plans provide the comprehensive vision and the general framework under which any management action should be planned and implemented to be most effective (e.g. Wolf Recovery Plans in US, USFWS 1982, 1987, 1989; Wolf in Europe, Boitani 2000). Indeed, Action/Management Plans are often highly detailed documents covering all aspects of a species' management, including zoning systems that allow for different management regimes and accounting for all human dimension programmes.

Another tool that is often used to support legal decisions is the Red List of Endangered Species that is issued annually by the IUCN (2000). All canid species are currently included and their threat to survival assessed according to a complex set of demographic and range size criteria. The Red Listing system is often applied at a national level and has entered many national law systems, thus becoming legally binding.

Community involvement and education

As conservation is often at the interface of habitat, animal populations and human activities, consideration for the human dimensions in wildlife management cannot be overemphasized. Large canid management is often more a socio-political issue than a biological one (Bath 1991) and solutions to most of the canid ecological and conservation problems are to be found in socio-economic and cultural systems (Machlis 1992). While the need for human dimension studies and management has long been appreciated by government agencies and conservation organizations, effective implementation of these approaches remains a challenge, especially outside North America. The human dimension was a large component of wolf restoration efforts in Yellowstone and great care was taken to study human values and attitudes towards wolves (Bath 1991; Bath and Buchanan 1989) and to manage public consultation and participation in the decision-making process. Human dimension analyses were crucial in successfully addressing the issues and concerns of many stakeholders, and finding the correct approach to working with them rather than against them.

In the Yukon, Minnesota, and Wisconsin difficult choices on wolf management were made successfully when the local public, through representatives of various interest groups, were given the authority to design their own management plans. Conflict resolution techniques based on careful and professional management of the decision-making processes are often used to facilitate the more difficult conservation programmes. Alternative Dispute Resolution (ADR) is one of the several interesting techniques used in this field (Wondolleck *et al.* 1994) and variations on its approach have been used successfully to reach a consensus on management of the wolf in Minnesota and Wisconsin (L. D. Mech and A. P. Wydeven personal communication). Where public involvement has not been properly planned, wolf management plans have remained highly controversial and substantially inapplicable, as in Alaska (Stephenson *et al.* 1995). Ideally, canid conservation should be implemented through full collaborative management approaches: the complexities of the issues involved and the decisions to be taken make this confrontation ideal ground for more advanced experiments in public participation (Borrini-Feyerabend and Buchan 1997; Boitani 2003; Fritts *et al.* 2003).

The direct participation of local groups in the conservation decision-making process is not always possible, but local community involvement should remain a fundamental tool of any conservation project. An extensive programme of information and education of local people has been an integral part of the Ethiopian Wolf Conservation Programme (Sillero-Zubiri and Macdonald 1997) and it has been instrumental in raising public awareness of the main conservation issues as well as raising people's acceptance of the Ethiopian wolf. In Italy, the successful recovery of the wolf to the current population level was determined by a combination of factors, of which one of the most important was the public information and education campaign, carried out for several years and through a great variety of means (Boitani 1992). All these examples point to the need to make education and information primary tools in conservation, too often they remain relegated to a secondary role constrained by lack of funds and of professional competence. Levels of public participation can be increased when local groups obtain direct benefits from conservation, such as revenues from sport hunting. Ecotourism is often included among the tools available for generating economic returns for those who also suffer economic losses from coexistence with large carnivores, particularly wolves. However, though ecotourism is often feasible and a potentially significant resource, it can only be applied to a few select cases, as increasing the offer would reduce the share of potential revenues.

Conclusion

A holistic approach

The need for interaction of several dimensions is now widely accepted but there is no single recipe for all conservation issues (Clark *et al.* 1993, 2001b). Collaboration, integration and multidisciplinary approaches are increasingly common themes when complex conservation issues are discussed as for most projects on carnivores (Gittleman *et al.* 2001a; Macdonald 2001). This is particularly true for canids because keys to their conservation very often lie in the very intricate interface between the species' biology and human interests and attitudes, and the latter varies across regions, cultures, and times. A shift in focus from the single-species to the more inclusive ecosystem approach is of paramount importance if conservation is to be effective and have long-lasting results. Moreover, many canid species show a remarkable biological flexibility and adapt to local situations with often-unexpected behavioural and ecological solutions. For these species, it is especially true that every conservation problem must be treated as a special case. Case studies provide excellent opportunities to learn new solutions to local problems and to study potential responses of both animals and humans, but they should not be taken as protocols or guidelines for other situations. Numerous conservation tools are available to the conservationist, but they form a body of knowledge that has to be digested and used wisely in order to apply the right mix of tools to each conservation programme. While all levels are important, it is crucial to remember that people are the biggest threat to canids (Woodroffe 2001b) and that reducing the conflict with humans is the most challenging but also most urgent need for canid conservation.

This brief analysis of some of the tools available for canid conservation must be set within a forward-looking framework as the future of conservation lies in a holistic approach. The traditional approach of focusing on single species and populations was all the more conspicuous as the species involved were large and charismatic vertebrates; moreover canids have been a relatively easy target for this approach because of their widespread prominence in most human cultures. A single species approach might be justified when the species plays the role of keystone or umbrella species, but the operational usefulness of these concepts is still quite vague and canids do not fit clearly into either of these species descriptions (Linnell *et al.* 2000b). There is now great consensus that focusing on one species only is inadequate for reaching long-lasting conservation results and that conservation tools are most effective when applied within the context of an ecosystem approach. Hence, canid conservation has to adopt this strategy with renewed energy. Since lasting conservation of canid populations, especially the large ones, cannot be achieved at the spatial scale of most protected areas, the challenge is to adopt a broader perspective and expand projects in time and spatial scales, including all components of the ecological and socio-economic contexts. The Carpathian Large Carnivore Project currently active in Romania for wolf, bear (*Ursus spp.*) and lynx (*Lynx lynx*) conservation (www.clcp.ro) is one of the best examples of an integrated conservation development project (ICDP) involving canids and it shows the potential of adopting this approach. Clark *et al.* (2001) convincingly argue that carnivore conservation, perhaps more than other conservation projects, needs to be viewed as a complex system of decision making that requires an interdisciplinary approach and the involvement of several professional figures. Recognizing this challenge is the first step to mastering all other conservation tools.

Acknowledgements

Axel would like to thank Husky Energy Inc. and the Wildlife Preservation Trust Canada for its funding, leadership and logistical support.

PART II

Case studies

CHAPTER 8

Arctic foxes

Consequences of resource predictability in the Arctic fox—two life history strategies

Anders Angerbjörn, Pall Hersteinsson, and Magnus Tannerfeldt

An adult Arctic fox *Alopex lagopus* female of the 'white' polar morph in summer fur. The Arctic fox has very thick and soft winter fur with two distinct colour morphs, 'blue' and 'white'. In winter, the 'white' morph is almost pure white, while in summer it is brown dorsally and light grey to white on it's underside. The 'blue' moults from chocolate brown in summer to lighter brown tinged with blue sheer in winter (see Fig. 8.1) © M. Tannerfeldt.

Introduction

Considerable interspecific variations in life history characteristics are found among mammals. These include age at maturity, maximum life span, fecundity, gestation length, rates of mortality, and natal dispersal (Roff 1992). These variations are, in many cases, species-specific which makes it hard to separate adaptation to different habitats (Southwood 1977) from long-term evolutionary changes or phylogenetic constraints (Stearns 1992).

Considering the consequences of environmental variability, characteristics of individual organisms, for example, natal dispersal, territoriality, and reproductive success, can influence population and community dynamics as well as geographical range, genetics and risk of population extinction (Lomnicki 1988). By intraspecific comparisons between populations living in contrasting habitats, the selective forces specifically connected to the ecological factors that differ between contrasted habitats can be separated from the general selection on a species. The Arctic fox (*Alopex lagopus*) exhibits considerable intraspecific variation in litter size, where rodent and non-rodent eating Arctic fox populations, respectively, are reported to have different reproductive strategies (Braestrup 1941; Tannerfeldt and Angerbjörn 1998) (Fig. 8.2). The relationship between food availability and life history strategy is crucial to many questions on predator population dynamics. In some cases there are even distinct intrapopulation differences in Arctic fox territorial behaviour, apparently linked to differences in food predictability and dispersion (Eide 2002).

We have examined life history characteristics of two Arctic fox populations, a relatively stable one in Iceland and a fluctuating one in Sweden, and we compare them in relation to abundance and fluctuations of their main food resources.

Resource distribution in time and space

Fluctuating food resources

In the Arctic fox's Holarctic range, productivity in inland areas is low but food resources can be extremely abundant in small patches and during short time periods. The dominant pattern in these resource fluctuations is determined by lemming (*Lemmus* and *Dicrostonyx* spp.) and vole (*Clethrionomys* and *Microtus* spp.) population peaks. These prey are superabundant every 3–5 years but are otherwise scarce (e.g. Hansson and Henttonen 1985; Stenseth and Ims 1993). However, there is considerable variation around this mean periodicity (Hanski *et al.* 1993), making these rodents an unpredictable resource for Arctic foxes. Population sizes of Arctic foxes in lemming areas are regulated by lemming numbers through variation in the recruitment of fox cubs each year (e.g. Macpherson 1969; Hersteinsson *et al.* 1989; Strand *et al.* 1999). Lemming population fluctuations are generally synchronized over large areas (Angerbjörn *et al.* 2001). Some avian predators, such as the snowy owl (*Nyctea scandiaca*), may respond to scarcity of lemmings by moving hundreds or even thousands of kilometres. Although Arctic foxes have been recorded as dispersing over similarly great distances (Eberhardt and Hanson 1978; Garrott and Eberhardt 1987), movements of that magnitude are probably very risky for them. Arctic foxes are short-lived in the wild and cannot expect to experience more than one rodent peak event (Hiruki and Stirling 1989; Tannerfeldt and Angerbjörn 1996). In Sweden, lemmings go through classic fluctuations and the Swedish Arctic fox population is a typical example of the strong relationship between Arctic foxes and lemmings (e.g. Angerbjörn *et al.* 1995).

Stable food resources

In other areas, Arctic foxes are sustained by more stable food resources. This occurs, for example, at bird cliffs and along ice-free coastlines, where food is available in all seasons although there may be substantial seasonal variability in both food abundance and type of prey available (Hersteinsson and Macdonald 1982; Prestrud 1992b). Under these circumstances, food resource levels are more predictable from year to year than in rodent areas. This is exemplified by Iceland where there are no lemmings or voles and the non-cyclic wood mouse (*Apodemus sylvaticus*) is the only rodent available to Arctic foxes. Iceland can roughly be divided into two main habitat types, coastal and inland, with prey availability generally higher in coastal habitats where the foxes feed on anything from tidal invertebrates to seal carcasses, although seabirds make up the most important constituent of the diet (Hersteinsson and Macdonald 1996). Inter-annual variation in food abundance is insignificant and irregular rather than cyclic, with a 10-year population cycle of the rock ptarmigan (*Lagopus mutus*) the only known exception (Gudmundsson 1960; Nielsen 1999). However, the ptarmigan is of minor importance as food to Arctic foxes in coastal habitats and only in northern and northeastern inland habitats do Arctic foxes show a numerical response, albeit slight, to ptarmigan fluctuations (Hersteinsson 1984).

Life history strategies

Litter size

Fluctuations in essential resources cause a strong selection pressure on the ability to adjust parental investment accordingly. Amongst canids, variance in female pre-birth investment is adjusted by litter size (Geffen *et al.* 1996). However, recruitment to the winter population can be divided in three phases: the number of litters produced, litter sizes, and cub survival rates. These are related to food supply for several canid species, but the mechanisms are largely unknown (e.g. Englund 1970; Bronson 1989). Lindström (1989) supplied large amounts of additional food to wild red foxes (*Vulpes vulpes*) in southern Sweden, causing an increase in the number of litters born. Also, Angerbjörn *et al.* (1991) conducted a field experiment with winter feeding of Arctic foxes and observed an increase in number of litters and litter size at food manipulated dens. Thus, in the Arctic fox, the number of litters and litter sizes seem to be regulated by food supply during winter and spring, through female condition (Hall 1989; Angerbjörn *et al.* 1991).

The Arctic fox has the largest known litter size in the order Carnivora, up to 19 young, and litter size is highly variable. Comparing data on placental scars and weaned litter sizes from a large number of populations (Tannerfeldt and Angerbjörn 1998), the range of litter sizes was shown to be larger in unpredictable than in predictable environments. Further, animals from unpredictable environments had more placental scars and larger weaned litters than those from predictable environments. Reviewing published Arctic fox studies, Tannerfeldt and Angerbjörn (1998) found that none of the several hundred observed 'non-rodent' litters comprised more than 12 cubs. This is well illustrated by a comparison between the fluctuating Arctic fox population in Sweden and the stable population in Iceland (Fig. 8.2). Some areas in Iceland, such as bird-cliffs, are consistently highly productive but the maximum number of placental scars found was only 10 ($N = 1048$; Hersteinsson 1990 and unpublished data). This is only half of the maximum scar counts in rodent areas (Macpherson 1969). We suggest, therefore, that energetic limitations explain intra-population differences in mean litter sizes, but that there is a genetic component behind the observed differences between populations related to food predictability. According to the 'jackpot hypothesis', populations with unpredictable food resources generally have larger litter sizes (Fig. 8.3).

Cub survival

Boutin (1990) reviewed field experiments involving food additions, and found a positive effect on juvenile survival for 9 out of the 12 mammal species, none of them carnivores. The relationship between food availability and cub survival was examined in a field experiment on Arctic foxes in northern Sweden (Tannerfeldt *et al.* 1994). Supplementary feeding at

Figure 8.1 An adult Arctic fox (*Alopex lagopus*) of the 'blue' colour morph in winter fur. Originally Braestrup (1941) connected non-rodent eating foxes in Greenland to the 'Blue' colour morph, more common along coasts, and rodent eating foxes to the 'White' colour morph, more common in inland areas © M. Tannerfeldt.

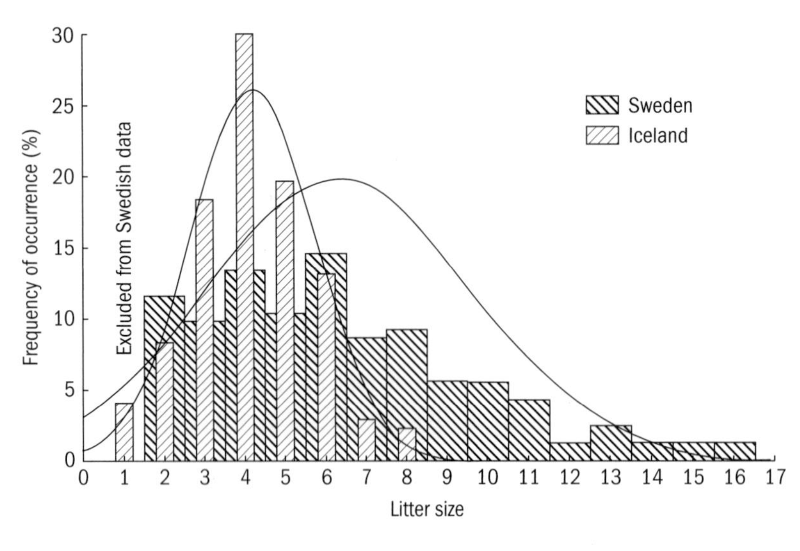

Figure 8.2 The frequencies of litter size at weaning in a habitat with unpredictable food resources (Sweden: Mean 6.3, S.D.±3.3, N=164) and in a habitat with stable food resources (Iceland: Mean 4.2, S.D.±1.5, N=309).

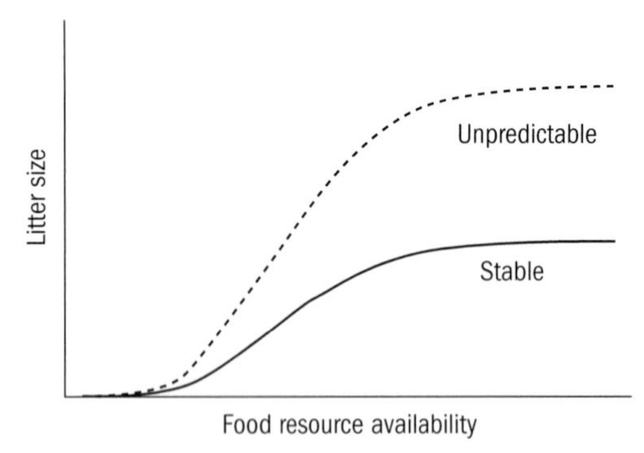

Figure 8.3 A conceptual model on the reaction norm for litter size in stable versus unpredictable environment against food resource availability (Tannerfeldt and Angerbjörn 1998).

dens during a year with medium food availability increased cub survival rates. Juvenile survival from weaning and over the next 6 weeks that year was 42% for non-fed and 92% for supplementary fed animals. During years with low food availability and no artificial feeding, no cubs survived at all (Angerbjörn and Tannerfeldt unpublished data). Only 8% of all juveniles (both fed and non-fed) survived from weaning until the first breeding season. Without artificial feeding, survival during the first year varied from 0% at low food availability to 12% at medium food availability (Tannerfeldt *et al.* 1994). The study was performed during a 20-year period without lemming peaks (Angerbjörn *et al.* 2001). Juvenile survival may be much higher during a lemming peak year.

In Iceland, on the other hand, Hersteinsson (1984 and unpublished data) found by analysing size of litters captured at breeding dens by foxhunters, and placental scars counts in their mothers, that survival from birth to 6 weeks of life was 85%, and 70% from early pregnancy to about 10 weeks of age (Fig. 8.4). This indicates that juvenile survival in Iceland is routinely similar to that during peak lemming years in Sweden. Excluding humans, the only predator on Arctic foxes in Iceland is the white-tailed sea-eagle (*Haliaeetus albicilla*), and mortality due to humans was excluded from the analysis. The hunting by people of Arctic foxes at breeding dens in Iceland peaks when cubs are 5–7 weeks old, and mostly occurs after they have reached 4 weeks of age (Hersteinsson 1988; see also sample sizes in Fig. 8.4).

Some authors have found high rates of predation on Arctic fox cubs. Garrott and Eberhardt (1982), for example, found dead cubs at 19% of the occupied breeding dens in their study area in northern Alaska, and attributed 65% of the registered deaths to predators. Macpherson (1969) suggested siblicide as a major cause of juvenile mortality, but we agree with Arvidson and Angerbjörn (1987) and Sklepkovych (1989) that there is no direct evidence of infanticide or siblicide even if young foxes feed on dead siblings.

Adult survival

In Iceland, hunters were the main cause of death, both in winter and at breeding dens in summer

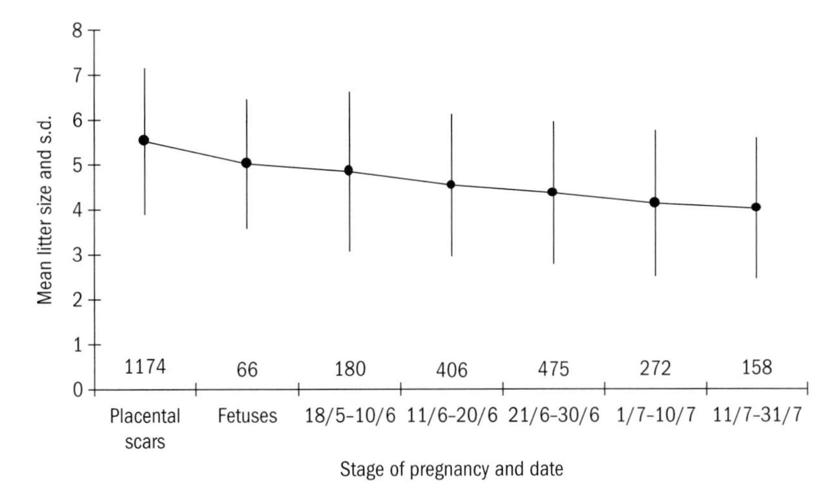

Figure 8.4 Arctic foxes are hunted in all seasons in Iceland, also at dens in summer. Dissections of vixens and data obtained from foxhunters can thus be used to track changes in mean litter size with time from early pregnancy (placental scars) through late pregnancy (fetuses) and at different periods from birth to late July (size of litters captured by foxhunters when hunting Arctic foxes at breeding dens in summer). The observed fall in mean litter size suggests that mortality from early pregnancy to weaning in early July is only about 30% (Hersteinsson 1984, 1988, unpublished). Mean date of birth is believed to be around 15 May. Numbers show sample sizes (no. of litters) and vertical lines show standard deviation.

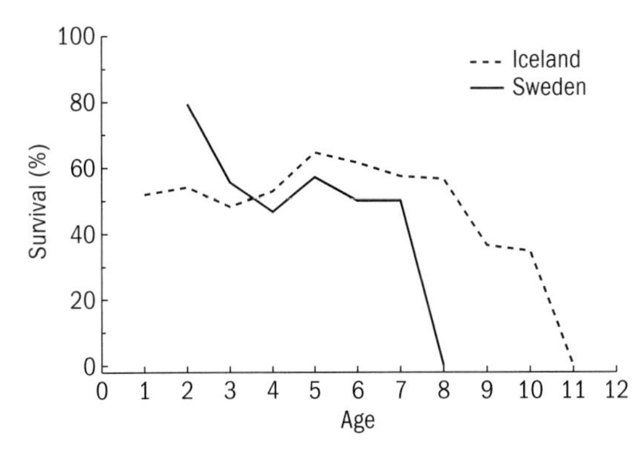

Figure 8.5 Pooled age-specific survival for Icelandic Arctic foxes born in 1985-90 (based on 1699 foxes born in these years and killed after the age of 4 months; Hersteinsson, unpublished data) and Swedish Arctic foxes born 1986-2001 (followed from birth to death, or from the time they immigrated to the study area assumed to have been 1.5 years old, n = 34, Tannerfeldt and Angerbjörn 1996, unpublished data).

(Hersteinsson 1992). An age-specific life table for foxes born in 1985–90 and killed after the age of 4 months, suggests that survival is sharply reduced at around 8 years of age with a maximum life expectancy of 11–12 years (Fig. 8.5).

In Sweden, some animals were followed from birth to death, others from the time they first settled in the area (assumed to have been at 1.5 yrs old, Tannerfeldt and Angerbjörn 1996). There, all adult mortality occurred during winter, from October to May, but the causes of death could not be established. There is a striking difference between Sweden and Iceland in maximum life expectancy with no fox exceeding 7 years of age in Sweden. There was no significant difference in longevity between the sexes in either population.

Natal dispersal

Natal dispersal appears to be a discrete event taking place mostly during the first or second year of a fox's life. Most Arctic foxes settle close to their natal areas. In Iceland dispersal distances were almost normally distributed with more than half of the foxes dispersing 10–30 km (Fig. 8.6). In Sweden, on the other hand, 50% of the foxes settled within 10 km of their natal den, mostly in an adjacent territory. However, 3 of 16 foxes tagged in Sweden as cubs dispersed 80–220 km from their natal den to successful reproduction. Thus in Sweden two strategies could be detected—going far or staying close. So, although the mean dispersal distance did not vary between

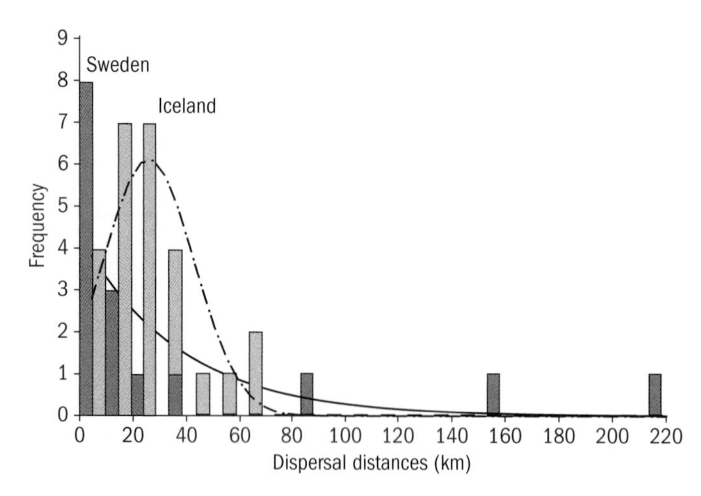

Figure 8.6 Dispersal distances of juvenile foxes in Iceland and Sweden. Two of the Icelandic foxes were females tagged as yearling non-breeders and recaptured as breeders the following year. All others were ear-tagged as cubs. Foxes were either observed to have settled down as adults or for some individuals in Iceland killed at ≥12 months of age. A normal distribution was fitted to the Icelandic data (Kolmogorov–Smirnov $d = 0.07, p > 0.05$) but not to the Swedish data ($d = 0.34, p < 0.05$) where instead a log-normal distribution was fitted ($d = 0.07, p > 0.05$).

Sweden and Iceland, the variance in dispersal was much larger in Sweden (Levene $F_{(1, 40)} = 12.3$, $p < 0.01$). The prevalences of the alternative dispersal strategies in Sweden were associated with the population phase of small rodents. We compared immigrating settlers (dispersing from at least 50 km away) with those born in the area (dispersing <10 km), that is, local settlers (Tannerfeldt and Angerbjörn 1996). Of 22 Arctic foxes, more immigrants (8) than locals (1) settled in the study area in pre-saturation years (increasing rodent populations). In saturation years (peak or decreasing rodent populations), the opposite was true (three immigrants compared with five locals: Fisher's exact test, 2-tailed $p < 0.05$). Further, immigrant females had a higher lifetime reproductive success (LRS) than did local females, whereas there appeared to be no difference in LRS between immigrant and local males (assuming monogamy—Tannerfeldt and Angerbjörn 1996).

There was no difference between sexes in dispersal distances (Sweden: $t = 1.50$, d.f. = 13, $p = 0.16$; Iceland: $t = 1.21$, d.f. = 24, $p = 0.24$). The dispersal strategy observed in areas where rodents are the mainstay of the diet is probably related to the large fluctuations in food resources that occur there. A disperser during decreasing rodent abundance in its natal area would be successful if it settled in an area where prey density was increasing or high, thus vastly improving the odds of it reproduced successfully. Further, a high quality individual has, by definition, a higher probability than average to survive long-distance dispersal and thus to gain large reproductive success—the 'jackpot'. It is likely that dispersal is a state-dependent life history characteristic of individuals. Thus weaker individuals might be predicted to avoid the risks of long-range dispersal and stay in the vicinity of their natal site although that entails a low probability of breeding until the next rodent peak. Strong individuals, on the other hand, might be predicted to capitalize on their superior fitness to disperse long distances in search of a resource-rich habitat (Tannerfeldt and Angerbjörn 1996).

Social organization

The basic social unit of the Arctic fox is the breeding pair. Both parents take an active part in rearing the cubs. For the first 3 weeks after birth, while the cubs are mostly dependent on milk, the female rarely leaves the den for any length of time and the male brings most of the food on which the female feeds during what is, for her, an energetically demanding period. As meat becomes a larger constituent of the cubs' diet the roles of the parents converge and the female takes an active part in hunting and provisioning the cubs.

In Sweden, detailed observations during the breeding season revealed that in 46 of the 55 reproductive events (84%) foxes reproduced in pairs (Table 8.1; Angerbjörn and Tannerfeldt, unpublished data). There were only two cases with a confirmed extra

Table 8.1 Arctic fox group composition during the breeding season in Iceland and Sweden. Percentage of groups against all reproductive events

Country	Habitat or pop. phase	Pre-weaning %					Post-weaning %					
		N	Pairs only	Single female	Extra adult	Cooperative breeding	*N*	Pairs only	Singles female	Extra adult	Split litter	Cooperative breeding
Iceland (stable food resources)	Coastal	11	36		64		10	100				
Sweden (fluctuating food resources)	Crash		4									
	Low		24	2				23				4
	Increase		55	4	4	4		45	4	4	8	8
	High		2			4						6
	Total	55	84	5	4	7	53	72	4	4	8	17

adult (one female and one of unknown sex) and another three instances of females without a helping male. In nine cases fox families reproduced cooperatively in the same den. In five of these cases two families merged post-weaning. For example, in one case a lone female with cubs joined her mother in a neighbouring den and territory after an attack by a red fox when one cub was killed. In another case three females, one male and three litters with a total of 27 cubs were together in a single den. The adults came from neighbouring territories and were related: mother and her two yearling daughters and yearling son (B. Elmhagen, unpublished data). In four cases a pair split the litter in two dens and reared the cubs separately.

In protected areas in Iceland, non-breeding females are frequently found within a breeding pair's territory and these appear to be mostly yearling offspring of the pair (Table 8.1; Hersteinsson 1984, 1999, unpublished data; Hersteinsson and Macdonald 1982). This appears to be less common where and when intensive foxhunting takes place, insofar as foxhunters rarely observe more than the pair at breeding dens.

In coastal habitats in Iceland the breeding pair maintains a territory throughout the year. Overlap between territories is generally small and borders between neighbouring territories are frequently maintained from year to year (Hersteinsson and Macdonald 1982; Hersteinsson 1984) although in some cases their exact location may shift by over 1 km (Hersteinsson, unpublished data). Scent marking by urination is much more frequent in areas of territory overlap than elsewhere and in addition the foxes use vocalizations and tail displays to advertise their presence on or near territory borders, which they patrol at least daily (Hersteinsson 1984, Frommolt *et al.* 2003). Thus they appear to expend much more time and energy in territorial defence and maintenance in Iceland than in Sweden. As food availability is stable from year to year, all territories are occupied each year where foxhunting is prohibited and there appears to be minimal scope for territory reduction. In the Hornstrandir Nature Reserve in northwest Iceland, mean territory size was estimated in 1999 at 12.1–13.5 km^2 or 4.0–4.5 km of coastline per territory and the number of occupied territories appears to be stable at 43–48 territories (Hersteinsson 1999; Hersteinsson *et al.* 2000). This is comparable to the 12.5 $\pm$ 5.3 km^2 (s.d.) mean size of territories and 7.2 $\pm$ 2.9 km (s.d.) length of coastline per territory in an earlier study elsewhere in northwestern Iceland, (Hersteinsson and Macdonald 1982).

In Sweden, on the other hand, where the population size and number of territories fluctuate widely from year to year, summer territories of the foxes varied between 17 and 31 km^2 with a mean of 24.6 km^2 (Angerbjörn *et al.* 1997). For Sweden we have no information regarding winter home ranges, but adults tend to stay in their summer ranges (Fig. 8.7). Territorial boundaries remain basically the same from year to year, although the location of the breeding den may change.

Discussion

By comparing Arctic foxes in the unpredictable habitat of northern Sweden and those in Iceland, where food availability is stable and predictable from year to year, we have pinpointed some important differences in selective forces on their reproductive and social strategies.

First, litter size at gestation and birth varies significantly between the two environments. The Swedish study area was characterized by the unpredictable availability of lemmings and other rodents, which were the mainstay of the foxes' diet. Fluctuations in the rodent population were translated into a numerical response by the Arctic fox, with a delay of about 1 year (Angerbjörn *et al.* 1999). Consequently, by the time rodent numbers are recovering from a population trough, the Arctic fox population is still small and intraspecific competition for resources is low. Thus, a superabundance of food becomes available and the foxes channel surplus energy into reproduction. Reproductive output appears to be limited only by the vixen's physiological constraints with regard to maximum litter size during late pregnancy and lactation. In Iceland, on the other hand, the Arctic fox population is adapted to a high population density that varies little from year to year, with severe competition for territories. Furthermore, while food availability is stable, it is generally lower than in peak rodent years in Sweden. Consequently litter sizes at gestation and birth are much smaller than in Sweden (Fig. 8.2). Even in Svalbard, with stable food availability, where individuals had a body fat content up to 40% in winter, maximum litter sizes were still low (Prestrud 1992a; Prestrud and Nilssen 1995). Litter size is then adjusted to the degree of predictability in resource variation. Within each reaction norm litter sizes are adjusted through a number of plastic traits, influenced by nutritional constraints and including reduced ovulation rates, prenatal losses, and litter size reduction during the lactation period. Because of these mechanisms of pre-natal litter size regulation, large litters are only produced when resources are abundant, and reproductive costs can be kept small. If reproductive costs are small, maximum litter sizes should be larger in strongly fluctuating and unpredictable environments than in stable environments (Fig. 8.3).

Figure 8.7 An adult Arctic fox *Alopex lagopus* at a den. Arctic Foxes in Sweden often stay at their dens during the winter © L. Liljemark.

Second, cub survival varies considerably in Sweden, depending on the abundance of rodents (Tannerfeldt *et al.* 1994). In years of low food availability very few cubs survive while survival is high in rodent peak years. In contrast, cub survival is high and inter-annually stable in Iceland, that is, about 70% from early pregnancy to the age of about 10 weeks (excluding mortality due to hunters).

Third, there is a difference in maximum life span between the two populations that could be related to breeding strategy, where a jackpot strategy has a cost in life span (Promislow and Harvey 1990).

Fourth, dispersal patterns vary between habitats. In Sweden, where fluctuations in rodent abundance are synchronized over large areas, young foxes seem either to stay close to their natal area or disperse over long distances. Dispersal over intermediate distances appears to be the least preferred strategy, perhaps because it is likely to lead to places where food availability differs little from those in the natal area, and thus the benefits of dispersal are unlikely to outweigh the risks. The heavy risk-taking associated with long-distance dispersal might make this option advantageous for only the highest quality individuals; indeed, immigrants to our Swedish study area had higher lifetime reproductive success than did non-dispersers (Tannerfeldt and Angerbjörn 1996). In contrast, Icelandic Arctic foxes showed a normal distribution of dispersal distances, most of them settling at intermediate distances of 10–30 km from the natal area. In conditions where there is a stable and predictable food supply with little prospect of significantly improved conditions with increased distance from the natal area, we deduce there would be selection against risky long-distance dispersal through areas densely populated by territorial foxes.

Fifth, social organization differed between the two populations. While a breeding pair was the core social unit in both areas, non-breeding yearling vixens were commonly found on their parents' territories in early summer in areas where foxes are protected in Iceland. We suggest that young vixens in Iceland use their natal territory as a base for exploration of their surroundings as they search for a vacant territory or a mate. As food availability of foxes is generally patchy in Iceland, particularly in coastal habitats, the minimum territory that can sustain a breeding pair and their cubs of the year, will frequently also support an extra adult, in agreement with the Resource Dispersion Hypothesis (Hersteinsson and Macdonald 1982; Kruuk and Macdonald 1985; Eide 2002). In contrast, non-breeding yearling females were rarely found on their natal territories in Sweden but were frequently found on adjacent territories. This is probably due to the presence of non-occupied territories during an increase phase of the population cycle. Thus the foxes on neighbouring territories may be closely related, facilitating the complex social structure occasionally observed.

Conclusions

Intraspecific variation in reproductive and social strategies of Arctic foxes in Sweden and Iceland suggests that adaptations to different resource distributions in time and space have resulted in divergence in strategies between the two populations. In Sweden, where food availability fluctuates widely in time but less in space (except on a large geographical scale), the foxes have adopted the 'jackpot' strategy of enormous variation in reproductive output from year to year with much inter-annual variation in cub and juvenile survival, depending on food availability. They have also adopted two dispersal strategies. One involves only minimal dispersal and correspondingly limited risks, with the possibility of forming a larger social unit with close relatives during the breeding season—probably for predator defence. The other involves very long-distance dispersal, with attendant high risks and no possibility of forming a large social group during the first year of breeding but with the possibility of discovering an area with superabundant food.

In Iceland, on the other hand, where food availability is predictable in time and space, reproductive output is stable with small litter sizes, high cub survival, intermediate dispersal distances, and female yearlings frequently using their natal territories as a base while searching for a vacant territory or mate in the neighbourhood. While helping behaviour occurs (Hersteinsson 1984), food contributed by yearlings to their younger siblings is generally of minor importance (Strand *et al.* 2000).

We suggest that these different strategies—and particularly variation in litter size—have a genetic basis. Furthermore, we hypothesize that individuals in the two populations should have different competitive abilities. It seems likely that the Icelandic population has experienced strong selection for competitive ability whereas the Swedish population has not, in that it generally only breeds when food is abundant and population density is low. Our conclusions accord with an early suggestion by Braestrup (1941) that Arctic foxes in inland (lemming) and coastal (non-lemming) habitats had different life history strategies. The differences we report between the Icelandic and Swedish circumstances resonate with the notions of *r* and *K* selection (MacArthur and Wilson 1967), a concept which, despite its shortcomings in interspecific comparisons, is applicable to our intraspecific contrast (Charlesworth 1980). It remains to be seen whether the consequences of this contrast between predictable and unpredictable resources extend to the different regulatory mechanisms, but we hypothesize that regulation of the Icelandic population is density dependent, and that of the Swedish population is density independent.

Acknowledgement

We extend our warmest thanks to all devoted, hard-working field workers in Sweden and Iceland and to the many foxhunters in Iceland who provided information and carcasses for research. LIFE-Nature and WWF Sweden provided economic support to The Arctic Fox Project SEFALO (AA and MT). This Science Research Fund of Iceland and the Ministry of the Environment in Iceland provided funds for the Icelandic part of the study (PH). We are grateful for funding from Oscar and Lili Lamms Stiftelse, Carl Tryggers Stiftelse, Magnus Bergvalls Stiftelse and the Royal Swedish Academy of Science.

CHAPTER 9

Island foxes

The evolution, behavioural ecology, and conservation of island foxes

Gary W. Roemer

A yearling, male Santa Cruz island fox just *Urocyon littoralis santacruzae* prior to dispersal © G. Roemer.

The island fox is (*Urocyon littoralis*) endemic to the California Channel Islands, a continental archipelago located off the coast of the southwestern United States. A descendent of the mainland gray fox (*U. cinereo argenteus*), it is hypothesized that island foxes first colonized the three northern Channel Islands (Santa Cruz, Santa Rosa, and San Miguel) by chance over-water dispersal. Native Americans then transported foxes from these islands to three southern Channel Islands (Santa Catalina, San Clemente, and San Nicolas). Each island fox population is currently recognized as a distinct subspecies, and both the hypothesized colonization scheme and the current taxonomic classification are supported by morphological and genetic evidence.

An insular existence has had a profound influence on the evolution, ecology, and genetic structure of island foxes. A dwarf form of the gray fox, island foxes are the smallest canid in North America. Compared to mainland canids of similar size, island foxes have shorter dispersal distances (mean = 1.39 km, SD = 1.26, range = 0.16–3.58 km, n = 8), a smaller average home range size (mean annual home range = 0.55 km^2, SD = 0.2, n = 14) and higher population densities (2.4–15.9 foxes/km^2). Although they are distributed as socially monogamous pairs, island foxes are not completely monogamous. Extra-pair fertilizations (EPFs) accounted for 25% of all offspring whose parents were determined through paternity analysis. This relatively high rate of EPFs may be related to high population density and the proximity of suitable partners other than social mates. Finally, the genetic gradient among fox populations appears steeper than mainland populations suggesting that smaller dispersal distances on islands result in increased population structure.

An insular existence coupled with small population size may have also increased the vulnerability of the island fox to extinction. Over the past decade, five of the six subspecies have declined and two are extinct in the wild. Factors contributing to these declines include predation by golden eagles (*Aquila chrysaetos*), the introduction of canine distemper virus and predator control efforts aimed at controlling foxes to protect an endangered bird. A multi-faceted conservation strategy that includes the live-capture and removal of golden eagles, the vaccination of foxes against canine distemper virus, the eradication of feral herbivores and the captive propagation of island foxes is currently underway to avert the impending extinction of this endemic canid.

Introduction

The bow sliced through the calm waters of the bay and then quietly slid to a stop on the gently sloping beach. Limú was glad the journey was half over. The crossing from Santa Cruz Island to this sheltered bay at the west end of Santa Catalina Island had been rough. The swells were half as tall as his *tamal* was long, and more than once he had felt they would lose their cargo. Their safety was critical to the success of his trading effort. He was sure they would fetch a handsome price, perhaps a few steatite bowls, soapstone cookware coveted by all the tribes of the Channel Islands, or maybe several sea otter pelts. Yes, indeed, he was sure these gentle creatures with their beautiful cinnamon, white and grey coats, their large hazel eyes, and their inquisitive and playful nature would bring a handsome price. Limú, of the Chumash Indians of Santa Cruz Island was about to introduce an animal that the Gabrielino Indians of Santa Catalina Island had never seen. The Gabrielino were about to meet the island fox.

Although fictional, this scene may well depict how island foxes colonized the southern most Channel Islands, a continental archipelago located off the coast of southern California, USA (Fig. 9.1). Archaeological, ethnographic, morphological, and genetic evidence support the contention that foxes were brought to the southern Channel Islands (Santa Catalina, San Clemente, and San Nicolas) by Native Americans between 2,200 and 5,200 YBP (Collins 1991a,b, 1993; Wayne *et al.* 1991b; Vellanoweth 1998). Native Americans of the Channel Islands harvested foxes to make arrow-quivers, capes, and headdresses from their pelts, they ceremonially buried foxes, conducted an Island Fox Dance and most likely kept foxes as pets or semi-domesticates (Collins 1991b). Island foxes played a prominent role in the

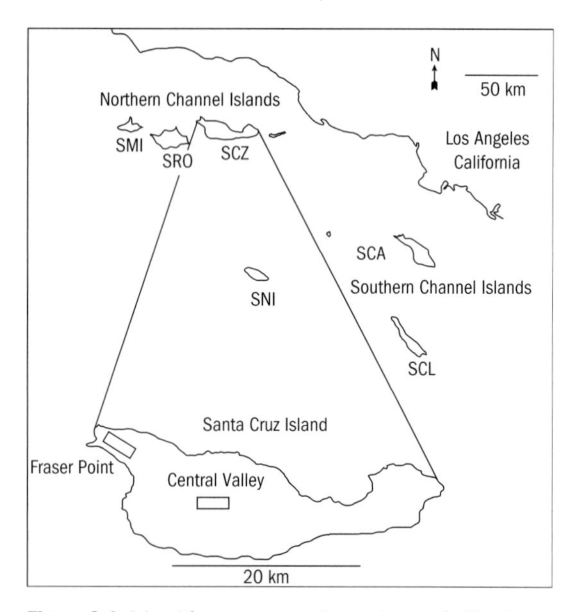

Figure 9.1 Island foxes occur on the six largest California Channel Islands. The three northern islands are Santa Cruz (SCZ), Santa Rosa (SRO), and San Miguel (SMI), and the three southern islands are Santa Catalina (SCA), San Clemente (SCL), and San Nicolas (SNI). The enlargement of Santa Cruz Island shows the relative placement and size of two trapping grids, Fraser Point and Central Valley, which were used to capture island foxes.

spiritual and personal lives of these island Americans. Fossil evidence dates the arrival of foxes to the northern Channel Islands (Santa Cruz, Santa Rosa, and San Miguel) much earlier, from 10,400 to 16,000 YBP (Orr 1968). Their actual colonization probably occurred between 18,000 and 40,000 years ago, when these northern islands were joined into one large island known as 'Santarosae' (Collins 1982, 1993; Johnson 1983). At its closest, Santarosae was a mere 6 km from the North American continent, having reached its maximum size 18,000–24,000 YBP. It is hypothesized that sometime during this period, mainland gray foxes, the progenitor of the island fox, colonized Santarosae by chance over-water dispersal, by either swimming or by rafting on floating debris (Collins 1982, 1993). As glaciers retreated and sea levels rose, Santarosae was subdivided into separate islands. Santa Cruz Island was formed first, some 11,500 YBP. Sea levels continued to rise separating the remaining land mass once again, approximately 9,500 YBP, to form Santa Rosa and San Miguel Islands. Native Americans then colonized the Channel Islands 9,000–10,000 YBP, and after establishment of an extensive trade route, transported foxes to the southern islands.

The island syndrome

In general, insular populations differ from their mainland counterparts in aspects of form, genetics, demography, and behaviour. They may be dwarfed or gigantic, they typically have lower levels of genetic diversity, occur at higher densities, have decreased dispersal tendencies, and reduced aggression (Stamps and Buechner 1985; Adler and Levins 1994; Burness *et al.* 2001). Island foxes are no exception. A dwarf form of the mainland gray fox, island foxes are the smallest North American canid, varying in body mass from 1.4 to 2.5 kg, roughly two-thirds the size of a mainland grey fox (Moore and Collins 1995; Roemer *et al.* in press). Further, each insular population differs in body mass and morphology (Collins 1982, 1993; Wayne *et al.* 1991; Roemer *et al.* in press). For example, Santa Catalina has the largest island foxes (*U. littoralis catalinae*), San Clemente's foxes (*U. littoralis clementae*) have the smallest craniums and San Miguel foxes (*U. littoralis littoralis*) have the shortest tails, owing to a reduction in the number of caudal vertebrae (Collins 1982; Moore and Collins 1995). Morphology is so distinctive that each fox population can be distinguished from the other solely on osteological traits. Using 29 cranial and mandibular characters measured from 2,207 island and gray fox specimens, Collins (1982, 1993) correctly classified 91% of all island fox specimens to their island of origin.

Insularity has also had a profound influence on genetic diversity and the phylogeography of the island fox. Founder events, genetic drift, and selection have played significant roles reducing phenotypic and genetic variation, and creating six genetically distinct populations. Island foxes contain about 35% of the genetic variation observed in mainland grey foxes, and the fox population on San Nicolas Island is one of the most genetically invariant wild populations known (Table 9.1). Genotype profiles generated from hypervariable minisatellite and microsatellite DNA were identical for all foxes assayed from San Nicolas (Gilbert *et al.* 1990; Goldstein *et al.* 1999). Of the five

Table 9.1 Genetic diversity at 19 microsatellite loci in the island fox and a mainland California population of gray fox

Population	*n*	Alleles/ locus	% Polymorphic	He	HW-He
San Miguel	17.8	1.7	47.4	0.106	0.155
	(0.8)	(0.2)		(0.045)	(0.048)
Santa Rosa	25.4	2.5	57.9	0.198	0.274
	(0.9)	(0.4)		(0.056)	(0.065)
Santa Cruz	22.4	2.3	57.9	0.209	0.284
	(1.1)	(0.3)		(0.047)	(0.058)
San Nicolas	26.4	1.0	0	0	0
	(0.9)	(0.0)		(0.0)	(0.0)
San Clemente	24.9	2.1	52.6	0.228	0.248
	(1.1)	(0.3)		(0.061)	(0.064)
Santa Catalina	25.4	2.5	89.5	0.341	0.405
	(1.1)	(0.2)		(0.046)	(0.043)
Grey fox	11.9	6.3	94.7	0.700	0.752
	(0.6)	(0.5)		(0.045)	(0.047)

Note: The San Nicolas island fox population is genetically invariant at all loci. *n*: mean sample size per locus (±SD); Alleles/locus: mean number of alleles per locus; % Polymorphic: the percentage of loci that had two or more alleles; He: direct count of heterozygosity and HW-He: heterozygosity assuming Hardy–Weinberg equilibrium in allele frequencies.

mitochondrial DNA (mtDNA) haplotypes found in island foxes, none is shared with a nearby mainland sample of gray foxes and all island fox populations share a unique restriction enzyme site, a synapomorphy that clusters the six populations into a single monophyletic clade (Wayne *et al.* 1991). Each island fox population contains population-specific restriction-fragment profiles (Gilbert *et al.* 1990). Genotypes generated from 19 microsatellite loci were used to correctly classify 181 out of 183 island/gray fox samples to their population of origin (Goldstein *et al.* 1999). The three northern island populations and the three southern island populations consistently cluster into two groups, and the Santa Rosa and San Miguel fox populations are more closely related than either is to Santa Cruz (Gilbert *et al.* 1990; Wayne *et al.* 1991; Goldstein *et al.* 1999). This phylogeographic structure supports the view that colonization followed by vicariant events created the northern island fox populations and that human-assisted dispersal aided the colonization of the southern islands. These morphologic and genetic differences also clearly justify the taxonomic classification of the island fox as a separate species (Wilson and Reeder 1993) and support the individual subspecific classifications of the six island fox populations (Hall 1981; Moore and Collins 1995).

Higher densities are predicted to be characteristic of island vertebrates (Adler and Levins 1994) and island foxes have some of the highest population densities of any canid. On Santa Cruz, densities from 7 to 8.1 foxes/km^2 have been recorded (Laughrin 1977; Roemer *et al.* 1994) and on San Clemente densities at three sites varied from 4.8 to 8 foxes/km^2 over a 10-year period (Roemer *et al.* 1994; Garcelon 1999; Roemer 1999). On San Miguel and San Nicolas Islands, the two smallest islands that harbour foxes, densities have varied from near zero to 16 foxes/km^2 (Coonan *et al.* 2000; Roemer 2000). Densities of mainland gray fox populations are typically much lower, averaging 1.2 to 2.1 foxes/km^2 across a range of studies (Fritzell and Haroldson 1982). Some variation in apparent density between species may arise from methodology. For example, on Santa Cruz Island, the density of island foxes in 1993 determined with a capture–recapture approach was 7.0 foxes/km^2 whereas density for the same population determined via home range size was approximately 35% lower or 4.5 foxes/km^2.

Regardless of methodology, densities of island foxes are high compared to mainland foxes and are probably a result of the small home ranges of island foxes. In mixed habitat on Santa Cruz Island, fox home ranges averaged between 0.25 and 0.33 km^2 (n = 12) (Crooks and Van Vuren 1996) and in coastal grassland seasonal home range size varied from 0.15 to 0.87 km^2 (n = 42), with a mean annual home range size of 0.55 km^2 (n = 14) (Fig. 9.2(a)—Roemer *et al.* 2001b). On San Clemente Island, home ranges are larger (mean = 0.77 km^2, n = 11) perhaps due to the lower productivity of this more southerly island (Thompson *et al.* 1998). On San Miguel, average home range size of five yearlings was 2.26 km^2 (range = 1.72–2.91 km^2) during a period of low density (T. Coonan personal communication) and on Santa Cruz fox home ranges expanded (range = 16–266%, n = 5) as territorial neighbours were killed by golden eagles, suggesting that density of foxes and the spatial distribution of neighbours may influence territory size (Roemer 1999; Roemer *et al.* 2001b). The small home range size observed in island foxes is related to their more insectivorous diet and to the high resource density common to insular ecosystems (Macdonald 1983; Stamps and Buechner 1985; Roemer *et al.* 2001b).

Insular species are predicted to have reduced aggression and a reduction in territoriality because of the increased costs of territory maintenance at high population densities (Stamps and Buechner 1985). These predictions stem from the difference in ecological conditions between insular and mainland systems. Insular systems typically have higher resource densities and lower levels of interspecific competition owing to a depauperate fauna. These conditions cause higher densities of both territory holders and non-territorial floaters. An increase in the number of floaters leads to increased defence costs for territory holders and ultimately to a reduction in territorial behaviour. This reduction in territoriality may be manifested as: (1) reduced territory size; (2) increased territory overlap; (3) acceptance of subordinates within the territory; and (4) reduced intraspecific aggression (Stamps and Buechner 1985). Contrary to these predictions, island foxes are distributed as socially monogamous pairs that defend discrete territories (Crooks and Van Vuren 1996; Roemer *et al.* 2001b). On Santa Cruz Island, home ranges of mated pairs overlapped (mean = 85%, SD = 0.05%) significantly more than those of neighbours (mean = 11%, SD = 0.13%) (Fig. 9.2(a)—Roemer *et al.* 2001b).

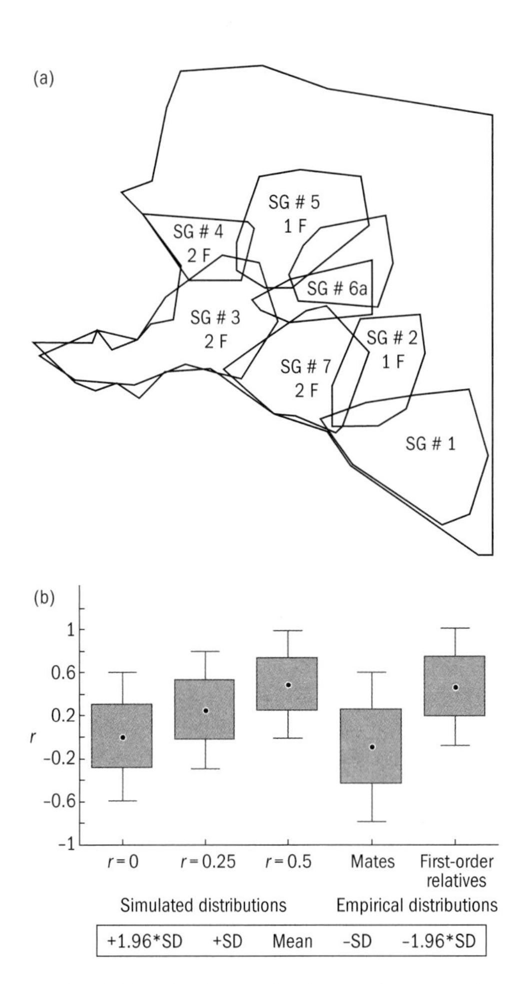

Figure 9.2 (a) The distribution of male, island fox territories at Fraser Point, Santa Cruz Island from April to August 1994. The letter and number (e.g. 1F) represent the number and sex of pups born in that season. (b) Empirical distributions (mean $\pm$ 1 SD and 1.96 SD) of the coefficient of relatedness (*r*) for first-order relatives and for mated pairs of island foxes. Social relationships among the foxes were determined from observations of spatial distribution or by paternity analysis using microsatellite loci. Simulated distributions of *r* for first-order relatives ($r = 0.5$), for distantly related relatives ($r = 0.25$), and for unrelated individuals ($r = 0$) are shown for comparison. (Modified from Roemer *et al.* 2001b).

Island foxes might also be predicted to be monogamous given their territorial nature and the general trend towards monogamy in the Canidae that increases with decreasing body size (Geffen *et al.* 1996; but see Sillero-Zubiri *et al.* 1996a). However, EPFs are not uncommon in island foxes, 4 (25%) of 16 pups whose parents were identified through paternity analysis were a result of EPFs (Roemer *et al.* 2001b). Three of the four EPFs were by the two largest male foxes in the study and bite wounds were observed on males only during the breeding season (Roemer 1999). These observations suggest that male–male competition for mates may be intense. Further, nearly all pairings occurred between unrelated male and female foxes, suggesting that foxes avoid inbreeding and practice mate choice (Fig. 9.2(b)—Roemer *et al.* 2001b).

Like other canids, island foxes display a relatively high degree of bi-parental care. Dependent young accompany parents on forays and have been observed foraging with their parents for insects in grassland, and for striped-shore crabs (*Pachygrapsus crassipes*) in intertidal habitats (G. Roemer personal observation). Additionally, both parents have been observed transferring artificially provided food to dependent pups, and vertebrate prey have been found outside of traps containing captured pups (Garcelon *et al.* 1999). It is also not uncommon for full-grown young to remain within their natal range into their second year, or for offspring to associate with their parents after gaining independence (Roemer 1999). Although helping by independent young may occur, it has not been observed. The average number of adult-sized foxes occupying a single territory on Santa Cruz Island was 3.38 (SD = 1.12, $n = 13$) (Roemer *et al.* 2001b).

Finally, because of small size and a finite border, dispersal opportunities on small islands are more limited than on the mainland. This reduction in dispersal distance is predicted to increase the viscosity of gene flow, creating greater population substructure compared with mainland populations (Roemer *et al.* 2001b). Similar to insular rodent populations (Sullivan 1977; Tamarin 1977), island foxes disperse less frequently and over shorter distances than mainland canids of similar size. On Santa Cruz, five juveniles (three males and two females) dispersed to areas within the study site (mean = 0.99 km, SD = 0.61) and one male dispersed 3.58 km from the study site. Thus, only one (17%) of six juvenile island foxes successfully dispersed from its natal area, moving a distance greater than two average home range diameters (average home range diameter = 0.84 km) (Roemer *et al.* 2001b). Tullar and Berchielli (1982, cf. Fritzell 1987) found that 63% of juvenile female and 73% of juvenile male gray foxes left their natal area. Dispersal distances of male and female island foxes were limited compared with those of mainland fox species. Eight island foxes on Santa Cruz, including six juveniles and two young adults that were probably subordinate offspring, dispersed an average of 1.39 km (SD = 1.26) with the longest dispersal distance recorded being less than 4 km. In Alabama, Nicholson *et al.* (1985) recorded a mean dispersal distance of 15 km (SD = 9.5) for three male gray foxes. The kit fox, only slightly larger than the island fox, has an average dispersal distance of 11.1 km (n = 47, range 1.7–31.5) (O'Farrell 1984). The longest recorded dispersal distance for a gray fox is 84 km (Sheldon 1953) and for the kit fox is 64 km (O'Neal 1985). The maximum beeline distance over which an island fox could disperse is 38 km, the total length of Santa Cruz Island. In addition, significant genetic subdivision was observed on Santa Cruz between two sampling sites separated by only 13 km, suggesting restricted gene flow (Nm = 1.6–2.5—Roemer *et al.* 2001b). Similar values of Nm are found between populations of mainland canids separated by several hundred kilometres (Mercure *et al.* 1993; Roy *et al.* 1994b).

Historic demography and the decline of the island fox

Prior to Laughrin's (1977, 1980) work in the early and mid-1970s, there had been no systematic attempts to quantify the abundance of island foxes. Laughrin (1977, 1980) estimated the density of foxes on all six islands by live trapping along road transects. Most populations were at moderately high density (1.2–4.3 foxes/km^2), except for San Nicolas (0.1–2.7 foxes/km^2) and Santa Catalina (0–0.8 foxes/km^2) that were at relatively lower densities (Laughrin 1980). Santa Catalina had apparently been at a low density throughout the 1970s whereas the San Nicolas population had declined between 1971 and 1977. Although feral cats were known competitors and suspected as a potential agent involved in the declines, the causes of the low numbers on Santa Catalina, or of the apparent decline on San Nicolas were unknown (Laughrin 1980).

In the 1980s, Kovach and Dow (1981, 1985) employed the first use of trapping grids to estimate island fox population density and size. Using a series of 12 small, trapping grids (30 traps), they trapped approximately 37% of San Nicolas Island. In 1981, density estimates varied from zero to 6.9 foxes/km^2 with an estimated population size of 110 foxes (Kovach and Dow 1981). In 1985, the estimate of population size increased to 520 foxes (Kovach and Dow 1985). These data implied that the fox population on San Nicolas had recovered from apparently low numbers in the 1970s.

A capture–recapture design incorporating large trapping grids (48–80 traps—see Fig. 9.1) has been used on San Clemente (1988–97 and 1999–2002), Santa Catalina (1989 and 1990), Santa Cruz (1993–99), San Miguel (1993–99), and San Nicolas Islands (2000–02) (Roemer *et al.* 1994, 2001a, 2002; Garcelon 1999; Roemer 1999, 2000; Coonan *et al.* 2000). Between the late 1980s and early 1990s, data from the capture–recapture studies coupled with anecdotal observations by island residents suggested that island fox populations on all the Channel Islands were at relatively high density. Depending on grid (habitat) and island, densities varied from 2.4 to 15.9 foxes/km^2 (Roemer *et al.* 1994; Coonan *et al.* 2000; Coonan 2003). Estimates of population size varied from approximately 350 adult foxes on San Miguel, the smallest island, to greater than 1300 adult foxes on Santa Cruz, the largest island (Roemer *et al.* 1994, 2001a). It was estimated that there were approximately 6400 adult foxes distributed among the six island populations (Roemer *et al.* 1994).

In the mid- to late-1990s, fox populations on the three northern islands underwent drastic population declines (Roemer 1999; Coonan *et al.* 2000; Roemer *et al.* 2001a). By 1998, mean fox density on San Miguel and Santa Cruz Islands had dropped to 0.8 ($\pm$1.0) foxes/km^2, capture success had decreased six-fold, from 25.7% (1993) to 4.3% (1998), and population size on both islands had plummeted. Only 15 adults were known to be alive on San Miguel in 1999 with an estimated 133 foxes remaining on Santa Cruz. The San Miguel and Santa Cruz fox populations were estimated to have a 50% probability of persistence within the next decade (Roemer *et al.* 2001a). Capture success in 1998 on nearby Santa Rosa Island was also low (4.8%) suggesting that fox

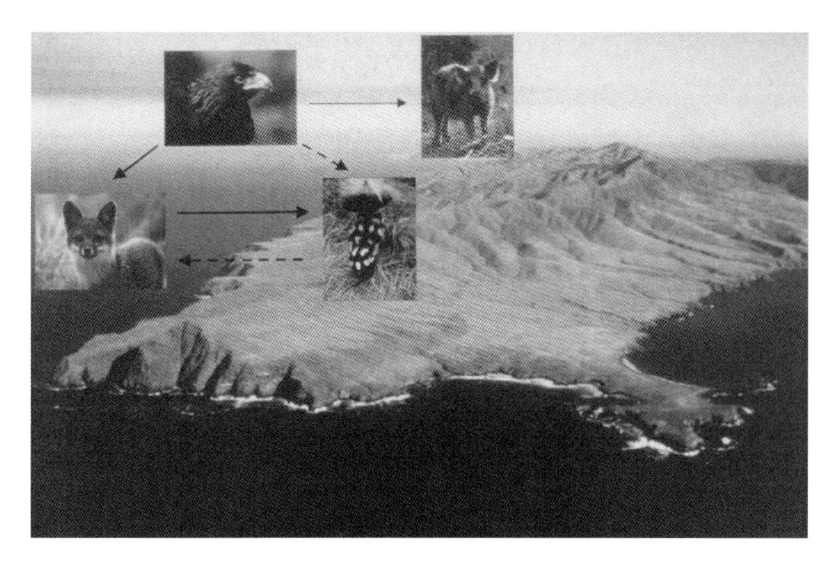

Figure 9.3 The introduction of feral pigs to Santa Cruz represented an abundant food source that enabled golden eagles to colonize the island. Golden eagles preyed on pigs, foxes and skunks, but predation pressure was greatest on the unwary fox. As foxes were driven to extinction, skunks were released from fox competition, and their numbers increased © G. Roemer.

populations had declined on all three northern Channel Islands. These data showed that the three subspecies on the northern Channel Islands were critically endangered and in need of immediate conservation action (Mace and Lande 1991; Coonan 2003; Roemer *et al.* 2001a, 2002).

Disease was initially suspected as a contributory agent, but further investigation proved that the most important proximate driver of the fox population declines was the presence of an exotic species, the feral pig (*Sus scrofa*) (Roemer *et al.* 2000a, 2001a, 2002). Pigs, by acting as an abundant food, enabled mainland golden eagles to colonize the northern Channel Islands and through hyperpredation caused the decline in the fox populations. Hyperpredation is a form of apparent competition whereby an introduced prey, well adapted to high predation pressure, indirectly facilitates the extinction of an indigenous prey by enabling a shared predator to increase in population size (Holt 1977; Courchamp *et al.* 1999). Pigs, by producing large numbers of piglets, sustained the eagle population and because of their high fecundity could cope with the increased levels of predation. In addition, as piglets mature, they eventually escape predation by growing beyond the size range that eagles typically prey upon (Roemer *et al.* 2002). Foxes, on the other hand, are small, active during the day, and produce relatively few young each year. Thus, predation by eagles had an asymmetrical effect on the more vulnerable fox, driving the fox populations toward extinction (Fig. 9.3).

The presence of pigs had further ramifications causing a wholesale reorganization of the island food web. Historically, island foxes were the largest terrestrial carnivores and were competitively dominant to the island spotted skunk (*Spilogale gracilis amphiala*) (Crooks and Van Vuren 1996; Roemer *et al.* 2002). Three to four times larger than an average skunk, a fox consumes nearly three times as much insects and small rodents. Prior to the arrival of eagles, capture success of foxes (28.3%, SD = 8.7%) was 35 times higher than that of skunks (0.8%, SD = 1.0%) on Santa Cruz Island (Roemer *et al.* 2002b). However, once fox populations declined, skunks were released from competition with foxes and subsequently increased. By 1999, skunk capture success had increased 17-fold to 13.9% (SD = 8.5%), and fox capture success had dropped to an all-time low of 4.3% (SD = 1.9%). These community-level dynamics were predicted by a mechanistic model parameterized with independent data sets, which also confirmed that pigs were the indirect driver of this food web transformation (Fig. 9.4). In this case, the presence of an exotic prey indirectly caused apparent competition to replace resource competition as the primary biotic force structuring this carnivore community (Roemer *et al.* 2002b).

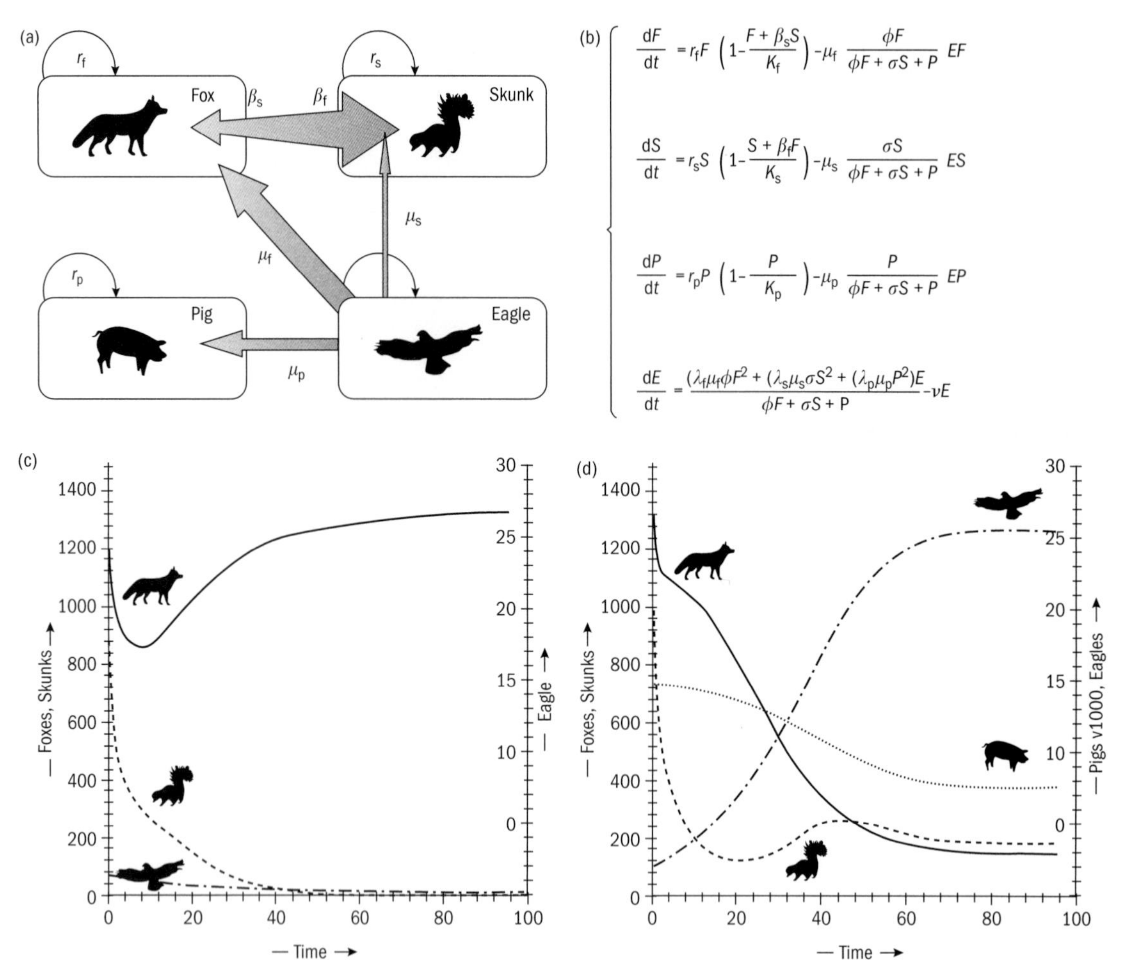

$$\frac{dF}{dt} = r_f F\left(1-\frac{F+\beta_s S}{K_f}\right) - \mu_f \frac{\phi F}{\phi F + \sigma S + P} EF$$

$$\frac{dS}{dt} = r_s S\left(1-\frac{S+\beta_f F}{K_s}\right) - \mu_s \frac{\sigma S}{\phi F + \sigma S + P} ES$$

$$\frac{dP}{dt} = r_p P\left(1-\frac{P}{K_p}\right) - \mu_p \frac{P}{\phi F + \sigma S + P} EP$$

$$\frac{dE}{dt} = \frac{(\lambda_f \mu_f \phi F^2 + (\lambda_s \mu_s \sigma S^2 + (\lambda_p \mu_p P^2)E}{\phi F + \sigma S + P} - \nu E$$

Figure 9.4 (a) Schematic representation, (b) corresponding set of equations, and (c–d) demographic relationships among island foxes, island spotted skunks, feral pigs, and golden eagles on Santa Cruz Island. (c) Without pigs, eagles are unable to colonize the island because of insufficient food. Foxes out-compete skunks and are the dominant terrestrial predator. (d) With pigs, eagles colonize the island and through hyperpredation drive the foxes toward extirpation. The decline in foxes releases skunks from fox competition. (From Roemer *et al.* 2002.)

Although pigs were unequivocally linked to the decline in foxes on the northern islands, it was further hypothesized that the ultimate cause of this interaction was a result of historic, human-induced perturbations to the islands, to the mainland, and to the surrounding marine environments (Roemer *et al.* 2001a). European agricultural practices together with overgrazing by introduced herbivores reduced vegetative cover and probably increased the vulnerability of foxes to a diurnal, avian predator. Environmental degradation of the marine environment then led to the extirpation of the bald eagle (*Haliaeetus leucocephalus*) from the Channel Islands by 1960 (Kiff 1980). Bald eagles are primarily piscivorous, forage over marine habitats, and are not a significant predator of the island fox. However, bald eagles are territorial, aggressive towards conspecifics and other raptors, and may have competed with golden eagles for nest sites (Roemer *et al.* 2001a). The extirpation of the bald eagle probably paved the way for colonization of the islands by golden eagles. Finally, increased urbanization along the southern California coast reduced golden eagle habitat possibly displacing them to new hunting grounds on the islands. This series of complex interactions may have ultimately allowed golden eagles to colonize the islands and drive island foxes toward extinction.

As the declines on the northern islands were in full swing, another catastrophe occurred when canine distemper virus was introduced to Santa Catalina Island between late 1998 and to mid-1999 (Timm *et al.* 2000). This epizootic caused an estimated 95% reduction in the fox population on the eastern 87% of Catalina Island (Timm *et al.* 2000). Luckily a suspected barrier to fox dispersal, and hence to the spread of the disease, was in place on the western end of the island in the form of the town of Two Harbors. The fox population on the remaining 13% of the island west of Two Harbours appeared to be unaffected with 49 individual foxes being captured in 137 trap nights (36% capture success) (Timm *et al.* 2000). Domestic dogs are hypothesized to have introduced the virus but the actual agent is unknown.

The fox population on San Clemente Island was thought to be in gradual decline since the early 1990s, with an estimated 20% chance of extinction in the next 100 years (Garcelon 1999; Roemer 1999; Roemer *et al.* 2001a). However, recent demographic modelling suggests that the population may be declining at a much higher rate (Roemer *et al.* 2000b). Deterministic estimates of intrinsic growth rate ($\lambda = 0.956$) imply a 4.4% decline in annual population size, or a decline of nearly 50% from 1988 ($N = 850$ foxes) to 2002 ($N = 457$ foxes).

This decline may have been exacerbated by an interesting endangered species conflict that is occurring on San Clemente Island (Roemer and Wayne 2003). Over the past decade, a monumental effort has been undertaken to prevent the extinction of the critically endangered San Clemente loggerhead shrike (*Lanius ludovicianus mearnsi*—Juola *et al.* 1997; USDA 1998). Erroneously classified as a separate subspecies in the 1930s (Miller 1931) based on a flawed method of systematic classification (Collister and Wicklum 1996), the San Clemente loggerhead shrike also has equivocal genetic distinction (Mundy *et al.* 1997a,b). As part of efforts to bring this 'subspecies' back from the brink of extinction, predator control measures have been instituted that include lethal removal, permanent removal to zoological institutions, and temporary containment of island foxes, an identified nest predator (Garcelon 1996; Cooper *et al.* 2001). Ten to thirty per cent of the fox population is placed in temporary confinement each year and/or repeatedly trapped on consecutive nights during the reproductive season (Cooper *et al.* 2001). Pregnant and lactating females are contained resulting in pups being born in captivity as well as dependent pups in the wild probably dying because of removal or confinement of their mother (Cooper *et al.* 2001). For example, in 1999, 49 foxes were held temporarily in small pens (~0.55 m^2) during the fox reproductive season. Of the foxes held captive, 20 females were suspected of having dependent pups still in the wild (Cooper *et al.* 2001). Adult female San Clemente island foxes wean an average of 1.25 (SE = 0.015) pups per reproductive event (Roemer 1999). Thus, an estimated 25 pups would be expected to have starved to death as a result of confinement of their mothers. The impact of these measures on the fox population has not been critically evaluated. Within the last decade, four island fox populations have experienced dramatic population declines and a fifth population is in significant decline. Only the San Nicolas fox population, the most genetically invariant population of all, is currently at high density (Roemer 2000).

Conservation of the island fox

Channel Islands National Park (CINP) has established several recovery actions to prevent the extinction of the northern Channel Island fox populations (Coonan 2001). In 1998, CINP established an Island Fox Conservation Working Group, a team of experts whose expertise was used to guide recovery actions on the northern Channel Islands (Coonan 2003). Following the recommendation of the Working Group, CINP contracted the Santa Cruz Predatory Bird Research Group to begin live-capture and removal of golden eagles. From November 1999 to June 2002, a total of 22 golden eagles, 20 adult or subadult golden eagles and 2 chicks, were live-captured and removed from Santa Cruz Island (Coonan 2003; B. Latta personal communication). In 1999, an island fox captive-breeding facility was

Figure 9.5 An inquistive Santa Cruz island fox (*Urocyon littoralis santacruzae*). This picture was taken by the author at a distance of ~3 m © G. Roemer.

established on San Miguel and a second facility was added on Santa Rosa in 2000. In June 2002, there were 28 foxes in captivity on San Miguel and 45 on Santa Rosa. Recently, each of these facilities has been divided into two separate facilities to safeguard against disease. Genetic information is being used to establish mated pairs (Gray *et al.* 2001) and the health of the captive populations is being monitored by veterinary examination and disease and parasite surveys (L. Munson and M. Willet personal communication). Releases are being planned for Santa Rosa Island in 2003, and demographic modelling is being used to guide future recovery efforts (Roemer *et al.* 2000b; Coonan 2003).

Monitoring of the Santa Cruz fox population is continuing (Fig. 9.5). The most recent estimate of the number of foxes on Santa Cruz is below 100. Eighty-two foxes were captured in an island-wide trapping effort, with six additional deaths of foxes owing to predation by golden eagles occurring in 2000–01 (Dennis *et al.* 2001). In cooperation with The Nature Conservancy, CINP has established a third captive breeding facility on Santa Cruz. Following recommended actions from the Island Fox Conservation Working Group, a total of 12 bald eagles have been released on Santa Cruz Island with another 12 birds planned for release over the next 4 years (60 bald eagles total) (Vallopi *et al.* 2000; Coonan 2002, 2003). A feral pig eradication effort is being planned and should be underway by 2003 (Coonan 2001).

The Institute for Wildlife Studies (IWS), funded by the Catalina Island Conservancy, has initiated a wild fox vaccination programme against the canine distemper virus on Santa Catalina, and is continuing monitoring efforts. It has also established a captive breeding facility (Timm *et al.* 2000). Unfortunately, during the construction of the captive facility, a decision was made to bring pregnant foxes into captivity in late spring 2001. Females gave birth in the facility and 12 of 18 pups subsequently died, owing to apparent stress-related abandonment by the females (Timm 2001). Currently there are 11 pairs of foxes in the facility that are adjusting well and captive reared foxes have been released in late 2001 (J. Floberg personal communication).

On San Clemente, the IWS, funded by the US Navy and in compliance with actions enforced by the US Fish and Wildlife Service, is continuing containment efforts of the island fox to protect the endangered San Clemente loggerhead shrike (Cooper *et al.* 2001). The fox population is declining. In 2000, monitoring of the San Nicolas island fox population was resumed (Roemer 2000) and is being continued by the IWS (G. Smith personal communication).

In response to the declines in fox populations, the IWS along with the Center for Biological Diversity petitioned the US Fish and Wildlife Service to list the three northern island fox populations and the Santa Catalina island fox population as endangered. Curiously, both the San Clemente and San Nicolas island fox populations, the only populations managed jointly by the US Navy and IWS, were excluded from this petition. The four populations are now being considered for listing as endangered under the Endangered Species Act in a proposed rule by the US Fish and Wildlife Service (USDI 2001).

The IUCN—Canid Specialist Group has undertaken an independent assessment of the status of the island fox (Roemer *et al.* in press). Using the IUCN Red List categories (IUCN 2001), the subspecies on

San Miguel and Santa Rosa Islands, *U. littoralis littoralis* and *U.littoralis santarosae*, are recommended listed as Extinct in the Wild. The subspecies on Santa Cruz and Santa Catalina Islands, *U. littoralis santacruzae* and *U. littoralis catalinae*, are recommended listed as Critically Endangered and the subspecies on San Clemente Island, *U. littoralis clementae*, is recommended listed as Endangered. The subspecies on San Nicolas Island, *U. littoralis dickeyi*, is recommended listed as Vulnerable (Roemer *et al.* in press). Further, it has been recommended that the entire species, not just four of the six subspecies, receive protection under the Endangered Species Act (Roemer 1999). The objective guidelines provided by the IUCN and used by Roemer *et al.* (in press) support this view.

Island foxes are located on islands that are managed by public or private resource agencies that are governed by state and federal laws that mandate the preservation of wildlife. Given that the island fox is unique in terms of its biological, scientific, and cultural qualities, the protection of this critically endangered canid is clearly a high priority.

Acknowledgements

Over the years many folks have assisted in the study of island foxes. To those who have ventured before me, to my colleagues, research assistants, volunteers, friends, and especially to those who still stand upon our shoulders—I thank you for your past, present and future hard work. Also, I would like to thank my Dad, it was through him that I was first introduced to all things natural, and to all island foxes, it was through them that I have realized how important all things natural truly are.

CHAPTER 10

Swift and kit foxes

Comparative ecology and conservation priorities of swift and kit foxes

Axel Moehrenschlager, Brian L. Cypher, Katherine Ralls, Rurik List, and Marsha A. Sovada

Kit fox *Vulpes macrotis* pups © C. Wrigert.

Leading causes for the endangerment of many canid populations are habitat loss, persecution and, in some circumstances, intraguild killing by other carnivores. Kit foxes (*Vulpes macrotis*) and swift foxes (*V. velox*) are affected by all of these factors as North America's grasslands and deserts are increasingly threatened, foxes are occasionally trapped or poisoned, and competition or predation by coyotes (*Canis latrans*) and red foxes (*Vulpes vulpes*) impacts their populations. Consequently, swift and kit fox populations are nationally or regionally threatened from Canada to Mexico.

Kit and swift foxes, which are currently classified as separate species, exhibit morphological and genetic differences. Traditionally, separate research priorities and action plans would be designated, and we questioned whether species differences would warrant this separation. Our objectives were to: (1) compare life history and ecological parameters of kit and swift foxes; (2) compare ecological factors or anthropogenic threats that impact species abundance and/or distribution; (3) recommend species-specific, topic-specific, or area-specific research priorities for kit and swift foxes. While variation in diet, home range use, den use, dispersal, and survival was high in space and time within species, we could not discern substantial differences in these parameters, or in population threats, between the species. Current data gaps illustrate that swift and kit fox conservation would benefit from increased research on: (1) defining habitat disturbance thresholds that lead to swift/kit fox exclusion; (2) understanding processes that determine the relative abundance of swift/kit foxes, coyotes, and red foxes in areas of sympatry; (3) population genetics; and (4) disease threats. Due to similarities in the ecology and population threats of swift and kit foxes, we propose that collaborative research and concerted action planning will maximize the efficiency of financial and political resources to develop applied conservation solutions for both species.

Introduction

Swift foxes and kit foxes occur in relatively flat, arid regions of North America. Historically, swift foxes occupied the mixed-grass and short-grass prairies from central Alberta to central Texas and eastern North Dakota to central Colorado (Fig. 10.1; Allardyce and Sovada 2003). Swift foxes are separated from kit foxes by the Rocky Mountains, but interbreeding does occur within a limited hybridization zone in New Mexico (Rohwer and Kilgore 1973; Mercure *et al.* 1993). Kit foxes range from southern Idaho and Oregon in the United States to northern Zacatecas and Nuevo León in northern Mexico (Fig. 10.1; McGrew 1979; Hall 1981; Moehrenschlager and List 1996).

Swift and kit foxes are phenotypically similar although kit foxes have slightly longer, less rounded ears (Dragoo *et al.* 1990) and weigh less than do swift foxes (Moehrenschlager *et al.* submitted). Although early morphometric comparisons and protein-electrophoresis suggested that these foxes constitute the same species (Ewer 1973; Clutton-Brock *et al.* 1976; Hall 1981; Dragoo *et al.* 1990; Wilson and Reeder 1993), more recent multivariate morphometric approaches (Stromberg and Boyce 1986) as well as mitochondrial DNA (mtDNA) restriction-site and sequence analyses (Mercure *et al.* 1993) have concluded that they are separate species. The San Joaquin kit fox (*V. macrotis mutica*) in southern California is topographically isolated from the main kit fox continuum and genetically distinct from other kit foxes

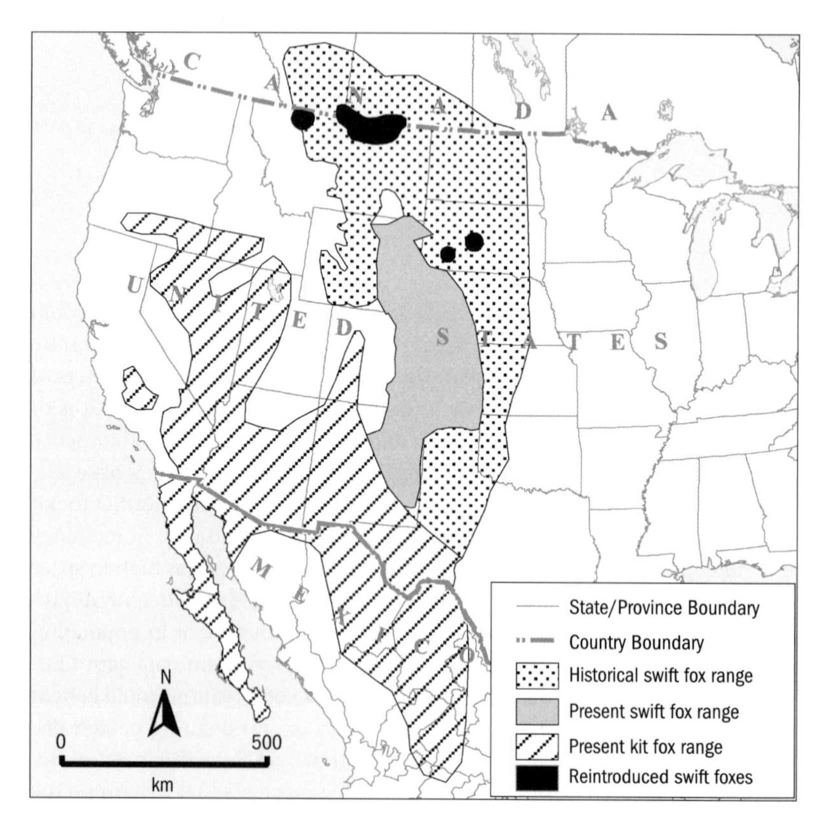

Figure 10.1 Distribution of swift and kit foxes in North America.

(Mercure *et al.* 1993). Kit foxes in northern Coahuila, Mexico, (*V. macrotis zinseri*) are also genetically unique, although they are closely allied with populations of kit foxes west of the Rocky Mountains in the United States (Maldonado *et al.* 1997). Subspecies classifications for swift foxes are likely unwarranted (Stromberg and Boyce 1986; Mercure *et al.* 1993). Whether one regards them as separate species or well-defined subspecies, it is clear that kit and swift foxes exhibit morphological and genetic differences and tend to inhabit different habitats.

Swift foxes are imperilled throughout their range and kit fox populations in California and Mexico are endangered. The swift fox was extirpated from Canada from the late 1930s until a reintroduction programme was initiated in 1983. Subsequently the population has grown (Moehrenschlager and Moehrenschlager 1999, 2001; Moehrenschlager *et al.* 2002) and, although the species is no longer considered extirpated, it is federally 'endangered' (COSEWIC 1998). Current estimates for the United States suggest that swift foxes are located in 39–42% of their historic range depending on how that is delineated (Sovada and Scheick 1999—nevertheless, a petition to declare swift foxes federally endangered (U.S. Fish and Wildlife Service 2001) was deemed unwarranted and an alternative threat designation has not been assigned. San Joaquin kit foxes are federally endangered in the United States and the kit fox is deemed vulnerable in Mexico.

To aid swift and kit fox conservation, our objectives are to: (1) compare life history and ecological parameters of swift and kit foxes; (2) compare ecological and anthropogenic factors that impact the abundance and/or distribution of these species; (3) recommend future species-, topic-, or area-specific research priorities.

Life history and ecology comparisons

Diet

Kit foxes are opportunistic foragers. Kit foxes consume a wide variety of prey items, although rodents and leporids usually constitute the bulk of the diet. In many locations, kangaroo rats (*Dipodomys* spp.) are preferred food (Fisher 1981; Cypher *et al.* 2000; Koopman *et al.* 2000). In other locations, ground squirrels (*Spermophilus* spp.), prairie dogs (*Cynomys* spp.), black-tailed jackrabbits (*Lepus californicus*), and cottontails (*Sylvilagus* spp.) are important prey (Table 10.1). Other frequently consumed items include pocket mice (*Perognathus* spp. and *Chaetodipus* spp.), deer mice (*Peromyscus maniculatus*), birds, and various insects.

Like kit foxes, swift foxes are opportunistic foragers, feeding primarily on a variety of mammals, but also birds, insects, plants, and carrion (Cutter 1958a; Kilgore 1969; Hines 1980; Uresk and Sharps 1981; Cameron 1984; Hines and Case 1991; Zimmerman 1998; Kitchen *et al.* 1999; Moehrenschlager *et al.* submitted; Sovada *et al.* 2001). Leporids have been reported as a primary prey item in several studies (Cutter 1958a; Kilgore 1969 [winter]; Cameron 1984; Zumbaugh *et al.* 1985). In South Dakota, mammals accounted for 49% of prey occurrences with black-tailed prairie dogs (*Cynomys ludovicianus*) as the primary prey item (Uresk and Sharps 1986). Sovada *et al.* (2001) in Kansas, and Hines and Case (1991) in Nebraska, found that murid rodents were the most frequently occurring prey of swift foxes. Several studies have reported a high frequency of insects, but insects likely constituted a small portion of biomass (Kilgore 1969). Birds and bird eggs have been identified as a food for swift foxes (Cutter 1958a; Kilgore 1969; Uresk and Sharps 1981; Moehrenschlager *et al.* submitted; Sovada *et al.* 2001). Prey species and their relative consumption may differ between seasons and areas, but prey classes are generally similar for swift and kit foxes.

Home range

Home range size of kit foxes ranges from 2 to 12 km^2 and swift fox home ranges vary from 8 to 43 km^2 (Table 10.1). Since swift fox home range sizes in Nebraska (Hines and Case 1991), Kansas (Sovada *et al.* 2001), and Alberta/Saskatchewan (Moehrenschlager *et al.* submitted) exceed those of kit foxes, this could suggest that swift foxes generally have larger home range sizes. However, differences in sample size, calculation method, and tracking duration blur such comparisons. Home range sizes do not only differ between areas or species but, indeed, also within areas over time. For example, adaptive kernel home ranges estimated for swift foxes on the Piñon

Table 10.1 Ecological and life-history parameters of swift and kit foxes

Species	Area	Yrs. data collected	Foxes radio-collared	#dens used/year	Diet	% Females with pups	Litter size
V. m. nevadensis	Toole Co., Utah	1951–80	Approx. 90 tagged[b,c]	–	Leporids, kangaroo rats, deer mice, birds[b]	–	Annual avg. 2.8–5.0[a,c]
V. m. nevadensis	Desert Experimental Range, Utah	1983	38	9.5, SD = 4.0	Kangaroo rats and pocket mice 34%, jackrabbits 20%, Jerusalem crickets 18%, birds 14%	100%	4.6
V. m. mutica	Buena Vista Valley, CA	1970–71	14	–	Kangaroo rats, cottontails, birds, insects	–	4
V. m. mutica	Camp Roberts, CA	1988–91	94	11.4–15.5[e]	Diurnal ground squirrels[a]	Adults 32%, juveniles 0%[d]	3.0 ± 0.28[d]
V. m. mutica	Carrizo Plain, CA	1989–91	38	~25[a]	Small rodents 40%, insects 35%, birds 9%[b]	0–57%[c]	–
V. m. mutica	Naval Petroleum Reserves, CA	1980–95	341 adults; 184 juveniles[a]	11.8 ± 0.5[b]	Leporids 44%, kangaroo rats 31%, leporids predominated until 1984, kangaroo rats after 1991[d]	Adults 61.1% ± 0.1; yearlings 18.2% ± 0.1[a]	Adults 3.8 ± 0.1; yearlings 2.5 ± 0.6[a]
V. m. mutica	Western Kern Co., CA	1989–93	103	–	Leporids 14%, kangaroo rats 62%, other rodents 18%, insects 22%	Adults 64%; yearlings 39%	Adults 3.7 ± 1.1 (SD); yearlings 4.5 ± 1.3 (SD)
V. m. neomexicana	Janos, Chihuahua	1994–96	7 : 4[c]	7–13[b]	Rodents 57.2% (prairie dogs 17.8%, kangaroo rats 14.2%, ground squirrels 10.8%, mice 14.4%), insects 18%, lagomorphs 12.2%[a]	75%[b]	3 ± 0.8[b]
V. velox	Beaver Co., OK	1965–66	0	–	Leporids, rodents, birds, insects	–	4.25, $n = 4$

Annual survival	Causes of death	Home range size	Genetic relationships	Dispersal	Disease	Source
–	Vehicles, predators, shooting[b]	–	–	–	Fleas, ticks[a,b]	a. Egoscue (1956) b. Egoscue (1962) c. Egoscue (1975)
–	–	3.1 km^2	–	Juvenile 24–64 km	Fleas, ticks, lice, tapeworms, nematodes	O'Neal *et al.* (1987)
–	Shooting, pup starvation, vehicles	2.6–5.2 km^2	–	–	Fleas, ticks	Morrell (1972)
Adults 0.53, juveniles 0.2[c]	Larger canids, mainly coyotes (75%)[c]	–	–	–	Rabies[b] Fleas[f] Parvovirus, distemper, toxoplamosis, leptospirosis, canine hepatitis[g]	a. Logan *et al.* (1992) b. White *et al.* (2000) c. Standley *et al.* (1992) d. Spencer *et al.* (1992) e. Reese *et al.* (1992) f. Spencer and Egoscue (1992) g. Standley and McCue (1997)
Adults 0.58–0.61; juveniles 0.21–0.41[d]	Larger canids, mainly coyotes (78%)[d]	11.6 ± 0.09 km[c] MCP	Pair members unrelated; female neighbours often related[e]	–	Fleas[a]	a. Ralls and White, unpublished b. White *et al.* (1995) c. White and Ralls (1993) d. Ralls and White (1995) e. Ralls *et al.* (2001)
Adults 0.44 ± 0.05; juveniles 0.14 ± 0.03[a]	Mainly predators, primarily coyotes; vehicles[a]	4.6 ± 0.4 km^2 MCP[e]	–	Juveniles 5.0 ± 0.9 miles; adults 3.0 ± 0.5 miles[c]; 0–52% juveniles dispersed, more ♂♂ than ♀♀[d]	Fleas[f] Parvovirus, canine hepatitis, distemper, tularemia, brucellosis, toxoplasmosis coccidioidomycosis[g]	a. Cypher *et al.* (2000) b. Koopman *et al.* (1998) c. Scrivner *et al.* (1987) d. List *et al.* (2003) e. Zoellick *et al.* (in press) f. Egoscue (1985) g. McCue and O'Farrell (1988)
Adults 0.56; juveniles 0.55	Mainly predators, primarily coyotes, also dogs and bobcats; vehicles	5.82 ± 0.45 km^2, 95% MCP	–	–	–	Spiegel (1996)
No collared foxes died during study but 3 lost their collars (mean tracking period $x = 252$ days)[a]	Non-collared foxes found dead: coyote = 2, car = 1, poisoned = 1, unknown 2[b]	10.98 ± 4.6 km^2 MCP[c]	–	–	–	a. List (1997) b. Personal observations c. List and Macdonald (2003)
–	Vehicles, shooting	–	–	–	Fleas, ticks, cestodes, nematodes	Kilgore (1969)

Table 10.1 (Continued)

Species	Area	Yrs. data collected	Foxes radio-collared	#dens used/year	Diet	% Females with pups	Litter size
V. velox	Piñon Canyon Maneuver Site; Las Animas Co., CO	1986–87[a] 1997–98[b,c] 1989–91[d]	23[a] 73[b] 99[c] 40 adults, 8 yearlings, 46 pups[d]	>20[a]	Leporids, insects, birds[a] Leporids, insects, rodents, berries[b]	44%–1998, 68%–1999[c]	3.4, $n = 5$[a] 2.1 ± 0.8 1997; 2.4 ± 1.2 1998[c] Trios 4.2 ± 0.7, $n = 5$; pairs 2.4 ± 0.3, $n = 13$[d]
V. velox	Wallace and Sherman Cos., KS	1996–97	41 adults, 25 juveniles	–	Murid rodents, leporids, insects, birds, seeds, berries[a]	44%	3.25, $n = 8$[b]
V. velox	Weld Co., CO; Central Plains Experimental Range and Pawnee National Grasslands	1994–97[a]	110[a]	–	Leporids, rodents, birds[b]	–	2.8, SD = 1.1, $n = 24$[a]; 3.6, $n = 16$[b]
V. velox	Albany Co., WY[b]	1996–97[a], 1996–99[b]	10[a], 56[b]	6 maximum[a]	Rodents, leporids, insects, birds[b]	79%[b]	4.6, SE = 0.36, $n = 25$[b]
V. velox	Western Kansas	1981–83	–	–	Leporids, rodents, birds insects, vegetation	–	–
V. velox	Shannon Co., SD	1978–79		–	Mammals (largely rodents and leporids) 49%, insects 27%, plants 13%, birds 6% [relative frequency][a]		4.0, $n = 5$[b]
V. velox	Blaine Co., MT	1996–97	9	–	Rodents, insects, vegetation, birds	–	–
V. velox	Hansford Co., TX	1953–56	–	–	Leporids, rodents, birds, insects		5.5, $n = 2$
V. velox	Alberta/ Saskatchewan	1989–92[a] 1994–98[b]	76[b]	<22[b]	Rodent 60%, insect 24%, bud 8%, leporid 7%, large mammal 2% [relative frequency][b]	85%[b]	3.8, $n = 29$[b]
V. velox	Wyoming	–	See citation		–	–	–

Annual survival	Causes of death	Home range size	Genetic relationships	Dispersal	Disease	Source
Adults 0.45; juveniles 0.13[z]; adults 0.64; juveniles 0.69 (n = 12);[b] adults 0.53 ± 0.05, pups 0.13 ± 0.13[d]	Coyotes (63%), eagles, badger, vehicle[a]; Coyotes, 80% of all mortalities[b,c]; Mainly predators 87%; primarily coyotes, raptors[d]	29.0 km^2 100%MCP: 22.8 km^2 95%ADK; n = 5[a] 7.6 km^2 SE = 0.5, 95% ADK[b]	–	Adults 11.9 km ± 8.8, n = 6; juveniles 12.6 ± 3.17, n = 8[c] Males 9.4 km ± 1.66 n = 8; Females 2.1 km ± 0.21 n = 6[d]	–	a. Andersen *et al.* (2001) b. Kitchen (1999) c. Schauster (2001) d. Covell (1992)
Adults 0.45; juveniles 0.37[c]	Coyotes, vehicles, poison[c]	15.9 km^2 SE = 1.6, 95% ADK[d]	–	Juveniles 14.7 km, SE = 4.8, n = 10	–	a. Sovada *et al.* (2001) b. Allardyce and Sovada (2003) c. Sovada *et al.* (1998) d. Sovada *et al.* (2003)
1995–0.57 (95%CI = 0.43–0.72) 1996–0.75 (0.59–0.9)[a]	Mainly coyotes, vehicles[a]	–	–	–	–	a. Roell (1999) b. Fitzgerald *et al.* (1983)
0.58[b]	Coyotes 45% Raptors, badger; n = 35[b]	11.7 km^2 SE = 1.3 95% ADK; 7.7 km^2, SE = 1.1 100% MCP[a]; See paper Home range given for each year and seasons[b]	–	–	Canine distemper[b]	a. Pechecek *et al.* (2000) b. Olson and Lindzey (2002) c. Olson *et al.* (2003)
–	–	–	–	–		Zumbaugh *et al.* (1985) a. Uresk and Sharps (1986) b. Hillman and Sharps (1978)
0.46 SD = 0.13	–	10.4 km^2 100% MCP; 12.3 km^2 95% ADK	–	–	–	Zimmerman (1998) Zimmerman *et al.* (2003)
0.46–0.64[b,c]	Coyotes, golden eagles, badger starvation, vehicles, poison, snaring	31.9 km^2 fixed kernel (n = 36)[c]		Juveniles[b]: ♂: 13.7 ± 3.8 km; ♀: 10.6 ± 4.6 km	–	a. Carbyn *et al.* (1994) Brechtel *et al.* (1993) b. Moehrenschlager (2000) c. Moehrenschlager and Macdonald (2003)
–	–	–	–	–	Sylvatic plague, canine distemper	a. Pybus and Williams (2002)

Canyon Maneuver site in Colorado averaged 22.8 km^2 from 1986 to 1987 (Andersen *et al.* 2003) and 7.6 km^2 from 1997 to 1998 (Kitchen *et al.* 1999).

Home ranges of adjacent kit fox family groups frequently overlap, but core areas of concentrated use are generally used exclusively by a given family (White and Ralls 1993; Spiegel 1996). Andersen *et al.* (2003) reported nearly total exclusion of a swift fox's core activity area to other same-sex individuals. Pechacek *et al.* (2000) and Sovada *et al.* (2003) found that adjacent family groups had minimal home range overlap and exclusive core areas. In Canada, Moehrenschlager *et al.* (submitted) reported swift fox home ranges overlapped by 77.1% among mates and 21.4% between neighbours. Despite variation in home range sizes in time and space, we find that intraspecific partitioning of space is similar between swift and kit foxes.

Den use

Swift and kit foxes use dens virtually every day of their lives. Both species use dens for escaping predators, avoiding temperature extremes and excessive water loss, diurnal resting cover, and raising young (Tannerfeldt *et al.* 2003; Cypher *et al.* 2003). Most dens are earthen, but man-made structures such as pipes, culverts, buildings, and rubble piles may also be used. Kit and swift foxes frequently enlarge rodent burrows and badger digs, but can readily dig their own dens (Cutter 1958b; Kilgore 1969; List and Cypher in press). Kit fox dens vary in the number of entrances and internal intricacy, but dens used for pup-rearing typically have multiple entrances (Egoscue 1962). Swift fox dens can have as many as 17 branches and up to 2 chambers at depths extending to 1 m (Kilgore 1969). An individual kit or swift fox will use multiple dens, and these dens typically are distributed throughout an individual's home range, which facilitates predator avoidance. From Mexico to Utah, the number of dens utilized by kit foxes per year vary from approximately 7 to 25. Similarly, individual swift foxes from Colorado to Alberta may utilize more than 20 dens per year (Table 10.1). Based on the available information, we cannot discern compelling differences between swift and kit foxes in terms of den structure or utilization.

Reproduction

Kit and swift foxes are monestrous and primarily monogamous with occasional polygyny (Egoscue 1962; Kilgore 1969; Ralls *et al.* 2001), as is common among small canids (Moehlman 1989). Pairs usually mate for life (Egoscue 1956). Kit foxes mate from mid-December to January and whelp from mid-February to mid March after a gestation of 49–55 days (Egoscue 1956; Zoellick *et al.* 1987). Litter size ranges from 1 to 7 and averages about 4 (Table 10.1). The proportion of kit fox females successfully breeding varies annually and appears to vary with food availability (White and Ralls 1993; Spiegel 1996; Cypher *et al.* 2000). Adult kit fox success can be 100% in some years (Table 10.1), but success among yearlings is generally lower even in optimal years. Pups emerge from dens at about 4 weeks, are weaned at about 8 weeks, and become independent at about 5–6 months (Morrell 1972; List and Cypher, in press). Pups, particularly females, may delay dispersal and remain in natal home ranges. Some of these individuals may assist parents in rearing the next litter of pups (O'Neal *et al.* 1987; List 1997; Koopman *et al.* 2000).

Figure 10.2 Kit fox *Vulpes macrotis* in Mexico © R. List.

The timing of swift fox breeding is dependent upon latitude, occurring December to January in Oklahoma (Kilgore 1969), January to February in Colorado (Scott-Brown *et al.* 1987; Covell 1992), February to early March in Nebraska (Hines 1980), and in March in Canada (Pruss 1994; Moehrenschlager 2003). Swift foxes may produce young in the year following their birth (Kilgore 1969; Table 10.1). The mean gestation period is 51 days (Schroeder 1985) after which litters with up to 8 kits (Moehrenschlager 2000), averaging 2.4–5.3 young (Covell 1992; Olson *et al.* 1997; Moehrenschlager 2000) are born. Pups open their eyes at 10–15 days, emerge from the natal den at about 4 weeks, and are weaned at 6–7 weeks of age (Kilgore 1969; Hines 1980). Like kit foxes, additional swift fox females are occasionally observed at den sites, and likely act as helpers in pup-raising (Kilgore 1969; Covell 1992; Olson *et al.* 1997; Sovada *et al.* 2003). In Colorado, litter sizes were greater for mated pairs that had helpers than pairs that did not. (Covell 1992). The only difference in reproduction that we can detect between swift and kit foxes is that swift foxes tend to breed later at northern latitudes; however, southern swift foxes such as those in Oklahoma breed from December to January like many kit fox populations.

Dispersal

Mean kit fox dispersal age on the Naval Petroleum Reserves in California was about 8 months, and many juveniles may not live long enough to reach dispersal age (Koopman *et al.* 1998). Koopman *et al.* (2000) found that 0–52% of pups dispersed, males were more likely to disperse than females, philopatric kit fox females were significantly heavier than dispersing females, 62.5% of pups died within 10 days of departure from natal home ranges, and survival tended to be higher for dispersing than philopatric males. Scrivner *et al.* (1987) found that juveniles and adults dispersed 8.0 ± 1.4 km and 4.8 ± 0.8 km, respectively. Although most dispersal movements probably are less than 10 km (Koopman *et al.* 1998), kit fox young in Utah dispersed as far as 64 km (O'Neal *et al.* 1987) and dispersals of over 100 km have been recorded in California (Scrivner *et al.* 1993).

Swift fox dispersal commences in August or September in Oklahoma (Kilgore 1969), September/October in southern Colorado (Covell 1992), October/November in Kansas (Sovada *et al.* 2003), and August in Canada (Pruss 1994; Moehrenschlager 2000). As many as 67% of juveniles in Canada still remained in natal home ranges during the 1.5 months that precede the subsequent breeding season. By comparison, all foxes that were 18 months or older had dispersed from natal territories (Moehrenschlager and Macdonald 2003). Although swift foxes that were translocated from Wyoming to Colorado had dispersal distances of 27.2 ± 14.2 km for adults and 19.3 ± 15.9 km for juveniles (Moehrenschlager 2000), average dispersal distances for naturally dispersing swift foxes in Colorado, Kansas, and Canada were less than 15 km (Table 10.1). As additional data become available, differences in swift and kit fox dispersal may emerge but these are not currently apparent.

Survival

Annual mortality rates for adult kit foxes range from 0.44 to 0.61 between areas and similarly mortality rates for swift foxes range from 0.47 to 0.63 (Table 10.1). Juvenile kit fox survival ranges from 0.14 to 0.55 and that of swift fox juveniles ranges from 0.13 to 0.69 (Table 10.1). Cypher *et al.* (2000) found that annual survival of kit foxes is variable and influenced by factors such as food availability and competitor abundance.

Mortality rates of both species are primarily attributable to predation or direct human causes and these are further discussed below; regardless of cause, survival rates are variable within study areas over time, among species, and, given this variability, are broadly similar between the fox species for both adult and juvenile cohorts.

Serum antibodies against a number of infectious diseases have been detected among kit and swift foxes (Table 10.1). However, there is no evidence that disease is an important mortality factor, although circumstantial evidence suggests that rabies could have contributed to a decline in one population of *V. macrotis mutica* (White *et al.* 2000). No significant disease outbreaks have been documented in swift fox populations to date; however, Olson (2000) reported deaths of two swift foxes to canine distemper.

Threats limiting distribution and abundance

Habitat availability

The kit fox inhabits the deserts and arid lands to the west and south of the shortgrass prairie occupied by the swift fox. The regions occupied by kit foxes generally receive less rainfall than do the areas occupied by swift foxes. Kit fox abundance varies greatly from year to year and changes in prey abundance are believed to be driven by widely variable rainfall (White and Ralls 1993; Ralls and Eberhardt 1997; White and Garrott 1997, 1999; Cypher *et al.* 2000).

Habitat types include desert scrub, chaparral, halophytic, and native and non-native grassland communities (McGrew 1979; O'Farrell 1987; Sheldon 1992). Kit foxes are found at elevations of 400–1900 m (List and Cypher 2002). They are less vulnerable to mortality from larger canids such as coyotes on flat or rolling grassland (Warrick and Cypher 1998), and Haight *et al.* (2002) included slope as a factor in classifying habitat quality (0–5% = good, 5–10% = fair).

The swift fox is predominately found on shortgrass and mixed-grass prairies in gently rolling or level terrain (Cutter 1958b; Kilgore 1969; Egoscue 1975; Hillman and Sharps 1978; Hines 1980). The conversion of native grassland prairies has been implicated as one of the most important factors for the contraction of the swift fox range (Hillman and Sharps 1978). However, swift foxes have adapted to regionally a variety of atypical habitats such as mixed agricultural areas (Kilgore 1969; Hines 1980; Sovada *et al.* 2003), sagebrush steppe, and shortgrass prairie transition (Olson *et al.* 1997).

Swift and kit foxes are prone to similar threats of loss, fragmentation, or degradation of habitats. Kit foxes in California can survive in urban environments (Cypher *et al.* 2003) and oil fields. Although Spiegel (1996) concluded that reduction and fragmentation of habitat due to oilfield-related construction and maintenance activities could lower carrying capacity for kit foxes, long-term studies by Cypher *et al.* (2000) found no evidence that oilfield activities at the Naval Petroleum Reserves in California were impacting kit fox populations. Effects of oilfields on swift foxes have not been studied. However, an experimental evaluation of oil pipeline development in Alberta found that swift fox den use was unaffected and that changes in radio-collared fox movements over time were similar between pipeline and control areas, but reproduction was apparently affected by site disturbance before development (Moehrenschlager and Macdonald 2003). In Canada, the oil and gas industry is expanding dramatically and previously isolated prairie areas are now targeted for exploration (Gauthier 2002). Associated road developments will potentially decrease the habitat carrying capacity, increase vehicle-caused swift fox mortalities, and impede population gene flow.

Although kit fox populations are thought to be stable or increasing in most western states of the United States, populations in California's San Joaquin Valley and Mexico are thought to be declining (List and Cypher in press). Habitat loss and degradation limit both distribution and abundance of kit foxes in these areas and are the main threats to their long-term survival (List and Cypher in press). The endangered San Joaquin kit fox, *V. macrotis mutica* once inhabited most of the San Joaquin Valley in California, but approximately 95% of the natural landscape has been replaced by irrigated agriculture, cities and towns, and industrial development resulting in declining kit fox populations (USFWS 1998). Habitat conversion to agriculture is slowing in the San Joaquin Valley, but habitat loss, fragmentation, and degradation due to urban and industrial development continues at a rapid rate (List and Cypher in press); thus the population continues to be threatened.

In the northeastern range of the kit fox in Mexico, only about 50% of endemic Mexican prairie dog (*Cynomys mexicanus*) towns remain, which kit foxes rely on heavily. In northwestern and northeastern Mexico, kit foxes inhabit prairie dog towns that are being converted to potato fields, and the road network in eastern Mexico is expanding, thereby increasing the risk that foxes will be killed by vehicles (List and Cypher in press; Cotera Correa 1996).

San Joaquin kit foxes use agricultural lands, particularly orchards, to a limited extent, but they are not able to utilize areas with irrigated agriculture. Similarly, swift foxes in Kansas occupy mixed agricultural areas that apply dry-land fallow farming practices (i.e. every other year crop fields remain fallow; Jackson and Choate 2000; Sovada *et al.* 2002). Survival rates between foxes in grassland and cropland sites were not significantly different suggesting that swift foxes may be able to adapt to such habitat in some cases (Sovada *et al.* 1998). However, an increasing trend towards crop irrigation in such areas could exclude swift foxes that have successfully adapted to dry-land farming practices.

Areas that are not lost to urban spread or irrigated for crop agriculture are not necessarily conducive to swift or kit fox populations. Most of the southern San Joaquin Valley in California was originally a desert vegetated by salt bush scrub with only a sparse cover of native annual grasses and forbs (Germano *et al.* 2001), but today only a few, fragmented patches remain. Even the remaining 'natural' lands have been invaded by exotic annual grasses and forbs (Germano *et al.* 2001). In the absence of grazing, these introduced plants can create an impenetrable thicket that has deleterious effects on kangaroo rats and other prey species of kit foxes (Germano *et al.* 2001). Thus, decisions to decrease or eliminate livestock grazing on conservation lands may have adverse effects on kit foxes. In the United States, the planting of tall, dense vegetation as a part of the United States Conservation Reserve Program, may also negatively impact swift foxes, which avoid these densely vegetated habitats (Allardyce and Sovada 2003; Moehrenschlager and Sovada in press).

Human-caused mortality

Direct anthropogenic sources of mortality for kit and swift foxes have decreased in significance, but still can be important in some locations. Both species were previously harvested throughout most of their range (Reid and Gannon 1928; Grinnell *et al.* 1937; Egoscue 1956, 1962; Johnson 1969), but harvest has declined in many areas and in California and Oregon is no longer permitted (Allardyce and Sovada 2002). Historically, many kit fox and swift foxes died from toxicants distributed for other predators, primarily wolves (*Canis lupus*) and coyotes (Grinnell 1914; Grinnell *et al.* 1937; Egoscue 1956; Young 1944, pp. 335–336). Rodent control programmes can result in the primary or secondary poisoning of kit foxes, but this source of mortality is now infrequent (Snow 1973). Vehicles have been and continue to be an important source of kit and swift fox mortality (Egoscue 1962; Sovada *et al.* 1998), and in some locations are responsible for over 10% of kit fox mortalities (Cypher *et al.* 2000). Other anthropogenic sources of mortality include illegal shooting (Morrell 1972; Cypher *et al.* 2000) and accidental death associated with agricultural and urban development (Knapp 1978; Sovada *et al.* 1998; Cypher unpublished data).

In Canada, swift foxes are legally protected against intentional killing, but radio-collared foxes have been poisoned or killed in coyote snares (Moehrenschlager 2000; Moehrenschlager *et al.* 2003). In the United States, swift foxes are legally protected in ten States, and are protected from harvest in seven of these. States that do provide harvest opportunities regulate these by season length and monitor harvest numbers annually. The 1972 Presidential ban on predator toxicant use (e.g. strychnine, compound 1080) on Federal lands may have contributed to swift fox recovery, but 1080 is currently being legalized in prairie areas of Saskatchewan, Canada, which will likely limit reintroduced swift fox populations (Moehrenschlager and Sovada in press).

Predation and competition

The situation where two predators compete for prey and the larger one kills the smaller species is known as intraguild predation. Theory suggests that intraguild predation can have major effects on the population dynamics of the smaller competitor (Holt and Polis 1997). Although coyote killing of swift and kit foxes is a well-known example of intraguild predation, subsequent effects on the fox populations are not entirely clear. Certainly, there is considerable overlap in both habitat and food use between coyotes and both fox species (Cypher *et al.* 1994; White *et al.* 1994, 1995; Kitchen *et al.* 1999). Some proportion of kit fox mortality due to coyotes appears to be additive (White and Garrott 1997; Cypher and Spencer 1998; Cypher *et al.* 2000). Simulation modelling suggests that adult kit fox mortality is independent of fox density but that juvenile fox mortality increases with fox density (White and Garrott 1999). Thus, density-dependent juvenile mortality by coyotes could help regulate kit fox population size, curtailing population growth at high fox densities but having less impact at low densities (White and Garrott 1999; Fig. 10.3).

Figure 10.3 Coyote *Canis latrans* © R. List.

The strength of interspecific competition likely varies with prey abundance (Creel *et al.* 2001), which could explain some of the variation in results among studies conducted at different times and places. For example, kit foxes in Chihuahua, Mexico, which prey on the world's largest remaining prairie dog population were rarely killed by coyotes (List and Macdonald 2003). Fox survival differences between Canada and Mexico were assessed relative to intra- and interspecific spatial patterns, dynamic interactions, diet, and predator avoidance habitats (Moehrenschlager and List 1996; Moehrenschlager *et al.* submitted). Although in Canada, coyotes and foxes were heavier than their Mexican counterparts, interspecific body size ratios were similar between the countries, suggesting similar energetic pressures for interspecific competition. Fox home ranges were significantly larger in Canada than Mexico, coyote home ranges were similar between countries, and ratios of coyote/fox home ranges were approximately four times larger in Mexico than Canada. Differences in interspecific home range size ratios and escape hole availability between countries were likely caused by differences in the abundance and diversity of prey. These factors probably caused observed differences in fox–coyote encounter rates and, consequently, fox mortalities between countries (Moehrenschlager *et al.* submitted).

Coyote control on the Naval Petroleum Reserves in California did not result in increased fox abundance or survival and coyote abundance was not consistently related to kit fox abundance (Cypher *et al.* 2000). Similarly, intensive coyote killing by landowners in Alberta and Saskatchewan apparently caused decreased coyote-caused mortality of swift foxes, but subsequent predation by golden eagles (*Aquila chrysaetos*) prevented fox survival rates from improving (Moehrenschlager *et al.* submitted).

The relationship between coyotes and kit foxes is complicated when red foxes, which are intermediate in size, are also present. In fact, red foxes may pose a greater threat to swift and kit foxes than coyotes. Red foxes have been able to displace smaller Arctic foxes (*Alopex lagopus*) in Scandinavia (Hersteinsson and Macdonald 1992; Cypher *et al.* 2001; Tannerfeldt *et al.* 2003).

Red foxes are known to kill kit foxes (Ralls and White 1995; Clark 2001) and swift foxes in Canada have greater dietary overlap with red foxes than coyotes (Moehrenschlager unpublished data), which may allow red foxes to competitively exclude swift foxes. Since red foxes tend to exist at higher densities than coyotes, the likelihood of red fox–swift fox encounters would be higher than coyote–swift fox encounters. Preliminary results from an experimental study examining the swift fox–red fox relationship suggest that red foxes can be a barrier to swift fox populations expanding into unoccupied, but suitable areas (M. A. Sovada unpublished data). Within one day, a family of red foxes in Saskatchewan took over a den that had been used by a swift fox for at least 2 weeks prior. Moreover, red fox dens in Alberta and Saskatchewan were significantly closer to human habitation than coyote dens while swift foxes dens were found at all distances (Moehrenschlager 2000). As coyotes avoid high human activity areas, red foxes may utilize these sites to begin their invasion of swift fox home ranges. While coyotes reduce swift fox numbers through direct, density-dependent killing within the swift fox range, red foxes could potentially exclude swift foxes through a combination of interference and exploitative competition. As such it is possible that coyotes sometimes have a beneficial effect on swift or kit foxes by excluding or reducing red foxes, since coyotes tend to exclude red foxes but not swift/kit foxes from their home ranges (White *et al.* 1994; Ralls and White 1995; Cypher *et al.* 2001).

Conclusions and recommendations

We compared swift and kit foxes to see if morphological and genetic differences might also be indicative of differences in life history, ecology, or population threats. We found that temporal variation in ecological parameters can be high as environmental conditions within respective study areas change seasonally or annually. Spatial variation also exists within study areas as swift or kit foxes utilize available habitats to different extents. Across the range of each species, the variability of these parameters is likely a product of temporal variability at individual sites, small-scale spatial variability at such areas, and large-scale geographic or climatic differences between regions. Yet, when this amount of variation for individual parameters is accounted for within each species, we cannot discern substantial differences between swift and kit foxes in diet, home range use, den use, reproduction, dispersal, or survival.

The distribution and abundance of kit and swift foxes has been depleted since Europeans began settling in North America. Three primary causes have left swift and kit foxes imperilled locally or nationally: (1) habitat loss and fragmentation; (2) human-caused mortality; and (3) predation and competition. These threats continue to persist today for both species although their causes, nature, and effects vary relative to historic times and differ between regions of the swift/kit fox complex.

We found that the threat of habitat loss due to cropland agriculture is slowing in Canada and the United States, but prairie dog towns and associated kit fox areas in northern Mexico continue to be lost at an alarming rate. Habitat fragmentation continues to be a universal threat as urbanization, industrial developments, and roads increasingly expand into swift and kit fox areas. Habitat degradation in prairie and desert regions is increasing as exotic species and changing agricultural practices limit the amount of suitable habitat for both species. The threat of human-caused mortality may be decreasing as legalized trapping is increasingly restricted throughout the range of swift and kit foxes. However, vehicle-caused mortality may be increasing as road networks increase in both prairie and grassland habitats. Incidental poisoning of northern swift foxes will likely increase in Canada as compound 1080 is regionally legalized. The threat that coyotes currently pose to swift foxes remains, but the threat of expanding red foxes on both swift and kit foxes is increasing.

Since swift foxes in Kansas and kit foxes in the San Joaquin Valley of California have shown an ability to adapt to some agricultural systems, while foxes in other regions have not, the question 'what levels of habitat loss, degradation, and fragmentation can swift and kit foxes tolerate?' arises. While individual foxes occasionally utilize human-modified habitats, one needs to understand which ratios of modified to pristine habitat, which degrees of connectivity, and what changes in habitat quality can swift and kit fox populations can tolerate.

We believe that the effect of habitat disturbance or other human-caused stressors on swift and kit foxes must be evaluated in terms of interactions with coyotes and red foxes. As red foxes adapt well to human habitation and farming, the question needs to be addressed to what extent they exclude swift and kit foxes from agricultural or natural areas that these smaller species could otherwise utilize. The fact that coyotes, whose killing of swift/kit foxes represents one of the strongest examples of intraguild pressure among carnivores, may be necessary to exclude red foxes suggests that scenarios ironically exist where too many or too few coyotes could lead to swift/kit fox exclusion. Determining where these equilibria fall and how they are affected by changing environmental conditions or human disturbance should be an area for future investigation. Scenarios may exist, for example, where the increase of a direct threat such as trapping or habitat fragmentation may have a net benefit for swift/kit foxes if, indirectly, the disadvantage to coyote and red fox populations is greater than the disadvantages this poses for swift/kit foxes. Habitat loss or fragmentation may decrease swift/kit fox carrying capacity as well as patch isolation and such habitat threats might be additive to those already posed by coyotes or red foxes. On the other hand, habitat alteration could favour swift/kit foxes in some cases; for example, kit foxes in Bakersfield, California have higher survival rates than conspecifics in surrounding natural areas because of predictable food availability and coyote exclusion in the city (Cypher *et al.* 2003).

Additional knowledge of population genetics is necessary to understand the relatedness, dispersal, and gene flow within fragmented swift and kit fox populations (e.g. Ralls *et al.* 2001; Schwartz *et al.* submitted), which may affect population viability and the potential spread of disease. Canine distemper has been noted among swift foxes (Olson 2000) and rabies among kit foxes (White *et al.* 2000). Both of these diseases can have devastating effects on endangered canid populations (Sillero-Zubiri and Macdonald 1998; Woodroffe and Ginsberg 1999a), but the sources and effects of these diseases in swift and kit fox populations are not understood.

Although regional variation in environmental conditions, demographics, and human-caused pressures are great throughout the swift/kit fox complex, this variability appears to be as great within the species as it is between them. Therefore, conservation solutions devised for one of these species will probably be relevant to the other. Common conservation planning for swift and kit foxes might be most effective to achieve financial, political, and ecological means that will sustain both species in the future.

Acknowledgements

A. Moehrenschlager would like to thank Husky Energy Inc. and the Wildlife Preservation Trust Canada for its funding, leadership and logistical support. A. Moehrenschlager and R. List are also grateful for the support of the People's Trust for Endangered Species through Wild CRU, Oxford.

CHAPTER 11

Blanford's foxes

Evolution, behavioural ecology, and conservation of Blanford's foxes

Eli Geffen

Blanford's fox *Vulpes cana* © C. and T. Stuart.

While the morphology typical of canids adapts them to endurance running, and consequently to life in open habitats, a few species have specialized to other habitats such as rainforests, or even to climbing trees or cliffs. The cliff-dwelling Blanford's fox (*Vulpes cana*) is one of these exceptional canid species. Previously, this species was thought to be confined within a restricted range of which the western-most extreme was Iran. However, its discovery in 1981 in Israel extended its range 1000 km further west, and was the first of series of records that revealed its presence throughout the Middle East. Following a decade of research on this previously almost unknown fox, I present here an overview of the Blanford's fox's phylogeny and taxonomic status, recent distributional changes, the behavioural ecology of the population studied in Israel, and its conservation status.

Morphological features

Blanford's fox is one of the smallest canids, weighing on average about 1 kg—a similar body mass to the fennec (*Vulpes zerda*). This delicate species has a bushy and very long tail (76% of body length; Geffen *et al.* 1992d). The body is brownish-grey, fading to pale yellow on the belly. The winter coat is soft and woolly with dense, black underwool. Its dorsal region is speckled with white-tipped hair (Fig. 11.1). The summer coat is less dense, paler, and the white-tipped hairs less apparent. Specimens from the eastern part of its distribution may be predominantly grey. A distinctive mid-dorsal black band extends from the nape of the neck caudally, becoming a mid-dorsal crest throughout the length of the tail. The tail is similar in colour to the body. The dark mid-dorsal band, which is a distinctive feature of the Israeli specimens, is less evident in specimens from Oman, although the black tail markings are equally developed (Harrison and Bates 1989). A distinctive dorsal black spot (violet gland) is present at the base of the tail, which usually has a black tip, although in some individuals the tip is white (4% in Israel and 26% in UAE; Smith *et al.* 2003b). The forefeet and hind feet are dorsally pale yellowish-white, while posteriorly they are dark grey. Unlike the other fox species in the Arabian deserts, the blackish pads of the feet and digits are hairless and the claws are cat-like, curved, sharp, and semi-retractile (Geffen *et al.* 1992d). The head is orange buff in colour, especially in the winter coat. The face is slender with a dark band extending from the upper part of the sharply pointed muzzle to the internal angle of the eyes. The ears are pale brown on both sides, with long white hairs along the antero-medial border, and intermediate in length between those of the red (*Vulpes vulpes*) and fennec foxes (Harrison and Bates 1991; Geffen *et al.* 1992d; Geffen 1994; Roberts 1977).

Compared with other small canids occurring in the Saharo-Arabian deserts, the relatively large and long tail of *Vulpes cana* is unique. Most canids are cursorial terrestrial carnivores, capable of prolonged trotting at a moderate speed (Taylor 1989), and adapted to long-distance travel over horizontal ground. Blanford's fox and the Arctic fox are the only canids known regularly to utilize both horizontal and vertical habitats. The only species that routinely climbs trees is the gray fox, *Urocyon cinereoargenteus* (Fritzell 1987). Large tails are typical of tree-dwelling carnivores such as stone martens (*Martens foina*) and ringtails (*Bassariscus astutus*). Jumping is usually an integral part of the locomotor pattern in fast-moving arboreal mammals and the large tail is probably an important counter balance during jumps and may function like a parachute (Taylor 1989). Mendelssohn *et al.* (1987) described the jumping ability of *Vulpes cana* as astonishing; captive individuals bounced from one wall to another or jumped to

Figure 11.1 Blanford's fox *Vulpes cana* in Ein Gedi, Israel © S. Kaufman.

the highest ledges (2–3 m) in their cage with remarkable ease and as part of their normal movements. Their small feet and naked pads provide sure footing even on the narrow ledges of a vertical wall. In the field, we observed foxes climbing vertical, crumbling cliffs by a series of jumps up the vertical sections. Their sharp, curved claws doubtless enhance traction on the more difficult vertical ascents.

Phylogeny and taxonomic status

Four fox species are common in the desert regions of the Middle East and northern Africa: Blanford's fox, the fennec fox, Ruppell's fox (*Vulpes rueppelli*), and the red fox (*Vulpes vulpes*). Recent phylogenetic analysis of mtDNA sequences shows that the former two taxa are a monophyletic clade distinct from the other fox-like canids (Fig. 11.2; Geffen *et al.* 1992e). However, the sequence divergence between the fennec fox and Blanford's fox is large, approximately 7%, indicating an ancient divergence as much as 3–4 million years ago (Ma). This divergence is coincident with the appearance of desert regions in the Middle East and northern Africa (Wickens 1984), and suggests that a fox-like progenitor entered these regions and diversified into two lineages. The fennec occupies a habitat in shifting sand dune environments whereas Blanford's fox is restricted to steep rocky slopes. Each species shows distinct morphologic adaptations for these habitats. For example, Blanford's fox has hairless feet adapted for climbing on bare rock and the fennec has furred pads for locomotion on shifting sand. Whereas other fox species usually occupy a range of habitats, the fennec and Blanford's fox, which are the smallest of all canids, show a strong affinity to a single, specific habitat. Their small size may be associated with their specialization to the arid, poorer quality habitats of the Arabian desert. Both Ruppell's fox and the red fox are 1.5–3 times larger than the Blanford's fox or the fennec, and do not persist in the poorer quality habitats where the latter two species occur (Mendelssohn *et al.* 1987; Harrison and Bates 1991).

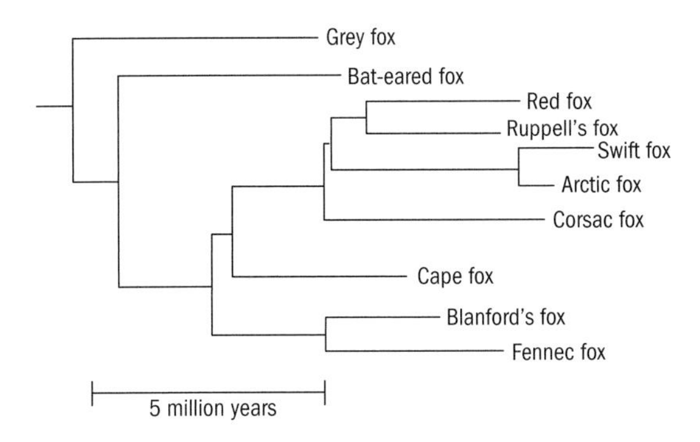

Figure 11.2 Phylogenetic tree generated by maximum-likelihood method (see text). Transition/transversion ratio = 5.0, Ln Likelihood = −1852.3. All branch lengths are significantly different from zero ($P < 0.05$). (Modified from Geffen *et al.* 1992e.)

Generic distinctions should be based on monophyletic groupings of taxa. The results of Geffen *et al.* (1992e) suggest that the fennec and Blanford's fox comprise a separate taxonomic entity since they define a consistent monophyletic group within the clade of the *Vulpes*-like foxes (Wayne *et al.* 1997). In fact, these analyses clearly indicate that the taxonomic division of the fox-like canids into *Alopex, Fennecus*, and *Vulpes* do not reflect genuine phylogenetic divisions within the clade; in evolutionary terms all vulpine-like species might more appropriately be classified within the genus *Vulpes*.

Distribution and abundance

Until recently the Blanford's fox was thought to be confined to mountainous regions in Iran (Lay 1967), Afghanistan (Blanford 1877; Hassinger 1973), Pakistan (Roberts 1977), and Turkistan (Bobrinskii *et al.* 1965; Novikov 1962; Fig. 11.3). The species was considered as one of the rarest predatory mammals in southwest Asia (Novikov 1962). It is seldom represented in scientific collections (Lay 1967), and trade in its fur is minimal compared with some other fox species (e.g. red and Arctic foxes; Ginsberg and Macdonald 1990). In 1981, it was discovered in Israel (Geffen *et al.* 1993; Ilany 1983), and since then in Egypt and Sinai (Geffen *et al.* 1993; Peters and Ršdel 1994), Jordan (Amr *et al.* 1996; Amr 2000), Oman (Harrison and Bates 1989), Saudi Arabia (Al Khalil 1993), and the United Arab Emirates (Stuart and Stuart 1995; Fig. 11.3). The current patchiness of its

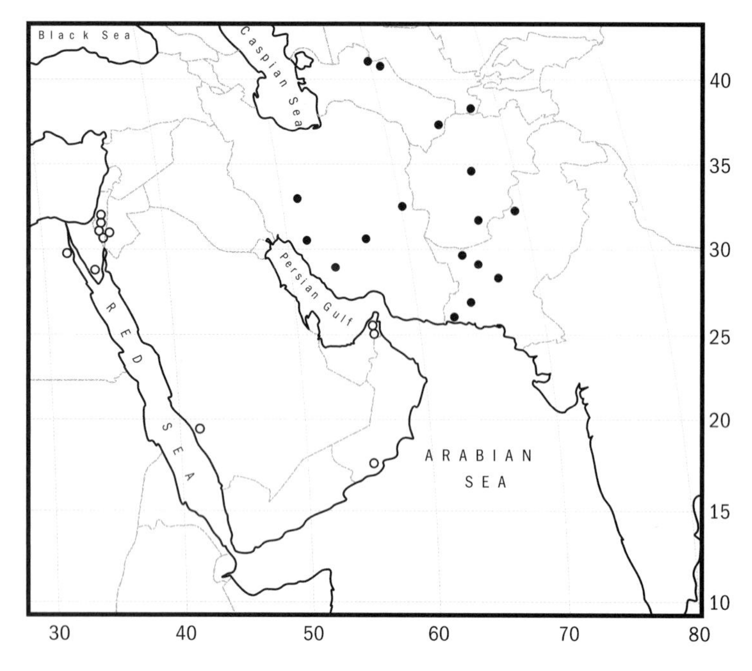

Figure 11.3 Distribution map for the Blanford's fox. Fill circles represent collection localities on the traditional range in southwestern Asia (before 1981). Empty circles represent collection localities after the discovery of the first individual in Israel (1981).

known distribution in the Middle-East probably reflects the inadequacy of records more than it does the species' real range. The elusive habits of Blanford's foxes, and their inaccessible habitat, make them hard to detect. It would not be surprising, therefore, to find this fox in northwest India or along the western Red Sea shore south to Ethiopia. These areas would certainly provide suitable habitat and climate. During the ice ages, when sea level was at a minimum, these foxes could have crossed overland to Africa via the Hormuz Straits and the Gulf of Suez. In fact, it is not clear whether the species radiated in to the Middle East or out from it into southwestern Asia. An ongoing phylogeographical study on the routes and time frames of the species' expansion into the Middle East and Africa using DNA from museum skins collected throughout the range and current molecular techniques may allow us to answer these questions.

The Blanford's fox is confined to mountainous regions (Lay 1967; Roberts 1977). Hassinger (1973) concluded that they are generally found below 2000 m in dry montane biotopes. All the records collected on the Persian Plateau are from foothills and mountains in the vicinity of lower plains and basins (Hassinger 1973; Roberts 1977). In that region, the habitat of Blanford's foxes comprises the slopes of rocky mountains with stony plains and patches of cultivation (Lay 1967; Roberts 1977). In the Middle East, this fox is confined to mountainous desert ranges, and inhabits steep, rocky slopes, canyons, and cliffs (Harrison and Bates 1989; Mendelssohn *et al.* 1987). In Israel, it is distributed along the western side of the Rift Valley and in the central Negev. Specimens from the central Negev were found in creeks that drain into the Rift Valley (Geffen *et al.* 1993). Apparently, *V. cana* can occur on various rock formations provided its other requirements are met. The distribution of Blanford's fox in the Arabian Desert is not limited by access to water (Geffen *et al.* 1992a). In Israel, it inhabits the driest and hottest regions. The densest population is found in the Judaean Desert at elevations of 100–350 m below sea level. This is in contrast to Roberts' (1977) remark that the species avoids low, warm valleys in Pakistan.

In the suitable habitat across southeastern Israel, extrapolations from the home-ranges of Blanford's foxes at two sites (Ein Gedi and Eilat) suggest their population density ranges between 0.8 and 1.0 individuals/km^2 (Geffen *et al.* 1992c). Comparable estimates for other sites are not available, but research in the United Arab Emirates estimated, based on catch per unit effort, suggests that Blanford's foxes are locally abundant in the north-eastern mountain range (Smith *et al.* 2003b). Although surveys in Israel and the United Arab Emirates indicate that Blanford's fox can be relatively common in its specific desert habitat, only about 10% of its known range has been surveyed. Additional surveys across its range are clearly necessary to provide a more comprehensive assessment of the distribution, abundance, and status of this canid.

Diet, energy expenditure, and habitat selection

Blanford's foxes in Israel are primarily insectivorous and frugivorous (Geffen *et al.* 1992b; Ilany 1983). Invertebrates are the major food, with beetles, grasshoppers, ants, and termites eaten most often (Geffen *et al.* 1992b). Plant foods consist mainly of the fruit of two species of caperbush, *Capparis cartilaginea* and *C. spinosa*. Fruits and plant material of Date Palm (*Phoenix dactylifera*), *Ochradenus baccatus*, *Fagonia mollis*, and various species of Gramineae are also eaten. Blanford's foxes in Pakistan are largely frugivorous, feeding on Russian olives (*Eleagnus bortensis*), melons, and grapes (Roberts 1977). In Israel, remains of vertebrates occurred in ca. 10% of fecal samples, and although neither seasonal nor individual differences were detected, the diet differed between the two sites examined (Geffen *et al.* 1992b).

Blanford's foxes almost always forage solitarily (92% of 463 observations; Geffen *et al.* 1992b). Mated pairs, which shared home ranges, differed significantly in the time of arrival at fruitful food patches and in the pattern of use of their home range (Geffen and Macdonald 1993).

The annual size of home ranges of Blanford's foxes in Israel was estimated at 0.5–2.0 km^2, with neither seasonal nor sexual differences (Geffen *et al.* 1992c). Dry creekbed was the most frequently visited habitat in all home ranges, and foxes spent significantly more time in this habitat than expected if their movements had been random. Home ranges (in km^2) at Ein Gedi, Israel encompassed an average ($\pm$ SD) of 63.44 $\pm$ 3.22% gravel scree, 3.63 $\pm$ 2.59% boulder scree, 28.38 $\pm$ 4.05% dry creekbed, and 4.54 $\pm$ 3.46% stream and spring. Average time ($\pm$ SD) spent by these foxes at Ein Gedi in gravel scree was 148.8 $\pm$ 109.8 min/night, 46.0 $\pm$ 63.8 min/night in boulder scree, 359.9 $\pm$ 141.9 min/night in dry creekbed, and 13.0 $\pm$ 27.9 min/night near a water source (Geffen *et al.* 1992c). Dry creekbed provided substantially more abundant prey for the foxes than did the other habitats, and sparse cover for invertebrates (Geffen *et al.* 1992b). Creekbed patches were used in proportion to their size, so that large patches were heavily used while small ones were rarely visited. Both the available area of creekbed in each range, and the area of creekbed patches used by the foxes, were independent of home range size. However, variance in size of home range was explained by the mean distance between the main denning area and the most frequently used patches of creekbed (Figs 11.4 and 11.5; Geffen *et al.* 1992c).

Daily energy expenditure of free-ranging Blanford's foxes near the Dead Sea was 0.63–0.65 kJ g^{-1} day^{-1}, with no significant seasonal difference (Geffen *et al.* 1992a). Mean rate of water intake in Ein Gedi was significantly higher in summer (0.11 $\pm$ 0.02 ml g^{-1} day^{-1}) than in winter (0.08 $\pm$ 0.01 ml g^{-1} day^{-1}). Geffen *et al.* (1992a) concluded that foxes maintained water and energy balances on a diet of invertebrates and fruits without drinking. Furthermore, this study

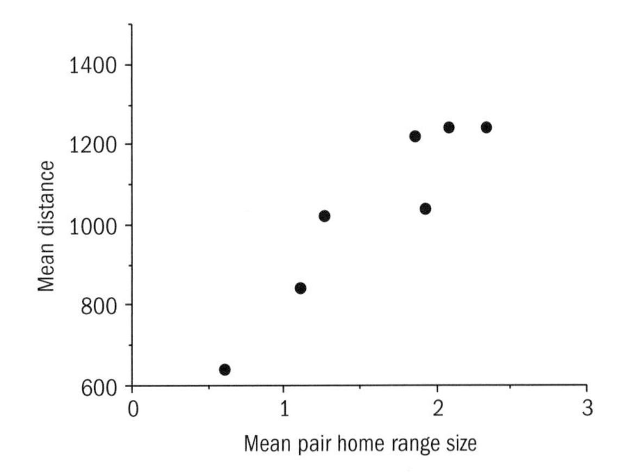

Figure 11.4 Mean distance (m) of the five most intensively used creekbed patches from the main denning area versus mean pair home range size (80% polygon; km^2). $r = 0.94$, $Z = 3.42$, $P < 0.001$. (Modified from Geffen *et al.* 1992c.)

Figure 11.5 Blanford's fox *Vulpes cana* © C. and T. Stuart.

suggested that Blanford's foxes foraged more for water than for energy because metabolic needs are met before water requirements when feeding on invertebrates. Blanford's foxes in Israel consume more fruit during the hot summer, which compensates for deficiencies in body water (Geffen *et al.* 1992a,b).

In Israel, Blanford's foxes are strictly nocturnal year-round. Geffen and Macdonald (1993) hypothesized that this activity pattern is an anti-predator response to diurnal raptors. The onset of activity is governed largely by light conditions, and closely follows sunset. Foxes were active *c.*8–9 h/night, independent of duration of darkness. Average distance (± SD) travelled per night was 9.3 ± 2.7 km, and size of nightly home range averaged 1.1 ± 0.7 km^2 (Geffen and Macdonald 1992). Significant seasonal or sexual differences in duration of activity, nightly distance traveled, or nightly home range were not detected. Except at their extremes, climatic conditions at night in the desert of Israel appeared to have little direct effect on the activity of Blanford's foxes (Geffen and Macdonald 1993).

Social organization and reproduction

Data from 11 radio-tracked Blanford's foxes studied over 2 years indicated that they were organized as strictly monogamous pairs in territories of *c.*1.6 km^2 that overlapped minimally (Fig. 11.6; Geffen and Macdonald 1992). Locations and configurations of home ranges were stable during that study. A shift in location of home range was observed only once following the death of a pair member. Three of five territories contained one, non-breeding yearling female during the mating season, but there was no evidence of polygyny (Geffen and Macdonald 1992). Monogamy may be advantageous in this species because the dispersion of their prey is such that to accommodate additional adults would generally require territorial expansion that would bring greater costs than benefits. The food resources in observed territories may be sufficient to support only a single breeding female.

Dens used by Blanford's foxes in Israel were usually on mountain slopes, and consisted of large rock and boulder piles or screes. The foxes appeared to use only natural cavities, and never dug burrows. Dens were used both for rearing young during spring and for day-time shelter throughout the year. During winter and spring, both members of a pair frequently occupied the same den, or adjacent dens at the same site, while during summer and autumn they often denned in separate locations. Changes in location of den from day to day were most common in summer and autumn (Geffen and Macdonald 1992).

Females are monoestrus and come into heat during January–February (in Israel). Gestation lasts

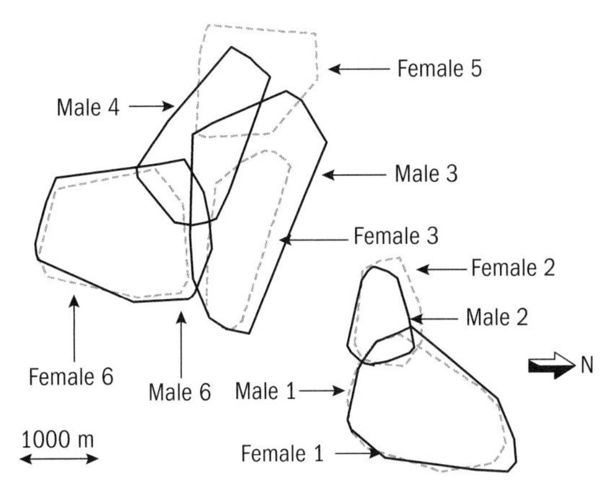

Figure 11.6 The spatial organization of home ranges (80% minimum polygon) of 10 Blanford's foxes at Ein Gedi during 1987. Solid lines represent home range boundaries of males and dashed lines of females. (Modified from Geffen and Macdonald 1992.)

c.50–60 days, and litter size is one to three pups. Females have 2–6 active teats, and the lactation period is 30–45 days. Neonates are born with soft black fur. Based on repeated measures of body mass of three young born in captivity, a neonate body mass of 29 g was estimated (Geffen 1994; Mendelssohn *et al.* 1987). At 3–4 months old, subadults weight 700–900 g. At *c*.2 months of age the young start to forage, accompanied by one of the parents, and at 3 months of age they start to forage alone. Juveniles have similar markings to the adult, but their coat is darker and more greyish. Sexual maturity is reached at 10–12 months of age (Geffen 1994). Offspring often remain on their natal home-range until autumn (October–November).

One to three cubs are born in spring (March). Young are entirely dependent upon their mother's milk for food and water until they begin to forage for themselves. Adult foxes have never been observed carrying food to the young and only one den was found with remains of prey at the entrance (Geffen and Macdonald 1992). As for other vulpine canids, there is no evidence that Blanford's foxes regurgitated to their young. Geffen and Macdonald (1992) had no indication that the male provides food either to the female or to the cubs, although they observed males grooming and accompanying 2–4 month old juveniles. Further, non-breeding adults were never observed to provide for the cubs in any way. Therefore, it appears that the direct contribution to survival of the young by any individual, other than the mother, is probably minimal. The issue of paternal and maternal investment is of special interest because previous observations have suggested that food is not provided to the female and cubs by the male. In such a system, polygyny is favoured because litters are small and paternal investment seems to be minimal. However, the fact that monogamy is the observed social system suggests that other, yet undetected energetic constraints may play a role. An ecophysiological study of energy expenditure in breeding pairs may elucidate why these foxes are monogamous.

Mortality, pathogens, and conservation

In Israel, old age or rabies were the primary causes of death (Geffen 1994). Maximum life span of Blanford's foxes in the Israeli populations was estimated as 4–5 years, but in captivity they may live longer (6 years; Geffen 1994). The potential predators of Blanford's foxes are leopards (*Panthera pardus*), red foxes, eagle owls (*Bubo bubo*), golden eagles (*Aquila chrysaetus*), and Bonelli's eagles (*Hieraeetus fasciatus*). Blanford's foxes were observed to flee from a red fox; the only known case of mortality from predation was probably attributable to a red fox. However, occasionally individuals will stand at a safe distance and bark at larger potential predators (e.g. leopard and human).

Blanford's foxes are threatened by rabies. During 1988–89, 11 Blanford's foxes were found dead in the two studied populations in Israel, of which two fresh carcasses tested positive for rabies (the remaining carasses were too decayed to yield results). All these individuals were marked adults. Considering that the Ein Gedi reserve may support only 7–10 breeding pairs, rabies could substantially diminish the population there. Rabies is a common disease throughout the Middle East, and in south-eastern Israel a few cases are reported every year. Oral rabies vaccination of wildlife is now being implemented in Israel. This may prove an important conservation measure for the seemingly fragmented, and thus vulnerable, populations of Blanford's foxes.

Only a single poisoning record of three Blanford's foxes and two red foxes is known from the United Arab Emirates. Probably poisoning is a rare cause of mortality in this species. Road kills also appear to be insignificant (none has been reported from Israel or the United Arab Emirates, and one from Saudi Arabia; Al Khalil 1993).

The Blanford's fox has been known for its luxurious coat and its pelts are occasionally sold in the bazaars of south-western Asia (e.g. Hassinger 1973; Roberts 1977). Records compiled by the Convention on International Trade in Endangered Species indicated that no Blanford's fox pelts were exported during 1983 and 1985–86; in 1980 and 1982 seven were exported; and in 1981 *c.*30 skins were exported from Afghanistan. However, in 1984, there is a clearly anomalous record of 519 Blanford's fox skins exported, mostly from Canada (Ginsberg and Macdonald 1990). In Israel, the species is completely protected by law, whereas in Jordan, Oman, and Saudi Arabia it is protected only within reserves. The issue of trapping is significant in this species because Blanford's foxes are not fearful of traps. Both Geffen *et al.* (1992d) and Smith *et al.* (2003b) emphasized that foxes at their study site readily entered box traps. In Israel, mean trapping success was 30.3% and 15.3%, and the mean number of recaptures during the study was 8.0 ± 4.9 and 6.7 ± 3.5 for Ein Gedi and Eilat, respectively. These data indicate a potential to harvest rapidly a local population using only a few box traps. Unlike many other fox species, for which intensive trapping effort and elaborate techniques are required in order to eradicate a population, the Blanford's fox could easily be extirpated locally using simple means.

Finally, the threat from habitat loss is limited as most of the area where this species occurs in Israel is a nature reserve. Political developments may change the status of the northern Judaean Desert. Human development along the Dead Sea coasts may also pose a considerable threat to that habitat. Similar concerns exist for the populations in the United Arab Emirates. Large scale quarrying in Wadi Siji (one of the capture sites in the UAE) and the construction of a road through that wadi is the most likely reason for the lower density of foxes in the area. The rugged arid mountain ranges throughout the distribution of the Blanford's fox are low impact, and the present overall concern about habitat loss is minor. However, specific localities where Blanford's fox density is high should be protected from ecological alterations (e.g. agriculture, grazing) and habitat loss.

CHAPTER 12

Red foxes

The behavioural ecology of red foxes in urban Bristol

Philip J. Baker and Stephen Harris

Red foxes *Vulpes vulpes*, adult sisters © J. M. Macdonald.

Introduction

The red fox (*Vulpes vulpes*) is the most widely distributed extant canid species, and is present in a broad range of habitats ranging from arctic tundra to deserts to city suburbs. Throughout its range, the general social system is a territorial breeding pair accompanied by up to eight subordinate individuals. However, there is substantial plasticity in social organization (e.g. Newsome 1995; Cavallini 1996), principally through variation in territory size, group size, and group structure (Table 12.1). Such differences represent the optimal response to changing ecological conditions (Macdonald 1983) and mortality rates (e.g. Macdonald and Carr 1989). Because of this variability, and because individuals do not forage cooperatively, the red fox represents one model for investigating mechanisms promoting group formation *per se*, and for the evolution of group living in this taxon specifically.

Several evolutionary steps are required to move from the ancestral system of intra-sexual territoriality shown by most solitary carnivores (Sandell 1989) to the system that red foxes exhibit today (Fig. 12.1). These include the evolution of paternal care, the existence of territories that subordinate animals can share without any *a priori* need for benefits from group living, an increase in the costs associated with dispersal and the evolution of mechanisms that enhance subordinates' inclusive fitness in the absence of direct reproduction. To date, research has tended to focus on two stages within this pathway. Resource-based hypotheses have emphasized the importance of spatio-temporal patterns of resource availability in leading to the formation of territories that can support subordinates at little cost to the breeding pair (e.g. Macdonald 1981; Carr and Macdonald 1986; Bacon *et al.* 1991a,b).

Table 12.1 Intraspecific variation in red fox social organization. Studies are ranked by increasing home range size

			Home range size (ha)			Group size			Subordinate sex ratio	
Reference	Country	Habitat	Mean	Range	*N*	Mean	Range	*N*	% ♂	% ♀
Baker *et al.* (1998, 2000)	England	Urban	18	8-28	8	4.5	2-10	8	51	49
Macdonald (1981)	England	Urban	45	19-72	7	4.4	2-5	5	0	100
Adkins and Stott (1988)	Canada	Urban	53	24-75	6	?	2-3	2	0	100
Meia and Weber (1995, 1996)	Switzerland	Rural	106	49-248	13	5.0	3-8	7	36	64
Kolb (1986)	Scotland	Urban	116	61-233	9	?	2-4	2	0	100
Mulder (1985)	Netherlands	Rural	153	105-200	56	?	2-4	56 foxes	0	100
Tsukada (1997)	Japan	Rural	~200	~100-300	14	?	2-5	7	50	50
Poulle *et al.* (1994)	France	Rural	203	48-376	6	?	2-5	6 adults	0	100
Reynolds and Tapper (1995)	England	Rural	245	70-360	8	2.1	1-3	8	0	100
Phillips and Catling (1991)	Australia	Urban	368	130-530	3	?	1-2	3	?	?
Cavallini (1992)	Japan	Mountain	494	357-631	5	3	–	1	0	100
Macdonald *et al.* (1999)	Saudi Arabia	Desert	852	?	31	3.0	2-4	9	9	91
Niewold (1980)	Netherlands	Rural	928	?	5	?	2-4	?	?	>90
Jones and Theberge (1982)	Canada	Tundra	1611	277-3420	7	?	–	–	–	–

Individual-fitness based hypotheses have emphasized the cost–benefit trade-off between remaining on a territory as a non-breeding subordinate versus dispersing and attempting to become the breeding individual in a non-natal group (e.g. Lindström 1986; Macdonald and Carr 1989). Occasionally these have been viewed as disparate entities (e.g. Blackwell and Bacon 1993). While it is the case that without sufficient resources subordinates could not survive on a territory, in the absence of fitness benefits selection would favour subordinates that disperse. Therefore, both processes are fundamental to the variability in social organization seen in the red fox. Consequently, to identify the ultimate and proximate mechanisms underlying this plasticity, data are required on individual reproductive success from populations under different ecological conditions monitored over time.

Urban foxes represent an ideal subject for such long-term studies since territories are small, group sizes are large, they are easily trapped (Baker *et al.* 2001a), they can be radio-tracked at close quarters, recovery rates of dead animals are high, and detailed information on individual foxes can be obtained from resident householders who regularly see the foxes in their garden. Therefore, it is possible to obtain detailed life histories of known individuals. In this chapter, we review the work undertaken in a long-term study in Bristol, England investigating the mechanisms and benefits of group formation in a population of urban red foxes.

The Bristol fox project

Bristol's foxes have been studied continuously since 1977. Early studies focused on population-level processes such as abundance and distribution (Harris 1981; Harris and Rayner 1986), demography (Harris and Smith 1987), and dispersal (Harris and Trewhella 1988; Trewhella and Harris 1988; Woollard and Harris 1990; Harris and White 1992), partly to develop a rabies contingency plan for Britain. Since the 1990s, the project has focused on individual patterns of behaviour within a small number of social groups in the north west of the city (e.g. Saunders *et al.* 1993, 1997; White and Harris 1994; White *et al.* 1996; Baker *et al.* 1998, 2000, 2001b).

A summary of the changes occurring in this population is given in Table 12.2. Prior to 1990, spring

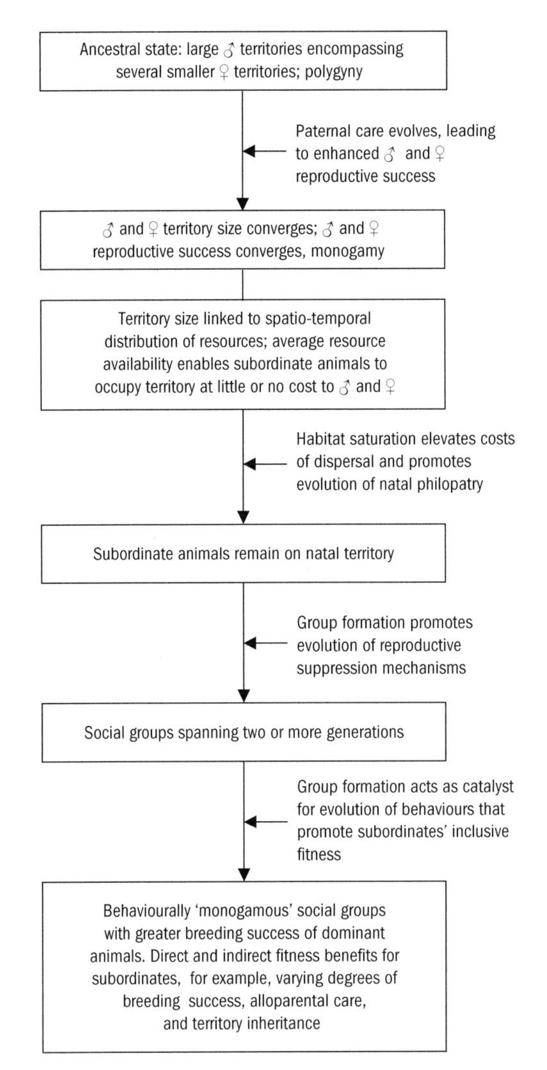

Figure 12.1. Potential pathway for the evolution of group living in the red fox. Male provisioning of females and cubs is a widespread feature of canid social systems and suggests one factor that may have promoted the convergence of male and female territory size, with males expending a substantially greater proportion of their investment on caring for the young of a single female. In the absence of any *a priori* benefits of group living, territories must be able to support subordinate animals without imposing costs on the dominant pair. Habitat saturation would increase the costs of dispersal, such that some individuals may opt to remain on their natal group. Such primitive groups may then act as a catalyst for the evolution of behaviours such as alloparental care, which would enhance the inclusive fitness of subordinate foxes.

density (i.e. adults and cubs) was approximately 20–30 foxes per km^2. Between 1990 and 1994, adult fox density increased dramatically, peaking at 37 foxes per km^2 in 1993; total fox density was highest in 1993 ($>$ 60 animals per km^2). These changes arose from increases in group size, the number of groups per unit area and the number of females breeding per group. In spring 1994, sarcoptic mange was detected in the population, and this triggered a dramatic decline in fox numbers; currently, fox density is approximately 10% of pre-mange numbers. For the purposes of this chapter, we will confine our discussion to the pre-mange period 1990–94, except where explicitly stated. During 1990–94, the study site comprised an area of 1.5 km^2 of predominantly medium-density housing; other habitat types include playing fields (21.3% of area), allotment gardens (2.1%), woodland (0.1%), and a cemetery (5.8%). During the mange outbreak, and up to the present day, the study area has been expanded to cover an area of approximately 9 km^2.

For each social group, we have attempted to obtain detailed information on the history of each fox born on the study site and those that moved into the study groups. Three general methods have been employed. A capture–mark–recapture programme has been used to obtain information on survival rates. Animals were cage-trapped (Baker *et al.* 2001a) and marked with ear tags. In particular, we tried to mark animals as cubs so that we could determine each fox's natal group. In addition to trapping, we requested that local householders supply information on foxes that they saw in their garden. To facilitate this process, full-grown foxes were fitted with radio-collars covered in coloured electrical insulating tape; in recent years, we have also marked foxes with unique combinations of colour-coded ear tags to help householders identify individuals.

Radio-tracking was used to quantify range utilization patterns and to delimit territorial boundaries. Daytime rest sites were used to monitor whether individuals had dispersed from a group. All animals reported dead were recovered for *post mortem* examination to obtain information on diet and female productivity rates, and to identify the fate of marked individuals. To maximize the number of foxes recovered, leaflets were delivered to every

Table 12.2 Summary of changes occurring in the Bristol fox population between 1980 and 1999. Reproduced with permission from Baker *et al.* (2001b)

	Spring 1980	Spring 1990	Spring 1993	Spring 1994	Summer 1995	Autumn 1995	Winter 1995	Summer 1998	Autumn 1999
Mean territory size (ha)	24	29	18	18	27	83	210	131	169
Group density (groups/km^2)	4.1	3.4	5.6	5.6	3.7	1.2	0.5	0.8	0.6
Mean group size (adults)	3.4	2.3	4.6	6.6	2.7	1.8	1.7	1.9	1.8
Adult density (adults/km^2)	13.9	7.8	25.8	37.0	10.0	2.2	0.9	1.5	1.1
Mean no. of females breeding per group	1.0	1.0	1.6	1.0	1.0	1.0	1.0	1.0	1.0
Mean emergent litter size	3.8	3.8	4.3	3.8	3.4	2.8	2.2	3.0	2.7
Cub density (cubs/km^2)	15.6	12.9	38.5	21.3	12.6	3.4	1.1	2.4	1.6
Total fox density (foxes/km^2)	29.5	20.7	64.3	58.3	22.6	5.6	2.0	3.9	2.7

house on the whole study site in several years requesting reports of foxes found dead. To date, we have ear tagged > 2350 foxes, of which approximately 280 have been radio-collared: we have recovered > 3200 foxes dead for *post mortem* examination, of which 1070 were tagged (46% of those tagged).

We have four goals in this chapter: (1) to summarize the behavioural ecology of this population; (2) to draw comparisons with other studies in other habitats; (3) to synthesize the information currently available for this species; and (4) to identify areas for further research. In particular, we address the following questions:

1. Does resource availability allow subordinate animals to exist on a territory at little or no cost to the dominant pair?
2. Does habitat saturation increase the costs of dispersal?
3. Do patterns of behaviour such as alloparental care and territory inheritance mechanisms enhance the fitness of subordinate animals?

Does resource availability allow subordinates to exist on a territory at little cost?

The Resource Dispersion Hypothesis (RDH: Macdonald 1981) suggests one mechanism by which subordinate animals are able to occupy a territory at little or no cost to the dominant pair. Specifically it proposes that dominant animals configure their territory in relation to their resource requirements during some critical limiting period so that, on average, outside the critical period, there are more resources than required by the breeding pair. Such territories would, therefore, be able to support subordinates under some circumstances.

Under this hypothesis, group size and territory size will generally not be linearly related, as group size is dependent on average resource availability, while territory size is correlated with the spatial distribution of resources during the critical period. Groups should also be more prevalent and larger in highly variable environments, as dominant animals would need to defend territories with greater average resource availability to compensate for the temporal variation (Carr and Macdonald 1986). Lastly, territory size should remain constant despite short-term variation in the pattern of resource availability; for example, if the period of lowest food availability to which territory size was geared was the animal's life, then territory size would be expected to remain constant within the lifetime of a dominant individual. However, group size may vary as resources fluctuate.

This hypothesis appears readily testable. For example, Kruuk and Macdonald (1985) proposed that removing a single key food patch during the critical period would make a territory untenable, and as a result the dominant pair would need to enlarge their territory. Yet, few such manipulation experiments have been undertaken, and there are several reasons for this. For example, territory sizes infer researchers must work over relatively large spatial scales (Table 12.1) and populations are often perturbed (e.g. Harris and Saunders 1993) so obtaining long-term data on individuals is problematical. But perhaps the most fundamental problems are identifying the important resource(s) and the critical period, and quantifying long-term patterns of resource availability, particularly during the critical period. Foxes take a wide variety of food types (e.g. Baker and Harris 2003) and will use many different localities for resting and breeding (e.g. Harris 1980; but see Lucherini *et al.* 1995). Such flexibility means that it is often difficult to identify which resource(s) need to be measured.

In urban areas, however, foxes are often reliant on food derived from human sources, mainly food deliberately supplied by householders; in Bristol, this comprised 60% of the diet by volume (Saunders *et al.* 1993). Temporal trends in the availability of food deliberately supplied was, therefore, quantifiable by questioning householders about their feeding practices. By questioning householders over a number of years, it was also possible to determine temporal trends in food availability throughout the period of study; these could then be compared to observed changes in group size and territory size (Baker *et al.* 2000).

Group size and territory size were not linearly related (Baker *et al.* 2000). There also appeared to have been a consistent increase in the amount of food supplied by householders in several territories, and this was matched by an increase in the size of some social groups. Other studies have also documented changes in group size but not territory size in relation to changes in food availability (von Schantz 1984a; Lindström 1989; Zabel and Taggart 1989; Meia and Weber 1995).

During the study, two territories divided to form two new groups. In both cases, females budded off part of their natal territory from their parent(s). For each of the new territories occupied by the parent(s), the amount of food available each week was broadly similar and approximately equal to the basic energetic requirements of a pair of foxes. However, the absolute number of houses supplying food varied markedly (range 14–49) (Baker *et al.* 2000). By contrast, Macdonald (1981) found that territories in Oxford, UK contained similar proportions of human-associated habitats and houses, but the amount of scraps provided per week varied considerably between territories (range 4.7–8.9 kg).

The results from Bristol are, therefore, only partially consistent with the RDH. Territories would not have been expected to divide within the lifetime of a dominant individual, as any reduction in territory size in response to a short-term increase in food availability would make the territory untenable in the long term: territory holders would experience elevated costs as they fought to increase their territory size in response to subsequent declines in food availability (von Schantz 1984b). We were not able to observe

Figure 12.2 Red fox, *Vulpes vulpes* © D. W. Macdonald.

whether these costs were realized in Bristol as the population subsequently declined because of an epizootic of sarcoptic mange (Baker *et al.* 2000, 2001b).

However, during this decline, dominant foxes increased territory size in the absence of any decline in food availability (Baker *et al.* 2000). At present, territories are still of a comparable size to those seen at the end of the mange epizootic (Table 12.2). Therefore, under some circumstances, dominant individuals may defend territories larger than the minimum required to meet their own requirements. This would suggest one other mechanism by which subordinates could survive on the territory of a dominant pair. Previously, this has been considered unlikely since, in the absence of concomitant benefits from subordinates, larger territories would impose additional defence costs (e.g. Blackwell and Bacon 1993). Yet non-minimal territories may not be atypical for a species that is susceptible to diseases such as rabies and sarcoptic mange (Chautan *et al.* 2000; Forchhammer and Asferg 2000), which is widely culled (Harris and Saunders 1993), and where territory size may be governed by interspecific processes (Sargeant *et al.* 1987). Non-minimal territories would always increase the food security (Carr and Macdonald 1986) of the dominant pair, and would be favoured where the benefits exceeded the costs of territorial defence. However, we currently know very little about the costs of territorial defence in this species, and how this varies with territory size and population density.

Although the RDH and non-minimal territories propose different mechanisms by which territories may support subordinates in the absence of cooperative benefits, they do not consider the fundamental question of whether subordinate animals exert a cost. For example, subordinate animals may enhance the likelihood of disease transmission. Subordinates may also increase the foraging burden on dominant animals through serial exploitation of food patches. Dominant animals would, therefore, need to expend more time and energy foraging, even where there was enough food to feed all the foxes on a territory, and this may in turn increase the mortality rate of dominant animals, for example, through decreasing physical condition and/or increasing exposure to certain mortality factors. To identify whether such costs exist, it would be necessary to compare the movement patterns and survival rates of dominant animals in the presence and absence of subordinates for constant levels of resource availability. Such comparisons may ultimately require group-manipulation experiments, where subordinate animals are removed temporarily, to control for differences between territories.

In summary, it is still unclear how resource availability interacts with territory defence economics and strategies to promote group living in this species. In particular, there is the need for experimental manipulations to test some of the predictions arising from the theoretical studies into the possible effects of spatio-temporal variation in resource availability. Urban fox populations may be particularly amenable to such field tests. Furthermore, there has been little focus on the costs of group living. However, all these aspects need to be addressed in populations under a range of ecological conditions.

Does habitat saturation increase the costs of dispersal?

The most plausible explanation for the evolution of natal philopatry in the red fox is habitat saturation. A saturated habitat would increase the costs of dispersal by reducing the opportunities for dispersing individuals to find vacant positions (Emlen 1982), although dispersers may be able to oust existing dominant animals and the success rate of this strategy may be dependent on individual quality rather than density. For example, in Bristol, some dispersing males managed to become the dominant animal in a group even where natal males were present (Baker *et al.* 1998). This would also facilitate out-breeding.

Generally, there is a paucity of quantified information concerning the costs and benefits of dispersal to individuals. The costs are generally believed to be associated with the act of dispersing itself, such as the risk associated with crossing unfamiliar terrain and through conflict with existing territory holders. These costs may, therefore, be manifested as elevated mortality rates, a decline in physical condition and risks of injury. Benefits are correlated with the attainment of breeding status in a non-natal group and are, therefore, associated with an increase in the number of dependent offspring produced.

Ultimately, however, the costs and benefits of dispersal should only be considered in comparison with the costs and benefits of natal philopatry: when the net benefits of dispersal exceed the net benefits of remaining, juveniles should disperse, and *vice versa*. Therefore, to compare these two strategies, all the elements of both dispersal and philopatry need to be quantified concurrently. For most studies this has proved logistically difficult. For example, our studies in Bristol have tended to focus on philopatric individuals since these, by definition, have remained on the study site. By contrast, although we have quantified many aspects relating to dispersal (e.g. Harris and Trewhella 1988; Trewhella *et al.* 1988; Woollard and Harris 1990; Harris and White 1992), we have not been able to quantify, for example, the mortality risk associated with extra-territorial movements and the rate with which dispersing animals become the dominant animal in a non-natal group. Recent advances in technology (e.g. satellite radio-tracking and DNA analysis) should facilitate such studies, as they will increase the amount of data that can be collected on individual life histories and movements. However, it is currently unclear how dispersal costs vary with mortality rates and population density, and under what conditions habitat saturation would promote natal philopatry.

Do alloparental care and territory inheritance enhance subordinate fitness?

Subordinate foxes perform a number of alloparental behaviours, including babysitting and provisioning nutritionally dependent young with food (Macdonald 1979a). For such behaviours to increase individual inclusive fitness, it is necessary that (1) cubs and alloparents are related and (2) the care given by alloparents increases cub survival.

Intra-group relatedness

Fox groups are generally characterized as comprising a monogamous breeding pair and related subordinates, typically offspring from previous years (e.g. Macdonald 1983). Under such a system, alloparents would generally be related to cubs by a factor of 0.5. In Bristol, all subordinates whose origins were known (i.e. tagged as cubs) were philopatric cubs or older females that had lost their dominant status (Baker *et al.* 1998), although we were not able to document the origins of all subordinates and some immigrating animals from outside the study site would not have been detected using our methodology. Niewold (1980) also found that subordinates were offspring retained from previous years, and this is consistent with a number of tagging studies (e.g. Trewhella *et al.* 1988). By contrast, in Zabel's (1986) study, subordinate animals were unrelated to the dominant pair.

Relatedness between helpers and cubs may also be complicated by variations in patterns of mating. Many studies have documented that more than one litter may be produced by groups where food availability permits (e.g. Macdonald 1979a; von Schantz 1984a; Zabel and Taggart 1989; Baker *et al.* 1998). Furthermore, it is often proposed that subordinates may regularly become pregnant as an insurance against the death of the dominant female (e.g. von Schantz 1981), but that this litter is typically lost through infanticide (see Harris and Smith 1987). Dominant males may find it hard to monopolize access to all females in their group where several females conceive. Under such circumstances, paternity assurance mechanisms such as mate-guarding and post-copulatory locks may evolve, and both are seen in foxes. The presence of such mechanisms indicates an evolutionary history of male–male competition for access to females. In addition, males are known to make extra-territorial movements during the breeding season (e.g. Zimen 1984; White *et al.* 1996). Such observations suggest that cuckoldry may be common in this species.

To investigate the pattern of paternity in Bristol's foxes, we collected tissue samples from cubs born during 1992–94 for DNA analysis. At this time, population density was very high and comprised equal numbers of subordinate males and females (Baker *et al.* 2000). Analyses of these sample with colleagues at the Institute of Zoology have shown a complex mating strategy, with dominant and subordinate males siring cubs both in their own and neighbouring groups, and that mixed paternity litters were present (Baker *et al.* in press). Such patterns of mating mean that, although we are confident that most subordinates were offspring from previous years, it was not easy to assess the degree of relatedness between adults in a social group and between helpers and cubs. Such assessments were further complicated by the pooled rearing of cubs, so that often it was not possible to determine even mother–offspring dyads. Similar maternity–paternity studies in other populations are also required to determine whether the pattern observed in Bristol is typical of the red fox, or whether the degree of cuckoldry is related to population density. If this is the case, then the retention of offspring from previous years does not necessarily imply that alloparent–cub relatedness is high.

Group size and cub survival

The effect of alloparental care on cub survival has typically been measured by comparing the number of adults with the number of cubs surviving to a chosen age (e.g. Moehlman 1979). Yet, such comparisons may be confounded if all adults do not help to the same degree. For example, von Schantz (1984a), and Meia and Weber (1996) both recorded adults that did not provision cubs to any degree, and this was also evident in some groups in Bristol. Therefore, a more appropriate measure would be, for example, the number of provisioning trips versus the number of surviving cubs.

In Bristol, such an analysis showed that the number of provisioning trips for each group was consistent, but that in larger groups each individual simply worked less hard. Consequently, there was no observable effect of alloparental care on cub survival (Baker *et al.* 1998), and it is yet to be shown that alloparental care increases cub survival in any study of red foxes, although Zabel and Taggart (1989) documented an increase in the number of cubs produced by each female in polygynous versus monogamous groups. The benefits of alloparental care are most likely to be evident where the costs of central-place foraging are high. This is likely to be the case where foraging for appropriate sized items is difficult and territory sizes are large. This may now be the case for Bristol's foxes, where territory sizes are approximately nine times larger than before the outbreak of mange (Baker *et al.* 2001b). Comparable data are, therefore, required for this post-mange population, as well as from other studies. Food-provisioning behaviour can also be easily manipulated by artificially supplying food that parents and alloparents can or cannot take back to the cubs.

Territory inheritance

In the absence of cooperative gains, a subordinate may benefit simply by remaining at home and waiting to inherit the territory from its same-sex parent (Lindström 1986). For such a strategy to be selected, the likelihood of a subordinate attaining dominant status must be greater by philopatry than by dispersing. An individual's decision to remain will, therefore, be partly dependent on the mortality rate of subordinate animals relative to the mortality rate of dominants. In Bristol, subordinate foxes appeared to have a significantly lower life expectancy (mean (±SD) age at death 2.1 ± 1.1 years) than dominants (4.5 ± 1.1 years) (Baker *et al.* 1998). Most subordinates, therefore, would not outlive their parents and, given that group sizes were large, most survivors would not inherit a territory either. Furthermore, the data we have suggest that the costs for those animals that dispersed were not extreme (Baker *et al.* 1998). It is not clear, therefore, why so many animals remained on their natal groups, although cub productivity in this population was high and subordinates may have benefited by being able to reproduce on their natal ranges; based on observational data (i.e. not DNA analysis), we estimated that 22 litters were produced in 15 group-years where subordinate females were present, an average of 1.5 litters per group (Baker *et al.* 1998). Furthermore, there was evidence of a relationship between fox numbers and food supplied by householders, with more food being supplied in response to the number of foxes remaining in a territory. Under such conditions, the costs of natal philopatry would have been further reduced.

Conclusions

It is apparent that we still understand very little about the complexity of social organization in the red fox and we need more long-term studies on individual populations, such as we have undertaken in Bristol. In this study, there was a clear relationship between average food availability and the size of some social groups, but the relationship between territory size and food availability was less clear (Fig. 12.3). Prior to the outbreak of mange, two territories divided in response to an increase in the amount of food supplied by householders, and following the outbreak of mange, territories increased in size despite no apparent decline in food availability. This flexibility in territory size within the lifetime of territory holders is generally not consistent with the RDH. In particular, we have shown that dominant animals will defend territories larger than required solely to meet their own requirements. Such a strategy would enhance the effects of habitat saturation, as it would constrain the number of territories that could be accommodated in a given area.

We could detect no effect of alloparental care on cub survival (Fig. 12.3). This may have arisen from the complex pattern of mating which affected relatedness between cubs and helpers, even though most helpers were philopatric offspring from previous years. The major benefit accruing to subordinates appeared to be the opportunities for direct breeding in their natal group. However, we have not yet been able to quantify the costs associated with competition for food within groups and the costs of dispersing in a high-density population.

Therefore, we believe that future research on red fox social systems should focus on (1) the costs of group living, (2) the costs of dispersal, (3) mating strategies, (4) the benefits of alloparental care, and (5) territory defence strategies in relation to resource availability. Most usefully, comparable data are needed from a broad spectrum of ecological conditions. From the perspective of the Bristol study, we are quantifying these factors following the outbreak of sarcoptic mange: current projects include the spatiotemporal patterns of food availability (R. Ansell personal communication), male mating behaviour (G. Iossa personal communication) and dispersal behaviour (C. Salsbury personal communication). Such data will enable us to compare the effects of food availability, population density, and mortality rates on the costs and benefits of group living in this species.

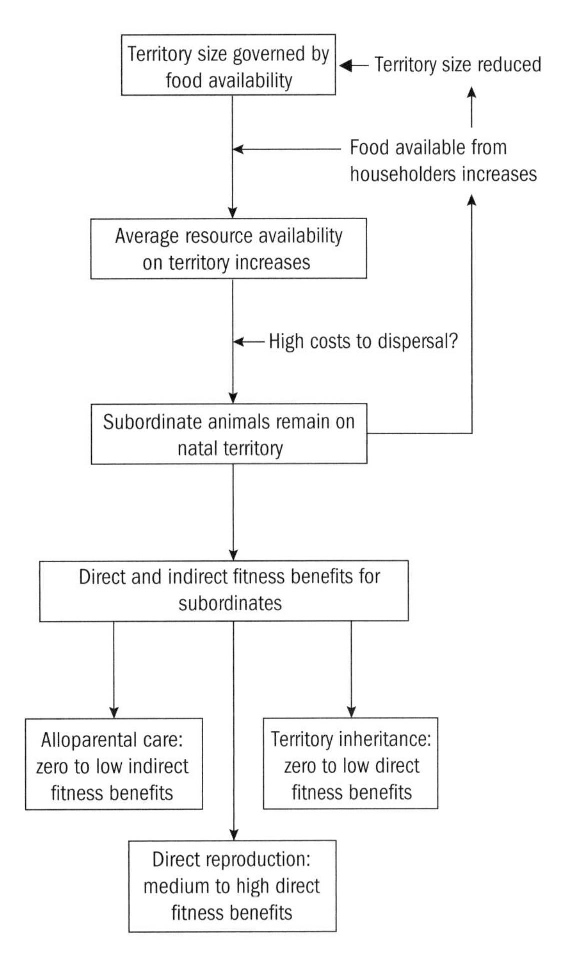

Figure 12.3 Summary of the mechanisms promoting group formation in Bristol's foxes. During 1990–94, there was a clear increase in the amount of food deliberately supplied by householders. This in turn led to an increase in group size and the number of social groups. Increased fox density also contributed to more householders feeding the foxes. We were not able to identify whether the costs of dispersal were a significant contributory factor in promoting the level of natal philopatry observed. Alloparental care did not appear to increase cub survival, and most subordinate animals would not have benefited from the inheritance of their natal territory. The major benefit to subordinate males and females appeared to be the opportunity for direct breeding.

Acknowledgements

The studies outlined in this chapter have been funded by the Australian Meat and Livestock Research and Development Corporation, the Dulverton Trust, the German Academic Exchange

Service, the International Fund for Animal Welfare, the Leverhulme Trust, the Ministry of Agriculture Fisheries and Food (now the Department for Environment, Food and Rural Affairs), the Natural Environment Research Council, The Royal Society, the Royal Society for the Prevention of Cruelty to Animals and the Science and Engineering Research Council. We would also like to thank *British Wildlife* for permission to reproduce Table 12.2, and the many members of the general public who have helped with this research.

CHAPTER 13

Raccoon dogs

Finnish and Japanese raccoon dogs—on the road to speciation?

Kaarina Kauhala and Midori Saeki

Raccoon dog *Nyctereutes procyonoides*. © Great Tanuki Club.

Japanese raccoon dogs (*Nyctereutes procyonoides viverrinus*) have been isolated from the populations (e.g. *N. p. ussuriensis*) on the mainland of Asia for about 12,000 years. Since the environment and climate of Japan differ greatly from that on the mainland, different selection pressures have affected the two populations. Several features of Finnish (*N. p. ussuriensis*, originally from SE Russia) and Japanese raccoon dogs were compared in order to evaluate the progress of the Japanese raccoon dog towards speciation. We reviewed the chromosome number, skull and tooth morphology, body size and weight, the ability to hibernate, reproduction, home ranges, habitat use and diet of Japanese and Finnish raccoon dogs.

Japanese raccoon dogs have fewer chromosomes than Finnish specimens, the difference being due to centric fusions. Skulls and teeth also separate well the specimens from the two provenances; the skull is larger and mandible is more robust in Finland, indicating that Finnish raccoon dogs are more carnivorous than Japanese raccoon dogs. The longer tooth rows and larger molars of Japanese raccoon dogs also suggest a more insectivorous/frugivorous diet in Japan. Since the climate is colder in Finland, Finnish raccoon dogs hibernate in winter and are able to gather large fat reserves in autumn. The volume of the stomach is larger and insulation of fur better in Finland than in Japan where raccoon dogs do not hibernate. Finnish raccoon dogs are also bigger than Japanese specimens. The raccoon dogs are monogamous in both areas and the male participates in pup rearing, but the litter size is larger in Finland; females save energy during winter lethargy and can thus invest more in reproduction. Differences in chromosome number and skull and tooth morphology suggest that the Japanese raccoon dog has adapted genetically to a new environment, a conclusion supported by aspects of their physiology and reproductive and behavioural biology. The species status of the Japanese raccoon dog should thus be seriously considered.

Introduction

Raccoon dogs in Japan have been isolated from those on the Asiatic mainland for 12,000 years or so. After the last glacial age (25,000–15,000 years ago), dominant forests of Japan changed drastically from boreal and temperate coniferous forests about 20,000 years ago to warm-temperate, evergreen broadleaf forests about 13,000–6000 years ago (Tsukada 1984; Matsuoka and Miyoshi 1998). The Japanese environment differs greatly from that on the mainland, raising the question of whether the Japanese raccoon dog is on its way to, or has achieved, species status. Our studies have focused on raccoon dogs in Finland (introduced from the Russian Far East only in the last century and known as the 'Ussuri raccoon dog') and Japan. In the Far East, the Ussuri raccoon dog lives in small broadleaf forests, shrub thickets, marshy areas, meadows, and on shores of lakes and streams, but avoids large coniferous forests (Novikov 1962; Stroganov 1969). As in the Far East, in Finland raccoon dogs prefer habitats with dense undergrowth along lakes and rivers, but they also inhabit coniferous and mixed forests (Kauhala 1996a). The climate is much colder in Finland than in Japan (excluding Hokkaido), the mean temperature of the year being 2–5°C in southern and central Finland. Winters are harsh, snow covers the earth from November to mid-April and the air temperature can fall below −20°C. The climate in the Russian Far East is similar to that in Finland, but winters can be even colder (Bartholomew and Son 1987). The climate of the Japanese Archipelago is diverse with the mean annual temperature ranging 5.9–22.4°C, however, the monthly mean temperature seldom falls below freezing point except for Hokkaido and high-altitude areas, and the annual precipitation averages 1852 mm (Observation Department 1992). Net Primary Productivity is higher in Japan than in Finland, for example, 15.5 ton/ha/year in Chiba in Honshu (Seino and Uchijima 1988) and 10 ton/ha/year in southern Finland. The differences between Finnish and Japanese raccoon dogs are likely to be attributable largely to the different selection pressures they have experienced by their ancestors during the past 12,000 years in Japan and in the Russian Far East. Adaptation to the Finnish environment during the past few decades may, however, also have affected the characteristics of Finnish raccoon dogs. To evaluate the progress of Japanese raccoon dogs towards speciation we compare here different features of Finnish and Japanese raccoon dogs.

Evolutional history

The natural range of the raccoon dog spans much of China, north-east Indochina, Korea, Amur, and Ussuri regions of Eastern Siberia, Mongolia and Japan (Ellerman and Morrison-Scott 1951; Ward and Wurster-Hill 1990; Henry Max, personal communication; Fig. 13.1). One sub-species, *N. p. ussuriensis* originally from south-east Russia has been introduced to eastern and northern Europe (Lavrov 1971), and now occurs in, for example, Finland, Russia, the Baltic States, Poland, and Germany (Mitchell-Jones *et al.* 1999). Five or six subspecies have been described including *N. p. viverrinus* in Japan (excluding Hokkaido) and *N. p. albus* in Hokkaido (Ellerman and Morrison-Scott 1951).

The forerunners of raccoon dogs were widely distributed throughout Europe and Asia during the mid-Pliocene (Ward *et al.* 1987), when *Nyctereutes*-species also occurred in Africa (Werdelin 2001). The earliest known *Nyctereutes* fossil from Africa (*c.*3.7 million years old) is from Northern Tanzania (Laetoli). *Nyctereutes donnezani* was found in Europe 4 million

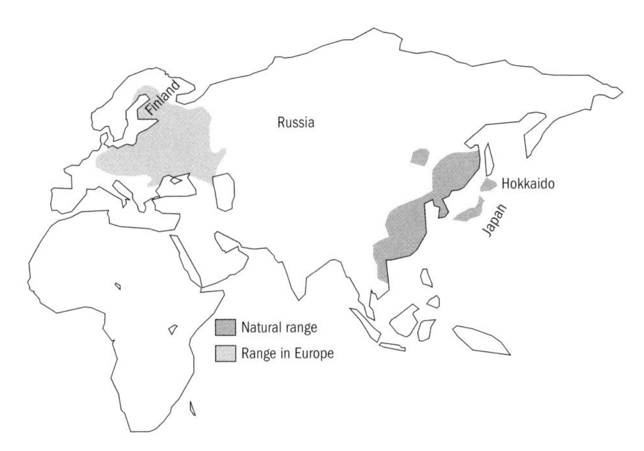

Figure 13.1 A map showing the natural range of the raccoon dog in the Far East and the range in Europe. (Modified from Ward and Nurster-Hill 1990.)

Latitude	Longitude	NPP	Home range	Reference
36.33	136.39	12.6	6.9	Fukue (1991)
33.14	129.66	17.3	10.3	Ikeda (1982)
31.41	130.51	17.6	30	Ikeda (1982)
35.31	139.4	15.8	30.7	Yamamoto (1993)
31.41	13.51	17.6	49	Ward and Wuster-Hill (1989)
39.48	140.13	11.1	59	Ward and Wuster-Hill (1989)
35.22	140.2	15.5	278	Saeki (2001)
35.53	138.1	12.2	609.5	Yamamoto *et al.* (1994)
		8	950	Kauhala
		10	700	Kauhala
		ton/ha/yr		

Peason correlation
$r = -0.767$ ($n = 10$)
$r = -0.980$ ($n = 8$)

NPP data from:
Seino and Uchijima (1988): Mesh maps of net primary productivity of natural vegetation of Japan (BCP-88-I-2-2). NIAES, Japan, p.131
NIAES: National Institute of Agro-Environmental Sciences

years ago (Kurtén 1968). *Nyctereutes megamastoides*, a large ancestor of raccoon dogs, lived in Europe and a similar form, *Nyctereutes sinensis* in China during the Pliocene and the early Pleistocene (Kurtén 1968; Ward *et al.* 1987). The distribution of the genus decreased during the Pleistocene; *N. megamastoides* became extinct, while *N. sinensis* became smaller, and the Later Pleistocene Chinese forms evolved to the modern species (Kurtén 1968).

The ancestors of the Japanese raccoon dog (*N. p. viverrinus*), which today inhabits Japan excluding Hokkaido, and the Ezo raccoon dog (*N. p. albus*), inhabiting Hokkaido, probably colonized Japan between 0.4 Ma and 12,000 years ago either through the Sakhalin or Korean peninsulas (Ward *et al.* 1987; Kawamura 1991; Dobson and Kawamura 1998). *N. viverrinus nipponicus* occurred in mid-Pleistocene in Japan, and was probably the link between *N. sinensis* and modern Japanese raccoon dogs (Shikama 1949; Y. Kawamura personal communication). When the Japan Sea opened about 12,000 years ago, Japanese raccoon dogs became isolated from the populations in the Asiatic mainland and adapted to a mild marine climate. These conditions differ greatly from those in SE Russia from where *N. p. ussuriensis* was translocated to Europe. Since more than 9000 raccoon dogs were introduced to European parts of Russia, founder effects are unlikely to have had a major effect on the gene pool of Finnish raccoon dogs.

Introduction to Europe

Motivated by the value of their fur, *N. p. ussuriensis* from SE Russia were introduced to European parts of the former Soviet Union during the first half of the twentieth century (Lavrov 1971). It spread rapidly (Fig. 13.2) and was first detected in Finland in the late 1930s (Siivonen 1958). During the 1940s, raccoon dogs were reported only occasionally in Finland, including Lapland, but they invaded in substantial numbers from the southeast during the 1950s and 1960s, and by the mid-1970s were widespread throughout the southern and central parts of the country (Helle and Kauhala 1991). The Finnish population peaked in the mid-1980s, and remained stable thereafter (Kauhala and Helle 1995). Raccoon dogs are now among the most numerous carnivores in Finland.

Winter lethargy

Raccoon dogs are unique among canids: they hibernate in areas where winters are harsh—like SE Russia and Finland—but remain active in areas

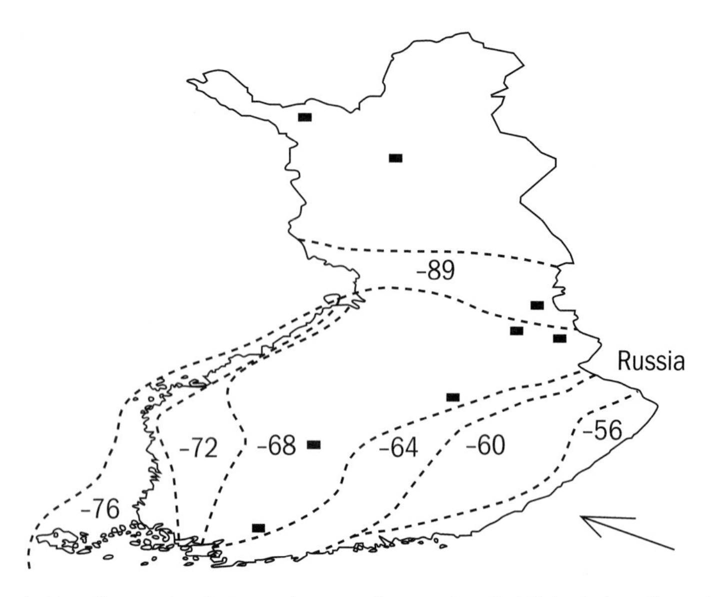

Figure 13.2 A map showing how the raccoon dog spread over southern and central Finland since the early 1950s. The arrow indicates the main route of colonization. The squares show the places of early observations of raccoon dogs during the 1930s and 1940s. (Siivonen 1958; Helle and Kauhala 1991)

with milder winters. Winter lethargy lasts usually from November until March in Finland, depending especially on snow conditions. During mild weather raccoon dogs may be active in mid-winter, especially in southern Finland. Soft, deep snow impedes the raccoon dog's movement, whereas they can move around when snow is scant or hard (and they follow roads and ski tracks in winter). Adults retreat to their dens earlier in autumn than do juveniles. The winter dens are usually in the middle of home ranges, although some raccoon dogs choose a winter den outside their normal home range (Kauhala *et al.* 1993a). An individual may use up to five different dens during winter.

Winter lethargy enables raccoon dogs to live at northern latitudes where almost no food would be available during winter. Badgers (*Meles meles*) are similarly inactive in winter, but the red fox (*Vulpes vulpes*) forages throughout. These traits reduce competition during the harshest season, but winter lethargy also determines the northern limit of the raccoon dog's distribution, as they need sufficient summer to gather enough fat before the winter to survive. Juveniles must first grow, and only after reaching adult body length, usually in October, can they accumulate fat reserves (Kauhala 1993). Adult raccoon dogs almost double their weight between early summer and late autumn; in late autumn, the mean fat content of individuals is 3.5 kg (43% of the total weight; Kauhala 1993). In addition to a thick layer of subcutaneous fat, fat reserves in the body cavity average 134 g (17–224 g, $n = 138$) and 84 g (0.5–329 g, $n = 675$) for adults and juveniles, respectively. The stomach volume of Finnish raccoon dogs is larger than that of Japanese specimens (160 and 75 ml, respectively), enabling them to eat more at a time (Korhonen *et al.* 1991); this may be connected to their ability to gather larger fat reserves. Also the insulation of fur is better in Finnish than in Japanese raccoon dogs; the amount of heat required to keep body temperature constant is in Japanese raccoon dogs 35–40% greater than in the Finnish raccoon dogs (Korhonen *et al.* 1991). In the mild, maritime climate of Japan, raccoon dogs do not show winter lethargy, and they have less dense fur and show no evidence of gathering the large fat reserves typical of Finnish (Ussuri) specimens.

Body size and weight

The mean body weight of adult Finnish raccoon dogs is 5.0 kg (3.1–7.3 kg, $n = 36$) in early summer (May–June) and 8.1 kg (3.9–12.4 kg, $n = 64$) in October when they are at their heaviest (Kauhala 1993, 1996b). The mean body weight of juveniles is 7.0 kg (3.3–9.7 kg, $n = 273$) in November when they are heaviest. The mean head and body length of adult raccoon dogs in Finland is 59.8 cm (51.5–70.5 cm, $n = 760$) for both sexes (Kauhala 1993, 1996b). Body weight/size thus differs between seasons and age groups but not between sexes.

Similarly, in Japan there is no sexual dimorphism in body weight or length. The mean body weight of adult Japanese raccoon dogs is 4.5 kg for both sexes (2.5–6.25 kg, $n = 64$), peaking at a mean of 5.0 kg in February and falling to 3.95 kg in May, and their mean head and body length is 56.7 cm ($n = 43$) (Kauhala and Saeki in press).

The smaller body size in Japan may indicate an adaptation to a milder climate, as mammals tend to be larger in colder areas (James 1970).

Reproduction

Methods in Finland

We examined 2647 female carcasses collected from hunters, including 599 $\geq$ 1 year (raccoon dogs reach sexual maturity at the age of *c.*10 months). To estimate the date of ovulation we examined the size and configuration of uterine horns and weighed and measured the foetuses. The number of *corpora lutea* was counted from histological sections and was considered an estimation of the number of ova, that is, fecundity. The embryonic and birth litter sizes were determined by examining the number of placenta and foetuses (spring sample) or placental scars (autumn sample; Helle and Kauhala 1995).

Time of ovulation in Finland

Females ($n = 109$) ovulated between 12 January and 18 April (Helle and Kauhala 1995; K. Kauhala, unpublished data), the mean date being 8 March (SD = 13.9 days). Only 2% of females ovulated in January and 25% before the 1 March, 69% by 15 March and 95% before the end of March. Since gestation lasts 2 months, most pups are born during the first half of May.

Females in SW Finland ovulated earlier than those in SE Finland; 46% of females ovulated by the 1 March and 85% by the 15 March in SW coast of Finland, the corresponding figures being 11% and 63% for SE Finland. This difference is due to the fact that climate is milder and winters shorter in SW Finland, and consequently, raccoon dogs become active earlier after their winter lethargy in SW Finland.

Fecundity and litter size in Finland

The mean number of *c. lutea*, of embryos and litter size at birth was higher in SW Finland than in NE Finland (Table 13.1; Helle and Kauhala 1995; K. Kauhala, unpublished data). The mean condition index (body weight/head-body length^2) of females in each area correlated positively with the mean number of *c. lutea* ($r = 0.94$, $p = 0.060$), the mean number of embryos ($r = 0.99$, $p = 0.013$) and the mean litter size ($r = 0.99$, $p = 0.002$).

The very high litter size in Finland probably is due to three features in the species's ecology: winter lethargy, omnivory, and monogamy. Because raccoon dogs are inactive in winter and have large fat reserves, they pass the winter in good condition regardless of the weather and food availability

Table 13.1 The number of *c. lutea*, embryos and pups at birth (mean ± SD) in southwestern and northeastern parts of the raccoon dog range in Finland

Area	*Corpora lutea*	Embryos	Litter size at birth
SW Finland	12.8 (±3.6)	10.4 (±3.1)	9.5 (±3.2)
NE Finland	10.4 (±2.6)	9.0 (±3.4)	7.0 (±2.6)
Mean	11.4 (±2.9)	9.6 (±2.5)	8.9 (±2.6)
Range	5–23	1–18	1–16
N	220	430	371

during winter and can thus invest heavily in reproduction (Kauhala 1996b). Because they are extremely omnivorous, they always find something to eat and the litter size is only slightly affected by varying food conditions. Finally, raccoon dogs are strictly monogamous in Finland and the female can rely on the help from the male in pup rearing. Litter size is similar (mean = 9) in SE Russia (Judin 1977).

Productivity of the population in Finland

The proportion of breeding females in the population averaged 80% (Helle and Kauhala 1995; K. Kauhala, unpublished data), being highest in SW Finland (85%) and lowest in NE Finland (67%). The productivity of the raccoon dog population was thus 7.2 pups/all adult females in the population. The productivity was highest in SW Finland (7.9) and lowest in NE Finland (4.7).

Climate seems to be the most important factor affecting productivity of the raccoon dog population in Finland (Kauhala and Helle 1995). The onset of spring determines the time of ovulation and, thus, the time when pups are born. In southern Finland pups are born earlier and have a longer time to grow and gather fat reserves before the onset of winter lethargy than do pups born in more northern areas. Therefore, young females are much heavier in late autumn in southern than in northern Finland and a larger proportion of them reproduces the following spring. Climate thus affects the proportion of reproducing females in the population, which is very important for the total productivity of raccoon dogs (Kauhala and Helle 1995).

Mean weight of raccoon dog pups at birth in Finland is 122 g (Kauhala 1996b). Mean litter weight is thus 1080 g, which is 21% of the mean weight of the female. These figures show that raccoon dog females invest heavily in reproduction, compared for instance with the red fox; the weight of a fox litter is 10–13% of the mean weight of the female (Kauhala 1996b).

Reproduction of the Japanese raccoon dog

The basic reproductive physiology of the raccoon dog is similar to that of other canids, and they too have a copulatory tie (Ikeda 1982). Testosterone levels in males peak in February–March in Japan, and progesterone levels in females coincide, even in the absence of males, suggesting that the raccoon dog is a monoestrous, seasonal, and spontaneous ovulator (Yoshioka *et al.* 1990). Raccoon dogs reach sexual maturity at 9–11 months, although yearling females tend to ovulate later than older females (M. Saeki, personal observation). Mean weight of 109 g at birth were reported in Japan (Ikeda 1983). The litter size of the Japanese raccoon dog is around 4 (2–5) (Okuzaki 1979; Ikeda 1983). Kinoshita and Yamamoto (1993) reported one female with 10 embryos in a sample of 43 female carcasses. In this urban environment the average number of *c. lutea* was 6.3 ($n = 6$, range 4–9+) from the carcasses, mostly road-kills, collected in Kawasaki-city.

Home ranges, mating system and parental care

Finnish raccoon dog

In Finland, there is no evidence of raccoon dogs living in groups, other than a pair and their pre-dispersal juveniles. Rather, raccoon dogs are monogamous, the pair sharing their home range and usually also moving together throughout the year (Kauhala *et al.* 1993a). The core areas of the home ranges of different pairs do not overlap, especially in the breeding season. In the autumn, peripheral areas may overlap, but non-paired adults were never located nearer than 500 m of each other. The maximum (95–100% utilization) harmonic mean home range size, revealed by radio-tracking (31 individuals) in southern Finland, was 184–950 ha and that of the core area (80–85% utilization) 80–340 ha, depending on the area and season (Kauhala *et al.* 1993a; K. Kauhala and K. Holmala unpublished data). The outer convex polygons were 187–700 ha. There were no significant annual or sexual differences in the average home range sizes, but the maximum home ranges, especially those of males, were larger in autumn than in summer. The home ranges of adult pairs were stable from year to year, only minor shifts occurring between seasons. In one case, following the death of its mate, an adult male moved to an area

6 km from the pairs' former range. The home ranges of juveniles in autumn were larger than those of adults, the maximum home ranges of juveniles averaging 15.7 km^2, core areas 5.4 km^2 and polygons 14.2 km^2.

Dispersal usually occurs between August and October, and we have no evidence of juveniles remaining in their natal range over winter. Greatest juvenile dispersal is >150 km, but most (79%) were found within 20 km from the marking location. None of the adults was discovered further than 10 km from the location at which they were tagged as adults (Kauhala and Helle 1994).

Both male and female participate in pup rearing; the male usually spent more time at the den with pups than did the female (Kauhala *et al.* 1998a). The female nurses the pups and forages for herself, because food items are small and are not usually carried to the den. While the female is foraging, the male stays with the pups. At night, males and females spent 61% and 50%, respectively, of their time at the den, but by day the comparable figures were 80% and 60%. Pups were seldom left alone during the first month of their life. At night, the parents took turns of babysitting: when one entered the den, the other left.

Observations in an enclosure revealed that a male spent more time with the pups outside the den than did the female, and also carried food (small fish) to them (Huttunen 2001). The female nursed the pups, but otherwise kept well away from them.

Japanese raccoon dog

Evidence from Japan suggests that there too the raccoon dog is basically monogamous, and pair bonds may endure through consecutive years (Yashiki 1987; Yachimori 1997). The mating period falls between February and April, and gestation lasts 61–63 days. Again, both parents tend the pups (Okuzaki 1979; Ikeda 1983; Yamamoto 1987), taking turns to attend the den for 30–50 days (Fukue 1991; Saeki 2001) although the male spends more time there (Saeki 2001). After weaning pups forage with their parents until dispersal starts in autumn.

Home-range sizes varying between 6.9 and 610 ha have been recorded (Ikeda *et al.* 1979; Ikeda 1982; Ward and Wuster-Hill 1989; Fukue 1991; Yamamoto 1993; Yamamoto *et al.* 1994). The most detailed study, by Saeki and Macdonald (submitted) was in countryside known as *satoyama* ('a landscape of core forests and the surroundings, which have been maintained through utilization and disturbance by local people with their daily life, self-sustainable agriculture, and other traditional industries' (Osumi and Fukamachi 2001)). There, seasonal home-range sizes were calculated by 95% MCP, 95% kernel estimates and range span for each animal in each season ($\bar{X} = 78.2 \pm 12.0$ ha for 95% MCP; $\bar{X} = 63.3 \pm 7.34$ ha for 95% kernel; $\bar{X} = 1533.6 \pm 110.6$ m for range span; $n = 54$). Descriptive statistics suggested that ranges were largest in the autumn (95% kernel estimate: Kruskal-Wallis Test, $H = 7.09$, DF = 3, $p = 0.069$), yearlings had significantly larger home ranges than did adults ($H = 6.11$, DF = 1, $p = 0.013$), and there were no apparent differences between sexes ($H = 0.67$, DF = 1, $p = 0.412$).

The mean percentage of overlap (95% and 100% MCPs) decreased in the order of pair, yearlings, adult-yearling and adults (Saeki and Macdonald submitted). There was no overlap between the core ranges of neighbouring breeding pairs, nor between the 50% springtime MCPs of non-pair adults or adults and yearlings. Mean percentages of overlap between yearlings were 27.7 ± 18.2 ($n = 4$) and 76.2 ± 0.64 ($n = 2$) within breeding pairs. In a subalpine area Yachimori (1997) reported similar patterns of range overlap.

In Japan, dispersal of young occurs from late autumn to the following early summer, and unlike the Finnish raccoon dogs, some individuals may stay or return to their natal areas (Saeki 2001).

Habitat use

Finnish raccoon dog

In southern Finland, the proportion of barren heath in the core areas of home ranges was greater in early summer than in midsummer or autumn, while the proportion of moist heath was higher in late summer (Kauhala 1996a). Raccoon dogs used lake shores more than expected in both seasons, particularly in early summer. This habitat provides frogs, lizards, and insects in early summer, and dense undergrowth for shelter. Furthermore, raccoon dogs sometimes flee into the water when chased by dogs or other larger predators. Rock piles, suitable for denning, are common on barren heath, which probably explains the heavy use of this habitat in early summer. In midsummer, when pups leave the dens, parents take them to meadows and abandoned fields where they can find insects, frogs, and strawberries. In late summer, raccoon dogs forage especially on moist heath where abundant berries, especially bilberries (*Vaccinium myrtillus*), ripen. Later in autumn, pine forests with abundant lingonberries (*Vaccinium vitis-idaea*) are favoured. Raccoon dogs also visit gardens in late summer and autumn, because their stomachs often contain cultivated berries and fruits during this season.

Japanese raccoon dog

The *satoyama* habitat where raccoon dogs were studied by Saeki and Macdonald (submitted) could be divided between the 'mountain type' typified by secondary forest and herbaceous areas, and the 'village type' comprised agricultural landscapes. In both, the least favoured habitat type was Japanese cedar (*Cryptomeria japonica*) plantation. Habitat selection differed with scale, in terms of preferences for the location of ranges, and preferences for the use of habitats within them. In village-type areas, home ranges were placed preferentially in rice fields and cropland, whereas the preference was less apparent in the placement of the ranges of raccoon dogs in the mountain-type areas. Within home ranges, rice fields were used less by the village type and preferred by the mountain type. Thus, the raccoon dog exhibited multi-scale habitat preferences in *satoyama*.

In urban habitats, raccoon dogs have been shown to inhabit areas, which have forest cover, varying from ≥20% in Tokyo (Nojima 1988) to ≥5% in Kawasaki-city (Yamamoto *et al.* 1995). Animals killed by traffic accidents were older in areas with ≥15% of forest cover than in areas with <15% of forest cover (Yamamoto *et al.* 1995), suggesting the importance of forest cover for long-term survival.

Diet

Finnish raccoon dog

Raccoon dogs are extremely omnivorous in Finland (Kauhala *et al.* 1993b,1998b, 1999). Voles and shrews are important prey in most areas. Fifty-six per cent of raccoon dog stomachs (n = 172) contained remains of mammals, 34% those of birds, 8% frogs or lizards, 20% fish, 51% invertebrates, 89% plants, and 49% carrion. Frogs, lizards, and invertebrates are eaten mainly in summer and autumn, reflecting their availability. Fish are consumed mainly in late winter, when raccoon dogs can find small fish discarded on the ice by fishermen. Berries and fruit are eaten frequently in late summer and autumn. Raccoon dogs also visit compost heaps; coffee beans, rubber bands, pieces of paper, etc. are often found in their stomachs.

Faeces were collected in early summer from small uninhabited islands and larger inhabited islands off the SW coast of Finland, and from the mainland of southern Finland. Birds, mainly female eider (*Somateria mollissima*), and fragments of egg shells were found more often in the diet in the small islands than in the larger islands or mainland (Table 13.2; Kauhala and Auniola 2001). It seems, however, unlikely that raccoon dogs affect eider numbers in the archipelago, since we estimated that they kill only a small proportion (1.2–3.5%) of brooding female eiders each year. Rodents, shrews, frogs, reptiles, and carrion are consumed most often on the mainland. The scarcity of frogs in the diet of raccoon dogs in the archipelago points to the conclusion that there are no frogs on the small islands; raccoon dogs frequently eat frogs, if they are available. Berries, especially crowberries (*Empetrum nigrum*), occurred more frequently in the faeces collected from the islands than those from the mainland.

Japanese raccoon dog

In Japan, racoon dog diet has been studied in several habitats but, notwithstanding regional and seasonal variations, all reveal a preponderance of small items, invertebrates and fruits, and if available, small vertebrates, such as rodents, birds, frogs, and fish.

Table 13.2 The percentage of faeces containing different food items on small uninhabited islands and on larger inhabited islands off the SW coast of Finland and on the mainland in southern Finland in early summer

Food item	Small islands	Large islands	Mainland
Rodents	45.4	34.1	53.8
Shrews	12.6	14.1	23.4
Hare	9.8	0.7	8.7
Waterfowl	66.7	28.1	2.1
Other birds	29.0	28.9	26.6
Eggs	39.9	11.1	15.2
Frogs and reptiles	8.2	10.4	42.4
Carrion	8.7	24.4	28.8
Invertebrates	86.9	77.0	69.6
Cereal	0.0	20.7	12.9
Berries	31.7	34.8	7.9
Other plants	75.4	82.2	35.3
Number of scats	183	135	850

Source: Original table from Kauhala and Auniola (2001).

In the subalpine zone (alt. 1600–2000 m), diet, expressed as percentage occurrence in faeces, included insects (90%), mainly Coleoptera, year around, artificial foods (58%) year around but less in summer, earthworms (58%) except from January to April, berries and seeds (49%) year around but less from January to April, and mammals (46%) from January to June (Yamamoto 1994).

In a mountainous area (alt. 500–1000 m), comparable figures were insects (78–100%), mainly Coleoptera in spring and summer, Orthoptera in autumn, and Hemiptera in winter, fruits (77–100% year around excluding May, 30%), Crustacea (*Geothelphusa dehaani*) (28–71%) from April to December, fish (9–27%) although none in July, August, and October, birds (8–21%) except in May, June, September, and October, small mammals (7–25%) except from July to September, and carrion, mainly sika deer (*Cervus nippon*) and serow (*Capricornis crispus*) (10–37%) from February to June (Sasaki and Kawabata 1994).

In the countryside (alt. 0–100 m), important food items appear to be insects (Orthoptera and Coleoptera) and earthworms throughout year with seasonally abundant food, such as persimmon fruit (*Diospyros kaki*) in autumn and early winter (Saeki 2001).

In urban and suburban areas, raccoon dogs feed predominantly around human dwellings, often on garbage, especially in winter and spring (Yamamoto 1991; Yamamoto and Kinoshita 1994). The percentage of occurrence in stomachs were garbage (72%) year around, insects (46%), mainly Coleoptera except winter, persimmon fruit (30%) in summer and autumn, earthworms (24%) except winter, birds (21%) in winter and spring, Myriapoda (11%) year around.

Skull and tooth morphology

Twenty two measurements of skull and tooth morphology were made on 65 skulls from Finland and 104 skulls from Honshu, Japan. Only adults ($\geq$1 year) were considered, and all teeth were measured, except for canines, which were used for ageing (Kauhala *et al.* 1998c).

Skulls of Finnish raccoon dogs are larger both absolutely and relative to body size than those of Japanese raccoon dogs. Mandible width and jaw height are the best absolute measurements for identifying the origin of the skulls (discriminant analysis: Eigenvalue = 19.6, Wilks' lambda = 0.049, $F = 60.7$, $p < 0.001$). Discriminant analysis resulted in 100% correct classification.

The skulls of the two provenances differ in shape, indeed all measurements in relation to skull size, except rostrum breadth, differed between populations (discriminant analysis resulted in 100% correct classification of skulls: Eigenvalue = 16.9, Wilks' lambda = 0.056, $F = 51.5$, $p < 0.001$). The mandibles of the Finnish raccoon dogs are more robust, and the jaw more powerful, than those of the Japanese form, which may indicate a more carnivorous niche in Finland. Japanese raccoon dogs have adapted to a different diet; they have a relatively longer rostrum and longer tooth rows than do Finnish specimens. Lower carnassials, and upper and lower post-carnassial molars, are larger in relation to skull size in Japan, the lower m2 being also absolutely longer in Japan. Japanese raccoon dogs thus have a larger grinding surface perhaps indicating a more insectivorous/frugivorous diet in *viverrinus* than in *ussuriensis*.

Sexual dimorphism is slight among both Japanese and Finnish raccoon dogs and all measurements overlapped between sexes, although males are larger on 5 skull measurements out of 22 in Japan and only one in Finland (Kauhala *et al.* 1998c). No sexual dimorphism exists in tooth measurements in Japan, whereas 2/27 dental measurements are dimorphic in Finland.

Chromosome number

Finnish raccoon dogs have 54 chromosomes (Mäkinen *et al.* 1986), while Japanese and Ezo raccoon dogs have 38 (Mäkinen *et al.* 1986; Ward *et al.* 1987; Wada *et al.* 1998). Finnish raccoon dogs have five metacentric and 21 acrocentric autosome pairs, while Japanese and Ezo have 13 metacentric and five acrocentric autosome pairs. The *nombre fondamental* (NF; the total number of euchromatic chromosome arms in the female) is the same (66) for animals of both provenances (Mäkinen *et al.* 1986). The number of B-chromosomes varies in both (within and between individuals), being 2–5 in *viverrinus* and 2–4 in *ussuriensis* (Mäkinen *et al.* 1986). The difference in chromosome number is probably due to Robertsonian translocations (centric fusions), which play a major role in karyotype evolution in other canids (Ward *et al.* 1987). Eight fusion events differentiated the karyotypes of the Japanese and Finnish raccoon dogs.

Conclusions

Since winters are mild in Japan, raccoon dogs have no need to be inactive in winter and thus they do not gather large fat reserves or hibernate. The body size is smaller, the insulation of fur poorer, and stomach volume smaller in Japan than in Finland, also suggesting genetic adaptation to a milder climate. Since Finnish (Ussuri) raccoon dogs save energy during winter by hibernating, they can produce larger litters than do Japanese raccoon dogs. Adaptation to different diets probably led to the differences in tooth and skull morphology. Finnish and Japanese raccoon dogs have thus adapted genetically to different climatic conditions and partly different diet, which is shown in their chromosome numbers, morphology, physiology, and reproductive biology.

The difference in chromosome numbers between raccoon dogs from Japan and the mainland is a result of Robertsonian translocations (centric fusions), which often occur during speciation; most differences in chromosome numbers between closely relates species of animals is the result of chromosomal fusions: two acrocentric chromosomes fuse into a single metacentric chromosome (Mayr 1976). Chromosomal reconstruction may be an important component of the speciation process. Polymorphism in chromosome number can, however, occur also within species. One well-known case is the house mouse (*Mus musculus*) in the Swiss Alps and in Scotland (Brooker 1982; Hauffe and Pialek 1997). Heterozygosity for chromosome structure will, however, usually result in decreased fertility. This means that if Japanese and Finnish raccoon dogs would meet and mate, the prediction is that their progeny would have decreased fertility or they would be sterile. The species status of the Japanese raccoon dog should thus be seriously considered.

CHAPTER 14

Bat-eared foxes

Bat-eared foxes 'insectivory' and luck: lessons from an extreme canid

Barbara Maas and David W. Macdonald

Bat-eared fox *Otocyon megalotis* © B Maas.

Introduction

What underlies interspecific variation in the behaviour of the wild Canidae? Answers focus on two factors, the impacts first of phylogeny and second of ecological adaptations (Macdonald and Sillero-Zubiri, Chapter 1, this volume). In both respects, the bat-eared fox, *Otocyon megalotis*, is a revealing element in the pattern of canid variation. First, phylogenetically, although bat-eared foxes probably arrived in Africa as recently as the Pliocene, the clade of which they are the sole survivors split off from other modern canids long ago (Petter 1964; van Valen 1964; Wang *et al.* Chapter 2, this volume). Views differ as to exactly when they diverged—molecular evidence puts them at the base of the fox clade or even lower (Geffen *et al.* 1992e; Wayne *et al.* 1997), whereas morphologists place them closest to the gray fox, *Urocyon*, clade (Tedford *et al.* 1997); either way, the last time they shared a common ancestor with any other modern canid was more than 6 million years ago (m.y.a.), and perhaps closer to 10–12 m.y.a. A comparably long, distinct evolutionary history is illustrated by the raccoon dog, *Nyctereutes procyonoides* (the subject of Chapter 13). Second, just as their ancestry is distinct, so too amongst modern canids the extent of their specialization on insectivory is extreme. Their dentition is unlike any other heterodont placental mammal because they possess between 1 and 4 pairs of extra molars (Coetze 1971; Clutton-Brock *et al.* 1976). Their teeth are small compared to those of other canids and, uniquely in the family, they have no carnassial shear. A sub-angular lobe at the insertion of the digastric muscle allows them to take 4–5 bites per second. Feeding on dung beetles, Scarabidae, links their fortunes to large ungulates (Malcolm 2001), but harvester termites, *Hodotermes mossambicus* or other termiter of the gonera *Macrotermes* or *Odentotermes*, are their most important food throughout their range (Nel 1978 1990; Nel and Mackie 1990; Wright 2004; Pacew 2000). Scarabidae, in particular, are conspicuously spatio-temporally heterogeneous in abundance within and between years—characteristics that are particularly relevant to one set of ideas seeking to explain inter- and intraspecific variation in sociality (Johnson *et al.* 2002; Macdonald *et al.* Chapter 4, this volume).

Against this background of phylogenetic and trophic extremism, we present a case study of bat-eared foxes in the Serengeti in order to shed light on the question, how does their behaviour differ from that of other canids, and why?

Study area and general methods

Between 1986 and 1990, one of us observed bat-eared foxes for a total of 2500 h in the Serengeti National Park, Tanzania (Maas 1993a). In the Serengeti, bat-eared foxes are common in open grassland and woodland boundaries but not short-grass plains (Lamprecht 1979; Malcolm 1986). Hendrichs (1972) recorded a density of 0.3–1.0 foxes/km^2 in the Serengeti, but as many as 9.2 foxes/km^2 in the breeding season, and 2.3 foxes/km^2 at other times have been recorded in Botswana (Berry 1978). Groups forage as a unit and have home ranges from less than 1 km^2 to more than 3 km^2 that are sometimes overlapping (Nel 1978; Lamprecht 1979; Malcolm 1986; Mackie and Nel 1989). In Laikipia, Kenya, neighbouring pairs occupied ranges which overlapped widely (c. 20%) and averaged 3.3 km^2 (Wright 2004). In the Seregenti's woodland boundary, and the open grasslands of southern and East Africa, insects are the primary food sources, with harvester termites (*H. mossambicus*) and beetles predominating, supplemented by smaller numbers of orthopterans, beetle larvae, and ants (Slater 1900; Shortridge 1934; Nel 1978; Lamprecht 1979; Berry 1981; Waser 1980; Stuart 1981; Malcolm 1986; Mackie 1988; Skinner and Smithers 1990).

The study area consisted of irregularly spaced *Acacia* trees in open bush country, forming part of the transitional boundary between *Acacia tortillas* woodland to the north and open grassland to the south. Rainfall is strongly seasonal with two peaks occurring during a rainy season from November to May (Sinclair 1979). Insect abundance is linked to rainfall (Waser 1980), and the onset of the rainy season in November is characterized by an explosion of insect activity. Harvester termite holes, vegetation, and ungulate droppings were surveyed during the dry season of 1987. A total of 2989 m^2 samples were collected in 25 m intervals along 72.5 km of parallel transects, spaced 100 m apart. Transects extended across bat-eared fox territories as well as similarly sized adjacent areas.

General demographic data (group size and composition, litter size, and mortality) were collected from 16 groups in 1986, 19 in 1987, 18 in 1988, and 13 in 1989 (of these, 12 were studied continuously). However, sample sizes are smaller for some measures, depending on when they were taken, because of the formation and disappearance of some groups over the course of each year, particularly in 1987 and 1988, when rabies broke out. In 1987, rabies struck when the young foxes were nutritionally independent and affected 95% of family groups, but in 1988, cubs were less than 2 months old when rabies infected 68% of groups.

These bat-eared foxes were individually recognizable, and were observed at close quarters from a vehicle by day through binoculars, and by night using an image intensifier. Following Altman (1974), scan samples of all group members in view were recorded in 1-min intervals during daylight and 2-min intervals at night, together with proximity data for all individuals in view. Focal groups were watched weekly during at least one morning (05:00–08:00 h) and one evening (17:00 to 19:30–20:30 h). Beginning on the fifth day before each full moon, 8–10 6-h night watches were carried out between 18:00 and 07:30 h, to provide data equivalent to one composite night each month for each group.

Group composition, dispersal, and philopatry

Bat-eared foxes in the Serengeti lived in family groups of one male and up to three females (Table 14.1), all of which invariably bred. Average breeding group size over 4 years was 2.44 (SE ± 0.09, $n = 18$), with 1.44 (±0.09) females per group. However, in 1986, each of the 16 territories was occupied by just one pair, whereas during the following 3 years (1987, 1988, and 1989) 44%, 67%, and 54% of groups, respectively, contained more than one breeding female. This inter-annual variation in group size was statistically significant ($\chi^2 = 9.24$, df = 3, $p < 0.002$). Of 65 breeding events recorded, only one took place in a group that contained an apparently non-breeding adult, a yearling male (see below).

At about 9 months old, both males and females dispersed, whereas a proportion of females displayed natal philopatry, and this was apparently the only mechanism to join a group with a surviving female. There was one case of permanent male philopatry when a non-dispersing male remained following his putative father's death. Of 136 cubs that survived to dispersal age, 105 (77%) emigrated, almost all (95%) of them at the start of the dry season in June. The number of cubs that dispersed from a group was positively correlated with the number surviving to dispersal age (Kendall correlation coefficient, 1986: $T = 0.7$, $z = 3.02$, $p < 0.01$, $n = 14$; 1987: $T = 0.74$, $z = 4.13$, $p < 0.0001$, $n = 18$; 1988: $T = 0.83$, $z = 4.46$, $p = 0.0001$, $n = 13$). At the time of dispersal, group size averaged 6 (±0.41, $n = 18$). Although variation between the 4 years 1986–89 did not reach statistical significance ($\chi^2 = 1.17$, df = 3, $p < 0.07$), group size at dispersal was smallest in 1989 (averaging 5.69 ± 0.38) when rabies struck while the cubs were small, and largest in 1987 (6.21 ± 0.43) when there was no rabies. Although adult group-members, and especially the male, were aggressively territorial to intruding conspecifics (which were young males in all 16 cases where they could be identified) around the breeding season, the period of dispersal was typified only through a decline in mutual-grooming and play which reversed as soon as the last disperser had left.

Table 14.1 Average ± SE number of breeding females per group

Year	Mean ± SE	N
1986	1	16
1987	2.5 ± 0.15	19
1988	2.7 ± 0.18	18
1989	2.5 ± 0.14	13

Most (93%) dispersers left the boundaries of their natal ranges, but seven known dispersers and one unidentified male formed four pairs that attempted neighbourhood settlement within one or both of their parents' territories. Of these, only one pair (both of whom could be identified) bred successfully for at least two subsequent seasons, displacing the female's natal group after a year. The newly formed pair had carved off a small part from the female's parental home range (SRI3, which had an extremely high termite density), and a large part of a neighbouring area (SR13B which, while not a breeding territory, had been occupied by a pair that had recently succumbed to disease). This sequence ended in one of five documented cases of adult dispersal. In this case, the occupants of SR13B—by then comprising the male, female and three almost fully grown cubs—finally took over SR13 after rabies had reduced the original group's membership from ten to just three. This followed a confrontation with the larger group in SR13B, whereupon SR13 settled in an unoccupied area about 1 km away; there, the surviving female was killed by a python, whereupon the male and their cub also disappeared. SR13B then took over almost exactly the original boundaries of SR13, and abandoned the peripheral area that had previously been their sanctuary.

Previously, one approximately 16-month-old subordinate female member of SRI3 disappeared after a period of escalating aggressive behaviour towards all other group members, especially her putative father, until the alpha-pair eventually drove her out.

In another instance, a male whose mate was killed by a vehicle apparently left the area and was not seen again. His territory was taken over by a newly formed pair the following year.

The only case of male natal philopatry involved a juvenile inheriting the territory at the end of the mating season in July, following the death then of his putative father; the male subsequently held his dead father's territory for three years. In another instance, a yearling male, after having left his family in October prior to the birth of the family's next litter, rejoined his group for 6 weeks after an absence of two. He then groomed and guarded the new cubs and defended them against predators (he, and one cub, disappeared during the same night).

Case histories suggested that the female recruits to groups were invariably born there and one-third were dominant amongst their siblings. Of 30 philopatric female recruits, 18 (60%) became the dominant or sole female, whereas 12 (40%) became subordinate members of a group. Ten (55.6%) of these eighteen heiresses became dominant over their siblings, while eight (44.4%) lived in pairs, cohabiting with their father and in one case, their brother during the next mating season. All eight of the latter were seen to be mounted by the male kin with which they cohabited. Territory inheritance by a daughter was never accompanied by her father's emigration. Although full copulation was never seen, males attempted to mount all the females in oestrus, regardless of whether or not they were their daughters or sisters. During 1988 and 1989, a total of eight pairs consisted of an adult male and one of his adult daughters. The average number of cubs raised from these apparently incestuous matings was 2.88 ($\pm$0.398), and was not significantly different to the number of cubs raised from potentially non-incestuous matings over the same period (2.5 $\pm$ 0.5; Mann–Whitney U-test: $p = 0.5$, $z = -0.67$, $U = 19.5$).

The dominant female's status was clearly defined by her close relationship with the male, rather than hostility to subordinates, and she was also the only female to urine-mark, either alone or in tandem with the male. The dominant female appeared to set the route when the group set out to forage in the evening. Of 30 females recruited into their natal groups, 18 inherited their mother's alpha status during the study, invariably following her death. The age at which these females acceded to dominance was 12.6 months (range 4–18, $n = 10$) in 1988 and 11.1 months (range 3–15, $N = 8$) in 1989; six females attained alpha status before they were 7 months old and before their siblings dispersed.

There was no relationship between recruitment and either numbers of cubs at emergence per family (Kendall correlation coefficient, 1986: $T = 0.2$, $z = 0.98$, $p = 0.3$, $n = 14$; 1987: $T = 0.28$, $z = 1.45$, $p = 0.15$, $n = 17$; 1988: $T = 0.2$, $z = 0.96$, $p = 0.3$, $n = 13$) or cub survival to dispersal age (1986: $T = 0.12$, $z = 0.61$, $p = 0.5$, $n = 14$; 1987: $T = 0.22$, $z = 1.16$, $p = 0.2$, $n = 17$; 1988: $T = -0.19$, $z = 0.91$, $p = 0.4$, $n = 13$).

Litter size, sex ratio, and reproductive success

Bat-eared foxes become sexually mature at 8–9 months of age and mate for life. Pair-bonding and mating take place from July to August with up to 10 copulations per day for several days (see also Rosenberg 1971), and with a copulatory tie lasting c. 4 min, followed by peculiar post-copulatory play (Le Clus 1971). Bat-eared foxes have one litter, of up to six cubs, per year, with births occurring from October to December (Nel *et al.* 1984), following a gestation period of 60–75 days.

To judge by the male's behaviour, in the Serengeti, all adult females in each group came into oestrus within a few days of each other, and they all bore and raised cubs, and lactated. Breeding dens contained different size classes of cubs, and the maximum number of distinct sizes always corresponded to the number of adult females.

Average litter size at emergence 8–12 days after parturition was 2.56 $\pm$ 0.13 ($n = 90$), but this average disguises substantial inter-annual variation. In 1986, when all territories were occupied solely by pairs, the single female per territory bore an average of 4.43 cubs ($\pm$0.36, $n = 14$ pairs), significantly more than the average litter size for 1987–89 (Fig. 14.1; Wilcoxon

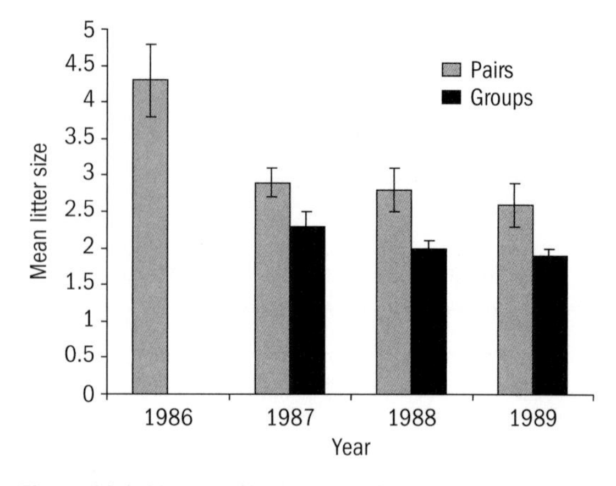

Figure 14.1 Mean ($\pm$SE) litter sizes for pairs and groups for the years 1986-89.

test: $p = 0.001$, $z = -3.24$, $n = 12$). Between 1987 and 1989, females living in pairs had smaller litters at emergence (averaging 2.7 ± 0.2) than in 1986 (Wilcoxon test: $p < 0.01$, $z = -2.66$, $n = 11$), but nonetheless their emerging litters were significantly larger than those of females living in groups of two or more (averaging 1.9 ± 0.1; Mann–Whitney *U*-test: $p < 0.01$, $z = -3.02$, $U = 203$, $n = 32$). However, despite the smaller per capita litter sizes of their members, groups with two or three adult females had a higher mean total number of cubs (Fig. 14.2).

Assuming that the single adult male associated with each group fathered all the cubs, then male reproductive success was highest in 1986 (averaging 4.43 cubs). During the years of lower productivity, from 1987 to 1989, males co-habiting with groups of females enjoyed a higher reproductive success (averaging 3.9 ± 0.4 cubs, $n = 15$) than did males in pairs (2.7 ± 0.2 cubs, $n = 17$; Mann–Whitney *U*-test: $p < 0.01$, $z = -2.61$, $U = 194.5$, $n = 31$).

There was a general shift from an unbiased offspring sex ratio in 1986 to a female-biased sex ratio over the following 3 years. Of a total of 198 cubs from litters of which all members were sexed at emergence, 58% were female (Binomial test: $n = 198$, $z = -2.06$, $p < 0.05$). However, this overview disguises a marked shift in sex ratio following emergence which swung from parity in 1986 (52% of 58 cubs were female), to a strong female bias between 1987 and 1989 (67% of 140).

Parental care and parent–offspring proximity

While the females gave birth inside the den, males were usually resting at one of the den entrances. Newborn cubs spent their first 10 days or so inside the natal den accompanied by their mother. Consequently, we know nothing of their number or sex ratio at birth. After 8–12 days, cubs appeared at the den entrance and began to explore the den area and, later, its immediate vicinity. Males were never seen to regurgitate to their mates or the cubs. However, following cub emergence, mothers spent increasing time away from the cubs whereas males took over all parental duties, with the obvious exception of nursing.

Males guarded, groomed, carried the cubs, and played with them. By day, males rested at one of the den entrances while females typically slept slightly further away, often hidden from the cubs' view by a small bush or tuft of grass. Thus, when the cubs emerged they would always encounter the male first. When sleeping outside the den, and until they were more than 3 months old, cubs invariably slept next to the male. Males played with and groomed the cubs assiduously, and from the age of 4 weeks cubs reciprocated; in contrast, until 3 months old, amicable socializing such as grooming and playing were rare between females and cubs. Females almost always rejected approaches by the cubs, walking away from them, sometimes snarling, growling, and even snapping at their young.

Figure 14.2 Nursing bat-eared fox *Otocyon megalotis* © B. Maas.

By night and while the cubs were young, males would remain at the den while females foraged. Occasionally males would leave the den together with the female and forage for 15–45 min. Thereafter, the male would remain at the den, or forage in its immediate vicinity, until the female returned 2.5–6 h later. Then, while the male foraged, the female would stay with the cubs for 15 min to 2 h until the male returned. Female guarding shifts were always shorter than those of males.

Once the cubs were 4 weeks old, males led them on their first foraging trips, during which they kept close to the male, which frequently indicated or passed items of food to them. Usually males would indicate food, be it carpets of foraging *Hodotermes*, or patches of active dung beetles. Occasionally, there were large dung beetles, which the male crushed in his mouth for the cubs. Both males and females accompanied the cubs on foraging trips once they were approximately 3–4.5 months old.

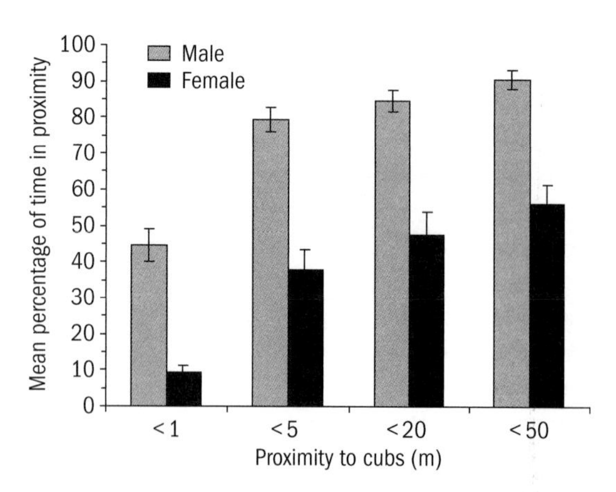

Figure 14.3 Mean (±SE) percentage time spent by males and females at <1, <5, <20, and <50 m proximity to the cubs.

Sex differences in parent–offspring proximity

Spatial geometry can reveal social structures (e.g. Macdonald *et al.* 2000), and this was so for the tendency of cubs to spend time (recorded during 2-min scan samples), at given proximities to male and female adults. Between weeks 3 and 12 (at weaning age), males spent a significantly greater proportion of time in all proximity ranges than did females (*t*-test, < 1 m: $t = 9.66$, $p < 0.0001$; < 5 m: $t = 6.26$, $p < 0.0001$; < 20 m: $t = 5.33$, $p < 0.001$; < 50 m: 6.61, $p < 0.0001$; Fig. 14.3), spending 90% of their time, on average, within 50 m of the cubs, compared to the females' average of 56%. This difference was comparably significant ($t > 5.03$, $p < 0.001$ in all cases) whether the foxes were members of a pair or a group, and whether, within groups, the females were dominant or subordinate. Males cohabiting with a group of females spent more time with the cubs ($t > 2.31$, $p < 0.05$ for all proximity ranges) than did those living as a pair, but there was no such distinction between females from groups and pairs. Within groups, subordinate females spent a significantly greater proportion of time in the 5 m ($t = -2.84$, $p < 0.05$) and 20 m ($t = -3.42, p < 0.01$) range of cubs than did dominant females. Time spent within 1 m of the cubs appeared to be exceptionally demanding, males were within this proximity for 45% of their time; for females, the equivalent figure was 10%.

Energetics of female reproduction

Female bat-eared foxes suckled their young for up to 14 weeks. In groups with more than one female present, only one female ever nursed cubs at a time, and when she did so all the cubs nursed. During periods of the day when all group members including females were present in the vicinity of the den together, while one female suckled the others would remain at a distance and lying down, rendering their teats unavailable. Suckling bouts were infrequent, typically occurring in the early evening prior to the female's departure from the den, and in the morning on her return, as well as during nightly guard shifts, and occasionally in the early afternoon. Suckling was usually initiated when the female summoned the cubs from the den or its vicinity with a soft whimpering cry. The average suckling bout lasted 3 min (191.11 ± 8.06 s, $n = 18$ females and 61 bouts). There were no significant differences between years in suckling bout length, for paired females (1986 versus 1987: $t = 0.22$, df = 5, $p > 0.05$), dominants (1987 versus 1988: $t = -0.32$, df = 8, $p > 0.05$), or subordinates (1987 versus 1988: $t = 0.19$, df = 11, $p > 0.05$). However, as revealed in Table 14.2, there were significant differences in suckling bout length between females categorized by group size and status

(ANOVA: df = 5, $p < 0.001$, $n = 29$). Thus, within groups, subordinates suckled for longer than did dominants ($t = -6.48$, df = 18, $p = 0.0001$), and female members of a pair suckled for longer than did the dominant female within a group ($t = -4.03$, df = 11, $p = 0.002$).

Sucking time may be proportional to milk intake. The product of sucking time and litter size gives a measure of the drain on the nursing female (referred to as 'cub time' on Table 14.2), and by this measure, too, subordinates invested more than did dominant females overall (Mann–Whitney U-test: $p = 0.03$, $z = -2.19$, $U = 79$, $n = 20$), and more than did paired females in 1987 (Mann–Whitney U-test: $p = 0.05$, $z = -2$, $U = 12$, $n = 8$) but not in 1986.

Both sucking bout length and cub time were significantly positively related to *H. mossambicus* foraging hole density in bat-eared fox territories (Kendall Rank correlation coefficient; suckling bout length: $p = 0.05$, $z = 1.96$, $T = 0.8$, $N = 8$; cub time: $p = 0.05$, $z = 1.96$, $T = 0.8$, $N = 8$).

What are the costs of reproduction to female bat-eared foxes? First, an estimate of the energetic cost of pregnancy can be obtained by relating litter size to maternal body weight (Gittleman 1986; Oftedal and Gittleman 1989). Second, an estimate of the energetic cost of nursing at peak lactation can be estimated from litter metabolic mass (LMM = litter size × cub weight$^{0.83}$) by multiplying LMM by 227, and an index of maternal energy investment is provided by the ratio of LMM to maternal metabolic mass ($W^{0.75}$) (Oftedal and Gittleman 1989). An LMM ratio indicates a high energy demand relative to maternal metabolism and, as argued by Oftedal (1984), indicates high expenditure of energy.

Based on the weights of five bat-eared foxes in the Serengeti, we use a maternal metabolic mass of $3.5^{0.75}$ kg. Neonate weight was estimated at 120 g

Table 14.2 Average sucking bout and cub time durations for paired females in 1986 and 1987, and for dominant (mother) and subordinate (daughter) females in groups in 1987 and 1988 (no focal observations were made on pairs in 1988)

Female status	Year	Mean sucking bout length	Mean cub time	*n*
Paired female	1986	178.22 ± 3.89	788.64 ± 52.39	5
	1987	184.35 ± 21.95	368.70 ± 43.9	2
Dominant	1987	146.02 ± 9.11	678.16 ± 84.87	5
	1988	150.18 ± 9.28	665.32 ± 134.43	5
Subordinate	1987	211.88 ± 16.11	999.50 ± 118.48	6
	1988	208.41 ± 9.59	1025.09 ± 149.91	7

Table 14.3 Comparative estimates of female energetic output during peak lactation in five canids. Developed from Oftedal (1984a), Gittleman and Oftedal (1987), and Oftedal and Gittleman (1989)

Species	Maternal weight (kg)	Litter size	Litter metabolic mass (kg$^{0.83}$)	Weight of young (kg) at peak lactation	Litter weight as % of maternal weight	Daily milk energy output (kcal)	Metabolic mass ratio
Bat-eared fox	3.5	4	3.29	0.79	13.7	747	1.29
Coyote	9.7	6	5.35	0.87	14.4	1200	0.97
Dhole	13.8	4.3	8.73	2.35	8.6	1970	1.22
Red fox	3.9	3.9	2.66	0.63	12.9	598	0.96
Grey fox	3.3	3.8	2.11	0.49	12.4	474	0.86

(Ewer 1973; Smithers 1983; Moehlman 1986) and cub weight at first consumption of solid food, when lactation demand is at its peak, was set at 790 g (following Smithers 1983). As a percentage of maternal metabolic mass in bat-eared foxes, litter weight at birth ranged between 8.7% (assuming a litter size of 2.6 as in 1987–89) and 13.7% (assuming a litter size of 4, following Gittleman 1989). In calculations by other authors (Oftedal 1984; Gittleman and Oftedal 1989; Oftedal and Gittleman 1989), only the arctic fox has a higher percentage gestational investment (litter weight is 16.2% of maternal metabolic mass). Metabolic mass ratio was estimated at 1.29, higher than estimates for four other canid species (Table 14.3), suggesting that suckling bat-eared fox litters place a high energy demand on the female relative to her metabolism.

Rabies and mortality

Mortality levels were significantly higher in 1987 and 1988 than in 1986 and 1989 ($G = 55.72$, df = 3, $p = 0.0001$), due to rabies outbreaks in those years. Rabies killed 85 (90%) of the 94 individually recognizable animals that died over the 4-year study, primarily in 1987 and 1988. The first rabies cases in 1987 were in early February when cubs born in 1986 were being weaned, but in 1988, rabies broke out in November, while the females were still suckling. Female mortality from rabies was significantly higher in 1988 (71% of 31 females died) than in 1987 (when 27% of 26 females died; $\chi^2 = 10.98$, df = 1, $p = 0.001$), but cub mortality was slightly lower (44% of 70 cubs in 1987 versus 36% of 61 cubs in 1988), although this difference was not statistically significant ($\chi^2 = 0.914$, df = 1, $p = 0.339$). Mortality in 1986 and 1989 was considerably lower for both cubs (approximately 3% mortality in both years, out of 62 and 41 cubs, respectively) and females (no mortality in 16 females in 1986 and approximately 11% in 20 females in 1989). No male mortality was recorded in 18 and 13 individuals in 1987 and 1989, respectively, but in 1986, 1(6%) of the 16 males died (from rabies) and in 1988, 5 of 18 males died, 3(16.7%) of them from rabies. Male mortality from rabies was significantly less than female mortality (1987: $\chi^2 = 8.69$, df = 1, $p = 0.003$; 1988: $\chi^2 = 13.47$, df = 1, $p < 0.0001$).

Deaths were diagnosed as rabies on the basis of the presence of inclusion bodies in the brain and tested positive using Indirect Fluorescent Antibody Technique, by the Tanzania Livestock Research Organization; the remainder were inferred.

Other causes of mortality were predation (three individuals) and road accidents (three individuals). Predators, all of which ate at least some of the foxes they killed, included martial eagles (*Polemaetus belicosus*), pythons (*Python sebae*), and spotted hyaenas (*Crocuta crocuta*). When a bat-eared fox was attacked by a martial eagle, the fox uttered an alarm call, whereupon five other members of its group charged towards the eagle. Together the foxes snapped and lunged at the bird, leaping into the air as it laboured to take-off. The foxes mobbed black-backed jackals ($n = 22$), golden jackals ($n = 12$), and spotted hyenas ($n = 45$), and occasionally aardwolves, mongooses, goshawk, and various snakes (total $n = 19$). An average of 2.8 ± 0.32 (range 1–7) foxes was involved in 22 mobbings of black-backed jackals, and 4.62 ± 0.36 (range 1–11) mobbed hyaenas, on average. These figures include instances where more foxes were available to mob the smaller predators, and thus the number participating can be interpreted as a measure of their assessment of risk. Groups of bat-eared foxes mobbing hyaenas were significantly larger than those mobbing jackals (Mann–Whitney *U*-test: $U = 1119$, $z = -3.56$, $N = 79$, $p < 0.001$); there were no differences in the size of mob harassing the two species of jackals ($U = 135$, $z = -0.111$, $n = 34$, $p = 0.9$). Occasionally, when the cubs were threatened by hyaenas, the dominant male fox attempted alone to distract and lead them away.

Resources and sociality

Bat-eared foxes were clearly territorial. Territories sketched onto maps averaged 0.62 km^2 (range 0.4–0.87, $n = 6$), and appeared to tessellate neatly. Although movements were not studied in detail, territorial borders appeared to be stable between years. There was a negative relationship between *Hodotermes* foraging hole density and territory size (Fig. 14.4) (Kendall's tau = −0.867, $z = -2.442$, $p = 0.01$).

Areas occupied by bat-eared fox territories were clustered in dispersed pockets. Both *H. mossambicus*

Figure 14.4 Bat-eared fox *Otocyon megalotis* foraging for *Hodotermes* © B. Maas.

and ungulate dung were significantly more abundant in these areas (*H. mossambicus* : $G = 485.7$, df = 1, $p = 0.0001$; ungulate droppings: $G = 33.9$, df = 1, $p = 0.0001$), and vegetation cover inside these areas was less dense. Because *Hodotermes* eat grass, grass height and vegetation cover were both negatively correlated with *Hodotermes* density (grass height: Kendall's $T = -0.491$, $z = 2.1$, $p < 0.05$; cover: $T = -0.855$, $z = -3.659$, $p < 0.001$).

Rainfall measured monthly at a rain gauge within the study area varied inter-annually. Rainfall in the whelping season was strongly correlated with scarabid abundance. The highest rainfall in a whelping season fell in 1986, when 107 mm fell in October, followed by 3 years of lower rainfall in this month (13, 64, and 4 mm in 1987–89, respectively).

When feeding on termite patches, group members fed closely together, but when feeding on beetles, beetle larvae, or grasshoppers, they foraged up to 200 m apart. Group members called each other to rich food patches with a low whistle.

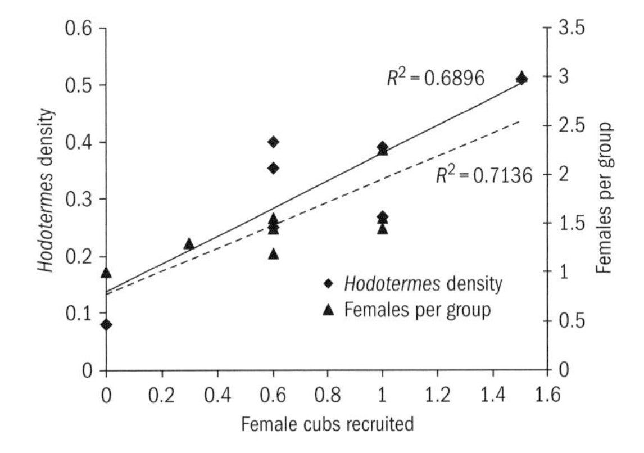

Figure 14.5 Relationship of philopatry (average number of female cubs recruited per family groups) with mean number of females per group and with average density per square metre of *Hodotermes* foraging holes within the group territory.

Group size, philopatry, and territory quality

There was no significant relationship between the number of adult females per group and *Hodotermes* hole density (Kendall correlation coefficient: $T = 0.15$, $z = 0.474$, $p = 0.6$, $n = 7$). However, this result may be confounded by high rabies mortality in 1987 and 1988, and a small sample size. Female philopatry was positively correlated to *Hodotermes* foraging hole density within each territory and to the average number of females in a group between 1987 and 1989 (Fig. 14.5).

Reproductive success and territory quality

Between 1986 and 1989, reproductive success (measured as average number of cubs per female at emergence) was positively correlated with *Hodotermes* foraging hole density measured in 1987, and this relationship persisted during 1987–89, years of lower October rainfall (although total annual rainfall was not particularly low in these years), with female reproductive success positively correlated with *Hodotermes* foraging hole density in 1987 (Kendall rank correlation coefficient: $p = 0.03$, $z = 2.16$, $T = 0.683$, $N = 7$). During this period, reproductive success varied negatively with the number of females per group (Kendall rank correlation coefficient: $p = 0.02$, $z = -2.54$, $T = -0.51$, $N = 14$). The cumulative consequences of these relationships are that male reproductive success was positively correlated with number of females per group (Fig. 14.6).

Social dynamics and territory quality

One obvious line of thought is that the females' reluctance to associate with cubs stemmed from the harrowing demands of foraging—which consumes their time and limits their energy. To shed light on this, one pair was selected—the female had spent less time close to her cubs than had any other in 1986—and in 1987, she was provided with commercial dry cat food, raisins, and water for 12 weeks after whelping. Despite the fact that this period of 1987 was, for most foxes, one of substantially scarcer resources than it had been in 1986, the provisioned female spent a significantly greater proportion of her time in close proximity to her cubs in 1987 (< 1 m: 11% versus 15%, $t = 4.37$, df $= 0.01$; <5 m: 41% versus 51%; $t = 13.26, p = 0.0001$). Furthermore, she was uniquely amicable towards them.

These observations are complemented by the predicament of a female whose mate died before the birth of her cubs in 1986. As a single mother, she spent significantly more time within 5–30 m of her cubs than her peers in 1986 did with theirs (<5 m: $t = 7.56$, $p < 0.0001$; <20 m: $t = 4.68$, $p < 0.01$; < 30 m: $t = 5.36$, $p < 0.001$); in contrast, she did not spend significantly more time within 1 m of her cubs than did other mothers (<1 m range: $t = 2.05$, $p > 0.05$). Indeed, the single mother spent a significantly lower proportion of time (24.2%) in the < 1 m range of her cubs than did pair-living males (39.7%; $t = 2.99$, $p < 0.01$), but otherwise her proximity scores did not differ from those of males that year. By 1987, that individual had recruited daughters to form a group along with her philopatric son and her behaviour towards her next litter of cubs was statistically indistinguishable to that of other females in her circumstances that year.

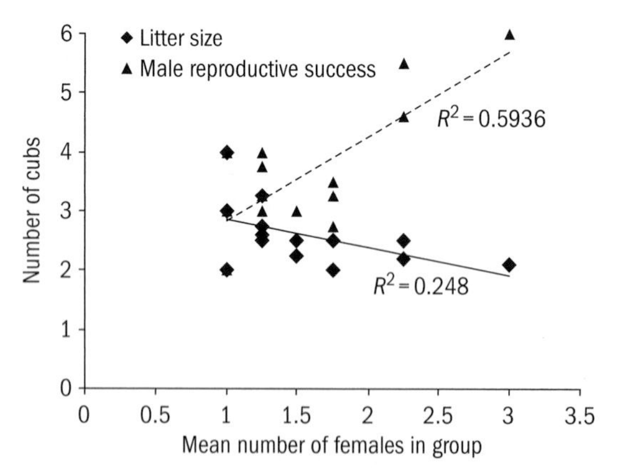

Figure 14.6 Relationship between number of females per group, litter size, and male reproductive success.

Discussion

Five questions of general theoretical interest emerge from these observations of one population of bat-eared foxes. In logical order these are, first, how do the energetics of insectivory and the foxes' life-history parameters interact to fashion their lifestyle? Second, how is that lifestyle moderated by patterns in the dispersion of their prey? Third, within the society that emerges from these foregoing considerations, what dictates their systems of mating and parental care and, fourth, sex ratio allocation? Finally, what lessons have emerged from this study with regard to the influences of phylogeny, resources and luck (as exemplified by the uncontrollable vicissitudes of rainfall and disease) on the biology of bat-eared foxes?

Energetics and insectivory

Body size is associated with various allometries and morphological constraints. In the case of canids, it is also associated with variation in social systems, running on a continuum from small species with female-biased groups and predominantly male dispersal to larger species with male-biased groups and predominantly female dispersal (e.g. Macdonald and Moehlman 1982; Creel and Macdonald 1995). As summarized by Macdonald and Sillero-Zubiri (Chapter 1, this volume), this continuum is explained in terms of the effect of body size on neonate development by Moehlman (1986, 1989) and Moehlman and Hofer (1997), whereas Geffen *et al.* (1996) favour instead resource-based explanations. Similarly, and with consequences reviewed by Macdonald *et al.* (Chapter 4, this volume) body size is linked via metabolic rate to energetic demands, leading Carbone *et al.* (1999) to argue that the relatively low availability of small prey limits the size of carnivores that they can sustain. In addition to the size-related allometry in metabolic demands imposed by surface-to-volume ratios, some species (e.g. African wild dog) have higher energy demands than predicted by their mass (Gorman *et al.* 199n).

In this context, no canid lives more exclusively on prey smaller than the insects to which bat-eared foxes are committed. While their gestation costs may be unexceptional, extrapolations from our observations of litter size and suckling bout lengths suggest that the energetic demands on lactating bat-eared foxes may be even higher than expected (high metabolic mass ratio and high estimated milk energy output at peak lactation). This situation may explain (1) their lower litter size following poor rainfall and (2) their parental behaviour.

We know of no other canid in which females are so determinedly stand-offish towards their emergent offspring, or in which males are so correspondingly paternally diligent, and we attribute both to energetic brinkmanship of the females. Because of the small and fiddly nature of insect prey, and bearing in mind that bat-eared foxes do not regurgitate, the only way to transport energy to cubs is in milk; the entire burden must be born by the female which, if Carbone *et al.* (1999) are correct, is already likely to face severe time constraints. Gittleman and Thompson (1989) have emphasized the necessity for rest and sleep for females during lactation (McNab 1987).

The extent to which male bat-eared foxes monopolize the task of parental care is at an extreme for canids (and indeed for mammals, Kleiman and Malcolm 1981). Komers and Brotherton (1997) suggest that paternal care is less relevant to the evolution of monogamy in mammals than in the dispersion of females. The effectiveness of paternal care in mammals is largely unquantified (but see captive studies by Wynne-Edwards and Lisha 1989; Cantoni and Brown 1997, and observation in the wild by Gubernick and Tejeri 2000). By spending most of their time with the cubs, males enable females to maximize their foraging time. Consequently, their own foraging time is curtailed, and studies elsewhere have revealed that they have smaller ranges than do lactating females (Mackie and Nel 1989) and spend less time actually feeding when accompanied by cubs (Nel 1990). In our study, males from multi-female groups spent a greater proportion of time close to cubs than did males from pairs, indicating advantages resulting from better food availability and/or labour-sharing between group members. The uniformity of maternal behaviour—irrespective of group size or litter size—suggests females have little room for manoeuvre, an interpretation supported by the atypical motherliness of the one female, given supplementary food and water. Although in a different ecosystem, Wright (2004) found that the proportion of bat-eared fox cubs nursing to 14 weeks increased with the proportion of time males spent at the den, which itself was related to measures of termite abundance with territory.

Communal care of the young may also allow females to conserve energy and share the burden of lactation (e.g. by increasing the interval between suckling) (Creel and Creel 1995). The costs to female canids of sharing a male (and therefore paternal care for the young) may often be primarily associated with the capture of prey and feeding of the young—two features not applicable to the natural history of these foxes, whereas the benefits of vigilance and guarding are non-depreciable.

Resource dispersion

Although they rarely eat mammalian prey (Lamprecht 1979; Skinner and Smithers 1990), bat-eared foxes are nonetheless dependent on mammals insofar as the distribution of their insect prey is heavily dependent upon the dung of large ungulates (Malcolm 2001). In the Serengeti, the presence, and grazing activity, of ungulates (and thus of their dung) is heterogeneous, and therefore so too is the dispersion of insects. Seasonal migrations bring the herds to the Serengeti for only the wet season, commencing in October (McNaughton 1989), primarily in search of phosphorus (Murray 2001). Once there, fine scale variations in soil and vegetation type cause them to graze patchily (Snaydon 1962; McNaughton and Banyikwa 1995), and the potential for insect communities to develop, sustained by their dung, is correspondingly patchy. The extent to which that potential is realized is determined by rainfall. As described above, when, as in 1986, rainfall is high between October and December, scarabid availability soars, whereas when the rains during these months following whelping are poor, as in 1987–89, so too is the supply of beetles and harvester termites' surface activity. Waser (1980) confirmed a direct link between rainfall and the availability of insect prey (see also Nel 1978, 1990). Locally, at least in some areas, numbers of bat-eared foxes can fluctuate from abundant to rare depending on rainfall and thus food availability (Waser 1980; Nel *et al.* 1984), and climate change in southern Africa has been invoked to explain the species' range extension (e.g. Stuart 1981; MacDonald 1982; Marais and Griffin 1993).

At the broad scale of their settlement throughout the ecosystem, the distribution of bat-eared foxes in the Serengeti was highly clustered into island communities, whose location matched the patchiness in insect availability driven by ungulate behaviour and thus, ultimately, interactions of soils, minerals, and vegetation. Families of bat-eared foxes occur in clusters in areas where harvester termites are present and are absent from areas lacking termites. Insofar as the territories in such clusters are occupied (a situation altered by epizootic disease), the prospects for dispersal may be limited—a situation analogous to that of an Ethiopian wolf facing dispersal from a mountain plateau (Sillero-Zubiri *et al.* Chapter 20, this volume).

At the fine scale of social and spatial arrangements within and between territories, the spatio-temporal heterogeneity in the dispersion of insects available to bat-eared foxes in the Serengeti creates almost exactly the conditions proposed by the Resource Dispersion Hypothesis (RDH) to facilitate social group formation (Macdonald 1983; Carr and Macdonald 1986; Macdonald *et al.* Chapter 4, this volume). Specifically, Macdonald and Carr (1989) argue that two ecological pressures will affect fundamentally the balance of costs and benefits of group membership: (1) the probability of successful dispersal and (2) the probability of accessing adequate resource security as a group-member (see Johnson *et al.* 2002). Watching the foxes, it appeared that food availability was patchy in that the foraging party would clearly make beeline journeys to particular foraging areas. Where resources are spatio-temporally heterogeneous, additional members may share the smallest territory required by the basic social unit (in canids generally a pair) at minimal cost, and if the risks of dispersal are high it may be advantageous to do so. That advantage may be enhanced if group membership brings sociological benefit—for example, through cooperation—and therefore the question of whether each additional group member is desirable (from the different perspectives of the candidate and the existing members) will be affected by the marginal advantage arising from its admission. If that marginal advantage is great, it may even pay to bear the costs of expanding the minimum necessary territory to accommodate the resource requirements of the new member (Kruuk and Macdonald 1985). Do these ideas help interpret the behaviour of the bat-eared foxes described here?

Assuming each *H. mossambicus* foraging hole is an equivalent measure of resource availability, all else being equal, each pair of foxes will require the same number of holes. RDH predicts that where these holes are widely dispersed, territories will be larger than where the holes are densely packed. As predicted, territory sizes were inversely related to the densities of *Hodotermes* foraging holes. Group size was not correlated to *H. mossambicus* foraging hole density (although this result may be confounded by the impact of rabies), but both the frequency of philopatry and reproductive success were. RDH predicts a correlation between group size and patch richness, but in this case *H. mossambicus* availability (including their often high renewal rate) was only one component of this richness—the other was scarabid availability.

On the basis of these results, we suggest that the following model might be tested by future study.

First, territory size is dictated by the availability of *H. mossambicus*, and territories are configured to sustain a pair of foxes under the likely worst conditions. We expect the surface abundance of *H. mossambicus* to increase following the grass growth stimulated by good rainy seasons and to wane as their food-stores deplete through prolonged drought. Between 1986 and 1989, bat-eared fox territories appeared stable. Whether territory sizes adapt to a fresh bottleneck each year, or are adapted over a sequence of years, the question arises of when is the bottleneck? We suggest that it is the dry season. At this time alternative prey are rare and the foxes depend almost entirely on *Hodotermes*. The availability of termites to foxes (i.e. the time they spend foraging above ground) is likely to depend on the supply of dried grass to be harvested, and this in turn depends on the rainfall during the preceding rainy season and current dry season.

Second, RDH predicts that group size will be determined by patch richness (e.g. the aggregate availability of *Hodotermes* and scarabids). Of these, variation in the availability of scarabids is conspicuous, and determined by wet (whelping) season rainfall. The substantially higher mean litter size in 1986, when October rainfall was high, suggests that cub survival is determined by the availability of scarabids and, following the rapid growth of grass, of surface active harvester termites (the mean litter size for pairs in 1986 was significantly higher than the mean litter size for pairs for 1987–89).

Whether these recruits remain in the group to become breeders during the following year will depend on the number of them surviving predation and disease, and whether the suggested dry season bottleneck in surface-feeding termites is sufficiently lenient to sustain them. We could not test the latter prediction readily because rainfall during the dry seasons within our intensive study was not particularly variable. However, remembering that in October 1986 all the foxes were settled in pairs, we predicted that the previous dry season must have imposed a stringent bottleneck. Indeed, rainfall in June 1986 was only 8 mm (well below the 29-year average of 34 mm), and this savage drought persisted until the end of September. Sinclair (1975) states that 23 mm of rain monthly is required to sustain grass growth, and termite surface activity is related to grass growth. The foregoing model, although speculative, suggests other ideas for future field test. Since the density of termite holes is seemingly not related to fox group size, whereas group size is negatively related to reproductive success, it is likely that the smaller litters produced by group-living females (in comparison to contemporary pair-living females) has a sociological explanation. Indeed, communal denning in Ethiopian wolves (*Canis simensis*) (Chapter 20, this volume), Arctic foxes (*Alopex lagopus*) (Chapter 8, this volume), and perhaps bush dogs (*Speothos venaticus*) (Macdonald 1996) and red foxes (*Vulpes vulpes*) (Macdonald 1987) have all been associated with reduced litter sizes.

In 1986, associated with high October–December rainfall, food availability appeared to be high during the time of lactation, in marked contrast to subsequent years when, by the time of emergence, litters were half the size of those in 1986. This raises the question of whether this litter-size reduction occurred pre- or postpartum, and in either case by what mechanism. The model proposed above would rationalize litter reduction as a consequence of reduced rainfall leading to a catastrophic delay in the appearance of scarabids, plausibly translating into a lowered capacity to generate milk—suckling bout lengths were longer in territories with high *Hodotermes* scores. The relevance of water is further supported by observations of nursing females licking dew from blades of grass.

We do not know when or how the litter reduction took place, but in other species, females may reduce their litter sizes both pre- and post-natally under suboptimal conditions (Harvey *et al.* 1988; Lindström 1988; Clutton-Brock *et al.* 1989; Packer *et al.* 1992). Smithers (1983) reports that bat-eared foxes characteristically carry 4–6 embryos (they have six teats). We have no evidence that during pregnancy the females can forecast the likely rainfall that will influence her food supply during the crucial months of lactation. To that extent, we turn to postpartum litter reduction. By what mechanism might this occur? Insofar as all emergent cubs looked healthy, general starvation seems implausible, raising the possibility of infanticide. Since the father almost never entered the den prior to cub emergence, and since litters were small in 1987–89, even in territories with only one breeding female, as well as in groups, the mother is a strong candidate. Insofar as litter reduction occurs, and if it is the means of sex ratio variation, it suggests selective infanticide. Two anecdotes are noteworthy: only six cubs can nurse simultaneously from one female, and even in groups the mothers suckled at different times; in only two cases did an aggregate litter in excess of six emerge from the den, and in both cases the 'surplus' cub disappeared within days. Intriguingly, Smither's (1983) notes that his pet bat-eared fox had only four teats (he was unaware that this was anomalous) and gave birth to litters of 4–6 cubs—on all four occasions when her litters were larger than four, but not otherwise, she ate any cubs in excess of four within days of their birth (and thus before investing substantially in nursing them).

Mating system and parental care

Amongst group-living canids, the generality is for reproduction to be the prerogative of only one, dominant female in each group, and for non-breeding subordinate females to act as alloparents (Macdonald *et al.* Chapter 4, this volume). Where more than one female does breed within a group, there are cases of communal denning and allosuckling, but again, the generality is for signs of hostility (sometimes fatal for the pups) between the mothers (Creel and Macdonald 1995). More broadly amongst carnivores, all or several females within groups of felids (e.g. Bertram 1979), hyaenids (e.g. Kruuk 1972), and some herpestids (e.g. Rood 1986) may reproduce communally—all families within the feliform branch of the Order. The bat-eared foxes of the Serengeti are thus a significant outlier in canid sociality in that all females bear cubs, den comunally, and allosuckle indiscriminately.

What selective pressures favour communal breeding by these foxes? Key observations are that (1) bat-eared foxes have six teats, (2) even in groups of three females, only one female at a time ever nursed the cubs, and (3) even the largest aggregate litter never exceeded six for more than 3 weeks of age. The critical question is why all three females in our groups produced about two cubs each, rather than the dominant producing six which were nursed by the other two. Are there reasons why (1) a dominant female should be unable to suppress the reproduction of her subordinates and (2) it would be advantageous for all females to breed?

Possibly, reproductive suppression and spontaneous lactation, may, phylogenetically, not be open to bat-eared foxes, as it seems that regurgitation is not. The only way for these foxes to transport nourishment to their cubs is as milk. A hypothetical dominant female might thus gain by allowing her female group-mates to bear cubs, which she then kills so that the bereaved mothers nurse her offspring. This did not happen (although it does in other canids, see Chapter 4). One possibility is that the act of infanticide is simply too risky—within the confines of a shared den, the subordinates might resist with the result that the dominant loses control and her own cubs suffer, and infanticide could proliferate to a tragedy of the commons. Several factors might erode the advantage to the dominant of monopolizing reproduction. Most obviously, the closeness of her relatedness to the subordinates' cubs (see below, inclusive fitness). A further clue may lie in the highly structured pattern whereby, even though two or three females were at the den, only one at a time would suckle the cubs; this may indicate optimization of either or both of milk yield and consumption. Most interestingly, subordinate females nursed for longer, and guarded more frequently, than did dominant females—perhaps reflecting that the dominant's status puts other drains on her time and energy, and perhaps indicating that she benefits from the greater assiduousness of her subordinate daughters and grand-daughters—benefits that might translate into her own lifetime reproductive success.

Benefits of cooperation

Our account reveals that potential benefits of group-formation in this population of bat-eared foxes include the added production of related cubs, matrilineal territory inheritance, collective nursing, grooming and general care of the cubs, corporate defence (and vigilance) against predators and defence of the territory, and huddling for warmth. How do these behaviours affect the balance of costs and benefits of increased group membership. On the one hand, females in groups produced smaller litters than did those in pairs. On the other hand, those in larger groups can share vigilance, maternal antibodies, and other duties, and if one is killed another is at hand to nurse her orphans. A further speculation is that communal nursing is a form of risk-sharing in the face of an uncertain prey supply and high mortality rate. The costs of tolerating additional group members—a factor emphasised by Macdonald and Carr (1989)—may be unusually low in bat-eared foxes. To a dominant female, the cost of sharing a male (who guards a den with equal assiduousness whether it contains one litter or two) or food (which is often highly renewable) may in some respects be low.

Territory inheritance and inbreeding

Territory inheritance can provide an advantage, from the perspectives of both parent and offspring, to natal philopatry. However, it is often supposed to involve a long wait, and to bring with it risks of inbreeding. In this case, the wait was short and the probability of inheritance high. Although the widespread evidence of multi-male mating in mammals (Wolff and Macdonald 2004) raises the expectation that extra-pair copulations will be found in bat-eared foxes, the frequent observations of fathers mounting their daughters in our study suggest—but do not prove—that father–daughter or sibling inbreeding may have been the norm in this population. In a different population, where bat-eared foxes invariably lived as pairs, Wright (2004) found confirming evidence of cubs fathered by extra-pair males in only 2 litters out of 14. He also noted that the close proximity maintained within each pair could have made philandering difficult. This may, of course, be different in circumstances where there is more than one female in a group.

Clearly, one factor that may contribute to cooperative behaviour of the bat-eared foxes is their genetic relatedness. In this context we report two systems. The most simple is the case where an unrelated males with a single female or with two sisters. In this case, the coefficient of relatedness between parents and cubs is 0.5, and between aunt and nieces/nephews is 0.25. However, in some years (1987–1989), some foxes lived in groups comprised of one adult male with two females, and a number of cubs. In these groups the male's relatedness to any observed cub can arise in two ways. Observations in those years suggested that for 80% of matings (12/15) the male was also the father of its mate, and therefore simultaneously the grandfather of the cub (in some groups this was true for all cubs, as both females were the daughters of the male). For these matings, the male/cub relatedness will exceed 0.50 (indeed, because the coefficient of relatedness between grandparent–grandoffspring is 0.25, the male/cub relatedness is estimated to be 0.75). The same applies to those cases where a female is related to her cub as its mother and through incestuous mating with her father. Furthermore, the coefficients of relatedness of communally nursing females to each others cubs may also be elevated—for example, that of a female to her sister's cub may be 0.5 (0.25 as their aunt and 0.25 through their shared father). In short, the relatedness between all adults in these groups, and the cubs, may be unusually high. A different point is that the evidence of the neighbourhood settlement suggests that the members of adjoining territories may also have kinship ties.

Sex-ratio allocation

Whereas sex ratio of cubs at emergence was equal in 1986 (the breeding season of peak food availability), a preponderance of daughters emerged during the failed rainy seasons on 1987–89 when food was less abundant. Departures from equal sex ratios among offspring can be explained when the determinants of fitness vary between the sexes (Clutton-Brock 1991). Thus, if traits affected by parental investment influence the fitness of one sex more than the other, it may pay parents to invest more in the sex expected to produce the most grandchildren per unit investment. That investment can occur during pregnancy or in the den or thereafter. Although we have no measures of differential parental investment in the sexes at any of these stages, we can observe its outcome in the sex ratio of cubs at emergence. Although the preponderance of daughters emerging in each year from 1987 to 1989 might be explained by factors beyond the control of the parents, for example, a disease afflicting males, we turn to two hypotheses that seek to explain such variation in terms of parental investment. First, Trivers and Willard (1973) hypothesized that for species with significant sexual dimorphism biased towards males, such that the costs of producing sons exceed those of daughters, only qualitatively superior mothers can afford to rear sons, while inferior mothers should produce daughters. Alternatively, the local resource competition hypothesis (Clarke 1978; Silk 1983; Julliard 2000) predicts that mothers in poor condition (in low quality territories) should produce offspring of the sex that is most likely to disperse from the natal area to reduce competition for resources in that locality. Alternatively, mothers in good condition (in high quality habitats) should produce the most philopatric sex. Both hypotheses make assumptions. The crucial ones for Trivers and Willard is that sons are more costly to produce, and that greater investment in them may reap rewards later. For local resource competition, the assumptions are that resources are spatially variable and that reproductive success better in some patches than in others; furthermore, dispersal rate should differ between males and females. All these conditions would seem to apply, although the modest sexual dimorphism characteristic of bat-eared foxes makes the extra cost (and value) of hefty sons the least clearly supported. Overall, and in the Serengeti, males (4.06 kg) are heavier than females (3.9 kg) (Gittleman 1983, 1989), although, in a sample from Botswana, females weighed marginally more than males (male: 4.03 kg, range 3.4–4.91, $n = 22$; female: 4.11 kg, 3.18–5.36, $n = 29$; Smithers 1971). The predictions of these two hypotheses are opposite. In breeding seasons when food was short, the foxes produced predominantly daughters—in accord with the offspring quality hypothesis and contrary to the local resource quality hypothesis. Amongst carnivores, this accords also with findings for badgers (*Meles meles*), but is opposite to those for Arctic foxes (Dugdale *et al.*, 2003; Goltsman *et al.* submitted). However, the interpretation of such results is seldom as simple as the beguilingly straightforward predictions of these two hypotheses might suggest. For example, in the case of the bat-eared foxes we have already established that opportunities for dispersal may have been low, which would alter the predictions of Julliard's hypothesis. Indeed, a diversity of complicating factors is already known to affect offspring sex ratio in polytocous species, such as maternal parity (African wild dogs, *Lycaon pictus*; Creel *et al.* 1998), maternal age and condition (coypu, *Myocastor coypus*; Gosling 1986b), and stress (golden hamsters, *Mesocricetus auratus*; Pratt *et al.* 1989).

Whatever the reason for the skewed sex ratio, how is it achieved? We speculate above that infanticide is a strong candidate, and raise the question of which individuals do the killing. Maternity analysis of the surviving cubs may suggest an answer.

Phylogeny, resources, and luck

Bat-eared foxes are survivors of an ancient lineage (currently represented by two subspecies (Coetzee 1977): *O. m. megalotis* (southern Africa), *O. m. virgatus* (East Africa)). The extent to which their ancient separation has bequeathed upon these foxes different constraints to those faced by other canids is unknown. There is evidence that they cannot regurgitate food, nor it seems can vulpine canids. There is no evidence of social suppression of female reproduction (common in both lupine and vulpine canids). The inverted U-posture of the tail is seemingly unique (Nel and Bester 1983). Nonetheless, there have been sufficient studies to reveal that, like other wild canids, bat-eared foxes display some intraspecific variation (Nel and Maas 2004; Wright 2004). That variation is doubtless rooted in regional differences in ecological circumstances and perhaps to the behaviour of the various species of termite on which they feed and, even within our study area, fox sociology varied between years and between groups within years. There is a plausible case that this can be attributed to variations in the dispersion of invertebrate prey, and the RDH suggests some predictions to test this. Nonetheless, our study reveals the enormous impact that an essentially chance event—infectious disease—can have on these predictions. It also reveals as crucial the question of whether—in the context of litter size—they can count to six. While the social lives of bat-eared foxes take a direction determined by their ancestry and local ecology, the outcome would appear to owe much to luck.

Acknowledgements

The thesis of which this work was a part was generously funded by the Max-Planck Institut für Verhaltensphysiology. We are grateful to Fran Tattersall and Paul Johnson for help in preparing this chapter.

CHAPTER 15

Patagonian foxes

Selection for introduced prey and conservation of culpeo and chilla foxes in Patagonia

Andrés J. Novaro, Martín C. Funes, and Jaime E. Jiménez

Culpeo fox *Pseudalopex culpaeus* © E C Montané.

Introduction

The culpeo (*Pseudalopex culpaeus*) and the South American grey fox or chilla (*P. griseus*) are closely related canids (Wayne *et al.* 1989a) that live in western and southern South America. The distributions of the culpeo and chilla overlap through most of the chilla range in Chile and western Argentina (Johnson *et al.* 1996; Fig. 15.1). Adult culpeos usually weigh 6–12 kg, adult chillas weigh 2–5 kg, and their sizes and size differences increase towards the south (Johnson and Franklin 1994a; Jiménez *et al.* 1995). Culpeo and chilla are opportunistic predators. Both canids feed primarily on small mammals but frequently consume introduced lagomorphs and livestock or its carrion wherever these have become abundant (i.e. Simonetti 1988; Johnson and Franklin 1994a; Jiménez *et al.* 1996; Novaro *et al.* 2000a). Additional foods are birds, lizards, insects, and fruits (Medel and Jaksic 1988).

The mechanisms that allow coexistence between culpeos and chillas have been the subject of debate (Fuentes and Jaksic 1979; Jiménez *et al.* 1995, 1996; Johnson *et al.* 1996). Fuentes and Jaksic (1979) argued that complementarity (Schoener 1974) in the use of trophic and spatial resources allows coexistence because they compensate high overlap in one niche dimension with low overlap in the other. Evidence from radio-tracking studies in two areas of sympatry in Chile, however, suggests that culpeos select habitats with higher prey densities and exclude chillas, which are thus confined to less productive habitats (Johnson and Franklin 1994b; Jiménez *et al.* 1996).

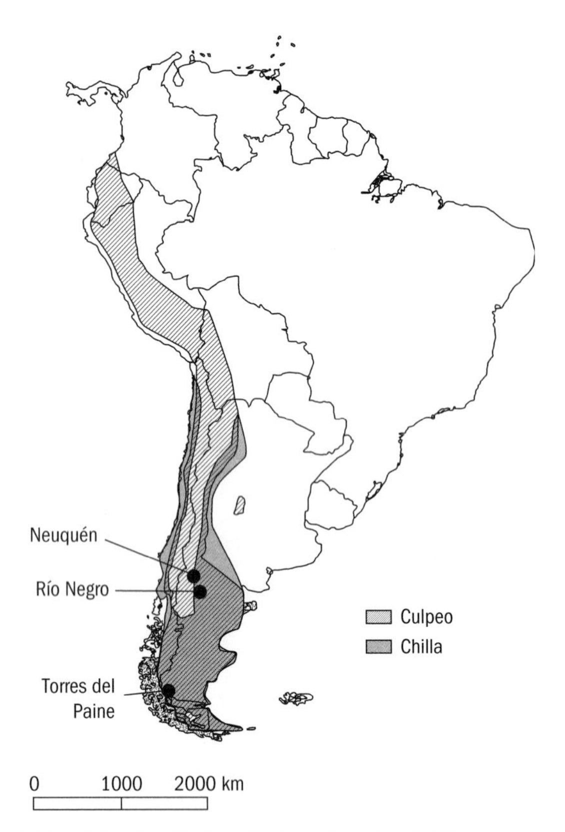

Figure 15.1 Distribution of culpeo (lines) and chilla (shaded) foxes in South America and location of study areas in Neuquén and Río Negro rangelands, Argentina, and Torres del Paine National Park, Chile.

Jiménez *et al.* (1996) expanded Fuentes and Jaksic's hypothesis by proposing that the presence of high-quality (large) prey such as lagomorphs may allow coexistence of culpeos and chillas in sympatry if the habitat is sufficiently complex to provide shelter for the smaller chilla from culpeos. Increasing body size differences between culpeos and chillas towards the southern portion of their range may also favour coexistence by permitting specialization in different food resources (Fuentes and Jaksic 1979; but see Jiménez *et al.* 1995). Unfortunately, the patterns of prey selection by culpeos and chillas are little known because most studies of their diets (reviewed by Medel and Jaksic 1988; Jiménez *et al.* 1996) did not include evaluations of prey availability or did not test for prey selection.

The culpeo and chilla are intensively hunted throughout Argentina and Chile because they are perceived as major predators of sheep, goats, and poultry (Bellati and von Thüngen 1990). In the Argentine provinces of Río Negro and Chubut, for example, control agencies paid US$10–25 between 1996 and 2001 for culpeo bounties, killing *c.*19,400 and 30,000 culpeos, respectively (Direcciones de Fauna of Río Negro and Chubut provinces unpublished data). In spite of their small size, chillas are perceived as lamb predators by many rural people in Argentina and Chile (Jiménez *et al.* 1996; Travaini *et al.* 2000b). Pressure to offer bounties for chillas in Argentine Patagonia has been mounting since the decline in fur demand in the 1990s led to increased chilla densities and anecdotal reports of chilla predation on sheep. The studies of the feeding ecology of both culpeos and chillas appear to contribute little to reducing this perceived conflict with people and, in particular, shed little light on what factors determine the incidence of predation on domestic species (Novaro *et al.* 2000a).

Here, our aims were to assess patterns of prey selection by culpeos and chillas in areas where the two species were sympatric and: (1) where sheep were abundant and the main wild prey, lagomorphs, had different densities; (2) where both canids were protected and sheep density was low. We use these comparisons to evaluate the competitive relationships between the culpeo and chilla and the factors that determine predation on livestock.

Our comparisons were based on two studies that reported data on culpeo and chilla food habits and a broad array of prey availability, and on unpublished information from one of these studies. In sheep rangelands of Neuquén, Argentina (Fig. 15.1), culpeo and chilla diets were studied from stomach contents provided by hunters (Novaro *et al.* 2000a). Culpeos and chillas selected strongly for European hares (*Lepus europaeus*) and culpeos also selected for sheep according to prey densities, whereas culpeos selected for hares and chillas for sheep carrion according to the biomasses of available foods. In this study, however, culpeos were sympatric with chillas only in the east of the study area, where hare densities were low (29.5 ± 11.6 hares/km^2). Only culpeos were present to the west, where hare densities were 54.3 ± 27.2 hares/km^2. Here, we compare culpeo and chilla prey selection in sympatry and culpeo prey selection between areas of allopatry and sympatry with chillas. Additionally, we compare data on diets from sympatric culpeo and chilla in a third area to the southeast (Río Negro; Fig. 15.1) where sheep were in similar numbers and hare were more abundant than in the allopatric Neuquén area (R. Cardon personal communication). We expected that different hare numbers would affect the trophic interactions between culpeos and chillas and their consumption of sheep, because population changes of principal prey can have significant effects on predator populations and their behaviour (Knick 1990; Poole 1994). In particular, considering the importance of hares in the diet of culpeos and chillas, we predicted that low hare numbers might be associated with increased overlap in their diets, presumably increasing competition for food, and increased predation on domestic livestock, and consequently increased animosity of farmers towards these canids. In addition to its focus on the highly practical question of depredations by these canids on domestic stock, this study brings together two highly topical areas of ecological theory, namely intra-guild competition and the ecosystem effects of introduced species (as reviewed, for example by Macdonald *et al.* 2001).

Johnson and Franklin (1994a) presented a comprehensive analysis of culpeo and chilla diets and prey abundances in Torres del Paine National Park in Chile (Fig. 15.1). Using data from faeces, these authors reported that culpeo and chilla diets differed in three main components: culpeos preyed more on hares, chillas consumed more carrion, and chillas were more omnivorous than culpeos, feeding more on arthropods and plants. Johnson and Franklin also reported that the differences in culpeo and chilla diets were associated with different prey availabilities in the habitats that both species used. Culpeo home ranges included habitats that had higher densities of hares, and chilla ranges included habitats where carrion was more abundant (Johnson and Franklin 1994b). A comparison of prey selection patterns between Torres del Paine and Neuquén sheds further light on prey selection by culpeos and chillas, and provides a contrast between circumstances where sheep were present in low numbers as opposed to densities typical of Patagonian ranches. A prerequisite for this comparison is a reconciliation of statistical methods used for the two studies.

Study areas

The Neuquén study area was located in northwestern Argentine Patagonia (40°S, 71°W), Province of Neuquén, on six sheep and cattle ranches encompassing a total area of 1420 km^2 (Fig. 15.1). Culpeo and chilla were sympatric on the two ranches to the east and only culpeos occurred on the other four ranches. The vegetation was characterized by a mixed steppe of grass and shrubs. Weather was dry and cold, with frosts throughout the year. Mean annual temperature was 11°C, and mean annual precipitation ranged from 28 to 75 cm on an east–west gradient and was concentrated during the winter. The Río Negro study area was located 150 km southeast of the Neuquén area, on small sheep ranches in the vicinity of the town of Comallo (41°S, 70°W). Vegetation and mean temperature were similar to Neuquén, but mean annual precipitation was 200 cm.

Torres del Paine National Park is located in the western foothills of the Andes Mountains in southern Chilean Patagonia (51°S, 73°W; Johnson and Franklin 1994a; Fig. 15.1). Seventy per cent of Torres del Paine was a dry steppe similar to the study area in Neuquén, but deciduous forest (*Nothofagus* spp.) patches were common in Torres del Paine. Mean annual precipitation (55 cm) was similar to that in Neuquén, but summers were wetter and mean annual temperature was lower in Torres del Paine (approximately 6°C). The mammal assemblage in Torres del Paine (Johnson *et al.* 1990) was similar to that in the Neuquén steppe, but guanacos (*Lama guanicoe*) were more abundant and sheep were rare in the park.

Methods

Neuquén and Río Negro rangelands

Food habits were determined through the analysis of stomach contents of 320 culpeo and 42 chilla killed by hunters in Neuquén between 1989 and 1994 and 18 culpeo and 19 chilla killed in Río Negro in 1989. The methodology is detailed in Novaro *et al.* (2000a). Prey items were identified as carrion if they were too large to have been killed by the foxes (e.g. cattle or horse) or when they contained larvae of Diptera. Young Sheep were considered as prey

(Bellati and von Thüngen 1990), although they were likely scavenged in many cases. This avoided underestimating potential predation on sheep.

We present results as per cent occurrence (number of times an item occurred as percentage of the total number of prey items in all stomachs) and as per cent mass of each item for stomach contents. We compared culpeo and chilla diets and dietary overlap between areas with different hare densities. Diets were compared between areas using log-linear analysis for frequencies (Zar 1996) and the von Mises test for continuous proportions (Stephens 1982; Maher and Brady 1986) for biomass consumed. We calculated food-niche overlap or diet similarity using Pianka's (1973) index: $O = \Sigma p_{xi} p_{yi} / (\Sigma p_{xi}^2 \Sigma p_{yi}^2)^{1/2}$, which ranges from 0 (complete dissimilarity) to 1 (similarity).

We estimated prey biomass in Neuquén as the product of prey density and mean body mass and assumed that prey biomass and density were acceptable combined estimators of prey availability (Jaksic *et al.* 1992). Prey activity patterns and habitat use may also be important components of prey availability but were not estimated. Density estimation methods for different prey and for carrion are described in Novaro *et al.* (2000a). Sheep availability was based on densities and body masses of sheep up to 1 year old for culpeos and up to 2 months old for chillas. Results are presented as mean density ± 1 SE. Diet and prey availability data were averaged throughout seasons.

To evaluate the role of prey body-size and density in selection by culpeo we analysed selectivity according to relative biomass and frequency of prey consumed and available. Both selectivity measures are necessary to assess selection of available prey. Prey selection was studied by comparing proportions of biomass of prey in stomachs to proportions of biomass available using overall MANOVA procedures based on an F-test (Girden 1992; PROC GLM, SAS Institute Inc. 1996). Prey selection also was studied by comparing frequencies of occurrence in diets to relative densities of each prey using a goodness-of-fit G-test (Zar 1996). When differences were significant ($p < 0.05$), we tested for selection or rejection of individual prey with individual MANOVA tests for each prey biomass (PROC GLM, SAS Institute Inc. 1996) and 95% Bonferroni confidence intervals for each prey frequency (Byers *et al.* 1984). Expected proportions were the proportions of biomass or density of each prey available to culpeos in allopatry and sympatry with chilla. Prey selection by sympatric chillas is reported in Novaro *et al.* (2000a).

Torres del Paine National Park

Johnson and Franklin (1994a) reported the per cent occurrence of prey in culpeo and chilla faeces. Per cent occurrences, however, do not represent the relative numbers of prey consumed, due to differential digestibility of prey of different sizes and types (Ackerman *et al.* 1984; Weaver 1993). Furthermore, the per cent biomass of prey consumed, needed to compare to the proportions of available prey to estimate selectivity (Novaro *et al.* 2000a), cannot be obtained directly from faecal samples as it can be from stomach contents. Therefore, we estimated the per cent biomass and number of prey consumed by culpeos and chillas using correction factors calculated by Lockie (1959) for the red fox, *Vulpes vulpes*, a canid that is intermediate in size between Patagonian culpeos and chillas (and comparable to values measured for jackals by Atkinson *et al.* 2002a). We assumed that per cent occurrence of prey in faeces was an acceptable approximation to per cent mass of undigested matter for each prey. Carrion consumption could not be estimated because correction factors were not available, so the per cent consumption of other items was overestimated in relation to percentages reported for culpeos and chillas from Neuquén and Río Negro. We did not calculate diet overlaps for corrected data from Torres del Paine because our diet correction for only some of the food items would have yielded overestimated overlap indices.

We calculated relative densities and biomass of the main prey in Torres del Paine from information on hare densities from Johnson and Franklin (1994a) and on sheep and upland goose (*Chloephaga picta*) densities from Iriarte *et al.* (1991). Some sheep occurred in Torres del Paine, mostly within chilla home ranges (Johnson and Franklin 1994a). We assumed a similar age structure and differential availability of sheep as for culpeos and chillas on Neuquén ranches. We calculated mean densities of cricetine rodents from Johnson and Franklin

(1994a)'s cricetine abundance data and mean body masses from Johnson *et al.* (1990). We divided cricetine abundances by the size of the trapping grids plus an extrapolated border strip of width equal to the mean distance moved by individuals (Seber 1982; Pearson *et al.* 1984) of a similar species assemblage in Neuquén (Novaro 1991; Corley *et al.* 1995).

Results

Culpeo and chilla diets

European hare and sheep comprised the majority of the biomass in the diet of culpeo and chilla in Neuquén, and hare comprised the majority of their diets in Río Negro and Torres del Paine (Tables 15.1 and 15.2). Culpeo diets differed significantly between areas of sympatry with chilla in Neuquén and Río Negro according to biomass ($Z = 5.01$, $df_1 = 9$, $df_2 = 729$, $P < 0.001$) and numbers of prey consumed ($\chi^2 = 10.7$, $df = 4$, $P < 0.05$). Chilla diets also differed significantly between areas of sympatry with culpeos in Neuquén and Río Negro according to biomass ($Z = 4.24$, $df_1 = 4$, $df_2 = 116$, $P < 0.001$) and numbers of prey consumed ($\chi^2 = 11.6$, $df = 4$, $P = 0.041$). The biomass and number of hares consumed by culpeos and chillas were less than half in Neuquén than in Río Negro, whereas sheep and carrion were consumed more frequently in Neuquén.

In spite of the similarity of sheep densities between areas of sympatry in Neuquén and Río Negro, culpeo and chilla diets in Río Negro were most similar to diets in Torres del Paine, where sheep were almost absent. The dominance of hare in the culpeo and chilla diets in Torres del Paine is more pronounced when the diets are presented as per cent biomass consumed as opposed to per cent occurrence in faeces. Conversely, the dominance of cricetines is emphasized when the numbers of prey consumed are reported (Table 15.2).

Contrary to our prediction, overlap between the diets of culpeos and chillas appeared higher where hares were more abundant. Overlap in Neuquén was 0.899 according to biomass and 0.950 according to numbers of prey consumed, and in Río Negro was 0.986 and 0.965, respectively (Table 15.1).

Culpeo and chilla prey selection

In Neuquén, both culpeos and chillas were selective in their food habits, but selectivity for certain prey differed according to whether biomass or densities of prey were considered. Prey selection by culpeos changed between areas of allopatry and sympatry with chillas (Tables 15.3 and 15.4). In spite of the lower hare density in the area of sympatry, in both areas culpeos consumed hares more than expected according to their biomass available, carrion less than expected, and cricetines in similar proportion

Figure 15.2 Chilla *Pseudalopex griseus* © R G del Solar.

Table 15.1 Culpeo and chilla diets in an area of sympatry (and low European-hare density) and culpeo diet in an adjacent area of allopatry (and intermediate hare density) in Neuquén rangelands, and culpeo and chilla diets in an area of sympatry (and high hare density) in Río Negro rangelands, Argentina

	Neuquén						Río Negro			
	Allopatric culpeo		Sympatric culpeo		Sympatric chilla		Sympatric culpeo		Sympatric chilla	
Prey type	%N	%B	%N[a]	%B[b]	%N[a]	%B[b]	%N[c]	%B[d]	%N[c]	%B[d]
Mammals										
Order Rodentia										
Cricetine	30.9	9.2	29.2	9.5	25.5	11.4	23.1	2.2	27.6	5.9
Ctenomys spp.	5.8	3.9	5.6	2.0			3.8	3.1	3.4	1.5
Caviidae	2.2	1.5	0.6	tr	3.6	6.5	3.8	7.8	3.4	5.4
Order Marsupialia										
Thylamys pusilla					3.6	0.1				
Order Edentata										
Chaetophractus and Zaedyus	2.4	1.1			5.5	0.8			6.9	10.8
Order Lagomorpha										
Lepus europaeus	29.0	56.6	20.5	35.1	16.4	14.6	57.7	85.2	41.4	71.6
Oryctolagus cuniculus	0.5	0.8								
Order Artiodactyla										
Sheep	12.7	14.8	17.4	34.4	20.0	40.3	3.8	0.4	3.4	3.9
Carrion[e]	8.3	10.5	11.2	15.7	9.1	21.0				
Unidentified mammals	2.9	0.2	4.3	1.3	9.1	0.5				
Total mammals	94.7	98.6	88.8	98.0	92.8	95.2	92.2	98.7	86.1	99.1
Birds										
Pterocnemia pennata			0.7	0.2						
Unidentified birds	3.6	1.3	6.8	1.8	1.8	2.5	7.7	1.4	10.3	1.0
Lizards	1.7	tr	3.7	tr	5.5	0.5			3.4	tr
Number of vertebrate food items	411		161		55		26		29	
Number of stomachs	239		81		42		18		19	
Percentage of stomachs with										
Invertebrates					17.6				47.4	
Schinus spp. seeds					2.9		16.7		21.1	

Note: %N = per cent occurrence, and %B = per cent biomass in stomachs.

Pianka indices of trophic overlap: [a] 0.950; [b] 0.899; [c] 0.965; [d] 0.986. [e] *Equus, Bos, Cervus, and Lama*, tr = trace, <0.05%.

Table 15.2 Per cent occurrence of prey in culpeo and chilla faeces (%O, from Johnson and Franklin 1994a) and per cent biomass (%B) and number of prey consumed (%N) in an area of high European-hare density in Torres del Paine National Park, Chile

	Culpeo			Chilla		
Prey type	%O	%B	%N	%O	%B	%N
Mammals						
Order Rodentia						
Cricetine	20.3	10.6	81.6	23.9	16.0	72.9
Order Lagomorpha						
Lepus europaeus	68.5	81.4	11.7	45.0	68.7	7.6
Order Carnivora						
Conepatus humboldti	0.8	1.1	1.1	0.5	0.9	0.5
Order Artiodactyla						
Sheep	0.8	0.8	0.4	6.4	8.7	2.5
Carrion (*Lama and Bos*)	2.2	–	–	13.8	–	–
Total mammals	92.6			89.6		
Birds						
Chloephaga picta	5.1	5.7	2.9	1.8	2.6	0.7
Unidentified birds	1.8	–	–	4.6	–	–
Frogs	0.0			0.2		
Lizards	0.5	0.3	2.2	3.8	3.1	15.6
Number of vertebrate food items	784			851		
Number of faeces	645			890		
Percentage of faeces with						
Beetles	2.0			41.8		
Scorpions	0.0			1.2		
Seeds	0.0			0.6		
Vegetation	2.8			7.0		
Berberis buxifolia	1.0			7.5		
Egg shells	0.7			3.4		

Note: %B and %N were calculated with correction factors from Lockie (1959). Missing percentages (–) are those that could not be calculated with correction factors.

to their availability. In the area of sympatry culpeos consumed sheep more than expected and in the area of allopatry they consumed sheep in similar proportion to its availability (Table 15.3). The frequencies of prey in the diets of culpeos also differed significantly from the relative densities of prey available in sympatry ($G = 403.9$, df $= 2$, $P < 0.001$) and allopatry ($G = 956.6$, df $= 3$, $P < 0.001$; Table 15.4). In both areas culpeos consumed hares and sheep significantly more than expected according to their densities and cricetines less than expected.

In Torres del Paine, culpeos and chillas were selective in their food habits according to numbers of prey consumed (Table 15.5) but their patterns of prey selection were different when we considered the biomass of prey (Table 15.6). The numbers of prey consumed by culpeos differed significantly from the relative densities of prey available ($G = 195.8$, df $= 3$, $P < 0.001$; Table 15.5). Culpeos consumed hares and geese significantly more than expected according to their densities and cricetines less than expected. According to biomass of prey,

Table 15.3 Prey selection by culpeos based on biomass consumed and available in areas of allopatry and sympatry with chillas in Neuquén, Argentina

	Prey	%EB	%B	Wilks' Lambda	df_1	df_2	*P* level
Sympatry (95)	Overall			0.133	3	31	0.0001
	Hare	18.0	37.0	0.221	1	54	0.0001
	Lamb	24.0	36.3	0.736	1	54	0.0323
	Carrion	41.2	16.6	0.362	1	54	0.0001
	Cricetines	9.8	10.1	0.995	1	54	0.9634
	Edentates	7.0	0.0	–	–	–	–
Allopatry (92)	Overall			0.015	4	188	0.0001
	Hare	28.7	61.4	0.141	1	192	0.0001
	Lamb	20.8	16.1	0.861	1	192	0.5472
	Carrion	35.8	11.4	0.382	1	192	0.0001
	Cricetines	8.6	10.0	0.958	1	192	0.7934
	Edentates	6.0	1.1	0.904	1	192	0.0182

Notes: Percentages of biomass available and in the diet are only for prey for which availability data were obtained. Percentages of prey in the diet (%B) were compared to percentages expected according to availability (%EB) using a MANOVA test. Per cent biomass of diet considered out of overall diet is indicated between parentheses.

Table 15.4 Prey selection by culpeos based on per cent prey occurrence in stomachs (%N) and density of prey in areas of sympatry and allopatry with chilla in Neuquén, Argentina

	Sympatry			Allopatry		
Prey type	Density (ind./km^2)	%EN[a]	%N ± BCI	Density (ind./km^2)	%EN[a]	%N ± BCI
Hare	29.5 ± 11.6	1.2 *M*	30.6 ± 11.1	54.3 ± 27.2	2.2 *M*	38.6 ± 7.1
Sheep	5.4 ± 0.5	0.2 *M*	25.9 ± 10.5	5.4 ± 0.5	0.2 *M*	16.9 ± 5.5
Cricetine rodents	2422.9 ± 1597	98.6 *L*	43.5 ± 11.9	2422.9 ± 1597	97.1 *L*	41.2 ± 7.2
Edentates	13.1 ± 6.9	0.5	0.0	13.1 ± 6.9	0.5 *M*	3.2 ± 2.6
Total prey items			108			308

Notes: Numbers added or subtracted from %N are 95% Bonferroni confidence intervals (BCI) for %N; expected percentages of prey consumed (%EN) were calculated from prey densities.

[a] Prey items are consumed significantly more (*M*) or less (*L*) than expected according to their availability if %EN are smaller than the lower limit or larger than the upper limit of each BCI, respectively ($P < 0.05$).

however, culpeos consumed most prey (except cricetines) in similar proportions to their availabilities (Table 15.6).

The numbers of prey consumed by chillas also differed significantly from the relative densities of prey available in Torres del Paine ($G = 394.6$, $df = 3$, $P < 0.001$; Table 15.5). Chillas consumed hares and sheep significantly more than expected according to their densities and cricetines less than expected. The chilla diet also differed from the biomass of available prey: chilla consumption of cricetines was *c*. 1/3 of that expected and hares and sheep were consumed in larger proportions than their availabilities (Table 15.6).

Table 15.5 Prey selection by culpeos and chillas based on per cent prey occurrence in faeces (%O, from Johnson and Franklin 1994a), numbers of prey consumed (%N), and density of prey in Torres del Paine

	Culpeo				Chilla			
Prey type	%O	Density (ind./km^2)	%EN[a]	%N ± BCI	%O	Density (ind./km^2)	%EN[a]	%N ± BCI
Hare	72.3	86.6 ± 30.3	3.64 *M*	12.1 ± 3.0	58.4	24.8 ± 7.7	1.07 *M*	9.1 ± 2.8
Sheep	0.8	0.2	0.01	0.5 ± 0.6	8.3	0.08	0.003 *M*	3.0 ± 1.7
Cricetines	21.4	2290 ± 1759	96.13 *L*	84.5 ± 3.3	31.0	2290 ± 1759	98.7 *L*	87.1 ± 3.3
C. picta	5.4	5.3	0.22 *M*	3.0 ± 1.6	2.3	5.3	0.23	0.8 ± 0.9
Total prey	742				656			

Notes: %N were calculated applying correction factors from Lockie (1959); numbers added or subtracted from %N are 95% Bonferroni confidence intervals (BCI) for %N; expected percentages of prey consumed (%EN) were calculated from prey densities (from Johnson and Franklin 1994a, and Iriarte *et al.* 1991).

[a] Prey items are consumed significantly more (*M*) or less (*L*) than expected according to their availability if %EN are smaller than the lower limit or larger than the upper limit of each BCI, respectively ($P < 0.05$).

Table 15.6 Prey selection by culpeos and chillas based on biomass of prey consumed and biomass of prey available in Torres del Paine

		Culpeo		Chilla	
Prey type	Weight (kg)	%B	%EB	%B	%EB
Hare	3.35	82.6	71.9	71.6	43.5
Sheep	25 and 5[a]	0.9	1.4	9.1	0.2
Cricetines	0.04	10.7	22.7	16.7	48.0
C. picta	3.0	5.8	3.9	2.7	8.3

Notes: Average body weight for cricetines are from Johnson *et al.* (1990), for hare from Johnson *et al.* (unpublished manuscript), and for sheep (lambs and 1-year-olds) from Novaro (unpublished data). Biomasses of prey consumed (% B in diet) were calculated applying correction factors from Lockie (1959) to data from Johnson and Franklin (1994a); expected percentages of biomass of prey available (%EB) were calculated from density data from Johnson and Franklin (1994a) and Iriarte *et al.* (1991).

[a] Body mass of sheep available to culpeos and chillas, respectively.

Discussion

Culpeo and chilla diets and prey selection

Diets and prey selection by culpeos and chillas in Patagonia were strongly affected by the local abundance of European hare and domestic sheep. In the Neuquén area where hare were scarce, which was also an area of sympatry with chillas, culpeos consumed more sheep than in the area of allopatry. In the Río Negro area of sympatry, where sheep were also present, culpeo and chilla consumed more hares than in either of the Neuquén areas. The changes in prey selection associated with hare abundance were different, however, depending on whether biomass or frequencies were used as the measure. Using biomasses as the yardstick, culpeos selected hares positively in both areas and selected sheep where hares were in low numbers (Table 15.3). When frequencies were the measure, hares and sheep were selected regardless of the abundance of hares. Conversely, in Torres del Paine, where hares were most abundant, consumption of hare by culpeos was not different from hare availability according to biomass. Thus, culpeos appear to select intensively for hares at intermediate and low hare densities, even in the presence of abundant sheep. Perhaps hare is a more profitable (Pyke *et al.* 1977) prey than sheep, due to greater vulnerability (Corbett and Newsome 1987). In the case of the chilla, strong selectivity for hares even at low hare densities (and when sheep and its carrion are abundant; Novaro *et al.* 2000a) suggests that hares are also a highly profitable prey. These results also agree with food preferences of other canids, which only take sheep when their preferred prey are scarce (i.e. red fox, Macdonald 1977a; coyote *Canis latrans*, Sacks and Neale 2002).

The low consumption of sheep in Río Negro in comparison with Neuquén may be related to differences in sheep management practices. In the study area in Río Negro sheep are raised in more labour-intensive, smaller ranches where humans proximity may deter predation. In the large ranches of the Neuquén study area, on the other hand, ranch hands check less frequently on the sheep. These differences may lead to lower predation on sheep or to lower carrion availability in Río Negro.

Our results suggest that spatial or temporal changes in hare densities could lead to increased predation on sheep and also on native prey in some areas in Patagonia. The increase in sheep predation is more likely in culpeos which, due to their larger size, are better able to kill lambs and even adult sheep than are chillas (Bellati and von Thüngen 1990). Our inability to distinguish between sheep that were scavenged or preyed upon prevents us from drawing conclusions about chilla predation on sheep. However, lower hare densities are likely to be associated with increased chilla consumption of sheep carrion and probable predation on lambs. In summary, areas and periods with low hare density may have increased conflicts between canids and humans in Patagonia.

The low selection for small cricetine rodents (according to both biomass and numbers) in relation to other prey suggests that these prey also may be less profitable than hares. This appears to be the case both in the presence of high (Neuquén) and low (Torres del Paine) numbers and biomass of sheep and its carrion as alternative foods. Low profitability of cricetines supports the conclusions of Jiménez *et al.* (1996) and Johnson and Franklin (1994a) that culpeos select habitats with higher densities of larger rodents or lagomorphs, but is at odds with their conclusion that cricetines would be selected by chillas. Chillas appear to be selecting strongly for hares, and may consume more cricetines than culpeos but not necessarily select for them in comparison with other prey.

Lower hare densities may not lead to increased competition for food between culpeo and chilla in Patagonia. The slightly lower indices of trophic overlap in Neuquén than in Río Negro were mainly due to higher consumption of sheep by culpeos and of carrion by chillas, as well as reduced consumption of hare by both. The ultimate explanation for the small change in trophic overlap, however, may be the overall high density of hare, livestock, and its carrion in Patagonian rangelands and the relatively low canid densities after several decades of intense hunting for fur (Novaro 1997b). These results support the overall conclusion by Johnson and Franklin (1994a) about the similarity of optimal diets by culpeo and chilla. The diet similarity index in Torres del Paine may be in fact much higher (0.94) than reported by these authors (0.14), apparently due to miscalculation of Pianka's index. In particular, our results indicate that selection for or against specific prey such as hares and cricetines is almost identical between these canids, even in the presence of larger prey such as sheep. As a consequence, the use of different habitat types reported by Johnson and Franklin is probably the result of exclusion of chillas by culpeos from more productive habitats and not a consequence of different selection patterns for habitat or food.

Coexistence and conservation of culpeos and chillas in Patagonia

Our findings suggest that in areas of sympatry in southern Chile and Argentina, diets and prey selection patterns of culpeo and chilla may be more similar than expected, especially when the numbers and biomass of prey consumed are considered. Overall, culpeo and chilla diets and prey selection differed mostly among areas with different prey densities and were strikingly similar between species in each area. Furthermore, this similarity in prey selection occurs in two areas where culpeo and chilla body sizes are most dissimilar (Novaro 1991; Johnson and Franklin 1994a). Thus, segregation of food resources through selection for different prey or for habitats with different prey availabilities are unlikely mechanisms to allow coexistence between these canids throughout their range. In summary, if the food component of the niche-complementary hypothesis does not allow segregation even under the most extreme body-size differences, habitat segregation may be the only mechanism promoting coexistence throughout their range.

Most of the habitats occupied today by culpeos and chillas, including the Patagonian steppe, have been highly modified by humans. Some of these modifications may promote local conditions that would alter the result of chilla displacement from productive habitats by culpeos. In the area of sympatry of the

Neuquén steppe, for example, culpeo and chilla home ranges overlap and they are frequently found in the same habitats, particularly where habitat structure is more homogeneous (Novaro and Funes unpublished data). Based on the similarity of prey selection patterns presented in this study, we conclude that the presence of introduced hares in Patagonia may be insufficient to explain the coexistence between culpeo and chilla. Conversely, three other human-related factors may contribute to promote or reduce coexistence in this region. First, the availability of large numbers of sheep (both as live prey and carrion) may help reduce interference between the two foxes, because food availability may be so high that aggressive interactions from culpeos towards chillas may occur only rarely. Additionally, hunting of culpeos to reduce sheep predation is much more intense than of chillas in Patagonian sheep ranges (Novaro 1997b). This source of mortality for culpeos may be sufficient to maintain culpeo numbers low at specific sites and thus allow the persistence of chillas even in homogeneous habitats. Finally, the introduction of sheep to Patagonia in the early 1900s was followed by the eradication of a larger carnivore, the puma (*Puma concolor*) from most of the steppe. The removal of pumas, which may have kept culpeo numbers low in many areas, might have affected chilla numbers negatively, as it has occurred with other carnivore guilds (Johnson *et al.* 1996, Linnell and Strand 2000). The current recolonization of much of Patagonia by pumas, probably due to a decline in sheep production, may lead to reduced culpeo densities and may allow higher densities or range expansions of chillas.

We therefore suggest that human disturbance through food supplementation or differential mortality may promote local (even within habitat) coexistence between culpeos and chillas in Patagonia, and likely in other areas of Chile and Argentina. However, the complexity of the interactions involving food availability, hunting by humans, and predation by larger carnivores determines that conditions for coexistence may depend on the local balance of these processes as well as on habitat complexity. As Johnson *et al.* (1996) point out, manipulation experiments (of food or mortality, even by closely monitoring removal or population reductions in removal conducted on sheep ranches) could provide additional understanding of the coexistence mechanisms between culpeos and chillas.

One implication of our results for managing canid–livestock conflicts in Patagonia is that predation on sheep and other domestic species may be more likely in areas where European hare densities are lower. Also, because hare numbers may fluctuate in Patagonia (Novaro *et al.* 2000a), predation on sheep may increase during hare declines. These predictions apply mostly to culpeo predation, but also may apply to predation by chillas on lambs and other small domestic animals. Preliminary data from Neuquén may confirm these predictions (Novaro and Funes, unpublished data). First, sheep and goat losses to culpeos in 1999 averaged 24% and 21%, respectively, for 12 families in the Chiquilihuin Mapuche-Indian land, where ranges are degraded and densities of wild prey (including hares) are low, whereas losses usually average 5–10% on large private ranches. Second, estimated predation on sheep in one of our study ranches increased from an annual average of 10% to 40% in 1995–96 after a hare decline. The implications of these conclusions are that predation control efforts (either by canid control and/or livestock protection) should be restricted temporally and spatially to areas and times that are more likely to experience high predation, and should not be applied indiscriminately, as is commonly done in Patagonia.

Another implication of our results is that strong selection for hares by culpeos and chillas may result in regulation of hare populations at low densities, reducing competition for pastures between hares, sheep, and native herbivores. An ongoing study of the effect of culpeo removal on hare population dynamics suggests that culpeo predation may help regulate hare numbers at low densities (Novaro *et al.* unpublished data). If these results are confirmed, Patagonian sheep ranchers may be better off by tolerating a certain level of canid predation on sheep, because their benefits from canid predation on hares may outweigh their losses due to occasional attacks on sheep (this parallels calculations by Macdonald *et al.* 2003 for the benefits to cereal farmers in the United Kingdom of tolerating red foxes which eat rabbits). Modelling and economic studies are needed to evaluate further the interactions between canids, hares, sheep, and pastures in the Patagonian steppe.

Acknowledgements

We thank O. Monsalvo, C. Rambeaud, and the Guardafaunas of Neuquén Province for assistance in the field, A. del Valle for logistical support, and the ranch owners and workers for their cooperation. S. Walker helped with statistical analyses. E. Donadio, C. Iudica, and M. Bongiorno assisted with diet analysis. Financial support was provided to A. J. N. by the Wildlife Conservation Society, Lincoln Park Zoo Scott Neotropic Fund, Argentinean Associations of Fur Traders (CIP), American Society of Mammalogists, and Sigma-Xi. A. J. N. was supported by fellowships from the Fulbright Commission, TCD and PSTC Programs at the University of Florida, and the University of Buenos Aires.

CHAPTER 16

Jackals

A comparative study of side-striped jackals in Zimbabwe: the influence of habitat and congeners

David W. Macdonald, Andrew J. Loveridge, and Robert P. D. Atkinson

Side-striped jackal *Canis adustus in* Hwange © A. J. Loveridge.

Introduction

The side-striped jackal (*Canis adustus*) is a medium-sized canid found throughout much of East and central Africa, excluding the equatorial zone, and extending as far south as Zimbabwe and northeastern South Africa. Despite its occurrence in about one-third of the continent—a distribution that, amongst the 12 African canid species, is second in expanse only to that of the golden jackal (*C. aureus*)—the side-striped jackal had been little studied. Between 1990 and 1997, we undertook research on side-striped jackals in Zimbabwe with the objectives of exploring intraspecific variation between two populations in contrasting environments (in one of which they were sympatric with black-backed jackals (*C. mesomelas*)), and of using this comparison to shed light on the epidemiology and control of jackal rabies.

Black-backed jackals occur in two disjunct areas, one in East Africa and the other in southern Africa (from southwestern Angola, southern Zimbabwe, and Mozambique southwards throughout South Africa). The two populations are separated by some 900 km (Coe and Skinner 1983).

The two jackal species diverged from a common ancestor only 2 million years ago (Wayne *et al.* 1989a). Their contemporary distributions suggest different habitat and climatic preferences; side-striped jackals occur throughout the moist savannah regions of tropical Africa, while black-backed jackals predominate in the drier regions of East Africa and the semi-desert regions of southern Africa. The kidneys of black-backs have a higher relative medullary thickness, an index of urine concentrating ability, an adaptation to arid conditions (Loveridge 1999). Nonetheless, both jackal species are widely sympatric in central and southern Zimbabwe, Mozambique, northern Botswana, and the northeastern region of South Africa (Fig. 16.1) and Van Valkenburgh (1988) estimates that they have coexisted, along with the golden jackal, in the Serengeti for at least 1.5 million years. Where they coexist, jackal species show some niche separation, with side-stripes using miombo woodland and moist valleys, golden jackals favouring open plains, and black-backs using *Acacia* woodland and scrub. However, all three species are adaptable. For example, black-backed jackals use open woodland where they are sympatric with golden jackals (Fuller *et al.* 1989), but in the absence of this, species in southern Africa use open grassland and most other habitats ranging from montane areas to desert

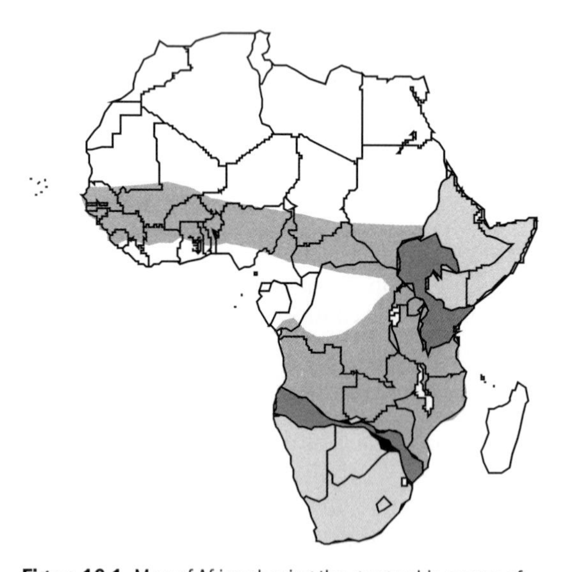

Figure 16.1 Map of Africa showing the geographic ranges of *C. mesomelas* (light grey shading) and *C. adustus* (medium grey shading) after Smithers (1983). The area where the two species of jackal are sympatric is shown in dark grey.

(Skinner and Smithers 1990). Allopatric side-stripes are similarly ubiquitous (Loveridge and Macdonald 2003). Rautenbach and Nel (1978) were the first to emphasize that similarities in habitat, body size, and activity patterns probably threw side-striped and black-backed jackals into competition.

Although both jackals are similar in size, the southern African black-backed jackals are generally smaller (Skinner and Smithers 1990). Side-striped jackals tend to have a longer, lighter skull and lighter dentition compared to the more robust skull and teeth of black-backed jackals (Skinner and Smithers 1990). Side-stripes have a larger relative molar grinding area, while black-backed jackals have a greater premolar cutting blade length. These characteristics suggest that side-stripes may be better adapted to omnivory (Van Valkenburgh 1991; Van Valkenburgh and Koepfli 1993). Both species are clearly omnivorous, with a varied diet that includes anything from fruit to insects and eggs to medium-sized mammals (Skinner and Smithers 1990). But black-backed jackals may be more inclined to hunt cooperatively, and hence to take larger mammalian prey (McKenzie 1990; and see later). Furthermore, black-backed jackals are considered vermin by livestock farmers whereas there is less evidence of side-stripes killing livestock (Van der Merwe 1953; Ansell 1960; Smithers and Wilson 1979) (Figs 16.2 and 16.3).

The behavioural ecology of all jackals, and indeed most medium-sized canids, is broadly similar. They are territorial, communicate occupancy—and much else—with olfactory and vocal signals, and live in social units developed around a breeding pair (Macdonald 1979c; Skinner and Smithers 1990). In southern Africa, mating occurs in the winter (mainly June–August), accompanied by increased vocalization and territoriality (Skead 1973; Bernard and Stuart 1992). Juveniles disperse annually although some may remain within their natal territory to act as helpers, a behaviour most fully documented for black-backed jackals in the Serengeti (Moehlman 1978, 1983).

Our research compared the natural history of side-striped jackals and black-backed jackals in two contrasting study areas in Zimbabwe. This chapter is a synthesis of results presented in the following papers: Atkinson *et al.* (2002a,b) Loveridge and Macdonald (2001, 2002, 2003), Rhodes *et al.* (1998), and Atkinson and Macdonald (in prep.).

Figure 16.2 Side-striped jackal *Canis adustus* © C. and T. Stuart.

Figure 16.3 Black-backed jackal *Canis mesomelas* © C. and T. Stuart.

Study areas

Highveld, North Zimbabwe

The Highveld study area (17° 45′S; 30° 30′E) encompassed 100 km^2 of commercial farmland, 40 km south-west of Harare, in West Mashonaland. The habitat comprised about 40% of each of *Brachystegia*-dominated miombo woodland and low-lying, seasonally flooded, natural grassland, both used by grazing animals. Planted grass leys, ploughed and fallow fields, and tobacco and maize crops, occupied the remaining 20%. Annual rainfall averaged 71–81 cm, with the great majority falling between November and March. Mean monthly temperatures were 13°C (June–July) to 21°C (October). We recorded 43 species of fruiting plants in the area, of which 24 were eaten by jackals, and of which at least one species was always available.

Black-backs were rare, large carnivores were absent, and wild ungulates were scarce. Springhare (*Pedetes capensis*) occurred at 0.06 ± 0.04/km, the most abundant rodents being multimammate mice (*Mastomys* spp.) and bushveld gerbils (*Tatera leucogaster*).

Hwange, southwestern Zimbabwe

The Hwange study area (18° 39′S; 27° 00′E) in Matabeleland lay within the 1000 km^2 Hwange Estate, northeast of Hwange National Park, and centred on the 4.9 km^2 Dete vlei. About 80% of the area was forest, dominated by Zimbabwe teak (*Baikiaea plurijuga*) and false mopane (*Guibortia coleosperma*); silver terminalia (*Terminalia sericea*) forming the ecotone between woodland natural grassland, which formed 18% of the study area; occasional small stands of camel thorn (*Acacia erioloba*) and hook-thorn (*Acacia fleckii*) bordered the low-lying, seasonally flooded grassland (vlei). There were three seasons: (1) wet season: late November to April or May ($<$10 and $>$370 mm per month); (2) cold, dry winter: June–August (2–26°C); and (3) hot, dry summer: September/October (17–31°C). There were no trees bearing edible wild fruits, and the only two fruiting shrubs, wild gooseberry (*Physalis augulata*) and tsamma melons (*Cucumis melo*), were available only in the wet season. Of six water holes along Dete vlei, four were permanently supplied with pumped ground water, and two were seasonal. Two safari camps and one hotel produced refuse, which was burnt and buried in open pits.

In addition to jackals there were four large carnivore species and high ungulate densities. Small- and medium-sized mammals included low densities of scrub hares (*Lepus saxatilis*) and abundant spring-hares (57.9/km^2), and low numbers of small rodents (following poor rains during 1994/95). Insects were widely abundant throughout the wet season, in grassland in March–May, and scarce in the dry season.

The Highveld and Hwange compared

Population density

In the Highveld, the resident population of territory holding adults was 20–30 side-striped jackals per 100 km^2 expanding to a breeding season peak of 80–120 per 100 km^2 (Rhodes *et al.* 1998). At Hwange, the population density of black-backed jackals was 53.9–79.1 per 100 km^2 expanding to 68.3–97.1 per 100 km^2 during the breeding season. Densities of side-stripes were more difficult to determine because this species was more cryptic than the black-back and not sighted on a regular basis. Based on approximately equal trapping success during the study for both species (11 side-stripes, 11 black-backs in 157 trap nights, with traps distributed evenly within the study site), we infer that approximately equal densities of the two species were present in the study area.

Spatial organization

Home ranges

In the Highveld there was no seasonal variation in range size, and annual ranges of side-stripes averaged 1210 ± 295 ha (938–1904 ha, n = 12). Individuals traversed about 14% of their ranges nightly and 33% over four consecutive nights (Rhodes *et al.* 1998). At Hwange, there were seasonal variations in range size, which were largest for both species during winter (Loveridge and Macdonald 2001). The ranges of side-stripes at Hwange averaged 154.5 ± 103.0 ha, almost an order of magnitude smaller than those on the Highveld, and significantly smaller than those of sympatric black-backs (210.8 ± 85.6 ha).

Pairs of side-striped jackals in the Highveld used their home range with a high degree of concordance, using the same areas with similar intensity and largely at the same times (concordance was least during the period of pup-rearing). The central core of each home range was used exclusively by its occupants, the periphery of the range overlapped widely with four or more neighbouring pairs (mean overlap between the territories of neighbouring pairs was 19% SD 13% — Atkinson and Macdonald in prep.). The territory of the single black-backed jackal tracked in the Highveld overlapped with those of its side-striped neighbours and appeared to be superimposed over parts of seven side-striped territories.

At Hwange, overlap between home ranges of the two species varied between the seasons ($F_{2,52} = 7.47$, $p < 0.001$), being greatest in the cold dry (mean 26 ± 18.6%, n = 18) and wet (24.7 ± 23.7%, n = 18) seasons, and lowest in the hot dry season (16.00 ± 22.5%, n = 19). However, there was no interspecific overlap between 50%-core home ranges (Loveridge and Macdonald 2001).

In both study areas, then, there was overlap between the ranges of neighbouring pairs. This is at odds with a perception of canid territories tessellating neatly, and requires explanation. Two non-exclusive possibilities are, first, that areas of overlap offer depleted food availability due to dual exploitation by neighbours, and thus merit less attention and less defence (as suggested by the Passive Range Exclusion hypothesis, Stewart *et al.* 1997); second, mortality may cause perturbation of the spatial organisation of the survivors (as suggested by Tuyttens and Macdonald 2000).

Social system

Side-striped jackals have often been considered as solitary (e.g. Rautenbach and Nel 1978), but in Hwange, and perhaps the Highveld, they formed social groups. Their tendency to use wooded terrain made it harder to confirm group sizes in side-stripes than in black-backs, and this was especially so in the Highveld where we could determine only that groups typically may have comprised a pair, two adult-sized non-breeders and pups of the year. However, at Hwange at least four of five side-striped territories included extra-pair members (up to seven in one case, two of which were known to be between 1 and 2 years of age). Two young adults that had taken up peripheral home ranges returned to their natal home ranges during the whelping season, and were apparently involved in the care of their younger siblings (as were at least three non-breeding adults in another territory (Loveridge and Macdonald 2001)).

At Hwange, of six black-backed jackal territories, four were occupied by a mated pair and 1–4 additional adults, while two contained only a pair. The extra-pair adults were seen tending and feeding pups of the breeding pairs. Breeding adults spent most time in each other's company, whereas non-breeding adults tended to be alone or associated with juveniles (Loveridge 1999).

Reproduction and dispersal

The reproductive biology of the two species is similar. At Hwange, both species mated between June and July, and most pups were born between September and October, and thus dependent on the denning site until at least early December, which corroborates findings in southern Africa (Ferguson *et al.* 1983; Bernard and Stuart 1992). Skinner and Smithers (1990) report mean litter sizes for side-stripes of 5.4 pups, but we estimate that only two per litter survive past 6 months (Rhodes *et al.* 1998); average lifespan is probably 3–4 years.

At Hwange, we documented the dispersal of 10 individuals, aged between 1 and 2 years, the majority of which dispersed between December and March. Side-stripes dispersed over a mean distance of 4.6 ± 3.51 km (three females, two males) and black-backs over 2.8 ± 2.05 km (two females, three males). Elsewhere black-backed jackals have been known to disperse up to 120 km (Ferguson *et al.* 1983). The longest dispersal recorded was by a side-striped jackal over 20 km.

In the Highveld, one side-striped jackal, an adult male, ranged over a large area before settling down to a much smaller area that was within its former home range. A young female stayed in the vicinity of her capture site for a month, then drifted slowly northwards, increasingly consolidating the use of the northern end of the drifting range. Three months after capture she had established a new permanent territory.

Movement patterns and habitat use

Activity patterns

Side-striped jackals in both study areas were almost exclusively nocturnal. At Hwange, this pattern was bigeminous (Fig. 16.4), as reported elsewhere for allopatric black-backed jackals (Ferguson *et al.* 1988) but at Hwange, while mostly nocturnal, they were also active during the morning and late afternoon. Peaks of activity were more pronounced in the side-stripes than the black-backs. In the Kalahari, black-back activity corresponds with that of their rodent prey (Ferguson *et al.* 1988) and Ethiopian wolves (*Canis simensis*) parallel the activity of their diurnal rodent prey (Sillero-Zubiri *et al.* 1995a). However, in Hwange, the main mammalian prey of black-backs was springhares, whose activity peaked between 21:00 and 03:00 h, apparently timed to avoid foraging jackals (see also Fenn and Macdonald 1995).

Movement patterns

Although canids are known to follow predictable routes (e.g. red fox: Doncaster and Macdonald 1997; Blanford's foxes *Vulpes cana*: Geffen and Macdonald 1993), our observations of side-striped jackals in the

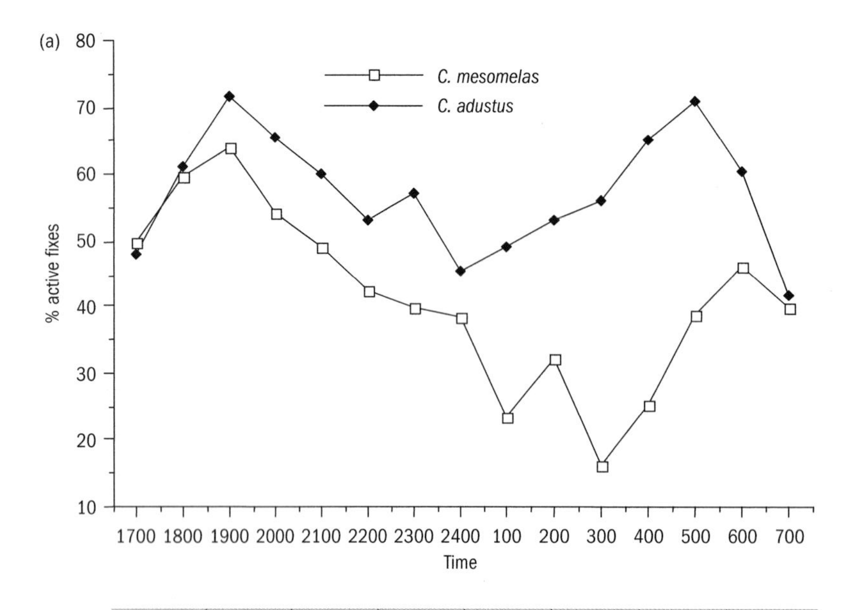

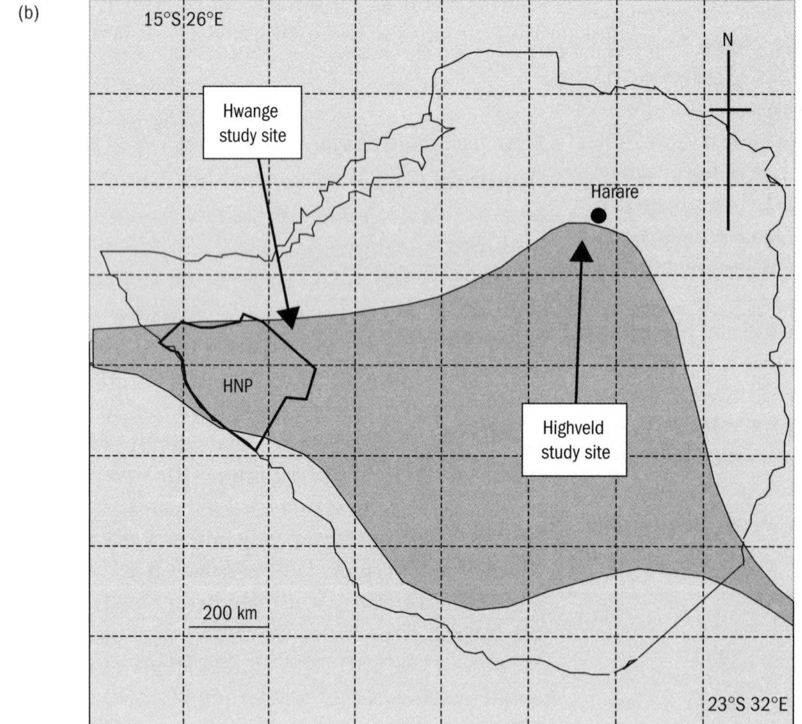

Figure 16.4 (a) Percentage of radio-telemetry relocations (fixes) spent active by *C. mesomelas* and *C. adustus* in the Hwange study site showing the bigeminous activity rhythm of both jackal species. (b) Map of Zimbabwe showing the two study sites. Dark grey shading denotes the area where *C. mesomelas* and *C. adustus* are sympatric. HNP=Hwange National Park.

Highveld revealed apparent disorder in nightly foraging trajectories and no evidence of between-night repetition in the order of visits to specific sites. If the jackals' movements were random, what was the nature of this randomness? To answer this, we sought to distinguish between two possibilities: Lévy flights (a class of random walk with fractal characteristics, Levy 1947) and Brownian walks. Levy flights are movements in two-dimensional space that are not described by the normal probability distribution with finite variance (i.e. tails which taper rapidly out); Levy statistics describe distributions with longer, power law tails. Extreme fluctuations are a characteristic of this distribution.

We concluded that three lines of evidence supported the hypothesis that jackals forage in a scale-free (fractal) manner (Atkinson *et al.* 2002b). First, the step-lengths taken by jackals do not yield a Gaussian (normal), but a power law distribution. Second, calculation from the detailed radio-tracking data yields a foraging trajectory dimension, D, for the points visited by each jackal, with a mean value of $D = 1.55 \pm 0.23$ (with no difference between males and females). Since this value of D lies clearly between 1 and 2, this non-integer dimension is strong evidence that jackals' movement patterns might be fractal or scale-free (see Mandelbrot 1982). Third, analysis of time series of displacement frequencies (the number of times they changed position per hour) of seven side-stripes exhibited long-range correlations with no characteristic time scale (i.e. a scale-free or fractal pattern). Lévy statistics may thus provide a suitable means of characterizing their movements.

Jackals are amongst the first mammals, and certainly the first carnivore, found to exhibit this pattern of behaviour (see Dicke and Burroughs 1988; Klafter *et al.* 1990; Viswanathan *et al.* 1996). Why should they move in this way? We argue that Levy flights are well suited to locating food resources in a complex and unpredictable environment, and scale-invariant search paths may well be a response to fractally distributed resources. In the competitive scramble to find resources, there may be selection pressure in favour of Lévy flights and against Brownian random movements, because Lévy flights are quicker to find new areas to exploit. As the one black-backed jackal encountered in the Highveld (see below) also exhibited Lévy foraging behaviour (as did raccoon dogs (*Nyctereutes procyonoides*) in two habitats (Saeki and Macdonald in prep.)), it is possible that it is a widespread response amongst similar animals with similar demands.

Habitat use

In the Highveld, the side-stripes used a range of habitats and preferentially utilized grassland. At Hwange, in contrast, the side-striped jackals used grassland less often than expected. While the two forms of grassland were botanically different, both were areas of high resource abundance in each of the respective study sites. Highveld grassland had high densities of rodents and seasonally abundant fruit, while Hwange grassland had high densities of springhares. In Hwange, at least, the open habitats had the additional advantage of high visibility for location of prey avoidance of predators. The Hwange grassland was the preferred habitat of black-backed jackals, which avoided woodland (Fig 16.5) and were generally averse to thick vegetation, as noted previously

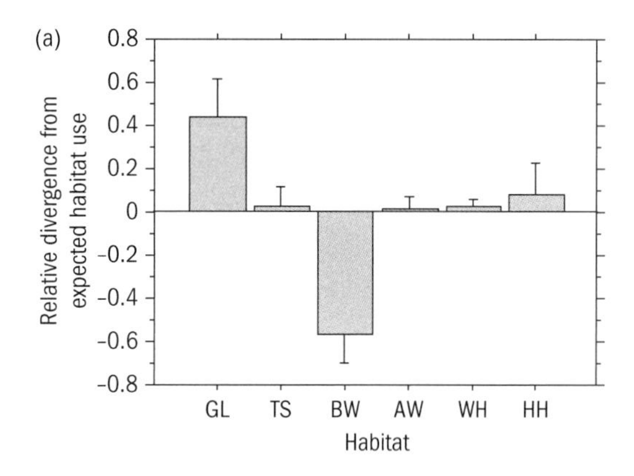

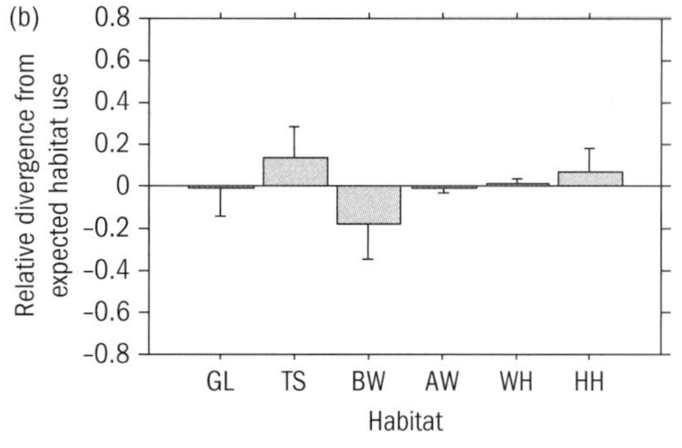

Figure 16.5 Mean divergence of actual from expected habitat use, based on habitat availability within home ranges, by (a) *C. mesomelas* and (b) *C. adustus* (Hwange). A value of zero represents expected habitat use, positive and negative values represent relative use of particular habitats to a greater or lesser extent than expected. Error bars show standard deviation. Habitat types are abbreviated as follows: GL = grassland; TS = *Terminalia sericea* scrub; BW = *Baikiaea* woodland; AW = *Acacia* woodland; WH = Waterhole; HH = Human habitation. (From Loveridge and Macdonald 2003.)

(Pienaar 1969; Skinner and Smithers 1990). In contrast, side-stripes used *Baikiaea* woodland and the *Terminalia* ecotone more than expected. Both species favoured human habitation, but side-striped jackals centred their home ranges on centres of human activity, benefiting from the regular, but low-quality, anthropogenic food resources. In short, as elsewhere, the side-stripes chose the more densely vegetated areas, whereas the black-backs occupied the open country (Loveridge and Macdonald 2002). This habitat use by side-stripes in Hwange contrasts with their habitat use in the Highveld area where this species uses grassland in preference to woodland (Atkinson *et al.* 2002b). The reason may be that competition with the black-backs in areas of sympatry causes this species to occupy habitat it would not normally use when allopatric. Indeed, in Hwange, we saw the black-backed jackals aggressively displace side-stripes from grassland on nearly all occasions that the two species were observed at the same time.

Atkinson and Macdonald's (submitted) further analysis of the home ranges in the Highveld revealed that the largest ranges were twice the area of the smaller ones. In accord with the null hypothesis that ranges were configured at random with respect to most habitat types, each habitat occupied a larger area, pro rata, of the larger ranges; in contrast, the representation of natural grassland—known to be the richest foraging ground—did not correlate with range size and its area had the lowest coefficient of variation between ranges. Furthermore, an index of habitat heterogeneity (which was higher where ranges comprised more, smaller patches, rather than fewer larger ones) was significantly negatively correlated with home range size. These results suggest, first, that home ranges may have been configured to encompass a roughly constant amount of important foraging habitat, in accordance with the Resource Dispersion Hypothesis (Macdonald 1983). Furthermore, it raises the question of why home range size should be less where patch size is smaller. Two possible answers, neither exclusive of the other, are first that smaller patches may have been richer than larger ones, or second that, insofar as food availability between patches is uncorrelated, as Carr and Macdonald (1986) predict more small patches will provide greater food security over time than will the same area comprised of fewer larger patches.

Food and foraging

Diet

Jackals are omnivorous and opportunistic foragers. In the Highveld, relative occurrence of fruit in the diet was 28%, the highest occurrence of any food type (Fig. 16.6(a,b)). In addition, fruit made up 30% of biomass in the diet, followed by 27% small mammals

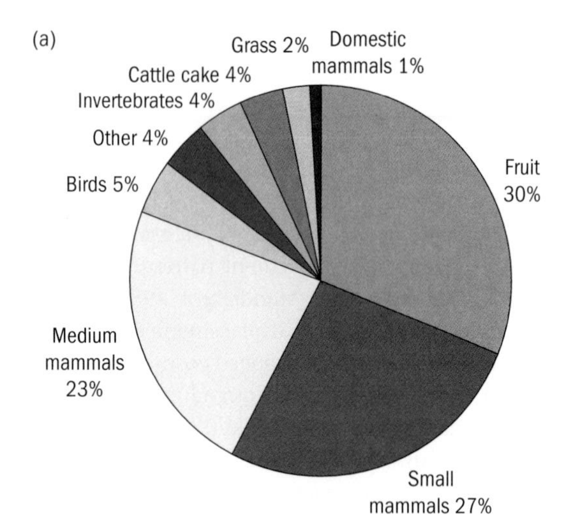

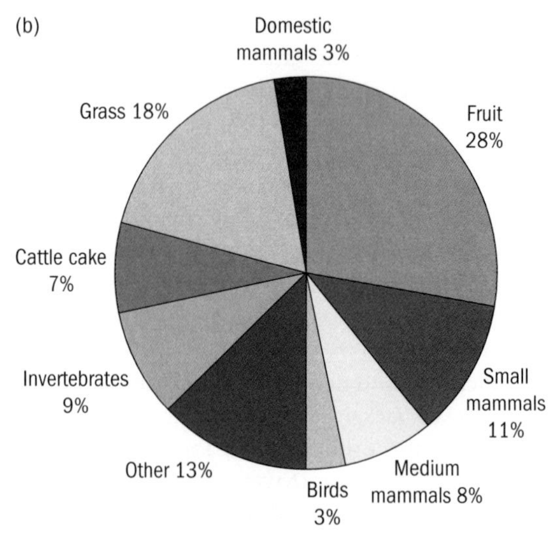

Figure 16.6 Composition of wild side-striped jackal (*C. adustus*) diet estimated by percentage of (a) fresh-weight biomass and (b) occurrences. The plot of percentage occurrence is derived by expressing the number of occurrences of each diet element as a percentage of all occurrences of all diet elements. (From Atkinson *et al.* 2002a.)

and 23% medium mammals (Atkinson *et al.* 2002a). Relative percentage occurrence was derived from the number of times food type occurred in the diet, that is, number of scats containing that item. Biomass was calculated from how much of food type was eaten, extrapolating from volumes of each item in jackal scats. Three species of mammal (multimammate mice (*Mastomys* spp), bushveld gerbil (*Tatera leucogaster*), and scrub hare (*Lepus saxatilus*)), and four species of fruit (mobola plum (*Parinari curatellifolia*), chocolate berry (*Vitex payos*), wild fig (*Ficus natalensis*), and waterberry (*Syzigium guineense*)) dominated these categories. Other items eaten included scavenged livestock, common duiker (*Sylvicapra grimmia*), and rock hyrax (*Procavia capenis*), invertebrates (including beetles, termites, giant millipedes (Diplopoda), ants (Hymenoptera: Formicidae), and sun spiders (Solifugae)), along with a miscellany of plastic and paper refuse, eggs, and kraal litter. Since Hwange was devoid of fruiting trees, the diet of side-stripes there was inevitably different. In Hwange, side-stripes 55% fresh weight biomass in the diet was of mammalian origin (Springhares (*Pedetes capensis*), rodents, and ungulate carrion), 30% was safari camp refuse, 7.9% arthropods (mostly Orthoptera), and only 2.8% was made up of fruit. In Hwange black-backs tended towards a more carnivorous diet with 83% of biomass ingested being of mammalian origin (64.5% of which was springhare). The only other food type of importance to this species were arthropods, largely coleopterans, which made up 13% of biomass ingested in the diet (Fig. 16.7). Certainly, in Hwange, black-backs were the more carnivorous of the two species of jackal (Loveridge and Macdonald 2003).

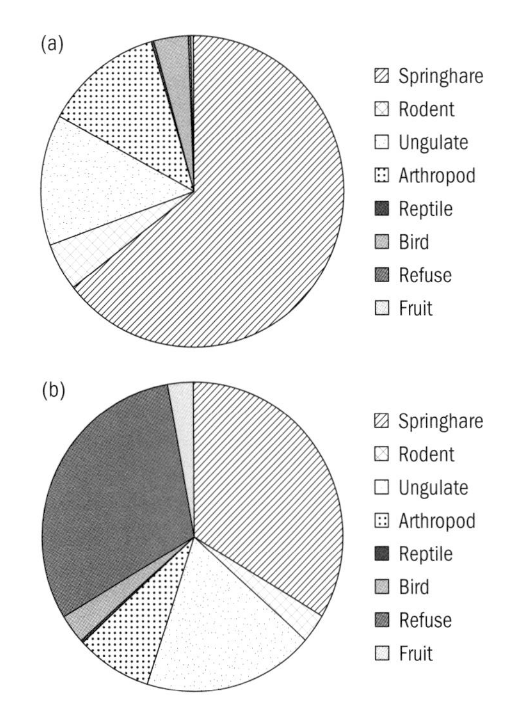

Figure 16.7 Comparison of the percentage fresh weight biomass in the diet of (a) black-backed jackals and (b) side-striped jackals in Hwange, Zimbabwe. (Adapted from Loveridge and Macdonald 2003.)

Opportunism in the side striped jackal was demonstrated in food preference tests on captive side-striped jackals (Atkinson *et al.* 2002a). They preferred animal prey to fruits (Table 16.1), but we could detect no particularly favoured species of mammal. We also tested termite alates, and they ranked between mammals and chocolate berry, their favourite fruit. The foods they preferred contained the highest levels of gross energy, carbohydrates, fats, and proteins and are highly digestible. On a monthly basis, small mammals were the single most important component of wild jackals' diet, comprising on average 22.6 ± 7.5% of fresh-weight biomass each month (range: 15–42.9%). Scrub hare (monthly mean ± SE = 10.1 ± 3.9%, range = 1.8–14%) and mobola plum (monthly mean ± SE = 11.4 ± 16%, range = 0.1–46.3%) were the second and third most important components, respectively.

However, despite their preference for high-energy animal prey, Highveld side-stripes did not spend more time in the habitats in which small mammals (their favoured food) were most abundant, nor did they eat more of them when they were most abundant. It seems, therefore, that although side-stripes prefer meat they do not actively seek small mammals, probably because there are always easier, if less palatable, things to find and eat (Atkinson *et al.* 2002a). Individual fruit species, however, were taken in proportion to their peak availability and jackal diet tracked fruit availability. Figure 16.8 shows the seasonal progression from waterberries (which were disliked by jackals in food choice trials), which contributed most to the percentage fresh-weight biomass of fruit in the diet in

Table 16.1 Choice order of food types by captive *C. adustus*

	Mouse	Springhare	Gerbil	Hare	Chocolate berry	Cattle cake	Mobola plum	Cape fig	Waterberry	Unripe fig
Mouse										
Springhare	=									
Gerbil	=	=								
Hare	=	=	=							
Chocolate berry	+	+	+	=						
Cattle cake	+	+	=	+	+					
Mobola plum	+	+	+	+	=	=				
Cape fig	+	+	+	+	=	+	+			
Waterberry	+	+	+	+	+	+	=	+		
Unripe fig	+	+	+	+	+	+	+	+	=	
Total wins	6	6	5	5	3	3	2	2	0	0

Notes: A '+' signifies a significant preference for column over row listed food types; '=' signifies no significant preference. There were insufficient data to include termites and ripe figs in the preference order.
From Atkinson *et al.* (2002a).

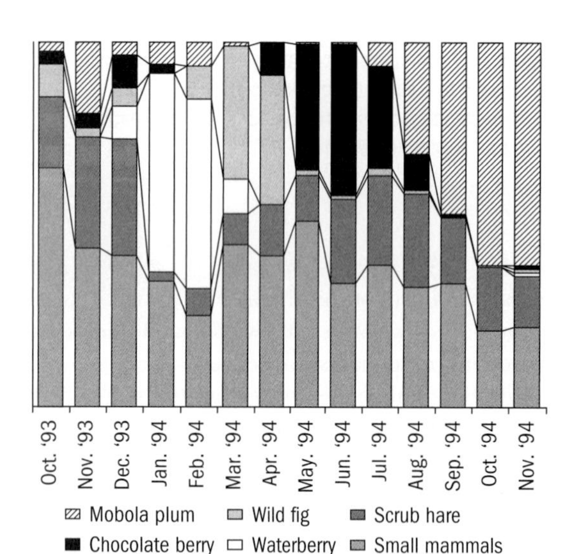

Figure 16.8 Seasonality in the overall abundance (percentage fresh-weight biomass) of the six most important foods in the diet of the wild jackals: small mammals, hares, mobola plum, waterberry, wild fig, and chocolate berry. Abundance is at its lowest during the drier months. (From Atkinson *et al.* 2002a.)

January and February, but were replaced by wild fig during March and April, followed by chocolate berry from May to July and mobola plum from August to November. In December, waterberries again began to dominate. Important fruit are most abundant during the wetter months, and mammals are eaten least during this time, further supporting the interpretation that the quest for fruit is the overwhelming determinant of foraging strategy for side-stripes in the Highveld.

Small mammals and scrub hares, in contrast, were taken throughout the year and formed a relatively constant proportion of the fresh-weight biomass of the jackals' diet, regardless of seasonal changes in their abundance.

Seasonality and dietary overlap in Hwange

Diet of the two jackal species in Hwange varied seasonally, with the widest diversity of food items eaten in the wet season (December–May) when resource abundance was high, and lowest in the hot dry season (September–November) when food availability (especially invertebrates) was at its lowest. Correspondingly dietary overlap between the species was greatest in the hot dry season and least during the wet season (Loveridge and Macdonald 2003). In general, side-striped jackals relied on a broader spectrum of food types than did black-backed jackals, leading to a greater index of dietary diversity (Levin's index: side-striped jackals $B = 3.56$–5.15, black-backed jackals $B = 2.86$–2.72, as calculated from relative

Table 16.2 Percentage occurrence (% occ), grams of biomass ingested (g), and % biomass ingested (%) by *C. mesomelas* and *C. adustus* over three seasons of the year, the wet season (December–May), the cold dry season (June–August), and the hot dry season (September–November)

		Springhare	Arthropod	Small vertebrate	Fruit	Ungulate	Refuse	*B*
Wet season								
adustus	% occ	7.4	40.7	3.7	11.1	7.4	29.6	**3.56**
(n = 15)	g	4.9	13.6	<0.1	0.2	6.9	3.5	
	%	16.8	46.6	0.3	0.6	23.6	12.0	**3.11**
mesomelas	% occ	22.2	50.6	9.8	9.9	6.2	1.2	**3.01**
(n = 51)	g	22.8	12.2	0.6	0.2	5.5	<0.1	
	%	55.5	29.4	1.4	0.5	13.3	0.2	**2.41**
Cold dry season								
adustus	% occ	21.4	29.0	9.0	13.1	4.8	22.9	**5.15**
(n = 64)	g	14.8	4.8	10.4	2.5	2.3	0.4	
	%	42	13.6	29.5	7.1	6.5	1.1	**3.38**
mesomelas	% occ	39.0	42.5	8.1	0.8	7.3	2.3	**2.86**
(n = 168)	g	40.1	8.8	3.5	0.02	7.5	0.3	
	%	66.6	14.6	5.8	0.03	12.5	0.5	**2.03**
Hot dry season								
adustus	% occ	16.8	32.9	3.1	5.4	9.0	24.0	**4.54**
(n = 107)	g	25.3	1.97	3.1	0.1	10.0	5.5	
	%	55.0	4.3	6.7	0.2	2.8	12.0	**2.69**
mesomelas	% occ	37.1	34.6	21.3	1.9	9.4	0.5	**3.72**
(n = 178)	g	28.7	2.5	3.1	<0.1	6.7	<0.1	
	%	69.7	6.06	7.5	0.2	16.3	0.2	**1.90**
Seasonal variance *C. adustus*		51.0	35.5	10.56	16.0	4.5	12.9	
Seasonal variance *C. mesomelas*		84.6	64	51.5	24.7	2.7	0.83	

Notes: The post rains period (April–May) is included in the wet season as these months are affected by the rains in previous months. Weights of biomass ingested are derived from volume of the dietary component in scats multiplied by a correction factor to give grams of food ingested. *B* is the niche breadth calculated for each species for each season using the Simpson index. Seasonal variance in the occurrence of each dietary item (standard deviation seasonal occurrence)^2 in the diet of each species is given in the final two rows.

From: Loveridge and Macdonald (2003).

percentage occurrence; side-striped jackals $B = 4.63–13.5$, black-backed jackals $B = 1.46–3.67$, calculated from percentage biomass ingested).

Estimates of dietary overlap (based on Pianka's index), using data for percentage occurrence and biomass ingested, were 0.84 and 0.76, respectively, which is high compared with other sympatric carnivores. Nonetheless, black-backs ate more springhares than did side-stripes, which in turn ate more safari camp refuse. Possibly the additional niche breadth created by human activity contributed to the coexistence of the two jackal species.

At Hwange, side-striped jackals had the greatest niche breadth in all seasons, perhaps reflecting a reliance on a greater diversity of food items. Seasonal variations in diet were most marked for springhares and insects (Table 16.2). Black-backed jackals had a lower niche breadth, especially during the cold dry season ($B = 1.46$), when the majority of biomass ingested was made up of springhare. Springhares formed the highest biomass ingested of any food item in both jackal species at all times of the year. For side-stripes, springhares made up the highest biomass during the hot dry season, and for black-backs during the cold dry season. Percentage occurrence of arthropod remains and average biomass ingested were highest in the wet season and lowest in the hot dry season in both species.

Cooperative hunting

Although both black-backed and golden jackals have been widely reported to hunt cooperatively (Wyman 1967; Lamprecht 1978a; Ferguson 1980; McKenzie 1990), evidence suggests that this increased success is equivocal (Macdonald *et al.* Chapter 4, this volume). Within social groups at Hwange, non-breeding adults generally do not associate with other adults (Loveridge 1999), so it is the breeding pair that has the opportunity to hunt cooperatively. We observed seven pair-hunting attempts for springhares, of which three were successful. In contrast, single black-backs were unsuccessful in all five hunts observed—a sample too small to be statistically relevant. More convincingly, the proportion of time for which members of a pair were in close proximity varied seasonally, being highest during the breeding season (June–July) and lowest when the female was preoccupied at the den (November–December). This pattern correlated with the percentage frequency of occurrence of springhares in the scats (Fig. 16.9).

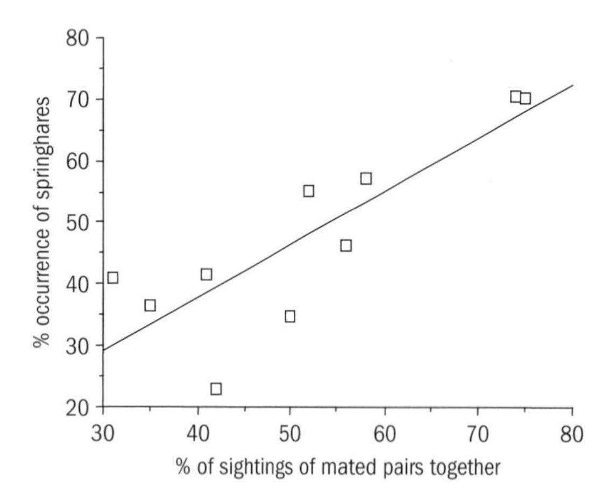

Figure 16.9 Percentage of time (sightings) that jackal pairs spent together (within 50 m of one another) and the percentage occurrence of springhare remains in scats. Peaks in percentage of sightings together in June and July correspond to the jackal breeding season. (From Loveridge 1999.)

Springhares are seemingly uniformly available throughout the year, so we suggest that their occurrence in the diet peaks when the jackals' breeding cycle enables members of a pair to hunt in company, whereas they are eaten least when insects are most abundant. An analogous situation occurs when coyotes exclusively exploit seasonal super-abundance of cicadas, while during the rest of the year rodents, leporids, and deer are their primary food (Cypher 1993; Cypher *et al.* 1994). Following the principle of least effort (Zipf 1949), black-backed jackals, and doubtless other canids, favour plentiful insect prey or fruit over elusive vertebrates.

Interspecific interactions and character displacement

Despite their smaller size, and contrary to expectation (cf. Johnson *et al.* 1996), black-backed jackals at Hwange almost always aggressively displaced side-striped jackals, and side-stripes avoided direct confrontation with them (Loveridge and Macdonald 2002). Although this relationship contradicts the generality of larger canids harassing smaller ones, it does resonate with Kingdon's (1997) remark that black-backed jackals are '. . . generally more aggressive than other jackal species'. This trait, which could be characterized as 'assertiveness', may be associated with the greater tendency of black-backed jackals to risk feeding alongside lion (*Panthera leo*) and spotted hyaena (*Crocuta crocuta*) (Estes 1967, 1991; Kingdon 1977; Mills 1990). We conclude that habitat partitioning at Hwange was mediated by aggressive exclusion of side-striped jackals from grassland by black-backed jackals.

Considering the evidence of interspecific antagonism and competition, one might have expected character displacement between these two similar species of jackal (e.g. Dayan *et al.* 1989). However, a comprehensive analysis (Van Valkenburgh and Wayne 1994) of cranial, dental, and mandibular traits of jackal species throughout Africa indicated, if anything, a tendency for sympatric jackals in East Africa to converge in size. The authors suggested that the large diversity of carnivores in East Africa limited the opportunity for character displacement. Consequently, East African black-backed jackals occupied a compressed niche and were less variable

Table 16.3 Residuals from GLM for *C. mesomelas*, for parameters showing significant interaction when testing the variation in sexual dimorphism between sympatry and allopatry (sex*location) ML = Mandible length, CBL = Candylo basal length, SKU = Skull length.

	Sympatric			Allopatric		
Variable	Male	Female	Difference in residual	Male	Female	Difference in residual
ML	0.69	−1.06	1.75	3.67	−2.65	6.32
CBL	2.48	−0.79	3.27	4.37	−4.17	8.54
SKU	0.36	−2.56	2.92	5.49	−3.62	9.11

in size and more sexually dimorphic than elsewhere in their range.

Prompted by this analysis, we measured 5 dimensions of a sample of 143 black-backs and 204 side-stripe skulls from southern Africa (Loveridge 1999). We found no evidence, for either species, of latitudinal corollaries of size or, amongst side-stripes, any degree of sexual dimorphism. We also found no evidence for character displacement in the size of either side-striped or black-backed jackals between areas of sympatry or allopatry in any of the parameters measured except a tentative indication that the widths of their zygomatic arches converged when sympatric. While we could discern no coherent pattern in the size ratios between male and female side-stripes between regions, the degree of sexual dimorphism in linear size and width of black-backed jackal skulls did differ (Table 16.3) between sympatric and allopatric populations. This mirrors Van Valkenburgh and Wayne's (1994) finding of an interaction between sexual dimorphism in Condylo-basal length and Mandibular length and locality for this species. In our sample, black-backed jackals appeared to be more dimorphic where allopatric (in Lowveld Zimbabwe, the Kalahari, and Eastern Cape) but less so when sympatric (Highveld Zimbabwe). The only anomaly is the Okavango Delta, where black-backs occur sympatrically with side-stripes but are nevertheless somewhat (although not significantly) dimorphic. In addition to any ecological or latitudinal explanation, this might be due to panmixia with the large and adjacent allopatric population in the Kalahari.

The increased sexual dimorphism in allopatric black-backed jackals was primarily due to increasing male and decreasing female size of skull parameters in comparison with populations that coexisted with

Table 16.4 Weights of jackals from the Highveld and Hwange study sites

	Weight (kg)	*n*	Weight range (kg)
Highveld			
C. adustus			
♀	7.45 ± 0.69	8	6.5–8.0
♂	9.56 ± 0.85	9	8.0–10.5
Hwange			
C. adustus			
♀	9.6 ± 0.40	6	9.0–10.0
♂	10.1 ± 0.60	4	9.3–10.5
Hwange			
C. mesomelas			
♀	6.6 ± 0.58	8	5.5–7.5
♂	7.6 ± 0.80	6	6.8–9.0

side-striped jackals. Overall, the largest males and smallest females occurred in the allopatric Eastern Cape population, while the smallest males and largest females occurred where sympatric with side-striped jackals in the Zimbabwe Highveld.

The adaptive significance of these changes probably lies in trophic adaptations rather than issues of bulk *per se* because the differences in dimorphism lie in skull morphology rather than overall size (Table 16.4). Van Valkenburgh and Wayne (1994) suggest that sympatry with golden jackals in East Africa explains the reduced sexual dimorphism in black-backs there (1.04) compared with South Africa (1.07). Dimorphism in our sample from the Eastern Cape (1.08) is close to theirs for South Africa as a whole, but where they are allopatric with side-striped jackals in Zimbabwe, we found black-backs to

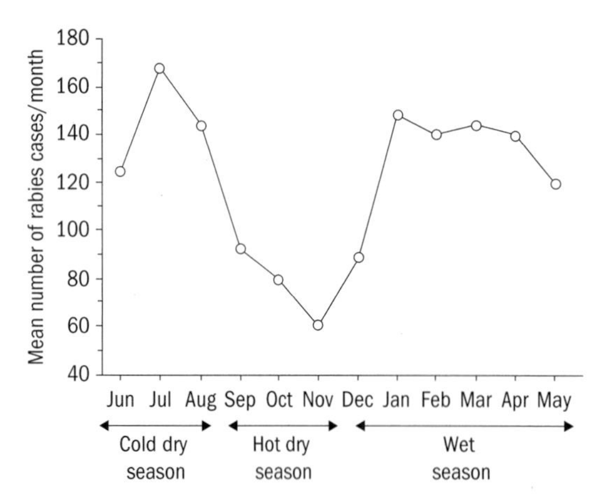

Figure 16.10 Mean monthly rabies incidence in jackals in Zimbabwe between 1950 and 1986 (modified from Foggin 1988). Peak rabies incidence coincides with the jackal mating season (usually around July in Zimbabwe), a marked decrease in rabies is experienced in the whelping season (around September) and while young pups are present at dens (during the hot dry season). (From Loveridge and Macdonald 2001.)

be barely dimorphic (1.01). Evidence for the difference in degree of dimorphism between areas of sympatry and allopatry within southern Africa is as compelling as that for the similar difference between East Africa and southern Africa, but cannot be explained with reference to golden jackals, which are absent from southern Africa. Notwithstanding some as yet unidentified factor at work in both areas where black-backs are sympatric with other jackals, a possible conclusion is that competition with side-striped jackals is as intense in southern Africa as is competition with other jackal species in East Africa. The relative contributions of side-striped and golden jackals, separately or together, to this impact on black-backs remains to be elucidated.

Rabies

Both side-striped and black-backed jackals are significant vectors of rabies in Zimbabwe, making up almost 25% of confirmed cases since 1950 (Fig. 16.10; Bingham *et al.* 1995). Jackals are the main vector responsible for transmission of the disease to domestic stock (Foggin 1988), and there is a significant association between the disease in jackals and cattle throughout southern Africa (Bingham and Foggin 1993; Swanepoel *et al.* 1993). The loss of livestock and the expense of vaccination make rabies economically significant in southern Africa. An important aim of our research was to contribute to the understanding and control of rabies.

Vector ecology and behaviour is an important aspect of rabies (Macdonald 1979), and their adaptations predispose all carnivores to being effective rabies vectors (Macdonald and Voigt 1985). Infection usually occurs though the injection of infected saliva through a bite wound, although mutual grooming can play a role. Jackals are extremely susceptible to rabies, succumbing to the disease after an incubation period of between 15 and 17 days when experimentally challenged (Foggin 1988; Bingham *et al.* 1995). The virus is present in saliva *c.*5 days before symptoms are patent. The rate of transmission is dependent on the rate at which individuals come into contact with one another and hence is linked to population density and behaviour.

Rabies has been continually present in Zimbabwe's dramatically expanding dog population for the last 40 years. However, there are also sporadic and spatially localized outbreaks of rabies in other species, including jackals. Side-striped jackals are commonly found on commercial farmland in Zimbabwe, as illustrated by our Highveld area, and the assumed continual contact between dogs and jackals along the communal land/commercial farmland boundaries is probably the primary route by which rabies enters the jackal population. This has led to the local supposition that wildlife was a reservoir for rabies. However, our field data, used in models developed from Anderson *et al.* (1981), suggest that jackals in Zimbabwe do not consistently occur at a sufficient density to support rabies endemically (Rhodes *et al.* 1998), a conclusion paralleled by research in the Serengeti (Cleaveland 1995; Cleaveland and Dye 1995). Nonetheless, jackal density in our Highveld area (*c.*1 per km^2) was close to the estimate of 1.4 jackals per km^2 required for disease maintenance, and Rhodes *et al.* (1998) suggest that rabies in a jackal community may persist for some time, thereby placing agricultural stock animals at risk. Where jackals exist at population densities of up to $2/km^2$, they could maintain the disease and possibly initiate front-like rabies epidemics, as have been observed around Harare in 1979–82 and 1990–95 (Foggin 1988; Bingham 1995).

Rabies and the natural history of jackals

At peak intensity, epidemics of jackal rabies in Zimbabwe may move up to 20 km/month (Foggin 1988) with peaks at intervals of 5–8 years (WHO 1992), within which seasonal highs in incidence correspond with seasonal peaks in jackal home range size, inter-territorial social interaction, and home range overlap during the mating season. A trough in rabies incidence corresponds to decreased contact between groups during the whelping season (Fig. 16.9, Loveridge and Macdonald 2001). At Hwange, but not in the Highveld, the home ranges of both side-striped and black-backed jackals expand in the mating season (June/July) from their minima during whelping (September–November) when the whole group's activity is focused at the den and, despite an increase in intra-group interactions, inter-group contact decreases. Births bolster the population of susceptible individuals, but pups have little contact with neighbours for the first 3 months of life.

At Hwange, we recorded 50 incidents of trespassing adult jackals, of which 21 were chased by territory holders; 10 (47.6%) of these chases resulted in physical contact and biting. Although this contact would have promoted rabies transmission, dispersal was localized, and seemed unlikely to lead to rapid spread of rabies. Dispersal is likely to be more common where vacancies are more available (e.g. Emlen 1991; Mumme and Koenig 1991), and hence where mortality is higher. This perturbation effect (Tuyttens and Macdonald 2000) may contribute to the rarity of rabies in undisturbed wildlife areas (Cumming 1982), and to its prevalence in the Highveld where jackals are frequently killed on commercial farmland.

Control

Our simulations suggest that if interactions with dogs ceased, the incidence of rabies in jackals could decline to zero. In contrast, if dogs continue to increase, and their vaccination does not improve (*c*.15–25% in the 1990s) jackal rabies is likely to increase, as is risk to livestock and humans. Historically, the control of jackal rabies in Zimbabwe has been based on killing jackals (Foggin 1988). Our results suggest instead a focused control and vaccination of dogs might be a more effective strategy. This might be complimented by an oral vaccination scheme to protect jackals living along the communal land-commercial farmland boundaries, targeting particularly the mating season (April–June) and period of pup recruitment (October–January).

Our study raises the possibility that realistic patterns of animal movement can be incorporated into spatial models of disease spread (Macdonald 1980c; Ball 1985; Smith and Harris 1991; Rhodes *et al.* 1998). The sprinkling of outbreaks of jackal (and fox) rabies ahead of main fronts, which is so characteristic of the disease, could conceivably be due to the virus 'freeing' the infected animal to make longer tracks across country. Even if infected, due to the power law nature of the step-lengths, extremely long steps are still less likely than shorter ones, and

Black backed jackals scavenging from an ungulate carcass. © M. G. L. Mills.

this may explain why the majority of rabid dogs, for example, do not make such journeys (Haig 1977). Furthermore, the consequences of Lévy-ranging behaviour extend to animal management and conservation. If Lévy foraging is a response to temporally and spatially unpredictable resources, then human activities that increase the proportion of such landscapes within an animal's territory could have dramatic effects on its movements.

Acknowledgements

At Hwange, we thank Touch the Wild safaris and the late Mr L. Reynolds. A. J. L. was funded by a Beit Trust Fellowship. We thank the farmers and their staff at Norton, where we were funded by the Overseas Development Administration (now DfID) and the Kapnec Trust. We are grateful to Mike Hoffmann, Tico McNutt, and Rolf Peters for comments.

CHAPTER 17

Coyotes

Coyotes in Yellowstone National Park: the influence of dominance on foraging, territoriality, and fitness

Eric M. Gese

A coyote *Canis latrans* pauses during its travels in Yellowstone National Park

Studies on the behavioural ecology of coyotes (*Canis latrans*) are inherently difficult due to their nocturnal and secretive habits. In Yellowstone National Park (YNP), Wyoming, the coyote population has not been subject to human persecution for several decades, allowing for direct observation of their behaviour, interactions among pack members, and how they deal with changes in their environment. From January 1991 to June 1993, over 2500 h of direct observation were collected on members of five resident packs, five transient individuals, and eight dispersing animals, in the Lamar River Valley of YNP. The presence of a dominance hierarchy within the resident packs greatly influenced access to food resources, individual fitness (i.e. mating opportunities, survival, and dispersal), and regulation of pack size. Alpha animals had the greatest access to ungulate carcasses in winter, diligently defended their territory against intruders, and consequently achieved a high degree of fitness in terms of acquiring all mating opportunities and reproductive success. Subordinate individuals (betas and pups) in the pack had less access to resources (mates and food), lower survival, higher dispersal rates, and thus reduced fitness as compared to alpha animals. Non-territorial coyotes (transients and dispersers) had even lower survival (mainly dispersing animals), no mating opportunities, and little access to ungulate carcasses during winter when resources were scarce. Being dominant and territorial was advantageous in coyote society by insuring access to mates, food, and space.

Introduction

The coyote, is an opportunistic, generalist predator that has expanded its distribution to most of North America and is probably one of the most widely researched canids. Yet, its typically nocturnal, secretive behaviour mean there have been only two studies—both in Grand Teton National Park, Wyoming—based on direct observation of wild coyotes (Camenzind 1978b; Bekoff and Wells 1986).

The coyote population in YNP has not been persecuted for several decades, and thus is tolerant of humans to an extent that has facilitated our studies of how coyotes deal with fluctuations in temperature, snow depth, snow-pack hardness, and food availability (e.g. Gese *et al.* 1996a–c). This chapter synthesizes the findings of over 2500 h of observation on coyotes in the Lamar River Valley, YNP, Wyoming (Gese *et al.* 1996a–c; Gese and Ruff 1997, 1998; Gese 2001b).

Study area

The study was conducted in a 70-km^2 area in the Lamar River Valley, YNP, Wyoming (Fig. 17.1;

Figure 17.1 Yellowstone National Park, Wyoming where the study was conducted.

44°52′N, 110°11′E), about 2000 m above sea level. Long, cold winters and short, cool summers characterize the climate in the valley (Dirks and Martner 1982; Houston 1982). Mean annual temperature and precipitation is 1.8°C and 31.7 cm, respectively, with most of the annual precipitation falling as snow (Dirks and Martner 1982; Houston 1982). Habitats included forest, mesic meadow, mesic shrub-meadow, riparian, grassland, sage-grassland, and road (see Gese *et al.* 1996a for habitat descriptions).

Predominant ungulate species included elk (*Cervus elaphus*), mule deer (*Odocoileus hemionus*), bison, (*Bison bison*), and bighorn sheep (*Ovis canadensis*). A few moose (*Alces alces*) and white-tailed deer (*Odocoileus virginianus*) inhabited the valley, and pronghorn antelope (*Antilocapra americana*) were present during summer. A major food source for coyotes during winter was elk carrion (Murie 1940; Houston 1978; Gese *et al.* 1996a). Small mammal species included microtines (*Microtus* spp.), mice (*Peromyscus* spp.), pocket gophers (*Thomomys talpoides*), and Uinta ground squirrels (*Spermophilus armatus*).

General methodology

The sampling design and methodologies for recording behavioural observations of coyotes were described in Gese *et al.* (1996a–c), Gese and Ruff (1997, 1998), and Gese (2001). In general, coyotes >5 months of age were captured with padded leg-hold traps with attached tranquilizer tabs, weighed, sexed, ear-tagged and radio-collared, and the vestigial first premolar of the lower jaw was extracted for ageing (Linhart and Knowlton 1967). Pups (8–12 weeks old) were captured at the den, ear-tagged, and surgically implanted with an intraperitoneal transmitter. We classified coyotes by age as pups (<12 months old), yearlings (12–24 months old), or adults (>24 months of age). Coyotes were also classified as residents or transients based upon their social interactions and affinity for one area (Bowen 1981; Gese *et al.* 1988). Members of a resident pack were further classified into different social classes, including alphas (dominant breeding adults), betas (adults and yearlings subordinate to the alphas but dominant over pups), or pups (young of the year subordinate to both alphas and betas), based upon the separate male and female dominance hierarchies observed in the pack (see Gese *et al.* 1996a–c for details on methodology).

Coyotes were observed with a 10–45× spotting scope from vantage points located throughout the valley during October–July; high grass (>1 m) precluded observation in August and September. We collected nocturnal observations using an 11× night-vision scope. Behavioural observations followed Gese *et al.* (1996a,b) in which we randomly sampled packs, and stratified individuals within each pack to allow for similar sampling of each sex and social class. We used focal-animal sampling (Lehner 1979; Martin and Bateson 1993), recording all behaviours for a single individual using a program on a notebook computer, or on a tape recorder and transcribed later. Whenever possible, we recorded the location at which behaviours (e.g. bed sites, dens, howling, scent-marking, predation, carcasses) occurred to the nearest 10-m grid intersection using the Universal Transverse Mercator (UTM) grid system on a 1 : 24,000 US Geological Survey topographic map. Snow depth, hardness, and layering were recorded every 1–2 days by excavation of a snow pit. Additional climate information was recorded at a permanent weather station within the study area. Available ungulate carcass biomass in the valley was estimated weekly (see Gese *et al.* 1996a). The sampling unit for all statistical tests was the individual coyote (Machlis *et al.* 1985). Statistical analyses of behaviours are described in Gese *et al.* (1996a–c) and used the software program SYSTAT (Wilkinson *et al.* 1992) following the recommendations in Steel and Torrie (1980), Sokal and Rohlf (1981), and Zar (1996).

Environmental conditions

The first winter (1990–91) of behavioural observations in YNP was mild, with little carcass biomass available to the coyotes in the valley (Fig. 17.2(a)). Maximum snow depth was 30 cm and the amount of known carcass biomass was <170 kg/wk. Coyotes were dependent upon small mammals, mostly voles, as their major food item during that winter. The second winter (1991–92) was characterized by deeper snow cover and higher carcass biomass (Fig. 17.2(b)). That winter had an early snowfall followed by a thaw, which re-froze into an ice layer on the ground and subsequently led to an early initiation of winter die-off of ungulates. Maximum snow depth was 46 cm, and known carcass biomass exceeded 200 kg/wk for 10 weeks. The third winter (1992–93) was similar to the second winter, with deep snow cover and high carcass biomass (Fig. 17.2(C)). Maximum snow depth was 63 cm, and for 6 weeks known carcass biomass was >200 kg/wk.

Social organization and dominance

From January 1991 to June 1993, we observed 49 resident coyotes from 5 packs for 2456 h and 5 transients for 51 h; 8 animals identified as dispersers were observed for 53 h. Of the 54 coyotes observed, 29 were males, 23 were females, and 2 unmarked coyotes were of unknown sex. We collared or implanted 31 coyotes with radio-transmitters, and 23 were unmarked but recognizable from physical characteristics. The coyotes in the Lamar River Valley were organized into relatively large packs (up to 10 individuals) with distinct territories (Fig. 17.3). These resident packs remained spatially stable, except in the last winter (1992–93) when the Soda Butte pack usurped a part of the Norris pack territory (Fig. 17.3(c); see Gese 1998 for details). Transient home ranges overlapped the resident territories. Territorial boundaries of resident packs were scent-marked and actively defended;

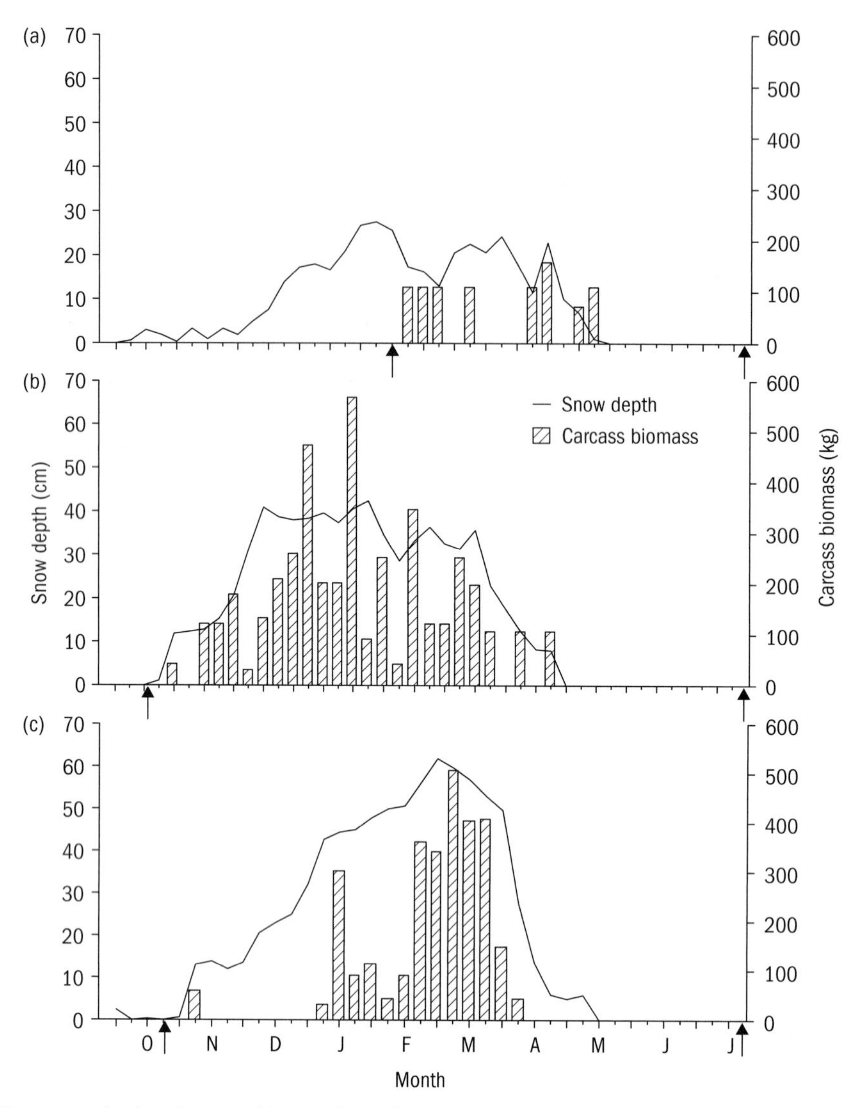

Figure 17.2 Mean snow depth and carcass biomass for each week during the winters of (a) 1990–91, (b) 1991–92, and (c) 1992–93 in the Lamar Valley, YNP, Wyoming. Arrows indicate the time span of data collection for each winter.

transient home ranges were not scent-marked or defended (Gese and Ruff 1997; Gese 2001). Each resident pack was comprised of an alpha pair and associated pack members, usually related offspring (Hatier 1995; Gese *et al.* 1996c). Associate animals that remained in the pack over winter usually helped feed and care for the offspring whelped by the alpha pair the subsequent spring (Hatier 1995). Dominance matrices for each pack demonstrated the presence of a social order or dominance hierarchy among both females and males (Gese *et al.* 1996c), similar to that described in a wolf pack (*Canis lupus*; Mech 1970). The presence of a dominance hierarchy in these packs played a major role on pack dynamics, foraging ecology, territorial maintenance, and ultimately individual fitness. The large packs we observed were probably a consequence of the combination of abundant prey biomass (Bekoff and Wells 1981; Geffen *et al.* 1996) and the lack of exploitation in the study area (Knowlton *et al.* 1999; Frank and Woodroffe 2001). For details on individuals observed and pack histories, see Gese *et al.* (1996a–c) (Fig. 17.4).

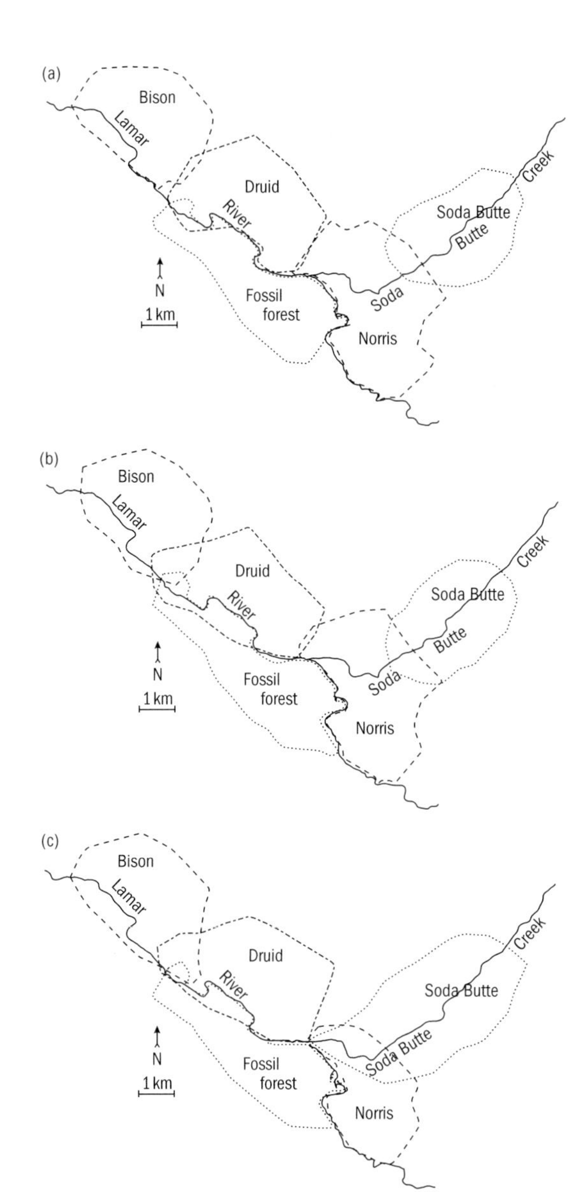

Figure 17.3 Spatial distribution and territorial boundaries of the five resident coyote packs occupying the Lamar River Valley in the winters of (a) 1990–91, (b) 1991–92, and (c) 1992–93, YNP, Wyoming.

Figure 17.4 The alpha male of the Soda Butte pack dominates the beta male (his 2-yr old son) at an elk calf (*Cervus elaphus*) the alpha pair just killed © E. M. Gese.

Behavioural activity budgets

The behavioural activity budgets of the coyotes in the Lamar River Valley changed throughout the year (Fig. 17.5). In the fall, coyotes spent much of their time travelling (60%) and hunting small mammals (13%). During winter, as snow depth increased and ungulate carcasses became available, coyotes travelled less (24%), hunted small mammals less (2%), and fed more on ungulate carcasses (2%) and rested (66%). During spring, the coyotes returned to travelling and hunting small mammals, with a corresponding decrease in the amount of time spent resting and feeding on ungulate carcasses. The ungulate carcasses that coyotes fed on during summer were mostly elk calves they killed, plus scavenging the remains of old carcasses from the previous winter. Transient coyotes showed similar proportions of activity as resident animals except for the amount of time spent feeding on carcasses. Members of resident coyote packs spent an average of 2% of their time feeding on carcasses, while transients spent only 0.3% feeding on carcasses ($t = 1.927$, $P = 0.056$). Transients, which were solitary animals, were at a disadvantage when attempting to obtain, feed on, or defend a carcass (Gese *et al.* 1996a). Bekoff and Wells (1981, 1982, 1986) reported similar changes in behavioural activity budgets of coyotes in Grand Teton National Park, Wyoming, in relation to social organization, and changes in snow depth and carcass availability.

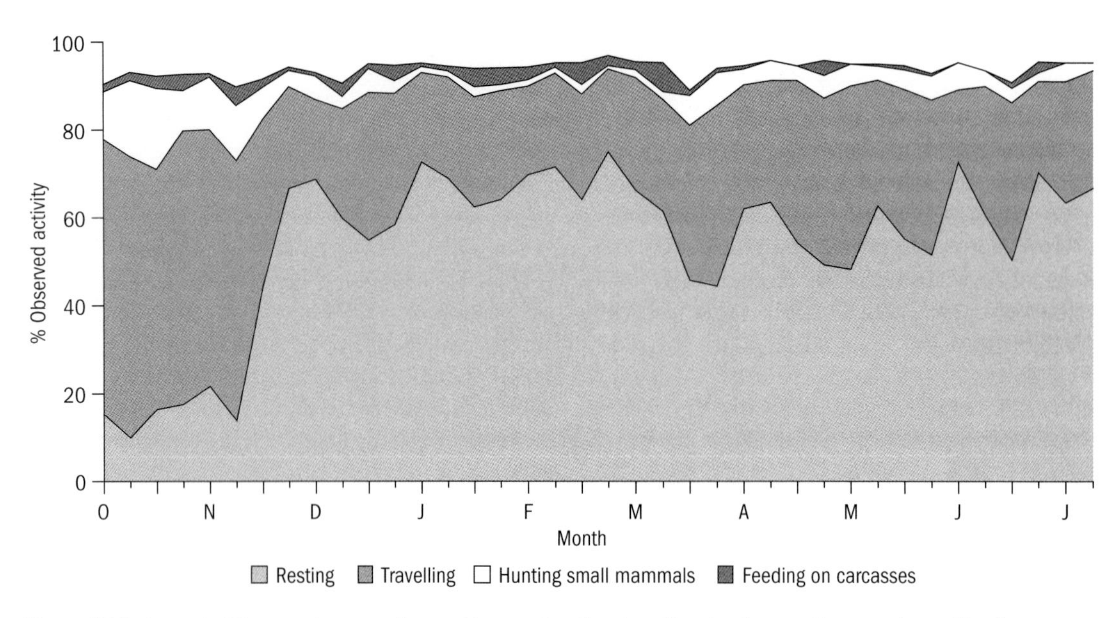

Figure 17.5 Amount of time coyotes were observed to spend resting, travelling, hunting small mammals, and feeding on carcasses each week during the three winters (1991–93) combined, YNP, Wyoming.

Foraging ecology

Coyotes hunted elk calves in early summer, while the calves were vulnerable during the first few weeks of life. Coyotes also hunted ground squirrels during summer when the squirrels emerged from hibernation. Voles were the principal small mammal food and constituted most of prey biomass ingested by coyotes year round. Even though large coyote packs existed, small mammals were always hunted by coyotes alone (Gese *et al.* 1996b). During the 2507 h of observation, we recorded 6433 prey detections of small mammals, 4439 attempts to capture prey, and 1545 captures of small mammals by coyotes. Many extrinsic and intrinsic factors influenced predation rates and capture success of small mammals by coyotes (Gese *et al.* 1996b). Habitat was a major factor influencing predation rates by coyotes on small mammals. Detection rates, attempt rates, and capture rates of small mammals by coyotes significantly varied among the various habitats (detection rate: $F = 39.82$, df $= 6, 1668$, $P < 0.001$; attempt rate: $F = 31.305$, df $= 6, 1668$, $P < 0.001$; capture rate: $F = 14.84$, df $= 6, 1668$, $P < 0.001$) with detection, attempt, and capture rates of small mammals being highest among mesic habitats (Table 17.1). Most *Microtus* species are associated with mesic habitats (Getz 1985). Dense vegetation also provides mechanical support for snow cover influencing the amount of subnivean space available at the ground surface for microtine passages (Spencer 1984). Coyotes readily exploited these habitats containing the highest prey densities and spent most of their time hunting small mammals in these habitats (Gese *et al.* 1996b).

Table 17.1 Influence of habitat type on detection, attempt, and capture rates (# prey/hour spent active) of small mammals by coyotes in the Lamar River Valley, YNP, Wyoming, 1991–93

Habitat type	Detection	Attempt	Capture
Shrub-meadow	8.0	5.4	1.8
Mesic meadow	7.3	5.2	1.6
Sage-grassland	4.6	3.1	1.0
Grassland	4.4	3.1	1.1
Riparian	2.2	1.3	0.5
Forest	1.4	1.0	0.4
Road	0.7	0.5	0.0

Table 17.2 Influence of snow depth on detection, attempt, and capture rates (# prey/hour spent active) of small mammals by coyotes in the Lamar River Valley, YNP, Wyoming, 1991–93

Snow depth	Detection	Attempt	Capture
None	5.8	3.5	1.7
Low (5–15 cm)	8.4	5.9	1.9
Moderate (16–25 cm)	5.0	3.2	1.0
Deep (26–40 cm)	3.7	2.6	0.9
Very deep (>40 cm)	3.4	2.4	0.5

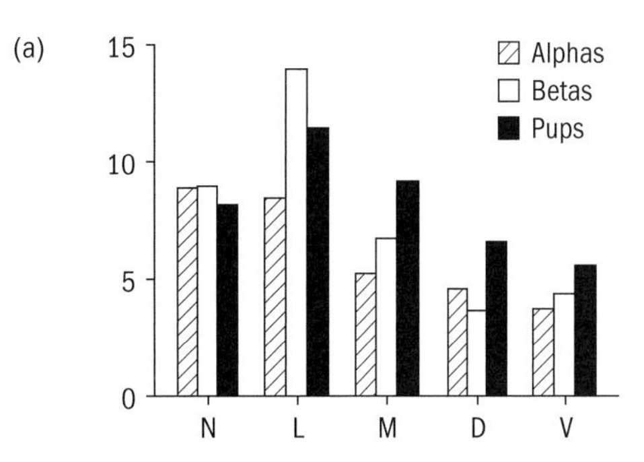

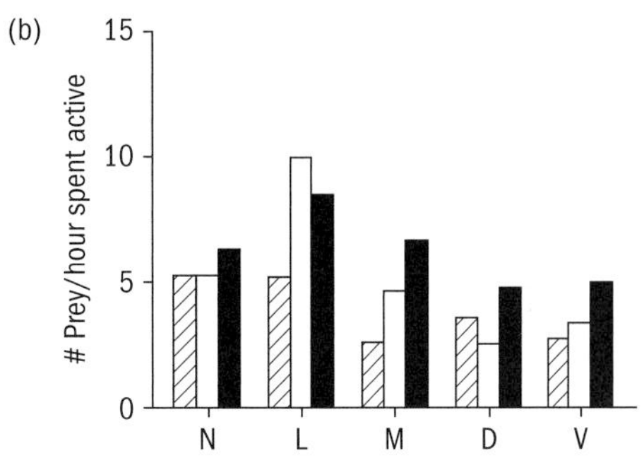

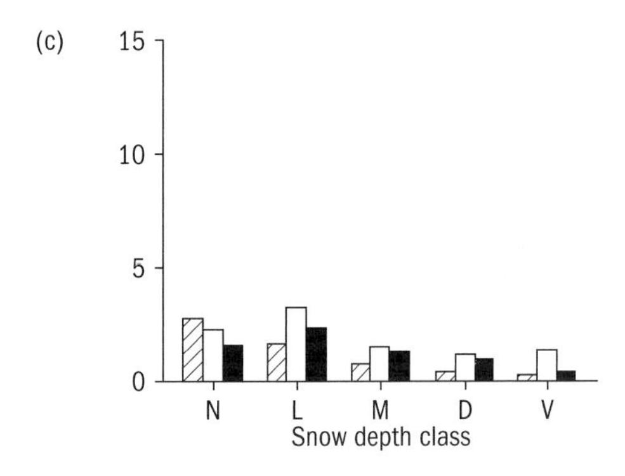

Figure 17.6 Rates (# prey/hour spent active) of small mammal (a) detection, (b) attempt, and (c) capture, for alpha, beta, and pup coyotes across varying snow depth classes in mesic-meadow habitat, YNP, Wyoming, 1991–93. Snow depth classes were: N (no snow), L (low, 5–15 cm), M (moderate, 16–25 cm), D (deep, 26–40 cm), and V (very deep, >40 cm).

Another important factor influencing predation on small mammals by coyotes was snow depth (Fig. 17.6). Snow depth was classed into none, low (5–15 cm), moderate (16–25 cm), deep (26–40 cm), and very deep (>40 cm). Detection rates, attempt rates, and capture rates of small mammals by coyotes varied among the different snow depth classes (detection rate: $F = 28.38$, df = 4, 1670, $P < 0.001$; attempt rate: $F = 24.35$, df = 4, 1670, $P < 0.001$; capture rate: $F = 15.26$, df = 4, 1670, $P < 0.001$) (Table 17.2; Fig. 17.6). Low snow cover actually increased prey detection rates, predation attempt rates, and capture rates of rodents by coyotes compared with bare ground. As snow depth increased, detection rates, attempt rates, and capture rates of small mammals by coyotes declined (Fig. 17.6).

Age and experience of the coyote was also a major factor influencing predation on small mammals. We found that even under the same environmental conditions (snow depth, habitat, snow-pack hardness, and wind speed), pups detected or showed that they detected more prey per hour than did older coyotes (Fig. 17.6). We believe that this higher detection rate by pups may have been due to increased responsiveness to an auditory cue (whether prey or not). It appeared that older coyotes may filter out irrelevant sounds from the environment and were more selective towards cues associated with prey (Gese *et al.* 1996b). Older coyotes also reduced the proportion of prey they attacked during adverse conditions, while pups continued to attack a high proportion of prey that they detected (possibly due to lack of experience). Alternatively, and more plausible, is that reduced access to ungulate carcasses (Gese *et al.* 1996a) may have forced pups to hunt small mammals under adverse conditions in order to survive and remain in the pack (Gese *et al.* 1996c).

During winter, the presence of a dominance hierarchy in the coyote packs dictated the level of resources acquired by individual members of the pack (Gese *et al.* 1996a). During winter as snow depth increased, access to small mammals (encounter, attempt, and capture rates) declined (Fig. 17.6; Gese *et al.* 1996b). However, as this snow cover limited access to the small mammal prey base by coyotes, it made foraging for plant material more difficult for ungulates (mainly elk). As winter progressed and the elk became nutritionally stressed, animals died due to malnutrition (Craighead *et al.* 1973; Houston 1978), or were weakened and killed by coyotes (Gese and Grothe 1995). Surprisingly, only 2–3 coyotes were needed to kill even an adult elk, but these elk were in extremely poor nutritional condition. Gese and Grothe (1995) reported several instances of coyote predation on elk and found that predation attempts on ungulates almost always involved the alpha pair (the alpha male was the main attacker) and the remainder of the pack did not participate in the attack, but were often observed to be watching the attack.

Once a kill had been made or an ungulate succumbed to winter stress, the resident pack would begin feeding on this resource. However, not all pack members fed equally (Gese *et al.* 1996a). Apparently, pups were restricted from feeding on the carcass by the older members of the pack (Fig. 17.7). The carcass was monopolized by the alpha pair first, then the higher ranking beta animals, then the lower ranking individuals, and lastly the pups (Fig. 17.7; Gese *et al.* 1996a). Even though these pups were the offspring of the alpha pair and usually related to the older betas in the pack, this restriction of access to the carcass indicated that the pups had to fend for themselves. Parent–offspring conflict (Trivers 1972, 1974), was apparent within these coyote packs as food resources became restricted during winter. In response to this resource partitioning, pups adopted a different foraging strategy and spent more time hunting small mammals even when conditions were poor (Fig. 17.7; Gese *et al.* 1996a,b; Fig. 17.8).

Evidence of resource partitioning in relation to social dominance has been found in other social carnivores. In the Namib Desert, spotted hyaenas (*Crocuta crocuta*) showed a linear dominance hierarchy when feeding on a carcass, in which subordinate animals eventually gained access to large carcasses, but not small carcasses (Tilson and Hamilton 1984).

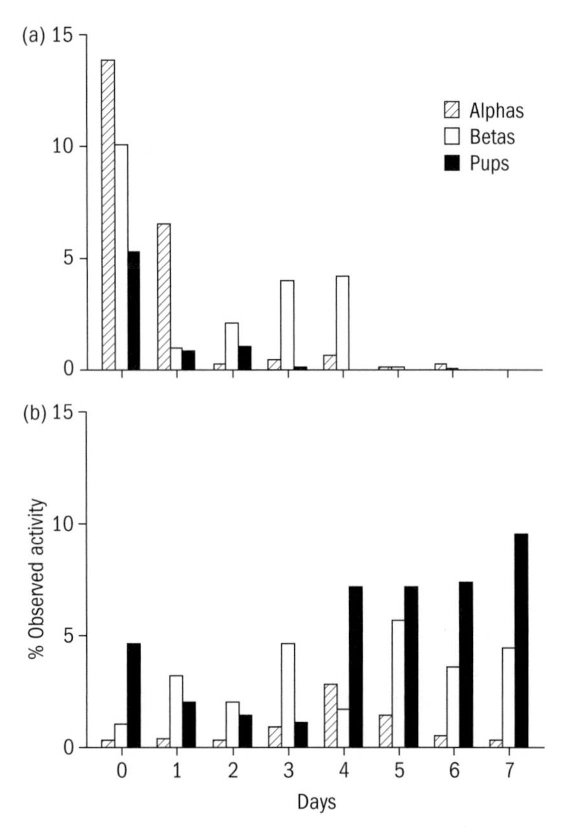

Figure 17.7 Amount of time alpha, beta, and pup coyotes were observed to spend (a) feeding on a carcass and (b) hunting small mammals, on the day (day 0) and the proceeding 7 days after an elk died or was killed by coyotes, YNP, Wyoming, 1991–93.

A correlation between social rank and feeding typified female spotted hyenas in the Masai Mara National Reserve in Kenya (Frank 1986), brown hyenas (*Hyaena brunnea*) (Owens and Owens 1978), and wolves (Zimen 1976), amongst others.

Influence of food availability on regulation of pack size

During our study, winter severity (mainly snow depth) determined ungulate carcass biomass, which in turn influenced coyote pack size as mediated by social dominance within the resident pack. Access to

Figure 17.8 Coyote *Canis latrans* pouncing through the snow to capture a vole underneath © E. M. Gese.

food resources during the winter bottleneck not only influenced coyote pack size, but also appeared to influence reproduction the subsequent spring. During the first winter, carcass biomass was low due to low snowfall. With limited food resources, competition for ungulate carcasses was high with access to those few carcasses determined by social rank within the pack (i.e. resource partitioning; Gese *et al.* 1996a). Subordinate individuals (i.e. low-ranking betas and pups) with limited access to ungulate carcasses attempted to compensate for this shortfall by hunting small mammals (Gese *et al.* 1996b). Those that could capture and subsist on small mammals often remained in the pack, but others that were less successful hunters of small prey dispersed (Gese *et al.* 1996c). With low prey biomass in the valley, coyote packs through the winter of 1990–91 remained small ($\bar{x}$ = 4.6 coyotes/pack in January) as pups from the previous year dispersed early (Gese *et al.* 1996a,c). Litter size (at den emergence) that spring (1991) averaged 5.0 pups/pack (Gese *et al.* 1996a). During the second winter (1991–92), increased snowfall resulted in an increase in available ungulate biomass in the form of winter kill. With more ungulate carcass biomass available, more of the pack had access to these resources and subsequently fewer individuals were forced to disperse (Gese *et al.* 1996c) and seek resources elsewhere, dispersal occurred later in winter, and pack size increased correspondingly ($\bar{x}$ = 5.8 coyotes/pack in January). Litter size increased to 7.8 pups/pack with one pack producing 2 litters (only the litter whelped by the alpha female survived beyond 4 months of age). During the final winter (1992–93), with similar high ungulate biomass in the valley, some coyotes did not disperse until late winter and pack size increased to 6.6 coyotes/pack (in January); litter size was not accurately determined that spring (Gese *et al.* 1996c).

The relationship between food abundance and regulation of canid populations has been documented (e.g. Zimen 1976; Keith 1983; Knowlton and Stoddart 1983; Fuller 1989; Fuller and Sievert 2001). Food abundance regulates coyote numbers by influencing reproduction, survival, dispersal, space-use patterns, and territory density (Todd *et al.* 1981; Todd and Keith 1983; Mills and Knowlton 1991; Knowlton *et al.* 1999). Coyote populations will increase and decrease with changes in food availability, particularly in areas with cyclic lagomorph populations. In areas where hares comprise a significant portion of the coyote diet, coyote numbers will rise and fall as snowshoe hare (*Lepus americanus*) or black-tailed jackrabbit (*Lepus californicus*) numbers change (Clark 1972; Todd *et al.* 1981; Knowlton and Stoddart 1992; O'Donoghue *et al.* 1997). The mechanisms for these responses are changes in ovulation rates and litter sizes, and changes in the percentage of adult and yearling coyotes that bred (Todd *et al.* 1981; Todd and Keith 1983). Food abundance also influences coyote numbers through its affect on dispersal of pups in winter (Gese *et al.* 1996c). In addition, food shortages can increase mortality rates, especially among juvenile coyotes as they disperse into unfamiliar areas (Knowlton *et al.* 1999).

Territorial maintenance and defence

The territory of an animal has been defined as the area that an animal will defend against individuals of the same species (Burt 1943; Mech 1970). Territoriality allows animals to exclude potential competitors from access to mates, food, space, and cover. Failure to defend the territory may have far-reaching consequences for the resident pack (e.g. Gese 1998). Canids use both direct and indirect mechanisms to maintain territorial boundaries, including scent-marking (Peters and Mech 1975; Camenzind 1978; Rothman and Mech 1979; Barrette and Messier 1980; Bowen and Cowan 1980; Wells and Bekoff 1981), howling (Harrington and Mech 1978a,b, 1979), and direct confrontation of intruders (Camenzind 1978; Bekoff and Wells 1986; Mech 1993, 1994). During this study, the importance of the presence of the dominance hierarchy in the resident packs was exemplified in the role pack members played in territory maintenance. Observations of the coyotes revealed that they defended their territorial borders both directly through confrontation of intruding animals, and indirectly via scent-marking and howling (Gese and Ruff 1997, 1998; Gese 2001). We found that the alpha pair of the pack was principally responsible for maintaining and defending the territory, with peak defence occurring during the breeding season.

Scent-marking

During observations of scent-marking behaviour, we recorded 3042 urinations, 451 defecations, 446 ground scratches, and 743 double-marks (Gese and Ruff 1997). Rates of urination, double-marking, and ground-scratching varied seasonally and among social classes (Fig. 17.9). Overall, alpha, beta, and pup

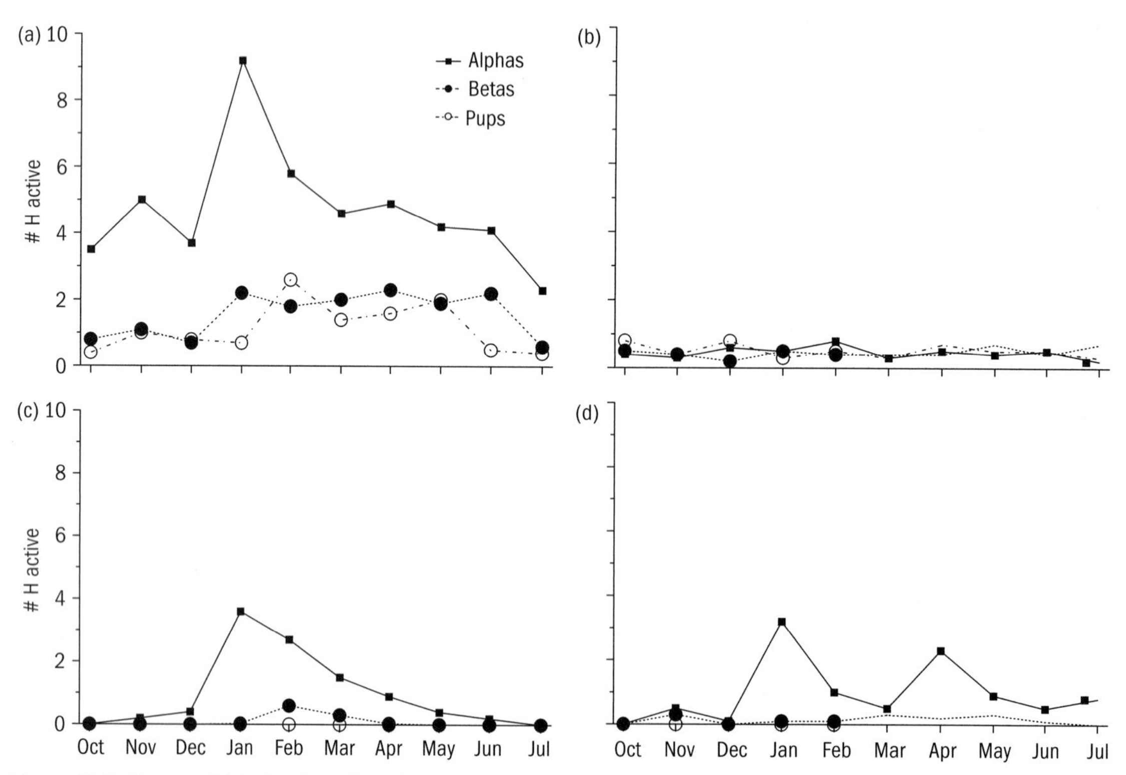

Figure 17.9 The rate of (a) urinations, (b) defecations, (c) double-marks and (d) ground scratches for alpha, beta and pup coyotes from October to July, YNP, Wyoming, 1991–93.

coyotes scent-marked at a rate of 5.1, 1.7, and 1.4 marks/h active, respectively. Double-marks were performed an average of 1.3, 0.1, and 0 marks/h active for alpha, beta, and pup coyotes, respectively. Scent-marking peaked during the breeding season (Fig. 17.9). We found that the alpha pair scent-marked the boundaries, using urinations, double-marking, and ground scratching, at a higher rate (6.0 marks/h) and frequency than in the core (2.7 marks/h) of their territory ($t = -3.039$, df = 82, $P = 0.003$). Beta coyotes participated to some degree in scent-marking, but not at the level of the alpha pair (Fig. 17.9; Gese and Ruff 1997). Pups seemed not to participate in scent-marking duties. Defecation rate was relatively constant all year (Fig. 17.9) and among social classes (0.5, 0.5, and 0.8 defecations/h for alphas, betas, and pups, respectively), and appeared to be relatively unimportant as a scent-marking signal (Gese and Ruff 1997). Asa *et al.* (1985) speculated that urine may be a better compound for scent-marking because faeces may not be as readily available for deposition as urine. Studies on the scent-marking of wolves (Peters and Mech 1975) and coyotes (Wells and Bekoff 1981) have reported similar results with territorial canids scent-marking more along the boundaries of their territory and dominant members scent-marking at higher rates than subordinates (see also Sillero-Zubiri and Macdonald 1998). Scent-marking increased during the breeding season when pair bonds are strengthened and breeding synchrony was initiated (Bekoff and Diamond 1976; Kennelly 1978). Scent-marking in dominant wolves changed seasonally and was correlated with changes in testosterone (Asa *et al.* 1990). Scent-marking by canids appears to influence demarcation of territorial boundaries and also provides internal information to members of the resident pack (Macdonald 1979a, 1985; Wells and Bekoff 1981). Scent-marks do not prevent animals from crossing territorial boundaries, but may serve as subtle repellents eliciting avoidance by potential intruders.

Howling

Another indirect means of territory maintenance that followed the same pattern as scent-marking was howling or long-range vocalizations. We recorded 517 howling events during the 2507 h of behavioural observations. Rates of howling varied seasonally and among the social classes (Fig. 17.10). The alpha pair spent more time howling (0.59%) and howled at a higher rate (0.33 howls/h) than both beta (0.15% and 0.10 howls/h) and pup (0.14% and 0.11 howls/h) coyotes (Gese and Ruff 1998). These alpha animals also howled at a greater frequency when near territorial boundaries (56% of howls) and howling rates peaked before and during the breeding season, then declined in the pup-rearing season (Fig. 17.10). In contrast, transient animals did not howl and appeared to maintain a 'low profile' and did not advertise their presence either through howling or scent-marking. Howling appeared to serve as a territorial spacing function that was mainly performed by the alpha pair. Research on howling among wolves and coyotes have found similar results with howling rates peaking during the breeding season, alpha members howling more frequently than subordinate individuals, and howling playing an important role in territory maintenance (Harrington and Mech 1978a,b, 1979, 1983; Walsh and Inglis 1989). Seasonal changes in howling rates among alpha animals may be related to increased pair-bond behaviour, hormonal changes, and territorial maintenance during the breeding season, with the decline possibly related to a reduced need to advertise their presence outside of the breeding season (Zimen 1976; Harrington and Mech 1978*a*; Gese and Ruff 1998).

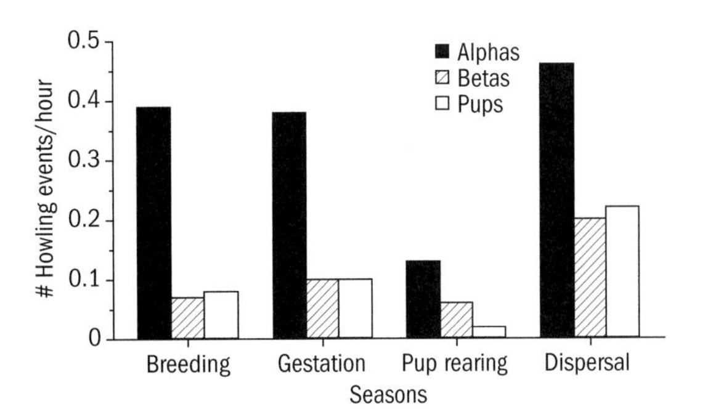

Figure 17.10 Howling rate for alpha, beta, and pup coyotes during the biological seasons, YNP, Wyoming 1991–93.

Direct defence

When intruding animals ignore indirect mechanisms of territory defence, canids must employ direct confrontation of intruders to reinforce territory boundaries (Camenzind 1978; Bekoff and Wells 1986; Mech 1993, 1994). Defence of a territory is usually a task undertaken by the dominant alpha pair (Mech 1970, 1993). We observed 112 instances of territorial defence by resident coyotes evicting trespassing animals (Gese 2001). These chases averaged 2.87 min in duration (range 0.3–26.8 min). Similar to the findings on howling and scent-marking rates, the alpha pair (mainly the alpha male) was most likely to be involved in territorial defence (87% of evictions). Beta coyotes were less likely to be involved (48% of the chases), while pups participated little in territorial defence (7% of the evictions). Pursuits of intruding coyotes terminated at the territory boundary and were followed by a robust session of howling and scent-marking at the border by the resident animal(s). Physical contact between the resident animals and intruders was observed, but consisted of ritualized displays of dominance and submission, with few serious injuries occurring. In contrast to the high mortality among wolves associated with a territorial trespass (e.g. Van Ballenberghe and Erickson 1973; Mech 1994), no intruding coyotes were killed during encounters with a resident pack. Intruders generally retreated from the resident territory quickly and often without any physical contact occurring between the residents and intruder(s). The group of coyotes pursuing an intruder or group of intruders usually had a numerical advantage over the group being chased (Gese 2001). Howling seems to serve as a long-distance warning to intruders, scent-marking as the visual and olfactory signal used at shorter distances, and direct confrontation if intruders ignored all the other territorial signals (Gese and Ruff 1997, 1998; Gese 2001).

Individual fitness

When we examined the benefits of a dominance hierarchy within the resident packs in terms of reproductive success and survival (i.e. fitness; Davies 1978), several key findings became evident. While the alpha coyotes have the risk of injury when confronting intruders or attacking large prey, they benefit greatly in terms of survival and reproduction (Gese 2001). We found that the alpha coyotes are the ones providing all of the reproductive output into the population with 93.7% of the alphas observed breeding and 66.1% of their pups being recruited into the population (Table 17.3). Even though one beta female produced a litter of pups, those pups did not survive to be recruited into the population (i.e. they all perished in <3 months). In addition, pup coyotes and dispersing coyotes had the lowest survival rates (0.64 and 0.13 annual survival rates for pups and dispersers, respectively). Most dispersing coyotes moved outside the park into areas where human persecution was more prevalent. Beta (0.96 annual survival) and alpha coyotes (0.91) had equal survival, but betas did not contribute to the reproductive effort (but may benefit

Table 17.3 Comparison of various reproductive, demographic, and foraging parameters between territorial and non-territorial coyotes, YNP, Wyoming, 1991–93

	% animals breeding	% pups surviving to 5 months	Annual survival rate	Annual dispersal rate	% time feeding on carcass	Small mammal capture rate (#/h)	Capture success (%)
Territorial							
Alphas (16)	93.7	66.1	0.91	0.03	2.7	2.3	38.2
Betas (31)	6.2	0	0.96	0.14	3.2	2.5	37.2
Pups (43)	0	–	0.64	0.30	0.6	2.2	27.6
Non-territorial							
Transients (5)	0	–	1.00	0.17	0.3	2.0	32.3
Dispersers (8)	0	–	0.13	B	0.4	0.6	22.0

Note: Numbers in parentheses are sample size for that cohort.

through inclusive fitness by helping related offspring; Hamilton 1964). Transient coyotes also had high survival, but again, produced no offspring (Table 17.3). In terms of dispersal rates, alpha coyotes rarely dispersed, while dispersal was much more common among betas, pups, and transients (Table 17.3). Alphas and betas had the greatest access to ungulate carcasses during winter, while pups, transients, and dispersers had little access to carcasses (Table 17.3). All cohorts of coyotes (alphas, betas, pups, and transients) were equally adept at capturing small mammals, while dispersing coyotes had the lowest success hunting small mammals (Table 17.3; Gese *et al.* 1996c). By defending a territory, the alpha pair benefited the most in terms of food resources, mating, space, and survival, when compared to other resident pack members (betas and pups) and non-territorial coyotes (transients and dispersers; Gese 2001). Essentially, within the coyote social system, the fitness of the alpha animals far exceeded all the other cohorts even when the risk of injury from territorial defence is considered (although the risk to the alphas seems almost non-existent).

In summary, in YNP, coyotes adapted to changes in prey abundance, availability, and vulnerability throughout the year, as well as changes in snow depth and temperature by modifying their behaviour, foraging strategies, and activity budgets. Differences in prey density within certain habitats were exploited by all coyotes as they spent more time hunting small mammals in habitats containing the highest reward. The presence of a dominance hierarchy in the resident pack, in conjunction with territoriality, allowed resident animals (particularly the alpha pair) more access to food, mates, and space and appeared to be evolutionary advantageous in coyote society.

Acknowledgements

I thank Pat Terletzky, Ed Schauster, Alden Whittaker, Alexa Calio, Melissa Pangraze, Lara Sox, Levon Yengoyan, Danny Rozen, Scott Grothe, Kezha Hatier, Jeanne Johnson, John Roach, and Valeria Vergara for field assistance; John Cary and John Coleman for computer programming; Robert Ruff for funding and logistical support; and David Macdonald, L. David Mech, and two anonymous referees for review of the manuscript. Data were collected while the author was a PhD student in the Department of Wildlife Ecology at the University of Wisconsin-Madison. Funding and support provided by the Department of Wildlife Ecology and the College of Agricultural and Life Sciences at the University of Wisconsin-Madison, National Park Service, Max McGraw Wildlife Foundation, US Fish and Wildlife Service, National Geographic Society, and the Hornocker Wildlife Research Center.

CHAPTER 18

Grey wolves—Isle Royale

Long-term population and predation dynamics of wolves on Isle Royale

John A. Vucetich and Rolf O. Peterson

Wolf *Canis lupus*, pack consumes a moose killed two days earlier in Isle Royale National park © R. O. Peterson.

The wolves (*Canis lupus*) of Isle Royale, an island (544 km^2) in Lake Superior (North America), have been studied with their primary prey, the moose (*Alces alces*), continuously and intensively since 1959. It is the longest study of such intensity in the world. The system is also importantly unique because on Isle Royale humans do not exploit wolves or moose, wolves are the only predator of moose, moose comprise an overwhelming majority of wolf prey, and the annual exchange of wolves and moose with the mainland is negligible. For this wolf–moose system, we present a chronology of research, general characteristics of the wolf population, and review some insights learned from studying the ecology of these wolves.

The wolves of Isle Royale

Chronology of wolves and wolf research on Isle Royale

Wolves first colonized Isle Royale National Park (Fig. 18.1) in the late 1940s—about 50 years after moose are thought to have first colonized the island. By 1930, moose probably exceeded 2000–3000 animals (4–6 moose/km^2; Peterson 1995b; see also Murie 1934). In 1934, a catastrophic, winter die-off reduced the moose population to a few hundred. In 1936, wildfire burned about 20% of the island, and subsequent moose population fluctuations during the next two decades were never documented. Another significant moose starvation event was

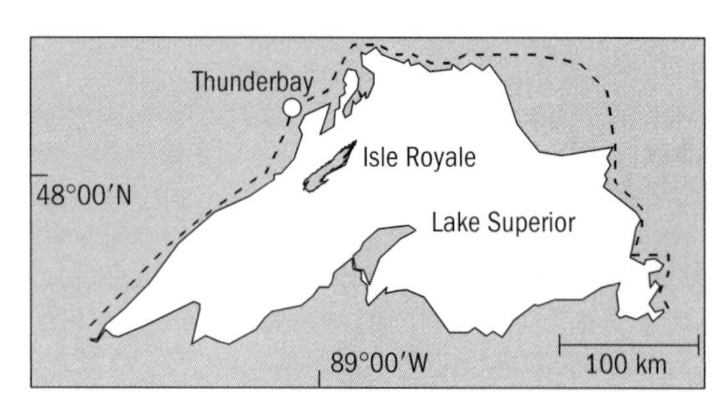

Figure 18.1 The location of Isle Royale National Park within Lake Superior, North America. The dotted line represents a highway.

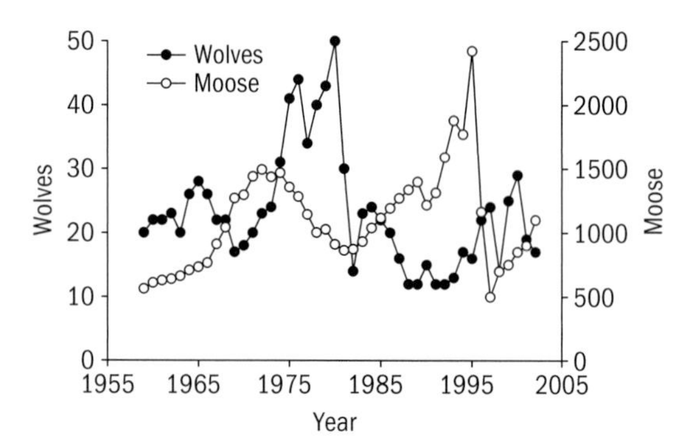

Figure 18.2 Population trajectories of wolves and moose on Isle Royale, 1959–2002. Each year the entire wolf population is counted from small aircraft (details in Peterson and Page 1988). The number of moose is estimated from aerial surveys (details in Peterson and Page 1993).

recorded in the late 1940s—about the time wolves arrived (Mech 1966).

Unable to stimulate federal sponsorship for Isle Royale wolf–moose research, Durward Allen moved and initiated what was envisioned to be a 10-year project in 1958 from Purdue University. During this 10-year period, Allen, graduate students, and post-doctoral investigators monitored wolf numbers annually, in addition to the ecology of moose, beaver (*Castor canadensis*), red fox (*Vulpes vulpes*), snowshoe hare (*Lepus americanus*), and deer mice (*Peromyscus maniculatus*) (Allen 1979). Long-term data sets, some extending back to 1959, now include wolf and moose population size, wolf social and spatial organization, wolf vital rates and predation rates, and characteristics of moose prey. These have been chronicled in a series of scientific and popular publications (Mech 1966; Jordan *et al.* 1967; Wolfe and Allen 1973; Peterson 1977; Allen 1979; Peterson and Page 1988; Peterson *et al.* 1998; Peterson 1995, 1999; Peterson and Vucetich 2001; Vucetich and Peterson 2002).

In addition, the ecology of lone wolves and small, non-territorial packs was reviewed by Thurber and Peterson (1993). Genetic characteristics of the wolf population were presented by Wayne *et al.* (1991) and Lehman *et al.* (1992). Evidence of occasional movement of wolves between the island and the mainland appears in Wolfe and Allen (1973) and Peterson (1979), but Wayne *et al.* (1991a) demonstrate that the entire wolf population in the late 1980s descended from a single maternal ancestor, probably the founding female.

In the late 1950s and early 1960s, wolves and moose both increased slowly (Fig. 18.2).

At the time, this relative stability was thought to characterize how the inclusion of predation led to a balance of nature (Allen and Mech 1963). During the late 1960s, the moose population nearly doubled from ~760 (1.4/km^2) to ~1400 (2.7/km^2). A series of winters with above average snowfall (1969–72) coincided with an end to rapid growth in the moose population. From 1969 to 1980, the wolf population nearly tripled from 17 to 50 wolves (from 31 to 92 wolves per 1000 km^2), during which time the moose population had risen, peaked, and began to fall again. Immediately after reaching this all-time peak, the population crashed in just 2 years to its lowest level ever, 14 wolves in 1982. These dramatic fluctuations inspire a very different interpretation of the balance of nature (Peterson 1995, 1999). The crash appears to have been caused by canine parvovirus, an infectious disease, and increased rates of wolves killing wolves in territorial disputes. Canine parvovirus was inadvertently introduced to Isle Royale by humans or their pet dogs, despite attempts to protect Isle Royale with the United States Wilderness Act (1963), which mandates 'wilderness to [be] affected primarily by the forces of nature, with the imprint of man's work substantially unnoticeable'. Since the crash, the number of wolves per moose has been substantially lower than before the crash, wolf extinction has at times seemed imminent, and the influence of inbreeding depression remains uncertain (Peterson *et al.* 1998). While wolves were at low density through the 1980s and early 1990s, the moose population increased to over 2000 animals (4 moose/km^2). Three-quarters of the moose starved to death in the severe winter of 1995–96, and aftershocks of this prey decline were seen in the wolf population through the close of the 1990s (Peterson and Vucetich 2001).

General characteristics of the Isle Royale wolf population

The wolf population is typically comprised of three or four packs. Each pack is typically comprised of 3–8 wolves, of which 2 or 3 are typically pups. In a typical year, one in six Isle Royale wolves lives as a loner or a member of a non-territorial pair. Although annual mortality rate varies substantially among years, one of every four or five wolves dies in a typical year (Fig. 18.3; Peterson *et al.* 1998). Two-thirds of all Isle Royale wolves die before the age of 5 years. Most deaths are probably associated with inter-pack strife or starvation. High and variable mortality rates are matched by similarly high and variable recruitment rates (Fig. 18.3; Peterson *et al.* 1998).

Associated with the high mortality rates is a dynamic social structure. On average, every 3 years one or more packs is dissolved, and usually replaced by a new pack in less than a year (Peterson and Page 1988; Peterson *et al.* 1998). For example, between 1980 and 1982, three old packs dissolved, and three new packs formed. Nevertheless, cases of long-term stability, such as a female wolf that led the west pack for 9 years (1987–95), also exist.

The Isle Royale packs are typically arranged linearly along the long axis of the island (Fig. 18.4). Wolves living in each of the pack territories experience vastly different densities of their primary prey, moose. Moose density at the east end of Isle Royale is typically twice that at the west end, and almost 10 times that of the middle third of Isle Royale (Fig. 18.4). Differences in moose density are attributable to differences in vegetation (Brander *et al.* 1990; McLaren and Janke 1996).

Depending on pack size and prey abundance, an Isle Royale wolf pack typically kills one moose every 4–10 days during winter (Thurber and Peterson 1993; Vucetich *et al.* 2002). Isle Royale wolves have some preference for killing calves and old moose (>9 yrs; Peterson 1977). Adult, wolf-killed moose frequently suffer from arthritis, jaw necrosis, or starvation (Peterson 1977). Depending on pack size and prey abundance, wolves may consume nearly all of a moose (including hide and bone marrow), or may consume only internal organs and some of the muscle tissue. Remaining tissue is typically consumed by scavenging foxes and ravens (*Corvus corax*).

The consequences of Isle Royale's isolation

Islands have long attracted the attention of evolutionary scientists (e.g. Wallace 1869; Kaneshiro 1988; Otte 1989; Roughgarden 1995; Sato *et al.* 2001). Case studies of island populations have also contributed importantly to understanding community ecology (Simberloff and Wilson 1969, Diamond 1975, Ricklefs and Bermingham 2001). Case studies of island populations, including the wolves and moose of Isle Royale,

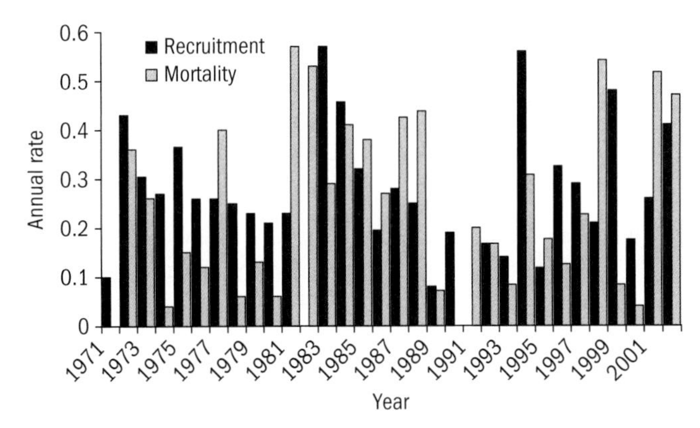

Figure 18.3 The annual rates of mortality and recruitment for the Isle Royale wolf population from 1971 to 2002 are relatively high and variable. The median mortality is 0.21 (interquartile range = (0.08, 0.40), coefficient of variation = 72%), and the median recruitment is 0.26 (interquartile range = (0.19, 0.32), coefficient of variation = 53%).

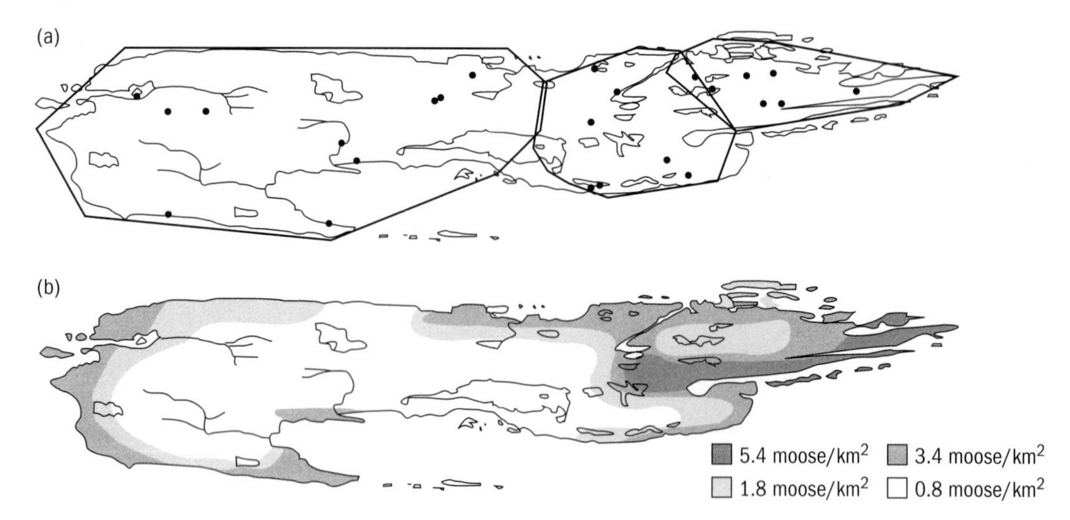

Figure 18.4 (a) Boundaries of the wolf packs on Isle Royale National Park in 2002. The dots represent the locations of moose kills made by each pack during a 6-week period in January and February of 2002. Winter kill rate data are based on these observations. In most years the population is comprised of three packs arranged along the long axis of the island. (b) Spatial variation in moose density on Isle Royale in 2002. Although absolute values change annually, the relative spatial pattern depicted here is representative of most years.

Table 18.1 Selected island case studies that have made distinctive contributions to various topics of population biology

Population regulation
Red deer on the Isle of Rhum (Albon *et al.* 2000), feral sheep of Soay Island (Coulson *et al.* 2001a), spiders on Gulf of California islands (Polis *et al.* 1998), community effects of predator removals (Terbough *et al.* 2001)
Predation
Wolf–moose interaction on Isle Royale (Vucetich *et al.* 2002), Mustelid–microtine interactions on islands in Fennoscandia (Heikkilae and Hanski 1994), fox–marten–hare interactions on islands in the northern Baltic (Marcstroem *et al.* 1989)
Competition
Arctic and snowshoe hare competition on islands off the coast of Newfoundland (Barta *et al.* 1989), Hermit crab competition on the San Juan Archipelago, Washington, USA (Abrams 1987), pararge butterfly competition on Madeira (Jones *et al.* 1998)
Extinction
Lizard populations in the Bahamas (Schoener *et al.* 2001), bird populations on islands off shore from Great Britain (Pimm *et al.* 1993; see also Vucetich *et al.* 2000), birds on Barro Colorado Island (Robinson 1999)
Population genetics
Sparrows of Mandarte Island (Keller 1998), *Peromyscus* on islands in inland lakes (e.g. Vucetich *et al.* 2001), finches of Galápagos (Grant and Grant 1992)

have also contributed uniquely to understanding several aspects of population biology (Table 18.1).

The value of the Isle Royale case study is importantly attributable to 25-km of open water (average annual temp ~4°C) that isolates Isle Royale from the mainland (Fig. 18.1). Although wolves would cross an ice bridge, they form only rarely, and access to the shoreline is limited by the town of Thunder Bay and a lakeshore highway. Because of this cold-water barrier, immigration and emigration have a negligible effect on demography of Isle Royale wolves and moose.

For most studies of demography, however, immigration and emigration represent substantial obstacles. When empirical observation is the basis for developing theory, the treatment of emigration and immigration often is *ad hoc* (e.g. Vucetich and Creel 1999), or heavily reliant on inference (e.g. Stacey and Taper 1992). Conversely, empirical validation of mechanistically based theory is limited by the estimation of emigration and immigration in real populations. Because the Isle Royale case study is unencumbered by these challenges, observations are more easily related to population biology theory (e.g. Vucetich *et al.* 1997; Eberhardt 1998).

The value of the Isle Royale case study is also facilitated by a synergy between observation and theory that is unencumbered by the complexity of ecological interactions associated with most other systems. Most generally, the species richness of Isle Royale mammals is only about one-third of that for nearby mainland areas. More specifically, potential prey such as white-tailed deer (*Odocoileus virginianus*) and potential competitors such as black bear (*Ursus americanus*) and coyote (*Canis latrans*) are absent from Isle Royale. Moreover, humans do not harvest either wolves or moose on Isle Royale. Thus, Isle Royale wolves and moose essentially represent a single-predator–single-prey system and can be adequately represented as a food chain. However, beaver is a minor component of summer diet for Isle Royale wolves (i.e. ~15%; Thurber and Peterson 1993). Elsewhere, wolves are embedded in substantially more complex food webs (see fig. 1 in Smith *et al.* 2003a; see also Polis and Strong 1996).

Another favourable property of the Isle Royale system is the number of individuals that comprise the populations of wolves and their prey. On average, Isle Royale is inhabited by 22 wolves and 1200 moose. At 544 km^2, Isle Royale is large enough to support a population of wolves, but small enough to permit complete annual censuses of the wolf population. Isle Royale is also small enough to permit annual surveys that include counting approximately 20% of the moose on Isle Royale (Peterson 1977). Our understanding of wolf–moose dynamics would be diminished, if Isle Royale were half or twice its size, or half or twice its distance from the mainland.

Summary of key contributions to science

Predation

Much predation research has focused on assessing how predation rates are affected by prey density, to the exclusion of other biotic and abiotic factors. This narrow scope reflects the canonization of early predation research that assessed only prey density, which seems to have slowed the acquisition of insights beyond those revealed by these early studies (e.g. Holling 1959; Rosenzweig and MacArthur 1963).

Research on wolves illustrates this historical interpretation of predation research. Most focus has been placed on the influence of prey density (e.g. Dale *et al.* 1994; Eberhardt 1998; Hayes and Harested 2000). However, despite vague appreciation that predation rates increase and carcass utilization decreases with deep snow cover (Nelson and Mech 1986; Fuller 1991; DelGiudice 1998), we have a poor understanding of how this important abiotic factor affects wolf–prey dynamics. Despite its potential importance, the effect of wolf density on predation rate is also poorly understood (Abrams and Ginzburg 2000; see also Yodzis 1994).

Since 1971, per capita kill rates (kills per wolf per month) have been estimated for a 45-day period each winter for each pack on Isle Royale. These observations indicate that estimated prey abundance on Isle Royale explains only 17% of the variation in the estimated per capita kill rates (Vucetich *et al.* 2002). However, a model that predicts kill rate from the ratio of prey to predators, a so-called ratio-dependent model (Akcakaya *et al.* 1995), explains 34% of the variation in kill rates, and outperformed models depending only on prey density as well as other models that depend on both predator and prey density (Vucetich *et al.* 2002). Also, a ratio-dependent kill rate model, modified to include the influence of seasonal snowfall, explains a total of 45% of the variation in per capita kill rate. Finally, when data from all packs each year are averaged, the per cent variation explained increases to 69% (Vucetich and Peterson unpublished result). Thus, an important component of variation in kill rates arises from variation among individual packs within a population (Fig. 18.5).

Figure 18.5 Wolf *Canis lupus*, pack in unsuccessful chase of moose, Isle Royale National Park © R. O. Peterson.

In contrast to our results, Messier (1994) reports that moose density explains 53% of the variation in per capita kill rates, and concludes that wolf predation is therefore a well-understood process. Several considerations suggest that this result and interpretation are misleading: (1) Messier's analysis is based on data from numerous short-term studies conducted across North America. Because spatial variation is not generally interchangeable with temporal variation, his analysis has limited relevance for temporal predation dynamics. (2) The explanatory power of Messier's analysis is artificially inflated because it relies heavily on data representing averages collected over several years from a single location. For example, Isle Royale is represented by 5 of the 14 data points in Messier's analysis. Each data point is an average of 5 years of data, and each year of data is an average of at least three packs. If these five data points were replaced by the >55 points that they represent, moose density would explain only 19% of the variation in kill rate. Although the inclusion of multiple observations from a single pack may represent pseudoreplication, this does not nullify the revelation that a substantial portion of variation in kill rate is unexplained and probably attributable to unexplained variation among packs within a population, and among years and within a single pack. Insights from Isle Royale suggest that wolf predation is more complex and less well-understood than has been suggested (cf. Messier 1994, p. 486).

Trophic cascades

Trophic cascades are the indirect effect of predator populations on plant populations, via direct influences on herbivore populations. Because the first well-documented trophic cascades were from marine (Paine 1966) and aquatic (Carpenter and Kitchell 1988) systems, and because of certain properties and assumptions of food chain models, trophic cascades are thought to be more common in aquatic systems than in terrestrial systems (Strong 1992; see also Chase 2000).

McLaren and Peterson (1994) reported that the dynamics of balsam fir (*Abies balsamea*) on Isle Royale were more closely linked to wolf–moose interactions than to seasonal weather patterns. This observation of predator dynamics impacting plant population dynamics was significant because such a pattern, manifest over several decades, had not been previously detected in a terrestrial system of long-lived vertebrates. Since then, numerous investigators have reported top-down effects in terrestrial ecosystems (reviewed by Schmitz *et al.* 2000; Chase 2000; see also Terbough *et al.* 2001).

Beyond assessing whether a community is or is not regulated by top-down processes, more recent trophic cascade research aims to: (1) compare the relative strengths of top-down and bottom-up processes (e.g. Polis *et al.* 1998); (2) assess top-down processes across different scales of space and time (Holt 2000; Power 2000); (3) understand the community characteristics that promote strong trophic cascades (Chase 2000); and (4) understand the relationship between the frequency of a trophic cascades and what portion of the community is affected (Polis *et al.* 2000).

A reassessment of the Isle Royale case study is needed to further understand (1) and (2). Preliminary re-analysis indicates that per capita rates of prey capture (a bottom-up process) explains only about 22% of the variation in wolf growth rate (Vucetich and Peterson in review). From the perspective of wolf growth rate, the rate of prey capture summarizes the influence of bottom-up processes. To the extent that this is true, top-down process on Isle Royale would appear very influential—perhaps 3–4 times more dominant that bottom-up processes ($(1-R^2)/R^2 = (1-0.22)/0.22 = 3.5$).

Another re-analysis indicates that substantially more variation in moose growth is explained by tree-ring growth of balsam fir (i.e. primary winter forage and a bottom-up process) than wolf abundance (i.e. the top-down process) (Vucetich and Peterson 2004). Importantly, weather variables explained more variation than did either of these biotic variables. Also, of the models examined, the most parsimonious explain only about half the variation in moose growth rate.

Multi-annual fluctuations

Populations, within and among taxa, exhibit a range of dynamical types: largely stable, eruptive, aperiodic multiannual fluctuations (MAF), and strong cycles with nearly constant periodicity. Population biologists have long been interested in understanding the mechanisms responsible for each dynamical type. The wolves and moose of Isle Royale clearly exhibit multiple consecutive years of increase, followed by multiple consecutive years of decrease (hereafter, MAF). These dynamics have been characterized as being cyclic (e.g. Peterson *et al.* 1984; McLaren and Peterson 1994; Post *et al.* 2002). This possibility is intriguing because most of our understanding of cycles is derived primarily from species with much smaller body size (e.g. hares, lemmings, *Synaptomys spp.*, and forest Lepidoptera) and much shorter cycle periods (10 years). However, it may be important to distinguish aperiodic MAF from MAF with nearly constant periods (hereafter, cycles). The distinction is important because the potential set of mechanisms that give rise to MAF may not be identical to those giving rise to cycles. If the Isle Royale system is cyclic, then the period is approximately 23 years, and we have observed approximately 1.8 cycles. This is hardly adequate for distinguishing between MAF and cycles.

Attempts to demonstrate cyclicity in the Isle Royale data distract from the value of assessing potential mechanisms that underlie the observed MAF. For example, the observed MAF may be the result of predator–prey *interactions*. If so, it would be important to discern whether observed MAF represent deterministic Lotka–Volterra processes, arising from destabilizing stochastic processes, or if they arise from a dynamic age structure of the moose population and wolves' limited ability to prey on prime-aged moose (3–9 years old).

In contrast to predator–prey interactions, moose might exhibit MAF independent of their interactions with wolves. Further, the MAF of wolves may arise merely as they track the fluctuating moose population. More specifically, moose MAF could represent delayed density dependence arising from intrinsic processes such as intraspecific competition or maternal effects (Berryman and Chen 1999; Keech *et al.* 2000). Alternatively, it is possible that MAF in moose arise from interactions with parasites, as has been considered for other vertebrate species (e.g. Moss and Watson 1995; Ives and Murray 1997).

Although assessing the constancy of the period in wolf–moose dynamics would be important, it seems unanswerable in the absence of a couple centuries of data or excessive reliance on inference. Fortunately, the pursuit of discerning the relative contributions of the above-mentioned processes would likely be feasible and profitable.

Foraging economics and the evolutionary maintenance of wolf sociality

Sociality is a conspicuous feature of wolves. A popular notion is that wolves live in large groups because it is required for capturing their usually large prey. However, observations from Isle Royale suggest that a single wolf can routinely capture moose, one of the largest species that wolves prey upon (Thurber and Peterson 1993). Nevertheless, group hunting may be favoured in wolves, like some other social species, because it confers increased foraging efficiency (Giraldeau and Caraco 2000). The positive relationship between average pack size and average prey size has been interpreted to support this claim (Nudds 1978; see Fig. 18.6). Observations from Isle Royale suggest that the per capita rate of prey capture decreases with pack size (Thurber and Peterson 1993; see also Schmidt and Mech 1997; Hayes *et al.* 2000). Such observations have been interpreted to mean that foraging economics do not favour sociality, and that kin selection is the sole selective force favouring sociality.

These ideas parallel the development of ideas related to understanding sociality in other large carnivores (e.g. Packer *et al.* 1990; Caro 1994; Creel and Creel 1995; Packer and Caro 1997). Unfortunately, most studies inadequately account for processes such as: (1) how foraging costs change with group size, (2) the instability of optimal group sizes (Giraldeau and Caraco 2000), and (3) per capita rates of food loss due to scavenging (by other species) for different group sizes. While some studies account for some of these factors (e.g. Carbone *et al.* 1997), no study to date has attempted to account for them all.

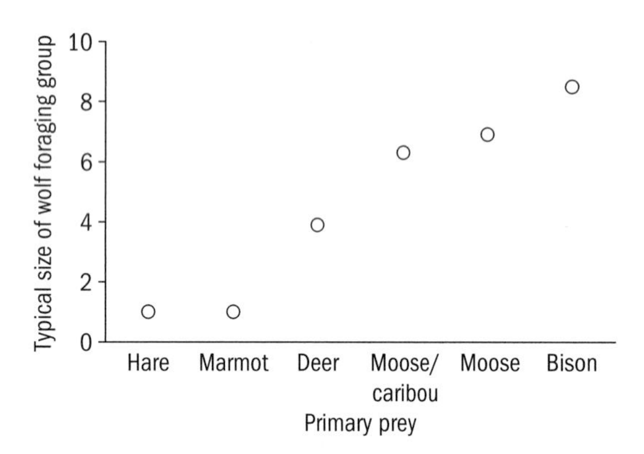

Figure 18.6 The relationship between prey type and size of wolf foraging group. Prey are ordered along the *x*-axis in order of size. Data are from Peterson *et al.* (1984), Carbyn *et al.* (1993), Messier (1994), Schmidt and Mech (1997), Hayes *et al.* (2000) and Schaller (2000) and represent averages for a study population or grand averages from several study populations. Although pack size tends to increase with prey size, variation in pack size for a given prey size is large. For example, pack sizes range from 2 to >20 for wolves that forage primarily moose and for wolves preying primarily on deer. Both increasing tendency and large variation are predicted by foraging theory based on rate maximization if the influence of raven scavenging is considered (Vucetich *et al.* 2004).

We recently used a combination of empirical observations and physiological modelling to estimate the net per capita rate of prey capture for Isle Royale wolves living in different-sized packs (Vucetich *et al.* 2004). This analysis suggests that the wolves living in larger packs capture less prey on a per capita basis than wolves in smaller packs. However, we also predicted the per capita rate of food loss due to scavenging for wolves living in different-sized packs. While feeding on a carcass, wolves may routinely lose 2–20 kg of prey per day to scavenging ravens. For rates of loss as low as 5 kg per day per moose carcass, the relationship between net rate of food intake and pack size is positive. Thus, large group size in wolves is favoured by social foraging benefits, because greater food-sharing costs in a larger pack are offset by smaller losses to scavengers.

Our analysis also indicates that because smaller prey are consumed faster, the rate of loss to scavengers is less, and wolves may afford to live in smaller packs when they forage on smaller prey. This is consistent with the observation that pack size tends to increase with prey size (Fig 18.6; see also Nudds 1978). However, our analysis also indicates that a wide range of pack sizes might form for any given prey size. This is consistent with observations that pack sizes for deer-killing wolves can be as large as 22 (Mech 2000a).

Kin selection certainly would seem to favour sociality in wolves (and other social carnivores). Another foraging theory, the resource dispersion hypothesis (see Macdonald *et al.*, Chapter 4, this volume), may also favour sociality in wolves. However, kin selection and the resource dispersion hypothesis do not appear to be the only selective force favouring sociality. Foraging economics also appears to favouring sociality among wolves, and perhaps other large, social carnivores.

Extinction risk and wolf sociality

Assessments of population viability and extinction risk have become a common pursuit in conservation research and a nearly ubiquitous component of managing endangered populations. An unresolved challenge seems to be assessing the accuracy and utility of analyses that are routinely limited by uncertainties in parameter estimates and model structure. Some conservation scientists seem optimistic about their value (e.g. Lindenmayer *et al.* 1993; Brook *et al.* 2000; see also www.cbsg.org/phvalist.htm), and others pessimistic (e.g. Beissinger and Westphal 1998; Ludwig 1999; White 2000). Evaluation of the accuracy and utility of viability assessments requires additional research to understand the consequences of ignoring potentially important processes or factors (e.g. age structure, density dependence, species interactions, and genetic processes).

Although most viability models ignore the influence of social structure, it may commonly affect population dynamics. The Isle Royale case study has been used to understand better how social structure affects extinction risk dynamics (Vucetich *et al.* 1997). Demographic data from the Isle Royale wolf population was used to construct a population viability model where each simulated wolf belonged to a pack and experienced age-specific mortality rates. The number of packs in the population depended on moose abundance, and recruitment was based on the number of packs in the population. The most important and general result of this analysis is that sociality may increase the population size required to eliminate demographic stochasticity as an important risk factor. The mechanism underlying this process is that the number of breeding units is equal to the number of packs, not the number of females. For example, the Isle Royale population has been small (e.g. 14 in 1982) and divided into just three packs, and much larger (e.g. 45 in 1976) but still comprised of the same number of breeding units (i.e. 3 packs). Since the publication of this chapter, further insights have been developed on how sociality and other behaviours affect extinction risk dynamics (e.g. Legendre *et al.* 1999; Reed 1999; Vucetich and Creel 1999; Courchamp *et al.* 2000; Courchamp and Macdonald 2001).

Evidence for inbreeding depression

Inbreeding depression is a decline in fitness due to inbreeding or genetic deterioration. Our understanding of inbreeding depression is based largely on theory (e.g. Vucetich and Waite 1999), laboratory experiments (e.g. Lacy *et al.* 1996), and captive zoo populations (e.g. Ballou 1997). Unfortunately, this understanding is limited by the simplifying assumptions that characterize theory and the artificial conditions that characterized laboratory and captive populations (e.g. Sheffer *et al.* 1997). Opportunities to examine inbreeding depression in unmanipulated populations (especially of vertebrate species) are rare and considered valuable (e.g. Wildt *et al.* 1987; Slate *et al.* 2000). The wolf population of Isle Royale has potential to provide insights on the nature of inbreeding depression. However, some obstacles prevent realizing any insights.

The Isle Royale wolf population was founded about 13 generations ago in the late 1940s by wolves that crossed an ice bridge connecting Isle Royale to the mainland (Mech 1966), and since has been completely isolated from the other wolf populations. Molecular studies suggest that all Isle Royale wolves have descended from a single female (Wayne *et al.* 1991). Demographic models indicate that the Isle Royale population has an effective population size (N_e) of approximately 3.8, and is expected to lose 13% ($= 1/2N_e$) of its genetic diversity each generation (which is 4.2 years (Peterson *et al.* 1998)). This rate of inbreeding is comparable to repeated matings among first cousins. Molecular analyses of wolf genetic diversity corroborate these high rates of loss (Wayne *et al.* 1991). In 2002, the Isle Royale population was expected to have only ~18% of the founding population's diversity. Such rapid losses of large amounts of genetic diversity generally increase the risk of inbreeding depression (Ehiobu *et al.* 1989).

Direct evidence (i.e. Laikre and Ryman 1991) and indirect evidence (i.e. Smith *et al.* 1997) suggest that wolves are generally vulnerable to fitness loss in response to high rates of inbreeding. However, the potential for inbreeding to reduce fitness is highly variable among taxa (Crnokrak and Roff 1999) and among populations within a taxa (Lacy *et al.* 1996; Vucetich and Waite 1999). Wolves clearly illustrated this principle: Two captive, inbred populations of wolves have failed to show any fitness losses (Kalinowski *et al.* 1999). Nevertheless, these populations were less inbred, and the statistical power of analyses may have been weak (Kalinowski *et al.* 1999).

These general uncertainties about inbreeding depression limit assessments of inbreeding depression for Isle Royale wolves. Nevertheless, several observations are suggestive of inbreeding depression. For example, the number of wolves per moose greater than 9 years old has been substantially lower ever since the wolf population crash from 50 in 1980 to 12 in 1982 (5.7 ± 0.3 SE versus 18.2 ± 3.9; see also fig. 5 in Peterson *et al.* 1998). This could represent a manifestation of inbreeding depression via reduced abilities to capture prey or convert captured prey into wolf recruitment, or an overall reduction in wolf survival, independent of prey abundance. However, it remains unclear whether ecological processes alone can explain the reduced number of wolves per old moose.

In 2000, we recovered the skeleton of a dead wolf with two grossly asymmetrical vertebrae (Fig. 18.7). Although the deformities appear to be developmental abnormalities, it is unclear whether the deformity: (1) has a genetic or environmental basis, (2) occurs with greater frequency in the Isle Royale population than in non-inbred populations, and (3) led to reduced fitness. In 2003, we discovered a wolf carcass with its two middle toe pads fused in both front feet—a condition common in certain inbred breeds of domestic dog.

Successful studies of inbreeding depression in free-ranging populations are generally based on comparisons among populations (e.g. Wildt *et al.* 1987) or individuals (e.g. Slate *et al.* 2000) that exhibit varying levels of inbreeding. In the absence of such a comparison, inbreeding depression in free-ranging populations is difficult to assess. The greatest obstacle for such comparisons involving the Isle Royale population is eliminating the possibility that differences in fitness are attributable to ecological factors.

Contributions to conservation

The conservation and recovery of wolf populations is actively pursued in numerous regions of North America and Europe. Although conservation research is generally assumed to be an important component of successful conservation, this assumption is rarely scrutinized. In this section, we assess the possible influence of the Isle Royale case study on wolf conservation.

Since human-caused mortality has been a primary cause of endangerment or remains a potential threat, wolf conservation might be facilitated by better understanding the extent to which human-caused mortality is additive or compensatory with other causes of wolf mortality. Because little is known about this process, it may be useful to know that mortality rates in the Isle Royale wolf population, which has never been exploited, can be high and

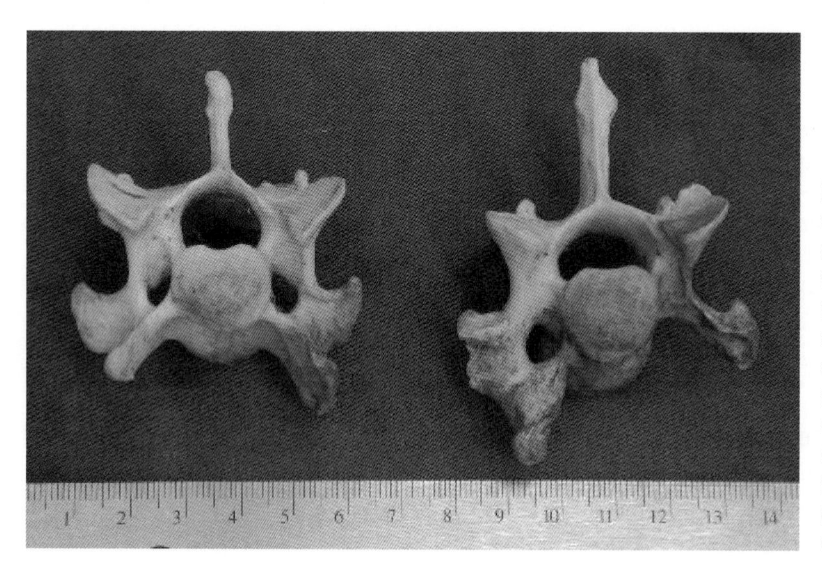

Figure 18.7 Gross asymmetry in a vertebrae from an Isle Royale wolf (#3529). This animal was born in the mid-1990s. Of the skeletal remains collected from approximately 35 Isle Royale wolves between 1959 and 2001, this is the only observed skeletal deformity. Nevertheless, the cause (environmental or genetic), frequency of occurrence, and fitness consequences of such deformities are unknown.

variable (Fig. 18.3). From this, one might infer that exploitation is not necessarily the cause of high and variable mortality rates in exploited wolf populations. Such an inference is, however, limited because Isle Royale may not be representative of unexploited wolf populations. Although this observation provides some perspective, its value for guiding conservation is limited.

Wolf conservation and recovery has also been concerned with understanding how many wolves and how large an area are required for population viability (e.g. Fritts and Carbyn 1995). The Isle Royale case study illustrates the possibility that a small, isolated population can persist, at least for several decades. This observation is also quite limited, because extinction and genetic drift are highly variable processes (Vucetich and Waite 1999), and a single case example may not be representative. Moreover, the requirements for long-term persistence are likely to differ from those for short-term persistence.

Wolf conservation is often justified by the notion that top predators, including wolves, are keystone species, and have a substantial influence on the ecosystems they inhabit. The Isle Royale case study provides a scientific basis for justifying this claim (McLaren and Peterson 1994) (Fig. 18.8).

Several issues have been critical to wolf conservation for which the Isle Royale case study has contributed little or no insight. These issues include: (1) taxonomic relationships among historical populations and recovering populations (e.g. Wilson *et al.* 2000), (2) biological details of how to translocate and release wolves into a new environment (e.g. hard and soft releases; see also van Manen *et al.* 2000), and (3) the amount of gene flow required to maintain natural population genetical processes (e.g. Forbes and Boyd 1997).

Perhaps the most important factor determining the success of wolf recovery and conservation has been the relationship between humans and wolves. The attitudes of North American humans towards wolves began to change in the late 1960s and early 1970s (Dunlap 1988). During this time, the Isle Royale case study was a prominent example to the general public of the value of wolves. *National Geographic* published *Wolves versus moose on Isle Royale* in its February 1963 issue, with a follow-up article in 1985 (i.e. Eliot 1985). Also during this time, two nationally broadcast films featured the Isle Royale case example (*Wolf Men* (1969) and *Death of a Legend* (1970)). The Isle Royale case study continues to heighten awareness of wolves for thousands of people through visitation to Isle Royale National Park, participation in EarthWatch expeditions (www.earthwatch.org/expeditions/peterson.html), wide distribution of annual reports, a popular book accounting the Isle Royale case study (Peterson 1995), and a web page (Vucetich and Peterson 2002).

The positive impact that Isle Royale wolves have had on the general public may also be reflected in the general public's interest in and support for wolf research on Isle Royale. National wire services

Figure 18.8 After killing a moose calf (*Alces alces*) in Isle Royale National Park, these two wolves (*Canis lupus*) successfully despatched the calf's mother. © R. O. Peterson.

consistently report the results of the annual wolf censuses on Isle Royale. Broad public support is also reflected in continuous financial support from a diverse array of non-governmental agencies and individuals, such as the National Rifle Association and Defenders of Wildlife. In fact, public interest alone forced Department of Interior officials to abandon an effort to scuttle the study in 1983 during the Reagan administration.

The Isle Royale case example may, however, also generate attitudes among the public that inadvertently hinder wolf recovery. First, the isolation and wilderness designation of Isle Royale probably contributes to the erroneous conception that wolves are restricted to wilderness areas far from where people live (cf. Haight *et al.* 1998; Mech 1995). This misconception may complicate the management of wolf–human conflicts. Second, the Isle Royale case study may also provide the general public with the idea that recovery of unexploited wolf populations will result in a 'balance of nature'. This is valuable, unless the general public does not appreciate that a 'balance of nature' may include periods of boom and bust for populations of both wolves and their prey.

An important, but difficult to quantify, contribution of the Isle Royale case study is professional training of people who have actively contributed to wolf conservation (and several of whom, are authors in this book). Specifically, D. Mech, whose PhD is based on the Isle Royale wolf population, has been a global leader in wolf conservation for four decades. R. Peterson also earned a PhD based on these wolves, and has contributed to the assessment and monitoring of wolf conservation in Alaska, the Great Lakes, and Scandinavia. D. W. Smith, studied the wolves of Isle Royale as a student for over a decade, and since 1996 has led the Yellowstone Wolf Project. M. K. Phillips was a field assistant for the Isle Royale wolf project in 1981, and has since directed reintroduction efforts of red wolves in North Carolina, and grey wolves in Yellowstone and the Southern Rocky Mountains. Finally, J. A. Vucetich began studying Isle Royale wolves in 1989, and has since contributed to monitoring and assessing wolf conservation in Michigan, in Algonquin Provincial Park, and for the Mexican Wolf Recovery Program.

The Isle Royale case study seems to have contributed to wolf conservation and recovery. However, the educational impact and inspiration of the Isle Royale story on the general public and researchers alike may have been more important than the scientific insight it has offered. Conservation scientists should consider the generality of this circumstance, and conduct their research with appropriate concern for its educational and inspirational impact on professionals, students, and the public.

Acknowledgements

Major support of these studies was received from the EarthWatch Institute, National Geographic Society, U.S. National Park Service, (DEB- and the U.S. National Science Foundation 9903671) plus numerous private donations. Any opinions, findings, and conclusions or recommendations expressed here are those of the authors and do not necessarily reflect the views of the National Science Foundation.

CHAPTER 19

Grey wolves—Yellowstone

Extermination and recovery of red wolf and grey wolf in the contiguous United States

Michael K. Phillips, Edward E. Bangs, L. David Mech, Brian T. Kelly, and Buddy B. Fazio

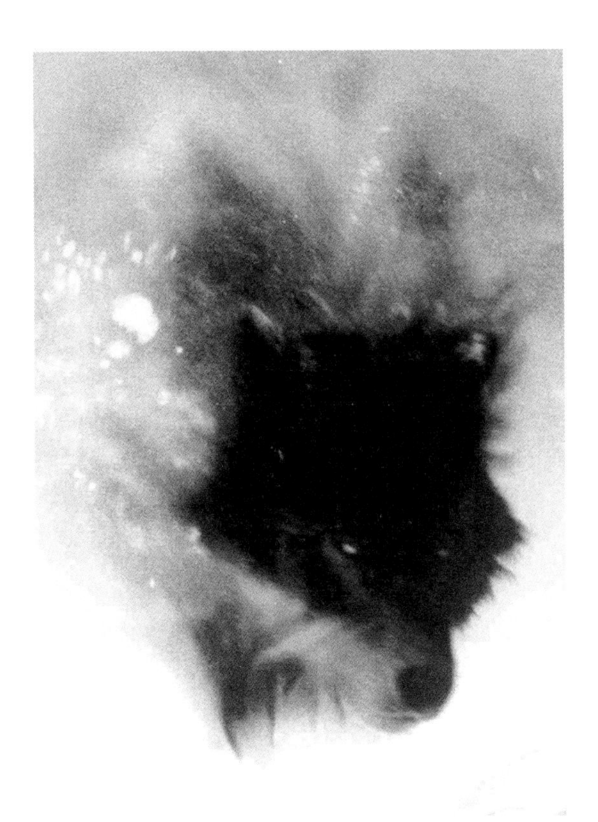

Grey wolf *Canis lupus* in snow © M. K. Phillips. Black wolf reintroduced to Yellowstone National Park © National Park Service.

As recently as 150 years ago, the grey wolf (*Canis lupus*) was distributed throughout the contiguous United States (US), except for the southeastern US from central Texas to the Atlantic coast where the red wolf (*Canis rufus*) occurred (Young and Goldman 1944; Nowak 1983). Conflict with agricultural interests resulted in government-supported eradication campaigns beginning in colonial Massachusetts in 1630 (Young and Goldman 1944; McIntyre 1995). Over the next 300 years, the campaigns were extended throughout the US resulting in the near extermination of both species. In recent decades, efforts to recover the red and grey wolf were carried out. This case study summarizes extermination and recovery efforts for both species in the contiguous US.

Wolf extermination

Historically, wolves were the most widely distributed large mammals in North America (Fig. 19.1). Together the two species probably numbered several hundred thousand individuals, and they occurred wherever large ungulates were found. Tolerant of environmental extremes, wolves inhabited areas from 15°N latitude (i.e. central Mexico) to the North Pole (Hall 1981; Nowak 1995). Wolf distribution was greatly reduced as a result of long-term extermination efforts that began as Europeans settled in North America. Conflict between the agrarian colonists and wolves prompted the establishment of bounties as early as 1630 (McIntyre 1995). Eventually wolf extermination became the policy of the federal government. Persecution reached a zenith in the late 1800s and early 1900s when the wolf's natural prey (i.e. bison (*Bison bison*), elk (*Cervus elaphus*), and deer (*Odocoileus* spp.)) had been greatly reduced due to unregulated exploitation (Schmidt 1978; US Fish and Wildlife Service 1987a). Bison were also killed as part of federal efforts to force Indians to submit to the reservation system (Isenberg 1992).

In the presence of reduced prey populations and expanded production of livestock, wolves increasingly depredated on the latter. In response the federal government and private citizens intensified control efforts. In 1915, the US Congress began funding a wolf control programme and assigned the mission of implementing it to the US Biological Survey. The goal was the 'absolute extermination' of the wolf, and poisoning was the main method used (McIntyre 1995).

By the 1930s, the numbers and distribution of wolves were reduced throughout the contiguous US, and by the 1940s wolves were almost absent (Young and Goldman 1944; Young 1970; Brown 1983; Nowak 1983). In the early 1950s, government trappers turned to northern Mexico and the few wolves from there that dispersed to the US. This influx was eliminated by the end of that decade (McIntyre 1995) when wolf numbers were at an all-time low. Then, less than 1000 wolves persisted in the remote regions of the Gulf Coast (red wolves) and the forests of northeastern Minnesota (grey wolves) (Fig. 19.1). Additionally, probably less than 20 grey wolves inhabited Isle Royale National Park, a 546 km^2 (210 miles^2) island in Lake Superior located about 32 km (20 miles) from the Minnesota mainland (Fig.19.1) (Stenlund 1955; Mech 1966; Peterson 1977; Fuller *et al.* 1992a; Thiel 1993). From the 1950s through the 1970s, studies provided insights into wolf ecology (Stenlund 1955; Pimlott 1967; Mech 1966; Mech and Frenzel 1971; Van Ballenberghe 1972; Peterson 1977) and helped foster a public desire to conserve the species.

Wolf recovery

In 1973, the Endangered Species Act (ESA) was passed (Public Law No. 93-205, as amended). This law provided significant protection for wolves and mechanisms for recovering both species. The first list of endangered species under this law included the red wolf, eastern timber wolf (*C. lupus lycaon*) and the Northern Rocky Mountain wolf (*C. lupus irremotus*) (US Fish and Wildlife Service 1974). In April 1976, the Mexican wolf (*C. lupus baileyi*) was listed as endangered (US Fish and Wildlife Service 1976a). In June 1976, *C. lupus monstrabilis* was listed as endangered (US Fish and Wildlife Service 1976b). In 1978, the US Fish and Wildlife Service (Service) combined the subspecific listings for the grey wolf and reclassified it at the species level (i.e. *C. lupus*) as 'endangered' throughout the contiguous US and Mexico, except for Minnesota where the species was reclassified to 'threatened' (Nowak 1978). Shortly after the wolves were listed, the Service began developing recovery plans.

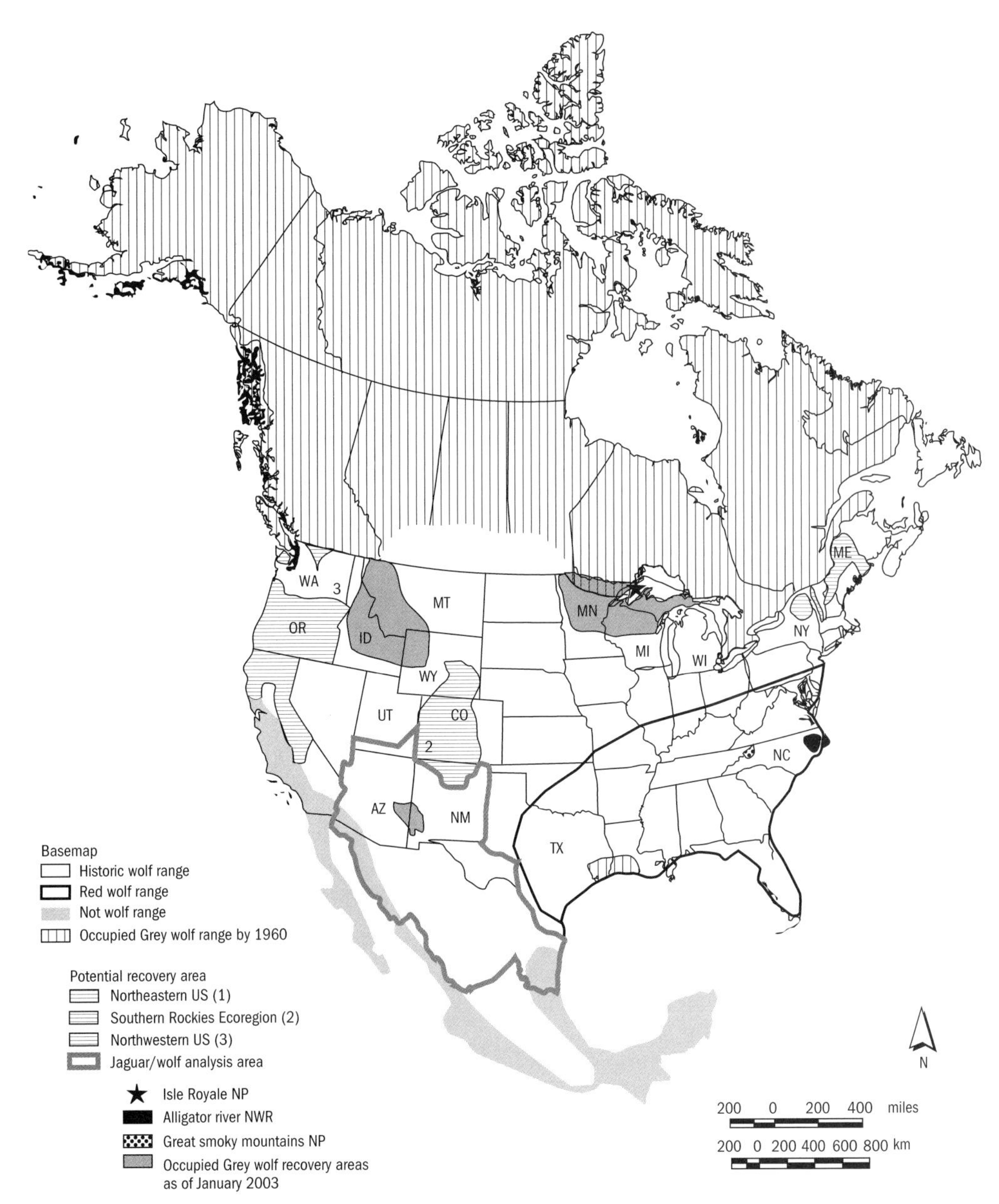

Figure 19.1 Areas that are relevant to the extermination and recovery of the red wolf and grey wolf in North America. The historic range that is portrayed for both species is from Nowak (1995). These recovery programmes were based on recognized wolf subspecies before Nowak's (1995) changes.

Red wolf

The decline of the red wolf was recognized in the 1960s (McCarley 1962). In addition to persecution by humans, the species was threatened by hybridization with coyotes (*Canis latrans*) (McCarley 1962; Nowak 1972, 1979). In 1973, a captive breeding programme was established at the Point Defiance Zoological Gardens, Tacoma, Washington. From a founding stock of 14 wolves, by December 2002, the captive population included 160 animals maintained at 32 facilities. By 1980, the red wolf was considered extinct in the wild (McCarley and Carley 1979; US Fish and Wildlife Service 1984).

The origins of the red wolf have been debated since the 1960s. Some authorities have considered the red wolf to be a full species (Nowak 1992, 2002), while others have considered that it might be a subspecies of the grey wolf (Lawrence and Bossert 1967; Phillips and Henry 1992) or a hybrid resulting from interbreedings of grey wolves and coyotes (Mech 1970; Wayne and Jenks 1991; Roy *et al.* 1996). The debate led to challenges to the integrity of the red wolf recovery programme (Gittleman and Pimm 1991) and was used by the American Sheep Industry as a rationale on which to petition the Secretary of Interior to remove the species from the list of endangered and threatened wildlife. The Service determined that the petition did not present substantial information to warrant delisting (Henry 1997). Recent genetics work suggests that the red wolf and eastern timber wolf share a close taxonomic relationship that justifies classification as a separate species, *Canis lycaon* (Wilson *et al.* 2000). However, Nowak (2002) presents morphological and distribution data that counter this claim and support the current taxonomic separation between the red wolf and the eastern timber wolf. The Service currently recognizes the red wolf as a valid species distinct from the grey wolf and coyote.

Recovery efforts

A recovery plan was finalized in 1984 that established the foundation for reintroducing up to 15 wolves for five consecutive years to the Alligator River National Wildlife Refuge (ARNWR) in northeastern North Carolina (Fig. 19.1) (US Fish and Wildlife Service 1984). The final plan called for the released wolves and their offspring to be designated as members of experimental-nonessential populations per section 10(j) of the ESA (Parker *et al.* 1986). Such a designation allows the Service to relax the restrictions of the ESA to facilitate wolf management (Parker and Phillips 1991). The ARNWR reintroduction is notable for several reasons, including its position as the first attempt ever to restore a carnivore species that was extinct in the wild. Wolves selected for release were taken from the Service's certified captive-breeding stock. Age, health, genetics, reproductive history, behaviour, and physical traits representative of the species were considered in the selection process. From 1987 through 2002, 85 red wolves were released on 38 occasions. During the fall of 2001, the last free-ranging red wolf that had been born in captivity died at the age of 13. After that the wild population consisted entirely of wild-born animals. By June 2003, free-ranging red wolves had given birth to >300 pups over four generations, and the population included approximately 100 red wolves in 20 packs across 6912 km^2 (2700 miles^2) of the restoration area. This area is composed of 60% private land and 40% public land that include three national wildlife refuges.

A revised red wolf recovery plan (US Fish and Wildlife Service 1989) called for additional reintroduction projects and indicated that for the foreseeable future it would not be feasible to downlist (change species' classification from endangered to threatened) or delist (remove species from list of threatened and endangered species) the red wolf (Table 19.1). In 1991, a second reintroduction project

Figure 19.2 Red wolf *Canis rufus* © M. S. Murri.

Table 19.1 Federal recovery criteria for the red wolf and the grey wolf

Recovery programme	Recovery area	Criteria for downlisting[a]	Criteria for delisting[b]
Red wolf	Southeastern United States	None[c]	None[c]
Eastern timber wolf	Minnesota, Michigan, Wisconsin	≥ 100 wolves inhabiting Michigan or Wisconsin for three consecutive years[d]	Assurance that the Minnesota population includes ≥1251 wolves Establishment of a second population outside of Minnesota and Isle Royale National Park If the second population is 160 km (100 miles) from the Minnesota population, it must consist of ≥200 wolves for at least 5 years (based on late winter counts) If the second population is within 160 km of the Minnesota population, then it must consist of ≥100 wolves for at least 5 consecutive years
Northern Rocky Mountain wolf	Montana, Wyoming, Idaho	10 breeding pairs[e] for 3 successive years in 2 of the recovery areas	≥30 breeding pairs comprising ≥300 wolves in a metapopulation with genetic exchange between subpopulations for 3 successive years
Mexican wolf	Southwestern United States	None[f]	None[f]

Notes:

[a] Downlisting refers to a classification change from endangered to threatened as per the federal ESA.

[b] Delisting refers to a classification change from threatened to removed from the list of endangered and threatened wildlife as per the federal ESA.

[c] The US Fish and Wildlife Service believes that establishment of 225 red wolves in the wild and maintenance of 325 animals in captivity would provide for preservation of the species. The Red Wolf Recovery Plan states that for the foreseeable future it is not feasible to either downlist or delist the species.

[d] Downlisting to threatened does not apply to the grey wolf in Minnesota, which was previously reclassified from endangered to threatened in 1978.

[e] A breeding pair is considered an adult male and an adult female wolf that have produced at least two pups that survived until December 31 of the year of their birth, during the previous breeding season.

[f] The Mexican Wolf Recovery Plan states that maintenance of a captive breeding programme and establishment of a population of >100 wolves would provide for maintenance of the subspecies. The plan further expresses no possibility for delisting the Mexican wolf.

was initiated in the Great Smoky Mountains National Park with the experimental release of one family (Fig. 19.1) (US Fish and Wildlife Service 1992a). Results suggested that restoration was feasible. The Service subsequently released 37 wolves from 1992 through 1996. Of these, 26 died or were recaptured after travelling outside the Park. Of 28 pups born in the wild and not removed, none survived the first year. Disease (canine parvo virus) was implicated in the death of many of the pups. Because of the inability of wolves to establish home ranges in the Park, low pup survival, and low winter prey availability, the Service terminated the project in 1998 (Henry 1998).

From 1987 through 1994 it seemed that the red wolf reintroduction project at ARNWR was succeeding (Phillips *et al.* 1996, 2003). During the mid-1990s, the prognosis changed as it became apparent that hybridization between red wolves and recently established coyotes was becoming increasingly common (Kelly and Phillips 2000). A comprehensive assessment (Kelly *et al.* 1999) facilitated development of an adaptive management plan to address the hybridization problem (Kelly 2000). The plan was implemented in April 1999 and called for hybridization to be eliminated or reduced through intensive fieldwork to euthanize or sterilize coyote and hybrids and promote the formation and maintenance of red wolf pairs. By 2002, the plan was beginning to show significant progress and the Service intends to continue its implementation. However, even if the effort at ARNWR ultimately proves successful, the ubiquitous distribution of coyotes indicates that hybridization with that species will remain the central challenge to red wolf recovery.

Problems of red wolf recovery

Red wolves and wild ungulates

Few conflicts with humans have arisen since the red wolves were released. White-tailed deer (*Odocoileus virginianus*) are abundant in northeastern North Carolina and hunter harvest has remained heavy since red wolves were reintroduced.

Red wolves and livestock

Very few depredations have been reported or documented for red wolves. For example, by June 2002, only three cases of minor depredations involving a pet or farm animals have been documented despite exhaustive efforts to investigate every complaint. No cases of wolf-induced loss of livestock have been reported for the recovery area.

The future of red wolf recovery

Despite the challenges arising from hybridization, the ARNWR restoration project is showing success in the presence of intensive management of wolves and coyotes. Overall the project illustrates that the values and successes of reintroduction efforts often have the potential to extend beyond the immediate preservation of the reintroduced species, to positively affect local citizens and communities, larger conservation efforts, and other imperiled species as well (Phillips 1990). A study done by Cornell University concluded that on average the ARNWR red wolf project generated an annual regional economic benefit of about $37.5 million due to increased tourism (Rosen 1997). Public opinion polls conducted as part of the Cornell study and by North Carolina State University (Quintal 1995) revealed that the majority of local residents strongly favoured red wolf recovery in northeastern North Carolina. Such support derives partly from the ecological effects generated by red wolves. Local landowners credit red wolf predation on raccoons (*Procyon lotor*) as benefiting populations of bobwhite quail (*Colinus virginianus*) and turkey (*Meleagris gallopavo*) by reducing nest predation by raccoons. Food habits data and observations by local landowners reveal that red wolf predation on coypu (*Myocaster coypu*) has the potential to reduce nutria damage to water control levees. For these reasons and others, Rosen (1997) predicted that the public would strongly support and benefit from efforts to reestablish red wolves elsewhere. It seems likely that the red wolf could be recovered through the reestablishment of additional populations via reintroduction of captive-born animals if not for the species' predilection to hybridize with coyotes.

The intensive management required to restore red wolves by minimizing hybridization with coyotes poses an important ethical question: Is it legitimate to disadvantage one species (the coyote) for the sake of another species (the red wolf)? This question was, of course, carefully considered when the adaptive management plan was developed. It was noted then that the coyote was non-native to the southeastern US and the coyote that was invading northeastern North Carolina represented a mix of genes from domestic dogs, western coyotes, and wolves (US Fish and Wildlife Service unpublished data). Consequently the Service determined that actively selecting against coyotes and hybrids, by capturing and sterilizing or euthanizing them, was a legitimate measure for restoring the red wolf.

Such management actions are not unique to red wolf recovery. Coyote control programmes have been included in efforts to conserve the imperiled San Joaquin kit fox (*Vulpes macrotis mutica*) (Cypher and Scrivner 1992) and restore the imperiled swift fox (*Vulpes velox*) (Kunkel *et al.* 2003). Eradication efforts directed at non-native trout (i.e. rainbow trout (*Oncorhynchus mykiss*) and brook trout (*Salvelinus fontinalis*)) are important components of efforts to restore populations of imperiled native trout such as Gila trout (*Oncorhynchus gilae*) (Propst *et al.* 1992) or several subspecies of cutthroat trout (*Oncorhynchus clarki*) (Gresswell 1991; Young and Harig 2001; New Mexico Department of Game and Fish 2002).

Grey wolf

Recovery of the grey wolf in the Western Great Lakes States (Minnesota, Wisconsin, Michigan)

The first recovery plan was written for the eastern timber wolf in May 1978 (US Fish and Wildlife Service 1978). A revised plan was finalized in 1992 and included two delisting criteria (Table 19.1) (US Fish and Wildlife Service 1992b). The recovery plan for the eastern timber wolf includes no goals or criteria for the wolf population on Isle Royale, because it is not considered important in the long-term survival of the species. The population on the island is small (i.e. it usually includes 12–25 animals and has never included more than 50 wolves) and is almost completely isolated from other wolf populations (Peterson *et al.* 1998).

Various surveys conducted from the late 1950s to 1973 indicate that the Minnesota population did not exceed 1000 animals during that time (Fuller *et al.* 1992). After federal protection in 1974, its increase began accelerating and, by January 2003, the Minnesota population included over 2500 animals (Minnesota Department of Natural Resources 2001; Refsnider 2003).

Wolves were considered extirpated from Wisconsin by 1960 (Thiel 1993). Until the mid-1970s occasional sightings were reported but there was no evidence of reproduction (Wisconsin Department of Natural Resources 1999). In response to persistent reports of wolves, population monitoring was initiated in 1979. By 1997, the Wisconsin wolf population had exceeded the criterion for downlisting to threatened (i.e. 80 or more wolves present for three successive years). By January 2003, the population included over 300 animals.

The last known breeding population of wolves in Michigan (outside of Isle Royale) occurred there in the mid-1950s. While numbers continued to decline through the 1970s, it is likely that wolves were never completely extirpated from the State (Michigan Department of Natural Resources 1997). During the late 1980s, reports of wolves in Michigan's Upper Peninsula (UP) began to increase. A pair produced pups there in 1991. Since then, the Michigan population has increased and spread throughout the UP with immigration occurring from Wisconsin, Minnesota, and Ontario. By 1997, the Michigan population had exceeded the threshold for downlisting to threatened. By January 2003, the population included over 300 animals.

Growth of wolf populations in the Great Lakes region prompted the recovery team to modify delisting criteria to consider wolves in Wisconsin and Michigan as a single population. The 1993–94 late winter count of the Wisconsin–Michigan population was the first to exceed 100 wolves. Subsequent late winter counts have all exceeded 100. Moreover, the Minnesota population has included ≥1251 wolves since at least the late 1980s. Consequently, by 1999, delisting criteria for the eastern timber wolf had been met.

Recovery of the grey wolf in the Northern Rocky Mountains

In 1974, an interagency team was formed and completed the Northern Rocky Mountain Wolf Recovery

Plan (US Fish and Wildlife Service 1980). Revisions to the plan (US Fish and Wildlife Service 1987b, 2002) focused recovery on northwestern Wyoming, western Montana, and central Idaho. These areas are characterized by large tracts of public land, healthy populations of native ungulates, and relatively few livestock. The plan indicated that about 300 wolves would inhabit the region at the time of recovery (Table 19.1). The Plan promoted natural recovery for Montana and Idaho, unless two packs had not become established in Idaho by 1992, at which time reintroduction would be considered. The Plan recognized that the most certain way to restore wolves to the Greater Yellowstone Ecosystem was by reintroducing animals to Yellowstone National Park (YNP).

By the 1970s, dispersing wolves from Canada were travelling through northwestern Montana, and by 1982 a pack used Glacier National Park (Ream and Mattson 1982). In 1986, the first litter of pups in over 50 years was born there (Ream *et al.* 1985, 1989). By January 2003, about 108 wolves inhabit northwestern Montana (US Fish and Wildlife Service *et al.* 2003).

In November 1991, the Service was directed by Congress to prepare an Environmental Impact Statement (EIS) on wolf reintroduction to YNP and central Idaho. The final EIS was published in April 1994 and recommended reintroducing about 15 wolves from Canada to each area every year for 3–5 years (US Fish and Wildlife Service 1994). The final EIS also recommended that released wolves and their offspring be designated as members of experimental-nonessential populations (Bangs 1994). By July 1994, the Secretary's of Interior and Agriculture had signed a 'Record of Decision' effecting the final EIS as the federal government's official policy.

By the end of 1994, several lawsuits had been filed by wolf proponents and opponents that questioned the application of the experimental-nonessential designation. In December 1997, a Wyoming federal judge determined that the designation had been illegally applied and ordered the Service to remove the reintroduced wolves and their offspring (US District Court, Court of Wyoming, Civil No. 94-CV-286-D (lead case), Civil No. 95-CV-027-D, Civil No. 95-CV-1015-D (consolidated)). Given the ramifications of his determination, the order was stayed on its execution pending appeal. The appeal was settled in January 2000 as the 10th Circuit Court of Appeals (Denver, Colorado) reversed the Wyoming court order (US Court of Appeals, Tenth Circuit, Nos. 97-8127, 98-8000, 98-8007, 98-8008, 98-8009, 98-8011).

In January 1995, 15 wolves from Alberta, Canada were released in Idaho. In January 1996, 20 wolves from British Columbia, Canada were released in Idaho (Bangs and Fritts 1996; Fritts *et al.* 1997). These animals spawned a population that by January 2003 included about 284 wolves (US Fish and Wildlife Service *et al.* 2003). During March 1995, 14 wolves from Alberta were released in YNP. In January 1996, 17 wolves from British Columbia were released in YNP (Phillips and Smith 1996). Furthermore, due to a wolf control action in northwestern Montana, 10 pups were placed in an acclimation pen in the Park in late 1996. These pups and three adults from an earlier reintroduction were released in the spring of 1997. By January 2003, this population included about 271 wolves (US Fish and Wildlife Service *et al.* 2003).

The reintroduced wolves adapted better than predicted. Only 2 years of reintroductions were required to ensure population establishment rather than 3–5 years of reintroductions as predicted (Fritts *et al.* 1997). Compared to predictions in the EIS, the wolves have produced more pups, survived at a higher rate, and caused fewer conflicts with humans (Phillips and Smith 1996; Bangs *et al.* 1998; Smith *et al.* 1999; Fritts *et al.* 2001). Additionally, over 100,000 visitors to YNP have observed wolves (YNP unpublished data) and public interest in recovery remains high.

Recovery of the Mexican grey wolf in the southwestern United States

Between 1977 and 1980, five wolves were captured in the Mexican states of Durango and Chihuahua. These four males and one pregnant female were transported to the Arizona-Sonora Desert Museum to establish a captive breeding programme. Shortly thereafter it was widely accepted that the Mexican wolf was extinct in the wild. In 1979, the Service formed a Mexican Wolf Recovery Team that finalized a binational recovery plan with Mexico by 1982 (US Fish and Wildlife Service 1982). While the plan contains no downlisting or delisting criteria (Table 19.1), a new plan will.

Given the absence of wild Mexican wolves, captive breeding is of central importance to recovery. By December 2002, the captive breeding programme included 230 animals maintained at 43 facilities in the US (30) and Mexico (13).

By 1997, the Service had completed a plan for reintroducing about 15 wolves every year for up to five consecutive years in the Blue Range Wolf Recovery Area (BRWRA) (US Fish and Wildlife Service 1996a). The plan called for designating reintroduced wolves and their offspring as members of an experimental-nonessential population (Parsons 1998). The BRWRA encompasses 17,752 km^2 (6854 miles^2) of the Gila National Forest in New Mexico and the Apache National Forest in New Mexico and Arizona. The reintroduction aims to restore about 100 wolves.

The Service began reintroductions by releasing 11 wolves in March 1998. From then until July 2002, the Service released another 63 wolves on 83 occasions. A comprehensive review of the reintroduction project was completed in June 2001 and recommended continuation of the project with modification (Paquet *et al.* 2001a). Reintroductions and management actions though April 2003 have resulted in the establishment of a population of Mexican wolves comprised of eight known packs, including two that formed naturally, of 22–37 individuals. Many of these packs are now producing pups every spring, and the Service has documented the production of one litter of second generation wild-born Mexican wolves.

Recovery of the grey wolf elsewhere

Other regions in the US possess suitable habitat for grey wolves. Recovery planning, however, has not been developed for these areas.

Northeastern United States

Recent studies show that suitable habitat and sufficient prey exist for wolves in the northeastern US from New York to Maine (Fig. 19.1) (Harrison and Chapin 1998; Mladenoff and Sickley 1998). These studies indicate that 1000 or more wolves could inhabit the region. However, Paquet *et al.* (2001b) conclude that while Adirondack State Park in New York contains sufficient habitat to support a small population of wolves, regional landscape conditions and development trends are not ideal for sustaining wolves over a long period. Mech (2001a) countered this conclusion by arguing that active management could resolve the shortcomings cited by Paquet *et al.* (2001b). While there is a remote possibility that wolves from Canada might recolonize the northeastern US, recovery will probably require reintroductions (Wydeven *et al.* 1998).

Recent genetic work suggests that the eastern timber wolf may be a separate species from the grey wolf and more closely related to the red wolf (Wilson *et al.* 2000). Interestingly, Nowak's (2002) reported that the red wolf's historic range extended into Maine. Verification of this will create new legal, policy, and management questions regarding recovery (Fascione *et al.* 2001). Moreover, there is concern that the eastern timber wolf, like the red wolf, readily hybridizes with coyotes thus complicating recovery efforts (Theberge and Theberge 1998, pp. 233–234, 250–262). Nonetheless, several non-governmental conservation organizations are advocating the wolf's return to the northeast. Public opinion surveys indicate strong support for the idea (Responsive Management 1996; Downs and Smith 1998).

Southern Rockies ecoregion

This ecoregion extends from southcentral Wyoming through western Colorado into north central New Mexico and contains 100,000 km^2 (39,000 miles^2) of public land that supports healthy populations of native ungulates (Fig. 19.1) (Shinneman *et al.* 2000). The ecoregion contains almost 1.5–1.8 times more public land than is available to wolves in the Yellowstone area (64,000 km^2 or 25,000 miles^2) and central Idaho (53,200 km^2 or 20,781 miles^2), and 6 times the amount of public land available to Mexican wolves in the BRWRA (i.e. 17,752 km^2 or 6854 miles^2). Moreover, the Southern Rockies contain 1.7–25 times more habitat than do other sites that have been considered for wolf recovery (Ferris *et al.* 1999). Extensive tracts of public land in the ecoregion are managed in a fashion that could facilitate wolf recovery (Shinneman *et al.* 2000; Carroll *et al.* 2003). For example, the ecoregion contains about 36,000 km^2 (14,000 miles^2) that are roadless and about 18,000 km^2 (7031 miles^2) that are legally designated or de facto wilderness.

A 1994 Congressionally mandated study concluded that the Colorado could support over 1000 wolves (Bennett 1994). Three additional studies also conclude that the Southern Rockies could support a self-sustaining population of wolves (Phillips *et al.* 2000; Southern Rockies Ecosystem Project 2000; Carroll *et al.* 2003). Recovery will require reintroductions as there is little chance for wolves to do so through natural recolonization (Carroll *et al.* 2003).

Some believe that wolf recovery in the Southern Rockies could be especially significant when considered against a continental perspective. Because the ecoregion is nearly equidistant from the Northern Rockies and the BRWRA, it is possible that a Southern Rockies population, through the production and movement of dispersers, would contribute to the establishment and maintenance of a wolf population that extended from Canada to Mexico. On the significance of restoring the wolf to the Southern Rockies, Mech (1999a) wrote: 'Ultimately, then, this restoration could connect the entire North American wolf population from Minnesota, Wisconsin, and Michigan through Canada and Alaska, down the Rocky Mountains into Mexico. It would be difficult to overestimate the biological and conservation value of this achievement'.

Several non-governmental conservation organizations are advocating the wolf's return to the Southern Rockies. Regional public opinion surveys indicate that there is strong support for the idea (Manfredo *et al.* 1994; Pate *et al.* 1996; Meadows 2001).

Northwestern US

A Congressionally mandated study of the feasibility of restoring wolves to Olympic National Park, Washington concluded that an estimated 56 wolves could survive within the Park boundaries (Fig. 19.1) (Ratti *et al.* 1999). The authors, however, observed a number of potential problems with restoration and urged that further consideration proceed cautiously. Carroll *et al.* (2001a) determined that there was good potential for restoring wolves to the Pacific Northwest (Fig. 19.1). They caution that current development trends may quickly obviate the accuracy of their results. Because of their proximity to wolf populations in British Columbia and Alberta, Washington state's North Cascades and Selkirk Mountains offer some but limited potential for natural wolf recolonization. Moreover, wolves from Montana and Idaho will likely recolonize eastern Oregon and Washington. One radio-collared wolf from Montana was found dead from unknown causes in eastern Washington (Fig. 19.3). In early 1999, a radio-collared female dispersed from Idaho

Figure 19.3 Radio-collared grey wolf *Canis lupus* © W. Campell.

to Oregon. She localized movements before being recaptured and returned to Idaho. Two other wolves from Idaho have been found dead in eastern Oregon. It is likely that additional wolves will similarly disperse there since they are capable of travelling great distances (up to 886 km (532 miles) straight line) (Fritts 1983).

Southwestern US and northern Mexico

In May 2002, Carroll *et al.* (2002) began a comprehensive assessment of potential habitat, landscape-level threats, and population viability for Mexican wolves (and jaguars (*Panthera onca*)) across the southwestern US and northern Mexico (Fig. 19.1). This area encompasses the majority of the estimated historic distribution of the Mexican wolf. Such a comprehensive conservation assessment has not been attempted previously due to challenges associated with gathering consistent habitat data over such a large region spanning two nations, and lack of tools to link population dynamics to mapped habitat data at this scale. To resolve the latter problem, the study will use the programme PATCH (Schumaker 1998), which provides a means of building biologically realistic regional-scale population models. Study results combined with previous work for the Southern Rockies ecoregion (Carroll *et al.* 2003) will be useful for developing a regional-scale strategy for recovering the wolf in the southwestern US and Mexico.

Proposed recovery future for the grey wolf

By the end of 2002, wolf numbers and distribution in the contiguous US were greater than during the previous several decades. At this time, >3700 wolves occupied about 4% of the species' historic range in the 48 contiguous states (Fig. 19.1). The wolf populations in the Great Lakes States and the Northern Rocky Mountains had exceeded delisting criteria since 1999 and 2002, respectively. The Mexican wolf reintroduction project in the BRWRA was showing signs of becoming firmly established.

In response to the species' improved conservation status and continued public interest in wolf restoration, the Service finalized a reclassification rule or national recovery strategy (Refsnider 2003) that established three distinct population segments (DPSs) for the grey wolf (i.e. areas supporting wolf populations that are somewhat separated from one another, are significant to the overall conservation of the species, and are considered separately under the ESA). The highlight of the strategy was the determination that the grey wolf would be delisted in the US, except for the southwestern portion of the country (Arizona, New Mexico, southern half of Utah and Colorado, and the western half of Oklahoma and Texas), based on the recovered populations in the Great Lakes States and the Northern Rocky Mountains. In the southwest, the wolf remains endangered, and the Service is required to develop recovery efforts that result in establishment of populations adequate for delisting.

The Service's strategy was comprehensive, complex, and controversial. It did not satisfy everyone, and litigation has resulted. Since about 95% of the species' historic range in the 48 states is unoccupied, wolf advocates argue that it is inappropriate for the strategy to not require the Service to restore more wolves to more places. Since wolf populations have significantly increased throughout select regions of the contiguous US, wolf opponents argue that it is appropriate for the strategy to not require the Service to restore more wolves to more places.

Disagreements over the specifics of the Service's strategy are to be expected and pivot on the federal government's responsibilities under the ESA. Unfortunately, the Act does not define the term recovery. On this matter the Service policy states: 'The goal of this process (recovery) is to restore listed species to a point where they are secure, self-sustaining components of their ecosystem and, thus, to allow delisting' (US Fish and Wildlife Service 1996b, p. 2). Some believe that recovery should be the establishment of functional densities of the species over a significant portion of suitable habitat within the species' historic range (Rohlf 1991; Tear *et al.* 1993; Shaffer and Stein 2000; Ninth Circuit Court of Appeal 2001). The ESA also includes the phrase 'significant portion of range' in the definitions for both an endangered species (any species which is in danger of extinction throughout all of a significant portion of its range) and threatened species (any species which is likely to become an endangered species within the foreseeable future throughout all or a significant portion of its range). Currently there is no accepted biological or legal standard for defining a 'significant portion of range'.

Problems of grey wolf recovery

As wolves become more common, conflicts with humans can increase in frequency, complexity, and seriousness. This is so because wolves prey on wild ungulates and sometimes on livestock, all of which are important resources for state wildlife managers, special interest groups, landowners, and private citizens.

Grey wolves and wild ungulates

To date, few conflicts have arisen over interactions between wolves and native, wild ungulates (cervids). In the western Great Lake states, where white-tailed deer are the primary prey, wolf predation will not usually negatively affect hunter harvest (Michigan Department of Natural Resources 1997; Wisconsin Department of Natural Resources 1999; Mech and Nelson 2000; Minnesota Department of Natural Resources 2001).

Figure 19.4 Grey wolf *Canis lupus* in Minnesota © D. W. Macdonald.

In the Northern Rockies, where elk and mule deer are also important prey and recovery has occurred in areas that support cougars (*Puma concolor*), bears (*Ursus* spp.), and coyotes, wolf predation, as one of many mortality factors affecting cervid survival, may negatively affect hunter harvests (Kunkel and Pletscher 1999). This issue was a central concern to the public regarding wolf recovery in the region. The EIS predicted that in the GYA (Greater Yellowstone Area) wolf predation may reduce elk and deer populations in some herds by 5–30% and 3–19%, respectively (US Fish and Wildlife Service 1994). Such reductions might prompt a reduction in the number of permits issued to hunters. To date, however, wolves in the GYA have apparently not effected a reduction in elk numbers (Smith *et al.* 2003a). However, wolves have most extensively preyed upon elk that inhabit YNK northern range. By winter 2002–03 this herd was estimated to include 12,000 animals which is 13% less than the 25-year average (1976–2001) of 13,890 (Smith *et al.* 2003a). We caution that the effects of wolf predation on elk cannot be fully understood after only a few years of study. Nonetheless, we believe that it is reasonable to expect lower cervid populations that remain low for extended periods where wolves, bears, cougars, coyotes, and humans vie for the same prey (Kunkel and Pletscher 1999) and where winter weather can greatly affect ungulate population dynamics (Mech *et al.* 2001).

The relationship between Mexican wolves and wild ungulates in the arid southwest is unclear. Clarity would require intensive monitoring of ungulates and reintroduced wolves and their wild-born offspring over an extended period.

Grey wolves and livestock

Conflicts between wolves and livestock or pets have occurred and they have been controversial and complex (Mech 1995, 1996, 1999a, 2001b; Clark *et al.* 1996; Mech *et al.* 1996, 2000; Phillips and Smith 1998). Even though wolf depredations are relatively uncommon, the public demands immediate and certain action when problems arise. For example, as the Minnesota wolf population increased during the last three decades, the number of wolves killed to resolve conflicts with livestock increased from 21 animals in 1980 to 216 in 2000 (Paul 2001). During those 21 years, 1875 wolves were killed. From 1987 through

July 2002, almost 125 wolves have been killed in control actions in the Northern Rockies. Interactions between wolves and livestock or pets have been a problem for the nascent population of Mexican wolves in the BRWRA. At a minimum, over half of the free-ranging Mexican wolves have been involved in such interactions (Paquet *et al.* 2001). Resolution of these conflicts is a common reason for Mexican wolves to be recaptured for re-release or permanent placement in captivity.

The frequency of wolf control belies the actual magnitude of the wolf–livestock problem. For example, only about 1% of farms in wolf range in Minnesota suffer verified wolf depredations (W. J. Paul, unpublished report, 1998 as cited by Mech *et al.* 2000). Similarly, in the Northern Rockies, average annual confirmed losses have been slight: 4 cattle and 28 sheep (and 4 dogs) in the GYA and 9 cattle and 29 sheep (and 2 dogs) in Idaho. These rates are one-third to one-half of the rates predicted in the EIS (US Fish and Wildlife Service 1994). Since 1987, in northwestern Montana, wolf depredations averaged six cattle and five sheep (and less than one dog) annually. In contrast, livestock producers in Montana annually reported losing annually an average of 142,000 sheep and 86,000 cattle to all causes between 1986 and 1991 (Bangs *et al.* 1995). While it is certain that far more livestock are lost to wolves than are verified (Roy and Dorrance 1976; Fritts 1982; Bangs *et al.* 2001), it is equally certain that wolf depredations have little effect on the economics of the livestock industry. Nonetheless, if not addressed quickly, wolf depredations can cause significant losses for individual producers and create great animosity towards wolf recovery. Many livestock producers have cooperated with recovery because they believe that wolf-induced problems will be resolved equitably. Monetary compensation for livestock losses has proven useful in this regard and for minimizing animosity towards wolves (Fischer 1989; Fischer *et al.* 1994).

The tension between promoting wolf survival and population expansion and killing wolves to resolve conflicts with humans has complicated wolf recovery. With the exception of lethal control, most approaches for resolving conflicts seem to be ineffective, cost-prohibitive, and/or logistically unwieldy when applied over a large scale (Cluff and Murray 1995; Mech *et al.* 1996). This reality prompted Mech (1995, p. 276) to observe that: 'Because wolf-taking by landowners or the public is the least expensive and most acceptable to people who do not regard the wolf as special, there will be greater local acceptance for wolf recovery in areas where such control is allowed. Thus, if wolf advocates could accept effective control, wolves could live in far more places'.

Conclusions

The conservation statuses of the red wolf and grey wolf have greatly improved since the 1950s when three centuries of intense persecution began to end as both species approached extinction in the contiguous US. This improvement is a direct result of science-based planning and implementation of recovery activities under the authority of and impetus provided by the ESA.

Progress notwithstanding, habitat loss continues to accelerate, further reducing the suitability of most areas to support wolves (Carroll *et al.* 2001, 2003; Paquet *et al.* 2001b). Moreover, coyotes are now firmly established throughout the US causing additional challenges to red wolf recovery and probably grey wolf recovery in the northeastern US. Nonetheless, significant credit is due to citizens, non-governmental conservation organizations, elected and appointed officials, state and tribal governments, livestock producers, and the federal government for recognizing the importance of recovering the red wolf and grey wolf, controversial but vitally important components of North America's natural heritage.

CHAPTER 20

Ethiopian wolves

Afroalpine ecology, solitary foraging, and intense sociality amongst Ethiopian wolves

Claudio Sillero-Zubiri, Jorgelina Marino, Dada Gottelli, and David W. Macdonald

Ethiopian wolf *Canis simensis* © C. Sillero.

Introduction

At *c.*20 kg, the Ethiopian wolf (*Canis simensis*) differs from such typical, medium-sized canids as the coyote (*Canis latrans*) in its unusually long legs and a long muzzle (Sillero-Zubiri and Gottelli 1994). Restricted to rodent-rich Afroalpine habitat within the Ethiopian highlands, its diurnal habits and distinctive coat render this species conspicuous. A bright tawny rufous fur, with a characteristic pattern of white marks, a thick black and white bushy tail, and broad, pointed ears result in a rather 'foxy' appearance. This, and its reliance upon small prey, misled early European naturalists to name this species the Simien fox. Uncertainty over its taxonomy led to an array of alternative vernacular names, including the Simien jackal, Abyssinian wolf, and *ky kebero* (which translates from Amharic to red jackal). Phylogenetic analysis based on mitochondrial DNA (mtDNA), however, have shown that Ethiopian wolves are indeed more closely related to grey wolves (*Canis lupus*) and coyotes than to any African canid (Gottelli *et al.* 1994). This is supported by their many similarities with wolves in biology and behaviour, and by the relative ease with which they hybridize with domestic dogs (Gottelli *et al.* 1994; Wayne and Gottelli 1997).

Unlike other medium- to large-sized canids, which typically are generalist predators and widely distributed (Ewer 1973; Macdonald 1992b), Ethiopian wolves combine conspicuous sociability with specialized, solitary foraging for a narrow range of Afroalpine rodents. Today, these wolves are confined to Afroalpine pockets in a handful of Ethiopian mountains, and total no more than 500 individuals, distributed in 7 small isolated populations (Fig. 20.1; Marino 2003). One of at least nineteen species of mammals restricted to the Afroalpine grasslands and heathlands of Ethiopia (Yalden and Largen 1992), Ethiopian wolves have only persisted in their fragmented habitat because of the sheer size of Ethiopia's mountain massif, which comprises 80% of Africa's land above 3000 m above sea level (a.s.l.) (Yalden 1983; Malcolm and Ashenafi 1997). The dominant herbivores in these high altitudes are rodents, adapted to the extreme diurnal temperature fluctuations, and these are the main prey of Ethiopian wolves (Sillero-Zubiri and Gottelli 1995a; Sillero-Zubiri *et al.* 1995a,b). As top predators of the Afroalpine ecosystem, Ethiopian wolves attain densities as high as 1.2 animals/km^2 in prime habitats (Sillero-Zubiri and Gottelli 1995b) and adult wolves have no known predators except man.

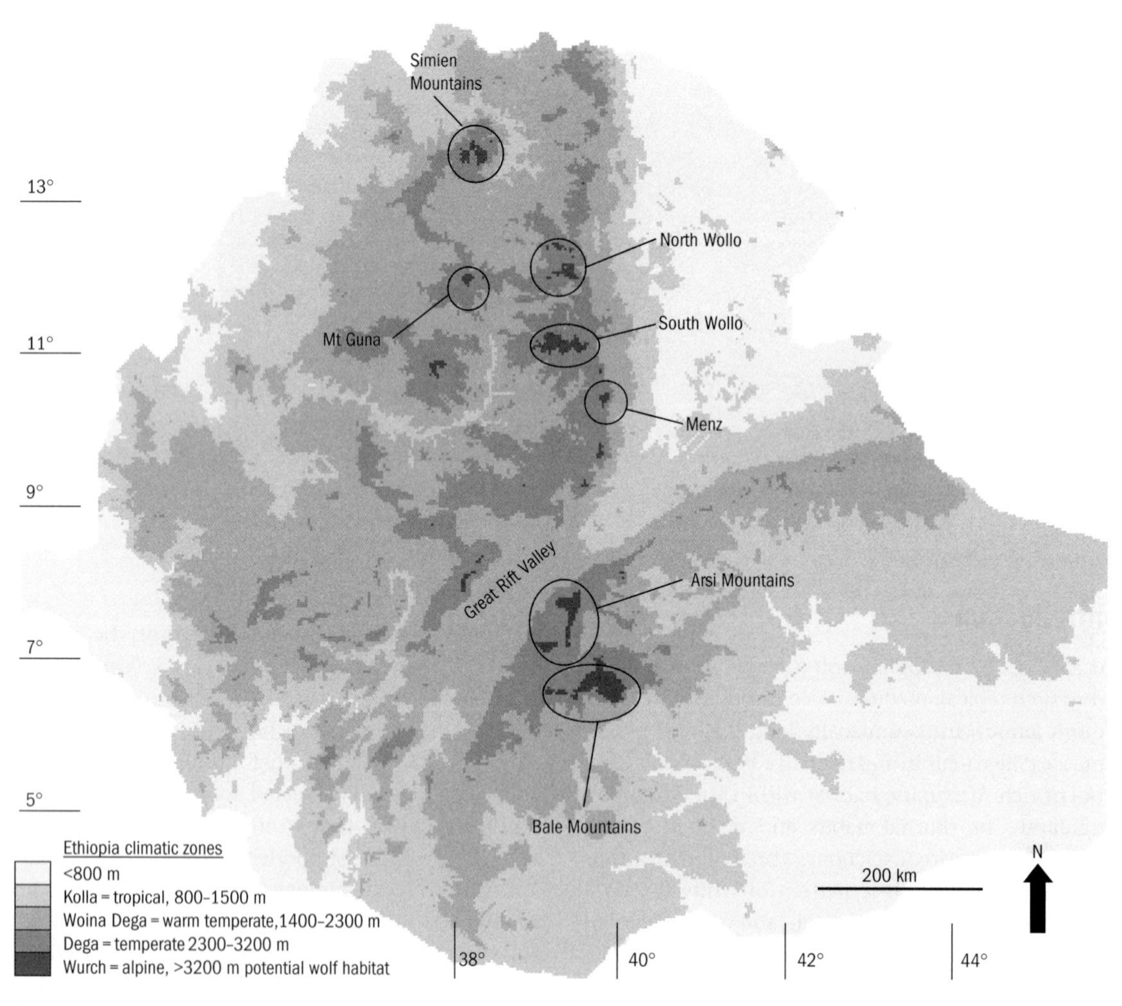

Figure 20.1 Distribution of extant wolf populations, with altitudinal climatic zones. The Afroalpine habitats shown in darker grey broadly correspond with the area of potential suitable habital for wolves.

The species may have originated from a grey wolf/coyote-like ancestor that invaded Ethiopia during the Pleistocene. During this period, Europe and Africa were connected by land bridges and alpine habitats formed a continuum that extended through Eastern Europe, the Middle East and Northeast Africa (Kingdon 1990). During the last glacial period (70,000–10,000 years BP) the African tropics were generally cooler and drier than at present (Bonnefille *et al.* 1990) and the hypothetical ancestor may have been pre-adapted to the cold-temperate Ethiopian highlands. There, the incoming canid specialized on the small mammals, particularly molerats (Rhyzomidae) and grass rats (Muridae) that filled the niche of the large grazing ungulates characteristic of the African plains.

The end of the Pleistocene brought a change in the climate, and the extensive Ethiopian Afroalpine steppes shrunk to their present state, reducing the habitat available to Ethiopian wolves by an order of magnitude (Gottelli and Sillero-Zubiri 1992; Gottelli *et al.* in press). Analyses of microsatellite and mtDNA have suggested that small population sizes may have characterized the evolution of Ethiopian wolves. The low number of mtDNA haplotypes and the low sequence divergence between them reflect a recent evolution of Ethiopian wolves; coalescence was estimated at just over 100,000 years ago (Gottelli *et al.* in press). Indeed, the Ethiopian wolf appears to have the most limited genetic variability at the population level of any extant canid (Wayne and Gottelli 1997).

An instructive comparison is with the cheetah (*Acinonyx jubatus*), whose range may also have undergone a dramatic contraction at the end of the Pleistocene. The current cheetah population is estimated at 15,000 (cf. 500 Ethiopian wolves) and the control region of their mtDNA includes at least eight mutations among the most distinct haplotypes (Freeman *et al.* 2001) compared with only three mutations in the control region of the most distinct haplotypes of the Ethiopian wolf.

The genetic patterns observed in modern Ethiopian wolves seems to be a relict of a late Pleistocene expansion into Afroalpine habitats; fragmentation and genetic drift over the last 10,000 years resulted in local loss of genetic variability, but the species as a whole conserved its genetic variability. Ironically, the specialization on Afroalpine rodents that was once the basis of the species' success is now the force that constrains Ethiopian wolves to a fragmented habitat (Kingdon 1990; Yalden and Largen 1992), and heightens the risk of local extinctions in the face of stochasticity and anthropogenic factors (Sillero-Zubiri and Macdonald 1997).

Our field studies of Ethiopian wolves began in 1988, with a focus in the Bale Mountains. Conservation and research activities continue in Bale and recently expanded to other populations in Ethiopia. In the following sections we draw on data presented in Sillero-Zubiri (1994), Sillero-Zubiri and Gottelli (1995a,b), Sillero-Zubiri and Macdonald (1998), Sillero-Zubiri *et al.* (1995a,b, 1996a,b, 1998).

Study area

We established three study areas: Web Valley (3450 m a.s.l.), Sanetti Plateau (4000 m), and Tullu Deemtu (3800–4300 m) in the central massif of the Bale Mountains National Park (BMNP) of southern Ethiopia (7°S, 42°E). BMNP is the largest realm of Afroalpine habitat in Africa, spanning over 1000 km^2 and harbouring over half of the global Ethiopian wolf population. The first two study areas represented typical open-grassland Afroalpine habitat and sustained the highest wolf densities (*c.*1.2 wolf/km^2). Tullu Deemtu was characterized by *Helichrysum* dwarf-scrub, also a common habitat type, which sustained a much lower wolf density (*c.*0.25 wolf/km^2).

The solitary wolf as the top predator of the Afroalpine rodent community

Diet

The diet of Ethiopian wolves was studied by scat analysis (689 droppings) and 946 h of watching focal animals that yielded 811 attempts to kill prey, of which 361 corresponded to successful kills/feeds (Sillero-Zubiri and Gottelli 1995a). Rodents accounted for 96% of all prey occurrences in droppings and 97% by volume of undigested faecal material. Wolf prey included six rodent species, Starck's hare (*Lepus starki*), cattle, birds, insects, and undigested sedge leaves, *Carex monostachya*. Giant molerat (*Tachyoryctes macrocephalus*—mean weight 618 g) was the main component in the overall diet (36% of total prey occurrences) and was present in 69% of all faecal samples, whereas diurnal rats *Arvicanthis blicki*, *Lophuromys melanonyx*, and *Otomys typus* (respective mean weight 126, 94, and 100 g) together accounted for 59% of occurrences and appeared in 78% of the samples. These four species together accounted for 86% of prey occurrences and no significant differences were found for these main four prey items between months or between dry and wet seasons.

Direct observations indicated a higher incidence of large prey (hare, rock hyrax *Procavia capensis capillosa*, birds, lambs, and antelopes) than suggested by scat analysis. Of all feeding instances observed, 69% were grass rats while giant molerat kills accounted for 22% of all successful attempts. Giant molerats formed the bulk of the prey by weight (40%), while diurnal rats were second (23%), although taken more often. Carrion, hares, hyraxes, and birds contributed the remaining 36.5% of the total prey weight, of which 12% was scavenged from livestock carcasses.

The diet was broadly similar at the three sites, with giant molerat as the single most important food item. In areas where this species is absent or rare, it is often replaced by the common molerat (*Tachyoryctes splendens*). For instance, in Bale's northern montane grasslands, common molerats constituted 32% of all animals eaten (Malcolm 1997), and in Menz, central Ethiopia, 31% of occurrences—17% by volume—in the wolf diet (Ashenafi 2001). Recent analysis of faeces from all other wolf populations revealed a diet amply dominated by rodents, even where molerats were absent (Marino 2004). This study found livestock remains at very low frequency in the northern highlands, where livestock abundance was high and rodent abundance low.

Foraging behaviour

During 946 h of focal observation away from dens, wolves spent 43% of their time foraging. They foraged solitarily throughout the day, travelling widely at a walk or trot, covering large areas of their home range. Peaks of foraging activity were synchronized with the activity of diurnal rodents above the ground. The wolves used various hunting strategies: molerats were commonly stalked, while zigzag and hole-checks were aimed at grass rats. Although foraging wolves were mostly observed alone, their daily hunting ranges overlapped considerably. Of 35 occasions in which more than one wolf was present during kills involving rats, only 23% were within 10 m. In the remaining observations, wolves shared the same foraging area, but did not appear to interfere with each other's foraging attempts or prey captures. Occasionally small packs hunted hares, antelope calves, and sheep. In 12 of 20 attempts to catch hares, 2–4 wolves hunted simultaneously. In the northern grasslands, wolves have been observed in packs of 3–4 animals hunting reedbuck *Redunca redunca* ($n = 3$) and a mountain nyala calf *Tragelaphus buxtoni* (Sillero-Zubiri personal observation) (Fig. 20.2).

Figure 20.2 Ethiopian wolves with pups © C. Sillero.

Rodents as predictors of Ethiopian wolf distribution

The role of the Afroalpine rodent community in limiting the distribution of Ethiopian wolves was studied in Bale by looking at the relationship between wolf abundance, and the species composition, relative abundance, and activity pattern of the rodent community in various habitats (Sillero-Zubiri *et al.* 1995a,b). Combined biomass of all diurnal rodents and hares in the Afroalpine grassland habitats was estimated at 24 kg/ha in Sanetti and 26 kg/ha in Web, with giant molerats contributing about half of this biomass (we assumed that an average molerat weighed 618 g (n = 11), and occurred at a biomass of 10–25 kg/ha, with patches of up to 55 kg/ha); hares averaged 2250 g (n = 4), giving a projected biomass of 0.4–0.7 kg/ha. Although hares were more conspicuous on the ground than were rodents, they accounted for only a small fraction of the total potential prey biomass.

Indices of giant molerat biomass for *Helichrysum* dwarf-scrub and ericaceous belt were only 1/5 and 1/150, respectively, of those in Afroalpine grasslands. Positive correlations between wolf density and molerat abundance in four areas (Tullu Deemtu, Sanetti, Web, and the ericaceous belt) suggested that molerats were a vital determinant of wolf presence (Sillero-Zubiri *et al.* 1995b). Because they are roughly six times the weight of any other rodent, hunting *T. macrocephalus* is likely to be considerably more efficient than hunting a smaller species. Nonetheless, the positive correlation between wolf abundance and an index of biomass of smaller rodents showed that the giant molerat was not the only determinant of wolf distribution. The biomass index for grass rats (in kilograms per 100 transect snap-trap nights) was highest on Sanetti Plateau, followed, in order, by Web Valley, montane grasslands, the ericaceous belt, and Tullu Deemtu (Table 20.1). *Arvicanthis blicki* and *L. melanonyx* were the most numerous species in Afroalpine grasslands. Ethiopian wolf density, measured both from observation and road counts, correlated positively with the total biomass index and the biomass index for diurnal species, but not for nocturnal species. Also a positive correlation was detected between rodent burrows and wolf signs (droppings or diggings) along habitat assessment transects. A similar correlation was found between wolf signs and the average index of fresh giant molerat signs (Sillero-Zubiri *et al.* 1995a,b).

Large mammal densities in the Afroalpine grasslands are low and, anyway, they might be largely unavailable to the wolves. Rodents were the most abundant, conveniently sized prey, and easiest to

Table 20.1 Ethiopian wolf density (individuals/km^2) and biomass index, weighted for sub-habitat area, for diurnal and nocturnal snap-trapped rodent prey

	Web Valley	Sanetti Plateau	Montane Grassland	Tullu Deemtu	Ericaceous Belt
Biomass index					
Diurnal rats	2.7	2.9	1.6	0.4	0.4
Nocturnal rats	1.8	2.1	1.2	1.4	1.7
Total	4.4	5.0	2.8	1.8	2.1
Ethiopian wolf density					
Road counts	1.0	1.2	0.1	0.2	0.1
Observation	1.2	1.2	0.3	0.2	0.1
Pack home ranges	6.5 ± 2.1	5.5 ± 1.3	7.4	13.4 ± 2.0	–
Group sizes	6.7 ± 0.7	4.9 ± 0.3	4.5 ± 0.3	2.6 ± 0.4	–

Note: The biomass index represents the biomass (kg) contributed per 100 trap nights using data from all months. Mean weights used as follows: *Arvicanthis blicki*: 126 g; *Lophuromis melanonyx*: 94 g; *L. flavopunctatus*: 49 g; *Stenocephalemys griseicauda*: 101.5 g; *S. albocaudata*: 129.5 g; *Otomys typus*: 100 g. Home ranges (km^2 ± SD) were estimated as average 100% minimum convex polygons of wolf packs in Bale between 1988 and 1991. Group size is the average number of adult and subadults (mean ± SE) in a pack.

catch. Their availability was more predictable, insofar as their abundance was closely associated to different habitat types. The predictability of the rodent prey may be one selective pressure favouring pack territoriality (Sillero-Zubiri 1994).

Wolf packs carve out the precious suitable habitat available

Pack home ranges and rodent biomass

While the Afroalpine rodent fauna constitutes a very rich and predictable source of food, the availability of Afroalpine habitats is limited by its geographical distribution. In Bale, all areas supporting a substantial rodent biomass were occupied by resident wolf packs.

Wolves were organized into discrete groups, and their composition was spatially and temporally stable. Groups were composed of 2–13 adults and subadults (>1 year old). Average group size for all 14 known packs in Web and Sanetti between 1988 and 1992 was 5.9 ± 0.5 (mean ± SE), with Web packs significantly larger than Sanetti's (Table 20.1). Tullu Deemtu packs were notably smaller and averaged 2.6 ± 0.4.

Home ranges of resident wolves overlapped almost completely with other pack members and entirely contained the home ranges of pups and juveniles (81–87% intragroup annual home range overlap between adult–adult and adult–subadult dyads, n = 4 packs). Home ranges of individual residents ranged between 2.0 and 15.0 km^2 (n = 92) and most of this variability was attributed to habitat type (Table 20.1). For instance, combined home ranges (i.e. pack home ranges estimated as minimum convex polygons) in Afroalpine grasslands averaged 6.5 km^2 ± 2.1 and 5.5 km^2 ± 1.3 (Web and Sanetti, respectively, n = 7 packs), while in *Helichrysum* dwarf-scrub, home ranges where twice as large and explained by the different density of prey species (Table 20.1). On the other hand, the home ranges of three non-resident *floater* females (sensu Sillero-Zubiri *et al.* 1996a) in Web overlapped widely with other packs and ranged from 8.5 to 18.7 km^2, their mean range being significantly larger than those of resident dominant females.

Small ranges, particularly those recorded in the grasslands and herbaceous communities of Web and Sanetti, are a reflection of the great density of the food resources available in some Afroalpine habitats. The ranges observed are among the smallest, and density among the highest reported for all eight *Canis* species (reviewed in Ginsberg and Macdonald 1990). Established relationships between metabolic rate, body weight, and size of home range in mammals (McNab 1963; Harestad and Bunnell 1979) would predict home ranges of 42 km^2, nearly eight times the mean values observed in Web and Sanetti.

Home range size was correlated with group size in the Afroalpine grasslands of Bale, and territories were enlarged whenever a reduction in group size in a neighbouring pack allowed it, which is indicative of an expansionist strategy (Kruuk and Macdonald 1985; Macdonald *et al.* Chapter 4, this volume). Under intense competition for rodent-rich grasslands, pack group size determine the outcome of territorial boundary clashes and the maintenance of a high quality range may be the greatest advantage of group-living (Sillero-Zubiri and Macdonald 1998).

Marking and territoriality

Studies of scent-marking behaviour and inter-pack aggression in Ethiopian wolf packs provided detailed evidence of territoriality (Sillero-Zubiri and Macdonald 1998). Movements and activity at the periphery of ranges was characterized by 'border patrols' during which groups of pack members of both sexes trot and walk along the territory boundary. In 167 km of border patrols totalling 68 h, 1208 scent marks were deposited at an overall rate of 7.2/km. Raised-leg urinations were the most frequently deposited scent mark (4.7/km), followed by ground scratching (2.3/km). Defecations and squat urinations during border patrols were rare (0.23/km and 0.04/km, respectively). Scent-marking rates were highest along or near territory boundaries (mean number of scent-marks deposited per kilometre significantly greater ($F_{(1,313)}$ = 6.40, P = 0.012) during patrols than at other times), where wolves vigorously over-marked neighbours' scent-marks. Most direct encounters between neighbouring wolves at territory borders were aggressive and involved repeated chases (102 out of 119 encounters) and the larger group was most likely to win (the larger group won in 77% of cases, whereas victorious and defeated groups were the same size in 15% of encounters).

In Bale, Ethiopian wolf packs occur at saturation density, in a system of highly stable tessellated territories (Sillero-Zubiri and Gottelli 1995b). Frequent scent-marking, inter-pack encounters and aversion to strangers' marks probably constrain each pack to its territory, while positive feedback keeps each territory boundary adequately marked. A further function of scent-marking may be to indicate sexual and social status. Wolves in Bale are seasonal breeders and in any given year, mating was synchronized to a period of 1–3 weeks in the latter part of the rainy season (August–October), suggesting that a social mechanism triggered mating (Sillero-Zubiri *et al.* 1998). Scent-marking might allow females to monitor their reproductive condition reciprocally, and synchronize their oestrus. On the other hand, neighbouring packs' males may gather information on the receptivity of females. While border encounters occurred throughout the year, peak intrusion pressure coincided with the mating season. Fifty out of 169 observed encounters, between wolves of neighbouring packs, consisted of territorial intrusions by small groups of neighbouring males, attracted by a receptive resident female. Highly seasonal mating may be connected to the occurrence of a philandering mating system in Ethiopian wolves (Sillero-Zubiri 1994; Sillero-Zubiri *et al.* 1996a).

Philopatry and the risk of inbreeding

Dispersal and philopatry

Lack of suitable habitat places a tight constraint on dispersal in Ethiopian wolves. In Bale, immigration was rare, births and deaths predominated over transfers between packs, and all pack members were potentially close kin. With kin of opposite sex residing in the same group, natal philopatry provided the potential for inbreeding, in a situation of severely limited dispersal opportunities (Sillero-Zubiri 1994).

Although dispersal was rare, that which occurred was sex biased. In Web and Sanetti, 63% of females dispersed at, or shortly before, sexual maturity at 2 years, some becoming floaters (Sillero-Zubiri *et al.* 1996a). Males did not disperse and were recruited into multi-male philopatric packs (pack fission, however, acts as short-distance male dispersal—see below). The population sex-ratio of adults was biased towards males at 1.9 : 1 ± 0.07 (SE), with the mean pack sex-ratio of adults at 2.6 : 1 ± 0.2.

Female breeding slots are the most coveted

Adaptive explanations of single-sex dispersal include avoidance of reproductive competition—for breeding status or resources (Dobson 1982)—or of inbreeding (Harvey and Ralls 1986; Wolff 1992). In Ethiopian wolves, observations of same-sex aggression prior to female dispersal support the competition-for-breeding-status hypothesis. In Bale, only the dominant female in each pack bred, indicating a high level of reproductive competition. Each breeding female was clearly dominant over her daughters. During the study period 1988–92, in all packs with more than one subordinate female, the lowest ranking female left the group at 18–28 months old. Fourteen subordinate females emigrated or disappeared from focal packs, whereas only four entered a different pack and two returned to their natal group, suggesting that approximately 57% of dispersing females either died or failed to find residence in the study population. Of the 14 females known prior to dispersal, 10 settled as floaters next to their natal territory.

No new packs were formed during the 4-year study. An apparent attempt by a subordinate female to split a pack—suggesting that fission could be a mechanism for pack formation—ended when the subordinate's litter succumbed, probably killed by the dominant female. During this period, female ascendancy to breeding status, either by immigration or inheritance, only occurred after the death of a dominant and so the chances of a female ever securing a breeding position were low. Five out of ten dominant females retained their breeding position throughout 4 years of observation, whereas the remainder maintained that role till they died. Breeding openings occurred at an average of 0.12 ± 0.09 opportunities for a subordinate female per year per pack. During contests for a breeding position, resident females appeared to have an advantage over floaters (three breeding females were replaced by their daughters after their deaths) (Fig. 20.3).

In late 1991, a rabies epidemic decimated the population in Bale and resulted in the disintegration of three out of five packs in Web, causing the sudden

Figure 20.3 Dominant and subordinate females with pups © C. Sillero.

opening of potential breeding opportunities (Sillero-Zubiri *et al.* 1996b). Rather than forming smaller, new breeding units, the surviving packs maintained their social cohesion and readjusted their territorial boundaries to occupy the habitat available (Marino 2004). Only 5 years after the die-off did new packs started to form, in one case by the fission of a large pack, and twice by the grouping of dispersing individuals, mostly from neighbouring packs. The successful establishment of new groups may depend not only on the availability of high-quality territories—virtually constrained by the 'expansionism' of the surviving packs—but also on the presence of a sufficient number of helpers to ensure the defence of a new territory and/or successful reproduction (Marino 2004). Despite the potential breeding vacancies opened to subordinate females after the population reduction, social factors involved in the formation of new breeding units led to inverse density dependent population growth at low densities (Marino 2004).

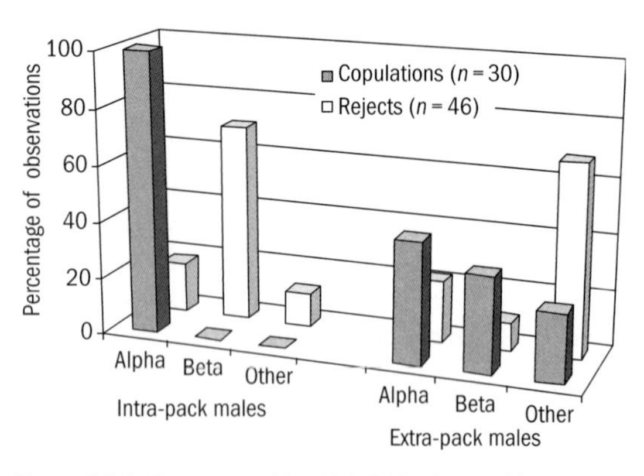

Figure 20.4 Frequency with which Ethiopian wolf females were observed in sexual encounters with males from their packs or neighbouring packs.

Extra pack copulations and multiple paternity

During the short mating season, the dominant female exercised choice in accepting when and with which male she mated. Of 30 observed instances of mating that involved copulation, only 9 (30%) took place with males from the female's pack, whereas the other 21 (70%) involved males from other packs (Fig. 20.4). Within packs, females copulated only with the dominant male, and rejected all mating attempts by lower ranking males. In contrast, mate choice with regard to male status was not apparent when a female courted and mated with outside males. Microsatellite DNA analysis confirmed the occurrence of multiplepaternity in two litters (Gottelli *et al.* 1994). Multiple paternity and male excursions into neighbouring territories during the mating season suggest that this is an important tactic for male wolves. Cuckoldry and multiple paternity may rival male philopatry and female-biased dispersal in importance as an out-breeding mechanism in a situation where habitat constraints impede dispersal. An alternative, but non-exclusive, explanation may be the prevention of infanticide from neighbours who, in this competitive milieu, could benefit by killing the offspring of neighbouring packs (Sillero-Zubiri 1994; Wolff and Macdonald 2004).

Cooperative breeding

Role of helpers at the den

All wolves that stayed in the natal pack helped to rear the litter of the dominant female, guarding the den, chasing potential predators, and regurgitating or carrying rodent prey to feed the pups (Sillero-Zubiri 1994). Given the high degree of relatedness among group members, subordinate wolves may increase the indirect component of their inclusive fitness by acting as helpers (Creel and Waser 1991) or, if competition within the group is intense, subordinates may be induced to help as a *payment* for remaining in the territory. Ethiopian wolf males helped throughout their lives and never dispersed; the dominant male at least shared paternity whereas subordinate males appeared generally to have no probability of fathering the pups, nevertheless they still helped. Subordinate females helped more intensely than did males for 1 or 2 years before dispersing or inheriting the breeding position. The balance of costs and benefits to all participants in a cooperative breeding system have been widely debated (e.g. Macdonald and Carr 1989; Emlen 1991; Solomon and French 1997). One aspect of the debate is whether groups containing many helpers deliver more food and care to the young than do smaller groups.

The development of the young was divisible into three broad stages (Sillero-Zubiri 1994). First, early denning (birth to 4 weeks), when the pups are confined to the den and are entirely dependent on milk. Second, mixed nutritional dependency (week 5 to week 11), when milk is supplemented by solid foods such as rodents provisioned by all pack members until pups are completely weaned. Third, post-weaning dependency (week 12 to 6 months), when the pups subsist almost entirely on solid foods supplied by breeders and non-breeding helpers. Juveniles were considered independent after 6 months, when they ceased receiving appreciable quantities of food from adults. A juvenile became subadult and was *recruited* at 1 year of age (Fig. 20.5).

Although the mother and putative father spent more time at the den on average than did other wolves, some non-breeders spent more time at the den than did the breeders themselves. The proportion of time pups were left unattended declined significantly as the number of helpers in the pack increased. Pack size may thus influence anti-predator behaviour, because babysitters were active in deterring and chasing potential predators. Unattended young might be taken by spotted hyaenas (*Crocuta crocuta*), domestic dogs, honey badgers (*Melivora capensis*) and eagles (*Aquilla verreauxi, Aquilla rapax*). However, there was no evidence that increases in pack size resulted in measurable increases in number of pups at any age.

Figure 20.5 Three month old pups © J. Marino.

Observations were made of nine Ethiopian wolf packs during the breeding season to quantify the amount of solid food provisioned to pups. Non-maternal food provisioning for 17 litters constituted 478 of 713 feedings (67%) observed other than nursing. Independent of the number of donors, there were significant differences in the rate of food provisioning (contributions per hour) by individuals of different breeding status, sex or age ($F_{(5,119)} = 9.08$, $P < 0.0001$; Table 20.2). Breeders contributed significantly more food than did non-breeders, and females more than males. The contributions by breeding females were greater than those by any other wolves (up to 0.3 contributions per hour). Dominant males made the second largest contribution, and non-breeding males contributed on average the least food. When the net contribution rate was considered (i.e. food items contributed minus items eaten by the individual helper), breeding females were still the most generous individuals, followed by subadult females, which contributed more than any other non-breeder. Subadult males on the other hand, did not always provision the packs' offspring.

The prediction of a positive correlation between the total amount of food delivered to the pups, and the number of non-breeding helpers present was not supported, since the presence of helpers did not increase feeding frequency at the den ($r_s = 0.18$, $n = 7$, $P > 0.05$). However, while the total food-provisioning rate did not increase significantly with the number of contributors to the den, the share

Table 20.2 Individual contributions of Ethiopian wolves to cooperative pup-care in relation to reproductive status, sex, and age during 2115 h of den observations

	Breeders (Mean ± SD)		Non-breeders (Mean ± SD)	
Behaviour/age	Male	Female	Males	Females
Babysitting				
Visits per hour				
Adults	1.9 ± 0.6	2.6 ± 0.7	1.1 ± 0.7	1.3 ± 1.2
Subadults			1.3 ± 0.7	1.6 ± 1.1
Percentage of time				
Adults	17.3 ± 8.6	23.6 ± 11.4	11.4 ± 9.5	11.4 ± 9.5
Subadults			8.6 ± 8.4	11.1 ± 8.8
Grooming rate				
Adults	0.04 ± 0.05	0.06 ± 0.06	0.02 ± 0.03	0.04 ± 0.08
Subadults			0.03 ± 0.04	0.03 ± 0.05
Food provisioning				
Hourly total food contribution				
Adults	0.06 ± 0.05	0.12 ± 0.09	0.03 ± 0.03	0.04 ± 0.06
Subadults			0.03 ± 0.03	0.06 ± 0.07
Hourly net food contribution				
Adults	0.05 ± 0.04	0.10 ± 0.08	0.02 ± 0.03	0.04 ± 0.04
Subadults			0.02 ± 0.04	0.06 ± 0.04

Note: Sample size was 17 breeding males, 18 breeding females, 49 non-breeding adult males, 11 non-breeding adult females, 16 subadult males, and 12 subadult females. Babysitting measured as the percentage of observation time in which individuals of a given category were present within 200 m of the den. Feeding measured in number of solid food items (i.e. whole rodents or regurgitations) contributed per hour.

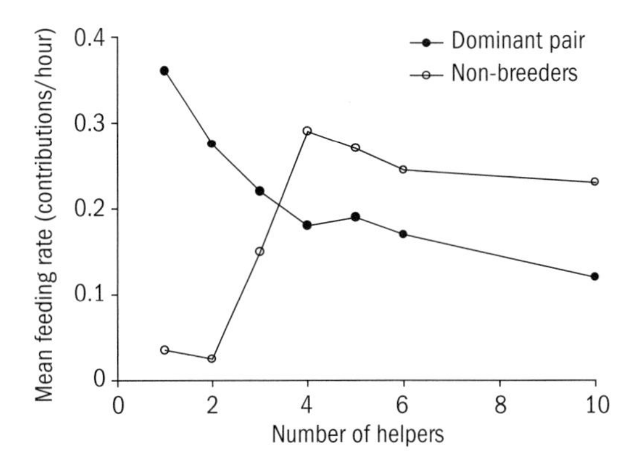

Figure 20.6 Rate of feeding pups (contribution of solid foods per hour) by the dominant pair in relation to the number of non-breeder helpers.

contributed by non-breeding helpers did do so. Food contributions by non-breeders were accompanied by reduced parental input in pup rearing—reducing food contributions by the dominant male ($r_s = -0.63$, $n = 7$, $P > 0.05$) and female ($r_s = -0.85$, $n = 7$, $P < 0.05$)—and hence a reduction in energy expenditure by the breeding pair (Fig. 20.6). The hypothesis that the number of non-breeding helpers enhances the reproductive output of the group was also not supported. The survival of pups at emergence was not correlated with the number of non-breeding helpers ($r_s = -0.26$, $n = 20$, $P > 0.05$), nor was survival at whelping ($r_s = -0.28$, $n = 20$, $P > 0.05$). Similarly, there was no significant correlation with survival at 6 months, 1 or 2 years and the number of non-breeding helpers. The litters observed, therefore, provide little evidence that helpers' feeding contributions *per se* influence the indirect fitness of helpers.

Allosuckling

The most extreme manifestation of cooperative care by Ethiopian wolves involved nursing the offspring of the dominant female, or *allosuckling* (Sillero-Zubiri 1994). Of the 20 successful breeding attempts observed, 8 dens had a subordinate female acting as allosuckler. Allosucklers were 2 years old or older, and often were closely related to the breeder (daughter or younger sibling). At least two allosucklers showed signs of pregnancy (or pseudo-pregnancy), but both either lost or deserted their own offspring before suckling the dominant female's. In at least two cases where an allosuckler was present, two females were seemingly pregnant. One might therefore have expected double the average litter size of pups from single females (5.1 ± 1.2 SD, $n = 12$, range = 2–6) to emerge. In contrast, significantly fewer pups than expected emerged ($t = 4.88$; $df = 16$; $P = 0.0002$), indeed, barely half the litter expected of a single female (2.6 ± 1.1 SD, $n = 8$, range = 1–4). Our evidence was that these few pups were invariably the offspring of the dominant female. The presence of an allosuckler was associated with distinct social unease in the pack and evident tension between the dominant and subordinate females. Female aggression inside the den may have an influence on pup mortality prior to emergence. In one case where litter size at parturition was known *a posteriori* from placental scars as five, only two pups had emerged from the den (Sillero-Zubiri 1994).

Allosuckling obviously has the potential to confer benefits to infants, and reduce the mother's energetic costs (Oftedal and Gittleman 1989). Mean suckling bout rate between weeks 4 and 13 was 0.26 ± 0.03 (SE) bouts per hour at dens with a single nursing female ($n = 9$). Assisted mothers suckled at a similar frequency ($n = 7$), but pups with access to two lactating females were suckled significantly more often, at 0.43 ± 0.05 bouts per hour ($t = 2.78$, $df = 60$, $P = 0.007$). Suckling was undertaken by only one female at a time. The suckling bouts of unassisted females were not only longer, but also involved more pups per event. Pups that nursed from two females may receive relatively more milk than those in larger litters with a single female, insofar as the share of female nursing time is a measure of milk flow. For those dens in which pups were produced, reproductive success at whelping (12 weeks) was variable but typically high. Mean whelping success at 4 months was 3.55 ± 0.47 (SE) pups for 20 litters, but dropped to 2.8 ± 0.45 (SE) at independence (6 months), and 2.0 ± 0.37 at 1 year of age ($n = 14$). Of 10 litters whose survival was monitored for at least 2 years, 7 produced an adult, at an average of 1.0 ± 0.25 per litter. Most den mortality and pre-whelping mortality was due to the mother's death ($n = 5$). Pre-independence mortality was mostly due to disease and starvation (Sillero-Zubiri 1994), while there was no evidence of losses to predation. Considering all packs that bred successfully ($n = 20$), the number of pups emerging from the den was not significantly correlated with the number of adults and subadults at the den ($r_s = -0.26$). Subsequent to emergence, pups whose mother was assisted by an allosuckler received a higher energetic input *per capita* until weaning (weeks 4–18) and enjoyed better survival than did those nursed by their mother alone.

Dominant females apparently benefited from allosuckling by sharing the costs of lactation, and thus lowering their *per capita* suckling frequency, without affecting a reduction in the pups' overall milk intake.

The foregoing results raise several interesting puzzles, which we hope the continuing research of our team will resolve. First, to the extent that the allosucklers do indeed make a long-term contribution to pup survival, this is initially disguised by the counter-intuitive earlier effect of litter reduction. Although the helpers in general, and allosucklers in particular, appeared to work assiduously for the well-being of the pups, and notwithstanding the rather large size of our data set, demonstrating any survival benefit was at best difficult. Perhaps such benefits are conditional upon circumstances. One intriguing speculation is that males nursed by two females do well: one such male grew up to acquire the dominant male position in his pack, another became dominant male in a pack with six adult males, and three survived a rabies epizootic in which nearly all other pack members perished (Sillero-Zubiri *et al.* 1996b). Additionally, the benefit of the presence of an allosuckler in a pack of Ethiopian wolves might be contingent on the availability of prey. In a good year, unassisted females may not need help, but allosuckler assistance might be important in harsh years. In a scenario where the chances of successful dispersal are very low, concentrating resources in fewer, fitter, individuals might raise their prospects of securing a dominant position, and eventually breeding status.

The cost of specialization, a conservation challenge

The apparently sterile Afroalpine steppes of Ethiopia support a rodent biomass, which is spatially and temporally predictable, and higher than all other figures quoted for Africa, which may explain why the Ethiopian wolf is the only canid to specialize so completely on rodents (Sillero-Zubiri and Gottelli 1995a). Moreover, the rodents' distribution and diurnal activity concur with Ethiopian wolves' diurnal and solitary foraging habits, and their confinement to Afroalpine habitats over 3000 m.

Global warming during the last 10,000 years progressively confined the Afroalpine ecosystem to the highest mountains, and 60% of all Ethiopian land above 3000 m has been converted to farmland. Two small populations recently became extinct when suitable wolf range shrunk below 20 km^2 (Marino 2003). Still, seven wolf subpopulations survive—occupying nearly every Afroalpine range in the country (e.g. 50 km^2 of habitat remaining in Mount Guna harbour no more than 10–15 wolves).

Ethiopian wolves face threats that arise from their isolation, small size, and the increasing contact with humans and their domestic dogs. Wolf killings seem to have decreased recently (Marino 2003), whereas transmission of rabies remains the main threat with serious consequences for small populations (Haydon *et al.* 2002). Protective measures require the consolidation of the management of protected areas and active efforts to monitor and protect all remaining populations, backed up by the establishment of a population management programme (Sillero-Zubiri and Macdonald 1997). Low genetic diversity, and the lack of evidence that phylogeographic patterns represent adaptive variation, remove hesitation about translocating across subpopulations (Gottelli *et al.* in press). On the face of it, their small, fragmented populations are a poor omen for Ethiopian wolves, but their concentration in a few clearly defined sites, their charisma and, we hope, a fair understanding of their biology, lend hope that with unwavering commitment to a well-founded management plan they may survive.

CHAPTER 21

Dholes

The behavioural ecology of dholes in India

Arun B. Venkataraman and A. J. T. Johnsingh

Dhole *Cuon alpinus* © A. J. T. Johnsingh.

Introduction

Past research on large social canids has focused particularly on the grey wolf (*Canis lupus*) and the African hunting dog (*Lycaon pictus*), and has produced important information to underpin significant advances in the conservation and management of the two species. In contrast, research on the dhole or Asiatic wild dog (*Cuon alpinus*), though steadily increasing, has been less intense and has largely occurred in southern and Central India that comprise only a small portion of its distribution. Research has mainly focused on the ecology and behaviour of the species and for the present, apart from promoting public awareness of the species, has contributed little towards its conservation and management. For example, the status and distribution of the species in its range countries are still poorly understood, greatly constraining conservation and management. This is critical, as recent evidence indicates that the species may be endangered in a number of range countries and could face extinction if immediate action is not taken. This is of concern, as apart from being a key element in prey–predator communities across a number of ecosystems, the dhole is of scientific importance due to the cooperative behaviour involved in its hunting and breeding (making it a challenging model for the evolution and maintenance of cooperative behaviour).

Status, distribution, and threats

The dhole is a medium-sized social canid, inhabiting forested tracts in south and southeast Asia. The dhole's global distribution once encompassed continental Asia (roughly east of 70°E), but it is now extinct in Russia (D. Miquelle personal communication). Dholes occur on the islands of Java and Sumatra but are absent in Japan, Sri Lanka, or Borneo (Pocock 1936). The status of the dhole remains vague within southeast Asia. Viable populations may exist in northern Myanmar (C. Wemmer personal communication) where, despite sufficient vegetation cover, prey densities tend to be very low. Viable populations exist in Java and Sumatra (S. Hedges personal communication).

Presently the largest populations probably occur in southern India, and in Central Indian Highlands (Johnsingh 1985). Possibly the best of the populations occur within the predominantly dry deciduous forests of the Lower Nilgiri Plateau or Mysore Plateau (Johnsingh 1985). This accords with the general belief that dhole densities are positively correlated with prey densities, which tend to be high in the drier savannah woodlands and moist and dry deciduous forests (Venkataraman and Narendra Babu 2001).

Dholes were once found in the *terai* region in the Himalayan foothills, but, despite its suitable habitat and high prey densities, sightings there are now only sporadic (Johnsingh 1985). Dhole populations here may not have recovered from past carnivore control programmes, initiated by the British and pursued to a lesser degree after the independence of India till the implementation of the Indian Wildlife Act in 1972 (Rangarajan 1998). However, anecdotal information from naturalists and hunters (Singh 1998) suggest that dholes have always lived at very low densities in the Corbett and Dudhwa National Parks, in the *terai* region. Probably, disease or the high densities of tigers were responsible for the low densities of dholes in these habitats (Johnsingh 1985).

In the past, single or pairs of dholes have been seen in protected areas in Eastern Rajasthan, western India, formerly connected with the Central Highlands. These dholes may have been stragglers moving through cultivated land into the protected areas. There is no evidence of the formation of viable populations or packs within both Rajasthan and the *terai* region (Johnsingh 1985).

The status of dholes in the wetter forests of northeast India and Bangladesh is unknown though the species is thought to be very rare or extinct in most northeast Indian states (except for Meghalaya and Arunachal Pradesh) and the Chittagong hill tracts of Bangladesh (Johnsingh 1985). It is believed that prey depletion caused by rampant hunting (often due to the collapse of law and order through insurgency, e.g. Manas Tiger Reserve in Assam) and large-scale habitat loss through shifting cultivation and large scale illegal felling arising as a result of increasing incidents of encroachments may have further endangered the species in all seven states of northeast India. For instance, in Arunachal Pradesh, the strong tradition of hunting, lack of enforcement of wildlife laws within the protected area network, and carnivore–human conflict, have led to active persecution of dholes and other carnivore species (Narendra Babu and Venkataraman 2001). This is in contrast to other areas in India, where direct hunting of dholes is rare. The areas mentioned above are indicated in Fig. 21.1.

An overview of research on dholes

Accounts of dhole behaviour have sporadically appeared in the *Journal of the Bombay Natural History Society* since 1895 (Hood 1895), most of them based on observations in India, made by government officers out on hunting expeditions (e.g. see Burton 1925, 1940; Connell 1944).

The earliest scientific findings on dholes, published by Davidar (1973, 1975) provided the first accounts of the mating and breeding behaviour of the species and was followed by a review paper on dhole biology by Cohen (1977). A 3-month field study conducted by Fox and Johnsingh (1975) described the species'

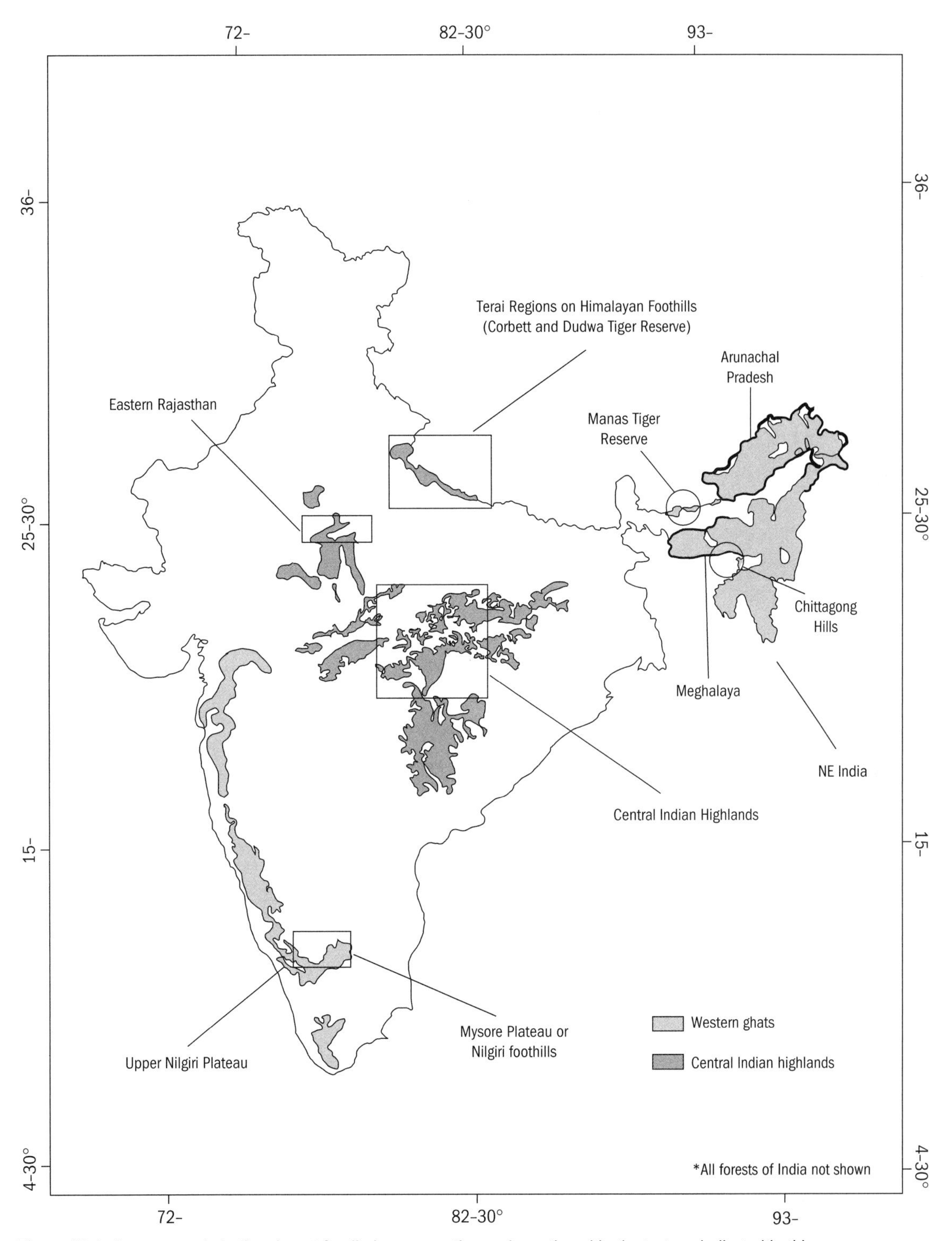

Figure 21.1 Forest areas in India relevant for dhole conservation and mentioned in the text are indicated in this map. The forest extent indicated in grey has been adapted from the State of Forest Report 1999, Forest Survey of India (Ministry of Environment and Forests), Dehra Dun, India.

feeding and hunting behaviour. The first extensive field study was carried out by Johnsingh (1982, 1983, 1992) on a single pack of dholes in the Bandipur Tiger Reserve, Mysore Plateau, southern India (Fig. 21.1) in 1976–78. The study provided detailed data on social and reproductive behaviour and feeding ecology. In the years 1990–95, field studies in the Mudumalai Wildlife Sanctuary, Mysore Plateau, southern India (Fig. 21.1), (Venkataraman *et al.* 1995; Venkataraman 1998) gave further insights into the foraging ecology and social biology of the species. Unless otherwise stated, the following account is based on these two studies.

Taxonomy and morphology

Systematic position

According to Thenius (1954), the genus *Cuon* is post-Pleistocene in origin, and related more closely to the extant jackals than to wolves. Earlier, Simpson (1945) had placed the dhole in the subfamily Simocyoninae of the family Canidae, together with the African wild dog and the bush dog (*Speothos venaticus*) of South America on the basis of shared anatomical features, most notably the reduction of the role of the crushing post-carnassial molars. *Lycaon* retains 42 teeth normally present in *Canis*, though these teeth are not very large, while in the other two members of this putative subfamily, these teeth are not merely small but some are absent (*Cuon* 40—III molar absent; *Speothos* 38) (Ewer 1973). Many have questioned Simpson's classification. Thenius (1954) noted that structural features of the vertebral column and the limb bones were quite different within the subfamily Simocyoninae, and suggested that similarities in dentition could be due to parallel or convergent evolution along phylogenetically distinct lines. Ewer (1973) argued that the characteristics of dhole dentition suggest a highly predatory habit, with diminished importance of vegetal matter in the diet and concluded that the features of dentition could have evolved independently in the three genera.

Clutton-Brock *et al.* (1976) provided further support for Thenius's view. They numerically analysed 90 morphological, ecological, and behavioural characteristics in 37 canid species. Their results indicated that when all of these characteristics were considered, *Cuon* was more similar to *Canis, Dusicyon*, and even *Alopex*, than to *Speothos* or *Lycaon*. When only skull and dental characters were considered, *Cuon* resembled *Speothos* and *Lycaon* most. According to Kleiman (1972) and Lorenz (1975), *Cuon, Lycaon*, and *Speothos* appear more closely related to other canid genera than to each other.

Simpson's (1945) association has also been questioned on behavioural grounds, *Cuon* and *Lycaon* sharing similar behavioural traits, but *Speothos* being quite distinct (Kleiman 1972). Further evidence of taxonomic distinctions between *Speothos, Cuon*, and *Lycaon* were provided by analysis of sequences from mitochondrial genes (cytochrome *b*, cytochrome *c* oxidase I, and cytochrome *c* oxidase II (Wayne *et al.* 1997). Here both *Lycaon* and *Cuon* were classified *Canis* like canids and *Speothos* within a clade consisting of another South American canid, the maned wolf (*Chrysocyon brachyrus*).

Earlier, two species of *Cuon*, the southern dhole (*Cuon javanicus*) and the northern dhole (*C. alpinus*), were distinguished by Mivart (1890) on the bases of body size and the second upper and lower molars. Ellerman and Morrison-Scott (1951), however, recognized 10 subspecies, later revised to 9 (Ellerman and Morrison-Scott 1966) or 11 according to Ginsberg and Macdonald (1990). Three subspecies recorded in India (Ellerman and Morrison-Scott 1966) include: *C. a. dukhunensis*, found south of the Ganga River; *C. a. primaevus*, seen in Kumaon, Nepal, Sikkim, and Bhutan; and *C. a. laniger* (Pocock 1936), occurring in Kashmir and Ladakh. *C. a. adustus*, a subspecies found in Myanmar, may range into adjacent parts of India. *C. a. infuscus* is the other subspecies found in Myanmar (Johnsingh 1985).

General description

On average, an adult male dhole weighs 18 kg, stands around 50 cm at the shoulder and is 130 cm long (including the 40–45 cm long tail). Females are slightly lighter in build. Sexual dimorphism is not very distinct with no known quantitative anatomical differences, and sexing them from a distance in the field is difficult.

Figure 21.2 Dholes *Cuon alpinus* in Bandipur Tiger Reserve trying to smell out the author in a tree © A. J. T. Johnsingh.

The genus *Cuon* is distinguished from *Canis* by:

(1) more rounded ears and proportionately shorter muzzle;
(2) the line of the face viewed sideways being slightly convex, that of *Canis* being straight or concave;
(3) having only two true molars on each side of the lower jaw, instead of the three typical of the genus *Canis*;

Foraging ecology

Prey

The prey of dholes varies from place to place. In Central India, where Brander (1927) worked, prey species were sambar (*Cervus unicolor*), chital (*Axis axis*), swamp deer (*Cervus duvauceli*), nilgai (*Boselaphus tragocamelus*), gaur (*Bos gaurus*), blackbuck (*Antilopus cervicapra*), and wild pig (*Sus scrofa*). Other animals occasionally hunted by dholes are Nilgiri langur (*Semnopithecus johnii*) (from scats—Adams 1949), Nilgiri tahr (*Hemitragus hylocrius*) (Williams 1971; Rice 1986), and cattle calves. Some ground-dwelling birds such as grey jungle fowl (*Gallus sonneratii*) may also be taken opportunistically (Davidar 1975).

Dholes are also known to feed on smaller prey, including rodents, and even on longicorn beetles (*Dorysthenes rostratus*). Scats consisting almost entirely of fresh grass blades and minimal animal remains have also indicated that dholes, like most other carnivores, consume grass (Johnsingh 1983). A dhole was once observed eating tender leaves of *Lantana camara*. Vegetal matter may serve as an anti-irritant, protecting the alimentary tract from bone fragments and may scour parasitic worms. (The possibility that dholes may eat grass and other vegetal matter with the intention of compensating for the lack of minerals in an all-meat diet appears dim, since the vegetal matter is defecated with little apparent change.)

Prey preference and cooperative hunting

Prey preference

Scat analysis from two packs inhabiting Mudumalai showed that one pack preferred chital over other prey species while the other favoured both chital and sambar. Preferences were assessed in a chi-square goodness of fit test, comparing the observed frequencies of occurrence of hair of each prey species in faeces, and biomass consumed, with that expected from its density and biomass within the study area (see Part A in Table 21.1). Biomass of each species consumed was extrapolated using differential digestibility corrections derived for the wolf (Floyd *et al.* 1978). The afore-mentioned preferences applied in terms of both number and biomass of prey consumed. The latter pack also ate cattle, which were utilized less than expected from their availability (Part A, Table 21.1). In contrast to the former pack, entirely contained within the protected area, this pack's home range included fringe zones of the protected areas, and a few areas outside, where cattle were abundant. Over 80% of chital killed by the former pack were less than a year old, while chital less than a year old constituted 61% of chital preyed upon by the latter pack (Part B, Table 21.1). Johnsingh (1983) also reported that chital fawns comprised a high proportion of chital killed. Among adult chital, Mudumalai packs preferred males over females in the ratio 2.6 : 1 (Venkataraman *et al.* 1995). Johnsingh (1983) used the population sex ratio (84 : 100) as obtained during the peak rutting season and concluded that predation on males was not significantly different from expected. Patel (1993), however, used the average sex ratio obtained by Johnsingh throughout the study (68 : 100) and found predation on males was significantly greater than expected.

Table 21.1 Prey preferences and preferences for age classes for two packs in the Mudumalai Sanctuary

Prey species (age class)	Remains in scat (%)		Biomass consumed (%)		Density individuals/km^2 ± 95% CI		Biomass (kg/km^2)	
	Pack A	Pack B	Pack A	Pack B	Pack A	Pack B	Pack A	Pack B
A. Prey preferences[a]								
Chital	70.4	41.0	61.1	33.4	16.0 ± 3.4	19.4 ± 5.2	841	1082
Sambar	21.7	23.3	32.4	36.1	7.9 ± 1.6	1.8 ± 0.9	1206	281
Cattle	4.3	15.2	5.8	25.6	3.1	50.0	543	8750

B. Preferences for age classes of chital and sambar[b]

	Individuals in population (%)	No. of individuals in kills (%)
Pack A		
Chital (<1 yr)	14.1	81
Chital (>1 yr)	85.9	19
Sambar (<1 yr)	13.5	76
Sambar (>1 yr)	86.5	24
Pack B		
Chital (<1 yr)	11.9	61
Chital (>1 yr)	85.1	39
Sambar (<1 yr)	15.8	67
Sambar (>1 yr)	84.2	33

Notes:

[a] Expected values were obtained from the density and biomass of the three prey species and observed values were compared with these in a chi-square goodness of fit test.

[b] Expected values were obtained from the proportion of each species' age classes in the population (Varman and Sukumar 1993, unpublished data) and observed values were compared with these in a chi-square goodness of fit test.

Dholes killed more sambar males than expected and preference for adult chital males could be a possible effect of heavy antlers serving as handicaps during escapes (Johnsingh 1983). Both packs in Mudumalai preferred sambar that were less than a year in age (Venkataraman *et al.* 1995) (Part B, Table 21.1). In the Nagarahole National Park, Karnataka, southern India (Fig. 21.3), dholes preferred medium-sized prey, such as chital (Karanth and Sunquist 1995).

It therefore appears that dhole packs prefer the young of medium-sized prey such as chital. In the three protected areas mentioned above, chital offer an

Figure 21.3 The study areas, Bandipur Tiger Reserve and Mudumalai Wildlife Sanctuary are located in the Mysore Plateau (foothills of Nilgiris), southern India. The contiguous Nagarahole National Park, where some research has been carried out on dholes, is also shown.

abundant and predictable source of food (see below). Furthermore, chital do not defend themselves or their young from attacks by dholes as the sambar do by getting into the waterhole (Johnsingh 1983). Indeed the lack of any direct anti-predator behaviour (apart from group vigilance and other benefits of herding) is striking. The above factors probably make chital, where available, a favoured prey species. This is true for only the foothills of the Nilgiris and the drier forests of Central India where chital are abundant—they are absent from the wetter forests of northeast India or the high altitude forests of Western Ghats.

Cooperative hunting

Members in dhole packs hunt cooperatively, particularly while hunting larger prey such as adult chital or sambar. However, in 92% of 48 occasions, the chase ended within 500 m of its starting point, and flushing out of small prey such as chital and sambar fawns from bushes was observed (Johnsingh 1983). While chasing prey for long distances, pack members are often observed to 'cut corners' reducing the distance between prey and the pack (Venkataraman *et al.* 1995). Relationships between pack size and hunting success or energetics have yet to be determined. There was no correlation between body weight of prey killed and pack size for packs in either Bandipur or Mudumalai (Venkataraman *et al.* 1995), but Bandipur kill data (Johnsingh 1983) indicated a negative correlation between per capita food consumption and pack size (Venkataraman *et al.* 1995).

Kills involving two dholes pulling down a chital fawn shared by all pack members were seen in Mudumalai (Venkataraman *et al.* 1995). In Bandipur, even three dholes were successful in killing a sambar fawn and two chital does (Johnsingh 1982). A significant proportion (20.4%) of the diet of a pack in Mudumalai comprised black-naped hare (*Lepus nigricollis*), an animal that can be captured by a single dhole (Venkataraman *et al.* 1995).

Home range

Dhole packs' home ranges are likely to vary as a function of prey density, composition, and distribution, all of which are influenced by habitat types. Estimates for dhole home ranges exist for the deciduous forests of Nagarhole, Mudumalai, and Bandipur. In Nagarhole, the home range for a single pack was estimated at 27.5 (adaptive kernel estimator), 23.4 (minimum convex polygon estimator), and 27.4 km^2 (harmonic mean estimator) (number of locations = 138) (Karanth and Sunquist 2000). In Mudumalai, the home range areas for packs were estimated at 83.3 and 54.2 km^2, respectively, using the minimum convex polygon method (number of locations = 276,103) (Venkataraman *et al.* 1995) Fig. 21.3). In Bandipur, the home range was estimated at 40 km^2 (Johnsingh 1982).

Venkataraman *et al.* (1995) analysed changes in home range areas across seasons, pack sizes, and years of study. For one pack, home range did not vary as a function of pack size, which varied between 4 and 10 over 5 study years. Furthermore, the home range area for the same pack remained constant throughout the study, although the dry season range was, on average, less than the wet season range (Venkataraman *et al.* 1995) (Fig. 21.3). The dry season coincided with the period when pups were less than 3 months of age, but even when a pack did not litter, the range was smaller in the dry season than in the wet season. Perhaps water scarcity confined the pack and its prey to areas with perennial water sources (Venkataraman *et al.* 1995).

Principal prey composition, distribution, and densities were similar for all three areas (and an additional area in Central India). Each had similar habitat, comprising dry deciduous forests interspersed with teak (*Tectona grandis*) or sal (*Shorea robusta*) plantations, and grassy meadows ideal for grazing herbivores (Neginhal 1974; Karanth and Sunquist 1992; Suresh *et al.* 1996) (see Table 21.2). There is little information on dhole home range or habitat utilization in other habitats (with other prey) such as the sub-tropical montane forests of the Nilgiri Hills, southern India and the wet evergreen forests of the Western Ghats and northeast India (Fig. 21.4).

Habitat utilization

In Mudumalai and Bandipur, there is evidence that dholes intensively use dry deciduous forests and savannah woodlands interspersed with meadows

Table 21.2 Dhole and principal prey densities for important dhole habitats in India

Area	Region	Chital density (km^2) (SE or %(CV))	Sambar density (km^2) (SE or %(CV))	Dhole density (km^2)	Habitat type
Mudumalai Wildlife Sanctuary	Nilgiri foothills, southern India	20.7 (5.7)	7.7 (1.3)	0.14	Dry and moist deciduous, thorn forests
Bandipur National Park and Tiger Reserve	Nilgiri foothills, southern India	48.4 (40%)	12.5 (24%)	0.09	Dry deciduous, thorn forests
Nagarhole National Park	Nilgiri foothills, southern India	38.1 (3.72)	1.5 (0.54)	Not available	Dry and moist deciduous
Pench Tiger Reserve	Chotta Nagpur Plateau, central India	51.3 (3.02)	9.66 (0.76)	0.3	Dry deciduous, thorn forests

Source: Prey densities for Mudumalai were from Arivazhagan and Sukumar (unpublished data), for Bandipur Varma *et al.* (unpublished data), for Pench and Nagarahole (Karanth and Nichols 2000). Dhole densities for Mudumalai and Bandipur were from Venkataraman and Narendra Babu (2001) and for Pench Johnsingh and Bhaskar (unpublished data).

Figure 21.4 Dhole *Cuon alpinus* after a dip in a puddle © A. J. T. Johnsingh.

with high chital density (Venkataraman and Narendra Babu 2001). Chital herds prefer open meadows located within these savannah woodlands (Sharath Chandra and Gadgil 1975) and live at low densities within closed forest, in contrast to sambar, which tend to remain in closed forests (Varma and Sukumar 1995). This results in a patchy distribution of chital herds as seen in the home range of a pack within Mudumalai (Fig. 21.5), where four patches can be discerned. The pack had a tendency to visit patches in a pre-determined sequence in one direction and, on reaching the last patch in the sequence,

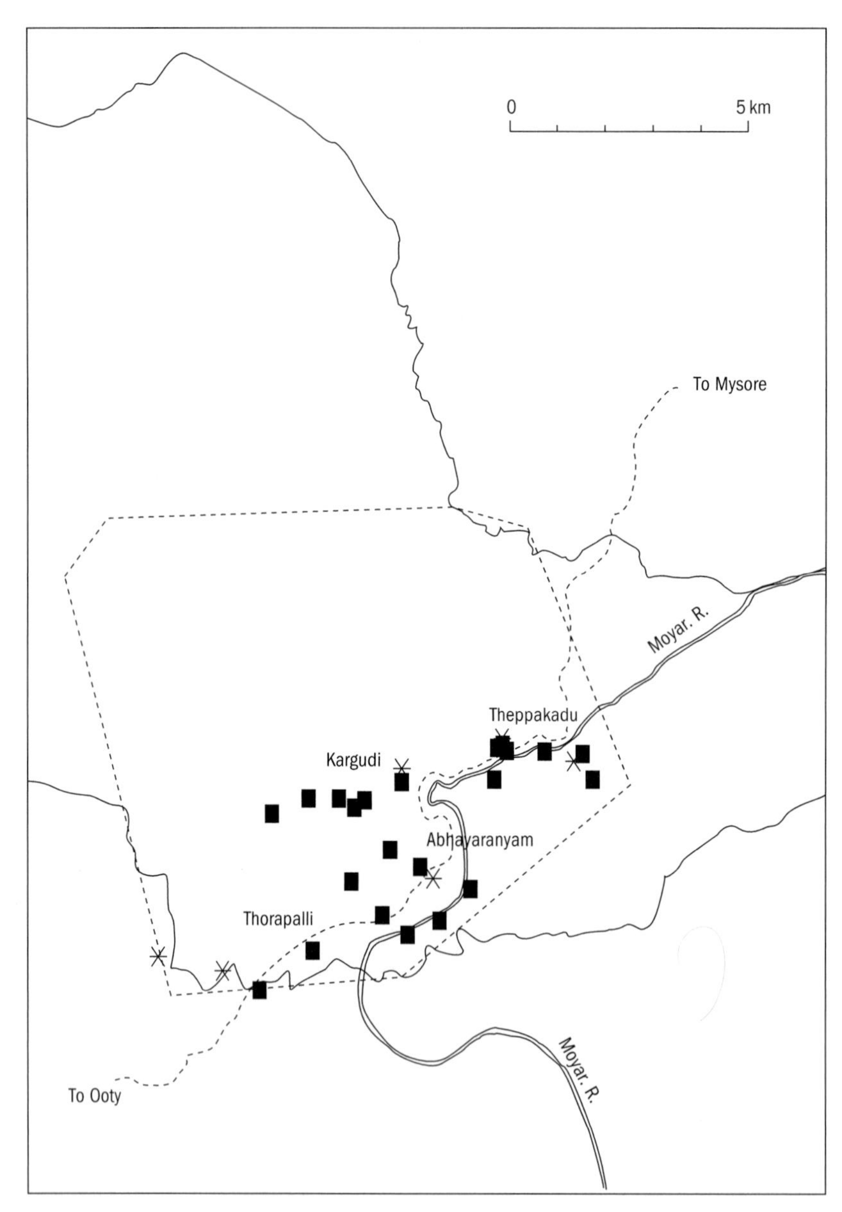

Figure 21.5 Map showing the major resource patches of the Kargudi pack. Each resource patch consists of chital herds (■) and resting sites of the dhole pack (⋇). The home range of the Pack (––––), as determined by the minimum convex polygon method, is also shown. Visits by the pack to the northern part of the home range were relatively rare.

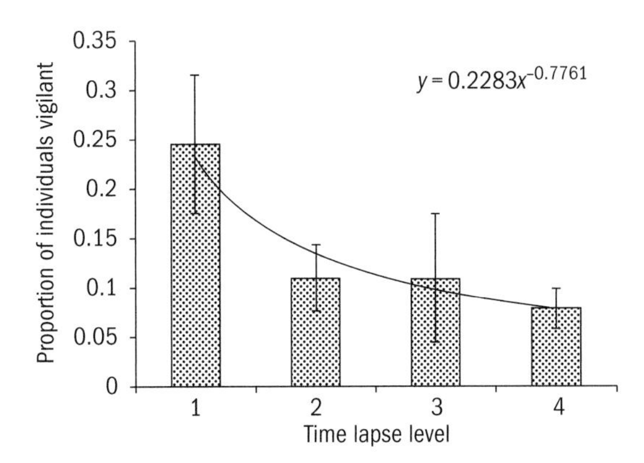

Figure 21.6 Mean proportion of chital individuals vigilant for time lapse levels 1–4. Each level represents a lapse of (1) <2 h, (2) 2–24 h, (3) 24–48 h, and (4) >48 h after a dhole pack visited and successfully or unsuccessfully hunted the herd under observation. Bars indicate standard errors.

returned to the starting patch, while other patches were visited in a random fashion. Residency time did not differ between patches and ranged from 1 to 7 days. Changes in the proportion of individuals vigilant in chital herds were monitored following hunts by dholes in a given patch, and revealed that the proportion of individuals vigilant was highest after a hunt (whether or not successful), and tapered off during the next two days (Fig. 21.6) (Venkataraman and Rao unpublished data). The enhanced vigilance of chital may thus cause a drop in dhole pack hunting success following a hunt and therefore packs may shift to fresh patches where vigilance is at base 'background' levels and hunting success is greater (Venkataraman *et al.* 1995).

Social organization

Over the period 1989–93, the number of adults in packs within Mudumalai varied from 4 to15 and the number of pups varied from 0 to 10 (mean ± s.d. for two packs, 4 ± 3.4 and 7.3 ± 1.9) (Venkataraman *et al.* 1995). In Bandipur, the number of adults and pups within a single pack varied from 5 to 11 (mean = 8.3) and 6 to 9, respectively (Johnsingh 1982). This pack was consistently male biased (Johnsingh 1982). A pack in Mudumalai was also male biased during most study years (Venkataraman *et al.* 1995). The male-biased sex ratio was caused by an almost 2-fold higher dispersal rate of females over males. The fate of dispersing females was unknown. Immigrations into packs were observed in two instances. In one case, three males joined a pack in Mudumalai (Venkataraman 1998) and in another instance, a female with two pups joined a pack within Bandipur (Johnsingh 1982).

Data were sufficient to assign dominance ranks for males, but not for the females, in a pack in Mudumalai. Using triad comparisons of rates of dominance and submission and rates of receiving both dominance and submissive acts (Tomback *et al.* 1989), ranking of five males indicated a linear dominance hierarchy. There was no correlation between dominance ranks and frequency of reproductive behaviour (sniffing anal region and mounting with and without penetration) displayed towards mature females (Venkataraman 1998). It must, however, be noted that actual copulation does involve a copulatory tie (Davidar 1973; Paulraj *et al.* 1992) and therefore mounting of females by males does not necessarily imply successful copulation.

Breeding biology

Only one female bred annually in each of the three packs studied in Bandipur and Mudumalai (Johnsingh 1982; Venkataraman 1998). However, for the two packs in Mudumalai, in 7 years when pack composition was definitively ascertained during the littering season, there were no other adult females in 5 of the years, and 4 and 1 in the remaining two years (Venkataraman 1998). Once, a sub-adult female (close to adulthood at 2 years) left the pack during the mating season but rejoined, not detectably pregnant, before the littering season commenced (Venkataraman 1998). Within dhole packs, in addition to the dominant pair, other adults and sub-adults were seen guarding, playing with, and regurgitating food to pups (Johnsingh 1982; Venkataraman 1998). Generally, there were no significant differences amongst pack members in the frequencies of seven behaviours displayed towards pups (playing with pups, accosting pups, being followed by pups, and regurgitating food) and at dens (entering dens, leaving dens, being at dens). However a single adult male, the putative 'uncle' of the dominant male, did display higher frequencies of pup care behaviours than others (Venkataraman 1998). Following a period when it was fully integrated within the pack, this individual paired up with a young female from the pack dispersing together (Venkataraman 1998). The dominant pair, most interestingly, was never seen guarding the den or provisioning food to the pups.

Discussion: the evolution and maintenance of sociality in dholes

The prevalence of sociality among pack-living canids has attracted much attention in the recent past. It has been hypothesized that the major selective forces for social carnivores include defence of territories or young (Packer *et al.* 1990), communal rearing of young (Malcom and Marten 1982), cooperative hunting (Alexander 1974; Bertram 1978b; Packer and Ruttan 1988; Creel and Creel 1995a), prey defence against kleptoparasitism from other predators (Lamprecht 1978b; Cooper 1991; Creel and Creel 1996), or a combination of these factors within a context in which resource dispersion is a fundamental ingredient (Macdonald 1983). Each of these theories is discussed in relevance to results discussed above.

Defence of territories or young

The Resource Dispersion Hypothesis (RDH) (Macdonald 1983) predicts that group living is facilitated where resources are dispersed such that the smallest defensible territory required by a primary pair (or other basic social unit) can also accommodate additional group members, conditions that may apply where food resources are patchily distributed and their abundance undergoes spatio-temporal variation. In the case of dholes, living on medium-to-large-sized deer, the conditions facilitating group formation might be the size of the prey (each carcass may provide a meal for several individuals), or the long flight distance of the prey or their herding behaviour (the territory size required to be able to hunt one deer may contain many deer) (Kruuk and Macdonald 1985; Carr and Macdonald 1986). Indeed, within Mudumalai, a dhole pack utilized chital herds within its home range, distributed in resource patches, sizes of which varied spatially and seasonally. Chital form large groups only from late April to September, in coincidence with the new flush of grass after the fire and rutting season. The availability of young animals also varied temporally as a consequence of asynchronous fawning, even though there is a rough peak from December to March corresponding to the rutting period from May to August (Johnsingh 1983). Furthermore, hunting success within patches varied due to heightened vigilance in herds subsequent to hunts, forcing the pack to move onto other patches where vigilance was at background levels. All these observations indicate that chital herds within the pack's home range closely follow the prey configuration and display the spatio-temporal variation described by the RDH (Venkataraman *et al.* 1995). However, as has been the case in other attempts to test this hypothesis (e.g. Da Silva *et al.* 1993), while the theory matches naturalistic observations, it has eluded critical test (Johnson *et al.* 2003).

Another prediction of the RDH is that, so long as patch richness and dispersion are not correlated, minimum home range areas and group size will be uncorrelated. For the one pack for which we have data, at Mudumalai, home range areas remained constant despite large fluctuations in pack size (Venkataraman *et al.* 1995). A third point is that RDH provides an explanation for why animals might form groups even in the absence of any cooperative benefit from doing so. In this case we cannot comment on the selective advantage of helping behaviour, and observe that the benefit of cooperative hunting is unknown; we have shown that dholes prefer to hunt fawns of both chital and sambar in the Mudumalai Sanctuary and the Bandipur National Park (Johnsingh 1983; Venkataraman *et al.* 1995) and these can clearly be killed by a single dog. However, the flushing activities of a larger pack may nonetheless increase their success at locating prey of all sizes (Johnsingh 1983).

Communal rearing of young

Among four parameters obtained from packs at the time of littering (number of adults in packs (mean ± s.d. = 6.5 ± 3.2, $n = 10$), number of sub-adults (3.7 ± 3.4, $n = 10$), age of the breeding female (4 ± 1.5, $n = 10$), and rainfall in the preceding breeding season (1015.5 ± 269.2, $n = 10$) thought to be highly correlated with chital fawning intensity) that were compared with litter sizes, only the age of the breeding female was significantly correlated with litter sizes in a bell-shaped relationship: as females get older, their litter sizes diminish (Venkataraman 1998). At this stage, there is no distinct evidence indicating that packs live in groups to accrue benefits from cooperative breeding (Venkataraman *et al.* 1998).

A likely consequence of the social system as described above is that there is scope for occasional matings by subordinate individuals, and thus of confused (Derix and Hoof 1995) or multiple paternity in litters, with associated consequences for communal care, infanticide risk, and kinship (Burke *et al.* 1989).

Cooperative hunting

The data currently available are sufficient to conclude only that even a small number of dholes are capable of killing prey, and that per capita consumption of meat may decline with increasing pack size. Much more data will be necessary to elucidate the foraging economics of cooperative hunting, and to test the proposal that a minimum of seven animals may be needed to optimize the efficiency of hunting for prey (Schoener 1971) and to feed a nursing bitch with a litter of 8–9 pups (Johnsingh 1982).

Defending prey against kleptoparitisitism

In eight separate interactions between dholes and large cats in the Bandipur and Mudumalai Sanctuary (Johnsingh 1982a; Venkataraman 1995), there was no evidence that interactions were initiated through dhole kills being parasitized by other large carnivores, though once, wild pigs (*Sus scrofa*) chased away dholes from a kill and scavenged (Johnsingh 1978a). The converse was, however, true. Eight dholes were once seen stealing the kill of a leopard (*Panthera pardus*). In other instances, packs of between 5 and 11 dholes spent considerable time (between 20 and 50 min) and effort chasing leopards from shared hunting areas (Venkataraman 1995). In a single instance when a pack of 10 dholes interacted with a tiger (*Panthera tigris*), mutual avoidance was observed (Venkataraman 1995). There was considerable overlap in the diets of all three large carnivores; competition with tigers may have been reduced by spatial and temporal partitioning of habitat and prey sex and size (Johnsingh 1983; Venkataraman 1995). Leopards and dholes, however, preyed on animals of the same weight class (Johnsingh 1983). Furthermore, leopards occasionally prey on dholes (Karanth and Sunquist 1995) and the dholes' tendency to chase leopards may be akin to mobbing—reducing the risk of unexpected attacks, especially on individuals separated from the main pack while hunting (Venkataraman 1995). Even though the minimum number of dholes needed to intimidate a leopard is unknown, as little as five dholes were seen engaging quite successfully in this behaviour.

Why do dholes live in packs?

In summary, of the possible explanations for group living in dholes, the consequences of their foraging for large, patchily dispersed ungulate prey would seem to be the most compelling on current evidence. In Mudumalai and Bandipur, although data are very scarce, it seems likely that dispersal is dangerous and constrained, perhaps tipping the trade-off between pack membership and emigration in favour of natal philopatry. Within the framework of pack-living set by resources, we have evidence that dholes cooperate in hunting, in the care of young, in attacking leopards, and these cooperative behaviour strengthen the advantage of pack living set by resources.

Acknowledgements

Dr. A. J. T. Johnsingh acknowledges the help he received from Mr. J. C. Daniel, Mrs. Dilnavaz Variava, Dr. Michael Fox, Dr. Madhav Gadgil, Dr. Ranjit Sinh and Mr. E. R. C. Davidar during the period he studied dholes. Arun would like to thank the Chief Wildlife Warden, Forest Department of Tamil Nadu for permission to carry out the dhole work; Prof. R. Sukumar for constant encouragement and financial support, the I.I.Sc. Field Station at Masinagudi, Mudumalai, for unstinting academic and logistic assistance; R. Arumugan for teachings on how to study dholes and being an unparalleled friend and companion in the forest; the Department of Science and Technology, Govt. of India, for a Young Scientist's Grant to carry out much of the research; the Wildlife Trust of India, New Delhi, for funding surveys in Northeast India; and last but not least, his wife Anu and son Ajay for sharing the travails and joys of exploring the life of dholes.

CHAPTER 22

African wild dogs

Demography and population dynamics of African wild dogs in three critical populations

Scott Creel, Michael G. L. Mills, and J. Weldon McNutt

African wild dog *Lycaon pictus*
© D. W. Macdonald.

African wild dogs (*Lycaon pictus*) are always found at low population densities, when compared to sympatric large carnivores. Consequently, most populations of wild dogs are small, and only a handful exceeds 500 individuals. Three of the largest remaining wild dog populations are found in Kruger National Park (South Africa), the Selous Game Reserve (Tanzania), and Northern Botswana. With a total of 1900–2500 individuals, these areas protect about one-third of the African wild dogs alive today, and the future of wild dog conservation rests in large part on these ecosystems.

Each of these populations has been studied in sufficient detail to provide good demographic data. In this chapter, we compare patterns of age-specific reproduction and survival among populations, and use Leslie matrices to investigate how wild dog demography affects population dynamics. These analyses are of interest for two basic reasons. First, the conservation of wild dogs is tightly linked to their fate in these populations, which creates immediate, practical interest in assessments of their demographic stability. Second, the long-term data sets compiled here reveal the demographic processes that drive the population dynamics of wild dogs. As Caswell (2001) noted, 'to know the survival probabilities and fertilities of every age class under a particular set of conditions is to possess a great deal of biological information about those circumstances. This information is most valuable when coupled with a comparative approach, in which the vital rates are measured under two or more different conditions'. Few studies have compared the demography of a large carnivore across several populations, simply because the data demands are so great (the data summarized here required 30 years of field work). Through quantitative comparisons, similarities and differences among populations can be identified, and these can be used to guide conservation actions.

In this chapter, we move between prospective and retrospective approaches to demography. Retrospective analyses identify the demographic variables that have, in the past, contributed most to observed variation among populations in the growth rate (λ). Prospective analysis identifies the vital rates that would make the strongest contribution to variation in population growth, if all variables were free to change. By combining the two approaches, we identify the demographic variables that have a strong impact on growth, and vary substantially among populations. From the perspective of identifying conservation priorities and evaluating management actions, these are the variables of greatest interest (Mills *et al.* 1999; de Kroon *et al.* 2000; Caswell 2001).

African wild dog ecology

African wild dogs are medium sized (18–28 kg) canids that live in highly cohesive packs holding from 2 to 28 adults, with a mean adult pack size of 4.8–8.9 across 5 ecosystems (Creel and Creel 2002). Wild dogs prey mainly on ungulates, focusing on wildebeest (*Connochaetes taurinus*), impala (*Aepyceros melampus*), kudu (*Strepsiceros tragelaphus*), gazelles (*Gazella thomsonii* and *grantii*), and warthogs (*Phacochoerus aethiopicus)* (Pienaar 1969; Fanshawe and Fitzgibbon 1993; Creel and Creel 1995; Fuller *et al.* 1995; McNutt 1995; Creel 2001). Differences in diet among ecosystems are strongly affected by the relative abundance of these species.

For wild dogs, cooperative breeding is nearly obligate. Most packs hold a single breeding female, though subordinates of both sexes sometimes produce offspring that are raised, particularly in large packs (Creel *et al.* 1997a; Girman *et al.* 1997; Creel and Creel 2002). Reproductive success is positively correlated with pack size, and few packs with less than four adults raise pups to independence (see below). The association between pack size and reproduction is driven by several factors. First, adults of both sexes provide alloparental care by guarding and feeding pups (Malcolm and Marten 1982). Second, hunting is more successful and energetically less costly in larger groups (Fanshawe and Fitzgibbon 1993; Creel 2001). Third, large packs are more successful at defending kills from scavengers, in ecosystems where loss of carcasses is common (Fanshawe and Fitzgibbon 1993; Creel and Creel 1996). Finally, large packs generally win aggressive clashes with other packs in the overlap zones of their home ranges (Creel and Creel 2002).

Factors affecting density and dynamics

Studies in several ecosystems have examined the factors that limit wild dog populations, and there are several recent reviews of this information (McNutt and Boggs 1997; Woodroffe *et al.* 1997; Creel and Creel 1998, 2002), which we summarize briefly. Like most large carnivores, wild dogs have disappeared from much of their historical range as human populations have expanded, and the dogs are now largely confined to protected areas and their peripheries. Wild dogs were actively destroyed by wildlife managers in most areas until the later part of the twentieth century, due to a perception that their method of killing prey was cruel, and that their cursorial hunting was disruptive for ungulate populations. Early in the 1970s, institutionalized culling of wild dogs came to an end, and they are now legally protected in the seven nations that hold substantial numbers (Fanshawe *et al.* 1991). Snaring and other human-caused deaths remain a substantial force of mortality in some populations (Drews 1995; Rasmussen 1996; Woodroffe and Ginsberg 1998). However, it is not clear that these problems affect wild dogs with greater force than they affect lions (*Panthera leo*) and spotted hyaenas (*Crocuta crocuta*), which generally have maintained thriving populations where wild dogs have declined or disappeared (Creel *et al.* 2001).

Wild dog population densities are invariably lower than the densities of sympatric large carnivores, typically by 1–2 orders of magnitude (Creel and Creel 1996). This basic pattern suggests that wild dogs differ from sympatric large carnivores in some fundamental aspect of their ecology. Data from several studies show that interspecific competition with spotted hyaenas and lions has strong effects on the density and distribution of wild dogs, both within and among ecosystems (Kruuk 1972; Frame and Frame 1981; Creel and Creel 1996; Mills and Gorman 1997; Gorman *et al.* 1998; Vucetich and Creel 1999; Creel *et al.* 2001). The impacts of competition on wild dogs are manifest through direct killing, interference competition at kills, loss of food, and exclusion from areas of high prey density. Although wild dogs do not attain high densities anywhere, their densities are lowest where spotted hyenas and lions are most common (Creel and Creel 1996; Mills and Gorman 1997; Creel *et al.* 2001). By holding wild dogs to low densities, interspecific competition plays a central role in making them vulnerable to extinction. Competition interacts with habitat fragmentation (Creel 2001). In the past, local numbers probably crashed when ecological conditions were unfavourable (just as they do now) but when conditions improved, recovery through recolonization or 'demographic rescue' was likely. Now, with increasing fragmentation and isolation, it is less likely that an area will quickly be recolonized when conditions are favourable (Fig 22.1).

Infectious diseases have also affected wild dog dynamics in several ecosystems. The literature on wild dogs often states that they are 'particularly sensitive to disease' (Fanshawe *et al.* 1991), or that infectious diseases have played 'a main role in the numerical and distributional decline of African wild dogs' (Kat *et al.* 1995). This idea is based mainly on data from the Serengeti ecosystem. There, wild dogs declined to local extinction while experiencing recurrent outbreaks of rabies and possibly canine distemper (Schaller 1972; Malcolm 1979; Gascoyne *et al.* 1993; Alexander and Appel 1994). The data from Serengeti clearly show that viral diseases can cause substantial mortality in wild dogs, and can contribute to a local extinction. However, the Serengeti population was probably vulnerable to extinction for other reasons. First, the population was small enough (less than 30 dogs) to be vulnerable to a knock-out blow, regardless of the cause (Ginsberg *et al.* 1995). Second, Serengeti dogs faced intense competition from larger carnivores (Frame 1986). Finally, Serengeti held a diverse suite of carnivores, many at high densities, that were known to carry rabies virus and/or canine distemper virus (CDV) (Maas 1993b; Alexander and Appel 1994; Alexander *et al.* 1994, 1995; Roelke-Parker *et al.* 1996). Under these conditions, it is expected that spillover transmission

Figure 22.1 Intra-guild hostility: lion (*Panthera leo*) kills African wild dog pup (*Lycaon pictus*) © M. G. L. Mills.

from high density species will endanger species living at lower density (Grenfell and Dobson 1995).

For these reasons, it might not be justified to generalize the conclusion that wild dogs are unusually vulnerable to diseases. Little is known about the regulatory role of diseases in other wild dog populations, but current data suggest that disease is not a major factor for all populations. Several dogs have died of infection with the bacterium *Bacillus anthracis*, in the Luangwa valley, Kruger National Park and Selous (Turnbull *et al.* 1991; Creel *et al.* 1995; van Heerden *et al.* 1995). In northern Botswana, 10 dogs in a pack of 12 died during an outbreak of canine distemper, which was confirmed as the cause of death for one wild dog (Alexander *et al.* 1996). Several packs disappeared at this time, but the population recovered quickly, much like the dynamics of Serengeti lions during an outbreak of CDV (Roelke-Parker *et al.* 1996). In Kruger and Selous there have not been detectable disease-related population declines over periods of 22 and 6 years, respectively (Reich 1981; van Heerden *et al.* 1995; Creel *et al.* 1997). Combining demography, serology, post-mortems, and veterinary examinations, van Heerden *et al.* (1995, p. 18) concluded that 'disease could not be incriminated as an important cause of death' in Kruger. In summary, current data are compatible with a wide range of views on the role of infectious disease in wild dog population dynamics. We suggest that the impact of viral diseases on wild dogs is similar to their impacts on other large carnivores, in which similar disease-driven declines are observed (Foggin 1988; Mills 1991, 1993; Maas 1993; Roelke-Parker *et al.* 1996). Local extinction triggered by disease has been described only for a wild dog population that was already precariously small—the fundamental problem rests with the ecological factors that keep population densities so low.

Study populations and methods

The three study areas share a few common features. All of the sites are relatively large, and all are situated in very large protected areas that are primarily woodland and wooded grassland. However, there are also clear differences among the populations (e.g. the prey base), and details for each area follow.

Northern Botswana

The data for Botswana come from an ongoing study that began in 1989. The 2600 km^2 (approx.) study area is situated at the northeastern corner of the Okavango Delta, a 14,000 km^2 freshwater alluvial fan with rich swamp vegetation at its centre and seasonal floodplains at its distal end. The study area lies at the end of the northern distributary of the delta, with habitats including recent and dormant seasonal floodplains, savannah woodlands, and shrublands. Floodplain habitats are a mosaic of small sinuate floodplains broken by wooded 'islands' ranging in size from a few square metres to several thousand hectares with *Acacia nigrescens, Croton megolobotrys*, and *Lonchocarpus capassa*, the predominant tree species. Woodlands are predominantly mopane and mopane shrub (*Colophospermum mopane*), Acacia (Thornveld) (*Acacia erioloba*), and Kalahari Bushveld (*Terminalia* spp., *Combretum* spp., *Acacia* spp.). The study area includes the northeastern section of the Moremi Wildlife Reserve and adjacent Wildlife Management Areas. Pastoralists are resident on portions of the study area.

The study population has ranged between 6 and 13 neighbouring wild dog packs per year and demographic characteristics of the sample are considered to be representative of the northern Botswana population, currently estimated at 700–850 individuals. Pack size averaged 10.4 adults and yearlings (range = 2–30, $N = 88$ breeding packs). Density estimates for wild dogs in northern Botswana range from a high of 35 per 1000 km^2 in the areas associated with northern Botswana's wetlands (the Okavango Delta including the study area and the Kwando-Linyanti River system in the north) to a low of 5 per 1000 km^2 in the drier habitats across most of northern Botswana. All individuals in the study population are identified using unique pelage markings and (excepting some immigrants) are known from birth. Reproduction in the population is seasonal (May–August). Population characteristics (age and sex of surviving adults) are summarized annually for all packs in June, the modal birth month. Age in years of individuals is based on the population's modal birth month. Whelping dates were known to be within a period of 2–3 weeks based on timing of observed mating and/or estimated by den attendance and growth/size of pups.

The primary prey species for wild dogs in the study area is impala, constituting 85% of their diet. Additional prey species include in order of frequency, kudu, red lechwe (*Kobus leche*), reedbuck (*Redunca arundinum*), steenbok (*Raphicerus campestris*), common duiker (*Sylvicapra grimmia*), and warthog. Calves of larger ungulate species are also occasionally and opportunistically preyed upon by Botswana's wild dogs including, wildebeest, tsessebe (*Damaliscus lunatus*), and buffalo (*Syncerus caffer*).

The Selous game reserve

In Selous, the data come from a study conducted from 1991 to 1997 on a 2600 km^2 site in the Northern Sector of the 43,600 km^2 Selous Game Reserve, which lies at the core of the 78,650 km^2 Selous ecosystem. Like most of southern Tanzania, Selous is dominated by miombo woodland, which is defined by two thornless deciduous trees, *Brachystegia* and *Julbernardia*. In Selous, miombo grades into chippya woodland, a lower and more open-canopied woodland that is dominated by *Combretum, Terminalia*, and *Pterocarpus*. Northern Selous also has substantial areas of thorn woodland, dominated by *Terminalia spinosa* and *Acacia drepanolobium*, and seasonally flooded plains dominated by grasses growing to more than 2 m (*Sporobolus pyramidalis, Setaria sphacelata, Andropogon gayanus*). Small areas of alkaline hardpan soil support short grass year round, usually with scattered palms. Most of the study area is moderate to dense woodland or long grass (though long grass areas are little used by wild dogs: Creel and Creel 2002). Approximately 80% of the study area is used for low-volume safari hunting, with the remainder used for low-volume photo-tourism.

Seven packs typically occupied the study site. Excluding pups less than a year old, the density of known individuals on the study site ranged from 35 adults and yearlings per 1000 km^2 to 46 adults and yearlings per 1000 km^2, averaging 38 per 1000 km^2. Including pups, mean density was 57 dogs per 1000 km^2. In other areas of the reserve, we used photographs and sightings to estimate densities from 16 to 24 adults and yearlings per 1000 km^2. Excluding pups and yearlings, pack size averaged 8.9 adults (range = 2–24). Including dogs of all ages, pack size averaged 18.9 (range 2–52). Methods of recording demographic data were almost identical to those described above for Botswana and below for Kruger, yielding 1068 annual records on the survival and reproduction of 365 individuals, with each dog contributing a mean of 2.9 annual records to the data set. (see Creel and Creel 2002 for details). Demographic data were tallied on June 15 each year, at the beginning of the breeding season (June–September, excepting two litters). Most dogs were of known age because they were first identified as pups or yearlings. For dogs identified as adults when we located a new pack, we estimated age on the basis of toothwear and pelage. Our analyses were not affected by the exclusion of dogs whose ages were estimated. For all three populations, our estimates of age-specific fecundity are limited to females, so data on survival were also restricted to females. Limited genetic data show that subordinate males sometimes father offspring (Girman *et al.* 1997; Creel and Creel 2002), so that paternity could not be assigned to the alpha male with complete confidence.

In systematic hunting data collected between 1991 and 1994 (N = 404 kills, 905 hunts), 17 species were hunted and 10 were killed (Creel and Creel 1995; Creel 2001). Wildebeest and impala were most commonly hunted (36% and 40% of hunts, respectively) and killed (54% and 29% of kills, respectively). Packs smaller than the median size preferred impala, while packs larger than the median preferred wildebeest (Creel and Creel 2002). In terms of mass killed, wildebeest was the most important prey, followed by impala: these two species formed the great bulk of the diet, with all other prey combined accounting for less than 10% of the mass killed.

Kruger National Park

The Kruger data were collected from 1989 until the present from an ongoing study in the 4280 km^2 Southern District. Six major habitats have been described (Mills and Gorman 1997); lowland sour bushveld, an open tree savannah dominated by *Terminalia sericea* and *Dichrostachys cineria* with a dense and tall grass layer, Malelane mountain bushveld, broken country with a heterogeneous vegetation, but where *Combretum apiculatum* is omnipresent, varying from moderate to dense bush savannah, *Combretum/Terminalia* bushveld, a relatively dense bush savannah with a moderate to dense short grass layer, *Acacia* thickets, dense woody vegetation particularly along the two perennial rivers, *Sclerocarya birrea/Acacia nigrescens* savannah, open tree savannah plains with a moderate shrub and dense grass layer, and the undulating Lebombo Hills, another heterogeneous dense to moderate bushveld area with north/south running ridges and bottomlands.

Demographic statistics are given for 1 January each year, the midpoint in the annual breeding cycle, and include all age groups. Eight to twelve packs inhabited the study area at varying densities of 19–39 dogs per 1000 km^2 in different years. The Southern District had a higher density of wild dogs than the Central and Northern Districts, where densities fluctuated between 4.4–18.6 and 2.8–22.5 dogs per 1000 km^2, respectively. Mean pack size was 10.4 (range 2–36, $N = 91$ packs). All individuals in the study population were identified by coat patterns and most were known from birth or as large pups.

Impala were the dominant prey species for wild dogs whether expressed as the percentage of kills (73.2%) or as the percentage of biomass in the diet (81%), with greater kudu as the secondarily important species (5.4% of kills and 8.1% of biomass), followed by common duiker (8.9% of kills and 4.4% of biomass), steenbok (8.9% of kills and 2.5% of biomass), bushbuck and reedbuck (each 1.8% of kills and 2.5% of biomass). Kruger has a more developed infrastructure and human presence than the other two study areas, and is used for photo-tourism, with active management of some animal populations.

Age-specific survival

For all three populations, we calculated rates of apparent survival (restricted to females) by determining whether an individual remained in the population from one breeding season to the next (exact dates give above). Our monitoring was sufficiently intense that we rarely encountered unknown wild dogs, and these appearances generally occurred under circumstances suggesting that the new dogs were immigrants (single-sex groups, moving widely: see McNutt 1996b; Creel and Creel 2002). When essentially all of the individuals in an area are resighted frequently, rates of apparent survival will be good estimates of true survival. To account for dogs that dispersed off the study site but survived, we modified the raw survival rates by assuming that the number of unknown immigrants of a given estimated age was equal to the number of undetected emigrants of that age. We also assumed that any dominant dog that disappeared was dead, because no alphas were ever known to disperse. The effect of these modifications on estimated survival rates was minor.

Patterns of survival differed strikingly among the three populations. In Kruger, survival was very poor for the first 2 years of a dog's life (Fig 22.2). Annual survival for pups was 0.35 (exact binomial CI: 0.29–0.42), and yearling survival was 0.45 (CI: 0.34–0.57). Altogether, only 16% of newborns survived to adulthood at 2 years of age. After reaching adulthood, annual survival was good, remaining above 0.72 for all ages except 5-year-olds, for which the sample size was not large ($N = 16$). There was no evidence for senescence in the Kruger survival rates. The survivorship curve for Kruger females falls below that of Botswana and Selous, and is far more concave than the curve for Selous (Fig. 22.3).

In Botswana, only half of all newborns survived their first year (mean annual pup survival = 0.48, CI = 0.42–0.54), though the survival rate for pups was 37% better than in Kruger. Botswana yearlings survived well (mean = 0.74, CI = 0.62–0.79), and actually had a higher probability of survival than any adult age class in the Botswana population (Fig. 22.2). The rate of survival from birth to adulthood was 35%, more than double the rate in Kruger, but 80% lower than in Selous (see below). Adult survival in Botswana ranged from 0.40 to 0.67 (0.50–0.67 if we exclude age 7, for which $N = 5$), less than the survival rate of adults in Kruger, with little or no evidence of senescence (Fig. 22.3).

Survival in Selous was better than in the other populations, particularly for young dogs. Pups survived at a rate of 0.75 (CI: 0.66–0.84), which exceeds the rates in Botswana and Kruger by 1.56 × and 2.33 ×, respectively. Selous yearlings also fared well, with an annual survival rate of 0.84 (CI = 0.73–0.91), which exceeds yearling survival in Botswana by 1.14 × and in Kruger by 1.87 ×. Adult survival in Selous showed evidence of senescence unlike the

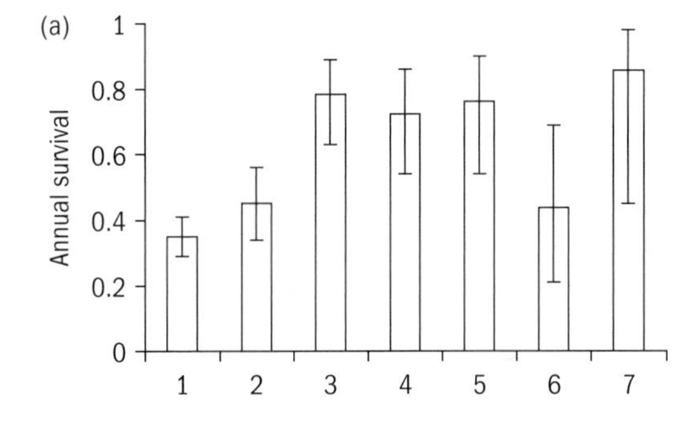

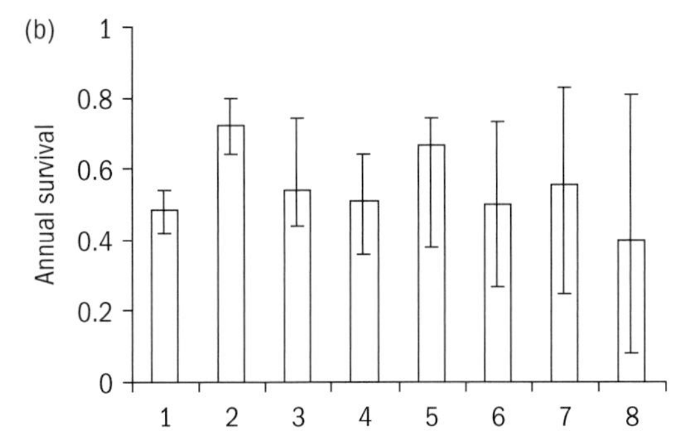

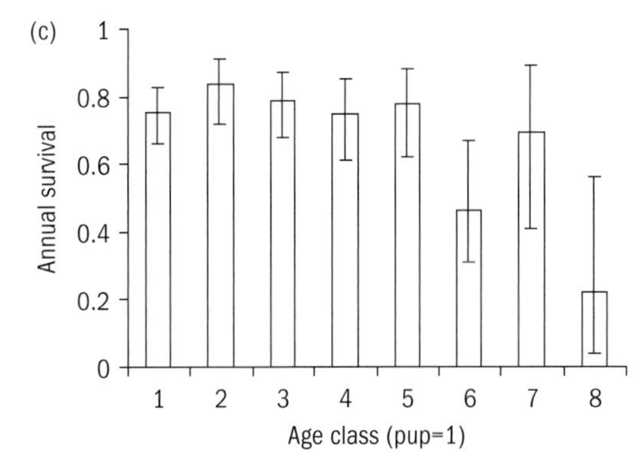

Figure 22.2 Age-specific annual survival rates for female wild dogs in (a) Kruger, (b) Northern Botswana, and (C) Selous. Error bars show 95% exact binomial confidence limits.

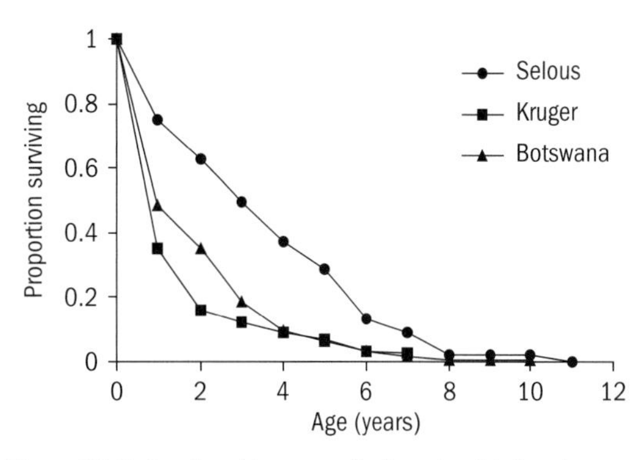

Figure 22.3 Survivorship curves for female wild dogs in Kruger, Northern Botswana, and Selous.

other two populations. Mean annual survival for 2-year-old to 4-year-old females was 0.77 ($N = 145$), while mean annual survival of dogs five or older was significantly lower (mean = 0.50, $N = 74$, $P = 0.02$, Fisher's exact test).

Examining the survivorship curves (Figs 22.2 and 22.3), survival to adulthood was much better in Selous (63%) than in Botswana (35%), which was in turn much better than in Kruger (16%). Differences among populations in adult survival were less pronounced. Survival was best for adults in Kruger and for young adults (2–4 years) in Selous, intermediate for adults in Botswana, and low for old adults (5+ years) in Selous.

Reproduction and age-specific fecundity

In all three populations, we knew when pups were born in most packs, through observations of mating, gestation, and pup size. The mass of a wild dog litter constitutes a greater proportion of the female's body mass than in any other canid, and this makes pregnant females very conspicuous (e.g. fig. 9.2 in Creel and Creel 2002). By identifying pregnant females, we could assign the mother's age to most litters, to determine age-specific fecundity. Knowing when litters were born, we counted them as soon as possible after emergence from the den. Wild dog pups remain underground for most of the first few weeks of life, so it is likely that one or more pups occasionally died before we made a count. However, the constraints on counting pups were the same for all three studies, so the data can be directly compared, though we cannot rule out the possibility that pre-count mortality varied among populations. Mean litter size differed substantially among the three populations (Fig. 22.4: $F_{2,148} = 4.91$, $P = 0.009$). Selous females had the smallest litters, with a mean of 7.5 ± 0.56 pups (SE, $N = 38$). Litter sizes were larger in Botswana (10.1 ± 0.37 pups, $N = 57$) and Kruger (9.4 ± 0.70 pups, $N = 57$).

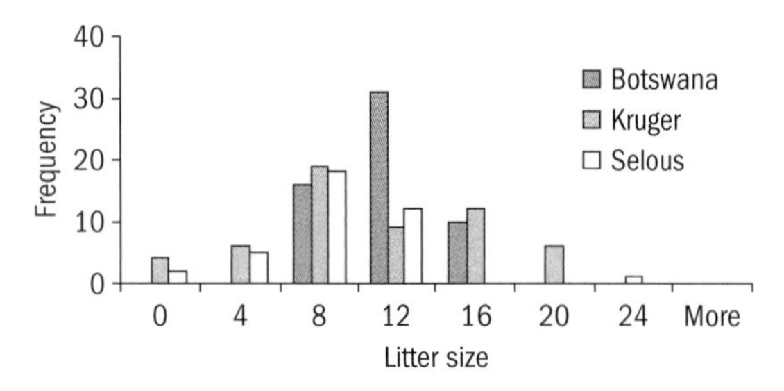

Figure 22.4 The frequency distribution of litter sizes for wild dogs in three population. The labels on the *X*-axis give the lower limit of each litter size category.

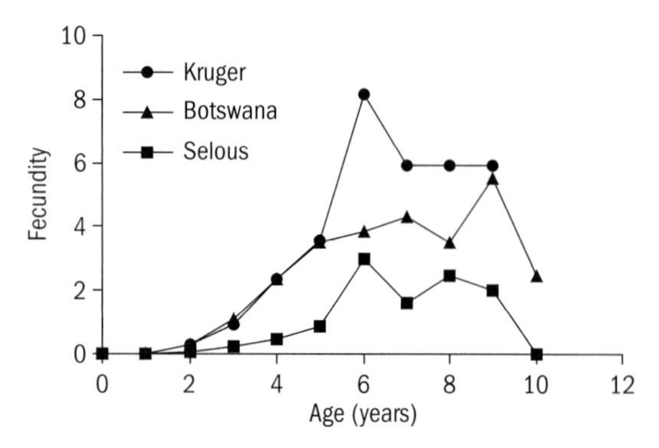

Figure 22.5 Age-specific fecundity curves for female wild dogs in three populations. Fecundity is measured as litter size divided by two, which must be taken into account when comparing this figure to other data on reproduction. For Kruger, data were pooled across ages 7–9 (these age classes were collapsed in our life table analyses to maintain a sample size of 5).

Age and reproduction

Age-specific fecundity curves for the three populations are shown in Fig. 22.5. The fecundity curve is highest in Kruger, lowest in Selous, and intermediate in Botswana. These differences in fecundity arise for three reasons. First, they parallel the differences in mean litter size just described: high fecundity in Kruger is partly due to the fact that litters are large among those females that breed. In a solitary or monogamous breeding system, this might completely explain differences in the fecundity curves, but it is important to recall that wild dogs are cooperative breeders. The majority of subordinate females does not breed (Malcolm and Marten 1982; Creel *et al.* 1997), and female dominance is positively correlated with age (Creel and Creel 2002). This breeding system, combined with differences among populations in the survivorship curves, has a strong effect on the fecundity curves. In Kruger and Botswana, poor survival among young age classes means that very few females reach ages older than 4. Consequently, most females that are 4 or older have no older pack mates of the same sex, and they are likely to be dominant. In Kruger and Botswana, the fecundity of females of age 4 or older is high partly because few are non-breeding subordinates. In Selous, with substantially better survival through early adulthood, it is more likely that females will have an older, dominant packmate of the same sex, and thus will not breed.

Third and last, the shapes of the fecundity curves are affected by the relationship between litter size and the age of the breeding female (Fig. 22.6). Here, we are not considering variation in the proportion of females that breed (as in point two), but are restricting the data to only those females that produce litters. In all three populations, litter sizes tend to be small for young females producing their first litters, and rise as female age increases. In Kruger, this rise continues unabated to the oldest age classes. In Selous, litter size decreases significantly among older females (Creel and Creel 2002), and Botswana shows a similar pattern. The tendency for litter sizes to decline among old breeders is based on relatively few observations, because a very small proportion of females survive to these ages (Fig. 22.3). Regardless of this pattern, the tendency for litter size to rise with age among young and prime-aged adults is shared by all three populations, with an effect on the shape of the fecundity curves (Fig. 22.5).

Pack size and reproduction

Wild dogs are communal hunters and cooperative breeders. As pack size increases, foraging success improves in several ways. Larger packs kill larger prey, chase prey over shorter distances, are more likely to make a kill, and make more multiple kills (Fanshawe and Fitzgibbon 1993; Creel and Creel 1995; Fuller *et al.* 1995; Creel 2001). In large packs there are more non-breeders, who help to raise offspring by guarding them from predation, regurgitating meat to pups at the den, regurgitating meat to

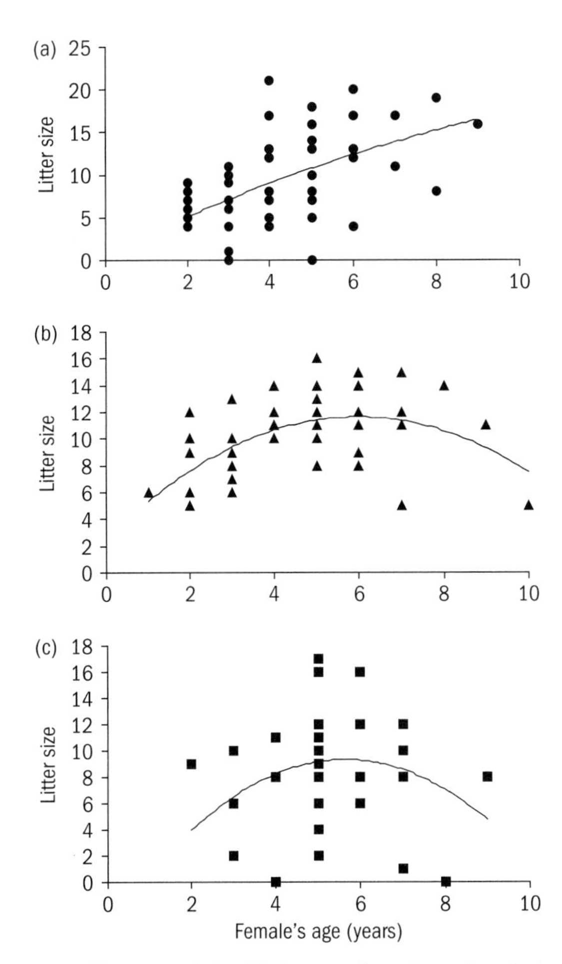

Figure 22.6 The relationship between litter size and mother's age, among reproductive females only, for female wild dogs in three populations—(a) Kruger, (b) Botswana, and (c) Selous. In each case, the curve is a second order polynomial fit by least squares.

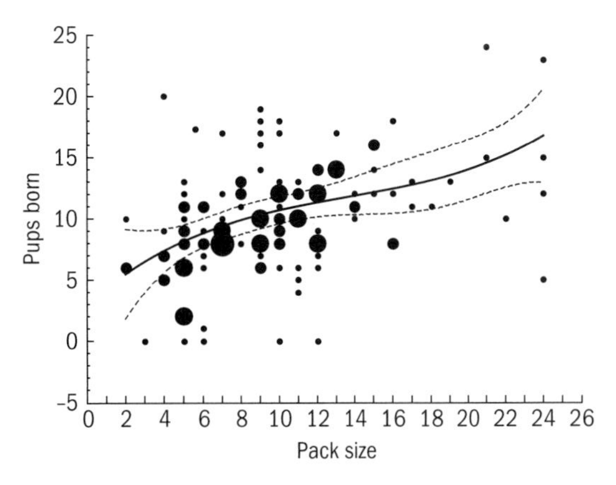

Figure 22.7 The relationship between pack size (adults and yearlings) at the onset of the breeding season and the annual number of pups born, for three wild dog populations. Point size is proportional to the number of identical observations. The curve is a third order polynomial with 95% confidence limits.

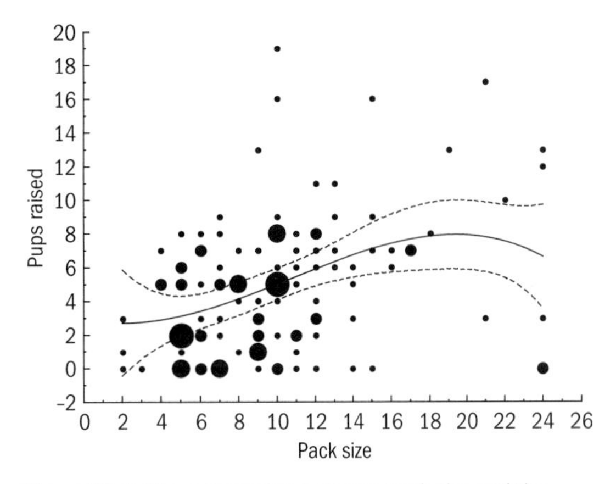

Figure 22.8 The relationship between pack size and the number of pups raised to 1 year for three wild dog populations. Point size is proportional to the number of identical observations. The curve is a third order polynomial with 95% confidence limits.

the alpha female during gestation and lactation, and giving priority of access to carcasses to the pups once they have left the den (Malcolm and Marten 1982; McNutt and Boggs 1997). If these benefits outweigh any costs of grouping, then reproductive success should be related to the number of adults in a pack. Two analyses show that this is indeed the case (Figs 22.7 and 22.8). First, the number of pups born increases as the number of adult pack members increases (Fig. 22.7: $F_{1,129} = 26.04$, $P < 0.0001$, $r^2 = 0.17$, $b = 0.41 \pm 0.08$ SE). Obviously, this relationship does not depend on the care given to pups after their birth, so it is probably dependent mainly on the benefits of communal hunting. The number of pups raised to 1 year also increases as pack size increases, an association that is stronger than the previous one (Fig. 22.8: $F_{1,113} = 71.52$, $P < 0.001$, $r^2 = 0.39$, $b = 0.62 \pm 0.07$). This relationship depends both on communal hunting and allo-parental behaviour.

Figure 22.9 A large litter of African wild dog pups (*Lycaon pictus*) in Kruger National Park © M. G. L. Mills.

In summary, reproduction is affected by age and pack size in all three populations. In general, reproductive success is better for older females in larger packs, though litter size might decrease for very old females in Botswana and Selous. Mean reproductive success is greater in Botswana and Kruger than in Selous, and is particularly high for old females in Kruger (Fig. 22.9).

Life-history trade-offs

Several properties of our demographic data suggest demographic compensation or life-history trade-offs consistent with the principle of allocation, which states that resources directed towards improving one vital rate must be diverted away from another vital rate. The three populations have similar growth rates, with λ slightly greater than one in each case (see below). This suggests that ecological constraints on growth are similar in the three ecosystems we studied. Assuming that λ is fixed over short time spans, an increase in one vital rate must be offset by a decrease in another rate. Three patterns in the age-specific vital rates might be the result of such life-history trade-offs. First, differences among populations suggest a trade-off between the number of pups produced and their survival. Litter sizes are largest in Kruger, where pup survival is lowest. Pups survive at a much higher rate in Selous, where litters are significantly smaller.

Second, there is an apparent trade-off between reproduction and adult survival. The survivorship curve for Selous lies well above the survivorship curves for Botswana and Kruger. If this difference was not offset by lower fecundity in Selous, one would predict a much greater difference in the growth rates than we actually observed. Finally, there may be a trade-off between juvenile survival and adult survival. This is suggested by the survivorship curves, particularly for Kruger, where very poor survival to adulthood is at least partially offset by good adult survival, with little sign of senescence.

Deterministic population growth rates and elasticities

Deterministic growth rates

We used our data on age-specific survival and fecundity to build a Leslie matrix for each population, and determined the annual growth rate (λ) from the dominant eigenvalue of the Leslie matrix. We confirmed these population growth rates by traditional life-table analysis in Excel, using Euler's equation to calculate the intrinsic rate of increase, $r\ (= \ln \lambda)$.

Given the differences in demography among populations, the deterministic growth rates were surprisingly similar in Kruger ($\lambda = 1.003$) and Botswana ($\lambda = 1.000$), both very close to zero growth. The

Selous population showed slightly positive deterministic growth ($\lambda = 1.038$, or 3.8% annual growth). "It should be pointed out that the selous population dataset derives from six years of observations, and therefore is less likely to reflect demographic variations than the comparitively longer periods of study (12 yr) in the other two populations. Estimates of demographic variance will depend both on sample size (which are not necessarily tied to duration) and also on the simple duration of observations." Overall, the demographic data suggest that these populations are more or less maintaining their current size (but see below for discussion of variance).

However, these growth rates fall in the region near $\lambda = 1$ that is of particular concern for modellers of population dynamics and viability. Botswana and Kruger, in particular, fall almost perfectly on the line between systematic growth and systematic decline. The slightest of shifts in mean rates of survival or reproduction would push λ across this line. Given the variances in our estimates of survivorship and fecundity, it is almost certain that these populations vary between periods of growth and decline. This is well illustrated in the Kruger population where censuses have revealed the total population to vary as follows: 357 in 1989, 434 in 1995, and 177 in 2000 (Davies 2000). Because the mean of a stochastically varying growth rate is smaller than the underlying deterministic growth rate (Dennis *et al.* 1991), realized growth rates could easily be negative under the demography we found for these wild dog populations, even with no change in patterns of survival and fecundity. To investigate dynamics in more detail, we used stochastic population projections (see below).

Impacts of vital rates on population growth

To determine which vital rates have the greatest potential impact on improving population growth, we calculated elasticities, using the shareware POPTOOLS. Elasticities measure the response of λ to changes in a single vital rate, with the magnitude of the change scaled to the mean of that vital rate. In other words, elasticity gives the impact on λ that would be obtained by changing each vital rate by a fixed proportion of its current mean, holding all other vital rates constant (Caswell 2001).

Figure 22.10 shows the elasticity of λ to changes in age-specific survivorship and fecundity. Despite the differences in demography among the three populations, the vital rates with the greatest effect on population growth are the same. For all three populations, improvements in the survival of pups and yearlings would have the greatest effect on growth, followed by improvements in the survival of young adults. The elasticity of λ to changes in fecundity were comparatively small. This result accords with the observation that λ is largest in Selous, where pups and yearlings had substantially better survival rates than in Botswana and Kruger.

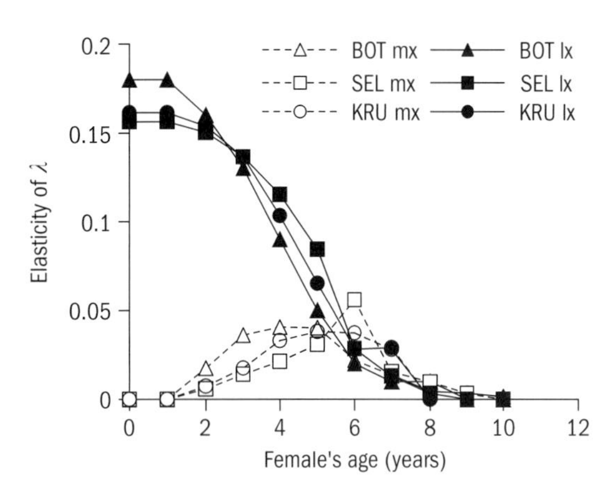

Figure 22.10 The elasticity of population growth (λ) to changes in age-specific survivorship or fecundity, in three wild dog populations. Population growth responds most strongly to changes in the survival of juveniles and young adults.

This result is also interesting because juvenile survival shows substantial variation within and among populations (see Fig. 22.2, also Creel and Creel 2002). There has been some debate over the value of elasticities to guide conservation, in part because the vital rates with the greatest elasticity may not show much variation in the real world (Mills *et al.* 1999; Caswell 2001; Ehrlen *et al.* 2001). For wild dogs this concern does not apply because the vital rate with the largest elasticity, pup survival, showed substantial variation among populations, with a 2.3-fold difference among population means. Within a single population (in Serengeti National Park) pup survival ranged from 0.24 ± 0.11 to 0.83 ± 0.17 over two decades (Frame *et al.* 1979; Malcolm and Marten 1982; Burrows *et al.* 1994). In accord with our results, the Serengeti population attained its highest density during a period of unusually good pup survival.

Where direct conservation action is envisioned, our results show that actions aimed at improving the survival of pups and yearlings will be most effective (Creel and Creel 2002). (This result does not agree with the IUCN African Wild Dog Action Plan (Woodroffe *et al.* 1997) which used the generalized PVA program VORTEX to conclude that adult survival has a greater effect on extinction risk than juvenile survival.) The best permissible methods of improving subadult survival will probably vary, depending on the specific factors that kill young dogs at a given time in a given place. Where direct action is considered necessary, a range of possibilities exists, particularly to improve the survival of pups. These might include: provision of food during gestation, lactation, or denning; actions to reduce the presence of lions and spotted hyaenas near den sites; oral vaccination of pups (several diseases affecting wild dogs are more likely to kill pups than adults).

Stochastic projections of population dynamics

The analysis above is retrospective, using empirical data to measure past growth rates and identify the vital rates with the strongest effect on growth. In this section we take a prospective approach, using the Leslie matrices to run stochastic population projections for each population. The purpose of these projections is to determine how differences in the populations' demography translate into future population dynamics and extinction risk. The stochastic Leslie matrix projections were implemented in Excel, using the shareware POPTOOLS. Briefly, Leslie matrix projections work as follows. At each time step (year), the population is tracked as a vector with the number of individuals in each age class. This vector is premultiplied by the Leslie matrix to give a new vector with the number of individuals in each age class at the next time step. This multiplication implements 1 year of age-structured population growth. We made survival stochastic by drawing the number of survivors from each age class at each time step from a binomial distribution with a mean determined from the Leslie matrix, and variance determined by the number of individuals in that age class at that time step. We made reproduction stochastic by drawing fecundities from a normal distribution with a mean determined by the Leslie matrix, and variance calculated directly for each population. The variance in fecundity was pooled across ages within each population, due to sample size limitation.

We set the initial age structure equal to the observed age distribution for each population, rather than the stable age distribution (see Caughley 1994 for discussion). To facilitate comparison among populations, we set the initial population size to 880 (the estimated size of the Selous population) for all three populations, but broad results were very similar for smaller initial population sizes. We projected dynamics 200 years *only* to allow differences among

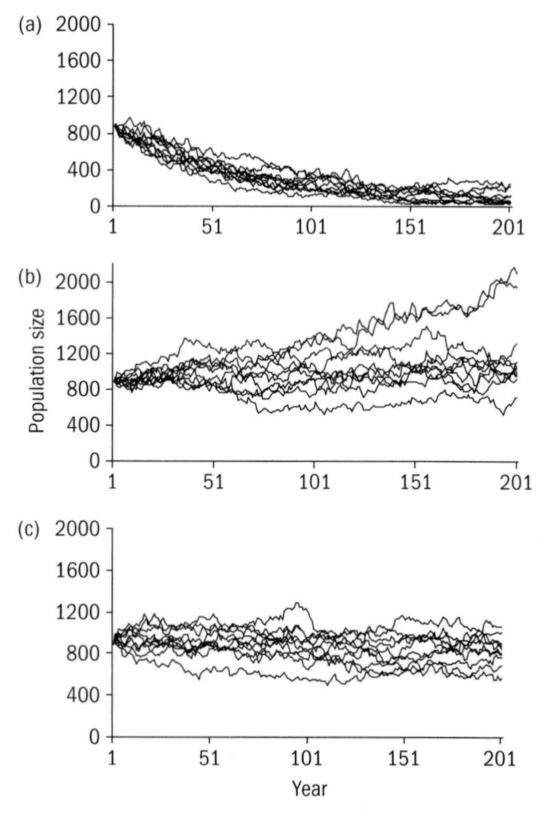

Figure 22.11 Examples of stochastic population projections for three populations with the demography of wild dogs in (a) Kruger, (b) Botswana, and (c) Selous, each with an initial population size of 880.

populations to become apparent, and we emphasize that it would be unwise to assume that the demography we observed in the field will remain unchanged for two centuries. The intention of these projections is to assess how dynamics are affected by differences in demography, not to determine long-term probabilities of local extinction. After running 100 projections per population, we used the linear regression of projected population size on years to determine the stochastic λ for each projection, and calculated a mean stochastic λ for each population.

Examples of projected dynamics for the three populations are shown in Fig. 22.11. Here, the differences among the three population are striking. Incorporating demographic stochasticity, the Selous (stochastic $\lambda = 1.000$) and Botswana (stochastic $\lambda = 1.000$) populations show no systematic growth or decline, though variation in the population trajectories is substantially greater for Botswana. In contrast, projections for the Kruger population show a clear tendency to decline, with stochastic $\lambda = 0.986$. Again, we do not mean to imply that the Kruger population will follow such a trajectory. With a change in population density, the demography may shift, and we do not know if growth is density dependent. However, we can conclude that the *current* demography of wild dogs in Kruger leads to dynamics that are more prone to collapse.

What processes drive this difference among populations? The answer is best seen by examining how the proportion of the dogs in each age class varies through time. Figure 22.12 shows a representative projection of age classes for each population. In Kruger, a large proportion of the population is young, due to high fecundity and a steep survivorship curve. However, the number of young individuals varies dramatically, jumping up in response to the survival of females into old, highly fecund age classes. The

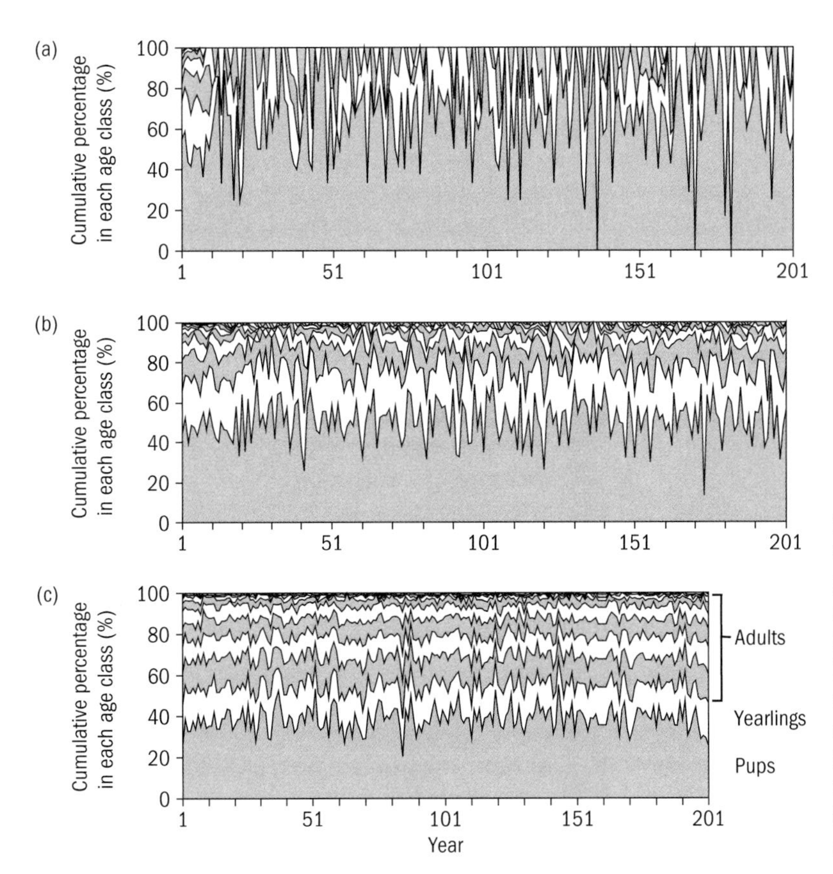

Figure 22.12 Changes in age structure during stochastic projections for each of three wild dog populations— (a) Kruger, (b) Botswana, and (c) Selous. In each panel, the bottom area (shaded) shows the proportion of the population made up of newborns. The area just above (white) shows yearlings, and the areas above show adults, separated into 1-year age classes.

probability of females surviving to these highly fecund age classes is low, but when it happens, the impact is large. Consequently, the age structure of Kruger is volatile in comparison to Botswana and Selous (Fig. 22.12). The greater this demographic variance, the larger will be the difference between the deterministic and stochastic growth rates (Dennis *et al.* 1991) as has been revealed in the Kruger population censuses (Davies 2000). This variation is amplified by large variation in litter size in Kruger, relative to Selous and Botswana (Fig. 22.4). Our projections suggest that the Kruger population maintains a deterministic λ near one, despite poor subadult survival, due to pulses of reproduction by the rare females who survive beyond 5 years (Fig. 22.12). The Selous and Botswana populations, with better survival and lower fecundity, are dynamically less volatile.

Conclusions

1. Patterns of age-specific survival and fecundity differ substantially for wild dogs in Kruger, Selous, and Northern Botswana.
2. Survival to adulthood is low (16%) in Kruger, high (63%) in Selous, and intermediate (35%) in Botswana.
3. Offsetting these differences, fecundity is highest in Kruger, intermediate in Botswana, and lowest in Selous. These differences are partially driven by differences in litter size, which is lower in Selous than in the other two populations. The differences among populations suggest demographic compensation or life-history trade-offs.
4. Despite differences in demography, the three populations have similar deterministic rates of population growth (Botswana: $\lambda = 1.000$, Kruger: $\lambda = 1.003$, Selous: $\lambda = 1.038$).
5. Kruger and Botswana have deterministic population growth rates sitting exactly on the line between systematic growth and systematic decline.
6. Stochastic population projections suggest that the demography of wild dogs in Kruger (very poor juvenile survival offset by high fecundity among old females) produces volatile population dynamics. High levels of demographic stochasticity in Kruger cause the stochastic growth rate to fall well below the deterministic growth rate. The demography of Selous dogs (good survival and lower fecundity) is more stable. Botswana is intermediate, as in most other aspects of our analysis.

Population dynamics are most affected by juvenile survival in all three populations, despite their differences in demography. Along with strong impacts on population growth, juvenile survival shows substantial variation within and among populations. Thus, juvenile survival should be a focal point for any direct conservation actions.

Acknowledgements

JWM is grateful to the Office of the President of Botswana and the Department of Wildlife and National Parks for permission to conduct research in Botswana. Most of the fieldwork represented in this paper was supported by the Frankfurt Zoological Society – Help for Threatened Wildlife. We are also grateful to the Woodland Park Zoo, Conservation International, ZSP, the Denver Zoo, the Folger Foundation, and numerous private donors. SC is grateful to Nancy Creel for all of her work on the Selous wild dog project, to the Tanzania Wildlife Division and the management of the Selous Game Reserve for support and permission to condict the research, and to the Frankfurt Zoological Society, National Science Foundation and Smithsonian Institution for grants. Gus Mills thanks SAN Parks, the Endangered wildlife Trust and The Tony and Lisette Lewis Foundation South Africa for support and Harriet Davies for help with data analysis.

PART III

A Conservation perspective

CHAPTER 23

Conservation

From theory to practice, without bluster

David W. Macdonald and Claudio Sillero-Zubiri

Ethiopian wolf *Canis simensis* © C. Sillero.

Do we know enough to be able to conserve canids and, as a related matter, to resolve—or at least manage—the conflicts that they face with people? If not, what must be discovered? The captivating processes, patterns, and details revealed in the preceding chapters of this book, and the exhilarating desire to know more than they fire in our imaginations, make it plain that the quarter of a century or so of research reported here enables scientists to answer many questions, and to ask many more. But do the answers in hand include those necessary to underpin conservation action? And, of the new generation of questions—however interesting they may be—which, if any, are prerequisites to that action? That we ask these questions does not indicate that we are victims of a utilitarian bigotry that spurns non-applied research—on the contrary, the scholarly unravelling of natural history is, in our view, a pursuit of the highest merit (and one which we will argue, below, is more important than generally credited as a force for canid conservation). However, conservation is a pressing matter, and so it is not a philistine action to draw a distinction between knowledge that is interesting and that which is also immediately useful, and to ask whether we have enough of the latter to solve the current problems of canid conservation. One measure of the answer can be found in *Canids: foxes, wolves, jackals, and dogs. Status survey and conservation action plan* (Sillero-Zubiri *et al.* in press).

The Canid Action Plan is the product of the deliberations of the Canid Specialist Group (CSG), itself one of more than 120 groups of specialists with a taxonomic focus on conservation under the aegis of the Species Survival Commission (SSC) of the IUCN—the World Conservation Union (initials originally abbreviating the International Union for the Conservation of Nature—www.iucn.org). The SSC uses a classification based on evaluated risk to assign a conservation status to each species (IUCN 2001). The CSG has classified each of the 36 living species of wild canid; their statuses (together with an interpretation of the categories) are given in Table 23.1. As its name implies, The Canid Action Plan aspires to

Table 23.1 Red List assessments of all canid species, using version 3.1 of the Red List categories and criteria (IUCN 2001)

Species	Red List 2004	Species	Red List 2004
Falklands wolf	EX	Coyote	LC
		Crab-eating fox	LC
Darwin's fox	CR C2a(ii)	Culpeo	LC
Island fox	CR A2be + 3e	Golden jackal	LC
Red wolf	CR D	Gray fox	LC
		Grey wolf	LC
African wild dog	EN C2b	Indian fox	LC
Dhole	EN C2a(i)	Kit fox	LC
Ethiopian wolf	EN C2a(i), D	Pampas fox	LC
		Raccoon dog	LC
Blanford's fox	VU C1	Red fox	LC
Bush dog	VU C2a(i)	Side-striped jackal	LC
Dingo	VU A1e	Swift fox	LC
Maned wolf	NT	Tibetan fox	LC
Arctic fox	LC	Fennec fox	DD
Bat-eared fox	LC	Hoary fox	DD
Black-backed jackal	LC	Pale fox	DD
Cape fox	LC	Rüppell's fox	DD
Chilla	LC	Sechuran fox	DD
Corsac fox	LC	Short-eared dog	DD

Notes:

Categories for the IUCN Red List of threathened species that apply to the Canidae. For a detailed treatment and definitions of the criteria refer to IUCN (2001) and www.redlist.org.

Extinct (EX): there is no reasonable doubt that the last individual of the taxon has died.

Critically Endangered (CR): the best available evidence indicates that the taxon meets any of the criteria A to E for Critically Endangered, and it is therefore considered to be facing an extremely high risk of extinction in the wild.

Endangered (EN): the best available evidence indicates that the taxon meets any of the criteria A to E for Endangered, and it is therefore considered to be facing a very high risk of extinction in the wild.

Vulnerable (VU): the best available evidence indicates that the taxon meets any of the criteria A to E for Vulnerable, and it is therefore considered to be facing a high risk of extinction in the wild.

Near Threatened (NT): the taxon has been evaluated against the criteria but does not qualify for any of the above now, but is close to qualifying for or is likely to qualify for a threatened category in the near future.

Least Concern (LC): the taxon has been evaluated against the criteria and does not qualify for any of the above. Widespread and abundant taxa are included in this category.

Data Deficient (DD): there is inadequate information to make a direct, or indirect, assessment of its risk of extinction based on its distribution and/or population status of the taxon. DD is not a category of threat, the taxon may be well studied, and its biology well known, but appropriate data on abundance and/or distribution are lacking.

identify important actions and to plan for their implementation. One way of doing this has been to canvas the views of the international community of canid specialists on the latest knowledge and status of each species, the threats they face, the questions that must be answered and the actions that must be taken to ameliorate these threats. One way to answer the question with which we opened this chapter—do we know enough (in terms of research findings) to conserve canids or, if not, what is it still necessary to discover—is to analyse the priorities submitted to the CSG by these specialists who represent a sort of Delphic circle of the well-informed.

Before summarizing these priorities, we should be plain about what this exercise is not: it is not an unbiased, statistically systematic sample, stratified by, for example, region, taxonomy, or nationality, or experience of the specialist. The number and nature of proposals submitted to the CSG is clearly biased by the energy and preoccupations of the individuals submitting them and the degree of threat, rarity, or perceived charisma of the species involved. Furthermore, we are keenly aware that most of the people submitting proposals were mainly biologists with aspirations in research, and that the whole action planning process focuses on species—we would have expected very different proposals had more development experts been involved, and had the action planning process been a system or area-based one. Nonetheless, we explore what lessons can be learnt from this process as it exists. The CSG maintains a list of 100 members and currently endorses more than 50 projects, it is structured into 7 regional and 13 thematic working groups (see www.canids.org), all of whom were invited to participate in the construction of the Canid Action Plan, which includes contributions from more than 90 specialists and has been reviewed by a further 80. Furthermore, 240 specialists from 38 countries gathered in Oxford in September 2001 at the Canid Biology and Conservation Conference, of which one explicit aim was to develop the Canid Action Plan through a series of workshops. We therefore conclude that two interesting topics can be explored on the basis of the priorities submitted by these specialists—first, we can gain a sense of the types of knowledge that are judged still to be lacking from the canid conservationist's armory

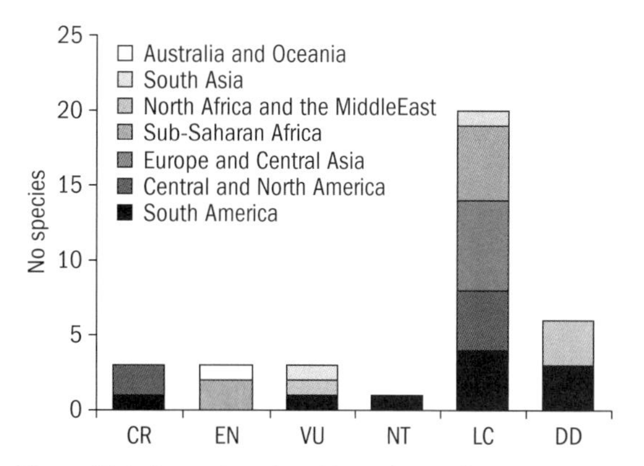

Figure 23.1 Proportion of canid taxa from different regions classified in the different IUCN Red List categories of threat (Sillero-Zubiri *et al.* in press).

and, second, we can learn something of the preoccupations and thought processes of the contributing specialists (and perhaps some strengths and weaknesses of the action planning process).

From the classification exercise undertaken by the CSG and reported by Sillero-Zubiri *et al.* (in press), a profile of the status of wild canids emerges (Fig. 23.1). We know that one canid species became extinct only relatively recently. The last Falklands wolf, also known as Malvinas fox (*Dusicyon australis*), disappeared in 1876, only a few years after Charles Darwin had warned of the species imminent demise (Darwin 1845). Of the 36 extant canid taxa 9 (25%) are listed as threatened (3 Critically Endangered, 3 Endangered, and 3 Vulnerable), whereas 1 is considered Near Threatened. The majority (56%) of species were considered safe and listed as Least Concern (20 species), and a further six (17%) were listed as Data Deficient, since there was insufficient information to make an informed assessment. The nine threatened canids are distributed in six geographic regions, so no obvious distribution pattern is apparent, although all three Critically Endangered species are located in the Americas. The most obvious trait shared by threatened canids is restricted distribution. Two species occur in islands (and, de facto, perhaps we should consider Ethiopian wolves in this category too), a fourth is found in a very small range.

The abundance of these submitted projects makes it clear that at least many of the specialists believe that more needs to be discovered if science is adequately to

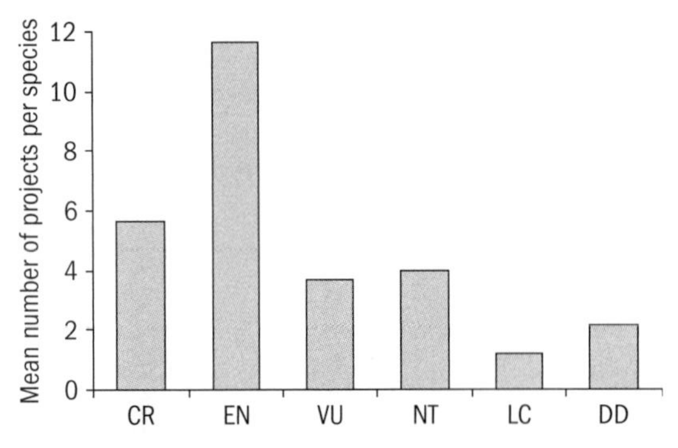

Figure 23.2 Mean number of projects and actions proposed in the Canid Action Plan for each species in relation to their category of threat ($n = 120$; Sillero-Zubiri *et al.* in press).

Table 23.2 Number of projects and actions proposed in the Canid Action Plan assigned to the main 10 themes identified (Sillero-Zubiri *et al.* in press)

Type of project	Number
Distribution, abundance, and monitoring	31
Taxonomy and genetics	11
Basic biology and field techniques	14
Community ecology	3
Disease and epidemiology	9
Human dimension	9
Planning, lobbying, and evaluation	12
Protected areas and active management	16
Captive breeding	7
Education	8

underpin canid conservation. However, 11 (30%) species were not named in any project; one might deduce that the consensus is that enough is known about these species already, but two other explanations should be born in mind. First, these species may include some that simply have no advocate within the specialist community, and some of these may be so poorly known that threats to them can only be guessed. Second, the preoccupations of the most vociferous specialists may lie with species where the conservation issue is endangerment through rarity, rather than management of conflict (a topic to which science may have more still to offer, as we will argue below). The distributions of threat categories for the species for which projects were submitted are given on Fig. 23.2. The 9 threatened species account for 60% of projects, averaging 6.6 projects per species, whereas the remaining 27 species are considered in an average by 1.6 projects.

Although some proposed projects encompass several topics, and some blend into actions, most can readily be assigned to one of ten research themes which clearly indicate the topics where the specialists believe more knowledge is needed (Table 23.2). These themes are enumerated in the following paragraphs (this summary is based on 120 projects and actions allocated to a single category, although it was sometimes difficult to decide where one project ended and another began). The examples that illustrate these themes are far from exhaustive. In particular, we will return below to the emphasis on research rather than actions.

The main themes

Distribution, abundance, and monitoring

By far the largest category encompassed research projects intended to answer simple questions about the whereabouts and abundance of less known species. In this context, an obvious example is the Darwin's fox (*Pseudalopex fulvipes*), previously known from only Chiloé Island, and recently discovered from an isolated mainland population in Nahuelbuta National Park. Clearly it is important to search for additional populations on mainland Chile, targeting the remaining dense, virgin forests between the Maullin and Nahuelbuta Mountains.

A more defuse problem is posed by the bush dog, for which proposals focus on questionnaire surveys throughout their historical range and on field surveys in protected areas throughout Paraguay and Brazil, using 1 km^2 grids incorporating camera traps, baited tracking stations, hair traps and faeces for molecular testing. The same techniques are proposed for field surveys for the Sechuran fox (*P. sechurae*) in southern Ecuador, and in northern and central Peru. Similarly, the status of the short-eared dog (*Atelocynus microtis*) is essentially unknown, and range-wide surveys are highlighted as a priority for this species and for the conservation of kit foxes (*Vulpes mutica*), with particular focus on Mexico and the Californian range of the San Joaquin subspecies (*V. macrotis mutica*), which is threatened by habitat loss and degradation, rodenticides, and expanding populations of non-native red foxes (see below). Similarly extensive surveys of the maned wolf and the dhole are called for.

Knowledge of some species is highly skewed by one or a few intensive studies. The Ethiopian wolf (*Canis simensis*) characterizes a situation where one population, that in the Bale Mountains National Park, is well known (but is sufficiently frail, and so exceptionally informative, that it needs continued monitoring), whereas there are other populations whose statuses are much less clear. In this case there is a need to continue the monitoring of wolves initiated by the Ethiopian Wolf Conservation Programme in several fragmentary populations and to initiate it in Delanta, Aboi Gara, Abuna Josef, and Akista (see Marino 2003). Similarly, whereas the status of African wild dogs (*Lycaon pictus*) in South Africa, Botswana, Zimbabwe, and Tanzania is rather well documented, the status of the potentially key population in Mozambique linking East and South African populations is unknown, and virtually nothing is known of the remnant West African populations and Angola.

Most specialists nominated priorities in terms of species whereas others did so by region. For example, the distributions of the desert foxes in North Africa and the Middle East, the pale, Rüppell's and fennec foxes (*Vulpes pallida, V. ruepellii and V. zerda*) are unknown in any detail, and particular attention is drawn to the central Sahara desert and to surveying the Rüppell's fox and Blanford's fox in southwest Saudi Arabia, Yemen, and Oman. In this region, the grey wolf is under intense hunting pressure, and surveys are called for in Iran, Iraq, and Syria, and throughout the southern Arabian peninsula in southern Saudi Arabia, Oman, and Yemen. Blanford's fox was discovered in the Middle East only 20 years ago, and more recently in Egypt—now the priority is to survey for this species in Africa (specifically in eastern Egypt, eastern Sudan, Eritrea, and Ethiopia)—as with other species, it is proposed that this may best be accomplished using a wide array of field techniques.

Several of these surveys propose that the data be used for predictive modelling and population habitat viability analysis—an example being analysis of the restoration potential and population viability of the Mexican wolf (*C. lupus baileyi*) in the southwestern United States and northern Mexico. This approach is focused on the crucial question of how much habitat is enough to ensure population viability and eventual species recovery.

Taxonomy and genetics

Knowledge of where canids occur presupposes being able to identify them, and some uncertainties about taxonomic status remain. Studies are needed to determine the taxonomy of *Canis aureus lupaster*, proposed by some to be a small wolf, rather than a large jackal. A combination of surveys—focusing initially in Egypt and Libya—and genetic analysis is needed. Similarly, the taxonomic status of the Cozumel Island gray fox (*Urocyon cinereoargenteus*) in Mexico is uncertain. Recently the major conservation genetic issues have focused on untangling the taxonomy of North American wolves (*Canis lupus*) and coyotes (*C. latrans*) and hence determining the classification of red wolves (*C. rufus*) and their legislative status. Hybridization with dogs is an issue for Ethiopian wolves, hence the call for genetic screening of all populations, together with the long-term maintenance of genetic variability in small, isolated wolf populations, and for which inbreeding among packs of highly related animals is likely. Identifying patterns in the spatial distributions of genes in a population is an important step to prioritizing areas.

Basic biology and techniques

Few would dissent from the view that conservation is more likely to be effective when underpinned by sound science. Approximately one-tenth of the proposals we compiled boiled down to either scientific background or development of techniques. Sometimes the driving question was clearly focused. For example, the only known mainland sanctuary for Darwin's foxes is Nahuelbuta National Park, but the species occurs beyond the park's borders in highly fragmented forest used for cattle grazing and wood extraction, and frequented by numerous people and their dogs, along with a high density of culpeos (*Pseudalopex culpaeus*), a potential predator of the Darwin's fox. A comparison of the populations inside and outside the park may reveal differences in reproductive success and population processes (although it might prove harder to answer the question of what to do about any emerging differences).

However, many other proposals sought to study the basic ecology of poorly known species. For example, priority was attached to a general study of bush dogs in the Mbaracayú Forest Biosphere Reserve, Paraguay, of the maned wolf as an element of mammal communities in a habitat mosaic at the pampa/forest boundary in Noel Kempff Mercado National Park in Bolivia, and of the short-eared dogs in Cocha Cashu Biological Station and the Alto Purus Reserved Zone, in southeastern Peru. Similarly broad studies were proposed for the isolated Ethiopian wolf populations in northern Ethiopia, and for the dhole (*Cuon alpinus*). While all these proposals focus on the species and places where least is known, another category of proposals seeks to capitalize on the high cost-effectiveness that can characterize longitudinal studies of well-known populations. Such efforts, many represented in the case studies in Chapters 8–22 have made the greatest contribution to scientific understanding of processes that are fundamental to conservation planning. For example, projects called for monitoring population and pack dynamics of wild dogs in Kruger National Park, South Africa and Ethiopian wolves in the Bale Mountains, both already the subject of case studies in this book.

A different category of fundamental research is directed at developing techniques that will have practical application. For example, microsatellite markers and DNA extraction methods from non-invasively collected samples (faeces or hair) to monitor released island foxes or, using capture-mark-recapture logic, to monitor inaccessible populations of Ethiopian wolves. The same techniques are prioritized for dholes, along with research into call-based survey methods, and into the practicalities of translocation.

Community ecology

An emerging fundamental of canid ecology is intra-guild competition and, in one guise or another, this was nominated as a priority in three projects (interspecific competition beyond the Canidae is also important). It is well established that coyotes kill San Joaquin kit foxes, as they do swift foxes (*V. velox*) restored to Canada (chapter 10)—although the abundance of coyotes may be indirectly determined by human intervention, the sympatry of these conflicting species is essentially natural. However, red foxes (*V. vulpes*) also threaten these small prairie foxes and whereas these larger congeners occur naturally in the Canadian prairies (again, notwithstanding the effects of human interventions on their numbers) they are an introduced species to California. This distinction may be held as ethically important when it comes to deciding on appropriate interventions to foster the San Joaquin kit fox, and a study of their interactions with red foxes is flagged as a priority. On the other side of the world, red foxes may be the main cause for local extinction of the Rüppell's fox in Israel, where Rüppell's foxes were the most abundant vulpine in the Negev Desert up until the 1960s. Thereafter an increase in human populations and associated agricultural developments probably explains increases in red fox numbers. This, in turn, coincided incriminatingly with a sharp decline in Rüppell's foxes.

Intra-guild competition is by no means restricted to interactions between canids, and one example identified as a priority for research was the impact of tigers (*Panthera tigris*) on dholes. Similarly, predation by golden eagles (*Aquila chrysaetos*) is probably responsible for the recent catastrophic declines of five of the six populations of island foxes (*Urocyon littoralis*) on the California Channel Islands—in total from over 6000 to less than 1200 individuals in a decade (Roemer *et al.* 2001a, 2002). The presence of feral pigs enabled eagles to colonize the islands, increase in population size, and over-exploit the fox (foxes from San Miguel and Santa Rosa Islands now exist only in captivity). Proposals initially hinged upon live-capture and translocation of golden eagles, the reintroduction of bald eagles as a potential deterrent to golden eagles, the establishment of four captive breeding facilities, and the eradication of the pigs. However, several golden eagles have persistently evaded capture, leading reluctantly to the proposal that they should be shot.

Dholes also provide a harrowing example of another feature of community ecology that abuts problematically on canid conservation, namely the dilemma of endangered predators killing endangered prey—the classic case being the rare dholes of Java preying upon banteng (*Bos javanicus*), a rare wild cow. The San Clemente island fox (*U.l. clementae*) provides another perplexing example. The foxes kill or disturb the Critically

Endangered San Clemente loggerhead shrike (*Lanius ludovicianus meamsi*), and so to protect the shrike the foxes are trapped (Cooper *et al.* 2001). Such clashes of conservation interest are increasingly common (e.g. African wild dogs disadvantaged by spotted hyaenas *Crocuta crocuta* and lions *Panthera leo*—Carbone *et al.* 1999; Creel and Creel 2002). Intervention in the structure and processes of (at least superficially) natural communities raise perplexing questions of ethics and priorities of an ilk that is likely to become as familiar as it is unwelcome (Macdonald 2001).

Disease

A notable development of this decade is that infectious disease has emerged as a topical issue in conservation (Macdonald 1996a). A dramatic instance is that on Santa Catalina an epizootic of canine distemper virus reduced the island fox population there by nearly 90% (Chapter 9, this volume)—current proposals suggest experiments with vaccine and continued disease surveillance. A closely parallel case is the decimation of the Mednyi Island subspecies of Arctic fox (*Alopex lagopus*) by mange introduced by dogs (Goltsman *et al.* 1996). The threat of canid diseases is recognized by the modern acceptance that it is good practice that anybody handling rare canids should routinely screen for antibodies indicative of zoonoses (e.g. Woodroffe *et al.* 1997). Following the debate on vaccinating African wild dogs against rabies (reviewed by Woodroffe 1997; 2001a), there have been interesting trials on vaccine efficacy in that species (Chapter 6, this volume). One proposal is that similar trials might usefully be explored on dholes, in anticipation of rabies or distemper threatening protected populations. Although disease screening might now be seen as an element of all canid field conservation projects, several proposed projects identified situations where it is a central priority. For example, domestic dogs are a common presence in and around Nahuelbuta National Park, the only protected mainland habitat of the Darwin's fox. Priority is therefore attached to identifying which canine diseases are present in the area and their prevalence. Similar fears underlie proposals to establish a domestic dog vaccination programme where small-eared dogs occur, and to evaluate the threat of rabies to Sechuran foxes. The threat of rabies to Ethiopian wolves was demonstrated all too dramatically in 1991 (Sillero-Zubiri *et al.* 1996b). Priority is therefore attached to protecting wolves from disease by vaccinating domestic dogs against rabies and canine distemper within the wolf's range (and against rabies alone in a buffer zone around the wolf's range). Research is needed into the feasibility of oral vaccination of domestic dogs, and into the feasibility of oral vaccination of the wolves themselves; exercises that require cost–benefit analyses.

Disease-oriented project priorities span the practical to the exploratory. For example, there is an immediate need to build models for rabies spread in southeastern Finland near the Russian border and to prevent the disease from spreading from Russia to Finland. A more 'blue skies' proposal is that to determine the variation of the MHC gene complex within and among Ethiopian wolf populations—this being a potential indicator of their ability to respond to disease.

Human dimension

Wherever the relationship lies on the spectrum from exploitation to conflict, a human dimension lies behind almost all canid conservation problems. Thus biologists working with the Sechuran fox argue that its future lies in introducing a 'carnivore conservation ethic' to people in rural areas of northern Peru and southern Ecuador. This involves, on the one hand, reducing predation by the foxes on poultry and, on the other, persuading people not to make amulets out of pieces of fox. Little is known of Tibetan foxes (*V. ferrilata*), other than their furs regularly turn out at local markets in Tibet; market surveys may help elucidate the extent of trade and its possible impact on wild populations. However, prejudices cannot be changed until they are first identified and second—if possible—debunked. That is why questionnaire surveys of the attitudes of local people to maned wolves were nominated as a priority. Furthermore, some problems cannot simply be talked out of existence—they require some sort of action—hence the project to design, produce, and position road signs to protect Ethiopian wolves, and the proposal to study how the impacts of urban developments and highways on San Joaquin kit foxes can be mitigated. In the case of dholes, the perceived need is to evaluate the relative merits and the feasibilities of compensation and insurance schemes (and improved husbandry techniques) for each country in the species' range.

One ideal is that people formerly hostile to canids come to value them. This aspiration is behind the proposal that ecotourism is developed around African wild dogs that either appear in, or are introduced to, game farms in the northern regions of South Africa. A combination of practices is needed to minimize the damage done by the dogs to commercially valuable game and to maximize the revenue generated by the dogs themselves.

Planning and evaluation

Time spent in reconnaissance is seldom wasted, or so runs the aphorism. Twelve of the projects submitted to the Canid Action Plan explicitly sought funds in order to develop a plan—in two cases this planning complimented an evaluation of progress under existing species-specific action plans (for the African wild dog—Woodroffe *et al.* 1997—and the Ethiopian wolf—Sillero-Zubiri *et al.* 1997). A third case concerns the dhole, for which a species action plan is in preparation, and the authors have highlighted the need for a study to develop a framework to guide the prioritization of dhole populations and conservation action, drawing attention specifically to whether genetic or ecological criteria might have primacy in ranking priorities. A separate, but linked, proposal is to review the current legal protection of dholes in each of the species' range states. The clearest statement of need for funding in order to plan was made in a proposal to update the US Fish and Wildlife Service's Mexican wolf recovery plan, authorized in 1982 and now obsolete. The need is to assemble the logistical, fiscal, and intellectual resources to develop a new recovery plan.

Finding funds to take action is often so preoccupying that conservationists may too infrequently have the option of reviewing their effectiveness. However this is central to the proposal to investigate the feasibility, effectiveness, and repeatability of an adaptive management plan originally formulated to reduce (or prevent) hybridization between red wolves (*Canis rufus*) and coyotes in northeastern North Carolina. From 1987 through 1994 efforts to restore a population of red wolves to the Alligator River National Wildlife Refuge were successful. During the mid-1990s hybridization between red wolves and coyotes became increasingly common. In response the US Fish and Wildlife Service developed an adaptive management plan that called for hybridization to be eliminated or reduced by killing or sterilizing coyotes and hybrids. Implementation of the plan began in April 1999 and before the situation was further complicated by genetic revelations suggesting the existence of *Canis lycaon* (Nowak 2002). Now an assessment of the plan's successes is needed to determine the likelihood of restoring the red wolf, given that coyotes are widespread throughout its historic range, and taking account of the changing consensus on systematics.

Protected areas and active management

Protected areas are important to several endangered canids. A top priority for the Ethiopian wolf is ratification that Bale Mountains National Park is gazetted by the Ethiopian parliament, and, hopefully, listed as a UNESCO World Heritage Site. Building on this foundation, there is a need for funds to develop an infrastructure for tourists, and the promotion of tourism. As illustrated by the restoration of red deer (*Cervus elaphus*) in Italy to facilitate the recovery of wolves, a basic predatory need is that a protected area must encompass sufficient prey, which is why priorities for dhole conservation list not only the creation of protected areas but also the fostering of prey populations therein.

Protected areas may not be sufficiently large to contain, and thus to protect, their occupants, as emphasized by Woodroffe (2001b). Various things can be done about this. First, some protection can be extended beyond the protected area—for example there is a need to revise the federal rules governing management of the Mexican wolves that travel outside the Blue Range Wolf Recovery Area in southeastern Arizona and southwestern New Mexico. Currently federal rules require wolves that wander outside the restoration area to be captured and returned or placed in captivity. This clearly thwarts the aspirations of both wolves and conservationists to encourage recolonization of suitable habitat outside the restoration area, and ignores the importance of genetic exchange between subpopulations. Local, state, and federal officials need to be lobbied to revise the rules. Similarly, in eastern Finland, 36 grey wolves close to the Russian border have been radio-tagged as part of a plan to protect them and thereby to encourage eastward recolonization into central and western Finland. Second, the effective size of a park can be increased by the creation of buffer zones around its periphery, as proposed for the Simien Mountains National Park to protect Ethiopian wolves. Third, parks can be linked through corridors. Dramatic cases that would aid African wild dogs include plans to develop transfrontier corridors between, for example, the Okavango in northern Botswana north through Namibia, and into southern Zambia; or through Kruger National Park, South Africa with southern Zimbabwe, and southwestern Mozambique. Similar corridors could link the Selous Game Reserve and Ruaha National Park in Tanzania and Selous Game Reserve and Niassa Game Reserve in northern Mozambique. A very different philosophy, tailored to situations where large protected areas are not an option, is the creation of managed metapopulations—an idea being applied to African wild dogs in South Africa. Management involves the challenge of simulating the processes of immigration and emigration in natural populations.

Captive breeding

The value of captive breeding to mammal conservation has been much debated (e.g. Kleiman 1989; Balmford *et al.* 1995), and some canid specialists see it as competing for funds with higher priority field projects. Two subspecies of island foxes are essentially extinct in the wild, and attempts to breed them in captivity have been only modestly successful. The remaining foxes on San Miguel were taken into captivity in 1999, since then their numbers have doubled from 14 to 28, but in 2002 only 3 of 10 pairs produced litters. Studies of hormonal profiles, mate selection, and microsatellite profiles (to avoid inbreeding) are all proposed. Variation in MHC alleles would be used as a step towards breeding for disease resistance. Meanwhile, techniques are being developed to obtain and store sperm and inseminate receptive female island foxes to improve captive propagation. Cryopreservation of genetic material has also been proposed for Ethiopian wolves, for which the establishment of a captive breeding population remains an unfulfilled priority.

One idea, proposed for Mexican wolves, is the need to establish facilities where captive-born canids can gain experience of the wild—in the case of the Mexican wolves an enclosure of some 1500 km^2 of wilderness is proposed. Attempts to train captive-bred canids to hunt and kill—such as wild dogs released in Namibia—have raised ethical questions (Scheepers and Venzke 1995).

Education

Rather than identifying topics on which lack of knowledge was a barrier, various proposals emphasized the importance of communicating existing information. Thus education as to the merits of conserving biodiversity in general, and canids in particular, is explicit in proposals to increase awareness of maned wolves by farmers, and of island foxes by boat-owners. In the southeastern United States the Red Wolf Coalition promotes red wolf recovery by fostering public–private partnerships, increasing public awareness, raising funds and other contributions, and acting as an advocate for the species. The Coalition's priorities include fund-raising for implementing education programmes and construction of a red wolf education centre in Columbia, North Carolina. Similar aspirations guide the Ethiopian Wolf Conservation Programme's plans to inform and lobby government officials about the status of the Ethiopian wolf, and to place information about this species within a proposed national syllabus for conservation education in Ethiopia's education system. The programme targets children, adults, and communities within and surrounding the Ethiopian wolf's ranges. Another comparable organization is the Wolf Forum for the Southern Rockies, which is seeking funds to promote educational initiatives designed to incorporate 'science-informed advocacy' into decisions about restoring grey wolves to the Southern Rockies Ecoregion in the United States (mostly western Colorado and northern New Mexico).

Recurring themes and unitary approaches

These ten categories of project proposal can be further broken down into those where the problem is essentially that we do not know where canids are, but fear the worst, and those where we know all too well where they are, because they are implicated in some sort of problem there.

Surveys and monitoring

First, ignorance of distribution and abundance (and even taxonomic status) makes it hard to be confident about the conservation status of many canids. In several cases—the bush dog and small-eared dog paramount amongst them—repeated but failed attempts to find them give grounds for pessimism. A similar situation applies to Saharan foxes, whose distribution and status can only be guessed from scant reports. However, it is difficult to know what to conclude from surveys that are sporadic and unsystematic. Happily, not only have various techniques been developed for surveying canids (reviewed by Gese in press; and Boitani *et al.* Chapter 7, this volume), but there is an immense literature on the monitoring of mammals in general (e.g. Wilson *et al.* 1996; Macdonald *et al.* 1998; Buckland 2001). Taken together, these make it feasible to identify the prerequisites and phases of a professionally robust survey project. These include a statistically appropriate sampling regime, stratified as necessary and, where it is important to detect differences in space or time, meticulous attention to statistical power. Failure to attend to these starting points can squander time and money, and cause misleading outcomes. Working at different spatial scales on different species under different circumstances will dictate which methods are most effective; cost will also impose constraints, but false economies are to be fiercely resisted, and raise the spectre of squandered effort. A hierarchy of methods is available, from questionnaire surveys through spotlight, spoor, and sign counts, to scent-stations, fur-traps, call-ins, and camera traps to trap-mark-and recapture surveys. Some can be combined, and the optimal combination and pattern of deployment requires detailed study of a copious and technical literature. Increasingly, there are innovative approaches to surveying rare mammals (e.g. Carbone *et al.* 2001), sophisticated molecular techniques to reduce ambiguities in identification using faeces and hair (Paxinos *et al.* 1997; Kohn *et al.* 1999), and approaches to use volunteers to spread the workload (Newman *et al.* 2003). Furthermore, distributional data can now be incorporated into spatially explicit models, incorporating Geographic Information Systems, that generate useful, testable extrapolations and have the added advantage of allowing biologists to make their worst mistakes in a virtual reality (e.g. Macdonald and Rushton 2003). Although the greatest number of projects submitted to the Canid Action Plan prioritized surveys of diverse canid species in far-flung corners of the globe, the principles governing such work, and the techniques open to them, suggest to us that a single outline project design—tuned to local circumstances—would fit them all (Table 23.3). A worthwhile and cost-effective survey requires excellence both in fieldcraft and quantitative methods; this combination of qualities is sadly not always apparent, perhaps because mammal surveys seem so easy and so obviously worthwhile—beguiling impressions, neither of which is correct.

Predatory problems

The overview of research projects submitted by specialists to the Canid Action Plan brings to mind an analogy to a broad point that we made in Chapter 1, namely that canids are intriguingly diverse yet remarkably similar. So too, because the locations are exotically diverse, and the circumstances so different in detail, when we learn of maned wolves, coyotes, Arctic foxes, red foxes, dholes, African wild dogs, and dingoes in conflict with farmers it is easy to be impressed with the diversity of canid problems in every corner of the globe. But the reality is that many of these are subtle variations on the same problem—the crab-eating fox killing poultry, the red fox killing a pet guinea pig, the coyote killing a lamb, and the African wild dog killing a cow are all actors in different productions of the same drama—although they act out the sequence on stages set in different theatres: respectively, an Amazonian village,

Table 23.3 Despite the diversity of species, peoples, regions, and problems involved, many species-based canid conservation problems (and perhaps those on some other taxa too) boil down to variations on just three themes; projects intended to address each of these three themes generally progresses along a predictable trajectory (see main text). Having this in mind while designing a new project may increase the likelihood of a productive outcome and this table provides, in effect, a tick-list of topics requiring attention from the outset. The flow of this table complements the processes in Fig. 23.3, where the goal is progressive reduction of negative impacts, with attention to all four elements of the Conservation Quartet at each stage of the hoped for progression of population recovery towards sustainable management. Because this table is set out in two dimensions the phases (Research, Community involvement, Education and Implementation) are listed in sequence—the reality is that they are often interactive and parts of each may occur in parallel.

		Project category		
Phase	Task	Survey	Predatory conflict	Infectious disease
Preparatory	Define	Define the problem, demonstrate need for project, and specify goals for solution		
	Positioning	Review positioning of project within Conservation Quartet and phase of Species Recovery Paradigm; evaluate scope within Impacts Reduction Scheme		
	Formulation	Formulate questions to be asked, hypotheses to be tested and predictions refined, and measures to be made		
	Permits	Determine law and land policy issues affecting species and secure permit to operate		
	Argue justification of	Choice of scale—(a) spatial (b) temporal General approach to problem Operational methodology and materials Quantitative methods—(a) sampling design (b) statistical power		
	Funding	Prepare realistic budget Secure funds Demonstrate funds are sufficient to achieve useful goals; if not, postpone or abandon project		
Research	Operational phases	Preparatory modelling	Define target of predation: (a) livestock, (b) game species, and (c) threatened prey species	Establish necessary zoological/veterinary collaboration
		Define threats Recruit/ train staff	Identify stakeholder groups Quantify perceptions held by each group	Undertake pilot prevalence survey Undertake epizootiological/ecological research to identify processes
		Undertake survey	Verify/refute perceptions: (a) use existing data, (b) identify critical unknowns, (c) gather new data	Develop simulation models of: (a) likely spread and (b) control options
		Enter data Primary analysis	Identify residual issues Identify and test mitigation/ control approaches	Field test of feasibility and effectiveness of vaccination / population control approaches
		Extrapolation (GIS)	Cycle through Impacts Reduction Scheme	
Community involvement		Determine local leadership capacity	Identify stakeholders (e.g. individuals, agencies, NGOs, organizations, networks)	If domestic dogs are involved, develop extension programme
		Identify local supporters	Create and nurture stakeholder coalition group	Enroll community participation on dog health programme

Table 23.3 (Continued)

Phase	Task	Project category: Survey	Predatory conflict	Infectious disease
		Enlist local help monitoring and enforcing law	Evaluate existing local approaches to deal with problem	
Education (and awareness)	Targets	Identify target groups for outreach, and their different preoccupations		
	Methods	Identify appropriate modalities for reaching each target group (e.g. formal, informal, outreach, media)		
	Implement	Produce educational materials and implement education campaigns		
	Evaluate	Evaluate impact of education programme and adapt accordingly		
Implementation	Strategic planning	Design recovery plan/ sustainable management plan	Produce management plan, with attention to: (a) all stakeholders, (b) interdisciplinarity, (c) ethics, and (d) finance	Produce disease management plan
	Implement	Protected area support, enforcement, restoration	Consider non-lethal or lethal control programme Implement actions	Vaccination or lethal control Implement actions
	Evaluate	Long-term monitoring	Quantify changes in predator losses	Quantify prevalence of pathogens following intervention
	Review/ adapt	Adaptive management	Cycle repeatedly through the process, systematically re-evaluating the questions posed in the preparatory phase.	
	Disseminate	Disseminate results in suitable media/forum (always ensuring underpinning science is published in peer-reviewed journals)		

a British suburban garden, a Texan ranch, or a Zimbabwean mopane woodland. In each case, the action by the canid is opportunistic predation on domestic stock, and the reaction is for the stock-owner to experience anger and to incur, at least perceived, material loss. Consequently, as all canids are opportunistic predators, all canids are to a greater or lesser extent in conflict with some group of people concerned with livestock husbandry. Another set of conflicts differ only in one detail from the foregoing, and this set arises from competition over wild prey, generally the quarry of hunters or harvesters. Again there is the illusion of diversity, the grey wolf killing moose (*Alces alces*) in Alaska, the red fox killing grey partidges (*Perdix perdix*) on British cereal farms, the Arctic fox in Iceland killing the eider ducks (*Somateria mollissima*) whose feathers would have been harvested for quilt-making, or the African wild dog killing a valuable sable antelope (*Hippotragus niger*) on a game ranch, denying the rancher both breeding stock and a trophy fee. Sometimes the prey is not human quarry, but has perceived resource value to another group of stakeholders in wildlife, for example conservationists; the red foxes killing avocets (*Recurvirostra avosetta*) or capercaillie (*Tetrao urogallus*) on, respectively, wetland or upland nature reserves in Britain, the Californian island foxes killing San Clemente shrikes, the Javan dholes (known there as ajags) killing rare banteng wild cattle (*Bos javanicus*). All these are examples of just one category of problem (fully analysed in Chapter 5, this volume), again the result of opportunistic predation but this time with the group that perceives the loss being those concerned with husbanding a wild prey species. There is one difference in detail between these two categories, and it is that the wildness of the prey generally makes them harder to count and to manage and, because they are part of a natural ecosystem, the task of measuring the consequences is vastly greater, as too is the task of mitigating them. However, the latter are mere differences in practical detail—the overarching point is that all the diverse examples given in the preceding paragraph actually boil down to just one problem: predation by canids on prey that are considered a resource by a group of people who therefore believe they endure material loss due to canids.

We might have feared that with 36 species of wild canid there would be at least 36*xy* different canid problems worldwide (where *x* is the number of different species killed by canids that are considered in some sense as the property of *y* different groups of people), but this distils to a less daunting prospect with the realization that they are all merely variations of only one problem (perhaps with two facets, if we allow the rather weak distinction between domestic versus wild prey). Furthermore, the reality that there is really only one type of problem makes it less surprising to discover that there is remarkable uniformity in the trajectories followed during attempts to deal with the many variants of this problem. Indeed, this trajectory leads predictably, in turn, to a rather limited, and again ubiquitous, set of possible solutions and outcomes. If we are correct in this, then there will be great gains to be made in the efficiency with which canid conflict issues are tackled by appreciating the wider canvas on which they are all drawn. With this hope in mind, we will consider some common denominators in a controversy caused by canid predation.

The trajectory commonly starts when a group of conservationists, generally warmly disposed towards canids (and often seeing themselves as having moral advantage) encounters a group of people whose livelihood or recreation hinges on fostering the numbers of potential canid prey (and often seeing themselves as aggrieved). This encounter tends to begin with a strained tenor that owes much to the probability that these two groups of people (i.e. stakeholders) may have different traits; while it may be politically unfashionable to draw attention to their differences, it is not patronizing to do so: for example, the conservationists may be knowledgeable in biological principles without being experienced or adept in practical matters—characteristics associated with a formal tertiary education of a type generally associated with higher socioeconomic class and urban experiences; the group hostile to canids may be formally unfamiliar with biological principles (which is not to say they do not grasp them intuitively), while nonetheless being knowledgeable naturalists with a high level of relevant practical skills built on generations of traditional, first hand, and largely rural experience. As a glib short-hand we will refer to these two groups as the pro- and the anti-canids, and the foregoing are just two of several potential caricatures of these protagonists. Of course, like all groups of people they will have many and varied qualities; amongst these, it is likely that the canids enhance the income and aspirations of the pro-canids, and detract from those of the anti-canids. The likely outcome of early encounters between these two groups is that the pro-canids will conclude that the anti-canids exaggerate the problems caused by canids, which, the anti-canids will conclude, are underestimated by the pro-canids. Both are likely to be proven correct in these conclusions by the next phase. This involves a study to quantify the match between reality and perception—the likely outcome is that the damage caused by the local canid is much less than had been perceived by the antis, but is more significant than hoped by the pros. It is also likely that the damage will not be inflicted uniformly by all individuals of the local wild canid population, nor that the costs will be born equally by all individuals of the local human community—thereby, for both species, raising awkward distinctions between individual and populational experiences. These are complicated issues and will have provoked a lot of discussion between the two groups, leading to the refreshing realization amongst some members of each group that some members of the other group are knowledgeable, well-intentioned, and likable. These elements begin to work together, increasingly respecting each other's knowledge and sharing, and thus becoming competent in, each other's worlds. They become eager to shed their old names of pro and anti, and prefer instead to be known by names like middle-way group, or forum or coalition. Having acknowledged the (albeit smaller than claimed) reality of canid damage, the next phase is to reduce it. Initially, this will be by working with the antis to tighten up their husbandry, and relax their prejudices. Damage, and conflict, are thereby reduced, and may be further reduced by importation of *avant guarde* ideas based on science. Indeed, research projects may be set up to develop these innovations further in the local context. However, there comes a point at which the conflict between wild canid and people is at a temporarily irreducible minimum—it can get smaller in only two ways, one is to increase the tolerance of the aggrieved people, and the other is to make a technical breakthrough that reduces their loss (both of which we will return to below).

Why have we devoted space to the foregoing caricature of the trajectory of a canid predation issue? On a broader scale, having between us devoted 50-man years to studies of carnivores, and particularly canids, and having watched many hundreds of projects come and go, we conclude that it would save a lot of time, money, and anguish if those setting out on the next generation of such projects acknowledged the unitary nature of the problem, anticipated the foregoing trajectory, and planned for each of the phases at the outset. At a more specific level, and noting that one of us grew up on a farm in Argentina, while the other currently runs a small farm in England (from which a red fox has killed 18 hens in the last month), we acknowledge the inconvenient reality that canids can be a bloody nuisance, and this is a reality that has some interesting, and politically awkward consequences that future canid conservationists might consider.

Having arrived at the point where damage caused by a canid is deemed irreducible, how is society to treat the people bearing the residual cost? In the short-term, there seem to be only three options. First, to allow some canids to be removed (probably killed), but only insofar as this demonstrably reduces the problem. Second, to compensate them for the loss. Third, for them to tolerate the loss.

Another major issue in canid conservation is intraguild competition. Is this merely another variant of the predation problem? After all, in essence it involves a bigger canid killing a smaller one, and this is problematic insofar as conservationists may value the disadvantaged species. Furthermore, as with the predation category, different groups of stakeholders are likely to hold views on the desired balance of different canid species and the means of achieving it. Since the issue boils down to intervening to influence the effect of one wild mammal on another, there may be small conceptual difference between the case where the issue is that coyotes kill game birds (or rare waders) and the case where coyotes kill swift foxes. In each case a section of the human community happens to be sufficiently enthusiastic about game birds (or rare waders) or about swift foxes, that it deems it legitimate to foster their survival by killing coyotes. Because, in the case of intra-guild killing, both species are members of the same family the potential technical solutions may be restricted (e.g. a repellent for one is likely to repel the other too), and some awkward ethical issues associated with the influence of fashion and inconsistency in conservation decision-making may come into unusually clear relief, but logically it seems that intra-guild competition (culminating in hyper-predation) is merely a complicated subset of predation issues.

Disease dilemmas

A distinct canid conservation issue is their involvement with infectious disease (fully explored in Chapter 6, this volume). Once again, despite the diversity of canid species and of their pathogens, there have been sufficient field studies of canid infectious diseases to reveal that this variety too distils down to just one or two genres of problem which, from a wider perspective are remarkably uniform, differing only between a rare canid paradigm and an abundant canid paradigm. In the latter case the trajectory has classically begun with an outbreak of zoonotic disease in domestic or agricultural stock, and led to the assumption by public health officials, generally veterinarians, that wild canids are the reservoir source. Consequently, schemes are implemented to kill wild canids (generally raising attendant issues of non-target victims) in the expectation of thereby reducing contact rate to a level below that at which the epizootic dies out.

Conservationists generally take either or both of two positions: one is to challenge the assumption that wild canids are the reservoir and second, whether they are or not, to predict that blanket killing is unlikely to work. One outcome has been that the resulting debate leads to field data, subsequently incorporated into epidemiological models, that suggest the reservoir is more likely to have been peri-domestic than sylvatic; in this case conservationists conclude that the priority is to tighten up management, surveillance, and vaccination of domestic animals (generally dogs), while government officials tend to continue plans to kill wild canids on the basis of it being better to be safe than sorry. Whether or not the wild canid is demonstrated to be the reservoir, the contemporary version of this story involves the conservationists arguing that killing them may prove counter-productive, through perturbation effects (Tuyttens and Macdonald 2000), and thus advocating vaccination schemes as an alternative. This, in turn, causes the veterinarians to call for research on the efficacy, and non-target effects, of the vaccine. The rare-canid paradigm is different insofar as the rare canid may be imperilled by infectious disease in an abundant vector that may be either a domestic or wild species. The relative numbers of potential vectors to victims thus shift the balance of practicalities when it comes to vaccination (or killing) that can be resolved only by cost-versus-efficacy studies.

Emergent questions

As captured in the aphorism that people may face problems when unable to see the wood for the trees, one purpose of this chapter has been to identify from the plethora of research projects and priorities (the trees), some salient features (the woods) of canid conservation. In summary, almost all of the topics identified as priorities by contributors to the Canid Action Plan can be encompassed in just three research topics, and can all be tackled within scarcely more research paradigms. These three priorities represent a quasi-independent validation of the structure of this book (in that it was determined before the Canid Action Plan was begun, far less written): the two categories of problem—predatory conflict and disease—are the topics of Chapter 5 and 6, and the problem of tracking down elusive canids is likely to be eased by the novel techniques that are the topic of Chapter 7. Identification of the three themes explored in Table 23.2 necessitates some caveats. First, we do not intend this as a gauntlet thrown down solely to provoke readers to think of projects that fall outside these three—there are many! For example, over-exploitation might occur without conflict, and studying the spatial requirements of a species can be vital to conservation planning, and might become necessary outside the context of any of the three themes, as might reintroduction techniques or captive-breeding (although any or all of the latter might become necessary within projects driven by conflict). Furthermore, we stress again that our review has grown out of the action planning process, which has a species emphasis—very different project designs might have grown from ecosystem or process-based approaches (although note that canid species may be particularly important as flagships in the latter approaches, and as flagships perhaps we should consider intense management areas specifically targeting endangered canids, akin to those for, say, rhinos). Indeed, a different starting point might have shifted the balance of projects from research to implementation (see the Conservation Quartet, below). Even were that true, it might emerge that the common denominators and shared trajectories of development-based projects could usefully be identified much as we have done here from a species-based perspective—indeed, one recommendation from this review is that species- and process-based should be better integrated. Furthermore, devising strategies for managing the transition from research-based to action-based projects (see Recovery Paradigm, Fig. 23.3) is a priority, as too is making explicit the essential measures of success for the latter. This involves such tricky issues as capacity-building and transfer of ownership. Despite these complications, three broad issues emerge from the Main Themes reviewed above, and can be phrased as questions.

First, is canid conservation substantially different from that of any other taxon? Second, what is the relevance of education? Third, do we know enough?

Are the requirements of canid conservation different?

Decisions that affect conservation are not based solely on science, but we think that science should underpin the framework within which they are made. That is, conclusions should be based on evidence, and evidence should be tested in a generally

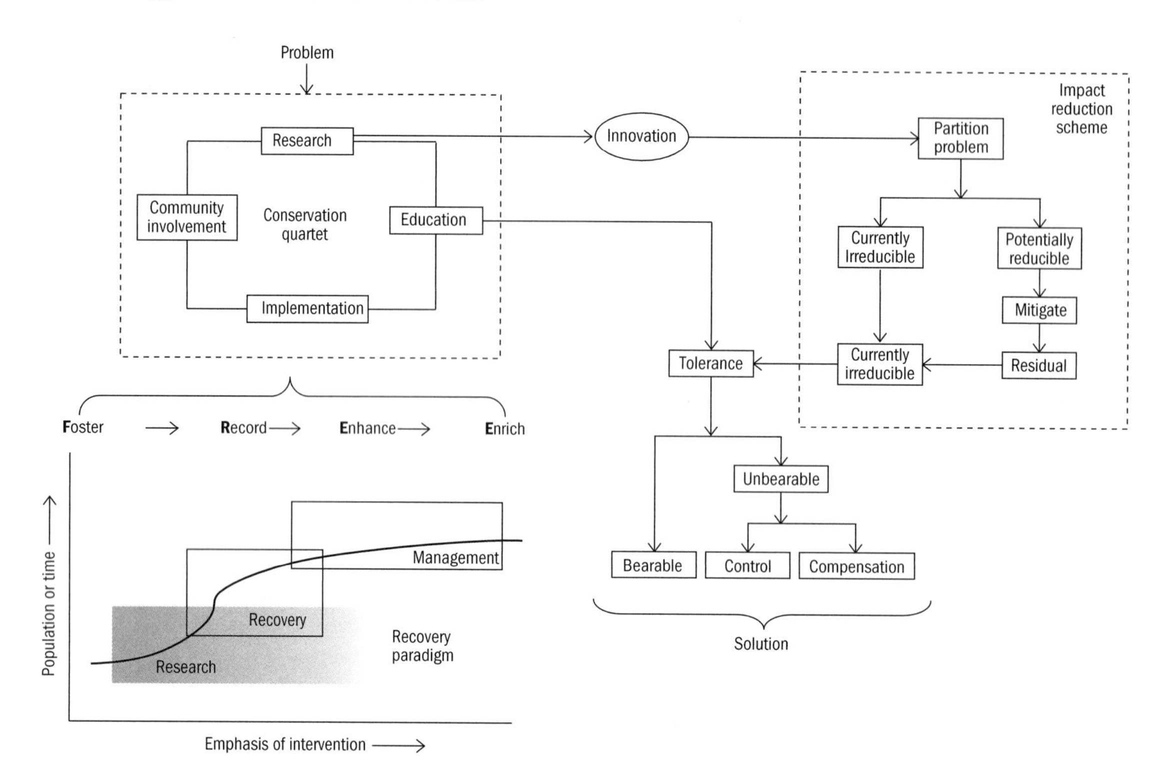

Figure 23.3 Canid (and other) conservation problems, and their solutions, can occur anywhere along a continuum of species recovery (a progression embodied for rare species in the acronym FREE), and at each point along this continuum projects might embrace any or all elements of the Conservation Quartet (Macdonald *et al.* 2002a). The emphasis of these elements may shift from research towards sustainable management, as time passes and, generally and hopefully, as a population recovers (Strachan in press). The progression will be different for species for which the problem is their rarity or their abundance: for the former the abscissa might normally be recovering population size, where for problematic abundant species it might be time; in both cases while the greatest intervention might be during the recovery phase, and the greatest emphasis on research at the outset, both research and management will have some role throughout. In parallel to this progression from endangerment or nuisance to sustainable management, there is a linked and iterative process flowing from problem to solution. Developing the notion that problems can be partitioned between reducible and irreducible elements (see Biodiversity Impacts Compensation Scheme in Macdonald 2001), and the balance between these will shift as currently intractable elements are rendered reducible by new innovation (itself engendered by the Research component of the quartet). At least from a species-based perspective, many conservation programmes can be approached by a rather small number of paradigms (see Table 23.3), and an essential element is to mitigate the reducible problems, thereby minimizing the current level of conflict. Conflict can be partitioned into that which is bearable (more or less willingly) by the afflicted stakeholders, and that which is unbearable. The extend to which these stakeholders will bear a conservation cost (such as predation) will depend on their tolerance which, in turn can be heavily affected by the education element of the quartet. Tolerance is affected by value, which is not merely financial, and may be attributed to both a species or a process of which it is a part. Two options are relevant to the unbearable component of current conflict: either to control (generally to kill) the problematic creature, or to compensate the aggrieved stakeholder. Each option raises questions explored in the text, and which can be partly answered by research of the types outlined in Table 23.3. In this diagram there could be overlap in the actions represented by the 'mitigation' and 'control' boxes, but these might loosely be partitioned as non-lethal and lethal interventions, respectively (and both, along with compensation, are facets of Implementation in the Quartet). In this figure the flow is from Problem to Solution—the process is iterative—the prevailing situation at the end of one pass through becoming the starting point for the next pass through; most of the right-hand side of the scheme represents operational processes exploded from the Implementation component of the quartet. As with all such schema, the reality is that every box interacts with every other, creating a web of links (e.g. access to compensation might be contingent on improved animal husbandry—a form of mitigation)—what we have shown with simplified arrows is the predominant flow that reveals what we regard as the salient lessons.

quantitative manner by questioning, hypothesizing, and predicting. With this in mind, how should conservation projects be constructed? Macdonald *et al.* (2000a) suggest both strategic and operational answers. Strategically, they propose that complete projects embrace a Conservation Quartet of four interdependent ingredients (Fig. 23.3): research (to crack the problem and identify the solution), education (to inform society, and influence opinion), community (to involve the stakeholders), and implementation (to get the job done). Operationally, projects with all four components could be targeted at any of the four courses of action that typify (sometimes in sequence) species conservation, and which are memorable in the acronym FREE: standing for Foster, Record, Enhance, and Enrich. A minimal goal is almost always to foster that which remains, and this necessitates recording its status (and monitoring how this changes). More ambitious than protecting the *status quo* is to enhance a species' circumstances, and perhaps even to restore creatures back into faunal communities from which we have removed them in the past. Against these generalities, are there peculiarities about canids that make the problem of conserving them different, and therefore call for a unique type of project? That question was phrased slightly more broadly by authors in Gittleman *et al.* (2001b) and answered specifically and affirmatively by both Ginsberg (2001) and Macdonald (2001). They noted that carnivores are distinguished by their lofty position in ecological pyramids, and that they inconvenienced and irritated people by eating their stock and their quarry (indeed, the larger ones may also attack and even eat people), and by being particularly involved in zoonotic disease. While people are particularly susceptible to the beauty and fascination of carnivores, they are also universally inflamed to wrath, even loathing, of them. Indeed, Kruuk (2002) argues that the strong emotions that typify human perceptions of carnivores have an innate basis. Although the Canidae is only one of the nine families comprising the Carnivora, they face the full spectrum of such problems characteristic of the Order as a whole—and provide many of the most extreme examples. In short, the problems posed by canid conservation are different in kind to those posed by many other orders of mammals, but no more so than is generally true for carnivores as a whole. That said, we can see no reason why projects in carnivore conservation should be approached differently to those on any other taxon—all four components of the Conservation Quartet remain relevant (and would also do so if the approach were process—rather than species-based) and should be thought out in advance, and the undeniable charisma of the subjects, and urgency of their plight, does not exempt fieldworkers from the exacting standards of logic and quantification necessary to sustain an evidenced-based argument.

What is the relevance of education?

Nobody can deny that many people perceive themselves to be in conflict with canids and while the perception may be exaggerated and the reality may be reducible, we accept that there may be a residual real and irreducible conflict. So what options are open? One is to find new ways to ameliorate the conflict, and we return below to the need for research. The other is to change people's perceptions, enhancing by various means the value they attribute to the canids in question, and this aim lies behind the various education projects mentioned above. Education can alleviate both the reducible and irreducible components of conflict (Fig. 23.3). Most obviously, where the perception of damage caused by canids is exaggerated, and where the actual damage is increased through failure to implement current knowledge, education can disabuse people of their prejudices and improve their competence. Less obviously, education about the rarity of a species, or even just the fascination of its lifestyle and the functioning of its ecosystem can radically affect the tolerance with which people view even irreducible conflict with canids. This may be a reality that is insufficiently acknowledged by those assigning priority to conservation projects: the possibility that the marginal impact on canid conservation of an extra banknote spent on education using existing knowledge may exceed the impact of that money being spent in the quest for new knowledge requires close scrutiny. We do not mean to be narrow-minded by proposing that for many conservation problems we know enough already: the emphasis can move towards the implementation segment of the quartet.

Our proposition that education will improve tolerance, by recalibrating the perceived costs and benefits that give a measure of conflict, implies that education affects value (a notion only partly reducible to money). Quantifying the value of nature, and specifically wild canids, is highly problematic and in this context is discussed by Macdonald *et al.* (2000a) and Macdonald (2001). Most tangibly, and perhaps least importantly, by combining biological and economical knowledge, it may be possible to debunk distorted perceptions (e.g. Macdonald *et al.* 2003 illustrate that a cereal farmer shooting a red fox in Britain may be costing himself a cash loss for the corn that will then be eaten by rabbits that the fox would otherwise have killed). Similarly, the mountain shepherd who sees an Ethiopian wolf as the basis of a hard currency tourist operation that benefits his community (and thus, crucially, himself) may re-evaluate his judgement of it as an occasional predator of lambs. Less tangibly, and perhaps most important, is the value people attribute to the existence of wild canids and the natural processes of which they are a part. For nature, as for art, knowledge is the route to understanding, and understanding is the basis of appreciation and thus value. Attaching value to an individual red fox is one of several reasons why shooting the one (mentioned above) that recently killed 18 of our chickens is rather low on our list of solutions. There is abundant evidence, although more anecdotal than systematic, that education about wild canids radically affects the value that people (including those in conflict with them) give them (reviewed in Sillero-Zubiri and Laurenson 2001). Of course, this puts a premium on the skills of the researcher as communicator, and raises worries about audience fatigue (even stunning natural history films are no longer novel in the developed world).

What sort of education best fosters conservation? Obviously, the answer is appropriate education, but what is appropriate may differ widely between the developed and developing world.

The fact that people may value the existence of nature reminds us that both value and nature are changeable. Many of the problems faced by many canid species arise in versions of nature that are no longer pristine. Indeed, almost by definition conservation problems are man-made, so it is temping to say that where modern man operates nature is no longer wholly natural. Outside protected areas people have a huge impact on wildlife, and even inside protected areas the mix of species is often heavily, if indirectly, influenced by anthropogenic factors. In short, for a substantial part of the world, a reversion to wilderness and *laissez faire* conservation is not an option. In consequence, the reality, and it requires a more sophisticated understanding than is current, is that much of conservation is directed by fashion. This is a big and provocative topic, but it has immediate relevance to canid conservation, and examples are legion but unnoticed. In Britain's Peak District National Nature Reserve, the statutory nature conservation agency subsidizes a gamekeeper to kill red foxes in association with agri-environmental schemes to foster ground-nesting lapwings (*Vanellus vanellus*). The story is more complicated than we have space to explore but, in essence, the conservation fashion (and it is self-evidently 'right' in the eyes of most stakeholders) is to value ground-nesting birds over small terrestrial carnivores—yet the fox is just as wild, just as natural and just as dependent on the man-made landscape as is the lapwing; the inclination to favour foxes over lapwings is close to whimsical, as is that to favour the San Clemente loggerhead shrike over the San Clemente island fox. Fashions aside, it is inarguable that humans routinely shape nature to their preferences, and we need to be much less coy about acknowledging that conservation too, is often an intervention. To conserve swift foxes probably necessitates killing coyotes, and conserving imperilled populations of Arctic foxes may involve killing red foxes. If they are to withstand the scrutiny of ethicists and logicians, many conservation decisions will require much more open-minded thought than has been customary. Seldom has this man-made aspect of nature, and the dilemmas it causes, been more vividly illustrated than by the notion of meta-population management as illustrated by African wild dogs in South Africa.

Because no sufficiently large piece of land remains in South Africa to support a self-sustaining, unmanaged population of wild dogs, the idea developed of coordinating the management of a constellation of smaller, satellite reserves according to current beliefs on the functioning of natural meta-populations. This would involve intervening to effect the emigration and immigration of individuals between these sub-populations, and thereby to obviate adverse effects of crowding and inbreeding (Mills *et al.* 1998). This intriguing concept not only brings into stark relief the degree of human involvement that will be needed if the distribution of African wild dogs in South Africa is to be restored, it also offers a challenging conservation mind-game. At what level of management does a reserve become a zoo, and how big does an enclosure have to be before it is 'natural enough' to qualify as a satellite within the meta-population? These questions are as pertinent to twenty-first century canid conservation as they are intractable—the relevant factors are widely interdisciplinary, bedevilled by complex trade-offs between largely incommensurable values, and any answers are certain to be matters of judgement rather than precision. What should be the criteria for membership of the meta-population club? It seems likely that the answer should be to do with having sufficient area and natural resources to sustain natural processes in the same way as a natural component of a meta-population might (were that properly understood). The natural process in question cannot merely be successful reproduction—otherwise a couple of wild dogs breeding in a cage might qualify—but perhaps reproduction based on an unmanaged prey population might qualify. However, exactly what is an unmanaged prey population—clearly releasing impala into a paddock full of wild dogs would not pass muster, but is a 'natural' predator–prey system conceivable for more than a short-timescale in an area supporting only one or two, or three packs of wild dogs? In any event, it is inconceivable that the mix of either prey or predator species in any reserve in Africa smaller than, say, Tanzania's Selous could be unaffected by management interventions. It seems clear that four biggish fenced reserves should qualify—Hluhluwe Umfolozi, Madikwe, Pilansberg, and Venetia—but what of smaller ones? The question is loaded because there is significant money at stake—smaller reserves could doubtless enhance their ecotourist revenue if they could boast the presence of wild dogs. However, what would these smaller reserves do with the surplus pups they bred, and indeed what will even the larger reserves do when they eventually fill up with wild dogs? The whole point is that there is nowhere else big enough to release them—the demand for zoos and research might soon be sated, and these options, together with the likely end point of simply killing surplus individuals each brings its own debate. By the way, almost exactly the same point was repeatedly missed by those who for decades have been advocating the introduction of grey wolves to the Scottish island of Rhum where they could feed on the abundant red deer, whose numbers are currently limited by riflemen. It is true that the wolves could probably fulfil the role of the riflemen, but when their numbers grew the only alternative to catastrophic oscillations in predators and prey (and the likely extinction of the wolves) would be to kill the surplus wolves. There is therefore a trade-off between the pleasure of retaining wild canids in a man-made world and the need for management to replace the natural processes we have usurped—that management is likely to involve killing, or at least sterilizing, some of the creatures we have sought to promote—a distasteful reality that may not be acceptable to a wide public, at least not unless they are exposed to a sophisticated education in ecological processes. Furthermore, do not take false comfort in believing the wild dogs of South Africa are an extreme case—intensive interventive management (based on meta-population thinking) is set to become increasingly necessary at least for big predators—we predict tigers in India will be a particularly high-profile example. That said, despite the gloom that attends discussion of small populations, one might take heart at the survival of Isle Royale's wolves (Chapter 18, this volume).

Do we know enough?

The enthralling case studies and comprehensive reviews in this book illustrate the depth of knowledge about wild canids. The similarity between the biology of different canids and the types of conservation problems they face make it unlikely that some unexpected feature of their basic biology will be revealed that crucially changes the prospects for their conservation. On the other hand, for species after species it is clear that their conservation hits a brick wall—a barrier that might be moved if we knew more. The nature of the barrier seems familiar—in the case of conflict over predation, the need is to find methods of preventing canids eating certain prey, and in the case of disease, the problem is to prevent them contracting or communicating infection. The priorities would therefore seem to be for ingenious, and probably non-lethal, methods of control and for the development of vaccination strategies. Both are categories of work that are likely to repay large-scale, carefully controlled field trials. Planning these, and cracking other conservation problems, will doubtless be aided by research into diverse aspects of canid biology, including their spatial requirements and population processes.

Finally, a prominent category of project proposed by contributors to the Canid Action Plan was for basic field studies of little known species. While the value of fundamental research for the sake of scholarship is something, as mentioned in our opening paragraph, that we would not contest, in the context of conservation funding is it not an indulgence to pursue basic research on species for which there is no more than an unspecified suspicion of threat? Perhaps counter-intuitively, there may be a strong conservation case for such research, but making that case comes with a ferociously strong caveat. First, it is arguable that one of the greatest contributions biological research has made to conservation is through its influence on the priorities of a wide public. In the developed world at least, discoveries of the intricacies of animal behaviour, revealed through enthralling films, photographs, and books have captivated and informed an ever wider public, radically changing their values, opinions and, ultimately, their politics. Conservation has to be paid for one way or another, and willingness to pay will be directly influenced by public understanding and appreciation of the value of biodiversity. Different arguments may have force in the developing world, where the best hope for conservation may lie in convincing people of the value of canids as part of the process that enables the ecosystem on which their livelihoods depend to function. Furthermore, few people make a greater impact in practice on the local protection of a species than does a highly motivated and fascinating field researcher—so the function of a scientist as a value-added warden-ambassador should not be ignored. Second, there is a tendency, perhaps on the increase, for bluster in conservation biology. All too often one reads that studying some aspect of a species' biology is 'crucial' for their conservation, without any plausible case being advanced as to why this is so. Worst of all, the non-existent case that conservation necessitated the study may then be used to justify any tawdriness in the resulting science. Conservation science will be blighted by such bluster, and by half-baked theory masquerading as a stepping-stone to practice. Far better to admit that while a carefully crafted study of an intriguing and unknown creature may have only tangential bearing on its conservation, it may produce results that help secure the commitment to conservation of a generation of decision-makers and tax-payers. Only the very best science is likely to enthral a wide public and therefore, perhaps paradoxically, if this genre of background investigations of canid private lives is truly to be crucial to conservation, the rigor demanded of it should be relentlessly high.

Red fox, *Vulpes vulpes* © D. W. Macdonald.

References

Abrams PA (1987). An analysis of competitive interactions between 3 hermit crab species. *Oecologia*, **72**, 233–247.

Abrams PA and Ginzburg LR (2000). The nature of predation: prey dependent, ratio dependent or neither? *Trends in Ecology and Evolution*, **15**, 337–341.

Ackerman BB, Lindzey FG and Hempker TP (1984). Cougar food habits in southern Utah. *Journal of Wildlife Management*, **48**, 147–155.

Adams EGP (1949). Jungle memories. Part IV. Wild dogs and wolves, etc. *Journal of the Bombay Natural History Society*, **48**, 645–655.

Adams JR, Leonard JA and Waits LP (2003). Widespread occurrence of a domestic dog mitochondrial DNA haplotype in southeastern US coyotes. *Molecular Ecology*, **12**, 541–546.

Adkins CA and Stott P (1998). Home ranges, movements and habitat associations of red foxes *Vulpes vulpes* in suburban Toronto, Ontario, Canada. *Journal of Zoology*, **244**, 335–346.

Adler GH and Levins R (1994). The island syndrome in rodent populations. *The Quarterly Review of Biology*, **69**, 473–490.

Aggarwal RK, Ramadevi J and Singh L (2003). Ancient origin and evolution of the Indian wolf: evidence from mitochondrial DNA typing of wolves from Trans-Himalayan region and Pennisular India. *Genome Biology*, **4**, P6.

Aguilar A, Roemer G, Debenham S, Binns M, Garcelon D, Wayne RK and High MHC (in press). Diversity maintained by balancing selection in an otherwise genetically monomorphic mammal. *Proceedings of the National Academy of Sciences*.

Akçakaya HR, Arditi R, Ginzburg LR (1995). Ratio-dependent predation: an abstraction that works. *Ecology*, **76**, 995–1004.

Al Khalil AD (1993). Ecological review and the distribution of Blanford's fox, *Vulpes cana*. In W Býttiker and F Krupp (eds), *Fauna of Saudi Arabia*, pp. 390–396. Natural History Museum, Basel, Switzerland.

Albon SD, Coulson TN, Brown D, Guinness FE, Pemberton JM and Clutton-Brock TH (2000). Temporal changes in key factors and key age groups influencing the population dynamics of female red deer. *Journal of Animal Ecology*, **69**, 1099–1110.

Alencar JE (1959). Calazar canino. Contribução para o estudo epidemiológico do calazar no Brasil. Tese Impr. Official Fortaleza, Ceará, Brazil.

Alencar JE (1961). Profilaxia do Calazar no Ceará, Brazil. *Revista do Instituto de Medicina Tropical de Sao Paulo*, **3**, 175–180.

Alexander RD (1974). The evolution of social behaviour. *Annual Review of Ecology and Systematics*, **5**, 325–383.

Alexander KA and Appel MJG (1994). African wild dogs (*Lycaon pictus*) endangered by a canine distemper epizootic among domestic dogs near the Masai Mara National Reserve, Kenya. *Journal of Wildlife Diseases*, **30**, 481–485.

Alexander KA, Kat PW, Wayne, RK and Fuller TK (1994). Serologic survey of selected canine pathogens among free-ranging jackals in Kenya. *Journal of Wildlife Diseases*, **30**, 486–491.

Alexander KA, Kat PW, Frank LG, Holekamp KE, Smale L, House C and Appel MJG (1995). Evidence of canine distemper virus infection among free-ranging spotted hyenas (*Crocuta crocuta*) in the Masai Mara, Kenya. *Journal of Zoo and Wildlife Medicine*, **26**, 201–206.

Alexander KA, Kat PW, Munson LA, Kalake A and Appel MJG (1996). Canine distemper-related mortality among wild dogs (*Lycaon pictus*) in Chobe National Park, Botswana. *Journal of Zoo and Wildlife Medicine*, **27**, 426–427.

Allardyce D and Sovada MA (2003). A review: ecology, historical distribution and status of swift foxes in North America. In M Sovada and L Carbyn (eds), *Swift fox conservation in a changing world*, pp. 3–18. Canadian Plains Research Center, University of Regina, Saskatchewan, Canada.

Allen DL (1979). *Wolves of Minong: their vital role in a wild community*. Houghton Mifflin Co., Boston, MA, USA.

Allen DL and Mech LD (1963). Wolves versus moose on Isle Royale. *National Geographic*, **123**, 200–219.

Allen LR and Sparkes EC (2001). The effect of dingo control on sheep and beef cattle in Queensland. *Journal of Applied Ecology*, **38**, 76–87.

Altmann J (1974). Observational study of animal behaviour: sampling methods. *Behaviour*, **49**, 227–265.

Amr ZS (2000). *Mammals of Jordan*. United Nations Environment Program, Amman.

Amr ZS, Kalishaw G, Yosef M, Chilcot BJ and Al-Budari A (1996). Carnivores of Dana Nature Reserve, Jordan. *Zoology in the Middle East*, **13**, 5–16.

Andelt WF (1982). Behavioural ecology of coyotes on Welder Wildlife Refuge, south Texas. Ph.D. dissertation, Colorado State University, Fort Collins, CO, USA.

Andelt WF (1985). Behavioral ecology of coyotes *Canis latrans* in south Texas, USA. *Wildlife Monographs*, **94**, 5–45.

Andelt WF (1987). Coyote predation. In M Novak, JA Baker, ME Obbard and B Malloch (eds), *Wild fur-bearer management and conservation in North America*, pp. 128–140. Ontario Ministry of Natural Resources and the Ontario Trappers Association, Ontario, Canada.

Andelt WF (1992). Effectiveness of livestock guarding dogs for reducing predation on domestic sheep. *Wildlife Society Bulletin*, **20**, 55–62.

Andelt WF (1999). Relative effectiveness of guarding-dog breeds to deter predation on domestic sheep in Colorado. *Wildlife Society Bulletin*, **27**, 706–714.

Andelt WF (2001). Livestock guard dogs, llamas and donkeys. Fact Sheet No. 1.218. Colorado State University Cooperative Extension, Fort Collins, CO, USA www.ext.colostate.edu/PUBS/LIVESTK/01218.html.

Andersen DE, Laurion TR, Cary JR, Sikes RS, McLeod MA and Gese EM (2003). Aspects of swift fox ecology in southeastern Colorado. In M Sovada and L Carbyn (eds), *Ecology and conservation of swift foxes in a changing world*, pp. 139–147. Canadian Plains Research Center, University of Regina, Saskatchewan, Canada.

Anderson RM and May RM (1991). *Infectious diseases of humans: dynamics and control*. Oxford University Press, Oxford.

Anderson RM, Jackson HC, May RM and Smith ADM (1981). Population dynamics of fox rabies in Europe. *Nature*, **289**, 765–771.

Andersone Z, Lucchini V, Randi E and Ozolins J (2002). Hybridisation between wolves and dogs in Latvia as documented using mitochondrial and microsatellite DNA markers. *Mammalian Biology*, **67**, 79–90.

Angerbjörn A, Arvidson B, Norén E and Strömgren L (1991). The effects of winter food on reproduction in the arctic fox (*Alopex lagopus*): a field experiment. *Journal of Animal Ecology*, **60**, 705–714.

Angerbjörn A, Tannerfeldt M, Bjärvall A, Ericson M, From J and Norén E (1995). Dynamics of the arctic fox population in Sweden. *Annales Zoologici Fennici*, **32**, 55–68.

Angerbjörn A, Ströman J and Becker D (1997). Home range pattern in arctic foxes. *Journal of Wildlife Research*, **2**, 9–14.

Angerbjörn A, Tannerfeldt M and Erlinge S (1999). Predator-prey relationship: arctic foxes and lemmings. *Journal of Animal Ecology*, **68**, 34–49.

Angerbjörn A, Lundberg H and Tannerfeldt M (2001). Geographical and temporal patterns of lemming population dynamics in Fennoscandia. *Ecography*, **24**, 298–308.

Angerbjörn A, Hersteinsson P and Tannerfeldt M (in press). Arctic fox (*Alopex lagopus*). In C Sillero-Zubiri, M Hoffmann and DW Macdonald (eds), *Canids: foxes, wolves, jackals and dogs. Status survey and conservation action plan*. IUCN/SSC Canid Specialist Group, Gland, Switzerland, and Cambridge, UK.

Ansell WFH (1960). *Mammals of Northern Rhodesia*. The Government Printer. Lusaka, Zambia.

Aquino R and Puertas P (1997). Observations of *Speothos venaticus* (Canidae: Carnivora) in its natural habitat in Peruvian Amazonia. *Zeitschrift für Saugetierkunde*, **62**, 117–118.

Arditi R and Ginzburg LR (1989). Coupling in predator-prey dynamics: ratio-dependence. *Journal of Theoretical Biology*, **139**, 311–326.

Arjo, WM and Pletscher DH (1999). Behavioural responses of coyotes to wolf recolonisation in northwestern Montana. *Canadian Journal of Zoology*, **77**, 1919–1927.

Armesto JJ, Rozzi R, Miranda P and Sabag C (1987). Plant/frugivore interaction in South American temperate forests. *Revista Chilena de Historia Natural*, **60**, 321–336.

Artois M (1997). Managing wildlife in the "Old World": a veterinary perspective. *Reproduction Feritility and Development*, **9**, 17–25.

Artois M, Delahay R, Guberti V and Cheeseman C (2001). Control of infectious diseases of wildlife in Europe. *Veterinary Journal*, **162**, 141–152.

Arvidson BE and Angerbjörn A (1987). Kannibalism bland fjällrävsvalpar? *Viltnytt*, **24**, 27–33.

Asa CS (1992). Contraceptive development and its application to captive and free-ranging wildlife. *Proceedings of the American Association of Zoological Parks and Aquariums*, 71–75.

Asa CS (1996a). Physiological and social aspects of reproduction of the wolf and their implications for contraception. In L Carbyn, SH Fritts and DR Seip (eds), *Ecology and conservation of wolves in a changing world*, pp. 283–286. Canadian Circumpolar Institute, Edmonton, Alberta, Canada.

Asa CS (1996b). The effects of contraceptives on behaviour. In US Seal, ED Plotka and PN Cohn (eds), *Contraception in wildlife, Book 1*, pp. 157–170. Edwin Mellen Press, Lewiston, NY, USA.

Asa CS (1998). Dogs (Canidae). In E Knobil and JD Neill (eds), *Encyclopedia of reproduction*, pp. 80–87. Academic Press, NY, USA.

Asa CS and Cossíos ED (in press). Sechura fox (*Pseudalopex sechurae*). In C Sillero-Zubiri, M Hoffmann and DW Macdonald (eds), *Canids: foxes, wolves, jackals and dogs. Status survey and conservation action plan*. IUCN/SSC Canid Specialist Group, Gland, Switzerland, and Cambridge, UK.

Asa CS and Porton I (1991). Concerns and prospects for contraception in carnivores. *Proceedings of the American Association of Zoo Veterinarians*, 298–303.

Asa CS and Valdespino C (1998). Canid reproductive biology: integration of proximate mechanisms and ultimate causes. *American Zoology*, **38**, 251–259.

Asa CS and Wallace MP (1990). Diet and activity pattern of the Sechuran desert fox (*Dusicyon sechurae*). *Journal of Mammalogy*, **71**, 69–72.

Asa CS, Mech LD and Seal US (1985). The use of urine, faeces, and anal-gland secretions in scent-marking by a captive wolf, *Canis lupus*, pack. *Animal Behaviour*, **33**, 1034–1036.

Asa CS, Mech LD, Seal US and Plotka ED (1990). The influence of social and endocrine factors on urine-marking by captive wolves (*Canis lupus*). *Hormones and Behavior*, **24**, 497–509.

Asa CS, Porton I, Plotka ED and Baker A (1996). Contraception. In DG Kleiman, ME Allen, KV Thompson and S Lumpkin (eds) *Wild mammals in captivity, Vol. 1: principles and techniques of captive management, Sect. A: captive propagation*, pp. 451–467. University of Chicago Press, Chicago, IL, USA.

Asa CS, Valdespino C and Cuzin F (in press). Fennec fox (*Vulpes zerda*). In C Sillero-Zubiri, M Hoffmann and DW Macdonald (eds), *Canids: foxes, wolves, jackals and dogs. Status survey and conservation action plan*. IUCN/SSC Canid Specialist Group, Gland, Switzerland, and Cambridge, UK.

Ashenafi ZT (2001). Common property resource management of an Afro-alpine habitat: supporting a population of the critically endangered Ethiopian wolf (*Canis simensis*), Ph.D. dissertation, Durrell Institute of Conservation and Ecology, University of Kent, Kent, UK.

Atkinson RPD (1997a). Side-striped jackal *Canis adustus*. In G Mills and L Hes (eds), *The complete book of southern African mammals*, pp. 197. Struik Publishers, Cape Town, South Africa.

Atkinson RPD (1997b). The ecology of the side-striped jackal (*Canis adustus* Sundevall), a vector of rabies in Zimbabwe. D. Phil. dissertation, University of Oxford, Oxford, UK.

Atkinson RPD, Macdonald DW and Kamizola R (2002a). Dietary opportunism in side-striped jackals *Canis adustus*. *Journal of Zoology*, **257**, 129–139.

Atkinson RPD, Rhodes CJ, Macdonald DW and Anderson RM (2002b). Scale-free dynamics in the movement patterns of jackals. *Oikos*, **98**, 134–140.

Atkinson RPD and Macdonald DMW (submitted). Territoriality in side-striped Jackals, *Canis adustus*, Sundevall: The influence of habitat structure. *J. Animal Ecology*.

Audet AM, Robbins CB and Lariviere S (2002). *Alopex lagopus. Mammalian Species*, **713**, 1–10.

Avise JC (1994). *Molecular markers, natural history and evolution*. Chapman and Hall, New York.

Avise JC (2000). *Phylogeography. The history and formation of species*. Harvard University Press, Cambridge, MA, USA.

Ayala JK and Noss A (2000). Censo por transectas en el Chaco Boliviano: limitaciones biológicas y sociales de la metodología. In E Cabrera, C Mercolli and R Resquin (eds), *Manejo de fauna silvestre en Amazonia y Latinoamérica*. Ricor Grafic S.A., Asunción, Paraguay.

Bacon PJ (1985). *Population dynamics of rabies in wildlife*. Academic Press, London, UK.

Bacon PJ and Blackwell PG (1993). *A comparison of models for the evolution of group living*. Research Report. Department of Probability and Statistics, University of Sheffield, UK.

Bacon PJ and Macdonald DW (1980). To control rabies: vaccinate foxes. *New Scientist*, **87**, 640–645.

Bacon PJ, Ball F and Blackwell P (1991a). Analysis of a model of group territoriality based on the Resource Dispersion Hypothesis. *Journal of Theoretical Biology*, **148**, 433–444.

Bacon PJ, Ball F and Blackwell P (1991b). A model for territory and group formation in a heterogeneous habitat. *Journal of Theoretical Biology*, **148**, 445–468.

Badridze J, Gurielidze Z, Todua GS, Badridze N and Butkhuzi L (1992). *The reintroduction of captive-raised large mammals into their natural habitat: problems and method*. Insitute of Zoology of the Academy of Sciences, Tibilisi, Georgia, USA.

Baer GM and Wandeler AI (1987). Rabies virus. In MJ Appel ed *Virus infections of carnivores*. Elsevier, Amsterdam, Netherlands.

Bailey EP (1992). Red foxes, *Vulpes vulpes*, as biological control agents for introduced Arctic foxes, *Alopex lagopus*, on Alaskan islands. *Canadian Field-Naturalist*, **106**, 200–205.

Bailey EP (1993). Introduction of foxes to Alaskan islands: history effects on avifauna, and eradication. Resource Publication No 193. U.S. Fish and Wildlife Service, Washington DC, USA.

Baker PJ and Harris S (2003). A review of the diet of red foxes in Britain and a preliminary assessment of its impact as a predator. In F Tattersall and W Manley (eds), *Conservation and conflict: mammals and farming in Britain*, pp. 120–140. Occasional Publication of the Linnean Society, Westbury Publishing, Otley, West Yorkshire, UK.

Baker PJ, Funk SM, Harris S and White PCL (2000). Flexible spatial organization of urban foxes, *Vulpes vulpes*, before and during an outbreak of sarcoptic mange. *Animal Behaviour*, **59**, 127–146.

Baker PJ, Robertson CPJ, Funk SM and Harris S (1998). Potential fitness benefits of group living in the red fox, *Vulpes vulpes. Animal Behaviour*, **56**, 1411–1424.

Baker PJ, Harris S, Robertson CPJ, Saunders G and White PCL (2001a). Differences in the capture rate of cage-trapped red foxes *Vulpes vulpes* and an evaluation of rabies control measures in Britain. *Journal of Applied Ecology*, **38**, 823–835.

Baker PJ, Newman T and Harris S (2001b). Bristol's foxes—40 years of change. *British Wildlife*, **12**, 411–417.

Baker PJ, Funk SM, Bruford MW and Harris S (in press). Polygynandry in a red fox population: implications for the evolution of group living in canids? *Behavioral Ecology*.

Baker RO and Timm RM (1998). Management of conflicts between urban coyotes and humans in southern California. In RO Baker and AC Crabb (eds), Proceedings of the 18th Vertebrate Pest Conference. University of California Davis.

Ball FG (1985). Spatial models for the spread and control of rabies in wildlife. pp. 197–222. In PJ Bacon (ed), *Population dynamics of rabies in wildlife*, Academic Press, London, UK.

Ballantyne EE and O'Donoghue SG (1954). Rabies control in Alberta. *Journal of the American Veterinary Medical Association*, **125**, 316–326.

Ballard WB, Whitman JS and Gardner CL (1987). Ecology of an exploited wolf population in south-central Alaska. *Wildlife Monographs*, **98**, 1–54.

Ballou JD (1997). Ancestral inbreeding only minimally affects inbreeding depression in mammalian populations. *Journal of Heredity*, **88**, 169–178.

Balmford A, Leader-Williams N, Green MJB (1995). Parks or arks: where to conserve threatened mammals. *Biodiversity and Conservation*, **4**, 595–607.

Bangs EE (1994). Establishment of a nonessential experimental population of grey wolves in Yellowstone National Park in Wyoming, Idaho, Montana central Idaho and southwestern Montana. *Federal Register*, **59**, 60252–60281.

Bangs EE and Fritts SH (1996). Reintroducing the grey wolf to central Idaho and Yellowstone National Park. *Wildlife Society Bulletin*, **24**, 402–413.

Bangs EE, Fritts SH, Harms DR, Fontaine JA, Jiminez MD, Brewster WG and Niemeyer CC (1995). Control of endangered grey wolves in Montana. In LN Carbyn, SH Fritts and DR Seip (eds), *Ecology and conservation of wolves in a changing world*, pp. 127–134. Occasional Publication No 35, Canadian Circumpolar Institute, Edmonton, Alberta, Canada.

Bangs EE, Fritts SH, Fontaine JA, Smith DW, Murphy KM, Mack CM and Niemeyer CC (1998). Status of grey wolf restoration in Montana, Idaho, and Wyoming. *Wildlife Society Bulletin*, **26**, 785–798.

Bangs EE, Fontaine J, Jiminez M, Meier T, Niemeyer C, Smith D, Murphy K, Guernsey D, Handegard L, Collinge M, Krischke R, Shivik J, Mack C, Babcock I, Asher V and

Domenici D (2001). Grey wolf restoration in the northwestern United States. *Endangered Species Update*, **18**, 147–152.

Barlow ND (1996). The ecology of wildlife disease control: simple models revisited. *Journal of Applied Ecology*, **33**, 303–314.

Barnes I, Matheus P, Shapiro B, Jensen D and Cooper A (2002). Dynamics of Pleistocene population extinctions in Beringian brown bears. *Science*, **295**, 2267–2270.

Barrette C and Messier F (1980). Scent-marking in free-ranging coyotes, *Canis latrans*. *Animal Behaviour*, **28**, 814–819.

Barta RM, Keith LB and Fitzgerald SM (1989). Demography of sympatric arctic and snowshoe hare populations: an experimental assessment of interspecific competition. *Canadian Journal of Zoology*, **67**, 2762–2775.

Bartholomew and Son (1987). *The Times atlas of the world*. Times Books Ltd., London, UK.

Barton NH and Hewitt GM (1989). Adaptation, speciation and hybrid zones. *Nature*, **341**, 497–503.

Bath AJ (1991). Public attitudes about wolf restoration in Yellowstone National Park. In R Keiter and MS Boyce (eds), *The Greater Yellowstone ecosystem: redefining America's wilderness heritage*, pp. 367–376. Yale University Press, New Haven, CT, USA.

Bath AJ and Buchanan T (1989). Attitudes of interest groups in Wyoming toward wolf restoration in Yellowstone National Park. *Wildlife Society Bulletin*, **17**, 519–525.

Bauman KL (2002). *Fennec fox (Vulpes zerda) North American Regional Studbook*, 2nd Edition. Saint Louis Zoological Park, St. Louis, Missouri, USA.

Beauchamp WD, Nudds TD and Clark RG (1996). Duck nest success declines with and without predator management. *Journal of Wildlife Management*, **60**, 258–264.

Beissinger SR and Westphal MI (1998). On the use of demographic models of population viability in endangered species management. *Journal of Wildlife Management*, **62**, 821–841.

Bekoff M (1978a). Scent-marking by free-ranging domestic dogs. *Biology of Behaviour*, **4**, 123–139.

Bekoff M ed (1978b). *Coyotes: biology, behavior, and management*. Academic Press, New York, USA.

Bekoff M and Diamond J (1976). Precopulatory and copulatory behavior in coyotes. *Journal of Mammalogy*, **57**, 372–375.

Bekoff M and Gese EM (2003). Coyote (*Canis latrans*). In GA Feldhamer, BC Thompson and JA Chapman (eds), *Wild mammals of North America: biology, management, and conservation*, pp. 467–481. Johns Hopkins University Press, Baltimore, USA.

Bekoff M and Wells MC (1980). The social ecology of coyotes. *Scientific American*, **242**, 130–151.

Bekoff M and Wells MC (1981). Behavioural budgeting by wild coyotes: the influence of food resources and social organization. *Animal Behaviour*, **29**, 794–801.

Bekoff M and Wells MC (1982). The behavioral ecology of coyotes: social organization, rearing patterns, space use, and resource defense. *Zeitschrift für Tierphysiologie*, **60**, 281–305.

Bekoff M and Wells MC (1986). Social ecology and behaviour of coyotes. *Advances in the study of Behavior*, **16**, 251–338.

Bellati J and von Thüngen J (1990). Lamb predation in Patagonian ranches. In LR Davis and RE Marsh (eds), *Proceedings 14th vertebrate pest conference*, pp. 263–268. University of California, Davis, CA, USA.

Belyaev DK and Trut LN (1975). Some genetic and endocrine effects of selection for domestication in silver foxes. In MW Fox (ed), *The wild canids. Their systematics, behavioural ecology and evolution*, pp. 416–426. Robert E. Krieger Publishing Co. Inc., Malabar, FL, USA.

Bennett LE (1994). Colorado grey wolf recovery: a biological feasibility study final report, 318 pp. U.S. Fish and Wildlife Service, Denver, CO, USA.

Bere RM (1955). The African wild dog. *Oryx*, **3**, 180–182.

Berg WE and Chesness RA (1978). Ecology of coyotes in northern Minnesota. In M Bekoff (ed), *Coyotes: biology, behavior, and management*, pp. 229–247. Academic Press, New York.

Berger JPB, Stacey L, Bellis MP and Johnson (2001). A mammalian predator-prey imbalance: grizzly bear and wolf extinction affect avian neotropical migrants. *Ecological Applications*, **11**, 947–960.

Bernard RTF and Stuart CT (1992). Correlates of diet and reproduction in the black-backed jackal. *South African Journal of Science*, **88**(5), 292–294.

Berry MPS (1978). Aspects of the ecology and behaviour of the bat-eared fox, *Otocyon megalotis* (Desmarest, 1822) in the upper Limpopo valley. M.Sc. dissertation, University of Pretoria, South Africa.

Berry MPS (1981). Stomach contents of bat-eared foxes, *Otocyon megalotis*, from the northern Transvaal. *South African Journal of Wildlife Research*, **11**, 28–30.

Berryman A and Chen X (1999). Population cycles: the relationship between cycle period and reproductive rate depends on the relative dominance of bottom-up or top-down control. *Oikos*, **87**, 589–593.

Berta A (1981). Evolution of large canids in South America. Anais II Congresso Latino-Americano de Paleontologia, pp. 835–845. Porto Alegre, Brazil.

Berta A (1982). *Cerdocyon thous*. *Mammalian Species*, **186**, 1–4.

Berta A (1984). The Pleistocene bush dog *Speothos pacivorus* (Canidae) from the Lagoa Santa caves, Brazil. *Journal of Mammalogy*, **65**, 549–559.

Berta, A (1986). *Atelocynus microtis*. *Mammalian Species*, **256**, 1–3.

Berta A (1987). Origin, diversification, and zoogeography of the South American Canidae. In BD Patterson and RM Timm (eds), *Studies in Neotropical mammalogy: essays in honor of Philip Hershkovitz. Fieldiana: Zoology*, **39**, 455–471.

Berta A (1988). Quaternary evolution and biogeography of the large South American Canidae (Mammalia: Carnivora). *University of California Publications of Geological Science*, **132**, 1–149.

Bertram BCR (1978a). Kin selection in lions and in evolution. In PPG Bateson and RA Hinde (eds), *Growing points in Ethology*, pp. 281–230. Cambridge University Press, London, UK.

Bertram BCR (1978b). Living in groups: predators and prey. In JR Krebs and NB Davies (eds), *Behavioural ecology: an evolutionary approach*, pp. 64–96. Blackwell Scientific Publications, Oxford, UK.

Bertram BCR (1979). Serengeti predators and their social systems. In PPG Bateson and RA Hinde (eds), *Growing points in ethology*, pp. 281–301. Cambridge University Press, Cambridge, UK.

Bertschinger HJ, Asa CS, Calle PP, Long JA, Bauman K, DeMatteo K, Jöchle W, Trigg TE and Human A (2001). Control of reproduction and sex related behaviour in exotic wild carnivores with the GnRH analogue deslorelin. *Journal of Reproduction and Fertility* (Supplement **57**), 275–283.

Bestelmeyer SV (2000). Solitary, reproductive and parental behavior of maned wolves (*Chrysocyon brachyurus*). M.Sc. dissertation, Colorado State University, Fort Collins, CO, USA.

Bester JL (1982). Die gedragsekologie en bestuur van die silwervos *Vulpes chama* (A. Smith, 1833) met spesiale verwysing na die Oranje-Vrystaat, M.Sc. dissertation, University of Pretoria, Pretoria, South Africa.

Bettini S and Gradoni L (1986). Canine leishmaniasis in the Mediterranean area and its implications for human leishmaniasis. *Insect Science Application*, **7**, 241–245.

Bibikov DI (1988). *Der wolf*. Die Neue Brehm-Bucherei. A. Ziemsen, Wittenberg Lutherstadt.

Bingham J (1995). Rabies in Zimbabwe. In J Bingham, GC Bishop and AA King (eds), *Proceedings of the third international conference of the southern and eastern African rabies group*, Editions Foundation Marcel Merieux, Lyon, France.

Bingham J and Foggin CM (1993). Jackal rabies in Zimbabwe. In G Thomson and A King (eds), *Rabies in southern and eastern Africa Onderstepoort Journal of Veterinary Research*, **60**, 365–366.

Bingham J, Kappeler A, Hill FWG, King AA, Perry BD and Foggin CM (1995). Efficacy of SAD (Berne) rabies vaccine given by the oral route in two species of jackal (*Canis mesomelas* and *Canis adustus*). *Journal of Wildlife Diseases*, **31**, 416–419.

Bingham J, Schumacher CL, Hill FWG and Aubert A (1999). Efficacy of SAG-2 oral rabies vaccine in two species of jackal (*Canis adustus* and *Canis mesomelas*). *Vaccine*, **17**, 551–558.

Birkhead TR and Nettleship DN (1995). Arctic fox influence on a seabird community in Labrador: a natural experiment. *Wilson Bulletin*, **107**, 397–412.

Blackwell P and Bacon PJ (1993). A critique of the Territory Inheritance Hypothesis. *Animal Behaviour*, **46**, 821–823.

Blanco JC, Reig S and de la Cuesta L (1992). Distribution, status, and conservation problems of the wolf *Canis lupus* in Spain. *Biological Conservation*, **60**, 73–80.

Blancou J, Aubert MFA and Artois M (1991). Fox rabies. In GM Baer (ed), *The natural history of rabies*, pp. 257–290. CRC press, Boca Raton, FL, USA.

Blanford WT (1877). Note on two species of Asiatic bears, the "mamh" of Balœchistan (*Ursus gedrosianus*) and *Ursus pruinosus* of Tibet, and on an apparently undescribed fox (*Vulpes canus*) from Balœchistan. *Journal of the Asiatic Society of Bengal*, **46**, 315–322.

Blejwas KM, Sacks BN, Jaeger MM and McCullough DR (2002). The effectiveness of selective removal of breeding coyotes in reducing sheep predation. *Journal of Wildlife Management*, **66**, 451–462.

Bobrinskii NA, Kuznetzov BA and Kuzyakin AP (1965). *Synopsis of the mammals of USSR*. Prosveshchenie, Moscow, Russia.

Boertje RD, Kelleyhouse DJ and Hayes RD (1995). Methods for reducing natural predation on moose in Alaska and Yukon: an evaluation. In LN Carbyn, SH Fritts and DR Seip (eds), *Ecology and conservation of wolves in a changing world*, pp. 505–513. Canadian Circumpolar Institute, Edmonton, Alberta.

Boissinot S and Boursot P (1997). Discordant phylogeographic patterns between Y chromosome and mitochondrial DNA in the house mouse: selection on the Y chromosome? *Genetics*, **146**, 1019–1034.

Boitani L (1983). Wolf and dog competition in Italy. *Acta Zoologica Fennica*, **174**, 259–264.

Boitani L (1984). Genetic considerations on wolf conservation in Italy. *Bollettino di Zoologia*, **51**, 367–373.

Boitani L (1992). Wolf research and conservation in Italy. *Biological Conservation*, **61**, 125–132.

Boitani L (1995). Ecological and cultural diversities in the evolution of wolf-human relationships. In LN Carbyn, SH Fritts and DR Seip (eds), *Ecology and conservation of wolves in a changing world*, pp. 3–11. Canadian Circumpolar Institute, Edmonton, Alberta, Canada.

Boitani L (2000). Action plan for the conservation of wolves in Europe. *Nature and Environment*, **113**, 1–86. (Council of Europe, Strasbourg).

Boitani L (2001). Hybridization and conservation of carnivores. In JL Gittleman, SM Funk, DW Macdonald and RK Wayne (eds), *Carnivore conservation*, pp. 123–144. Cambridge University Press, Cambridge, UK.

Boitani L (2003). Wolf conservation and recovery. In DL Mech and L Boitani (eds), *Wolves: ecology, behavior and conservation*, pp. 317–340. University of Chicago Press, Chicago, IL, USA.

Boitani L, Barrasso P and Grimod I (1984). Ranging behavior of the Red Fox in the Gran Paradiso National Park Italy. *Bollettino di Zoologia*, **51**, 275–284.

Boitani L, Corsi F, Reggiani G, Sinibaldi J and Trapanese P (1999). *A databank for the conservation and management of the African mammals*, p. 1150 Istituto di Ecologia Applicata, Roma, Italy.

Boitzov LV (1937). The arctic fox; biology, food habits, breeding (in Russian). *Transactions of the Arctic Institute, Leningrad*, **65**(7), 144.

Bomford M and O'Brien PH (1990). Sonic deterrents in animal damage control: a review of device tests and effectiveness. *Wildlife Society Bulletin*, **18**, 411–422.

Bonnefille R, Roeland JC and Guiot J (1990). Temperature and rainfall estimates for the past 40,000 years in equatorial Africa. *Nature*, **346**, 347–349.

Bookhout TA (ed), (1994). *Research and management techniques for wildlife and habitats*. The Wildlife Society, Bethesda, MD, USA.

Borrini-Feyerabend G and Buchan D (1997). *Beyond fences: seeking social sustainability in conservation*. IUCN Biodiversity Support Program. Gland, Switzerland. http://www.bsponline.org/publications/showhtml.

Bossart JL and Prowell DP (1998). Genetic estimates of population structure and gene flow: limitations, lessons and new directions. *Trends in Ecology and Evolution*, **13**, 202–206.

Bothma J du P (1966). Food of the silver fox *Vulpes chama*. *Zoologica Africana*, **2**, 205–221.

Boutin S (1990). Food supplementation experiments with terrestrial vertebrates: patterns, problems, and the future. *Canadian Journal of Zoology*, **68**, 203–220.

Bowen WD (1978). Social organisation of the coyote in relation to prey size. Ph.D. dissertation, University of British Columbia, Vancouver BC, Canada.

Bowen WD (1981). Variation in coyote social organisation: the influence of prey size. *Canadian Journal of Zoology*, **59**, 639–652.

Bowen WD (1982). Home range and spatial organization of coyotes in jasper national park, Alberta. *Journal of Wildlife Management*, **46**, 201–216.

Bowen WD and Cowan IM (1980). Scent marking in coyotes. *Canadian Journal of Zoology*, **58**, 473–480.

Boyce MS (2000). Modeling predator-prey dynamics. In L Boitani and TK Fuller (eds), *Research techniques in animal ecology*, pp. 253–287, Columbia University Press, New York, USA.

Boyle DB (1994). Disease and fertility in wildlife and feral populations: options for vaccine delivery using vectors. *Reproduction, Fertility and Development*, **6**, 393–400.

Bradley MP (1994). Experimental strategies for the development of an immunocontraceptive vaccine for the European Red Fox, *Vulpes vulpes*. In "immunological control of fertility: from gamete to gonads". *Reproduction Fertility and Development*, **6**, 307–317.

Brady CA (1979). Observations on the behaviour and ecology of the crab-eating fox (*Cerdocyon thous*). In JF Eisenberg (ed), *Vertebrate ecology in the northern neo-tropics*, pp. 161–171. Smithsonian Institution Press, Washington DC, USA.

Braestrup FW (1941). A study on the Arctic fox in Greenland. *Meddelelser om Grönland*, **131**, 1–102.

Brand CJ, Pybus MJ, Ballard WB and Peterson RO (1995). Infectious and parasitic diseases of the gray wolf and their potential effects on wolf populations in North America. In LN Carbyn, SH Fritts and DR Seip (eds), *Ecology and conservation of wolves in a changing world*, pp. 419–430. Canadian Circumpolar Institute, Edmonton, Canada.

Brand DJ (1993). *The influence of behaviour on the management of black-backed jackal*. Ph.D. dissertation, University of Stellenbosch, South Africa.

Brander AD (1927). *Wild animals in Central India*. Edward Arnold Co., London, UK.

Brander TA, Peterson RO and Risenhoover (1990). Balsam fir on Isle Royale: effects of moose herbivory and population density. *Ecology*, **71**, 155–164.

Breck S, Williamson R, Niemeyer C and Shivik JA (2002). Non-lethal radio activated guard for deterring wolf depredation in Idaho: summary and call for research. In RM Timm and RH Shmidt (eds), *Proceedings of the 20 Vertebrate Pest Conference*, pp. 223–226. University of California Davis, Davis, CA, USA.

Breitenmoser U (1998). Large predators in the Alps: the fall and rise of man's competitors. *Biological Conservation*, **83**, 279–289.

Brochier B, Aubert MFA, Pastoret PP, Masson E, Schon J, Lombard M, Chappuis G, Languet B and Desmettre Ph (1996). Field use of a vaccinia-rabies recombinant vaccine for the control of sylvatic rabies in Europe and North America. *Revue scientifique et technique OIE*, **15**, 947–965.

Bromley C and Gese EM (2001). Effects of sterilization on territory fidelity and maintenance, pair bonds, and survival rates of free-ranging coyotes. *Canadian Journal of Zoology*, **79**, 386–392.

Bronson FH (1989). *Mammalian reproductive biology*. Chicago University Press, Chicago, IL, USA.

Brook BW, O'Grady JJ, Chapman AP, Burgman MA, Akcakaya HR and Frankham R (2000). Predictive accuracy of population viability analysis in conservation biology. *Nature*, **404**, 385–387.

Brooker PC (1982). Robertsonian translocations in *Mus musculus* from N. E. Scotland and Orkney. *Heredity*, **48**, 305–309.

Brooks D (1992). Notes on group size, density, and habitat association of the Pampas fox (*Dusicyon gymnocercus*) in the Paraguayan Chaco. *Mammalia*, **56**, 314–316.

Brown CJ (1988). Scavenging raptors on farmlands. *African Wildlife*, **42**, 103–105.

Brown DE (1983). *The wolf in the southwest: the making of an endangered species*, 195 pp. University of Arizona Press, Tucson, AZ, USA.

Bruford MW and Wayne RK (1993). Microsatellites and their application to population genetic studies. *Current Biology*, **3**, 939–943.

Bruford MW, Chessman DJ, Coote T, Green HAA, Haines SA, O'Ryan C and Williams TR (1996). Microsatellites and their application to conservation genetics. In TB Smith and RK Wayne (eds), *Molecular genetic approaches in conservation*. Oxford University Press, New York, USA.

Bryant HN (1992). The carnivora of the lac pelletier lower fauna (Eocene: Duchesnean), Cypress Hills formation, Saskatchewan. *Journal of Paleontology*, **66**, 847–855.

Buckland ST (2001). *Introduction to distance sampling: estimating abundance of biological populations*. Oxford University Press, Oxford, UK.

Buckland ST, Anderson DR, Burnham KP and Laake JL (1993). *Distance sampling: estimating abundance of biological populations*. Chapman and Hall, London, UK.

Burke T, Davies NB, Bruford MW and Hatchwell BJ (1989). Parental care and mating behaviour of polyandrous dunnocks *Prunella* modularis related to paternity by DNA fingerprinting. *Nature*, **338**, 249–251.

Burke T, Hanotte O and Pijlen IV (1996). Microsatellite analysis in conservation genetics. In TB Smith and RK Wayne (eds), *Molecular genetic approaches in conservation*. Oxford University Press, New York, USA.

Burness GP, Diamond J and Flannery T (2001). Dinosaurs, dragons, and dwarfs: The evolution of maximal body size. *Proceedings of the National Academy of Sciences*, **98**, 14518–23.

Burrows R (1992). Rabies in wild dogs. *Nature*, **359**, 277.

Burrows R (1995). Demographic changes and social consequences in wild dogs, 1964–1992. In ARE Sinclair and P Arcese (eds), *Serengeti II. Dynamics, management and conservation of an ecosystem*, pp. 385–400. University of Chicago Press, Chicago, IL, USA.

Burrows R, Hofer H and East ML (1994). Demography, extinction and intervention in a small population: the case of the Serengeti wild dogs. *Proceedings of the Royal Society of London B*, **256**, 281–292.

Burt WH (1943). Territoriality and home range concepts as applied to mammals. *Journal of Mammalogy*, **24**, 346–352.

Burton RW (1925). Panther and wild dogs. *Journal of the Bombay Natural History Society*, **30**, 910–911.

Burton RW (1940). The Indian wild dog. *Journal of the Bombay Natural History Society*, **41**, 691–715.

Busby JR (1991). BIOCLIM—a bioclimatic analysis and prediction system. In CR Margules and MP Austin (eds), *Nature conservation: cost effective biological surveys and data analysis*, pp. 64–68. CSIRO, Melbourne, Australia.

Butler D (1994). Bid to protect wolves from genetic pollution. *Nature*, **370**, 497.

Byers CR, Steinhorst RK and Krausman PR (1984). Clarification of a technique for analysis of utilization-availability data. *Journal of Wildlife Management*, **48**, 1050–1053.

Cabrera A (1931). On some South American canine genera. *Journal of Mammalogy*, **12**, 54–66.

Cabrera A (1958). Catálogo de los mamíferos de América del Sur. *Revista del Museo Argentino de Ciencias Naturales "Bernadino Rivadavia"*, **4**, 1–307.

Cabrera A (1976). Regiones Fitogeográficas Argentinas. In *Enciclopedia Argentina de Agricultura y Jardinería, Tomo 2, Fascículo 1*. Editorial ACME, Buenos Aires, Argentina.

California Code of Regulations (1992). State of California Fish and Game Commission, Department of Fish and Game, Section 670.5.(b)(6)(F), pp. 279–280.

Camenzind FJ (1978a). Behavioural ecology of coyotes (*Canis latrans*) on the National Elk Refuge, Jackson, Wyoming. Ph.D. dissertation, University of Wyoming, Laramie, WY, USA.

Camenzind FJ (1978b). Behavioral ecology of coyotes on the National Elk Refuge, Jackson, Wyoming. In M Bekoff (ed), *Coyotes: biology, behavior, and management*, pp. 267–294. Academic Press, New York, USA.

Cameron MW (1984). The swift fox (*Vulpes velox*) on the Pawnee National Grassland: its food habits, population dynamics and ecology. M.Sc. dissertation, University of North Colorado, Greeley, CO, USA.

Campos CM and Ojeda RA (1996). Dispersal and germination of *Prosopis flexuosa* (Fabaceae) seeds by desert mammals in Argentina. *Journal of Arid Environments*, **35**, 707–714.

Canon SK (1995). Management of coyotes for pronghorn? In D Rollins, C Richardson, T Blankenship, K Canon and SE Henke (eds), *Coyotes in the Southwest: a compendium of our knowledge*, pp. 97–103. Texas Parks & Wildlife Department, Austin, TX, USA.

Cantoni D and Brown RE (1997). Paternal investment and reproductive success in the California mouse, *Peromyscus californicus*. *Animal Behaviour*, **54**, 377–386.

Carbone C, Du Toit JT and Gordon IJ (1997). Feeding success in African wild dogs: does kleptoparasitism by spotted hyenas influence hunting group size? *Journal of Animal Ecology*, **66**, 318–326.

Carbone C, Mace GM, Roberts SC and Macdonald DW (1999). Energetic constraints on the diet of terrestrial carnivores. *Nature London*, **402**, 286–288.

Carbone C, Christie S, Conforti K, Coulson T, Franklin N, Ginsberg JR, Griffiths M, Holden J, Kawanishi K, Kinnaird M, Laidlaw R, Lynam A, Macdonald DW, Martyr D, McDougal C, Nath L, Seidensticker OBTJ, Smith DJL, Sunquist M, Tilson R and Wan Shahruddin WN (2001). The use of photographic rates to estimate densities of tigers and other cryptic mammals. *Animal Conservation*, **4**, 75–79.

Carbyn LN (1982). Coyote population fluctuations and spatial distribution in relation to wolf territories in Riding Mountain National Park, Manitoba. *Canadian Field-Naturalist*, **96**, 176–183.

Carbyn LN, Oosenbrug SM and Anion DW (1993). *Wolves, bison, and the dynamics related to the peace-athabasca delta in Wood Buffalo National Park*. Canadian Circumpolar Institute, Alberta, Canada.

Carbyn L, Fritts SH and Seip DR (eds), (1995). *Ecology and conservation of wolves in a changing world*. Canadian Circumpolar Institute, Edmonton, Alberta, Canada.

Carmichael LE, Nagy JA, Larter NC and Strobeck C (2001). Prey specialization may influence patterns of gene flow in wolves of the Canadian Northwest. *Molecular Ecology*, **10**, 2787–2798.

Caro TM (1994). *Cheetahs of the Serengeti plains*. University of Chicago Press, Chicago, IL, USA.

Carpenter JW, Appel MJG, Erickson RC and Novilla MN (1976). Fatal vaccine-induced canine distemper virus infection in black-footed ferrets. *Journal of the American Veterinary Medical Association*, **169**, 961–964.

Carpenter MA, Appel MJG, Roelke PME, Munson L, Hofer H, East M and O'Brien SJ (1998). Genetic characterization of canine distemper virus in Serengeti carnivores. *Veterinary Immunology and Immunopathology*, **65**, 259–266.

Carpenter SR and Kitchell JF (1988). Consumer control of lake productivity. *Bioscience*, **38**, 764–769.

Carr GM and Macdonald DW (1986). The sociality of solitary foragers: a model based on resource dispersion. *Animal Behaviour*, **34**, 1540–1549.

Carroll C, Noss RF, Schumaker NH and Paquet PC (2001). Is the restoration of the wolf, wolverine, and grizzly bear to Oregon and California biologically feasible? In D Maehr, R Noss and J Larkin (eds), *Large mammal restoration: ecological and social implications*, pp. 24–46. Island Press, Washington DC, USA.

Carroll C, Phillips MK and Lopez Gonzalez CA (2002). *Spatial analysis of restoration potential and population viability of the Mexican wolf (Canis lupus baileyi) and jaguar (Panthera onca) in the southwestern United States and northern Mexico*, 9 pp. Klamath Center for Conservation Research, Orleans, CA, USA.

Carroll C, Phillips MK, Schumaker NH and Smith DW (2003). Impacts of landscape change on wolf restoration success: planning a reintroduction program using static and dynamic spatial models. *Conservation Biology*, **17**, 536–548.

Caswell H (2001). *Matrix population models: construction, analysis and interpretation*. Sinauer, Sunderland, MA, USA.

Caughley G (1994). Directions in conservation biology. *Journal of Animal Ecology*, **63**, 215–244.

Caughley G, Pech R and Grice D (1992). Effect of fertility control on a population's productivity. *Wildlife Research*, **19**, 623–627.

Cavalcanti SMC and Knowlton FF (1998). Evaluation of physical and behavioral traits of llamas associated with aggressiveness toward sheep-threatening canids. *Applied Animal Behaviour Science*, **61**, 143–158.

Cavallini P (1992). Ranging behavior of the red fox (*Vulpes vulpes*) in rural southern Japan. *Journal of Mammalogy*, **73**, 321–325.

Cavallini P (1995). Variation in the body size of the red fox. *Annales Zoologici Fennici*, **32**, 421–427.

Cavallini P (1996). Variation in the social system of the red fox. *Ethology, Ecology and Evolution*, **8**, 323–342.

Charlesworth B (1980). *Evolution in age-structured populations*. Cambridge University Press, Cambridge, UK.

Chase JM (2000). Are there real differences among aquatic and terrestrial food webs? *Trends in Ecology and Evolution*, **15**, 408–412.

Chautan M, Pontier D and Artois M (2000). Role of rabies in recent demographic changes in red fox (*Vulpes vulpes*) populations in Europe. *Mammalia*, **64**, 391–410.

Ciucci P and Boitani L (1998). Wolf and dog depredation on livestock in central Italy. *Wildlife Society Bulletin*, **26**, 504–514.

Ciucci P, Boitani L, Francisci F and Andreoli G (1997). Home range, activity and movements of a wolf pack in centreal Italy. *Journal of Zoology London*, **243**, 803–819.

Clark AB (1978). Sex ratio and local resource competition in a prosimian primate. *Science*, **201**, 163–165.

Clark FW (1972). Influence of jackrabbit density on coyote population change. *Journal of Wildlife Management*, **36**, 343–356.

Clark HO Jr. (2001). Endangered San Joaquin kit fox and non-native red fox: interspecific competitive interactions. Masters thesis, California State University, Fresno, CA, USA.

Clark JA (1993). The endangered species act: its history, provisions, and effectiveness. In TW Clark, RP Reading and AL Clarke (eds), *Endangered species recovery*, pp. 19–46. Island Press, Washington DC, USA.

Clark TW, Reading RP and Clarke AL (1993). Synthesis. In TW Clark, RP Reading and AL Clarke (eds), *Endangered species recovery*, pp. 417–431. Island Press, Washington DC, USA.

Clark TW, Curlee AP and Reading RP (1996). Crafting effective solutions to the large carnivore conservation problem. *Conservation Biology*, **10**, 940–948.

Clark TW, Mattson DJ, Reading RP and Miller BJ (2001a). Some approaches and solutions. In JL Gittleman, SM Funk, DW Macdonald and RK Wayne (eds), *Carnivore conservation*, pp. 223–240. Cambridge University Press, Cambridge, UK.

Clark TW, Mattson DJ, Reading RP and Miller BJ (2001b). Interdisciplinary problem solving in carnivore conservation: an introduction. In JL Gittleman, SM Funk, DW Macdonald and RK Wayne (eds), *Carnivore conservation*, pp. 223–240. Cambridge University Press, Cambridge, UK.

Clark WR and Fritzell EK (1992). A review of population dynamics of furbearers. In DR McCullough and RH Barrett (eds), *Wildlife 2001: populations*, pp. 899–910. Elsevier Science Publishers, Barking, Essex, UK.

Cleaveland S and Dye C (1995). Maintenance of a microparasite infecting several host species: rabies in the Serengeti. *Parasitology*, **111**, S33–S47.

Cleaveland S, Appel MGA, Chalmers WSK, Chillingworth C, Kaare M and Dye C (2000). Serological and demographic evidence for domestic dogs as a source of canine distemper virus infection for Serengeti wildlife. *Veterinary Microbiology*, **72**, 217–227.

Cleaveland S, Hess GR, Dobson AP, Laurenson MK, McCallum HI, Roberts MG and Woodroffe R (2002). The role of pathogens in biological conservation. In PJ Hudson, A Rizzoli, BT Grenfell, H Heesterbeek and AP Dobson (eds), *The ecology of wildlife diseases*, pp. 139–150. Oxford University Press, Oxford, UK.

Cleaveland S, Kaare M, Tiringa P, Mlengeya T and Barrat J (2003). A dog rabies vaccination campaign in rural Africa: impact on the incidence of dog rabies and human dog-bite injuries. *Vaccine*, **21**(17–18), 1965–1973.

Cleaveland SC (1995). Rabies in the Serengeti: the role of domestic dogs and wildlife in maintenance of the disease. In J Bingham, GC Bishop and AA King (eds), *Proceedings of the third international conference of the southern and eastern African rabies group*. Editions Foundation Marcel Merieux. Lyon, France.

Cluff HD and Murray DL (1995). Review of wolf control methods in North America. In LN Carbyn, SH Fritts and DR Seip (eds), *Ecology and conservation of wolves in a changing world*, pp. 491–504. Canadian Circumpolar Institute, Edmonton, Alberta, Canada.

Clutton-Brock J, Corbet GB and Hill M (1976). A review of the family Canidae with a classification by numerical methods. *Bulletin of the British Museum of Natural History and Zoology*, **29**, 119–199.

Clutton-Brock TH (1991). *The evolution of parental care*. Princeton University Press, Princeton, NJ, USA.

Clutton-Brock TH, Albon SD and Guinness FE (1989). Fitness costs of gestation and lactation in wild mammals. *Nature*, **337**, 260–261.

Coe MJ and Skinner JD (1993). Connections, disjunctions and endemism in the eastern and southern African mammal faunas. *Transactions of the Royal Society of South Africa*, **48**, 233–255.

Coetzee CG (1977). Order carnivora. In J Meester and HW Setzer (eds), *The mammals of Africa: an identification manual*, Part 8, pp. 3–4. Smithsonian Institution Press, Washington DC, USA.

Cohen JA (1977). A review of the biology of the dhole or Asiatic wild dog (*Cuon alpinus* Pallas). *Animal Regulation Studies*, **1**, 141–158.

Coleman PG and Dye C (1996). Immunization coverage required to prevent outbreaks of dog rabies. *Vaccine*, **14**, 185–186.

Collie JS and Spencer PD (1994). Modeling predator-prey dynamics in a fluctuating environment. *Canadian Journal of Fisheries and Aquatic Sciences*, **51**, 2665–2672.

Collins PW (1982). *Origin and differentiation of the island fox: a study of evolution in insular populations*, 303pp. M.A. dissertation, University of California, Santa Barbara, CA, USA.

Collins PW (1991a). Interaction between island foxes (*Urocyon littoralis*) and Indians on islands off the coast of Southern California: I. Morphologic and archaeological evidence of human assisted dispersal. *Journal of Ethnobiology*, **11**, 51–81.

Collins PW (1991b). Interaction between island foxes (*Urocyon littoralis*) and native Americans on islands off the coast of Southern California: II. Ethnographic, archaeo-logical and historical evidence. *Journal of Ethnobiology*, **11**, 205–229.

Collins PW (1993). Taxonomic and biogeographic relationships of the island fox (*Urocyon littoralis*) and gray fox (*U. cinereoargenteus*) from Western North America. In FG Hochberg (ed), *Third California islands symposium: recent advances in research on the California islands*, Santa Barbara Museum of Natural History, Santa Barbara, CA, USA.

Collister DM and Wicklum D (1996). Intraspecific variation in loggerhead shrikes: sexual dimorphism and implication for subspecies classification. *The Auk*, **113**, 221–223.

National Research Council (U.S.) Committee on wolf and Grizzly Bear Populations in Alaska 1997, wolves, bears and their prey in Alaska: biological and social challenges in wildlife management. National Academy Press, Washington DC, USA.

Concannon P, Altszuler N, Hampshire J, Butler WR and Hansel W (1980). Growth hormone, prolactin, and cortisol in dogs developing mammary nodules and an acromegaly-like appearance during treatment with medroxyprogesterone acetate. *Endocrinology*, **106**, 1173– 1177.

Connell W (1944). Wild dogs attacking tiger. *Journal of the Bombay Natural History Society*, **44**, 468–470.

Conner MC, Labisky RF and Progulske DR (1983). Scent-station indices as measures of population abundance for bobcats, raccoons, gray foxes, and oposums. *Wildlife Society Bulletin*, **11**, 146–152.

Conner MM, Jaeger M, Weller TJ and McCullough DR (1998). Impact of coyote removal on sheep depredation. *Journal of Wildlife Management*, **62**, 690–699.

Conover MR (2002). *Resolving human-wildlife conflicts*. Lewis Publishers, Boca Raton, FL, USA.

Conover MR and Kessler KK (1994). Diminisher producer participation in an aversive conditioning program to reduce coyote predation on sheep. *Wildlife Society Bulletin*, **22**, 229–233.

Coonan TJ (2003). Recovery strategy for island foxes (*Urocyon littoralis*) on the northern Channel Islands, Finalreport, Channel Islands National Park, Ventura CA, USA.

Coonan TJ, Schwemm CA, Roemer GW and Austin G (2000). Population decline of island foxes (*Urocyon littoralis*) on San Miguel Island. In DR Browne, KL Mitchell and HW Chaney (eds), *Proceedings of the fifth channel islands symposium*, pp. 289–297. US Department of the Interior, Minerals Management Service, Pacific OCS Region, Camarillo, CA, USA.

Coonan T (2002). *Findings of the island fox conservation working group*. Unpublished report, National Parks Service, Ventura, CA, USA.

Cooper DM, Kershmer EL, Schmidt GA and Garcelon DK (2001). San Clemente loggerhead shrike predator research and management program—(2000). *Final Report*, U.S. Navy, Natural Resources Mangement Branch, Southwest Division, Naval Facilities Engineering Command, San Diego, CA, USA. Institute for Wildlife Studies, Arcata, CA, USA.

Cooper G, Amos W, Hoffman D and Rubinsztein DC (1996). Network analysis of human Y microsatellite haplotypes. *Human Molecular Genetics*, **5**, 1759–1766.

Cooper SM (1991). Optimal group size: the need for lions to defend their kills against loss to spotted hyaena. *African Journal of Ecology*, **63**, 215–244.

Coppinger R and Coppinger L (2002). *Dogs: a new understanding of canine origin, behavior and evolution*. University of Chicago Press, Chicago, IL, USA.

Corbett L (1995). *The dingo in Australia and Asia*. University of New South Wales Press Ltd, Sydney, Australia.

Corbett LK (1988). Social dynamics of a captive dingo pack: population by dominant female infanticide. *Ethology*, **78**, 177–178.

Corbett LK (in press). Dingo (*Canis lupus dingo*). In C Sillero-Zubiri, M Hoffmann and DW Macdonald (eds), Canids: foxes, wolves, jackals and dogs. Status survey and conservation action plan. IUCN/SSC Canid Specialist Group, Gland, Switzerland, and Cambridge, UK.

Corbett LK and Newsome AE (1987). The feeding ecology of the dingo III. Dietary relationships with widely fluctuating prey populations in arid Australia: an hypothesis of alternation of predation. *Oecologia*, **74**, 215–227.

Corley JC, Fernandez GF, Capurro AF, Novaro AJ, Funes MC and Travaini A (1995). Selection of cricetine prey by the culpeo fox in Patagonia: a differential prey vulnerability hypothesis. *Mammalia*, **59**, 315–325.

Corsi F, Dupre' E and Boitani L (1999). A large-scale model of wolf distribution in Italy for conservation planning. *Conservation Biology*, **13**(1), 150–159.

Corsi F, De Leeuw J and Skidmore A (2000). Modeling species distribution with GIS. In L Boitani and TK Fuller (eds),

Research techniques in animal ecology, pp. 389–434. Columbia University Press, New York, USA.

Costa CHN and Courtenay O (submitted). A new record of the hoary fox *Pseudalopex vetulus* in north Brazil. *Mammalia*.

Cotera Correa M (1996). Untersuchungen zur ökologis-chen Anpassung des Wüstenfuchses *Vulpes macrotis zinseri* B, 105pp. In Nuevo León, Mexiko. Ph.D. dissertation, Ludwig-Maximilians-Universität München, Germany.

Coulson T, Catchpole EA, Albon SD, *et al.* (2001a). Age, sex, density, winter weather, and population crashes in Soay sheep. *Science*, **292**, 1528–1531.

Coulson T, Mace GM, Hudson E and Possingham H (2001b).The use and abuse of population viability analysis. *Trends in Ecology and Evolution*, **16**(5), 219–221.

Courchamp F and Macdonald DW (2001). Crucial importance of pack size in the African wild dog Lycaon pictus. *Animal Conservation*, **4**, 169–174.

Courchamp F and Sugihara G (1999). Biological control of alien predator populations to protect native island prey species from extinction. *Ecological Applications*, **9**, 112–123.

Courchamp F, Langlais M and Sugihara G (1999). Control of rabbits to protect island birds from cat predation. *Biological Conservations*, **89**, 219–225.

Courchamp F, Grenfell BT, Clutton-Brock TH (2000). Impact of natural enemies on obligately cooperative breeders. *Oikos*, **91**, 311–322.

Courchamp F, Rasmussen GSA and Macdonald DW (2002). Small pack size imposes a trade-off between hunting and pup-guarding in the painted hunting dog *Lycaon pictus*. *Behavioral Ecology*, **13**, 20–27.

Courtenay O (1998). *The epidemiology and control of canine visceral leishmaniasis in Amazon Brazil*, Ph.D. dissertation, University of London, London, UK.

Courtenay O and Maffei L (in press). Crab-eating fox (*Cerdocyon thous*). In C Sillero-Zubiri, M Hoffmann and DW Macdonald (eds), Canids: foxes, wolves, jackals and dogs. Status survey and conservation action plan. IUCN/SSC Canid Specialist Group, Gland, Switzerland, and Cambridge, UK.

Courtenay O, Macdonald DW, Lainson R, Shaw JJ and Dye C (1994). Epidemiology of canine leishmaniasis: a comparative serological study of dogs and foxes in Amazon Brazil. *Parasitology*, **109**, 273–279.

Courtenay O, Santana EW, Johnson PJ, Vasconcelos IAB and Vasconcelos AW (1996). Visceral leishmaniasis in the hoary zorro *Dusicyon vetulus*: a case of mistaken identity. *Transactions of the Royal Society of Tropical Medicine and Hygiene*, **90**, 498–502.

Courtenay O, Quinnell RJ and Chalmers WSK (2001). Contact rates between wild and domestic canids: no evidence of parvovirus or canine distemper virus in crab-eating foxes. *Veterinary Microbiology*, **81**, 9–19.

Courtenay O, Quinnell RJ, Garcez LM, Shaw JJ and Dye C (2002a). Infectiousness in a cohort of Brazilian dogs: why culling fails to control visceral leishmaniasis in areas of high transmission. *Journal of Infectious Diseases*, **186**, 1314–1320.

Courtenay O, Quinnell RJ, Garcez LM and Dye C (2002b). Low infectiousness of a wildlife host of *Leishmania infantum*: the crab-eating fox is not important for transmission. *Parasitology*, **125**, 407–414.

Courtenay O, Macdonald DW, Gillingham S and Dias R (submitted). The insectivorous canid of Latin America: Social management and bi-parental care in a family of hoary foxes, *Pseudalopex vetulus*. *Journal of Zoology London*.

Covell DF (1992). Ecology of the swift fox (*Vulpes velox*) in southeastern Colorado, 111 pp. M.Sc. dissertation, University of Wisconsin-Madison, Madison, WI, USA.

Cozza K, Fico R, Battistini ML and Rogers E (1996). The damage conservation interface illustrated by predation on domestic livestock in central Italy. *Biological Conservation*, **78**, 329–336.

Craighead JJ, Craighead FC Jr, Ruff RL and O'Gara BW (1973). Home ranges and activity patterns of nonmigratory elk of the Madison drainage herd as determined by biotelemetry. *Wildlife Monographs*, **33**, 1–50.

Crandall KA, Bininda-Emonds ORP, Mace GM and Wayne RK (2000). Considering evolutionary processes in conservation biology. *Trends in Ecology and Evolution*, **15**, 290–295.

Cravino JL, Calvar ME, Berruti MA, Fontana NA and Poetti JC (1997). American southern cone foxes: predators or prey? An Uruguayan study case. *Journal of Wildlife Research*, **2**, 107–114.

Creel S (1997). Cooperative hunting and group size: assumptions and currencies. *Animal Behaviour*, **54**, 1319–1324.

Creel S (2001). Cooperative hunting and sociality in African wild dogs, *Lycaon pictus*, In LA Dugatkin (ed), *Model systems in behavioral ecology*, pp. 466–490. Princeton University Press, Princeton, NJ, USA.

Creel SR and Creel NM (1991). Energetics, reproductive suppression and obligate communal breeding in carnivores. *Behavioural Ecology and Sociobiology*, **28**, 263–270.

Creel S and Creel NM (1995). Communal hunting and pack size in African wild dogs. *Lycaon pictus, Animal Behaviour*, **50**, 1325–1339.

Creel S and Creel NM (1996). Limitations of African wild dogs by competition with larger carnivores. *Conservation Biology*, **10**, 526–538.

Creel S and Creel NM (1998). Six ecological factors that may limit African wild dogs. *Lycaon pictus, Animal Conservation*, **1**, 1–9.

Creel S and Creel NM (2002). *The African wild dog: behavior, ecology and conservation*, Princeton University Press, Princeton, NJ, USA.

Creel SR and Macdonald DW (1995). Sociality, group size, and reproductive suppression among carnivores. *Advances in the Study of Behaviour*, **24**, 203–257.

Creel SR and Waser PM (1991). Failures of reproductive suppression in dwarf mongooses *Helogale parvula*: accident or adaptation? *Behavioural Ecology*, **2**, 7–15.

Creel SR and Waser PM (1994). Inclusive fitness and reproductive strategies in dwarf mongooses. *Behavioural Ecology*, **5**, 339–348.

Creel S, Creel NM, Matovelo JA, Mtambo MMA, Batamuzi EK and Cooper JE (1995). The effects of anthrax on endangered African wild dogs (*Lycaon pictus*). *Journal of Zoology London*, **236**, 199–209.

Creel SR, Creel NM and Monfort SL (1996). Radiocollaring and stress hormones in African wild dogs. *Conservation Biology*, **10**, 1–6.

Creel S, Creel NM, Mills MGL and Monfort SL (1997a). Rank and reproduction in cooperatively breeding African wild dogs: behavioral and endocrine correlates. *Behavioral Ecology*, **8**, 298–306.

Creel S, Creel NM, Munson L, Sanderlin D and Appel MJG (1997b). Serosurvey for selected viral diseases and demography of African wild dogs in Tanzania. *Journal of Wildlife Diseases*, **33**, 823–832.

Creel S, Spong G and Creel N (2001). Interspecific competition and the population biology of extinction-prone carnivores. In JL Gittleman, SM Funk, DW Macdonald and RK Wayne (eds), *Carnivore conservation*, pp. 35–60. Cambridge University Press, Cambridge, UK.

Creel SR, Monfort SL, Wildt DE and Waser PM (1991). Spontaneous lactation is an adaptive result of pseudo-regnancy. *Nature*, **351**, 660–662.

Crespo JA (1971). Ecologia del zorro gris, *Dusicyon gymnocerus antiguus* en la Provincia de La Pampa. Revista Museo Argentino de Ciencias Naturales "B. Rivadavia" Ecologia, **5**, 147–205.

Crespo JA and de Carlo JM (1963). Estudio ecológico de una población de zorros colorados, *Dusicyon culpaeus culpaeus* (Molina) en el oeste de la provincia de Neuquén. Revista Museo Argentino de Ciencias Naturales "B. Rivadavia", *Ecologia*, **1**, 1–55.

Crnokrak P and Roff DA (1999). Inbreeding depression in the wild. *Heredity*, **83**, 260–270.

Crooks KR and Van Vuren D (1995). Resource utilization by two insular endemic mammalian carnivores, the island fox and island spotted skunk. *Oecologia*, **104**, 301–307.

Crooks KR and Van VD (1996). Spatial organization of the island fox (*Urocyon littoralis*) on Santa Cruz Island, California. *Journal of Mammalogy*, **77**, 801–806.

Crusafont-Pairó M (1950). El primer representante del género Canis en el Pontiense eurasiatico (Canis cipio nova sp.). *Boletin de la Real Sociedad Española de Historia Natural (Geologia)*, 48.

Cumming DHM (1982). A case history of the spread of rabies in an African country. *South African Journal of Science*, **78**, 443–447.

Cutter WL (1958a). Food habits of the swift fox in northern Texas. *Journal of Mammalogy*, **39**, 527–532.

Cutter WL (1958b). Denning of the swift fox in northern Texas. *Journal of Mammalogy*, **39**, 70–74.

Cypher BL (1993). Food item use by three sympatric canids in southern Illinois. *Transactions of the Illinois State Academy of Science*, **86**, 139–144.

Cypher BL and Scrivner JH (1992). Coyote control to protect the endangered San Joaquin kit foxes at the Naval Petroleum Reserves, California. *Proceedings Vertebrate Pest Conference*, **15**, 42–47.

Cypher BL and Spencer KA (1998). Competitive interactions between coyote and San Joaquin kit foxes. *Journal of Mammalogy*, **79**, 204–214.

Cypher BL, Spencer KA and Scrivener JH (1994). Food-item use by coyotes at Naval Petroleum Reserves in California. *The Southwestern Naturalist*, **39**, 91–95.

Cypher BL, Warrick GD, Otten MRM, O'Farrell TP, Berry WH, Harris CE, Kato TT, McCue PM, Scrivner JH and Zoellick BW (2000). Population dynamics of San Joaquin kit foxes at the Naval Petroleum Reserves in California. *Wildlife Monographs*, **145**, 1–43.

Cypher BL, Clark HO, Jr, Kelly PA, Van Horn Job C, Warrick GW and Williams DF (2001). Interspecific interactions among wild canids: implications for the conservation of endangered San Joaquin kit foxes. *Endangered Species Update*, **18**, 171–174.

Cypher BL, Kelly PA and Williams DF (2003). Factors influencing populations of endangered San Joaquin kit foxes. In MA Sovada and L Carbyn (eds), *Swift fox conservation in a changing world*, pp. 125–137. Canadian Plains Research Center, University of Regina, Saskatchewan, Canada.

Dahier T (2002). Année 2001: Bilan des dommages sur les troupeax domestiques. *L'Infoloups*, **10**, 11.

Dale BW, Adams LG and Bowyer RT (1994). Functional response of wolves preying on barren-ground caribou in a multiple-prey ecosystem. *Journal of Animal Ecology*, **63**, 644–652.

Dalén L, Götherström A, Tannerfeldt M and Angerbjörn A (2002). Is the endangered Fennoscandian arctic fox (*Alopex lagopus*) population genetically isolated? *Biological Conservation*, **105**, 171–178.

Dalponte JC (1997). Diet of the hoary fox, *Lycalopex vetulus*, in Mato Grosso, Brazil. *Mammalia*, **61**, 537–546.

Dalponte J and Courtenay O (in press). Hoary fox (*Pseudalopex vetulus*). In C Sillero-Zubiri, M Hoffmann and DW Macdonald (eds), *Canids: foxes, wolves, jackals and dogs. Status survey and conservation action plan*. IUCN/SSC Canid Specialist Group, Gland, Switzerland, and Cambridge, UK.

Darwin C (1845). Journal of researchers into the natural history and geology of the countries visited during the voyage of the H.M.S 'Beagle' round the world, 2nd Edition. John Murray, London, UK.

Da-Silva J, Woodroffe R and Macdonald DW (1993). Habitat, food availability and group territoriality in European badgers. *Oecologia*, **95**, 558–564.

David J, Andral L and Artois M (1982). Computer simulation model of the epi-enzootic disease of vulpine rabies. *Ecological Modelling*, **15**, 107–125.

Davidar ERC (1973). Dhole or Indian wild dog (*Cuon alpinus*) mating. *Journal of the Bombay Natural History Society*, **70**, 373–374.

Davidar ERC (1975). Observations at the dens of the dhole or Indian wild dog (*Cuon alpinus* Pallas). *Journal of the Bombay Natural History Society*, **71**, 373–374.

Davies HT (1998). The suitability of Matusadona National Park, Zimbabwe, as a site for wild dog (*Lycaon pictus*) introduction. M.Sc. dissertation, University of Zimbabwe, Harare, Zimbabwe.

Davies HT (2000). The 1999/2000 wild dog photographic survey. *Unpublished Scientific Report* 3/2000. South African National Parks, Skukuza, South Africa.

Davies NB (1978). Ecological questions about territorial behaviour. In JR Krebs and NB Davies (eds), *Behavioural ecology: an evolutionary approach*, pp. 317–350. Blackwell Scientific Publications, Oxford, UK.

Davies NB, Hartley IR, Hatchwell BJ, Desrochers A, Skeer J and Nebel D (1995). The polygynandrous mating system of the alpine accentor *Prunella collaris*. I. Ecological causes and reproductive conflicts. *Animal Behaviour*, **49**, 769–788.

Dawkins R (1979). Twelve misunderstandings of kin selection. *Zietschrift fur Tierpsychologie*, **51**, 184–200.

Dayan T, Tchernov E, Yom TY and Simberloff D (1989). Ecological character displacement in Saharo-Arabian *Vulpes*: outfoxing Bergmann's rule. *Oikos*, **55**, 263–272.

Dayan T, Simberloff D, Tchernov E and Yom TY (1992). Canine carnassials: character displacement in the wolves, jackals and foxes of Israel. *Biological Journal of the Linnean Society*, **45**, 315–331.

de Kroon H, van Groenendal J and Ehrlen J (2000). Elasticities: a review of methods and model limitations. *Ecology*, **81**, 617–618.

Deane LM (1956). Leishmaniose visceral no Brasil. Estudos sôbre reservatórios e transmissores realizados no Estado do Ceará. Serviço Nacional de Educação Sanitária, Rio de Janeiro, Brazil.

Deane LM and Deane MP (1954a). Encontro de *Leishmanias* nas vísceras e na pele de uma raposa, em zona endêmica de calazar, nos arredores de Sobral, Ceará. *O Hospital*, **45**, 419–421.

Deane MP and Deane LM (1954b). Infecção experimental do *Phlebotomus longipalpis* em raposa (*Lycolopex vetulus*), naturalmente parasitada pela *Leishmania donovani*. *O Hospital*, **46**, 651–653.

Deane LM and Deane MP (1955). Observações preliminares sôbre a importância comparativa do homem, do cão e da raposa (*Lycalopex vetulus*) como reservatórios da *Leishmania donovani*, em área endêmica de Calazar, no Ceará. *O Hospital*, **48**, 79–98.

Defler TR and Santacruz A (1994). A capture of and some notes on *Atelocynus microtis* (Sclater, 1883) (Carnivora: Canidae) in the Colombian Amazon. *Trianea*, **5**, 417–419.

DelGuidice GD (1998). Surplus killing of white-tailed deer by wolves in northcentral Minnesota. *Journal of Mammalogy*, **79**, 227–236.

DelGiudice GD, Stone J, Mech LD and Seal US (1992). Sampling considerations involved with monitoring the nutritional status of gray wolves *Canis lupus* via biochemical analysis of snow urine. *Transactions of the Congress International Union of Game Biologists*, **18**, 35–38.

Delpietro HA, Gury DF, Larghi OP, Mena SC and Abramo L (1997). Monoclonal antibody characterization of rabies virus strains isolated in the River Plate Basin. *Journal of Veterinary Medicine Series B*, **44**, 477–483.

Dennis B, Munholland PL and Scott JM (1991). Estimation of growth and extinction parameters for endangered species. *Ecological Monographs*, **61**, 115–142.

Dennis M, Randall K, Schmidt G and Garcelon D (2001). *Island fox (Urocyon littoralis santacruzae) distribution, abundance and survival on Santa Cruz Island, California. Progress report: May through October 2001*. Institute for Wildlife Studies, Arcata, CA, USA.

U.S. Department of the Interior (2001). 50 CFR Part 17, Endangered and threatened wildlife and plants; Listing the San Miguel Island fox, Santa Rosa Island fox, Santa Cruz Island fox and Santa Catalina Island fox as endangered. *Federal Register*, **66**(237), 63654–63665.

Derix RRW and van Hoof J (1995). Male and female partner preferences in a captive wolf pack (*Canis lupus*): specificity versus spread of sexual attention. *Behaviour*, **132**, 127–149.

Despain DG (1990). *Yellowstone vegetation: consequences of environment and history in a natural setting*. Roberts Rinehart Publishers, Boulder, CO, USA.

Diamond JM (1975). Assembly of species communities. In ML Cody and JM Diamond (eds), *Ecology and evolution of communities*, pp. 342–444. Harvard University Press, Cambridge, MA, USA.

Dicke M and Burroughs PA (1988). Using fractal dimensions for characterising tortuosity of animal trails. *Physiological Entomology*, 393–398.

Dietz JM (1984). Ecology and social organization of the maned wolf (*Chrysocyon brachyurus*). *Smithsonian Contribution to Zoology*, **392**, 1–51.

Dietz JM (1985). *Chrysocyon brachyurus*. *Mammalian Species*, **234**, 1–4.

Dirks RA and Martner BE (1982). The climate of Yellowstone and Grand Teton National Parks. *National Park Service Occassional Paper No. 6*.

Dobson AP (1988). Restoring island ecosystems: the potential of parasites to control introduced mammals. *Conservation Biology*, **2**, 31–39.

Dobson FS (1982). Competition for males and predominant juvenile male dispersal in mammals. *Animal Behaviour*, **30**, 1183–1192.

Dobson M and Kawamura Y (1998). Origin of the Japanese land mammal fauna: allocation of extant species to historically-based categories. *Quaternary Research*, **37**, 385–395.

Dolf G, Schläpfer J, Gaillard C, Randi E, Lucchini V, Breitenmoser U and Stahlberger-Saitbekova N (2000). Differentiation of the Italian wolf and the domestic dog based on microsatellite analysis. *Genetics Selection Evolution*, **32**, 533–541.

Doncaster CP and Macdonald DW (1991). Drifting territoriality in the red fox *Vulpes vulpes*, *Journal of Animal Ecology*, **60**, 423–440.

Doncaster CP and Macdonald DW (1997). Activity patterns and interactions of red foxes (*Vulpes vulpes*) in Oxford city. *Journal of Zoology London*, **241**, 73–87.

Doncaster CP and Woodroffe RB (1993). Den site can determine shape and size of badger territories: implications for group living. *Oikos*, **66**, 88–93.

Donovan ML, Rabe DL Jr and Olson CE (1987). Use of geographic information system to develop habitat suitability models. *Wildlife Society Bulletin*, **15**, 574–579.

Doolan SP and Macdonald DW (1999). Co-operative rearing by slender-tailed meerkats (*Suricata suricatta*) in the Southern Kalahari. *Ethology*, **105**, 851–866.

Dorst J and Dandelot P (1970). *A field guide to the larger mammals of Africa*. Collins, London, UK.

Downs G and Smith L (1998). *Summary report of the Maine wolf attitude survey*. Center for Research and Evaluation, University of Maine, Orono, Maine, USA.

Dragesco-Joffé A (1993). *La vie sauvage au Sahara*. Delachaux et Niestlé, Lausanne, Switzerland.

Dragoo JW, Choate JR, Yates TL and O'Farrell TP (1990). Evolutionary and taxonomic relationships among North American arid land foxes. *Journal of Mammalogy*, **71**, 318–332.

Dragoo, JW and Wayne RK (2003). Systematics and population genetics of kit and Swift Foxes. In MA Sovada, and L Carnyn (eds), *Ecology and Conservation of Swift Foxes in a Changing World*, pp. 207–222. Canadian Plains Research Center, University of Regina, Saskatchewan, Canada.

Drews C (1995). Road kills by public traffic in Mikumi National Park, Tanzania, with a note on baboon mortality. *African Journal of Ecology*, **33**, 89–100.

Dunlap TR (1988). *Saving America's Wildlife*. Princeton University Press, Princeton, NJ, USA.

Durán J, Cattan P and Jañez J (1985). The gray fox *Canis griseus* (Gray) in Chilean Patagonia (Southern Chile). *Biological Conservation*, **34**, 141–148.

Durbin LS (1998). Individuality in the whistle call of the Asiatic wild dog *Cuon alpinus*. *Bioacoustics*, **9**, 197–206.

Durbin LS, Hedges S, Duckworth W and Venkataraman AB (in press). Dhole (*Cuon alpinus*). In C Sillero-Zubiri, M Hoffmann and DW Macdonald (eds), *Canids: foxes, wolves, jackals and dogs. Status survey and conservation action plan*. IUCN/SSC Canid Specialist Group, Gland, Switzerland, and Cambridge, UK.

Durchfeld B, Baumgartner W, Herbst W and Brahm R (1990). Vaccine-associated canine distemper infection in a litter of African hunting dogs (*Lycaon pictus*). *Zentralblatt für Veterinrmedizin B*, **37**, 203–212.

Dye C (1996). The logic of visceral leishmaniasis control. *American Journal of Tropical Medicine and Hygiene*, **55**, 125–130.

Eberhardt LL (1998). Applying difference equations to wolf predation. *Canadian Journal of Zoology*, **76**, 380–386.

Eberhardt LE and Hanson WC (1978). Long-distance movements of arctic foxes tagged in northern Alaska. *Canadian Field-Naturalist*, **92**, 386–389.

Eberhardt LE, Hanson WC, Bengtson JL, Garrott RA and Hanson EE (1982). Arctic fox home range characteristics in an oil-development area. *Journal of Wildlife Management*, **46**, 183–190.

Eberhardt LE, Garrott RA and Hanson WC (1983). Winter movements of arctic foxes, *Alopex lagopus*, in a petroleum development area. *Canadian Field-Naturalist*, **97**, 66–70.

Edwards SV and Hedrick PW (1998). Evolution and ecology of MHC molecules: from genomics to sexual selection. *Trends in Ecology & Evolution*, **13**, 305–311.

Egoscue HJ (1956). Preliminary studies of the kit fox in Utah. *Journal of Mammalogy*, **37**, 351–357.

Egoscue HJ (1962). Ecology and life history of the kit fox in Tooele County, Utah. *Ecology*, **43**, 481–497.

Egoscue HJ (1975). Population dynamics of the kit fox in western Utah. *Bulletin of the Southern California Academy of Sciences*, **74**, 122–177.

Egoscue HJ (1979). *Vulpes velox, Mammalian Species*, **122**, 1–5.

Egoscue HJ (1985). Kit fox flea relationships on the Naval Petroleum Reserves, Kern County, California. *Southern California Academy of Sciences Bulletin*, **84**, 127–132.

Ehiobu NG, Goddard ME and Taylor JF (1989). Effect of rate of inbreeding on inbreeding depression in Drosophila melanogaster. *Theoretical and Applied Genetics*, **77**, 123–127.

Ehrlen J, van Groenendal J and de Kroon H (2001). Reliability of elasticity analysis: reply to Mills et al. *Conservation Biology*, **15**, 278–280.

Eide NE (2002). Spatial ecology of arctic foxes. Relations to resource distribution, and spatiotemporal dynamics in prey abundance. Ph.D. dissertation. Agricultural University of Norway, Ås, Norway.

Eisenberg JF, O'Connell MA and August PV (1979). Density, productivity and distribution of mammals in two Venezuelan habitats. In JF Eisenberg (ed), *Vertebrate ecology in the northern neotropics*, pp. 187–207. Smithsonian Institution Press, Washington DC, USA.

Ellegren H (1999). Inbreeding and relatedness in Scandinavian grey wolves *Canis lupus*, *Hereditas*, **130**, 239–244.

Ellegren H, Savolainen P and Rosen B (1996). The genetical history of an isolated population of the endangered grey wolf *Canis lupus*: a study of nuclear and mitochondrial polymorphisms. *Philosophical Transactions of the Royal Society of London Series B Biological Sciences*, **351**, 1661–1669.

Ellerman JR and Morrison-Scott TCS (1951). *Checklist of Palaearctic and Indian mammals*. British Museum of Natural History, London, UK.

Ellerman JR and Morrison-Scott TCS (1966). Checklist of Palaearctic and Indian mammals, pp. 1758–1946. *2nd Edition*. British Museum (Natural History), London, UK.

Elliot JL (1985). Isle Royale-Northwoods Park Primeval. *National Geographic*, **167**, 534–550.

Elmhagen B, Tannerfeldt M, Verucci P and Angerbjörn A (2000). The arctic fox (*Alopex lagopus*): an opportunistic specialist. *Journal of Zoology London*, **251**, 139–149.

Elzinga CL, Salzer DW, Willoughby JW and Gibbs JP (2001). Monitoring plant and animal populations. Blackwell Science, Malden, MA, USA.

Emerson BC, Paradis E and Thebaud C (2001). Revealing the demographic histories of species using DNA sequences. *Trends in Ecology & Evolution*, **16**, 707–716.

Emlen ST (1982). The evolution of helping. I. An ecological constraints model. *American Naturalist*, **119**, 29–39.

Emlen ST (1991). Evolution of cooperative breeding in birds and mammals. In JR Krebs and NB Davies (eds), *Behavioural ecology. An evolutionary approach*, pp. 301–337. *3rd Edition*. Blackwell Scientific Publications, Oxford, UK.

Enck JW and Brown TL (2002). New Yorkers' attitudes toward restoring wolves to the Adirondack Park. *Wildlife Society Bulletin*, **30**, 16–28.

Englund J (1970). Some aspects of reproduction and mortality rates in Swedish foxes (*Vulpes vulpes*), 1961–63 and 1966–69. *Viltrevy [Swedish Wildlife]*, **8**, 1–82.

Erickson DW (1982). Estimating and using furbearer harvest information. In GC Sanderson (ed), *Midwest furbearer management*, pp. 53–65. Proceedings of the Symposium of the 43rd Midwest Fish and Wildlife Conference, Wichita, KS, USA.

Estelle VB, Mabee TJ and Farmer AH (1996). Effectiveness of predator exclosures for pectoral sandpiper nests in Alaska. *Journal of Field Ornithology*, **67**, 447–452.

Estes RD (1967). Predators and scavengers. *Natural History*, **76**, 20–29, 38–47.

Estes RD (1991). *The behaviour guide to African mammals, including hoofed mammals, carnivores and primates*, University of California Press, Berkley and Los Angeles, CA, USA.

Estes RD and Goddard J (1967). Prey selection and hunting behavior of the African hunting dog. *Journal of Wildlife Management*, **31**, 52–70.

Ewer RF (1973). *The Carnivores*, Cornell University Press, Ithaca, NY, USA.

Fain SR, DeLisle LA, Taylor BF, LeMay JP and Jarrell P (1995). Nuclear DNA fingerprint and mitochondrial DNA sequence variation in captive populations of Mexican gray wolves. *Unpublished report*. US Fish and Wildlife Service, Albuquerque, NM, USA.

Fanshawe JH and Fitzgibbon CD (1993). Factors influencing the hunting success of an African wild dog pack. *Animal Behaviour*, **45**, 479–490.

Fanshawe JH, Frame LH and Ginsberg JR (1991). The wild dog—Africa's vanishing carnivore. *Oryx*, **25**, 137–146.

Fanshawe JH, Ginsberg JR, Sillero-Zubiri C and Woodroffe R (1997). The status and distribution of remaining wild dog populations. In R Woodroffe, JR Ginsberg and DW Macdonald (eds), *The African wild dog-status survey and conservation action plan*, pp. 11–57. IUCN Canid Specialist Group. IUCN, Gland, Switzerland and Cambridge, UK.

Fascione N, Osborn LGI, Kendrot SR and Paquet PC (2001). *Canis soupus*: eastern wolf genetics and its implications for wolf recovery in the northeastern United States. *Endangered Species Update*, **18**, 159–163.

Fearneyhough MG, Wilson PJ and Clark KA (1998). Results of an oral vaccination program for coyotes. *Journal of the American Veterinary Medical Association*, **212**, 498–502.

Federoff NE and Nowak RM (1998). Cranial and dental abnormalities of the endangered red wolf *Canis rufus*. *Acta Theriologica*, **43**, 293–300.

Fedriani JM and Kohn MH (2001). Genotyping faeces links individuals to their diet. *Ecology Letters*, **4**, 477–483.

Feeney S (1999). *Comparative osteology, myology, and loco-motor specializations of the fore and hind limbs of the North American foxes* Vulpes vulpes *and* Urocyon cinereoargenteus. Ph.D. dissertation. University of Massachusetts, Amherst, MA, USA.

Fekadu M, Endeshaw T, Alemu W, Bogale Y, Teshager T and Olson JG (1996). Possible human-to-human transmission of rabies in Ethiopia. *Ethiopian Medical Journal*, **34**, 123–127.

Fenn MGP and Macdonald DW (1995). Use of middens by red foxes: risk reverses rhythms of rats. *Journal of Mammalogy*, **76**, 1130–1136.

Fentress JC and Ryon J (1982). A long-term study of distributed pup feeding in captive wolves. In FH Harrington and PC Paquet (eds), *Wolves of the World*, pp. 238–261. Noyes Publications, Park Ridge, NJ, USA.

Ferguson JWH (1980). *Die ecologie van die rooijakkals, Canis mesomelas Schreber 1778 met spesiale verwysing na bewegings en sociale organisaise*. M.Sc. dissertation, University of Pretoria, Pretoria, South Africa.

Ferguson JWH, Nel JAJ and De Wet MJ (1983). Social organization and movement patterns of black-backed jackals *Canis mesomelas* in South Africa. *Journal of Zoology London*, **199**, 487–502.

Ferguson JWH, Galpin JS and De Wet MJ (1988). Factors affecting the activity patterns of black-backed jackals *Canis mesomelas, Journal of Zoology London*, **214**, 55–69.

Ferguson WW (1981). The systematic position of *Canis aureus lupaster* (Carnivora: Canidae) and the occurrence of *Canis lupus* in north Africa, Egypt and Sinai. *Mammalia*, **45**, 459–465.

Ferrell RE, Morizot DC, Horn J and Carley CJ (1978). Biochemical markers in species endangered by introgression: the red wolf. *Biochemical Genetics*, **18**, 39–49.

Ferris RM, Shaffer M, Fascione N, Ring-Erickson G and Davidson D (1999). *Places for wolves: a blueprint for restoration and long-term recovery in the lower 48 states*, 31 pp. Defenders of Wildlife, Washington DC, USA.

Fieberg J and Ellner SP (2000). When is it meaningful to estimate an extinction probability? *Ecology*, **81(7)**, 2040–2047.

Fischer H (1989). Restoring the wolf: defenders launches a compensation fund. *Defenders*, **64**, 9–36.

Fischer H, Snape B and Hudson W (1994). Building economic incentives into the Endangered Species Act. *Endangered Species Technical Bulletin*, **19**, 4–5.

Fisher JL (1981). Kit fox diet in south-central Arizona. M.Sc. dissertation, University of Arizona, Tucson, AZ, USA.

Fisher RA, Putt W and Hackel E (1976). An investigation of the products of 53 gene loci in three species of wild Canidae: Canis lupus, Canis latrans, and Canis familiaris. *Biochemical Genetics*, **14**, 963–974.

Fitzgerald JP, Loy RR and Cameron M (1983). Status of swift fox on the Pawnee National Grassland, Colorado. 21 pp. Unpublished manuscript. University of Northern Colorado, CO, USA.

Fleming P, Corbett L, Harden R and Thomson P (2001). Managing the impacts of dingoes and other wild dogs. Bureau of Rural Sciences, Canberra, Australia.

Floyd TJ, Mech LD and Jordan PA (1978). Relating wolf scat content to prey consumed. *Journal of Wildlife Management*, **42**, 528–532.

Flynn JJ and Nedbal MA (1998). Phylogeny of the Carnivora (Mammalia): congruence vs. incompatibility among multiple data sets. *Molecular Phylogenetics and Evolution*, **9**, 414–426.

Foggin CM (1988). *Rabies and rabies-related viruses in Zimbabwe: historical, virological and ecological aspects*. Ph.D. dissertation, University of Zimbabwe, Harare, Zimbabwe.

Follman EH (1973). Comparative ecology and behavior of red and gray foxes. Ph.D. dissertation, Southern Illinois University, Carbondale, IL, USA.

Foran DR, Minta SC and Heinemeyer KS (1997). DNA-based analysis of hair to identify species and individuals for population research and monitoring. *Wildlife Society Bulletin*, **25**, 840–847.

Forbes SH and Boyd DK (1996). Genetic variation of naturally colonizing wolves in the Central Rocky Mountains. *Conservation Biology*, **10**, 1082–1090.

Forbes SH and Boyd DK (1997). Genetic structure and migration in native and reintroduced Rocky Mountain wolf populations. *Conservation Biology*, **11**, 1226–1234.

Forchhammer MC and Asferg T (2000). Invading parasites cause a structural shift in red fox dynamics. *Proceedings of the Royal Society of London B*, **267**, 779–786.

Fox MW (1971). *Behaviour of wolves, dogs and related canids*. Cape, London.

Fox MW (1984). *The whistling hunters. Field studies of the Asiatic wild dog (Cuon alpinus)*. State University of New York Press, Albany, NY, USA.

Fox MW and Johnsingh AJT (1975). Hunting and feeding in the wild dog. *Journal of the Bombay Natural History Society*, **72**, 321–326.

Frafjord K and Prestrud P (1992). Home ranges and movements of arctic foxes *Alopex lagopus* in Svalbard. *Polar Biology*, **12**, 519–526.

Frafjord K, Becker D and Angerbjorn A (1989). Interactions between Arctic and red foxes in Scandinavia: predation and aggression. *Arctic*, **42**, 354–356.

Frame GW (1986). Carnivore competition and resource use in the Serengeti ecosystem of Tanzania. Ph.D. dissertation, Utah State University, Logan, UT, USA.

Frame GW and Frame LH (1981). *Swift and enduring: cheetahs and wild dogs of the Serengeti*. E. P. Dutton, New York, USA.

Frame LH, Malcolm JR, Frame GW and van Lawick H (1979). Social organization of African wild dogs (*Lycaon pictus*) on the Serengeti Plains. *Zeitschrift für Tierpsychologie*, **50**, 225–249.

Frank DW, Kirton KT, Murchison TE, Quinlan WJ, Coleman ME, Gilbertson TJ, Feenstra ES and Kimbell FA (1979). Mammary tumors and serum hormones in the bitch treated with medroxyprogesterone acetate or progesterone for four years. *Fertility and Sterility*, **31**, 340–346.

Frank LG (1986). Social organization of the spotted hyaena *Crocuta crocuta*. II. Dominance and reproduction. *Animal Behaviour*, **34**, 1510–1527.

Frank LG and Woodroffe R (2001). Behaviour of carnivores in exploited and controlled populations. In JL Gittleman, SM Funk, DW Macdonald and RK Wayne (eds), *Carnivore conservation*, pp 419–442. Cambridge University Press, Cambridge, UK.

Frankham R, Ballou JD and Briscoe DA (2002). *Introduction to conservation genetics*. Cambridge University Press, Cambridge, UK.

Franklin IR (1980). Evolutionary changes in small populations. In ME Soulé and BA Wilcox (eds), *Conservation biology. An evolutionary-ecological approach*, pp. 135–149. Sinauer Associates, Sunderland, MA, USA.

Franklin IR and Frankham R (1998). How large must populations be to retain evolutionary potential. *Animal Conservation*, **1**, 69–70.

Frati F, Hartl GB, Lovari S, Delibes M and Markov G (1998). Quaternary radiation and genetic structure of the red fox *Vulpes vulpes* in the Mediterranean Basin, as revealed by allozymes and mitochondrial DNA. *Journal of Zoology*, **245**, 43–51.

Frati F, Lovari S and Hartl GB (2000). Does protection from hunting favour genetic uniformity in the red fox? *Zeitschrift für Saugetierkunde*, **65**, 76–83.

Fredrickson R and Hedrick P (2002). Body size in endangered Mexican wolves: effects of inbreeding and cross-lineage matings. *Animal Conservation*, **5**, 39–43.

Freeman AR, Machuch D, Mckeown S, Walzer C, Mcconnell DJ and Bradley DG (2001). Sequence variation in the mitochondrial DNA control region of wild African cheetahs (*Acynonyx jubatus*). *Heredity*, **86**, 355–362.

Fritts SH (1982). Wolf depredation on livestock in Minnesota. *Research Report* 145, United States Fish and Wildlife Service, Fort Snelling, MN, USA.

Fritts SH (1983). Record dispersal by a wolf from Minnesota. *Journal of Mammalogy*, **64**, 166–167.

Fritts SH and Mech LD (1981). Dynamics, movements, and feeding ecology of a newly protected wolf population in northwestern Minnesota. *Wildlife Monographs* No. 80.

Fritts SH and Carbyn LN (1995). Population viability nature reserves and the outlook for gray wolf conservation in North America. *Restoration Ecology*, **3**, 26–38.

Fritts SH and Paul WJ (1989). Interactions of wolves and dogs in Minnesota. *Wildlife Society Bulletin*, **17**, 121–123.

Fritts SH, Paul WJ and Mech LD (1984). Movement of translocated wolves in Minnesota. *Journal of Wildlife Management*, **48**, 709–721.

Fritts SH, Paul WJ and Mech LD (1985). Can relocated wolves survive? *Wildlife Society Bulletin*, **13**, 459–463.

Fritts SH, Bangs EE, Fontaine JA, Johnson MR, Phillips MK, Koch ED and Gunson JR (1997). Planning and implementing a reintroduction of wolves to Yellowstone National Park and central Idaho. *Restoration Ecology*, **5**, 7–27.

Fritts SH, Mack CM, Smith DW, Murphy KM, Phillips MK, Jimenez MD, Bangs EE, Fontaine JA, Niemeyer CC, Brewster WG and Kaminski TJ (2001). Outcomes of hard and soft releases of wolves in central Idaho and the Greater

Yellowstone Area. In DS Maehr, RF Noss and JL Larkin (eds), *Large mammal restoration; ecological and sociological challenges in the 21st century*, pp. 125–147. Island Press. Washington DC, USA.

Fritts SH and Mech DL (1981). Dynamics, movements, and feeding ecology of a newly protected wolf population in northwestern Minnesota. *Wildlife Monographs*, **80**, 1–80.

Fritts SH, Stephenson RO, Hayes RD and Boitani L (2003). Wolves and humans. In LD Mech and L Boitani (eds), *Wolves: ecology, behavior and conservation*, pp. 289–316. University of Chicago Press, Chicago, IL, USA.

Fritzell EK (1987). Gray fox and island gray fox. In M Novak, JA Baker, ME Obbard and B Malloch (eds), *Wild furbearer management and conservation in North America*, pp. 408–420. Ministry of Natural Resources, Ontario, Canada.

Fritzell EK and Haroldson KJ (1982). *Urocyon cinereoargenteus*, *Mammalian Species*, **189**, 1–8.

Frommolt KH, Kruchenkova EP and Russig H (1997). Individuality of territorial barking in Arctic foxes, *Alopex lagopus* (L. 1758). *Zeitschrift fuer Saeugetierkunde*, **62**, 66–70.

Frommolt K-H, Goltsman ME and Macdonald DW (2003). Barking foxes, *Alopex lagopus*: field experiments in individual recognition in a territorial mammal. *Animal Behaviour*, **65**, 509–518.

Fuentes ER and Jaksic FM (1979). Latitudinal size variation of Chilean foxes: tests of alternative hypotheses. *Ecology*, **60**, 43–47.

Fukue Y (1991). Utilisation pattern of home range and parental care of the raccoon dogs at the Kanazawa University Campus. M.Sc. dissertation, Kanazawa University, Japan (In Japanese.)

Fuller TK (1989). Population dynamics of wolves in north-central Minnesota. *Wildlife Monographs*, **105**, 1–41.

Fuller TK (1991). Effect of snow depth on wolf activity and prey selection in north central Minnesota. *Canadian Journal of Zoology*, **69**, 283–287.

Fuller TK and Kat PW (1990). Movements, activities, and prey relationships of African wild dogs (*Lycaon pictus*) near Aitong, southwestern Kenya. *African Journal of Ecology*, **28**, 330–350.

Fuller TK and Sampson BA (1988). Evaluation of a simulated howling survey for wolves. *Journal of Wildlife of Management*, **52**, 60–63.

Fuller TK and Sievert PR (2001). Carnivore demography and the consequences of changes in prey availability. In JL Gittleman, SM Funk, DW Macdonald and RK Wayne eds *Carnivore conservation*, pp. 163–178. Cambridge University Press, Cambridge, UK.

Fuller TK, Biknevicius AR, Kat PW, Van VB and Wayne RK (1989). The ecology of three sympatric jackal species in the Rift Valley of Kenya. *African Journal of Ecology*, **27**, 313–324.

Fuller TK, Berg WE, Radde GL, Lenarz MS and Joselyn GB (1992a). A history and current estimate of wolf distribution and numbers in north central Minnesota. *Wildlife Society Bulletin*, **20**, 42–54.

Fuller TK, Kat PW, Bulger JB, Maddock AH, Ginsberg JR, Burrows R, McNutt JW and Mills MGL (1992b). Population dynamics of African wild dogs. In DR McCullough and RH Barrett (eds), *Wildlife 2001: populations*, pp. 1125–1139. Elsevier Applied Science, New York, USA.

Fuller TK, Mills MGL, Borner M, Laurenson K and Kat PW (1992c). Long distance dispersal by African wild dogs in East and South Africa. *Journal of African Zoology*, **106**, 535–537.

Fuller TK, Nicholls TH and Kat PW (1995). Prey and estimated food consumption of African wild dogs in Kenya. *South African Journal of Wildlife Research*, **25**, 106–110.

Funes MC and Novaro AJ (1999). Rol de la fauna silvestre en la economía del poblador rural, provincia del Neuquén, Argentina. *Revista Argentina de Producción Animal*, **19**, 265–271.

Funk SM, Fiorello CV, Cleaveland S and Gompper ME (2001). The role of disease in carnivore ecology and conservation. In JL Gittleman, SM Funk, DW Macdonald and RK Wayne, eds *Carnivore conservation*, pp. 442–466. Cambridge University Press, Cambridge, UK.

Gangloff L (1972). Breeding fennec foxes *Fennecus zerda* at Strasbourg Zoo. *International Zoo Yearbook*, **12**, 115–116.

Garcelon DK (1996). Development of a video monitoring system to observe nest sites of the San Clemente logger-head shrike on San Clemente Island. *Unpublished report*, U.S. Navy, Natural Resources Management Branch, SW Division, Naval Facilities Engineering Command, San Diego, CA, USA.

Garcelon DK (1999). Island fox population analysis and management recommendations, 56 pp. Draft report to the U.S. Navy. Institute for Wildlife Studies, Arcata, CA, USA.

Garcelon DK, Wayne RK and Gonzales BJ (1992). A sero-logical survey of the Island fox *Urocyon littoralis* on the Channel Islands, California. *Journal of Wildlife Diseases*, **28**, 223–229.

Garcelon DK, Roemer GW, Phillips RB and Coonan TJ (1999). Food provisioning by island foxes, *Urocyon littoralis*, to conspecifics caught in traps. *Southwest Naturalist*, **44**, 83–86.

García-Moreno J, Matocq MD, Roy MS, Geffen E and Wayne RK (1996). Relationships and genetic purity of the endangered Mexican wolf based on analysis of microsatellite loci. *Conservation Biology*, **10**, 376–389.

Garrott RA and Eberhardt LE (1982). Mortality of arctic fox pups in northern Alaska. *Journal of Mammalogy*, **63**, 173–174.

Garrott RA and Eberhardt LE (1987). Arctic fox. In M Novak, JA Baker, ME Obbard and B Malloch, (eds), *Wild furbearer management and conservation in north America*, pp.395–406. Ministry of Natural Resources, Ontario, Canada.

Gasaway WC, Boertje RD, Grangaard DV, Kelleyhouse DG, Stephenson RO and Larsen DG (1992). The role of predation in limiting moose at low densities in Alaska and Yukon and implications for conservation. *Wildlife Monographs*, **120**, 5–59.

Gascoyne SC, King AA, Laurenson MK, Borner M, Schildger B and Barrat J (1993a). Aspects of rabies infection and control in the conservation of the African wild dog (*Lycaon pictus*) in the Serengeti region, Tanzania. *Onderstepoort Journal of Veterinary Research*, **60**, 415–420.

Gascoyne SC, Laurenson MK, Lelo S and Borner M (1993b). Rabies in African wild dogs *Lycaon pictus* in the Serengeti region, Tanzania. *Journal of Wildlife Diseases*, **29**, 396–402.

Gaston AJ (1978). Demography of the jungle babbler Turgoides straitus. *Journal of Animal Ecology*, **47**, 845–879.

Gaston KJ (1991). How large is a species' geographic range? *Oikos*, **61**, 434–438.

Gauither DA and Licht DS (2003). The socio-economic context for swift fox conservation in the prairies of North America. In M Sovada and L Carbyn (eds), *The swift fox: ecology and conservation of swift foxes in a changing world*, pp. 19–28. Canadian Plains Research Center, University of Regina, Saskatchewan, Canada.

Gauthier-Pilters H (1962). Beobachtungen en Feneks (*Fennecus zerda* Zimm.). *Zietschrift für Tierpsychologie*, **19**, 440–464.

Gauthier-Pilters H (1967). The fennec. *African Wildlife*, **21**, 117–125.

Geffen E (1994). Blanford's fox, *Vulpes cana*. *Mammalian Species*, **462**, 1–4.

Geffen E and Macdonald DW (1992). Small size and monogamy: spatial organization of the Blanford's fox, *Vulpes cana*. *Animal Behavior*, **44**, 1123–1130.

Geffen E and Macdonald DW (1993). Activity and movement patterns of Blandford's foxes. *Journal of Mammalogy*, **74**, 455–463.

Geffen E, Dagan AA, Kam M, Hefner R and Nagy KA (1992a). Daily energy expenditure and water flux of free-living Blanford's foxes (*Vulpes cana*): a small desert carnivore. *Journal of Animal Ecology*, **61**, 611–617.

Geffen E, Hefner R, Macdonald DW and Ucko M (1992b). Diet and foraging behavior of the Blanford's fox, *Vulpes cana*, in Israel. *Journal of Mammalogy*, **73**, 395–402.

Geffen E, Hefner R, Macdonald DW and Ucko M (1992c). Habitat selection and home range in the Blanford's fox, *Vulpes cana*: compatibility with the Resource Dispersion Hypothesis. *Oecologia*, **91**, 75–81.

Geffen E, Hefner R, Macdonald DW and Ucko M (1992d). Morphological adaptations and seasonal weight changes in the Blanford's fox, *Vulpes cana*. *Journal of Arid Environments*, **23**, 287–292.

Geffen E, Mercure A, Girman DJ, Macdonald DW and Wayne RK (1992e). Phylogeny of the fox-like canids: analysis of mtDNA restriction fragment, site, and cytochrome b sequence data. *Journal of Zoology London*, **228**, 27–39.

Geffen E, Hefner R, Macdonald DW and Ucko M (1993). Biotope and distribution of the Blanford's fox. *Oryx*, **27**, 104–108.

Geffen E, Gompper ME, Gittleman JL, Luh HK, Macdonald DW and Wayne RK (1996). Size, life-history traits, and social organization in the Canidae: a reevaluation. *American Naturalist*, **147**, 140–160.

Germano DJ, Rathbun GB and Saslaw LR (2001). Managing exotic grasses and conserving declining species. *Wildlife Society Bulletin*, **29**, 551–559.

Gese EM (1998). Response of neighboring coyotes (*Canis latrans*) to social disruption in an adjacent pack. *Canadian Journal of Zoology*, **76**, 1960–1963.

Gese EM (2001a). Monitoring of terrestrial carnivores. In JL Gittleman, SM Funk, DW Macdonald and RK Wayne eds *Carnivore conservation*, pp. 372–396. Cambridge University Press, Cambridge, UK.

Gese EM (2001b). Territorial defense by coyotes (*Canis latrans*) in Yellowstone National Park, Wyoming: who, how, where, when, and why. *Canadian Journal of Zoology*, **79**, 980–987.

Gese EM (in press). Survey and census techniques for canids. In C Sillero-Zubiri, M Hoffmann and DW Macdonald (eds), *Canids: foxes, wolves, jackals and dogs. Status survey and conservation action plan*. IUCN/SSC Canid Specialist Group, Gland, Switzerland, and Cambridge, UK.

Gese EM and Bekoff M (in press). Coyote (*Canis latrans*). In C Sillero-Zubiri, M Hoffmann and DW Macdonald (eds), *Canids: foxes, wolves, jackals and dogs. Status survey and conservation action plan*. IUCN/SSC Canid Specialist Group, Gland, Switzerland, and Cambridge, UK.

Gese EM and Grothe S (1995). Analysis of coyote predation on deer and elk during winter in Yellowstone National Park, Wyoming. *American Midland Naturalist*, **133**, 36–43.

Gese EM and Mech LD (1991). Dispersal of wolves (*Canis lupus*) in northeastern Minnesota, 1969–1989. *Canadian Journal of Zoology*, **69**, 2946–2955.

Gese EM and Ruff RL (1997). Scent-marking by coyotes, *Canis latrans*: the influence of social and ecological factors. *Animal Behaviour*, **54**, 1155–1166.

Gese EM and Ruff RL (1998). Howling by coyotes (*Canis latrans*): variation among social classes, seasons, and pack sizes. *Canadian Journal of Zoology*, **76**, 1037–1043.

Gese EM, Rongstad OJ and Mytton WR (1988). Home range and habitat use of coyotes in southeastern Colorado. *Journal of Wildlife Management*, **52**, 640–646.

Gese EM, Rongstad OJ and Mytton WR (1989). Population dynamics of coyotes in Southeastern Colorado. *Journal of Wildlife Management*, **53**, 174–181.

Gese EM, Ruff RL and Crabtree RL (1996a). Foraging ecology of coyotes (*Canis latrans*): the influence of extrinsic factors and a dominance hierarchy. *Canadian Journal of Zoology*, **74**, 769–783.

Gese EM, Ruff RL and Crabtree RL (1996b). Intrinsic and extrinsic factors influencing coyote predation of small mammals in Yellowstone National Park. *Canadian Journal of Zoology*, **74**, 784–797.

Gese EM, Ruff RL and Crabtree RL (1996c). Social and nutritional factors influencing the dispersal of resident coyotes. *Animal Behaviour*, **52**, 1025–1043.

Gese EM, Schultz RD, Johnson MR, Williams ES, Crabtree RL and Ruff RL (1997). Serological survey for diseases in free-ranging coyotes (*Canis latrans*) in Yellowstone National Park, Wyoming. *Journal of Wildlife Diseases*, **33**, 47–56.

Getz LL (1985). Habitats. *Special Publication, American Society of Mammalogy*, **8**, 286–309.

Gier HT (1968). *Coyotes in Kansas*. Kansas Agricultural Experiment Station, Kansas State University, Manhattan, KS, USA.

Gilbert DA, Lehman N, O'Brien SJ and Wayne RK (1990). Genetic fingerprinting reflects population differentiation in the California channel island fox. *Nature*, **344**, 764–767.

Ginsberg JR (1994). Captive breeding, reintroduction, and the conservation of canids. In PJS Olney, G Mace and ATC Feistner (eds), *Creative conservation: interactive management of wild and captive animals*, pp. 365–383. Chapman and Hall, London, UK.

Ginsberg JR (2001). Setting priorities for carnivore conservation: what makes carnivores different? In JL Gittleman, SM Funk, DW Macdonald and RK Wayne (eds), *Carnivore conservation*, pp. 498–523. Cambridge University Press, Cambridge, UK.

Ginsberg JR and Macdonald DW (1990). *Foxes, wolves, jackals and dogs: an action plan for the conservation of canids*. IUCN, Gland, Switzerland & Cambridge, UK.

Ginsberg JR and Woodroffe R (1997). Extinction risks faced by remaining wild dog populations. In R Woodroffe, JR Ginsberg and DW Macdonald (eds), *The African wild dog-status survey and conservation action plan*, pp. 75–87. IUCN, Gland, Switzerland and Cambridge, UK.

Ginsberg JR, Alexander KA, Creel S, Kat PW, McNutt JW and Mills MGL (1995a). Handling and survivorship of African wild dog (Lycaon pictus) in five ecosystems. *Conservation Biology*, **9**, 665–674.

Ginsberg JR, Mace GM and Albon S (1995b). Local extinction in a small and declining population: Serengeti wild dogs. *Proceedings of the Royal Society of London, Series B*, **262**, 221–228.

Ginsburg L (1999). Order Carnivora, Chapter 10. In GE Rössner and K Heissig (eds), *The Miocene land Mammals of Europe*, pp. 109–148. Verlag Dr. Friedrich Pfeil, Munich, Germany.

Gipson PS (1975). Efficiency of trapping in capturing offending coyotes. *Wildlife Management*, **39**, 45–47.

Gipson PS, Ballard WB and Noval RM (1998). Famous North American wolves and the credibility of early wildlife literature. *Wildlife Society Bulletin*, **26**, 808–816.

Giraldeau L-A and Caraco T (2000). *Social foraging theory*. Princeton University Press, Princeton, NJ, USA.

Giraudoux P, Raoul F, Bardonnet K, Vuillaume P, Tourneux F, Cliquet F, Delattre P and Vuitton DA (2001). Alveolar echinococcosis: characteristics of a possible emergence and new perspectives in epidemiosurveillance. *Medicine et al maladies infectieuses*, **31**, 247–256.

Girden ER (1992). *ANOVA: repeated measures*. Sage Publications, Newbury Park, CA, USA.

Girman DJ and Wayne RK (1997). Genetic perspectives on wild dog conservation. In R Woodroffe, JR Ginsberg, and DW Macdonald (eds) *The African wild dog-tatus survey and conservation action plan*, pp. 7–10. IUCN, Gland, Switzerland and Cambridge, UK; Gittleman JL (1986). Carnivore life history patterns: allometric, phylogenetic and ecological associations. *American Naturalist*, **127**, 744–771.

Girman DJ, Kat PW, Mills MG, Ginsberg JR, Borner M, Wilson V, Fanshawe JH, Fitzgibbon C, Lau LM and Wayne RK (1993). Molecular genetic and morphological analyses of the African wild dog (*Lycaon pictus*). *Journal of Heredity*, **84**, 450–459.

Girman DJ, MGL Mills, E Geffen and RK Wayne (1997). A molecular genetic analysis of social structure, dispersal, and interpack relationships of the African wild dog (*Lycaon pictus*). *Behavioral Ecology and Sociobiology*, **40**, 187–198.

Girman DJ, Vilà C, Geffen E, Creel S, Mills MGL, McNutt JW, Ginsberg J, Kat PW, Mamiya KH and Wayne RK (2001). Patterns of population subdivision, gene flow and genetic variability in the African wild dog (*Lycaon pictus*). *Molecular Ecology*, **10**, 1703–1723.

Gittleman JL (1985). Functions of communal care in mammals. In PH Greenwood, PH Harvey and M Slatkin (eds), *Evolution: essays in honour of John Maynard Smith*, pp. 181–205. Cambridge University Press, Cambridge, UK.

Gittleman JL (1986). Carnivore life history patterns: allo-metric, phylogenetic, and ecological associations. *American Naturalist*, **127**, 744–777.

Gittleman JL (1989). Carnivore group living: comparative trends. In JL Gittleman ed. *Carnivore behavior, ecology, and evolution*, pp. 183–207. Chapman and Hall, London, UK.

Gittleman JL and Harvey PJ (1982). Carnivore home range size, metabolic needs and ecology. *Behavioural Ecology and Sociobiology*, **10**, 57–63.

Gittleman JL and Oftedal OT (1987). Comparative growth and lactation energetics in carnivores. *Sympozium on Zoological Society of London*, **57**, 41–77.

Gittleman JL and Pimm SL (1991). Crying wolf in North America. *Nature*, **351**, 524–525.

Gittleman JL and Thompson SD (1989). Energy allocation in mammalian reproduction. *American Zoologist*, **28**, 863–875.

Gittleman JL, Funk SM, Macdonald DW and Wayne RK, eds (2001a). *Carnivore conservation*. Cambridge University Press, Cambridge, UK.

Gittleman JL, Funk SM, Macdonald DW and Wayne RK (2001b). Why 'carnivore conservation'? In JL Gittleman, SM Funk, DW Macdonald and RK Wayne (eds), *Carnivore conservation*, pp.1–7. Cambridge University Press, Cambridge, UK.

Golani I and Keller AA (1975). A longitudinal field study of the behavior of a pair of golden jackals. In MW Fox (ed), *The wild canids: their systematics, behavioral ecology and evolution*, pp. 303–335. Van Nostrand, Reinhold, NY, USA.

Goldstein DB and Pollock DD (1997). Launching microsatellites: a review of mutation processes and methods of phylogenetic inference. *Journal of Heredity*, **88**, 335–342.

Goldstein DB, Roemer GW, Smith DA, Reich DE, Bergman A and Wayne RK (1999). The use of microsatellite variation to infer population structure and demographic history in a natural model system. *Genetics*, **151**, 797–801.

Goldstein PZ, Salle R, Amato G and Vogler AP (2000). Conservation genetics at the species boundary. *Conservation Biology*, **14**, 120–131.

Golightly RTJ and Ohmart RD (1984). Water economy of two desert canids: coyote and kit fox. *Journal of Mammalogy*, **65**, 51–58.

Goltsman M, Kruchenkova EP and Macdonald DW (1996). The Mednyi Arctic foxes: treating a population imperiled by disease. *Oryx*, **30**, 251–258.

Goltsman M, Kruchenkova EP, Sergeev S, Johnson PJ and Macdonald DW (submitted). Effect of food availability on dispersal, and cub sex ratios in the insular population of Arctic fox, *Alopex lagopus semenovi*. *Ecology*.

González del Solar R and Rae J (in press). Chilla (*Pseudalopex griseus*). In C Sillero-Zubiri, M. Hoffman and DW Macdonald (eds), *Canids: foxes, wolves, jackals and dogs. Status survey* and *conservation action plan*. IUCN/SSC Canid Specialist Group, Gland, Switzerland, and Cambridge, UK.

González del Solar R, Puig S, Videla F and Roig V (1997). Diet composition of the South American grey fox, *Pseudalopex griseus* Gray 1836, in northeastern Mendoza. *Mammalia*, **61**, 617–621.

Gorman ML, Mills MG, Raath JP and Speakman JR (1998). High hunting costs make African wild dogs vulnerable to kleptoparasitism by hyaenas. *Nature*, **391**, 479–481.

Gosling LM (1986). Selective abortion of entire litters in the Coypu *Myocastor-Coypus* adaptive control of off-pring production in relation to quality and sex. *American Naturalist*, **127**(6), 772–795.

Goszczynski J (1999). Fox, raccoon dog and badger densities in north eastern Poland. *Acta Theriologica*, **44**, 413–420.

Gottelli D and Sillero-Zubiri C (1992). The Ethiopian wolf—an endangered endemic canid. *Oryx*, **26**, 205–214.

Gottelli D, Sillero-Zubiri C, Applebaum GD, Roy MS, Girman DJ, Garcia-Moreno J, Otsrander EA and Wayne RK (1994). Molecular genetics of the most endangered canid: the Ethiopian wolf, *Canis simensis*. *Molecular Ecology*, **3**, 301–312.

Gottelli D, Marino J, Sillero-Zubiri C, Fonk SM (in press). The effect of the last Glacial Age on speciation and population genetic structure of the endangered Ethiopian Wolf (*Canis simensis*). *Molecular Ecology*.

Grandy, Stallman B and Macdonald DW (2003). The science and sociology of hunting: shifting practices and perceptions in the United States and Great Britain. In *The State of Animals II*: 2003, DJ Salem and AN Rowan, (eds), Humane Society Press, Washington DC, pp. 107–130.

Grant PR and Grant BR (1992). Demography and the genetically effective sizes of two populations of Darwin's finches. *Ecology*, **73**, 766–784.

Gray MM, Roemer GW and Torres E (2001). Genetic assessment of relatedness among individuals in the island fox (*Urocyon littoralis*) captive breeding program. *Final report*. Channel Islands National Park, Ventura, CA, USA.

Greenwood RJ, Sargeant AB, Johnson DH, Cowardin LM and Shaffer TL (1995). Factors associated with duck nest success in the Prairie Pothole Region of Canada. *Wildlife Monographs*, 1–57.

Grenfell BT and Dobson AP (1995). *Ecology of infectious diseases in natural populations*. Cambridge University Press, Cambridge, UK.

Gresswell RE (1991). Use of antimycin for removal of brook trout from a tributary of Yellowstone Lake. *North American Journal of Fisheries Management*, **11**, 83–90.

Grinnell JD (1914). *The wolf hunter*. Grosset and Dunlap Publishers, New York, NY, USA.

Grinnell J, Dixon DS and Linsdale JM (1937). *Fur-bearing mammals of California*, Vol. 2. University of California Press, Berkeley, CA, USA.

Gubernick DG and Teferi T (2000). Adaptive significance of parental care in a monogamous mammal. *Proceedings of the Royal Society, London, Series B*, **267**, 147–150.

Gudmundsson F (1960). Some reflections on ptarmigan cycles in Iceland. *International Ornithological Congress*, **12**, 259–265.

Guilday JE (1962). Supernumerary molars of *Otocyon*. *Journal of Mammalogy*, **43**, 455–462.

Gustavson CR, Kelly DJ, Sweeney M and Garcia J (1976). Prey lithium aversions. I. Coyotes and wolves. *Behavioral Biology*, **17**, 61–72.

Gutierrez AP and Baumgaertner JU (1984). Multitrophic models of predator-prey energetics. II. A realistic model of plant-herbivore-parasitoid-predator interactions. *Canadian Entomologist*, **115**, 933–949.

Hackett D (1990). Predator problems on California's north coast: economic impacts. In GA Giusti, RM Timm and RH Schmidt (eds), *Predator Management in North Coastal California*, pp. 23–27. Hopland Field Station, Publ. 101.

Haig DA (1977). Rabies in animals. In C Caplan ed. *Rabies, the facts*. Oxford University Press, Oxford, UK.

Haight RG and Mech LD (1997). Computer simulation of vasectomy for wolf control. *Journal of Wildlife Management*, **61**, 1023–1031.

Haight RG, Mladenoff DJ and Wydeven AP (1998). Modeling disjunct gray wolf populations in semi-wild landscapes. *Conservation Biology*, **12**, 879–888.

Haight RG, Cypher B, Kelly PA, Phillips S, Possingham H, Ralls K, Starfield AM, White PJ and Williams D (2002). Optimizing habitat protection using demographic models of population viability. *Conservation Biology*, **16**, 1386–1397.

Hall A and Harwood J (1990). *The intervet guide to vaccinating wildlife*. Sea Mammal Research Unit, Cambridge, UK.

Hall ER (1981). The mammals of North America, *2nd Edition*, 1181 pp. John Wiley and Sons, New York, NY, USA.

Hall ER and Kelson KR (1959). *The mammals of North America*, Vol II. The Ronald Press, New York, USA.

Hall M (1989). *Parameters associated with cyclic populations of arctic fox* (Alopex lagopus) *near Eskimo Point, Northwest Territories: Morphometry, age, condition, seasonal and multiannual influences*. M.Sc. dissertation, Laurentian University, Ontario, Canada.

Hamilton WD (1964). The genetical evolution of social behaviour. *Journal of Theoretical Biology*, **7**, 1–52.

Hamilton WD (1967). Extraordinary sex ratios. *Science*, **156**, 477–488.

Hancock JM (1999). Microsatellites and other simple sequences: genomic context and mutational mechanisms. In DB Golstein and C Schlötterer (eds), *Micro-satellites. Evolution and applications*. pp. 1–9. Oxford University Press, Oxford, UK.

Hanotte O, Tawah CL, Bradley DG, Okomo M, Verjee Y, Ochieng J and Rege JEO (2000). Geographic distribution and frequency of a taurine *Bos taurus* and an indicine *Bos indicus* Y specific allele amongst sub-Saharan African cattle breeds. *Molecular Ecology*, **9**, 387–396.

Hanski I, Turchin P, Korpimäki E and Henttonen H (1993). Population oscillations of boreal rodents: regulation by mustelid predators leads to chaos. *Nature*, **364**, 232–235.

Hansson L and Henttonen H (1985). Gradients in density variations of small rodents: the importance of latitude and snow cover. *Oecologia*, **67**, 394–402.

Happold DCD (1987). *The mammals of Nigeria*. Oxford University Press, Oxford, UK.

Harden RH (1985). The ecology of the dingo *Canis familiaris dingo* in northeastern New South Wales Australia. 1. Movements and home range. *Australian Wildlife Research*, **12**, 25–38.

Harestad AS and Bunnell FL (1979). Home range and body weight—a reevaluation. *Ecology*, **60**, 389–402.

Harrington FH (1987). Aggressive howling in wolves. *Animal Behaviour*, **35**, 7–12.

Harrington FH and Mech LD (1978a). Howling at two Minnesota wolf pack summer homesites. *Canadian Journal of Zoology*, **56**, 2024–2028.

Harrington FH and Mech LD (1978b). Wolf vocalization. In RL Hall and HS Sharp (eds), *Wolf and man: evolution in parallel*, pp. 109–132. Academic Press, NY, USA.

Harrington FH and Mech LD (1979). Wolf howling and its role in territory maintenance. *Behaviour*, **68**, 207–249.

Harrington FH and Mech LD (1982). Patterns of homesite attendance in two Minnesota wolf packs. In FH Harrington and PC Paquet (eds), *Wolves of the world*, pp. 81–105. Noyes Publications, Park Ridge, NJ, USA.

Harrington FH and Mech LD (1983). Wolf pack spacing: howling as a territory-independent spacing mechanism in a territorial population. *Behavioral Ecology and Sociobiology*, **12**, 161–168.

Harrington FH and Paquet PC (1982). *Wolves of the world*. Noyes Publications, Park Ridge, NJ, USA.

Harrington FH, Mech LD and Fritts SH (1983). Pack size and wolf survival: their relationship under varying ecological conditions. *Behavioural Ecology and Sociobiology*, **13**, 19–26.

Harris S (1980). Home ranges and patterns of distribution of foxes (*Vulpes vulpes*) in an urban area, as revealed by radio tracking. In CJ Amlaner and DW Macdonald (eds), *A handbook on biotelemetry and radiotracking*, pp. 685–690. Pergamon Press, Oxford, UK.

Harris S (1981). An estimation of the number of foxes (*Vulpes vulpes*) in the city of Bristol, and some possible factors affecting their distribution. *Journal of Applied Ecology*, **18**, 455–465.

Harris S and Baker P (2001). *Urban foxes, 2nd Edition*. Whittet Books, Stowmarket, UK.

Harris S and Rayner JMV (1986). Urban fox (*Vulpes vulpes*) population estimates and habitat requirements in several British cities. *Journal of Animal Ecology*, **55**, 575–591.

Harris S and Saunders G (1993). The control of canid populations. In N Dunstone and ML Gorman (eds), *Mammals as predators. Symposia of the Zoological Society of London*, **65**, 441–464. Clarendon Press, Oxford, UK.

Harris S and Smith GC (1987). Demography of two urban fox (*Vulpes vulpes*) populations. *Journal of Applied Ecology*, **24**, 75–86.

Harris S and Trewhella WJ (1988). An analysis of some of the factors affecting dispersal in an urban fox (*Vulpes vulpes*) population. *Journal of Applied Ecology*, **25**, 409–422.

Harris S and White PCL (1992). Is reduced affiliative rather than increased agonistic behaviour associated with dispersal in red foxes? *Animal Behaviour*, **44**, 1085–1089.

Harrison DL and Bates PJJ (1989). Observations on two mammal species new to the Sultanate of Oman, *Vulpes cana* new record Blanford, 1877 (Carnivora: Canidae) and *Nycteris thebaica* new record Geoffroy, 1818 (Chiroptera: Nycteridae). *Bonner Zoologische Beitraege*, **40**, 73–77.

Harrison DL and Bates PJJ (1991). *The mammals of Arabia*. Harrison Zoological Museum Publication, Sevenoaks, Kent, UK.

Harrison DJ and Chapin TG (1998). Extent and connectivity of habitat for wolves in eastern North America. *Wildlife Society Bulletin*, **26**, 767–775.

Harrison RL (1997). A comparison of gray fox ecology between residential and undeveloped rural landscapes. *Journal of Wildlife Management*, **61**, 112–122.

Harrison RL (2003). Swift fox demography, movements, denning, and diet in New Mexico. Southwestern Naturalist, **48**, 261–273.

Harvey PH and Ralls K (1986). Do animals avoid incest? *Nature*, **320**, 575–576.

Harvey PH, Stenning MJ and Campbell B (1988). Factors influencing reproductive success in the pied fly-catcher. In TH Clutton-Brock (ed), *Reproductive success*, pp. 181–196. University of Chicago Press, Chicago, IL, USA.

Hassinger JD (1973). A survey of the mammals of Afghanistan. *Fieldiana: Zoology*, **60**, 1–195.

Hatier KG (1995). *Effects of helping behaviors on coyote packs in Yellowstone National Park, Wyoming*. M.S. dissertation, Montana State University, Bozeman, MO, USA.

Hauffe HC and Pialek J (1997). Evolution of the chromosomal races of *Mus musculus domesticus* in the Rhaetian Alps: the roles of whole-arm reciprocal translocation and zonal radiation. *Linnean Society Biological Journal*, **62**, 255–278.

Haydon DT, Laurenson MK and Sillero-Zubiri C (2002). Integrating epidemiology into population viability analysis: managing the risk posed by rabies and canine distemper to the Ethiopian wolf. *Conservation Biology*, **16**, 1372–1385.

Hayes FG (2000). The Brooksville 2 local fauna (Arikareean, Latest Oligocene): Hernando County, Florida. *Bulletin of Florida Museum of Natural History*, **43**, 1–47.

Hayes RD and Harestad AS (2000). Wolf functional response and regulation of moose in the Yukon. *Canadian Journal of Zoology*, **78**, 60–66.

Hayes RD, Baer AM, Wotschikowsky U and Harestad AS (2000). Kill rate by wolves on moose in the Yukon. *Canadian Journal of Zoology*, **78**, 49–59.

Hedrick PW (1994). Evolutionary genetics of the major histocompatibility complex. *American Naturalist*, **143**, 945–964.

Hedrick PW and Kalinowski ST (2000). Inbreeding depression in conservation biology. *Annual Review Ecology and Systematics*, **31**, 139–162.

Hedrick PW and Kim TJ (1999). Genetics of complex polymorphisms: parasites and maintenance of the major histocompatibility complex variation. In RS Singh and CB Krimbas (eds), *Evolutionary genetics: from molecules to morphology*, pp. 204–234. Cambridge University Press, Cambridge, UK.

Hedrick PW, Miller PS, Geffen E and Wayne R (1997). Genetic evaluation of the three captive Mexican wolf lineages. *Zoo Biology*, **16**, 47–69.

Hedrick PW, Lee RN and Parker KM (2000). Major histocompatibility complex (MHC) variation in the endangered Mexican wolf and related canids. *Heredity*, **85**, 617–624.

Hedrick PW, Lee RN and Garrigan D (2002). Major histocompatibility complex variation in red wolves: evidence for common ancestry with coyotes and balancing selection. *Molecular Ecology*, **11**, 1905–1913.

Heikkilae J, Below A and Hanski I (1994). Synchronous dynamics of microtine rodent populations on islands in Lake Inari in northern Fennoscandia, Evidence for regulation by mustelid predators. *Oikos*, **70**, 245–252.

Helle E and Kauhala K (1991). Distributional history and present status of the raccoon dog in Finland. *Holarctic Ecology*, **14**, 278–286.

Helle E and Kauhala K (1995). Reproduction in the raccoon dog in Finland. *Journal of Mammalogy*, **76**, 1036–1046.

Hemmer H (1989). *Domestication: the decline of environmental appreciation*. Cambridge University Press, Cambridge, UK.

Hendrichs H (1972). Beobachtungen und Untersuchungen zur Ökologie und Ethologie, insbesondere zur sozialen Organisation ostafrikanischer Säugetiere. *Zietschrift fur Tierpsychologie*, **30**, 146–189.

Henke SE (1997). Effects of modified live-virus canine distemper vaccines in gray foxes. *Journal of Wildlife Rehabilitation*, **20**, 3–7.

Henke SE and Bryant FC (1999). Effects of coyote removal on the faunal community in western Texas. *Journal of Wildlife Management*, **63**, 1066–1081.

Henry DJ (1996). *Red fox: the catlike canine*. Smithsonian Institution Press, Washington DC, USA.

Henry VG (1997). 90-day finding for a petition to delist the red wolf. *Federal Register*, **62**, 64799–64800.

Henry VG (1998). Notice of termination of the red wolf reintroduction project tin the Great Smoky Mountains National Park. *Federal Register*, **63**, 54151–54153.

Henshaw RE, Lockwood R, Shideler R and Stephenson RO (1979). Experimental release of captive wolves. In E Klinghammer (ed), *The behaviour and ecology of wolves*, pp. 319–345. Garland STPM Press, NY, USA.

Herrera EA and Macdonald DW (1987). *Group stability and the structure of a capybara population*, In Mammal population studies, pp. 115–130. Proceeding of the Symposia of the Zoological Society of London, 58.

Herrero S (1985). *Bear attacks: their causes and avoidance*. Nick Lyons Books, NY, USA.

Hersteinsson P (1984). *The behavioural ecology of the arctic fox* (Alopex lagopus) *in Iceland*. D.Phil. dissertation, University of Oxford, UK.

Hersteinsson P (1988). When is the denhunting of arctic foxes most successful? *Wildlife Management News (Iceland)*, **4**, 5–13. (In Icelandic with English summary.)

Hersteinsson P (1990). The fertility of Icelandic arctic fox vixens. *Wildlife Mangagement News (Iceland)*, **6**, 19–27. (In Icelandic with English summary.)

Hersteinsson P (1992). Demography of the arctic fox (*Alopex lagopus*) population in Iceland. In DR McCullough and RH Barrett (eds), *Wildlife 2001: populations*, pp. 954–964. Elsevier, London, UK.

Hersteinsson P (1999). *The Arctic foxes of Hornstrandir*, Ritverk sf, Reykjavik, Iceland.

Hersteinsson P and Macdonald DW (1982). Some comparisons between red and arctic foxes, *Vulpes vulpes* and *Alopex lagopus*, as revealed by radiotracking. *Symposium of Zoological Society of London*, **49**, 259–289.

Hersteinsson P and Macdonald DW (1992). Interspecific competition and the geographical distribution of red and arctic foxes (*Vulpes vulpes* and *Alopex lagopus*). *Oikos*, **64**, 505–515.

Hersteinsson P and Macdonald DW (1996). Diet of arctic foxes (*Alopex lagopus*) in Iceland. *Journal of Zoology*, **240**, 457–474.

Hersteinsson P, Angerbjörn A, Frafjord K and Kaikusalo A (1989). The arctic fox in Fennoscandia and Iceland: management problems. *Biological Conservation*, **49**, 67–81.

Hersteinsson P, Bjornsson Th, Unnsteinsdottir ER, Olafsdottir AH, Sigthorsdottir H and Eiriksson Th (2000). The arctic fox in Hornstrandir: number of dens occupied and dispersal of foxes out of the reserve. *Náttúrufrœdingurinn*, **69**, 131–142. (In Icelandic with English summary.)

Hewitt DA and England GCW (2001). Manipulation of canine fertility using in vitro culture techniques. *Journal of Reproductive Fertility* (Supplement), **57**, 111–125.

Heydon MJ and Reynolds JC (2000a). Demography of rural foxes (*Vulpes vulpes*) in relation to cull intensity in three contrasting regions of Britain. *Journal of Zoology London*, **251**, 265–276.

Heydon MJ and Reynolds JC (2000b). Fox (*Vulpes vulpes*) management in three contrasting regions of Britain, in relation to agricultural and sporting interests. *Journal of Zoology London*, **251**, 237–252.

Heydon MJ, Reynolds JC and Short MJ (2000). Variation in abundance of foxes (*Vulpes vulpes*) between three regions of rural Britain, in relation to landscape and other variables. *Journal of Zoology London*, **251**, 253–264.

Hillman CN and Sharps JC (1978). Return of the swift fox to the northern Great Plains. *Proceedings of the South Dakota Academy of Science*, **57**, 154–162.

Hines TD (1980). *An ecological study of Vulpes velox in Nebraska*. M.Sc. dissertation. University of Nebraska, Lincoln, Nebraska, USA.

Hines TD and Case RM (1991). Diet, home range, movements, and activity periods of swift fox in Nebraska. *Prairie Naturalist*, **23**, 131–138.

Hiruki LM and Stirling I (1989). Population-dynamics of the arctic fox, *Alopex-lagopus*, on Banks Island, Northwest Territories. *Canadian Field-Naturalist*, **103**, 380–387.

Hofmeyr M, Bingham J, Lane EP, Ide A and Nel L (2000). Rabies in African wild dogs (*Lycaon pictus*) in the Madikwe game reserve, South Africa. *Veterinary Record*, **146**, 50–52.

Hofreiter M, Serre D, Poinar HN, Kuch M and Pääbo S (2001). Ancient DNA. *Nature Review Genetics*, **2**, 353–359.

Hollander AD, Davis FW and Stoms DM (1994). Hierarchical representation of species distributions using maps, images and sighting data. In RI Miller (ed), *Mapping the diversity of nature*, pp. 71–88. Chapman and Hall, London, UK.

Holling CS (1959). The components of predation as revealed by a study of small mammal predation of the European pine sawfly. *Canadian Entomologist*, **91**, 293–320.

Holt RD (1977). Predation, apparent competition and the structure of prey communities. *Theory of Population Biology*, **12**, 197–229.

Holt RD (2000). Trophic cascades in terrestrial ecosystems. Reflections on Polis *et al. Trends in Ecology and Evolution*, **15**, 444–445.

Holt RD and Polis GA (1997). A theoretical framework for intraguild predation. *American Naturalist*, **149**, 745–764.

Honacki JH, Kinman KE and Koeppl JW (1982). *Mammal species of the world. A taxonomic and geographic reference*. Allen Press, Lawrence, KS, USA.

Hood R (1895). Wild dogs. *Journal of the Bombay Natural History Society*, **10**, 127–132.

Hoogland JL (1979). Aggression, ectoparasitism and other possible costs of prairie dog (Sciuridae: *Cynomys* spp.) coloniality. *Behaviour*, **69**, 1–35.

Houston DB (1978). Elk as winter-spring food for carnivores in northern Yellowstone National Park. *Journal of Applied Ecology*, **15**, 653–661.

Houston DB (1982). *The northern Yellowstone elk: ecology and management*. Macmillan, NY, USA.

Howard WE (1969). Relationship of wildlife to sheep husbandry in Patagonia, Argentina. Report on UNDP/SF/FAO/INTA Project 14.

Howard WE (1990). *Why lions need to be hunted*, pp. 66–68. Third Mountain Lion Workshop, Prescott, AZ, USA.

Hudson PJ, Rizzoli A, Grenfell BT, Heesterbeek H and Dobson AP (2002). *The ecology of wildlife diseases*. Oxford University Press, Oxford, UK.

Huey R (1969). Winter diet of the Peruvian desert fox. *Ecology*, **50**, 1089–1091.

Hughes AL and Nei M (1988). Pattern of nucleotide substitution at major histocompatibility complex class I loci reveals overdominant selection. *Nature*, **335**, 167–170.

Hughes AL and Nei M (1992). Maintenance of MHC polymorphism. *Nature*, **355**, 402–403.

Hughes AL, Hughes MK, Howell CY and Nei M (1994). Natural selection at the class II major histocompatibility complex loci in mammals. *Philosophical Transactions of the Royal Society of London Series B Biological Sciences*, **346**, 359–366.

Hunt RM Jr (1974). The auditory bulla in Carnivora: an anatomical basis for reappraisal of carnivore evolution. *Journal of Morphology*, **143**, 21–76.

Hutchins M and Conway WG (1995). Beyond Noah's Ark: the evolving role of modern zoological parks and aquariums in field conservation. *International Zoo Yearbook*, **34**, 117–130.

Huttunen M (2001). *Supikoiravanhempien ajankäyttö ja työnjako pentujen hoidossa*. [Time allocation and division of labour of raccoon dogs in pup rearing.] M.S. dissertation, University of Helsinki, Helsinki, Finland. (In Finnish.)

Ikeda H (1982). *Socio-ecological study on the raccoon dog,* Nyctereutes procyonoides viverrinus, *with reference to the habitat utilisation pattern*. Ph.D. dissertation, Kyushu University, Japan.

Ikeda H (1983). Development of young and parental care of raccoon dog, *Nyctereutes procyonoides viverrinus* Temminck, in captivity. *The Journal of the Mammalogical Society of Japan*, **9**, 229–236.

Ikeda H, Eguchi K and Ono Y (1979). Home range utilization of a raccoon dog, *Nyctereutes procyonoides viverrinus*, Temminck in a small islet in Western Kyushu. *Japanese Journal of Ecology*, **29**, 35–48.

Ilany G (1983). Blanford's fox, *Vulpes cana*, Blanford 1877, a new species to Israel. *Israel Journal of Zoology*, **32**, 150.

INDEC (2002). *Censo Nacional Agropecuario*. Instituto Nacional de Estadística y Censos, Buenos Aires, Argentina.

Insley H (1977). An estimate of the population density of the red fox (Vulpes vulpes) in the New Forest, Hampshire. *Journal of Zoology*, **183**, 549–553.

Inverarity JD (1895). The Indian wild dog (*Cyon dukhunensis*). *Journal of the Bombay Natural History Society*, **10**, 449–452.

Iriarte JA and Jaksic FM (1986). The fur trade in Chile: an overview of seventy-five years of export data (1910–1984). *Biological Conservation*, **38**, 243–253.

Iriarte JA, Johnson WE and Franklin WL (1991). Feeding ecology of the Patagonian puma in southernmost Chile. *Revista Chilena de Historia Natural*, **64**, 145–156.

Isenberg AC (1992). Toward a policy of destruction: buffaloes, law and the market, 1803–1183. *Great Plains Quarterly*, **12**, 227–241.

IUCN (2000). *2000 IUCN Red List of threatened species*. IUCN—The World Conservation Union, Gland, Switzerland and Cambridge, UK.

IUCN (2001). *IUCN Red List categories and criteria: version 3.1*. IUCN Species Survival Commission, Gland, Switzerland and Cambridge, UK.

Ives AR and Murray DL (1997). Can sublethal parasitism destabilize predator-prey population dynamics? A model of snowshoe hares predators and parasites. *Journal of Animal Ecology*, **66**, 265–278.

Jabara AG (1962). Induction of canine ovarian tumors by diethylstilbestrol and progesterone. *Australian Journal of Experimental Biology and Medicine*, **40**, 139–152.

Jackson VL and Choate JR (2000). Dens and den sites of the swift fox, *Vulpes velox*. *Southwestern Naturalist*, **45**, 212–220.

Jácomo ATA (1999). *Nicho alimentar do lobo guará (Chrysocyon brachyurus Illiger, 1811) no Parque Nacional das Emas*. M.Sc. dissertation. Universidade federal de Goiás, Goiás, Brazil.

Jaksic FM and Yáñez JL (1983). Rabbit and fox introductions in Tierra del Fuego: history and assessment of the attempts at biological control of the rabbit infestation. *Biological Conservation*, **26**, 367–374.

Jaksic FM, Schlatter P and Yáñez JL (1980). Feeding ecology of central Chilean foxes *Dusicyon culpaeus* and *D. griseus*. *Journal of Mammalogy*, **61**, 254–260.

Jaksic FM, Yáñez JL and Rau JR (1983). Trophic relations of the southernmost populations of *Dusicyon* in Chile. *Journal of Mammalogy*, **64**, 693–697.

Jaksic FM, Jiménez JE, Medel RG and Marquet PA (1990). Habitat and diet of Darwin's fox (*Pseudalopex fulvipes*) on the Chilean mainland. *Journal of Mammalogy*, **71**, 246–248.

Jaksic FM, Jiménez JE, Castro SA and Feinsinger P (1992). Numerical and functional response of predators to a long-term decline in mammalian prey at a semi-arid Neotropical site. *Oecologia*, **89**, 90–101.

James FC (1970). Geographic size variations in birds and its relationship to climate. *Ecology*, **51**, 365–390.

Janczak ME (1998). Finding of no significant impact (FONSI) for the proposed predator damage management to protect the federally endangered San Clemente loggerhead shrike on San Clemente Island, California. Department of the Navy, SW Division, Naval Facilities Engineering Command, San Diego, CA, USA.

Jedrzejewski W and Jedrzejewska B (1993). Predation of rodents in Bialowieza Primeval Forest, Poland. *Ecography*, **16**, 47–64.

Jeffreys AJ, Wilson V and Thein SL (1985). Hypervariable minisatellite regions in human DNA. *Nature*, **327**, 139–144.

Jenks SM and Wayne RK (1992). Problems and policy for species threatened by hybridization: the red wolf as a case study. In DR McCullough and RH Barrett (eds), *Wildlife 2001: populations*, pp. 237–251. Elsevier Science Publishers, London, UK.

Jennions MD and Macdonald DW (1994). Cooperative breeding in mammals. *Trends in Ecology and Evolution*, **9**, 89–93.

Jensen B (1966). Untersuchungen über Füchse und Fuchsbekämpfung im Zusammenhang mit der Tollwut (Rabies) in Dänemark. *Tagungsberichte*, **90**, 187–189.

Jessup DA, Boyce WM and Clarke RK (1991). Diseases shared by wild, exotic and domestic sheep. In LA Renecker and RJ Hudson (eds), *Wildlife production: conservation and sustainable development* pp. 438–445. University of Alaska, Fairbanks, AK, USA.

Jhala YV (1994). Predation on blackbuck by wolves in Velvadar National Park, Gujarat, India. *Conservation Biology*, **7**, 874–881.

Jhala YV and Sharma DK (1997). Childlifting by wolves in eastern Uttar Pradesh, India. *Journal of Wildlife Research*, **2**, 94–101.

Jhala YV and Moehlman PD (in press). Golden jackal (*Canis aureus*). In C Sillero-Zubiri, M Hoffmann, and DW Macdonald (eds), *Canids: foxes, wolves, jackals and dogs. Status survey and conservation action plan*. IUCN/SSC Canid Specialist Group, Gland, Switzerland, and Cambridge, UK.

Jiménez JE (1993). *Comparative ecology of Dusicyon foxes at the Chinchilla National Reserve in northeastern Chile*. M.Sc. dissertation. University of Florida, Gainesville, FL, USA.

Jiménez JE (2000). Viability of the endangered Darwin's fox (*Pseudalopex fulvipes*): assessing ecological factors in the last mainland population. Progress report for the Lincoln Park Zoo Neotropic Fund, Chicago, IL, USA.

Jiménez JE and McMahon E (in press). Darwin's fox (*Pseudalopex fulvipes*). In C Sillero-Zubiri, M Hoffmann, and DW Macdonald (eds), *Canids: foxes, wolves, jackals and dogs. Status survey and conservation action plan*, IUCN/SSC Canid Specialist Group, Gland, Switzerland and Cambridge, UK.

Jiménez J and Novaro AJ (in press). Culpeo (*Pseudalopex culpaeus*). In C Sillero-Zubiri, M Hoffmann and DW Macdonald (eds), *Canids: foxes, wolves, jackals and dogs. Status survey and conservation action plan*. IUCN/SSC Canid Specialist Group, Gland, Switzerland and Cambridge, UK.

Jimenez JE, Marquet PA, Medel RG and Jaksic FM (1990). Comparative ecology of Darwin's fox *Pseudalopex fulvipes* in minland and iland settings of southern Chile. *Revista Chilena de Historia Natural*, **63**, 177–186.

Jiménez JE, Yáñez JL, Tabilo EL and Jaksic FM (1995). Body size of Chilean foxes: a new pattern in light of new data. *Acta Theriologica*, **40**, 321–326.

Jiménez JE, Yáñez JL, Tabilo EL and Jaksic FM (1996). Niche complementarity of South American foxes: reanalysis and test of a hypothesis. *Revista Chilena de Historia Natural*, **69**, 113–123.

Jobling MA and Tyler-Smith C (2000). New uses for new haplotypes. The human Y chromosome, disease and fuction. *Trends in Genetics*, **16**, 356–362.

Johnsingh AJT (1978a). A wildboar (*Sus scrofa*) sharing wild dogs' (*Cuon alpinus*) kill. *Journal of the Bombay Natural History Society*, **75**, 211–212.

Johnsingh AJT (1978b). Some aspects of the ecology and behaviour of the Indian Fox-*Vulpes bengalensis* Shaw. *Journal of the Bombay Natural History Society*, **75**, 397–405.

Johnsingh AJT (1982). Reproductive and social behaviour of the dhole, *Cuon alpinus* Canidae. *Journal of Zoology London*, **198**, 443–463.

Johnsingh AJT (1983). Large mammal prey-predators in Bandipur. *Journal of the Bombay Natural History Society*, **80**, 1–57.

Johnsingh AJT (1985). Distribution and status of dhole *Cuon alpinus* Pallas, 1811 in South Asia. *Mammalia*, **49**, 203–208.

Johnsingh AJT (1992). Prey selection in three large sympatric carnivores in Bandipur. *Mammalia*, **56**, 517–526.

Johnsingh AJT and Jhala YV (in press). Indian fox (*Vulpes bengalensis*). In C Sillero-Zubiri, M Hoffmann, and DW Macdonald (eds), *Canids: foxes, wolves, jackals and dogs.*

Status survey and conservation action plan, IUCN/SSC Canid Specialist Group, Gland, Switzerland, and Cambridge, UK.

Johnson DDP, Macdonald DW, Kays R and Blackwell PG (2003). Response to Revilla, and Buckley and Ruxton: the resource dispersion hypothesis. *Trends in Ecology and Evolution*, **18**, 381–382.

Johnson DDP, Kays R, Blackwell PG and Macdonald DW (2002). Does the resource dispersion hypothesis explain group living? *Trends in Ecology and Evolution*, **17**, 563–570.

Johnson DDP, Macdonald DWM and Johnson P (submitted). Why do social carnivores have small home ranges? *Oikos*.

Johnson DL (1983). The California continental borderland: landbridges, watergaps and biotic dispersals. *Quarternary Coastlines*, **1983**, 481–527.

Johnson DR (1969). Returns of the American fur company, 1835–1839. *Journal of Mammalogy*, **50**, 836–839.

Johnson WE (1992). *Comparative ecology of the two sympatric South American foxes,* Dusicyon culpaeus *and* D. griseus. Ph.D. dissertation, Iowa State University, Omas, Iowa, USA.

Johnson WE (in press). Evaluating and predicting the impacts of exploitation and trade on canid populations. In C Sillero-Zubiri, M Hoffmann and DW Macdonald (eds), *Canids: foxes, wolves, jackals and dogs. Status survey and conservation action plan.* IUCN/SSC Canid Specialist Group, Gland, Switzerland, and Cambridge, UK.

Johnson WE and Franklin WL (1994a). Role of body size in the diets of sympatric gray and culpeo foxes. *Journal of Mammalogy*, **75**, 163–174.

Johnson WE and Franklin WL (1994b). Spatial resource partitioning by sympatric grey fox (*Dusicyon griseus*) and culpeo fox (*Dusicyon culpaeus*) in southern Chile. *Canadian Journal of Zoology*, **72**, 1788–1793.

Johnson WE and Franklin WL (1994c). Conservation implications of the South American grey fox (*Dusicyon griseus*) socioecology in the Patagonia of southern Chile. *Vida Silvestre Neotropical*, **3**, 16–23.

Johnson WE, Fuller TK and Franklin WL (1996). Sympatry in canids: a review and assessment. In JL Gittleman (ed), *Carnivore behavior, ecology and evolution*, pp. 189–218. Cornell University Press, Ithaca, NY, USA.

Johnson WE, Eizirik E and Lento GM (2001). The control, exploitation and conservation of carnivores. In JL Gittleman, SM Funk, DW Macdonald, and RK Wayne (eds), *Carnivore conservation*, pp. 192–220. Cambridge University Press, Cambridge, UK.

Jones DM and Theberge JB (1982). Summer home range and habitat utilisation of the red fox (*Vulpes vulpes*) in a tundra habitat, northwest British Columbia. *Canadian Journal of Zoology*, **60**, 807–812.

Jones E (1990). Physical characteristics and taxonomic status of wild canids *Canis familiaris* from the eastern highlands of Victoria Australia. *Australian Wildlife Research*, **17**, 69–82.

Jones MJ, Lace LA, Harrison EC and Stevens-Wood B (1998). Territorial behaviour in the speckled wood butterflies Pararge xiphia and P. aegeria of Madeira: a mechanism for interspecific competition. *Ecography*, **21**, 297–305.

Jordan PA, Shelton PC and Allen DL (1967). Numbers, turnover and social structure of the Isle Royale wolf population. *American Zoologist*, **7**, 233–252.

Jorde LB, Watkins WS, Bamshad ML, Dixon ME, Ricker CE, Seielstad MT and Batzer MA (2000). The distribution of human genetic diversity: a comparison of mitochondrial, autosomal and Y-chromosome data. *American Journal of Human Genetics*, **66**, 979–988.

Juarez KM and Marinho-Filho J (2002). Diet, habitat use, and home ranges of sympatric canids in central Brazil. *Journal of Mammalogy*, **83**, 925–933.

Judin VG (1977). *Enotovidnaja sobaka Primor'ja v Priamur'ja.* Nauka, Moskwa, Russia.

Julliard R (2000). Sex-specific dispersal in spatially varying environments leads to habitat-dependent evolutionarily stable offspring sex ratios. *Behavioral Ecology*, **11**, 421–428.

Juola FA, Everett WT and Koehler CE (1997). Final report: 1996 population and habitat survey of the loggerhead shrike on NALF San Clemente Island, California. Department of the Navy, SW Division, Naval Facilities Engineering Command, San Diego, CA, USA.

Kalinowski ST, Hedrick PW and Miller PS (1999). No inbreeding depression observed in Mexican and red wolf captive breeding programs. *Conservation Biology*, **13**, 1371–1377.

Kaneshiro KY (1988). Speciation in the Hawaiian Drosophila. Sexual selection appears to play an important role. *Bioscience*, **38**, 258–263.

Karanth KU and Nichols JD (2000). *Ecological status and conservation of tigers in India.* Final technical report to the Division of international conservation, US Fish and Wildlife Service, Washington and the Wildlife Conservation Society, New York, Centre for Wildlife Studies, Bangalore, India.

Karanth KU and Sunquist ME (1992). Population structure, density and biomass of large herbivores in the tropical forests of Nagarahole, India. *Journal of Tropical Ecology*, **8**, 21–35.

Karanth KU and Sunquist ME (1995). Prey selection by tiger, leopard and dhole in tropical forests. *Journal of Animal Ecology*, **64**, 439–450.

Karanth KU and Sunquist ME (2000). Behavioural correlates of predation by tiger (*Panthera tigris*), leopard (*Panthera pardus*) and dhole (*Cuon alpinus*) in Nagarahole, India. *Journal of Zoology London*, **250**, 255–265.

Kat PW, Alexander KA, Smith JS and Munson L (1995). Rabies and African wild dogs in Kenya. *Proceedings of the Royal Society of London B*, **262**, 229–233.

Kauhala K (1993). Growth, size and fat reserves of the raccoon dog in Finland. *Acta Theriologica*, **38**, 139–150.

Kauhala K (1996a). Habitat use of raccoon dogs, *Nyctereutes procyonoides*, in southern Finland. *Zeitschrift für Säugetierkunde*, **61**, 269–275.

Kauhala K (1996b). Reproductive strategies of the raccoon dog and the red fox in Finland. *Acta Theriologica*, **41**, 51–58.

Kauhala K and Auniola M (2001). Diet of raccoon dogs in summer in the Finnish archipelago. *Ecography*, **24**, 151–156.

Kauhala K and Helle E (1994). Supikoiran elinpiireistä ja yksiavioisuudesta Etelä-Suomessa [Home ranges and monogamy of the raccoon dog in southern Finland]. *Suomen Riista*, **40**, 32–41.

Kauhala K and Helle E (1995). Population ecology of the raccoon dog in Finland—a synthesis. *Wildlife Biology*, **1**, 3–9.

Kauhala K and Saeki M (in press). Raccon dog (*Nyctereutes procyonoides*). In C Sillero-Zubiri, M Hoffmann and DW Macdonald (eds), *Canids: foxes, wolves, jackals and dogs. Status survey and conservation action plan.* IUCN/SSC Canid Specialist Group, Gland, Switzerland and Cambridge, UK.

Kauhala K, Helle E and Taskinen K (1993a). Home range of the raccoon dog (*Nyctereutes procyonoides*) in southern Finland. *Journal of Zoology London*, **231**, 95–106.

Kauhala K, Kaunisto M and Helle E (1993b). Diet of the raccoon dog, *Nyctereutes procyonoides*, in Finland. *Zeitschrift für Säugetierkunde*, **58**, 129–136.

Kauhala K, Helle E and Pietilä H (1998a). Time allocation of male and female raccoon dogs to pup rearing at the den. *Acta Theriologica*, **43**, 301–310.

Kauhala K, Laukkanen P and von Rege I (1998b). Summer food composition and food niche overlap of the raccoon dog, red fox and badger in Finland. *Ecography*, **21**, 457–463.

Kauhala K, Viranta S, Kishimoto M, Helle E and Obara I (1998c). Skull and tooth morphology of Finnish and Japanese raccoon dogs. *Annales Zoologici Fennici*, **35**, 1–16.

Kauhala K, Laukkanen P and von Rége I (1999). Supikoiran, ketun ja mäyrän ravinnon koostumus ja riistan osuus ravinnossa alkukesällä (Summary: Summer food composition of the raccoon dog, red fox and badger in Finland, with reference to small game). *Suomen Riista*, **45**, 63–72.

Kawamura Y (1991). Quaternary mammalian faunas in the Japanese Islands. *Quaternary Research*, **30**, 213–220.

Kays RW and Gittleman JL (1995). Home range size and social behavior of kinkajous (*Potos flavus*) in the Republic of Panama. *Biotropica*, **27**, 530–534.

Keech MA, Bowyer RT, Ver Hoef JM, Boertje RD, Dale BW and Stephenson TR (2000). Life-history consequences of maternal condition in Alaskan moose. *Journal of Wildlife Management*, **64**, 450–462.

Keeler C (1975). Genetics of behaviour variations in color phases of the red fox. In MW Fox (ed), *The wild canids. Their systematics, behavioural ecology and evolution*, pp. 399–413. Robert E. Krieger Publishing Co. Inc., Malabar, FL, USA.

Keith LB (1983). *Population dynamics of wolves.* In LN Carbyn (ed), Wolves in Canada and Alaska—their status, biology, and management, pp. 66–77. Canadian Wildlife Service Report Series 45.

Keller LF (1998). Inbreeding and its fitness effects in an insular population of song sparrows (Melospiza melodia). *Evolution*, **52**, 240–250.

Kellert SR (1985). Public perceptions of predators, particularly the wolf and coyote. *Biological Conservation*, **31**, 167–189.

Kellert SR, Black M, Rush CR and Bath AJ (1996). Human culture and large carnivore conservation in North America. *Conservation Biology*, **10**, 977–990.

Kelly BT (2000). *Red wolf recovery program adaptive work plan FY00-02*, 15pp. U.S. Fish and Wildlife Service, Alligator River National Wildlife Refuge, Manteo, NC, USA.

Kelly BT and Phillips MK (2000). Red wolf. In R Reading and B Miller (eds), *Endangered animals: a reference guide to conflicting issues*, pp. 247–252. Greenwood Press, Westport, CN, USA.

Kelly BT, Miller PS and Seal US (eds), (1999). *Population and habitat viability assessment workshop for the red wolf (Canis rufus)*, 88pp. SSC/IUCN Conservation Breeding Specialist Group, Apple Valley, MN, USA.

Kelly BT, Beyer A and Phillips MK (in press). Red wolf (*Canis rufus*). In C Sillero-Zubiri, M Hoffmann and DW Macdonald (eds), *Canids: foxes, wolves, jackals and dogs. Status survey and conservation action plan*, IUCN/SSC Canid Specialist Group, Gland, Switzerland, and Cambridge, UK.

Kelly DW, Mustafa Z and Dye C (1997). Differential application of lambda-cyhalothrin to control the sandfly *Lutzomyia longipalpis*. *Medical and Veterinary Entomology*, **11**, 13–24.

Kennedy PK, Kennedy ML, Clarkson PL and Liepins IS (1991). Genetic variability in natural populations of the gray wolf, *Canis lupus*. *Canadian Journal of Zoology*, **69**, 1183–1188.

Kennelly JJ (1978). Coyote reproduction. In M Bekoff (ed), *Coyotes: biology, behavior, and management*, pp. 73–93. Academic Press, New York, USA.

Kiff LF (1980). Historical changes in resident populations of California Islands raptors. In DM Power ed. *The California Islands: proceedings of a multidisciplinary symposium*, pp. 651–673. Santa Barbara Museum of Natural History, Santa Barbara, CA, USA.

Kilgore DL (1969). An ecological study of the swift fox (*Vulpes velox*) in the Oklahoma Panhandle. *American Midland Naturalist*, **81**, 512–534.

Killick-Kendrick R (1985). Some epidemiological consequences of the evolutionary fit between leishmaniae and their phlebotomine vectors. *Bulletin de la Société de Pathologie Exotique*, **78**, 747–755.

Kingdon J (1977). *East African mammals: an atlas of evolution in Africa*, Vol. IIIA *(Carnivores)*, Academic Press, London, UK.

Kingdon J (1990). *Island Africa*. Collins, London, UK.

Kingdon J (1997). *The Kingdon field guide to African mammals.* Academic Press, London, UK.

Kinnear JE, Onus ML and Sumner NR (1998). Fox control and rock-wallaby population dynamics: II. An update. *Wildlife Research*, **25**, 81–88.

Kinosita A and Yamamoto Y (1993). Research on raccoon dog *Nyctereutes procyonoides viverrinus* TEMMICK in Kawasaki (II). (in Japanese) *Bulletin of the Kawasaki Municipal Science Museum for Youth*, **4**, 45–50.

Kitala P, McDermott J, Kyule M, Gathuma J, Perry B and Wandeler A (2001). Dog ecology and demography information to support the planning of rabies control in Machakos District, Kenya. *Acta Tropica*, **78**, 217–230.

Kitala PM, McDermott JJ, Kyule MN and Gathuma JM (2000). Community-based active surveillance for rabies in Machakos District, Kenya. *Preventive Veterinary Medicine*, **44**, 73–85.

Kitchen AM (1999). Resource partitioning between coyotes and swift foxes: space, time, and diet. M.Sc. Dissertation, Utah State University, Logan, UT, USA.

Kitchen AM, Gese EM and Schauster ER (1999). Resource partitioning between coyotes and swift foxes: space, time and diet. *Canadian Journal of Zoology*, **77**, 1645–1656.

Kitchen AM, Gese EM and Schauster ER (2000). Changes in coyote activity patterns due to reduced exposure to human persecution. *Canadian Journal of Zoology*, **78**, 853–857.

Klafter J, White BS and Levandowsky M (1990). Microzooplankton feeding behaviour and the Lévy walk. In W Alt and G Hoffmann (eds), *Biological motion. Lecture notes in biomathematics*. Springer-Verlag.

Kleiman DG (1972). Social behaviour of the maned wolf (*Chrysocyon brachyurus*) and bush dog (*Speothos venaticus*): a study in contrast. *Journal of Mammalogy*, **53**, 791–806.

Kleiman DG (1977). Monogamy in mammals. *The Quarterly Review of Biology*, **52**, 39–69.

Kleiman DG (1989). Reintroduction of captive mammals for conservation. *Bioscience*, **39**, 152–161.

Kleiman DG and Beck BB (1994). Criteria for reintroductions. In PJS Olney, GM Mace and ATC Feistner (eds), *Creative conservation: interactive management of wild and captive animals*, pp. 287–303. Chapman and Hall, London.

Kleiman DG and Eisenberg JF (1973). Comparisons of canid and felid social systems from an evolutionary perspective. *Animal Behaviour*, **21**, 637–659.

Kleiman DG and Malcolm JR (1981). The evolution of male parental investment in mammals. In DJ Gubernick and PH Klopfer (eds), *Parental care in mammals*, pp. 347–387, Plenum Press, New York, USA.

Knapp DK (1978). Effects of agricultural development in Kern County, California, on the San Joaquin kit fox in 1977. California Department of Fish and Game, Non-game Wildlife Investigations Final Report, Project E-1-1, Job V-1.21, Sacramento, CA, USA.

Knick ST (1990). Ecology of bobcats relative to exploitation and a prey decline in southeastern Idaho. *Wildlife Monographs*, **108**, 1–42.

Knobel DL, Du Toit JT and Bingham J (2002). Development of a bait and baiting system for delivery of oral rabies vaccine to free-ranging African wild dogs (*Lycaon pictus*). *Journal of Wildlife Diseases*, **38**, 352–362.

Knowlton FF (1972). Preliminary interpretations of coyote population mechanics with some management implications. *Journal of Wildlife Management*, **36**, 369–382.

Knowlton FF and Stoddart LC (1983). Coyote population mechanics: another look. In FL Bunnell, DS Eastman and JM Peek (eds), Natural regulation of wildlife populations. *Forest, Wildlife and Range Experiment Station*, pp. 93–111. Proceedings 14, University of Idaho, Moscow, USSR.

Knowlton FF and Stoddart LC (1992). Some observations from two coyote-prey studies. In A Boer (ed), *Ecology and management of the eastern coyote*, pp. 101–121. Wildlife Research Unit, University of New Brunswick, Fredericton, USA.

Knowlton FF, Gese EM and Jaeger MM (1999). Coyote depredation control: an interface between biology and management. *Journal of Range Management*, **52**, 398–412.

Koehler JK, Platz CC Jr, Waddell W, Jones MH and Behrns S (1998). Semen parameters and electron microscope observations of spermatozoa of the red wolf, *Canis rufus*. *Journal of Reproductive Fertility*, **114**, 95–101.

Kohn MH and Wayne RK (1997). Facts from feces revisited. *Trends in Ecology and Evolution*, **12**, 223–227.

Kohn MH, York EC, Kamradt DA, Haught G, Sauvajot RM and Wayne RK (1999). Estimating population size by genotyping faeces. *Proceedings of the Royal Society of London, Series B*, **266**, 657–663.

Kojola I and Kuittinen J (2002). Wolf attacks on dogs in Finland. *Wildlife Society Bulletin*, **30**, 498–501.

Kolb HH (1986). Some observations on the home ranges of vixens (*Vulpes vulpes*) in the suburbs of Edinburgh. *Journal of Zoology*, **210**, 636–639.

Kolenosky GB and Standfield RO (1975). Morphological and ecological variation among gray wolves (*Canis lupus*) of Ontario, Canada. In MW Fox (ed), *The Wild canids*, pp. 62–72. Van Nostrand Reinhold, New York, USA.

Komers PE and Brotherton PNM (1997). Female space use is the best predictor of monogamy in mammals. *Proceedings of the Royal Society, London, Series B*, **264**, 1261–1279.

Koopman ME, Scrivner JH and Kato TT (1998). Patterns of den use by San Joaquin kit foxes. *Journal of Wildlife Management*, **62**, 373–379.

Koopman ME, Cypher BL and Scrivner JH (2000). Dispersal patterns of San Joaquin kit foxes (*Vulpes macrotis mutica*). *Journal of Mammalogy*, **81**, 213–222.

Korhonen H, Mononen J and Harri M (1991). Evolutionary comparison of energy economy between Finnish and Japanese raccoon dogs. *Comparative Biochemistry and Physiology*, **100A**, 293–295.

Koskinen MT, Haugen TO and Primmer CR (2002). Contemporary fisherian life-history evolution in small salmonid populations. *Nature*, **419**, 826–830.

Kovach SD and Dow RJ (1981). Status and ecology of the island fox on San Nicolas Island, 1980, 33pp. Technical Memorandum TM-81-28. Pacific Missile Defense Center, Pt. Mugu, CA, USA.

Kovach SD and Dow RJ (1985). Island fox research on San Nicolas Island. Annual report, Department of the Navy, Pacific Missile Test Center, San Diego, CA, USA.

Kowalczyk R, Zalewski A, Jedrzejewska B and Jedrzejewski W (2000). Jenot—ni pies, ni borsuk (in Polish). *Lowiec Polski*, **11**, 1–20.

Kowalski K (1988). The food of the sand fox *Vulpes rueppellii* Schinz 1825 in the Egyptian Sahara. *Folia Biologica Cracow*, **36**, 89–94.

Krebs CJ (1999). *Ecological methodology*. Benjamin/Cummings, Menlo Park, CA, USA.

Krebs JR and Davies NB (1991). *Behavioural ecology: an evolutionary approach, 3rd Edition*. Blackwell Scientific Publications, Oxford, UK.

Krebs JR, Anderson R, Clutton-Brock T, Morrison I, Young D, Donelly C, Frost S and Woodroffe R (1997). *Bovine tuberculosis in cattle and badgers—an independent scientific review*. H.M.S.O, London, UK.

Kruchenkova EP, Goltsman M and Macdonald DW (1996). The Mednyi arctic foxes: treating a population imperilled by disease. *Oryx*, **30**, 251–258.

Kruchenkova EP, Goltsman M, Sergeev S and Macdonald DW (submitted). Ineffective helpers and a disinclination to disperse in two island populations of Arctic fox, *Alopex lagopus*.

Kruuk H (1972). *The spotted hyena*. University of Chicago Press, Chicago, IL, USA.

Kruuk H (1978). Spatial organisation and territorial behaviour of the European badger, *Meles meles*. *Journal of Zoology London*, **184**, 1–19.

Kruuk H (1980). *The effects of large carnivores on livestock and animal husbandry in Marsabit District, Kenya*. IPAL Report E-4. United Nations Environmental Programme, Nairobi, Kenya.

Kruuk H (2002). *Hunter and hunted*. Cambridge University Press, Cambridge, UK.

Kruuk H and Macdonald DW (1985). Group territories of carnivores: empires and enclaves. In RM Sibly and RH Smith (eds), *Behavioural ecology: ecological consequences of adaptive behaviour*, pp. 521–536. Blackwell Scientific Publications, Oxford, UK.

Kruuk H and Moorhouse A (1991). The spatial organisation of otters (*Lutra lutra*) in Shetland. *Journal of Zoology London*, **224**, 41–57.

Kruuk H and Parish T (1982). Factors affecting population density, group size and territory size of the European badger. *Journal of Zoology London*, **196**, 31–39.

Kühne W (1965). Communal food distribution and division of labour in African hunting dogs. *Nature*, **205**, 443–444.

Kunkel K and Pletscher DH (1999). Species-specific population dynamics of cervids in a multipredator ecosystem. *Journal of Wildlife Management*, **63**, 1082–1093.

Kunkel K, Honness K, Phillips M and Carbyn L (2003). Assessing restoration of swift fox in the northern Great Plains. In MA Sovada and L Carbyn (eds), *The swift fox: ecology and conservation of swift foxes in a changing world*, pp. 189–198. Canadian Plains Research Center, University of Regina, Saskatchewan, Canada.

Kurtén K (1968). *Pleistocene mammals of Europe*, p. 117. Weidenfeld and Nicolson, London, UK.

Lacy RC (1997). Importance of genetic variation to the viability of mammalian populations. *Journal of Mammalogy*, **78**, 320–335.

Lacy RC, Alaks G and Walsh A (1996). Hierarchical analysis of inbreeding depression in Peromyscus polionotus. *Evolution*, **50**, 2187–2200.

Lade JA, Murray ND, Marks CA and Robinson NA (1996). Microsatellite differentiation between Phillip island and mainland Australian populations of the red fox *Vulpes vulpes*. *Molecular Ecology*, **5**, 81–87.

Laikre L and Ryman N (1991). Inbreeding depression in a captive wolf (*Canis lupus*) population. *Conservation Biology*, **5**, 33–40.

Laikre L, Ryman N and Thompson EA (1993). Hereditary blindness in a captive wolf (*Canis lupus*) population: frequency reduction of a deleterious allele in relation to gene conservation. *Conservation Biology*, **7**, 592–601.

Lainson R (1988). Demographic changes and their influence on the epidemiology of the leishmaniases. In MW Service (ed), *Demography and vector-borne diseases* CRC Press, FL, USA.

Lainson R and Shaw JJ (1971). Epidemiological considerations of the leishmaniases, with particular reference to the New World. In AM Fallis (ed), *Ecology and physiology of parasites*, pp. 21–57. University of Toronto Press, Toronto, Canada.

Lainson R and Shaw JJ (1979). The role of animals in the epidemiology of South American leishmaniasis. In WHR Lumsden and DA Evans (eds), *Biology of the Kinetoplastida*, pp. 1–116. Vol. 2. Academic Press, London, UK.

Lainson R, Dye C, Shaw JJ, Macdonald D, Courtenay O, Souza AA and Silveira FT (1990). Amazonian visceral leishmaniasis: distribution of the vector *Lutzomyia longipalpis* (Lutz and Neiva) in relation to the fox *Cerdocyon thous* (L.) and the efficiency of this reservoir host as a source of infection. *Memorias do Instituto Oswaldo Cruz*, **85**, 135–137.

Lainson R, Shaw JJ and Lins ZC (1969). Leishmaniasis in Brazil. IV. The fox, *Cerdocyon thous* (L.) as a reservoir of *Leishmania donovani* in Pará state, Brazil. *Transactions of the Royal Society of Tropical Medicine and Hygiene*, **63**, 741–745.

Lainson R, Shaw JJ, Silveira FT and Braga RR (1987). American visceral leishmaniasis: on the origin of *Leishmania (Leishmania) chagasi*. *Transactions of the Royal Society of Tropical Medicine and Hygiene*, **81**, 517.

Lamprecht J (1978a). On diet, foraging behaviour and interspecific food competition of jackals in the Serengeti National Park, East Africa. *Zeitschrift für Säugetierkunde*, **43**, 210–223.

Lamprecht J (1978b). The relationship between food competition and foraging group size in some large carnivores. *Zeitschrift fur Tierpsychologie*, **46**, 337–343.

Lamprecht J (1979). Field observations on the behaviour and social system of the bat-eared fox *Otocyon megalotis* Desmarest. *Zeitschrift für Tierpsychologie*, **49**, 260–284.

Lande R (1988). Genetics and demography in biological conservation. *Science*, **241**, 1455–1460.

Lande R (1993). Risks of population extinction from demographic and environmental stochasticity and random catastrophes. *American Naturalist*, **142**, 911–927.

Landy JM (2000). Testing livestock guard donkeys in the Swiss Alps. *Carnivore Damage Prevention News*, **1**, 6–7, www.kora.unibe.ch.

Langguth A (1975). Ecology and evolution in the South American canids. In MW Fox (ed), *The wild canids*, pp. 192–206. Van Nostrand Reinhold Company, New York, USA.

Laughrin L (1977). *The island fox: a field study of its behavior and ecology*, 83pp. Ph.D. dissertation, University of California, Santa Barbara, CA, USA.

Laughrin L (1980). Populations and status of the island fox. In DM Power ed. *The California islands: proceedings of a multidisciplinary symposium*, pp. 745–749. Santa Barbara Museum of Natural History, Santa Barbara, CA, USA.

Laurenson K, Shiferaw F and Sillero-Zubiri C (1997). Disease, domestic dogs and the Ethiopian wolf: the current situation. In C Sillero-Zubiri and DW Macdonald (eds), *The Ethiopian wolf. Status survey and conservation action plan*, pp. 32–40. IUCN, Gland, Switzerland, and Cambridge, UK.

Laurenson MK, Sillero-Zubiri C, Thompson H, Shiferaw F and Malcolm JR (1998). Disease threats to endangered species: patterns of infection by canine pathogens in Ethiopian wolves (*Canis simensis*) and sympatric domestic dogs. *Animal Conservation*, **1**, 273–280.

Laurenson MK, Haydon DT and EWCP (in press). Guiding management decisions for disease control in endangered populations: lessons from a quantitative cost- effectiveness analysis of Ethiopian wolf protection. *Conservation Biology*.

Lavrov NP (1971). I togi introduktsii enotovidnoj sobaki (Npg) vothel'nye oblasti SSSR. *Trudy kafedry biologii MGZPI*, **29**, 101–166. (In Russian.)

Lawrence B and Bossert WH (1967). Multiple character analysis of *Canis lupus, latrans* and *familiaris* with a discussion of the relationship of *Canis niger*. *American Zoologist*, **7**, 223–232.

Lay DM (1967). A study of the mammals of Iran. *Fieldiana: Zoology*, **54**, 1–282.

Legendre S, Clobert J, Moeller AP and Sorci G (1999). Demographic stochasticity and social mating system in the process of extinction of small populations, the case of passerines introduced to New Zealand. *American Naturalist*, **153**, 449–463.

Lehman N and Wayne RK (1991). Analysis of coyote mitochondrial DNA genotype frequencies: estimation of effective number of alleles. *Genetics*, **128**, 405–416.

Lehman N, Clarkson P, Mech LD, Meier TJ and Wayne RK (1992). The use of DNA fingerprinting and mitochondrial DNA to study genetic relationships within and among wolf packs. *Behavioural, Ecology and Sociobiology*, **30**, 83–94.

Lehman N, Eisenhawer A, Hansen K, Mech LD, Peterson RO, Gogan PJP and Wayne RK (1991). Introgression of coyote mitochondrial DNA into sympatric North American gray wolf populations. *Evolution*, **45**, 104–119.

Lehner PN (1979). *Handbook of ethological methods*. Garland STPM Press, New York, USA.

Leite Pitman MRP and Williams RSR (in press). Short eared dog (*Atelocynus microtis*). In C Sillero-Zubiri, M Hoffmann and DW Macdonald (eds), *Canids: foxes, wolves, jackals and dogs. Status survey and conservation action plan*. IUCN/SSC Canid Specialist Group, Gland, Switzerland, and Cambridge, UK.

Lenain DM (2000). *Fox populations of a protected area in Saudi Arabia*. M.Phil. dissertation, University of Herefordshire, Hereford, UK.

Leonard JA, Wayne RK and Cooper A (2000). Population genetics of Ice age brown bears. *Proceedings of the National Academy of Sciences USA*, **97**, 1651–1654.

Leutenegger W (1982). Encephalization and obstetrics in primates with particular reference to human evolution. In E Armstron and D Falk (eds), *Primate brain evolution: methods and concept*, pp. 85–95. Plenum Press, New York, USA.

Leutenegger W and Cheverud J (1982). Correlates of sexual dimorphism in primates: ecological and size variables. *International Journal of Primatology*, **3**, 387–402.

Lever C (1994). *Naturalized animals: the ecology of successfully introduced species*. T & A.D. Poyser Ltd., London, UK.

Lévy P (1947). *Processus stochastiques et mouvement Brownian*. Gautier-Villars, Paris, France.

Lindenmayer DB, Clark TW, Lacy RC and Thomas VC (1993). Population viability analysis as a tool in wildlife conservation policy, with reference to Australia. *Environmental Management*, **17**, 745–758.

Lindsay IM and Macdonald DW (1986). Behaviour and ecology of the Ruppell's Fox *Vulpes rüppellii* in Oman. *Mammalia*, **50**, 461–474.

Lindström ER (1986). Territory inheritance and the evolution of group-living in carnivores. *Animal Behaviour*, **34**, 1825–1835.

Lindström ER (1988). Reproductive effort of the red fox, *Vulpes vulpes* and future supply of a fluctuating prey. *Oikos*, **52**, 115–119.

Lindström ER (1989). Food limitation and social regulation in a red fox population. *Holarctic Ecology*, **12**, 70–79.

Lindström ER (1992). Diet, reproduction, recruitment and growth of the red fox (*Vulpes vulpes*) in relation to population density—the sarcoptic mange event in Scandinavia. In DR McCullough and RH Barrett (eds), *Wildlife 2001: populations*, pp. 922–931. Elsevier Science Publishers, Barking, Essex, UK.

Lindström ER (1993). Group formation and group persistence: on realism, elegance and simplification. *Animal Behaviour*, **46**, 824–826.

Lindstrom ER and Hornfeldt B (1994). Vole cycles, snow depth and fox predation. *Oikos*, **70**, 156–160

Lingle S (2002). Coyote predation and habitat segregation of white-tailed deer and mule deer. *Ecology*, **83**, 2037–2048.

Linhart SB and Knowlton FF (1967). Determining age of coyotes by tooth cementum layers. *Journal of Wildlife Management*, **31**, 362–365.

Linhart SB and Knowlton FF (1975). Determining the relative abundance of coyotes by scent station lines. *Wildlife Society Bulletin*, **3**, 119–124.

Linhart SB and Robinson WB (1972). Some relative carnivore densities in areas under sustained coyote control. *Journal of Mammalogy*, **53**, 880–884.

Linhart SB, Sterner RT, Carrigan TC and Henne DR (1979). Komondor guard dogs reduce sheep losses to coyotes: a preliminary evaluation. *Journal of Range Management*, **32**, 238–241.

Linhart SB, Roberts JD and Dasels GJ (1982). Electric fencing reduces coyote predation on pastured sheep. *Journal of Range Management*, **35**, 276–281.

Linnell JDC and Strand O (2000). Interference interactions, co-existence and conservation of mammalian carnivores. *Diversity and Distributions*, **6**, 169–176.

Linnell JDC, Smith ME, Odden J, Kaczensky P and Swenson JE (1996). Carnivores and sheep farming Norway. Strategies for the reduction of carnivore-livestock conflicts: a review. *NINA Oppdragsmelding*, **443**, 1–118.

Linnell JDC, Aanes R and Swenson JE (1997). Translocation of carnivores as a method for managing problem animals: a review. *Biodiversity and Conservation*, **6**, 1245–1257.

Linnell JDC, Odden J, Smith ME, Aanes R and Swenson JE (1999). Large carnivores that kill livestock: do "problem individuals" really exist? *Wildlife Society Bulletin*, **27**, 698–705.

Linnell JDC *et al.* (2002a). The fear of wolves: a review of wolf attacks on humans. *NINA Oppdragsmelding*, **731**, 1–65.

Linnell JDC, Swenson JE and Andersen R (2000b). Conservation of biodiversity in Scandinavian boreal forests: large carnivores as flagships, umbrellas, indicators, or keystones? *Biodiversity and Conservation*, **9**, 857–868.

Linscombe G (1994). *U.S. fur harvest (1970–92) and fur value (1974–1992) statistics by state and region.* Louisiana Department of Wildlife and Fisheries, Baton Rouge, Louisiana, USA.

List R (1997). *Ecology of the kit fox* (Vulpes macrotis) *and coyote* (Canis latrans) *and the conservation of the prairie dog ecosystem in northern Mexico.* D.Phil. dissertation, University of Oxford, Oxford, UK.

List R and Cypher BL (in press). *Vulpes macrotis*. Black-backed jackal (*Canis mesomelas*). In C Sillero-Zubiri, M Hoffmann and DW Macdonald (eds), *Canids: foxes, wolves, jackals and dogs. Status survey and conservation action plan*, IUCN/SSC Canid Specialist Group, Gland, Switzerland, and Cambridge, UK.

List R and Jimenez Guzmán A (in press). *Vulpes macrotis*. In G Ceballo, G Oliva and H Arita (eds), *Atlas Mastozoológico de México*. CONABIO-UNAM, D.F., Mexico.

List R and Macdonald DW (2003). Home range and habitat use of the kit fox (*Vulpes macrotis*) in a prairie dog (*Cynomys ludovicianus*) complex. *Journal of Zoology London*, **259**, 1–5.

List R, Manzano-Fischer P and Macdonald DW (2003). Coyote and kit fox diets in prairie dog towns and adjacent grasslands in Mexico. In M Sovada and L Carbyn (eds), *The swift fox: ecology and conservation of swift foxes in a changing world*, pp. 183–188. Canadian Plains Research Center, University of Regina, Saskatchewan, Canada.

Lloyd HG (1980). *The red fox*. Batsford, London, UK.

Lockie JD (1959). The estimation of the food of foxes. *Journal of Wildlife Management*, **23**(2), 224–227.

Logan CG, Barry WH, Standley WG and Kato T (1992). Prey abundance and food habits of San Joaquin kit foxes (*Vulpes velox macrotis*) at Camp Roberts Army Guard training site, California. United States Department of Energy Topical Report No. EGG 10617-2158, National Technical Information Service, Springfield, Virginia, USA.

Lomnicki A (1988). *Population ecology of individuals*. Princeton University Press, Princeton, NJ, USA.

Lord RD (1961). A population study of the gray fox. *American Midland Naturalist*, **66**, 87–109.

Lorenz KZ (1954). *Man meets dog*. Methuen & Co., London, UK.

Lorenz KZ (1975). Foreward to (pp. VII–XII), In M Fox (ed), *The Wild canids*. Van Nostrand Reinhold, New York, NY, USA.

Lorenzini R and Fico R (1995). A genetic investigation of enzyme polymorphism shared by wolf and dog: suggestions for conservation of the wolf in Italy. *Acta Theriologica*, **3**, 101–110.

Loveridge AJ (1999). *Behavioural ecology and rabies transmission in sympatric southern African jackals*. D.Phil. dissertation. University of Oxford, Oxford, UK.

Loveridge AJ and Macdonald DW (2001). Seasonality in spatial organization and dispersal of sympatric jackals (*Canis mesomelas* and *C. adustus*): implications for rabies management. *Journal of Zoology London*, **253**, 101–111.

Loveridge AJ and Macdonald DW (2002). Habitat ecology of two sympatric species of jackals in Zimbabwe. *Journal of Mammalogy*, **83**, 599–607.

Loveridge AJ and Macdonald DW (2003). Niche separation in sympatric jackals (*Canis mesomelas* and *C. adustus*). *Journal of Zoology London*, **259**, 143–153.

Loveridge AJ and Nel JAJ (in press). Black-backed jackal (*Canis mesomelas*). In C Sillero-Zubiri, M Hoffmann and DW Macdonald (eds), *Canids: foxes, wolves, jackals and dogs. Status survey and conservation action plan*, IUCN/SSC Canid Specialist Group, Gland, Switzerland, and Cambridge, UK.

Lucchini V, Fabbri E, Marucco F, Ricci S, Boitani L and Randi E (2002). Non-invasive molecular tracking of colonizing wolf (*Canis lupus*) packs in the western Italian Alps. *Molecular Ecology*, **11**, 857–868.

Lucherini M, Lovari S and Crema G (1995). Habitat use and ranging behaviour of the red fox (*Vulpes vulpes*) in a Mediterranean rural area: is shelter availability a key factor? *Journal of Zoology*, **237**, 577–591.

Lucherini M, Pessino M, Farias AA (in press). Pampas fox (*Pseudalopex gymnocereus*). In C Sillero-Zubiri, M Hoffmann and DW Macdonald (eds), *Canids: foxes, wolves, jackals and dogs. Status survey and conservation action plan*, IUCN/SSC Canid Specialist Group, Gland, Switzerland, and Cambridge, UK.

Ludwig D (1999). Is it meaningful to estimate a probability of extinction? *Ecology*, **80**, 298–310.

Lyles AM and Dobson AP (1993). Infectious disease and intensive management: population dynamics, threatened hosts and their parasites. *Journal of Zoo and Wildlife Medicine*, **24**, 315–326.

Lynch CD (1975). The distribution of mammals in the Orange Free State, South Africa. *Navorsinge van die Nasionale Museum, Bloemfontein*, **3**, 109–139.

Lynch M and Lande R (1998). The critical effective size for a genetically secure population. *Animal Conservation*, **1**, 70–72.

Maas B (1993a). *The behavioural ecology and social organistion of the bat-eared fox in the Serengeti National Park, Tanzania*. D.Phil. dissertation, Cambridge, UK.

Maas B (1993b). Bat-eared fox behavioural ecology and the incidence of rabies in the Serengeti National Park. *Onderstepoort Journal of Veterinary Research*, **60**, 389–393.

MacArthur RH and Wilson EO (1967). *Theory of island biogeography*. Princeton University Press, Princeton, NJ, USA.

Macdonald DW (1976). Food caching by red foxes and some other carnivores. *Zeitschrift für Tierpsychologie*, **42**, 170–185.

Macdonald DW (1977a). On food preference in the red fox. *Mammal Review*, **7**, 7–23.

Macdonald DW (1977b). The behavioural ecology of the red fox. In C Kaplan (ed), *Rabies, the facts*, pp. 70–90. Oxford University Press, Oxford, UK.

Macdonald DW (1979a). "Helpers" in fox society. *Nature London*, **282**, 69–71.

Macdonald DW (1979b). Some observations and field experiments on the urine marking behaviour of the red fox, *Vulpes vulpes* L. *Zeitschrift für Tierphysiologie*, **51**, 1–22.

Macdonald DW (1979c). The flexible social system of the golden jackal, *Canis aureus*. *Behavioural Ecology and Sociobiology*, **5**, 17–38.

Macdonald DW (1980a). Social factors affecting reproduction amongst red foxes. In E Zimen (ed), *The red fox: symposium on behaviour and ecology*, pp. 131–183. W Junk, The Hague, Netherlands.

Macdonald DW (1980b). Pattern of scent marking with urine and feces amongst social communities. *Proceedings of the Royal Society*, **45**, 107–139.

Macdonald DW (1980c). *Rabies and wildlife: a biologist's perspective*. Oxford University Press, Oxford, UK.

Macdonald DW (1981). Resource dispersion and the social organisation of the red fox (*Vulpes vulpes*). In JA Chapman and D Pursley (eds), *The first international worldwide furbearer conference*, pp. 918–949. Frostburg, Maryland, USA.

Macdonald DW (1983). The ecology of carnivore social behaviour. *Nature*, **301**, 379–384.

Macdonald DW (1984). *The encyclopedia of mammals*. Facts on File, New York, USA.

Macdonald DW (1985). The carnivores: order Carnivora. In RE Brown and DW Macdonald (eds), *Social odours in mammals*, pp. 619–722. Oxford University Press, Oxford, UK.

Macdonald DW (1987). *Running with the fox*. Unwin Hymen, London, UK.

Macdonald DW (1992a). Cause of wild dog deaths. *Nature*, **360**, 633–634.

Macdonald DW (1992b). *The velvet claw: a natural history of the carnivores*. BBC Books, London, UK.

Macdonald DW (1993). Rabies and wildlife: a conservation problem? *Onderstepoort Journal of Veterinary Research*, **60**, 351–355.

Macdonald DW (1995). *Wildlife rabies: the implications for Britain*, pp. 33–48. Unresolved questions for the control of wildlife rabies: social perturbation and interspecific interactions. Rabies in a changing world. Proceedings of the British Small Animal Veterinary Association, Cheltenham, UK.

Macdonald DW (1996a). Dangerous liaisons and disease. *Nature*, **379**, 400–401.

Macdonald DW (1996b). Social behaviour of captive bush dogs (*Speothos venaticus*). *Journal of Zoology London*, **239**, 525–543.

Macdonald DW (2001). Postscript: science, compromise and tough choices. In JL Gittleman, SM Funk, DW Macdonald and RK Wayne (eds), *Carnivore conservation*, pp. 524–538. Cambridge University Press, Cambridge, UK.

Macdonald DW and Bacon PJ (1982). Fox society, contact rate and rabies epizootiology. *Comparative Immunology Microbiology and Infectious Diseases*, **5**, 247–256.

Macdonald DW and Baker SE (2004). Non-lethal control of fox predation: the potential of generalised aversion. *Animal Welfare*, **13**, 77–85.

Macdonald DW and Carr GM (1981). Foxes beware: you are back in fashion. *New Scientist*, **89**, 9–11.

Macdonald DW and Carr GM (1989). Food security and the rewards of tolerance. In V Standen and RA Foley (eds), *Comparative socioecology: the behavioural ecology of humans and other mammals*, pp. 75–99. *Special Publication of the British Ecological Society*, **8**, 75–99.

Macdonald DW and Carr GM (1995). Variation in dog society: between resource dispersion and social flux. In J Serpell (ed), *Biology of the domestic dog*, pp. 199–210. Cambridge University Press, Cambridge, UK.

Macdonald DW and Courtenay O (1993). Wild and domestic canids as reservoirs of American visceral leishmaniasis in Amazonia. In N Dunstone and ML Gorman (eds), Mammals as predators. *Proceeding of the Symposia of the Zoological Society of London*, **65**, 465–479.

Macdonald DW and Courtenay O (1996). Enduring social relationships in a population of crab-eating zorros, *Cerdocyon thous*, in Amazonian Brazil (Carnivora, Canidae). *Journal of Zoology London*, **239**, 329–355.

Macdonald DW and Johnson PJ (2000). Farmers and the custody of the countryside: trends in loss and conservation of non-productive habitats 1981–1998. *Biological Conservation*, **94**, 221–234.

Macdonald DW and Johnson DPP (2001). Dispersal in theory and practice: consequences for conservation biology. In J Clobert, E Danchin, AA Dhondt and JD Nichols (eds), *Dispersal*, pp. 358–372. Oxford University Press, Oxford, UK.

Macdonald DW and Moehlman PD (1982). Co-operation, altruism, and restraint in the reproduction of carnivores. In PPG Bateson and P Klopfer (eds), *Perspectives in ethology*, **5**, 433–467. Plenum Press, New York, USA.

Macdonald DW and Newdick MT (1982). The distribution and ecology of foxes, *Vulpes vulpes* (L.). In urban areas. In R Bornkamm, JA Lee and MRD Seaward (eds), *Urban ecology*, pp. 123–135. Blackwell Scientific Publications, Oxford, UK.

Macdonald DW and Rushton S (2003). Modelling space use and dispersal of mammals in real landscapes: a tool for conservation. *Journal of Biogeography*, **30**, 607–620.

Macdonald DW and Sillero-Zubiri C (2002). Large carnivores and conflict: lion conservation in context. In AJ Loveridge, T Lynam and DW Macdonald (eds), *Lion conservation research. Workshop 2: modelling conflict*, p. 18 Wildlife Conservation Research Unit, Oxford, UK.

Macdonald DW and Thom MD (2001). Alien carnivores: unwelcome experiments in ecological theory. In JL Gittleman, SM Funk, DW Macdonald and RK Wayne (eds), *Carnivore conservation*, pp. 93–122. Cambridge University Press, Cambridge, UK.

Macdonald DW and Voigt DR (1985). The biological basis of rabies models. In JP Bacon (ed), *Population dynamics of rabies in wildlife*. Academic Press, London, UK.

Macdonald DW, Artois M, Aubert M, Bishop DL, Ginsberg JR, King A, Kock N and Perry BD (1992). Cause of wild dog deaths. *Nature*, **360**, 633–634.

Macdonald DW, Barasso P and Boitani L (1980). Foxes, wolves and conservation in the Abruzzo Mountains, Italy. In E Zimen (ed), *The red fox, behaviour and ecology*, pp. 223–235. W Junk, The Hague, Netherlands.

Macdonald DW, Mace GM and Rushton SP (1998). *Proposals for future monitoring of British mammals*. Department for the Environment, Transport and the Regions, London, UK.

Macdonald DW, Courtenay O, Forbes S and Mathews F (1999). The red fox (*Vulpes vulpes*) in Saudi Arabia: loose-knit groupings in the absence of territoriality. *Journal of Zoology*, **249**, 383–391.

Macdonald DW, Mace GM and Rushton SP (2000a). British mammals: is there a radical future? In A Entwistle and N Dunstone (eds), *Priorities for the conservation of mammalian diversity: has the panda had its day?*, pp. 175–205. Cambridge University Press, Cambridge, UK.

Macdonald DW, Stewart PD, Stopka P and Yamaguchi N (2000b). Measuring the dynamics of mammalian societies: an ecologist's guide to ethological methods. In L Boitani and TK Fuller (eds), *Research techniques in animal ecology: controversies and consequences*, pp. 332–388. Columbia University Press. New York, USA.

Macdonald DW, Tattersall FH, Johnson PJ, Carbone C, Reynolds JC, Langbein J, Rushton SP and Shirley MDF (eds) (2000c). *Managing British mammals: case studies from the hunting debate*. Wildlife Conservation Research Unit, Oxford, UK.

Macdonald DW, Moorhouse TP, Enck JW and Tattersall FH (2002a). *Mammals. Handbook of Ecological Restoration*, **1**, 389–408.

Macdonald DW, Moorhouse TP and Enck JW (2002b). The ecological context: a species population perspective. In MR Perrow and AJ Davy (eds), *Handbook of ecological restoration. Vol. 1: Principles of restoration*. Cambridge University Press, Cambridge, UK.

Macdonald DW, Newman C, Dean J and Buesching CD (in press). Factors underlying the distribution of Eurasian badger *Meles meles* setts in a high-density area. *Oikos*.

Macdonald DW, Reynolds JC, Carbone C, Mathews F and Johnson PJ (2003). The bioeconomics of fox control. In FH Tattersall and WJ Manley (eds), *Conservation and conflict: mammals and farming in Britain*, pp. 220–236. Linnean Society Occasional Publication, Westbury Publishing, Yorkshire, UK.

Mace GM and Lande R (1991). Assessing extinction threats: toward a reevaluation of IUCN threatened species categories. *Conservation Biology*, **5**, 148–157.

Mace GM, Smith TB, Bruford MW and Wayne RK (1996). Molecular genetic approaches in conservation, an overview of the issues. In TB Smith and RK Wayne (eds), *Molecular genetic approaches in conservation*, pp. 1–12. Oxford University Press, Oxford, UK.

Machlis GE (1992). The contribution of sociology to biodiversity research and management. *Biological Conservation*, **62**, 161–170.

Machlis L, Dodd PWD and Fentress JC (1985). The pooling fallacy: problems arising when individuals contribute more than one observation to the data set. *Zeitschrift für Tierpsychologie*, **68**, 201–214.

Mackie AJ (1988). *Bat-eared foxes* Otocyon megalotis *as predators on termites Hodotermes mossambicus in the Orange Free State*. M.Sc. dissertation, University of Stellenbosch, South Africa.

Mackie AJ and Nel JAJ (1989). Habitat selection, home range use, and group size of bat-eared foxes in the Orange Free State. *South African Journal of Wildlife Research*, **19**, 135–139.

Mackowiak M, Maki J, Motes-Kreimeyer L, Harbin T and van Kampen K (1999). Vaccination of wildlife against rabies: successful use of a vectored vaccine obtained by recombinant technology. *Advances in Veterinary Medicine*, **41**, 571–583.

Macpherson A (1969). The dynamics of Canadian arctic fox populations. *Canadian Wildlife Service Report Serie*, **8**, 1–49.

Maddock AH and Mills MGL (1994). Population characteristics of African wild dogs *Lycaon pictus* in the eastern Transvaal Lowveld, South-Africa, as revealed through photographic records. *Biological Conservation*, **67**, 57–62.

Maffei L and Taber AB (in press). Area de acción, actividad y uso de hábitat del zorro patas negras, *Cerdocyon thous* Linnaeus, 1776 (Carnivora: Canidae) en un bosque seco. *Revista Chilena de Historia Natural*.

Maher DS and Brady JR (1986). Food habits of bobcats in Florida. *Journal of Mammalogy*, **67**, 133–138.

Mahi-Brown CA, Yanagimachi R, Nelson MC, Yanagimachi H and Palumbo N (1988). Ovarian histopathology of bitches immunized with porcine zonae pellucida. *American Journal of Reproductive Immunology and Microbiology*, **18**, 94–103.

Major JT and Sherburne JA (1987). Interspecific relationships of coyotes, bobcats, and red foxes in western Maine. *Journal of Wildlife Management*, **51**, 606–616.

Mäkinen A, Kuokkanen M-T and Valtonen M (1986). A chromosome-branding study in the finnish and the Japanese raccoon dog. *Hereditas*, **105**, 97–105.

Malcolm JR (1979). *Social organization and communal rearing in African wild dogs*. Ph.D. dissertation. Harvard University, Cambridge, MA, USA.

Malcolm JR (1986). Socio-ecology of bat-eared foxes (*Otocyon megalotis*). *Journal of Zoology London*, **208**, 457–467.

Malcolm JR (1997). The diet of the Ethiopian wolf (*Canis simensis* Rüeppell) from a grassland area of the Bale Mountains, Ethiopia. *African Journal of Ecology*, **35**, 162–164.

Malcolm JR (2001). [bat eared fox] p. 58. In DW Macdonald (ed), *New encyclopaedia of mammals*. Oxford University Press, Oxford, UK.

Malcolm JR and Ashenafi ZT (1997). Conservation of Afroalpine habitats. In C Sillero-Zubiri and DW Macdonald (eds), *The Ethiopian wolf: status survey and conservation action plan*, pp. 61–63. IUCN/SSC Canid Specialist Group, Gland, Switzerland and Cambridge, UK.

Malcom JR and Marten K (1982). Natural selection and the communal rearing of pups in African wild dogs *Lycaon pictus*. *Behavioural Ecology and Sociobiology*, **10**, 1–13.

Maldonado JE, Cotera M, Geffen E and Wayne RK (1997). Relationships of the endangered Mexican kit fox (*Vulpes macrotis zinseri*) to North American arid-land foxes based on mitochondrial DNA sequence data. *The Southwestern Naturalist*, **42**, 460–470.

Manakadan R and Rahmani AR (2000). Population and ecology of the Indian fox *Vulpes bengalensis* at Rollapadu Wildlife Sanctuary, Andhra Pradesh, India. *Journal of the Bombay Natural History Society*, **97**, 3–14.

Mandelbrot BB (1982). *The fractal geometry of nature*. WH Freeman, San Francisco, CA, USA.

Manfredo MJ, Bright AD, Pate J and Tischbein G (1994). *Colorado residents' attitudes and perceptions toward reintroduction of the grey wolf* (Canis lupus) *into Colorado*. Project Report No. 21. Human Dimensions in Natural Resources Unit, Colorado State University, Fort Collins, CO, USA.

Marais E and Griffin M (1993). Range extension in the bat-eared fox *Otocyon megalotis* in Namibia. *Madoqua*, **18**, 187–188.

Marcstroem V, Keith LB, Engren E and Cary JR (1989). Demographic responses of arctic hares (*Lepus timidus*) to experimental reductions of red foxes (*Vulpes vulpes*) and martens (*Martes martes*). *Canadian Journal of Zoology*, **67**, 658–668.

Marino J (2003). Threathened Ethiopian wolves persist in small isolated afroalpine enclaves. *Oryx*, **37**, 62–71.

Marino J (2004). *Spatial ecology of the Ethiopian wolf, canis simensis*. D.Phil. dissertation, Oxford University, Oxford, UK.

Marquet PA, Contreras LC, Torres-Murua JC, Silva SI and Jaksic FM (1993). Food habits of *Pseudalopex* foxes in the Atacama desert, pre-Andean ranges, and the high- Andean plateau of northermost Chile. *Mammalia*, **57**, 130–135.

Marshall LG (1977). Evolution of the carnivorous adaptive zone in South America. In MK Hecht, PC Goody and BM Hecht (eds), *Major patterns in vertebrate evolution*, pp. 709–721. Plenum, New York, USA.

Marshall LG, Butler RF, Drake RE, Curtis GH and Tedford RH (1979). Calibration of the great American interchange. *Science*, **204**, 272–279.

Martin LD (1989). Fossil history of the terrestrial Carnivora. In JL Gittleman (ed), *Carnivore behaviour, ecology, and evolution*, pp 536–568. Chapman and Hall, London, UK.

Martin P and Bateson P (1993). Measuring behavior: an introductory guide, 2nd Edition. Cambridge University Press, London, UK.

Martin R (1973). Trois nouvelles espèces de Caninae (Canidae, Carnivora) des gisements plio-villafranchiens d'Europe. *Documents des Laboratoires de Géologie de la Faculté des Sciences de Lyon*, **57**, 87–96.

Massoia E (1982). *Dusicyon gymnocercus lordi*. una nueva subespecie del "zorro gris grande" (Mammalia Carnívora Canidae). *Neotropica*, **28**, 147–152.

Matsuoka K and Miyoshi N (1998). Changes in broad-leaved evergreen forests since the advanced period of the last glacial age. In Y Yasuda and N Miyoshi (eds), *Vegetational history of Japanese archipelago*, pp. 224–236. Asakura-shoten, Tokyo, Japan. (In Japanese.)

Mauricio IL, Howard MK, Stothard JR and Miles MA (1999). Genomic diversity in the *Leishmania donovani* complex. *Parasitology*, **119**, 237–246.

Mauricio IL, Gaunt MW, Stothard JR and Miles MA (2001). Genetic typing and phylogeny of the *Leishmania donovani* complex by restriction analysis of PCR amplified gp63 intergenic regions. *Parasitology*, **122**, 393–403.

Matthee CA and Robinson TJ (1997). Mitochondrial DNA phylogeography and comparative cytogenetics of the springhare, *Pedetes capensis* (Mammalia: Rodentia). *Journal of Mammalian Evolution*, **4**, 53–73.

Mayr E (1976). *Populations, species and evolution. An abridgment of animal species and evolution*. The Belknap Press of Harvard University Press, Cambridge, UK.

McAdoo JK and DA Klebenow (1978). Predation on range sheep with no predator control. *Journal of Range Management*, **31**, 111–114.

McCarley H (1962). The taxonomic status of wild Canis (Canidae) in the southcentral United States. *Southwestern Naturalist*, **7**, 227–235.

McCarley H and Carley CJ (1979). Recent changes in the distribution and status of wild red wolves (*Canis rufus*). Endangered Species Report No. 4, U.S. Fish and Wildlife Service, Albuquerque, NM, USA.

McCormick AE (1983). Canine distemper in African hunting dogs (*Lycaon pictus*)—possibly vaccine induced. *Journal of Zoo Animal Medicine*, **14**, 66–71.

McCue PM and O'Farrell TP (1988). Serological survey for selected diseases in the endangered San Joaquin kit fox (*Vulpes macrotis mutica*). *Journal of Wildlife Diseases*, **24**, 274–281.

McGrew JC (1979). *Vulpes macrotis*. *Mammalian Species*, **123**, 1–6.

McIlroy JC, Cooper RJ, Gifford EJ, Green BF and Newgrain KW (1986). The effect on wild dogs, *Canis f. familiaris*, of 1080 poisoning campaigns in Kosciusko National Park, New South Wales. *Australian Wildlife Research*, **13**, 535–544.

McIntyre R ed. (1995). *War against the wolf: America's campaign to exterminate the wolf*. Voyageur Press, Stillwater, Minnesota, USA.

McKenzie AA (1990). *Co-operative hunting in the black-backed jackal* (Canis mesomelas) *Schreber*. Ph.D. dissertation, University of Pretoria, South Africa.

McLaren BE and Janke RA (1996). Seedbed and canopy cover effects on balsam fir seedling establishment in Isle Royale National Park. *Canadian Journal of Forest Research*, **26**, 782–793.

McLaren BE and Peterson RO (1994). Wolves, moose and tree rings on Isle Royale. *Science*, **266**, 1555–1558.

McMahon E (2002). *Status and conservation of the zorro chilote on Mainland Chile*. Progress report for SAG and CONAF, Temuco, Chile.

McNab BK (1963). Bioenergetics and the determination of home range size. *American Naturalist*, **97**, 133–140.

McNab BK (1987). Basal rate of metabolism, body size, and food habits in the order *Carnivora*. In JL Gittleman (ed), *Carnivore behaviour, ecology and evolution*, pp. 335–354. Chapman and Hall, London, UK.

McNaughton SJ (1989). Interactions of plants of the field layer with large herbivores. In PA Jewell and GMO Maloiy (eds), *The biology of large african mammals in their environment. Zoological Society of London Symposium*, **61**, 15–29. Clarendon Press, Oxford, UK.

McNaughton SJ and Banyikwa FF (1995). Plant communities and herbivory. In ARE Sinclair and P Arcese (eds), *Serengeti II: dynamics, management, and conservation of an ecosystem*, pp. 49–70. University of Chicago Press, Chicago, IL, USA and London, UK.

McNay ME (2000). A case-history of wolf-human encounters in Alaska and Canada *Wildlife Technical Bulletin*, **13**, 45. Alaska Department of Fish and Game.

McNutt JW (1995). *Sociality and dispersal in African wild dogs,* Lycaon pictus. Ph.D. dissertation, University of California, Davis, CA, USA.

McNutt JW (1996a). Adoption in African wild dogs, *Lycaon pictus*. *Journal of Zoology*, **240**, 163–173.

McNutt JW (1996b). Sex-biased dispersal in African wild dogs, *Lycaon pictus*. *Animal Behaviour*, **52**, 1067–1077.

McNutt JW and Boggs L (1997). *Running wild: dispelling the myths of the African wild dog*. Smithsonian Institution Press, Washington DC, USA.

Meadows LE and Knowlton FF (2000). Efficacy of guard llamas to reduce canine predation on domestic sheep. *Wildlife Society Bulletin*, **28**, 614–622.

Meadows R (2001). *Southern Rockies wildlife and wilderness survey report*, 121 pp. Decision Research, Washington DC, USA.

Mech LD (1966). *The wolves of Isle Royale*, 210 pp. Fauna Series No. 7. U.S. National Park Service, Washington DC, USA.

Mech LD (1970). *The wolf: the ecology and behavior of an endangered species*. Natural History Press, Doubleday, New York, USA.

Mech LD (1974). *Canis lupus*. *Mammalian Species*, **37**, 1–6.

Mech LD (1977a). Wolf-pack buffer zones as prey reservoirs. *Science*, **198**, 320–321.

Mech LD (1977b). Productivity, mortality and population trends of wolves on northeastern Minnesota. *Journal of Mammalogy*, **58**, 559–574.

Mech LD (1987). Age, season, distance, direction, and social aspects of wolf dispersal from a Minnesota pack. In BD Chepko-Sade and ZT Halpin eds *Mammalian dispersal patterns*, pp. 55–74. University of Chicago Press, Chicago, IL, USA.

Mech LD (1993). Details of a confrontation between two wild wolves. *Canadian Journal of Zoology*, **71**, 1900–1903.

Mech LD (1994). Buffer zones of territories of gray wolves as regions of intraspecific strife. *Journal of Mammalogy*, **75**, 199–202.

Mech LD (1995). The challenge and opportunity of recovering wolf populations. *Conservation Biology*, **9**, 270–278.

Mech LD (1996). A new era for carnivore conservation. *Wildlife Society Bulletin*, **24**, 397–401.

Mech LD (1999a). Estimated costs of maintaining a recovered wolf population in agricultural regions of Minnesota. *Wildlife Society Bulletin*, **26**, 817–822.

Mech LD (1999b). Alpha status, dominance, and division of labor in wolf packs. *Canadian Journal of Zoology*, **77**, 1196–1203.

Mech LD (2000a). A record large wolf, *Canis lupus*, pack in Minnesota. *Canadian Field Naturalist*, **114**, 504–505.

Mech LD (2000b). *Comments on proposed wolf reclassification rule*. Letter to US Fish and Wildlife Service, Fort Snelling, Minnesota. November 9.

Mech LD (2001a). Mech challenges study's pessimistic outlook. *International Wolf*, **11**, 8.

Mech LD (2001b). Managing Minnesota's recovered wolves. *Wildlife Society Bulletin*, **29**, 70–77.

Mech LD and Boitani L (eds) (2003). *Wolves: behavior, ecology and conservation*. University of Chicago Press, Chicago, IL, USA.

Mech LD and Boitani L (in press). Grey wolf (*Canis lupus*). In C Sillero-Zubiri, M Hoffmann and DW Macdonald (eds), *Canids: foxes, wolves, jackals and dogs. Status survey and conservation action plan*. IUCN/SSC Canid Specialist Group, Gland, Switzerland, and Cambridge, UK.

Mech LD and Frenzel LD eds (1971). Ecological studies of the timber wolf in northeastern Minnesota. U.S. Department of Agriculture Forest Service Research Paper NC-52. North Central Forest Experiment Station, St. Paul, Minnesota, USA.

Mech LD and Goyal SM (1995). Effects of canine parvovirus on gray wolves in Minnesota. *Journal of Wildlife Management*, **59**, 565–570.

Mech LD and Nelson ME (2000). Do wolves affect white-tailed buck harvest in northeastern Minnesota? *Journal of Wildlife Management*, **64**, 129–136.

Mech LD and Peters G (1977). The study of chemical communication in free-ranging mammals. In D Müller- Schwarze and MM Mozell (eds), *Chemical signals in vertebrates*, pp. 321–331. Plenum Press, New York, USA.

Mech LD, Adams LG, Meier TJ, Burch JW and Dale BW (1998). *The wolves of Denali*. University of Minnesota Press, Minneapolis, USA.

Mech LD, Fritts SH and Nelson ME (1996). Wolf management in the 21st century: from public input to sterilization. *Journal of Wildlife Research*, **1**, 195–198.

Mech LD, Harper EK, Meier TJ and Paul WJ (2000). Assessing factors that may predispose Minnesota farms to wolf depredation on cattle. *Wildlife Society Bulletin*, **28**, 623–629.

Mech LD, Smith DW, Murphy KM and MacNulty DR (2001). Winter severity and wolf predation on a formerly wolf-free elk herd. *Journal of Wildlife Management*, **65**, 998–1003.

Medel R and Jaksic FM (1988). Ecología de los canidos sudamericanos: una revisión. *Revista Chilena de Historia Natural*, **61**, 67–79.

Medel RG, Jimenez JE, Jaksic FM, Yanez JL and Armesto JJ (1990). Discovery of a continental population of the rare Darwin's fox *Dusicyon fulvipes* new-record Martin 1837 in Chile. *Biological Conservation*, **51**, 71–78.

Meher-Homji VM (1984). Udhagamandalam (Ootacamund): a biogeographic perspective. *Indian Geographic Journal*, **82**, 3889–3911.

Meia JS and Weber JM (1995). Home ranges and movements of red foxes in central Europe: stability despite environmental changes. *Canadian Journal of Zoology*, **73**, 1960–1966.

Meia JS and Weber JM (1996). Social organization of red foxes (*Vulpes vulpes*) in the Swiss Jura Mountains. *Zeitschrift für Saugetierkunde*, **61**, 257–268.

Mendelssohn H, Yom-Tov Y, Ilany G and Meninger D (1987). On the occurrence of Blanford's fox, *Vulpes cana* Blanford, 1877, in Israel and Sinai. *Mammalia*, **51**, 459–462.

Mercure A, Ralls K, Koepfli KP and Wayne RK (1993). Genetic subdivisions among small canids: mitochondrial DNA differentiation of swift, kit, and arctic foxes. *Evolution*, **47**, 1313–1328.

Meriggi A and Lovari S (1996). A review of wolf predation in southern Europe: does the wolf prefer wild prey to livestock? *Journal of Applied Ecology*, **33**, 1561–1571.

Mertens A, Promberger C and Gheorge P (2002). Testing and implementing the use of electric fences for night corrals in Romania. *Carnivore Damage Prevention News*, **5**, 2–5. www.kora.unibe.ch.

Messier F (1985). Solitary living and extraterritorial movements of wolves in relation to social status and prey abundance. *Canadian Journal of Zoology*, **63**, 239–245.

Messier F (1994). Ungulate population models with predation. A case study with the North American moose. *Ecology*, **75**, 478–488.

Messier F and Barrette C (1982). The social system of the coyote (*Canis latrans*) in a forested habitat. *Canadian Journal of Zoology*, **60**, 1743–1753.

Michigan Department of Natural Resources (1997). *Michigan grey wolf recovery and management plan*, 58 pp. Lansing, Michigan, USA.

Miller AH (1931). Systematic revision and natural history of the American shrikes (*Lanius*). *University of California Publications in Zoology*, **38**, 11–242.

Miller RI (ed) (1994). *Mapping the diversity of nature*. Chapman and Hall, London, UK.

Miller S and Rottmann J (1976). *Guía para el reconocimiento de mamíferos chilenos*. Editora Nacional Gabriela Mistral, Santiago, Chile.

Mills LS and Knowlton FF (1991). Coyote space use in relation to prey abundance. *Canadian Journal of Zoology*, **69**, 1516–1521.

Mills LS, Doak DF and Wisdom MJ (1999). Reliability of conservation actions based on elasticity analysis of matrix models. *Conservation Biology*, **13**, 815–829.

Mills LS, Citta JJ, Lair KP and Schwartz MK (2000). Estimating animal abundance using non-invasive sampling: promise and pitfalls. *Ecological Applications*, **10**, 283–294.

Mills MGL (1982). Factors affecting group size and territory size of the brown hyaena, *Hyaena brunnea*, in the southern Kalahari. *Zietschrift Tierpsychologic*, **48**, 113–141.

Mills MGL (1989a). *Kalahari hyaenas: the behavioural ecology of two species*. Unwin Hymen Press, London, UK.

Mills MGL (1989b). The comparative behavioural ecology of hyaenas: the importance of diet and food dispersion. In JL Gittleman (ed), *Carnivore behaviour, ecology and evolution*, pp. 125–142. Cornell University Press, Ithaca, NY, USA.

Mills MGL (1990). *Kalahari hyaenas: the comparative behavioural ecology of two species*. Unwin Hyman, London, UK.

Mills MGL (1993). Social systems and behaviour of the African wild dog *Lycaon pictus* and the spotted hyaena *Crocuta crocuta* with special reference to rabies. *Onderstepoort Journal of Veterinary Research*, **60**, 405–409.

Mills MGL and Gorman ML (1997). Factors affecting the density and distribution of wold dogs in the Kruger National Park. *Conservation Biology*, **11**, 1397–1406.

Mills MGL, Ellis S, Woodroffe R, Maddock A, Stander P, Rasmussen G, Pole A, Fletcher P, Bruford M, Wildt DE, Macdonald DW and Seal US (1998). *Population and habitat viability assessment for the African wild dog (Lycaon pictus) in Southern Africa*. IUCN/SSC Conservation Breeding Specialist Group, Apple Valley, MN, USA.

Minnesota Department of Natural Resources (2001). *Minnesota wolf management plan*, 80 pp. Minnesota Department of Natural Resources, St. Paul, MN, USA.

Minsky D (1980). Preventing fox predation at a least tern colony with an electric fence. *Journal of Field Ornithology*, **51**, 180–181.

Mitchell-Jones AJ, Amori G, Bogdanowicz W *et al.* (1999). *The atlas of European mammals*, pp. 320–321. Poyser Natural History, London, UK.

Mivart St G (1890). *Dogs, jackals, wolves and foxes: a monograph of the Canidae*. London, UK.

Mladenoff DJ and Sickley TA (1998). Assessing potential grey wolf restoration in the northeastern United States: a spatial prediction of favorable habitat and potential population levels. *Journal of Wildlife Management*, **62**, 1–10.

Mladenoff DJ, Sickley TA, Haight RG and Wydeven AP (1995). A regional landscape analysis and prediction of favorable gray wolf habitat in the northern Great Lakes region. *Conservation Biology*, **9**, 279–294.

Moehlman PD (1978). *Socioecology of silver-backed and golden jackals*. Serengeti Research Institute report no. 241. National Geographic Society, Washington DC, USA.

Moehlman PD (1979). Jackal helpers and pup survival. *Nature*, **277**, 382–383.

Moehlman PD (1983). Socioecology of silverbacked and golden jackals (*Canis mesomelas* and *Canis aureus*). In

JF Eisenberg and DG Kleiman (eds), *Recent advances in the study of mammalian behavior*, pp. 423–453. American Society of Mammalogists, Lawrence, KS, USA.

Moehlman PD (1986). Ecology of cooperation in canids. In DI Rubenstein and RW Wrangham (eds), *Ecological aspects of social evolution: birds and mammals*, pp. 64–86. Princeton University Press, Princeton, NJ, USA.

Moehlman PD (1989). Intraspecific variation in canid social systems. In JL Gittleman (ed), *Carnivore behavior, ecology, and evolution*, pp. 164–182. Cornell University Press, Ithaca, NY, USA.

Moehlman PD and Hofer H (1997). Cooperative breeding, reproductive suppression and body mass in canids. In NG Solomon and JA French (eds), *Cooperative breeding in mammals*, pp. 76–128. Cambridge University Press, Cambridge, UK.

Moehrenschlager A (2000). *Effects of ecological and human factors on the behaviour and population dynamics of reintroduced Canadian swift foxes* (Vulpes velox). D.Phil. dissertation, University of Oxford, Oxford, UK.

Moehrenschlager A and List R (1996). Comparative ecology of North American prairie foxes—conservation through collaboration. In DW Macdonald and FH Tattersall (eds), *The WildCRU review*, pp. 22–28. Wildlife Conservation and Research Unit, Oxford, UK.

Moehrenschlager A and Macdonald DW (2003). Movement and survival parameters of translocated and resident swift foxes. *Animal Conservation*, **6**, 199–206.

Moehrenschlager A and Moehrenschlager CAJ (2001). *Census of swift fox* (Vulpes velox) *in Canada and Northern Montana: 2000–2001*. Report to Alberta Environmental Protection, Edmonton, Alberta, Canada.

Moehrenschlager A, Macdonald DW and Moehrenschlager C (2003). Reducing capture-related injuries and radio-collaring effects on swift foxes (*Vulpes velox*). In M Sovada and L Carbyn (eds), *Swift fox conservation in a changing world*, pp. 107–113. Canadian Plains Research Center, University of Regina, Saskatchewan, Canada.

Moehrenschlager A and Somers M (in press). Canid reintroductions and metapopulation management. In C Sillero-Zubiri, M Hoffmann and DW Macdonald (eds), *Canids: foxes, wolves, jackals and dogs. Status survey and conservation action plan*, IUCN/SSC Canid Specialist Group, Gland, Switzerland, and Cambridge, UK.

Moehrenschlager A, List R and Macdonald DW (submitted). Intraguild killing depends on prey dynamics. Mexican kit foxes escape coyotes while Canadian swift foxes cannot.

Moehrenschlager A and Sovada MA (in press). *Vulpes velox*. In C Sillero-Zubiri, M Hoffmann and DW Macdonald (eds), *Canids: foxes, wolves, jackals and dogs. Status survey and conservation action plan*, IUCN/SSC Canid Specialist Group, Gland, Switzerland, and Cambridge, UK.

Moehrenschlager C and Moehrenschlager A (1999). Canadian swift fox (*Vulpes velox*) population assessment: Winter, 1999, 37 pp. Alberta Environmental Protection, Edmonton, Alberta, Canada.

Moen R and DelGiudice GD (1997). Simulating nitrogen metabolism and urinary urea nitrogen: creatinine ratios in ruminants. *Journal of Wildlife Management*, **61**, 881–894.

Momen H, Pacheco RS, Cupolillo E and Grimaldi G Jr (1993). Molecular evidence for the importation of Old World *Leishmania* into the Americas. *Biological Research*, **26**, 249–255.

Monfort SL, Wasser SK, Washburn KL, Burke BA, Brewer BA and Creel SR (1997). Steroid metabolism and validation of non-invasive endocrine monitoring in the African wild dog (*Lycaon pictus*). *Zoo Biology*, **16**, 533–548.

Montali RJ, Bartz CR, Teare JA, Allen JT, Appel MJG and Bush M (1983). Clinical trials with canine distemper vaccines in exotic carnivores. *Journal of the American Veterinary Medical Association*, **183**, 1163–1167.

Montgomery GG and Lubin D (1978). Social structure and food habits of crab-eating fox (*Cerdocyon thous*) in Venezuelan llanos. *Acta Científica Venezuelica*, **29**, 382–383.

Moore CM and Collins PW (1995). *Urocyon littoralis* (Baird, 1858). *Mammalian Species*, **489**, 1–7.

Moore GC and Parker GR (1992). Colonization by the eastern coyote (*Canis latrans*). In A Boer (ed), *Ecology and management of the eastern coyote*. Wildlife Research Unit, University of New Brunswick, Fredericton, Canada.

Moritz CC (1994). Defining "evolutionarily significant units" for conservation. *Trends in Ecology and Evolution*, **9**, 373–375.

Morrell S (1972). Life history of the San Joaquin kit fox. *California Fish and Game*, **58**, 162–174.

Morrison ML, Marcot BG and Mannan RW (1992). *Wildlife-habitat relationships: concepts and applications*. University of Wisconsin Press, Madison, Wisconsin, USA.

Moss R, Watson A and Parr R (1995). Experimental prevention of a population cycle in red grouse. *Ecology*, **77**, 1512–1530.

Motta-Junior JC, Lombardi JA and Talamoni SA (1994). Notes on crab-eating fox (*Dusicyon thous*) seed dispersal and food habits in southeastern Brazil. *Mammalia*, **58**, 156–159.

Motta-Júnior JC, Talamoni SA, Lombardi JA and Simokomaki K (1996). Diet of the maned wolf, *Chrysocyon brachyurus*, in central Brazil. *Journal of Zoology London*, **240**, 277–284.

Moutou F (1997). Dog vaccination around the Serengeti. *Oryx*, **31**, 14.

Mulder JL (1985). Spatial organization, movements and dispersal in a Dutch red fox (*Vulpes vulpes*) population: some preliminary results. *Revue d'Ecologie (Terre et la Vie)*, **40**, 133–138.

Mumme RL and Koenig WD (1991). Explanations for avian helping behaviour. *Trends in Ecology and Evolution*, **6**, 343–344.

Mundy NI, Winchell CS and Woodruff DS. (1997a). Genetic differences between the endangered San Clemente loggerhead shrike *Lanius ludovicianus mearnsi* and two neighboring subspecies demonstrated by mtDNA control region and cytochrome b sequence variation. *Molecular Ecology*, **6**, 29–37.

Mundy NI, Winchell CS, Burr T and Woodruff DS (1997b). Microsatellite variation and microevolution in the critically endangered San Clemente Island loggerhead shrike *(Lanius ludovicianus mearns). Proceedings of the Royal Society of London B*, **264**, 869–875.

Munthe K (1979). *The skeleton of the Borophaginae (Carnivora, Canidae): morphology and function*. Ph.D. dissertation, University California, Berkeley, CA, USA.

Munthe K (1989). The skeleton of the Borophaginae (Carnivora, Canidae), morphology and function. *University of California Publications Bulletin of Department of Geological Sciences*, **133**, 1–115.

Munthe K (1998). Canidae. In CM Janis, KM Scott, and LL Jacobs (eds), *Evolution of tertiary mammals of North America, Vol. 1: terrestrial carnivores, ungulates, and ungulatelike mammals*, pp. 124–143. Cambridge University Press, Cambridge, UK.

Murie A (1934). *The moose of Isle Royale*. Miscellaneous Publications of the Museum of Zoology, University of Michigan No. 25.

Murie A (1940). Ecology of the coyote in the Yellowstone. National Park Service, Fauna Series No. 4.

Murphy WJ, Eizirik E, O'Brien SJ, *et al*. (2001). Resolution of the early placental mammal radiation using Bayesian phylogenetics. *Science*, **294**, 2348–2351.

Murray M (2001). Serengeti migrations article. In *New encyclopaedia of mammals*, pp. 256–257 Oxford University Press, Oxford, UK.

Musiani M and Visalberghi E (2001). Effectiveness of fladry on wolves in captivity. *Wildlife Society Bulletin*, **29**, 91–98.

Myers N, Mittermeier RA, Mittermeier CG, da Fonseca GAB and Kent J (2000). Biodiversity hotspots for conservation priorities. *Nature*, **403**, 853–858.

Narendra Babu V and Venkataraman AB (2001). Dhole depredation and its consequences on the carnivore community in Arunachal Pradesh, Northeast India. *Proceedings of the Canid Biology and Conservation Conference, 2001*, Oxford University, Oxford, UK.

Nasimovich A and Isakov Y (eds) (1985). *Arctic fox, red fox and racoon dog: distribution of resources, ecology, use and conservation* (in Russian). Janka, Moscow, USSR.

Nass RD and Theade J (1988). Electric fences for reducing sheep losses to predators. *Journal of Range Management*, **41**, 251–252.

National Research Council (U.S.) Committee on Management of Wolf and Grizzly Bear Populations in Alaska (1997). *Wolves, bears, and their prey in Alaska: biological and social challenges in wildlife management*. National Academy Press.

Naumov NP, Gol'tsman ME, Kruchenkova EP, Ovsyanikov NG, Popov SV and Smirin VM (1981). Social behaviour of Arctic foxes on Menyi island. Factors determining space-time regime of activity (in Russian). In NP Naumov (ed), *Ecology, population structure and intraspecific communicative processes in Mammals*, pp. 31–75. Nauka, Moscow, USSR.

Neginhal SG (1974). *Project Tiger management plan for Bandipur Tiger Reserve, Karnataka state, India*, Ministry of Agriculture, New Delhi, India.

Nei M (1987). *Molecular evolutionary genetics*. Columbia University Press, New York, USA.

Nel JAJ (1978). Notes on the food and foraging behavior of the bat-eared fox, *Otocyon megalotis*. *Bulletin Carnegie Museum Natural History*, **6**, 132–137.

Nel JAJ (1990). Foraging and feeding by bat-eared foxes *Otocyon megalotis* in the southwestern Kalahari. *Koedoe*, **33**, 9–16.

Nel JAJ (1993). The bat-eared fox: a prime candidate for rabies vector? *Onderstepoort Journal of Veterinary Research*, **60**, 395–397.

Nel JAJ and Bester HM (1983). Communication in the southern bat-eared fox *Otocyon m. megalotis* (Desmarest, 1822). *Zeitschrift fur Saugetierkunde*, **48**, 277–290.

Nel JAJ and Mackie AJ (1990). Food and foraging behaviour of bat-eared foxes in the south-eastern Orange Free State. *South African Journal or Wildlife Research*, **20**, 162–166.

Nel JAJ, Mills MGL and van Aarde RJ (1984). Fluctuating group size in bat-eared foxes (*Otocyon m. megalotis*) in the southwestern Kalahari. *Journal of Zoology London*, **203**, 294–298.

Nelson ME and Mech LD (1986). Relationship between snow depth and gray wolf predation on white-tailed deer. *Journal of Wildlife Management*, **50**, 471–474.

New Mexico Department of Game and Fish (2002). *Long range plan for the management of Rio Grande cutthroat trout in New Mexico*. New Mexico Department of Game and Fish, Santa Fe, NM, USA.

Newman Bueshing CCD and Macdonald DW (2003). Validating mammal monitoring methods and assessing the performance of volunteers in wildlife conservation—*"Sed quis custodiet ipsos custodies". Biological Conservation*, **113**, 189–197.

Newman TJ, Baker PJ and Harris S (2002). Nutritional condition and survival of red foxes with sarcoptic mange. *Canadian Journal of Zoology*, **80**, 154–161.

Newsome AE (1995). Socio-ecological models for red fox populations subject to fertility control in Australia. *Annales Zoologica Fennici*, **32**, 99–110.

Newsome AE and Corbett LK (1985). The identity of the dingo *Canis familiaris dingo*. 3. The incidence of dingoes dogs and hybrids and their coat colors in remote and settled regions of Australia. *Australian Journal of Zoology*, **33**, 363–376.

Newsome AE, Parer I and Cattling PC (1989). Prolonged prey suppression by carnivores—predator removal experiments. *Oecologia*, **78**, 458–467.

Nicholson WS (1982). *An ecological study of the gray fox in east central Alabama*, M.Sc. dissertation. Auburn University, Auburn, Alabama, USA.

Nicholson WS, Hill EP and Briggs D (1985). Denning, pup-rearing, and dispersal in the gray fox in east-central Alabama. *Journal of Wildlife Management*, **49**, 33–37.

Nielsen OK (1999). Gyrfalcon predation on ptarmigan: numerical and functional responses. *Journal of Animal Ecology*, **68**, 1034–1050.

Niewold FJJ (1980). Aspects of the social structure of red fox populations: a summary. In E Zimen (ed), *The red fox: symposium on behaviour and ecology*, pp. 185–195. W Junk, The Hague, Netherlands.

Ninth Circuit Court of Appeals (2001). Defenders of Wildlife *et al.* v. Secretary of the Department of Interior. Nos. 99-56362 and 00-55496.

Nojima K (1988). *Environmental preference of the raccoon dogs in the urban Tokyo and Kanagawa Prefectures*. B. Sc. dissertation, Tokyo University of Agriculture and Technology, Japan. (In Japanese)

Noll-Banholzer U (1979). Water balance and kidney structure in the fennec. *Comparative Biochemistry and Physiology*, **62A**, 593–597.

Norton-Griffiths M, Herlocker D and Pennyquick L (1975). The patterns of rainfall in the Serengeti Ecosystem, Tanzania. *East African Wildlife Journal*, **13**, 347–374.

Novaro AJ (1991). *Feeding ecology and abundance of a harvested population of culpeo fox (Dusicyon culpaeus) in Patagonia*. M.S. dissertation, University of Florida, Gainesville, FL, USA.

Novaro AJ (1995). Sustainability of harvest of culpeo foxes in Patagonia. *Oryx*, **29**, 18–22.

Novaro AJ (1997a). *Pseudalopex culpaeus* Burmeister, 1856. *Mammalian Species*, **558**, 1–8.

Novaro AJ (1997b). *Source-sink dynamics induced by hunting: case study of culpeo foxes on rangelands in Patagonia, Argentina*. Ph.D. dissertation, University of Florida, Gainesville, FL, USA.

Novaro AJ, Funes MC and Walker RS (2000a). Ecological extinction of native prey of a carnivore assemblage in Argentine Patagonia. *Biological Conservation*, **92**, 25–33.

Novaro AJ, Funes MC, Rambeaud C and Monsalvo O (2000b). Calibración del índice de estaciones odoríferas para estimar tendencias poblacionales del zorro colorado (*Pseudalopex culpaeus*) en Patagonia. *Mastozoología Neotropical*, **7**, 81–88.

Novikov GA (1962). *Carnivorous mammals in the fauna of the U.S.S.R*, pp. 79–85. The Israel Program for Scientific Translations, Jerusalem, Israel.

Nowak RM (1972). The mysterious wolf of the south. *Natural History*, **81**, 51–53, 74–77.

Nowak RM (1978). Reclassification of the grey wolf in the United States and Mexico, with determination of critical habitat in Michigan and Minnesota. *Federal Register*, **43**, 9607–9615.

Nowak RM (1979). North American Quaternary *Canis*. *University of Kansas Museum of Natural History Monograph*, **6**, 1–154.

Nowak RM (1983). A perspective on the taxonomy of wolves in North America. In LN Carbyn (ed), *Wolves in Canada and Alaska: their status, biology, and management*, pp. 10–19. Canadian Wildlife Service Report Series No. 45.

Nowak RM (1992). The red wolf is not a hybrid. *Conservation Biology*, **6**, 593–595.

Nowak RM (1995). Hybridization: the double-edged threat. *Canid News*, **3**, 2–6.

Nowak RM (1995). Another look at wolf taxonomy. In LN Carbyn, SH Fritts and DR Seip (eds), *Ecology and conservation of wolves in a changing world: proceedings of the second North American symposium on wolves*, pp. 375–397. Canadian Circumpolar Institute, University of Alberta, Edmonton, Canada.

Nowak RM (1999). *Walker's mammals of the world, 6th Edition*. John Hopkins University Press, Baltimore, MD, USA.

Nowak RM (2002). The original status of wolves in eastern North America. *Southeastern Naturalist*, **1**, 95–130.

Nowak RM and Paradiso JL (1983). *Walker's mammals of the world, 4th Edition*. Vol. II. John Hopkins University Press, Baltimore, MD, USA.

Nudds TD (1978). Convergence of group size strategies by mammalian social carnivores. *American Naturalist*, **112**, 957–960.

Nunley GL (1995). Sheep and goat losses in relation to coyote damage management in Texas. In D Rollins, C Richardson, T Blankenship, K Canon and SE Henke (eds), *Coyotes in the Southwest: a compendium of our knowledge*, pp. 114–123. Texas Parks & Wildlife Department, Austin, Texas, USA.

Nyhus P, Fischer H, Madden F and Osofsky S (2003). Taking the bite out of wildlife damage. The challenges of wildlife compensation schemes. *Conservation in Practice*, **4**, 37–40.

Observations Department (1992). *Temperature, relative humidity, duration of sunshine, precipitation, frost and snow (normal values)(Av. 1961–90)*. Meteorological Agency. <http://www.stat.go.jp/>

O'Donoghue M, Boutin S, Krebs CJ and Hofer EJ. (1997). Numerical responses of coyotes and lynx to the snowshoe hare cycle. *Oikos*, **80**, 150–162.

O'Farrell TP (1984). Conservation of the endangered San Joaquin kit fox *Vulpes macrotis mutica*, on the Naval Petroleum Reserves, California. *Acta Zoologic Fennici*, **172**, 207–208.

O'Farrell TP (1987). Kit fox. In M Novak, JA Baker, ME Obbard, and B Malloch (eds), *Wild furbearer management and conservation in North America*, pp. 423–431. Ontario Trappers Association, North Bay, Ontario, Canada.

Oftedal OT (1984). Milk composition, milk yield and energy output at peak lactation: a comparative review. *Symposium of the Zoological Society, London*, **51**, 33–85.

Oftedal OT and Gittleman JL (1989). Energy output during reproduction in carnivores. In JL Gittleman (ed), *Carnivore behavior, ecology and evolution*, pp. 355–378. Chapman and Hall, London, UK.

Ogada MO, Woodroffe R, Oguge NO and Frank LG (2003). Limiting depredation by African carnivores: the role of livestock husbandry. *Conservation Biology*, **17**, 1521–1530.

O'Gara BW, Brawley KC, Munoz JR and Henne DR. (1983). Predation on domestic sheep on a western Montana ranch. *Wildlife Society Bulletin*, **11**, 253–264.

Ognev SI (1931). *Mammals of Eastern Europe and Northern Asia*, Vol. 2. Israel Program for Scientific Translations, Jerusalem, Israel.

Ojeda RA and Mares MA (1982). Conservation of South American mammals: Argentina as a paradigm. In MA Mares and H Genoways (eds), *Mammalian biology in South America. Pymatuning symposium of ecology*, pp. 505–521. University of Pittsburgh, Pittsburgh, Pennsylvania, USA.

Okoniewski JC and Chambers RE (1984). Coyote vocal response to an electric siren and human howling. *Journal Wildlife Management*, **48**, 217–222.

Okuzaki M (1979). Reproduction of raccoon dogs, *Nyctereutes procyonoides viverrinus* Temminck, in captivity. *Journal of Kagawa Nutrition College*, **10**, 99–103. (In Japanese.)

Olachea FV, Bellati JP, Suárez MC, Pueyo JM and Robles CA. (1981). Mortalidad perinatal de corderos en el oeste de la Provincia de Río Negro. *Revista de Medicina Veterinaria*, **62**, 128–134.

Olfermann E (1996). *Population ecology of the Rüppell's fox and the red fox in a semi-desert environment of Saudi Arabia*. Ph.D. dissertation, University of Bielefeld, Germany.

Olivier M and G Lust (1998). Two DNA sequences specific for the canine Y chromosome. *Animal Genetics*, **29**, 146–149.

Olivier M, Breen M, Binns MM and Lust G (1999). Localization and characterization of nucleotide sequences from the canine Y chromosome. *Chromosome Research*, **7**, 223–233.

Olrog CC and Lucero MM (1981). *Guide to the mammals of Argentina*. Fundacion Miguel Lillo, Tucuman, Argentina (in Spanish).

Olson TL (2000). *Population characteristics, habitat selection patterns, and diet of swift foxes in southeast Wyoming*. M.Sc. dissertation, University of Wyoming, Laramie, USA.

Olson TL, Dieni JS and Lindzey FG (1997). *Swift fox survey evaluation, productivity, and survivorship in southeast Wyoming*, 30pp. Wyoming Cooperative Fish and Wildlife Research Unit, Wyoming, USA.

Olson TL, Dieni JS, Lindzey FG and Anderson SH. (2003). Swift fox detection probability using tracking plate transects in Southeast. In M Sovada and L Carbyn (eds), *Swift fox conservation in a changing world*, pp. 93–98. Canadian Plains Research Center, University of Regina, Saskatchewan, Canada.

O'Neal GT (1985). *Behavioral ecology of the Nevada kit fox (Vulpes macrotis nevadensis)*. M.S. dissertation, Brigham Young University, Provo, Utah, USA.

O'Neal GT, Flinders JT and Clary WP (1987). Behavioral ecology of the Nevada kit fox (*Vulpes macrotis nevadensis*) on a managed desert rangeland. In HH Genoways (ed), *Current mammalogy*, pp. 443–481. Plenum Press, New York, NY, USA.

Orr PC (1968). *Prehistory of Santa Rosa Island*, 253pp. Santa Barbara Museum of Natural History, Santa Barbara, CA, USA.

Osgood WH (1943). The mammals of Chile. *Field Museum of Natural History, Zoological Series*, **30**, 1–268.

Osumi K and Fukamachi K (2001). *A note for considering satoyama*. (In Japanese.) <http://homepage.mac.com/hitou/satoyama/docs/osumi(2001)html>

Otte D (1989). Speciation in Hawaiian crickets. In D Otte and JA Endler (eds), *Speciation and its consequences*, pp. 482–526. Sinauer, Sunderland, Massachusetts, USA.

Ovsyanikov N and Poyarkov A (in press). Corsac fox (*Vulpes corsac*). In C Sillero-Zubiri, M Hoffmann and DW Macdonald eds *Canids: foxes, wolves, jackals and dogs. Status survey and conservation action plan*. IUCN/SSC Canid Specialist Group, Gland, Switzerland, and Cambridge, UK.

Ovsyanikov NG (1983). *Behaviour and social organization of the arctic fox*. Isd-vo TSNIL Glavochoti RF, Moscow, USSR. (In Russian.)

Owens MJ and Owens DD (1978). Feeding ecology and its influence on social organization in brown hyenas (*Hyaena brunnea*, Thunberg) of the central Kalahari Desert. *East African Wildlife Journal*, **16**, 113–135.

Packard JM, Mech LD and Seal US (1983). Social influences on reproduction in wolves. In LN Carbyn (ed), *Wolves in Canada and Alaska: their status, biology and management*, pp. 78–85. Canadian Wildlife Service, Edmonton, Canada.

Packard JM, Seal US, Mech LD and Plotka ED (1985). Causes of reproductive failure in two family groups of wolves, *Canis lupus*. *Zietschrift fuer Tierpsychologie*, **68**, 24–40.

Packard RL and Bowers JH (1970). Distributional notes on some foxes from western Texas and eastern New Mexico. *Southwestern Naturalist*, **14**, 450–451.

Packer C and Caro TM (1997). Foraging costs in social carnivores. *Animal Behaviour*, **54**, 1317–1318.

Packer C and Ruttan L (1988). The evolution of cooperative hunting. *American Naturalist*, **136**, 1–19.

Packer C, Scheel D and Pusey AE (1990). Why lions form groups; food is not enough. *American Naturalist*, **136**, 1–19.

Packer C, Lewis S and Pusey A (1992). A comparative analysis of non-offspring nursing. *Animal Behaviour*, **43**, 265–281.

Paine RT (1966). Food web complexity and species diversity. *American Naturalist*, **100**, 65–75.

Palomares F and Caro TM (1999). Interspecific killing among mammalian carnivores. *The American Naturalist*, **153**, 492–508.

Paquet PC (1992). Prey use strategies of sympatric wolves and coyotes in Riding Mountain National Park, Manitoba. *Journal of Mammalogy*, **73**, 337–343.

Paquet PC, Vucetich J, Phillips MK and Vucetich L (2001a). *Mexican wolf recovery: three-year program review and assessment*. Prepared by the IUCNSSC Conservation Breeding Specialist Group, Apple Valley, Minnesota for the U.S. Fish and Wildlife Service, Albuquerque, NM, USA.

Paquet PC, Strittholt JR, Staus NL, Wilson PJ, Grewal S and White BN (2001b). Feasibility of timber wolf reintroduction in Adirondack Park. In DS Maehr, RF Noss and JL Larkin (eds), *Large mammal restoration: ecological and social challenges in the 21st century*, 375pp. Island Press, Washington DC, USA.

Paradiso JL (1972). *Canis rufus*. *Mammalian Species*, **22**, 1–4.

Parker WT (1987). *A plan for reestablishing the red wolf on Aligator River National Wildlife Refuge, North Carolina*. Red Wolf Management Series Technical Report No. 1. USFWS, Atlanta, GE, USA.

Parker WT and Phillips MK (1991). Application of the experimental population designation to the recovery of endangered red wolves. *Wildlife Society Bulletin*, **19**, 73–79.

Parker WT, Jones MP and Poulos PG (1986). Determination of experimental population status for an introduced population of wolves in North Carolina—final rule. *Federal Register*, **51**, 41790–41796.

Parsons DR (1998). Establishment of a nonessential experimental population of the Mexican grey wolf in Arizona and New Mexico. *Federal Register*, **63**, 1752–1772.

Pastoret PP and Brochier B (1999). Epidemiology and control of fox rabies in Europe. *Vaccine*, **17**, 1750–1754.

Pate J, Manfredo MJ, Bight AD and Tischbein G (1996). Coloradan's attitudes toward reintroducing the grey wolf into Colorado. *Wildlife Society Bulletin*, **24**, 421–428.

Patel AH (1983). Are chital stags more vulnerable to dhole predation than does? *Journal of the Bombay Natural History Society*, **89**, 153–155.

Paul WJ (2001). Wolf depredation on livestock in Minnesota: annual update of statistics—2000, 13pp. U.S. Department of Agriculture, Grand Rapids, MN, USA.

Paulraj S, Sundarajan N, Manimozhi A and Walker S (1992). Reproduction of the Indian wild dog (*Cuon alpinus*) in captivity. *Zoo Biology*, **11**, 235–241.

Pauw A (2000). Parental care in a polygynous group of bat-eared foxes, *Otocyon megalotis* (*Carnivora: Canidae*). *African Zoology*, **35**, 139–145.

Paxinos E, McIntosh C, Ralls K and Fleischer R (1997). A non-invasive method for distinguishing among canid species: amplification and enzyme restriction of DNA from dung. *Molecular Ecology*, **6**, 483–486.

Pearson O, Martin S and Bellati J (1984). Demography and reproduction of the silky desert mouse (*Eligmodontia*) in Argentina. *Fieldiana Zoology*, n.s., **39**, 433–446.

Pechacek P, Lindzey FG and Anderson SH. (2000). Home range size and spatial organization of swift fox *Vulpes velox* (Say, 1823) in southeastern Wyoming. *Zeitschrift für Saügetierkunde*, **65**, 209–215.

Peres CA (1991). Observations on hunting by small-eared (*Atelocynus microtis*) and bush dogs (*Speothos venaticus*) in central-western Amazonia. *Mammalia*, **55**, 635–639.

Perry BD (1993). Dog ecology in eastern and southern Africa: implications for rabies control. *Onderstepoort Journal of Veterinary Research*, **60**, 429–436.

Peters G and Rodel R (1994). Blanford's fox in Africa. *Bonner Zoologische Beitraege*, **45**, 99–111.

Peters RP and Mech LD. (1975). Scent-marking in wolves. *American Scientist*, **63**, 628–637.

Peterson RO (1977). *Wolf ecology and prey relationships on Isle Royale*. National Park Service Scientific Monograph Series Number 11. U.S. Government Printing Office, Washington DC, USA.

Peterson RO (1979). Social rejection following mating of a subordinate wolf. *Journal of Mammalogy*, **60**, 219–221.

Peterson RO (1995a). Wolves as interspecific competitors in canid ecology. In LN Carbyn, SH Fritts, and DR Seip (eds), *Ecology and conservation of wolves in a changing world*, pp. 315–324. Canadian Circumpolar Institute, Edmonton, Alberta, Canada.

Peterson RO (1995b). *The wolves of Isle Royale: a broken balance*. Willow Creek Press, Minocqua, Wisconsin, USA.

Peterson RO (1999). Wolf-moose interaction on Isle Royale: The end of natural regulation? *Ecological Applications*, **9**, 10–16.

Peterson RO and Allen DL (1974). Snow conditions as a parameter in moose-wolf relationships. *Naturaliste Canadienne*, **101**, 481–492.

Peterson RO and Page RE (1988). The rise and fall of Isle Royale wolves, 1975–1986. *Journal of Mammalogy*, **69**, 89–99.

Peterson RO and Page RE (1993). Detection of moose in midwinter from fixed wing aircraft over dense forest cover. *Wildlife Society Bulletin*, **21**, 80–86.

Peterson RO and Vucetich JA (2001). *Ecological studies of wolves on Isle Royale, annual report 2000–2001*. Michigan Technological University, Houghton, Michigan, USA.

Peterson RO, Page RE and Dodge KM (1984a). Wolves moose and the allometry of population cycles. *Science*, **224**, 1350–1352.

Peterson RO, Woolington JD and Bailey TN (1984b). Wolves of the Kenai Peninsula, Alaska. *Wildlife Monographs*, **88**, 1–52.

Peterson RO, Thomas NJ, Thurber JM, Vucetich JA and Waite TA (1998). Population limitation and the wolves of Isle Royale. *Journal of Mammalogy*, **79**, 828–841.

Petter F (1957). La reproduction du fennec. *Mammalia*, **21**, 307–309.

Petter G (1964). Origine du genre *Otocyon* (Canidae africain de la sous-famille des otocyoninae). *Mammalia*, **28**, 330–344.

Phillips M and Catling PC (1991). Home range and activity patterns of red foxes in Nadgee Nature Reserve. *Wildlife Research*, **18**, 677–686.

Phillips MK (1990). Measures on the value and success of a reintroduction project: red wolf reintroduction in the Alligator River National Wildlife Refuge. *Endangered Species Update*, **8**, 24–26.

Phillips MK (1995). Conserving the red wolf. *Canid News*, **3**, 13–17.

Phillips MK and Henry VG (1992). Comments on red wolf taxonomy. *Conservation Biology*, **6**, 596–599.

Phillips MK and Smith DW (1996). *The wolves of Yellowstone*. Voyageur Press, Stillwater, Minnesota, USA.

Phillips MK and Smith DW (1998). Grey wolves and private landowners in the Greater Yellowstone Area. *Transaction of the North American Wildlife and Natural Resources Conference*, **63**, 443–450.

Phillips MK, Smith R, Henry VG and Lucash C (1995). Red wolf reintroduction program. In LN Carbyn, SH Fritts and DR Seip (eds), *Ecology and conservation of wolves in a changing world*, pp. 157–168. Occasional Publication No 35, Canadian Circumpolar Institute, Edmonton, Alberta, Canada.

Phillips MK, Fascione N, Miller P and Byers O (2000). *Wolves in the Southern Rockies: a population and habitat viability assessment*, 111pp. IUCN-SSC Conservation Breeding Specialist Group, Apple Valley, MN, USA.

Phillips MK, Henry VG and Kelly BT (2003). Restoration of the red wolf. In LD Mech and L Boitani (eds), *Wolves: behavior, ecology and conservation*, pp. 272–288. University of Chicago Press, Chicago, IL, USA.

Pía MV, López MS and Novaro AJ (2003). Effects of livestock on the feeding ecology of endemic culpeo foxes (*Pseudalopex culpaeus* smithersi) in central Argentina. *Revista Chilena de Historia Natural*, **76**, 313–321.

Pianka ER (1973). The structure of lizard communities. *Annual Review of Ecology and Systematics*, **4**, 53–74.

Pienaar U de V (1969). Predator-prey relationships amongst the larger mammals of Kruger National Park. *Koedoe*, **12**, 108–176.

Pilgrim KL, Boyd DK and Forbes SH (1998). Testing for wolf-coyote hybridization in the Rocky Mountains using mitochondrial DNA. *Journal of Wildlife Management*, **62**, 683–689.

Pimlott DH (1967). Wolf predation and ungulate populations. *American Zoologist*, **7**, 267–278.

Pimlott DH, Shannon JA and Kolenosky GB (1969). *The ecology of the timber wolf in Algonquin Provincial Park, Ontario*. Ontario Department of Lands and Forest Research Report (Wildl.) 87. 92 pp.

Pimm SL, Diamond J, Reed TM, Russell GJ and Verner J (1993). Times to extinction for small populations of large birds. *Proceedings of the National Academy of Sciences USA*, **90**, 10871–10875.

Pineda MH, Reimers TJ, Faulkner LC, Hopwood MC and Seidel GE Jr (1977). Azoospermia in dogs induced by injection of schlerosing agents into the caudae of the epididymides. *American Journal of Veterinary Research*, **38**, 831–838.

Pletscher DH, Ream RR, Boyd DK, Fairchild MW and Kunkel KE (1997). Population dynamics of a recolonizing wolf population. *Journal of Wildlife Management*, **61**, 459–465.

Poche RM, Evans SJ, Sultana P, Haque ME, Sterner R and Siddique MA (1987). Notes on the golden jackal (*Canis aureus*) in Bangladesh. *Mammalia*, **51**, 259–270.

Pocock RI (1936). The Asiatic wild dog or dhole (*Cuon javanicus*). *Proceedings of the Zoological Society of London*, **1936**, 33–65.

Pole AJ (2000). *The behaviour and ecology of African wild dogs, Lycaon pictus, in an environment with reduced competitor density*. Ph.D. dissertation, University of Aberdeen, Aberdeen, UK.

Polis GA (1999). Why are parts of the world green? Multiple factors control productivity and the distribution of biomass. *Oikos*, **86**, 3–15.

Polis GA and Strong DR (1996). Food web complexity and community dynamics. *American Naturalist*, **147**, 813–846.

Polis GA, Hurd SD, Jackson CT and Sanchez-Pinero F (1998). Multifactor population limitation. Variable spatial and temporal control of spiders on Gulf of California islands. *Ecology*, **79**, 490–502.

Polis GA, Sears ALW, Huxel GR, Strong DR and Maron J (2000). When is a trophic cascade a trophic cascade? *Trends in Ecology and Evolution*, **15**, 473–475.

Poole KG (1994). Characteristics of an unharvested lynx population during a snowshoe hare decline. *Journal of Wildlife Management*, **58**, 608–618.

Pople AR, Grigg GC, Cairns SC, Beard LA and Alexander P (2000). Trends in the numbers of red kangaroos and emus on either side of the South Australian dingo fence: evidence for predator regulation? *Wildlife Research*, **27**, 269–276.

Porton I (1983). Bush dog urine-marking: its role in pair formation and maintenance. *Animal Behaviour*, **31**, 1061–1069.

Posada D and Crandall KA (2001). Intraspecific gene genealogies: trees grafting into networks. *Trends in Ecology and Evolution*, **16**, 37–45.

Post ES, Stenseth NC, Peterson RO, Vucetich JA and Ellis AM (2002). Phase dependence and population cycles in a large mammal predator-prey system. *Ecology*, **83**, 2997–3002.

Potts, WK and Wakeland EK (1993). Evolution of MHC genetic diversity—A tale of incest, pestilence and sexual preference. *Trends Genetics*, **9**, 408–412.

Potts WK, Manning CJ and Wakeland EK (1991). Mating patterns in seminatural populations of mice influenced by MHC genotype. *Nature*, **352**, 619–621.

Poulle ML, Artois M and Roeder JJ (1994). Dynamics of spatial relationships among members of a fox group (*Vulpes vulpes*: Mammalia: Carnivora). *Journal of Zoology London*, **233**, 93–106.

Power ME (2000). What enables trophic cascades? Commentary on Polis *et al*. *Trends in Ecology and Evolution*, **15**, 443–444.

Pratt NC, Huck UW and Lisk RD (1989). Do pregnant hamsters react to stress by producing fewer males? *Animal Behaviour*, **37**, 155–157.

Prestrud P (1992a). *Arctic foxes in Svalbard: population ecology and rabies*. Ph.D. dissertation, Norsk Polarinstitutt, Oslo, Norway.

Prestrud P (1992b). Food habits and observations of the hunting behaviour of arctic foxes, *Alopex lagopus*, in Svalbard. *Canadian Field-Naturalist*, **106**, 225–236.

Prestrud P and Nilssen K (1995). Growth, size, and sexual dimorphism in arctic foxes. *Journal of Mammalogy*, **76**, 522–530.

Primm SA (1996). A pragmatic approach to grizzly bear conservation. *Conservation Biology*, **10**, 1026–1035.

Pritchard JK, Seielstad MT, Perez-Lezaun A and Feldman MW (1999). Population growth of human Y chromosomes: a study of Y chromosome microsatellites. *Molecular Biology Evolution*, **16**, 1791–1798.

Promislow DEL and Harvey PH (1990). Living fast and dying young: a comparative analysis of life-history variation among mammals. *Journal of Zoology London*, **229**, 417–437.

Propst DL, Stefferud JA and Turner PR (1992). Conservation and status of Gila trout, *Oncorhynchus gilae*. *Southwestern Naturalist*, **37**, 117–125.

Pruss SD (1994). *An observational natal den study of wild swift fox* (Vulpes velox) *on the Canadian Prairie*. M.Sc. dissertation, Department of Environmental Design, University of Calgary, Calgary, Canada.

Pulliam HR (1988). Sources, sinks and population regulation. *American Naturalist*, **132**, 652–661.

Pybus MJ and Williams ES (2003). Parasites and diseases of wild swift fox—a review, pp. 231–236. In M Sovada and L Carbyn (eds), *Swift fox conservation in a changing world.*

Canadian Plains Research Center, University of Regina, Saskatchewan, Canada.

Pyke GH, Pulliam HR and Charnov EL (1977). Optimal foraging: a selective review of theory and tests. *Quarterly Review of Biology*, **52**, 137–154.

Pyrah D (1984). Social distribution and population estimates of coyotes *Canis latrans* in North-Central Montana USA. *Journal of Wildlife Management*, **48**, 679–690.

Qiu Z-x and Tedford RH (1990). A Pliocene species of *Vulpes* from Yushe, Shanxi. *Vertebrata PalAsiatica*, **28**, 245–258.

Queller DC (1992). A general model for kin selection. *Evolution*, **46**, 376–380.

Queller DC, Strassman JE and Hughes CR (1993). Microsatellites and kinship. *Trends in Ecology and Evolution*, **8**, 285–288.

Quinnell RJ, Courtenay O, Garcez L and Dye C (1997). The epidemiology of canine leishmaniasis: transmission rates estimated from a cohort study in Amazonian Brazil. *Parasitology*, **115**, 143–156.

Quintal PKM (1995). *Public attitudes and beliefs about the red wolf and its recovery in North Carolina*. M.Sc. dissertation, North Carolina State University, Raleigh, NC, USA.

Rabb GB, Woolpy JH and Ginsberg BE (1967). Social relationships in a group of captive wolves. *American Zoologist*, **7**, 305–311.

Rabinovich JE, Capurro A, Folgarait P, Kitzberger P, Kramer G, Novaro A, Puppo M and Travaini A (1987). *Estado del conocimiento sobre doce especies de la fauna silvestre argentina de valor comercial*, pp. 6–7. Report to Federación Argentina de Comercialización e Industrial- ización de la Fauna, Buenos Aires, Argentina.

Ralls K and Eberhardt LL (1997). Assessment of abundance of San Joaquin kit foxes by spotlight surveys. *Journal of Mammalogy*, **78**, 65–73.

Ralls K and White PJ (1995). Predation on San Joaquin kit foxes by larger canids. *Journal of Mammalogy*, **76**, 723–729.

Ralls K, Harvey PH and Lyles AM (1986). Inbreeding in natural populations of birds and mammals. In E Soule (ed), *Conservation biology: the science of scarcity and diversity*, pp. 35–36. Sinauer Associates, Sunderland, MA, USA.

Ralls K, Pilgrim KL, White PJ, Paxinos EE, Schwartz MK and Fleischer RC (2001). Kinship, social relationships, and den sharing in kit foxes. *Journal of Mammalogy*, **82**, 858–866.

Randi E (1993). Effects of fragmentation and isolation on genetic variability of the Italian populations of wolf *Canis lupus* and brown bear *Ursus arctos*. *Acta Theriologica*, **38**, 113–120.

Randi E and Lucchini V (2002). Detecting rare introgression of domestic dog genes into wild wolf (*Canis lupus*) populations by Bayesian admixture analyses of microsatellite variation. *Conservation Genetics*, **3**, 31–45.

Randi E, Lucchini V and Francisci F (1993). Allozyme variability in the Italian wolf (*Canis lupus*) population. *Heredity*, **71**, 516–522.

Randi E, Francisci F and Lucchini V (1995). Mitochondrial DNA restriction-fragment-length monomorphism in the Italian wolf (*Canis lupus*) population. *Journal of Zoological Systematics & Evolutionary Research*, **33**, 97–100.

Randi E, Lucchini V, Christensen MF, Nadia M, Funk SM, Gaudenz D and Loeschcke V (2000). Mitochondrial DNA variability in Italian and East European wolves: detecting the consequences of small population size and hybridization. *Conservation Biology*, **14**, 464–473.

Rangrajan M (1998). *Studies in history*, pp 226–299. Sage Publications, New Delhi, India.

Ransom DJ, Rongstad OJ and Rusch DH (1987). Nesting ecology of Rio Grande turkeys. *Journal of Wildlife Management*, **51**, 435–439.

Rasmussen GSA (1996a). Predation on bat-eared foxes *Otocyon megalotis* by Cape hunting dogs *Lycaon pictus*. *Koedoe*, **39**, 127–129.

Rasmussen GSA (1996b). Highly endangered painted hunting dogs used as an excuse for stock loss. *Farmer (Zimbabwe)*, **66**, 30–31.

Rasmussen GSA (1999). Livestock predation by the painted hunting dog *Lycaon pictus* in a cattle ranching region of Zimbabwe: a case study. *Biological Conservation*, **88**, 133–139.

Ratter JA, Riveiro JF and Bridgewater S (1997). The Brazilian cerrado vegetation and threats to its biodiversity. *Annals of Botany*, **80**, 223–230.

Ratti JT, Weinstein M, Scott JM, Avsharian P, Gillesberg A, Miller CA, Szepanski MM and Bomar LK (1999). *Feasibility study on the reintroduction of grey wolves to the Olympic Peninsula*, 359 pp. Idaho Cooperative Research, University of Idaho, Moscow, USSR.

Rau JR, Martinez DR, Low JR and Tilleria MS (1995). Predation by gray foxes (*Pseudalopex griseus*) on cursorial, scansorial, and arboreal small mammals in a protected wildlife area of southern Chile. *Revista Chilena de Historia Natural*, **68**, 333–340.

Rautenbach I and Nel JA (1978). Co-existence in Transvaal Carnivora. *Bulletin of the Carnegie Museum of Natural History*, **6**, 138–145.

Ream RR and Mattson IU (1982). Wolf status in the northern Rockies. In FH Harrington and PC Paquet (eds), *Wolves of the world: perspectives on behavior, ecology, and management*, pp. 362–381. Noyes Publications, Park Ridge, NJ, USA.

Ream RR, Harris R, Smith J and Boyd D (1985). Movement patterns of a lone wolf, *Canis lupus*, in unoccupied wolf range, southeastern British Columbia. *Canadian Field- Naturalist*, **99**, 234–239.

Ream RR, Fairchild MW, Boyd DK and Blakesley A (1989). First wolf den in western United States in recent history. *Northwestern Naturalist*, **70**, 39–40.

Redford KH and Eisenberg JF (1992). *Mammals of the neotropics, the southern cone*, Vol. 2. The University of Chicago Press, Chicago, IL, USA.

Reed DH and Bryant EH (2000). Experimental tests of minimum viable population size. *Animal Conservation*, **3**, 7–14.

Reed JM (1999). The role of behavior in recent avian extinctions and endangerments. *Conservation Biology*, **13**, 232–241.

Reese EA, Standley WG and Berry WH (1992). *Habitat, soils, and den use of San Joaquin kit fox* (Vulpes velox macrotis) *at Camp Roberts Army National Guard Training Site, California*. United

States Department of Energy Topical Report No. EGG 10617–2156, National Technical Information Service, Springfield, VI, USA.

Reeve HK, Westneat DF, Noon WA, Sherman PW and Aquadro CF (1990). DNA fingerprinting reveals high levels of inbreeding in colonies of the eusocial naked mole rat. *Proceedings of the National Academy of Sciences USA*, **87**, 2496–2499.

Refsnider R (2003). Final rule to reclassifiy and remove the gray wolf from the list of endangered and threatened wildlife in portions of the conterminous United States. *Federal Register*, **68**, 15804–15875.

Reich A (1981a). *The behaviour and ecology of the African wild dog,* Lycaon pictus, *in the Kruger National Park*. Ph.D. dissertation, Yale University, New Haven, CN, USA.

Reich A (1981b). Sequential mobilization of bone marrow fat in the impala (*Aepyceros melampus*) and analyses of condition of wild dogs (*Lycaon pictus*). *Journal of Zoology London*, **194**, 409–419.

Reich DE, Wayne RK and Goldstein DB (1999). Genetic evidence for a recent origin by hybridization of red wolves. *Molecular Ecology*, **8**, 139–144.

Reid FA (1997). *A field guide to the mammals of central America and southeast Mexico*. Oxford University Press, Oxford, UK.

Reid R and Gannon CG (1928). Natural history notes on the journals of Alexander Henry. *North Dakota Historical Quarterly*, **2**, 168–201.

Responsive Management (1996). *Public opinion on and attitudes toward the reintroduction of the eastern timber wolf to Adirondack Park*. Report to Defenders of Wildlife, Washington DC, USA.

Reubel GH and Hinds L (2000). *The development of an immunocontraceptive vaccine for the control of fox populations in Australia* Annual report to Environment Australia, Biodiversity Group, Invasive Species Program: 21.

Reynolds JC (1999). The potential for exploiting conditioned taste aversion (CTA) in wildlife management. In DP Cowan and CJ Feare (eds), *Advances in vertebrate pest management*, pp. 267–282. Filander Verlag, Fürth, Germany.

Reynolds JC (2000). Conditioned taste aversion—the end of the line. *The Game Conservancy Review*, **31**, 57–60.

Reynolds JC and Tapper SC (1994). Are foxes on the increase? *The Game Conservancy Review*, **25**, 94–96.

Reynolds JC and Tapper SC (1995). The ecology of the red fox *Vulpes vulpes* in relation to small game in rural southern England. *Wildlife Biology*, **1**, 105–119.

Reynolds JC and Tapper SC (1996). Control of mammalian predators in game management and conservation. *Mammal Review*, **26**, 127–156.

Rhodes CJ, Atkinson RPD Anderson RM and Macdonald DW (1998). Rabies in Zimbabwe: reservoir dogs and the implications for disease control. *Philosophical Transactions of the Royal Society of London B*, **353**, 999–1010.

Rice CG (1986). Observations on predators and prey at Eravikulam National Park, Kerala. *Journal of the Bombay Natural History Society*, **83**, 283–305.

Richardson PRK (1987). Food consumption and seasonal variation in the diet of the aardwolf *Proteles cristatus* in southern Africa. *Zeitschrift fuer Saeugetierkunde*, **52**, 307–325.

Ricklefs RE and Bermingham E (2001). Nonequilibrium diversity dynamics of the lesser Antillean Avifauna. *Science*, **294**, 1522–1524.

Rigg R (2001). Livestock guarding dogs: their current use world wide. IUCN/SSC Canid Specialist Group Occasional Paper **1**, 1–133, www/canids.org/occassionalpaper/.

Riley GA and McBride RT (1972). *A survey of the red wolf* (Canis rufus). Scientific Wildlife Report No. 162, U.S. Fish and Wildlife Service, Washington DC, USA.

Roberts TJ (1977). *The mammals of Pakistan*. Ernest Benn Limited, London, UK.

Robinson NA and Marks CA (2001). Genetic structure and dispersal of red foxes (*Vulpes vulpes*) in urban Melbourne. *Australian Journal of Zoology*, **49**, 589–601.

Robinson WD (1999). Long-term changes in the Avifauna of Barro Colorado Island Panama a tropical forest isolate. *Conservation Biology*, **13**, 85–97.

Rodden M, Rodrigues F and Bestelmeyer S (in press). Maned wolf (*Chrysocyon brachyurus*). In C Sillero-Zubiri, M Hoffmann and DW Macdonald (eds), *Canids: foxes, wolves, jackals and dogs. Status survey and conservation action plan*. IUCN/SSC Canid Specialist Group, Gland, Switzerland, and Cambridge, UK.

Roelke-Parker ME, Munson L, Packer C, Kock R, Cleaveland S, Carpenter M, O'Brien SJ, Popischil A, Hofman-Lehman R, Lutz H, Mwamengele GLM, Mgasa MN, Machange GA, Summers BA and Appel MJG (1996). A canine distemper virus epidemic in Serengeti lions (*Panthera leo*). *Nature*, **379**, 441–445.

Roell BJ (1999). *Demography and spatial use of swift fox* (Vulpes velox) *in northeastern Colorado*. M.A. dissertation, University of Northern Colorado, Greeley, CO, USA.

Roemer GW (1999). *The ecology and conservation of the island fox* (Urocyon littoralis). Ph.D. dissertation, University of California, Los Angeles, CA, USA.

Roemer GW (2000). *Annual Progress Report 2000—Summary of the Demography of the San Nicolas Island Fox* (Urocyon littoralis dickeyi). A report submitted to the US Department of the Navy, Pacific Missile Test Center, Point Mugu, CA, USA.

Roemer GW and Wayne RK (2003). Conservation in conflict: the tale of two endangered species. *Conservation Biology*, **17**, 1251–1260.

Roemer GW, Garcelon DK, Coonan TJ and Schwemm C (1994). The use of capture-recapture methods for estimating, monitoring, and conserving island fox populations. In WL Halvorsen and GJ Maender (eds), *The fourth California Islands Symposium: update on the status of resources*, pp. 387–400. Santa Barbara Museum of Natural History, Santa Barbara, CA, USA.

Roemer GW, Coonan TJ, Garcelon DK, Starbird CH and McCall JW (2000a). Spatial and temporal variation in the seroprevalence of canine heartworm antigen in the island fox. *Journal of Wildlife Diseases*, **36**, 723–728.

Roemer GW, Miller PS, Laake J, Wilcox C and Coonan TJ (2000b). *Island fox demographic workshop report*. Final Report, Channel Islands National Park, Ventura, CA, USA.

Roemer GW, Coonan TJ, Garcelon DK, Bascompte J and Laughrin L (2001a). Feral pigs facilitate hyperpredation by golden eagles and indirectly cause the decline of the island fox. *Animal Conservation*, **4**, 307–318.

Roemer GW, Smith DA, Garcelon DK and Wayne RK (2001b). The behavioural ecology of the island fox (*Urocyon littoralis*). *Journal of Zoology London*, **255**, 1–14.

Roemer GW, Donlan CJ and Courchamp F (2002). Golden eagles, feral pigs, and insular carnivores: How exotic species turn native predators into prey. *Proceedings of the National Academy of Sciences USA*, **99**, 791–796.

Roemer GW, Coonan TJ, Munson L and Wayne RK. (in press). Island fox (*Urocyon littoralis*). In C Sillero-Zubiri, M Hoffmann and DW Macdonald (eds), *Canids: foxes, wolves, jackals and dogs. Status survey and conservation action plan*. IUCN/SSC Canid Specialist Group, Gland, Switzerland, and Cambridge, UK.

Roff DA (1992). *The evolution of life histories*. Chapman and Hall, New York, USA.

Rohlf DJ (1991). Six biological reasons why the Endangered Species Act doesn't work—and what to do about it. *Conservation Biology*, **5**, 273–282.

Rohwer SA and Kilgore DL (1973). Interbreeding in arid-land foxes, *Vulpes velox* and *V. macrotis*. *Systematic Zoology*, **22**, 157–166.

Rood JP (1986). Ecology and social evolution in the mongooses. In D Rubenstein and RW Wrangham (eds), *Ecological Aspects of Social Evolution*. Princeton University Press, Princeton, NY, USA.

Rook L (1992). "Canis" *monticinensis* sp. nov., a new Canidae (Carnivora, Mammalia) from the late Messinian of Italy. *Bolletino della Società Paleontologica Italiana*, **31**, 151–156.

Rook L (1994). The Plio-Pleistocene Old World *Canis* (Xenocyon) ex gr. *falconeri*. *Bolletino della Società Paleontologica Italiana*, **33**, 71–82.

Rook L and Torre D (1996a). The latest Villafranchian—early Galerian small dogs of the Mediterranean area. *Acta Zoologica Cracoviensia*, **39**, 427–434.

Rook L and Torre D (1996b). The wolf-event in western Europe and the beginning of the Late Villafranchian. Neues Jahrbuch für Geologie und Paläontologie Abhandlungen, 1996, 495–501.

Roscoe DE, Holste WC, Sorhage FE, Campbell C, Niezgoda M, Buchannan R, Diehl D, Rupprecht CE and Niu HS (1998). Efficacy of an oral vaccinia-rabies glycoprotein recombinant vaccine in controlling epidemic raccoon rabies in New Jersey. *Journal of Wildlife Diseases*, **34**, 752–763.

Rosen W (1997). *Red wolf recovery in northeastern North Carolina and the Great Smoky Mountains National Park: public attitudes and economic impacts*. College of Human Ecology, Cornell University, Ithaca, New York, USA.

Rosenberg H (1971). Breeding the bat-eared fox *Otocyon megalotis* at Utica Zoo. *International Zoo Yearbook*, **11**, 101–102.

Rosenzweig M and MacArthur RH (1963). Graphical representation and stability conditions of predator-prey interaction. *American Naturalist*, **97**, 217–223.

Rosevear DR (1974). *The carnivores of West Africa*. British Museum (Natural History), London, UK.

Roth JD (2003). Variability in marine resources affects Arctic fox population dynamics. *Journal of Animal Ecology*, **72**, 668–676.

Rothman RJ and Mech LD (1979). Scent-marking in lone wolves and newly formed pairs. *Animal Behaviour*, **27**, 750–760.

Roughgarden J (1995). *Anolis lizards of the caribbean: ecology, evolution, and plate tectonics*. Oxford University Press, Oxford, UK.

Roulin A and Heeb P (1999). The immunologial function of allosuckling. *Ecology Letters*, **2**, 319–324.

Rousset F and Raymond M (1997). Statistical analyses of population genetic data: new tools, old concepts. *Trends in Ecology and Evolution*, **12**, 313–317.

Rowe-Rowe DT (1982). Home range and movements of black-backed jackals in an African montane region. *South African Journal of Wildlife Research*, **12**, 79–84.

Roy LD and Dorrance MJ (1976). *Methods of investigating predation of domestic livestock*. Alberta Agriculture, Edmonton, Alberta, Canada.

Roy MS, Geffen E, Smith D, Östrander EA and Wayne RK (1994b). Patterns of differentiation and hybridization in North American wolflike canids, revealed by analysis of microsatellite loci. *Molecular Biology Evolution*, **11**, 553–570.

Roy MS, Girman DJ and Wayne RK (1994a). The use of museum specimens to reconstruct the genetic variability and relationships of extinct populations. *Experientia*, **50**, 551–557.

Roy MS, Geffen E, Smith D and Wayne RK (1996). Molecular genetics of pre-1940 red wolves. *Conservation Biology*, **10**, 1413–1424.

Ryman N, Jorde PE and Laikre L (1995). Supportive breeding and variance effective population size. *Conservation Biology*, **9**, 1619–1628.

Sacks BN and Neale JCC (2002). Foraging strategy of a generalist predator toward a special prey: coyote predation on sheep. *Ecological Applications*, **12**, 299–306.

Sacks BN, Blejwas KM and Jaeger MM (1999a). Relative vulnerability of coyotes to removal methods on a northern California ranch. *Journal of Wildlife Management*, **63**, 939–949.

Sacks BN, Jaeger MM, Neale JCC and McCullough DR (1999b). Territoriality and breeding status of coyotes relative to sheep predation. *Journal of Wildlife Management*, **63**, 593–605.

Sadlier RMFS (1982). Energy consumption and subsequent partitioning in lactating black-tailed deer. *Canadian Journal of Zoology*, **60**, 382–386.

Saeki M (2001). *Ecology and conservation of the raccoon dog (Nyctereutes procyonoides) in Japan*. D.Phil. dissertation, University of Oxford, Oxford, UK.

Saeki M and Macdonald DW (submitted). The behaviour of raccoon dogs (*Nyctereutes procyonoides*) in a mosaic habitat: habitat selections and movements.

Saleh MA and Basuony MI (1988). A contribution to the mammalogy of the Sinaï peninsula. *Mammalia*, **62**, 557–575.

Salvatori V, Vaglio LG, Meserve PL, Boitani L and Campanella A (19991). Spatial organization, activity, and social interactions of culpeo foxes (*Pseudalopex culpaeus*) in north-central Chile. *Journal of Mammalogy*, **80**, 980–985.

Sandell M (1989). The mating tactics and spacing patterns of solitary carnivores. In JL Gittleman (ed), *Carnivore behavior, ecology, and evolution*. Vol. 1, pp. 164–182. Cornell University Press, New York, USA.

Sanford E (1999). Regulation of keystone predation by small changes in ocean temperature. *Science*, **283**, 2095–2097.

Sargeant AB and Allen SH (1989). Observed interactions between coyotes and red foxes. *Journal of Mammalogy*, **70**, 631–633.

Sargeant AB, Allen SH and Hastings JO (1987). Spatial relation between sympatric coyotes and red foxes in North Dakota. *Journal of Wildlife Management*, **51**, 285–293.

Sargeant AB, Greenwood RJ, Sovada MA and Shaffer TL (1994). Distribution and abundance of predators that affect duck production—prairie pothole region. *United States Department of the Interior, Fish & Wildlife Service*, **194**, 96, Washington DC, USA.

Sargeant GA, Johnson DJ and Berg WE (1998). Interpreting carnivore scent-station surveys. *Journal of Wildlife Management*, **62**, 1235–1245.

SAS Institute Inc. (1996). *Language guide for personal computers*, Version 6.10 Edition. SAS Institute Inc., Cary, NC, USA.

Sasaki H and Kawabata M (1994). Food habits of the raccoon dog *Nyctereutes procyonoides viverrinus* in a mountainous area of Japan. *The Journal of the Mammalogical Society of Japan*, **19**, 1–8.

Sato A, Tichy H, O'Huigin C, Grant PR, Grant BR and Klein J (2001). On the Origin of Darwin's Finches. *Molecular Biology and Evolution*, **18**, 299–311.

Saunders G, White PCL, Harris S and Rayner JMV (1993). Urban foxes (*Vulpes vulpes*): food acquisition, time and energy budgeting of a generalized predator. In N Dunstone and ML Gorman (eds) *Mammals as predators. Symposia of the Zoological Society of London*, **65**, 215–234.

Saunders G, White PCL and Harris S (1997). Habitat utilisation by urban foxes (*Vulpes vulpes*) and the implications for rabies control. *Mammalia*, **61**, 497–510.

Saunders G, McIlroy J, Berghout M, Kay B, Gifford E, Perry R and van de Ven R (2002). The effects of induced sterility on the territorial behaviour and survival of foxes. *Journal of Applied Ecology*, **39**, 56–66.

Scandura M, Apollonio M and Mattioli L (2001). Recent recovery of the Italian wolf population: a genetic investigation using microsatellites. *Mammalian Biology*, **66**, 321–331.

Schaller GB (1967). *The deer and the tiger*. University of Chicago Press, Chicago, IL, USA.

Schaller G (1972). *The Serengeti lion*. University of Chicago Press, Chicago, IL, USA.

Schaller GB (1998). *Wildlife of the Tibetan Steppe*. University of Chicago Press, Chicago, IL, USA.

Schaller GB and Ginsberg JR (in press). Tibetan fox (*Vulpes ferrilata*). In C Sillero-Zubiri, M Hoffmann and DW Macdonald (eds), *Canids: foxes, wolves, jackals and dogs. Status survey and conservation action plan*. IUCN/SSC Canid Specialist Group, Gland, Switzerland, and Cambridge, UK.

Schauster ER (2001). *Swift fox* (Vulpes velox) *on the Pinon Canyon Maneuver Site, Colorado: population ecology and evaluation of survey methods*. M.Sc. dissertation, Utah State University, Logan, Utah, USA.

Schauster ER, Gese EM and Kitchen AM (2002). Population ecology of swift foxes (*Vulpes velox*) in southeastern Colorado. *Canadian Journal of Zoology*, **80**, 307–319.

Scheepers JL and Venzke KAE (1995). Attempts to reintroduce Africa wild dogs *Lycaon pictus* into Etosha National Park, Namibia. *South African Journal of Wildlife Research*, **25**, 138–140.

Schenkel R (1947). Expression studies of wolves. *Behaviour*, **1**, 81–129.

Schenkel R (1967). Submission: its features and function in the wolf and dog. *American Zoologist*, **7**, 319–329.

Schmidt JL (1978). Early management: intentional and otherwise. In JL Schmidt and DL Gilbert (eds), *Big game of North America: ecology and management*, pp. 257–270. Stackpole Books, Harrisburg, Pennsylvania, USA.

Schmidt PA and Mech LD (1997). Wolf pack size and food acquisition. *American Naturalist*, **150**, 513–517.

Schmitz OJ, Hambaeck PA and Beckerman AP (2000). Trophic cascades in terrestrial systems, a review of the effects of carnivore removals on plants. *American Naturalist*, **155**, 141–153.

Schoener TW (1971). Theory of feeding strategies. *Annual Review of Ecology and Systematics*, **2**, 364–404.

Schoener TW (1974). Resource partitioning in ecological communities. *Science*, **185**, 27–39.

Schoener TW, Spiller DA and Losos JB (2001). Predators increase the risk of catastrophic extinction of prey populations. *Nature*, **412**, 183–186.

Schroeder C (1985). *A preliminary management plan for securing swift fox reintroductions into Canada*. M.Sc. dissertation, Faculty of Environmental Design, The University of Calgary, Calgary, Alberta, Canada.

Schumaker NH (1998). *A user's guide to the PATCH model*. EPA/600/R-98/135. U.S. Environmental Protection Agency, Environmental Research Laboratory, Corvallis, OR, USA.

Schwartz M, Ralls K, Williams D and Fleischer RC (in press). Genetic variation and substructure of San Joaquin kit fox populations. *Molecular Ecology*.

Scott JM, Davis F, Csuti B *et al.* (1993). Gap analysis: a geographic approach to protection of biological diversity. *Wildlife Monography*, **123**, 1–41.

Scott JP and Fuller John L (1974). *Dog behavior: the genetic basis*. University of Chicago Press, Chicago, IL, USA.

Scott-Brown JM, Herrero S and Reynolds J (1987). Swift fox. In M Novak, JA Baker, ME Obbard and B Malloch (eds), *Wild furbearer conservation and management in North America*, pp. 432–441. Ontario Ministry of Natural Resources and the Ontario Trappers Association, Ontario, Canada.

Scrivner JH, O'Farrell TP and Kato TT (1987). *Dispersal of San Joaquin kit foxes, Vulpes macrotis mutica, on Naval Petroleum Reserve #1*, 33pp. Kern County, CA, USA.

Scrivner JH, O'Farrell TP and Hammer KL (1993). Summary and evaluation of the kit fox relocation program, Naval Petroleum Reserve #1, Santa Barbara Operations document. Kern County, California, USA.

Seal US (1975). Molecular approaches to taxonomic problems in the Canidae. In MW Fox (ed), *The wild canids: their systematics, behavioural ecology and evolution*, pp. 27–39. Van Nostrand Reinhold Company, New York, USA.

Seal US and Lacy RC (1998). *Florida panther Felis concolor conyi viability analysis and species survival plan*. Conservation Breeding Specialist Group (SSC/IUCN). Apple Valley, MN, USA.

Seber GAF (1982). *The estimation of animal abundance and related parameters*. Macmillan Publishing Co., New York, USA.

Seddon JM and Ellegren H (2002). MHC class II genes in European wolves: a comparison with dogs. *Immunogenetics*, **54**, 490–500.

Seddon PJ (1999). Persistence without intervention: assessing success in wildlife reintroductions. Trends in *Ecology and Evolution*, **14**, 1.

Seidensticker J and Lumpkin S (1992). Mountain lions don't stalk people? True or false. *Smithsonian*, **22**, 113–122.

Seielstad M, Bekele E, Ibrahim M, Toure A and Traore M (1999). A view of modern human origins from Y chromosome microsatellite variation. *Genome Research*, **9**, 558–567.

Seino H and Uchijima Z (1988). *Mesh maps of net primary productivity of natural vegetation of Japan (BCP-88-I-2-2)*, p.131. NIAES (National Institute of Agro-Environmental Sciences), Japan.

Shaffer ML and Stein B (2000). Safeguarding our precious heritage. In BA Stein, LS Kutner and JS Adams (eds), *Precious heritage: the status of biodiversity in the United States*, pp. 301–321. Oxford University Press, New York, NY, USA.

Sharathchandra HC and Gadgil M (1975). A year of Bandipur. *Journal of the Bombay Natural History Society*, **72**, 625–647.

Sharma DK, Maldonado JE, Jhala YV and Fleischer RC (2004). Ancient wolf lineages in India. *Proceedings of the Royal Society (Suppl.) Biology Letters*, **271**, S1–S4.

Sharps JC and Whitcher MF (1984). *Swift fox reintroduction techniques*. South Dakota Department of Game, Fish, and Parks, Rapid City, South Dakota, USA.

Sheffer RJ, Hedrick PW, Minckley WL and Velasco AL (1997). Fitness in the endangered Gila topminnow. *Conservation Biology*, **11**, 162–171.

Sheiff A and Baker JA (1987). Marketing and international fur markets. In M Novak, JA Baker, ME Obbard and B Malloch (eds), *Wild furbearer conservation and management in North America*, pp. 862–877. Ontario Ministry of Natural Resources and the Ontario Trappers Association, Ontario, Canada.

Sheldon JW (1992). *Wild dogs: the natural history of the nondomestic Canidae*. Academic Press, San Diego, CA, USA.

Sheldon WG (1953). Returns on banded red and gray foxes in New York state. *Journal of Mammalogy*, **30**, 236–246.

Shields WM, Templeton A and Davis S (1987). *Genetic assessment of the current captive breeding program for the Mexican wolf* (Canis lupus baileyi). Final contract report 516.6-73-13. Department of Game and Fish, Santa Fe, NM, USA.

Shikama T (1949). The Kuzuu ossuaries. *Science report of Tohoku University Ser 2 (Geology)*, **23**, 128–135. (In Japanese.)

Shinneman D, McClellan R and Smith R (2000). *The state of the Southern Rockies Ecoregion*, 137pp. Southern Rockies Ecosystem Project, Nederland, CO, USA.

Shivik JA and Martin DJ (2001). Aversive and disruptive stimulus applications for managing predation. *Proceedings of the Ninth Eastern Wildlife Damage Management Conference*, pp. 111–119.

Shivik JA, Jaeger MM and Barrett RH (1996). Coyote movements in relation to spatial distribution of sheep. *Journal of Wildlife Management*, **60**, 422–430.

Shivik JA, Asher V, Bradley L, Kunkle K, Phillips M, Breck S and Bangs E (2002). Electronic aversive conditioning for managing wolf predation. pp. 227–231 In RM Timm and RH Schmidt (eds), *Proceedings of the 20st Vertebrate Pest Conference*. University of California Davis, Davis, CA, USA.

Shivik JA, Treves A and Callahan M (2003). Non-lethal techniques: primary and secondary repellents for managing predation. *Conservation Biology*, **17**, 1531–1537.

Shortridge GC (1934). *The mammals of South West Africa*. Heinemann, London, UK.

Sidorov GN and Botvinkin AD (1987). The corsac fox *Vulpes corsac* in Southern Siberia Russian USSR. *Zoologicheskii Zhurnal*, **66**, 914–927.

Siivonen L (1958). Supikoiran varhaisimmasta historiasta Suomessa. [The early history of the raccoon dog in Finland.] *Suomen Riista*, **12**, 165–166. (In Finnish.)

Silk JB (1983). Local resource competition and facultative adjustment of sex ratios in relation to competitive abilities. *American Naturalist*, **121**, 56–66.

Sillero-Zubiri C (1994). *Behavioural ecology of the Ethiopian wolf, Canis simensis*. D.Phil. dissertation. University of Oxford, Oxford, UK.

Sillero-Zubiri C (in press). Pallid fox (*Vulpes pallida*). In C Sillero-Zubiri, M Hoffmann, and DW Macdonald (eds), *Canids: foxes, wolves, jackals and dogs. Status survey and conservation action plan*. IUCN/SSC Canid Specialist Group, Gland, Switzerland, and Cambridge, UK.

Sillero-Zubiri C and Gottelli D (1991). Ethiopia: domestic dogs may doom endangered jackal. *Wildlife Conservation*, **94**, 15.

Sillero-Zubiri C and Gottelli D (1994). *Canis simensis*. *Mammalian Species*, **485**, 1–6.

Sillero-Zubiri C and Gottelli D (1995a). Diet and feeding behavior of ethiopian wolves (*Canis simensis*). *Journal of Mammalogy*, **76**, 531–541.

Sillero-Zubiri C and Gottelli D (1995b). Spatial organization in the Ethiopian wolf *Canis simensis*—large packs and small stable home ranges. *Journal of Zoology*, **237**, 65–81.

Sillero-Zubiri C and Laurenson MK (2001). Interactions between carnivores and local communities: conflict or co-existence? In JL Gittleman, SM Funk, DW Macdonald, and

RK Wayne (eds), *Carnivore conservation*, pp. 282–312. Cambridge University Press, Cambridge, UK.

Sillero-Zubiri C and Macdonald DW (1997). *The Ethiopian wolf: status survey and conservation action plan*. IUCN/SSC Canid Specialist Group, Gland, Switzerland, and Cambridge, UK.

Sillero-Zubiri C and Macdonald DW (1998). Scent-marking and territorial behaviour of Ethiopian wolves *Canis simensis*. *Journal of Zoology London*, **245**, 351–361.

Sillero-Zubiri C, Tattersall FH and Macdonald DW (1995a). Bale mountains rodent communities and their relevance to the Ethiopian wolf (*Canis simensis*). *African Journal of Ecology*, **33**, 301–320.

Sillero-Zubiri C, Tattersall FH and Macdonald DW (1995b). Habitat selection and daily activity of giant molerats *Tachyoryctes macrocephalus*: significance to the Ethiopian wolf *Canis simensis* in the Afroalpine ecosystem. *Biological Conservation*, **72**, 77–84.

Sillero-Zubiri C, Gottelli D and Macdonald DW (1996a). Male philopatry, extra-pack copulations and inbreeding avoidance in Ethiopian wolves (*Canis simensis*). *Behavioral Ecology and Sociobiology*, **38**, 331–340.

Sillero-Zubiri C, King AA and Macdonald DW (1996b). Rabies and mortality in Ethiopian wolves (*Canis simensis*). *Journal of Wildlife Diseases*, **32**, 80–86.

Sillero-Zubiri C, Johnson PJ and Macdonald DW (1998). A hypothesis for breeding synchrony in Ethiopian wolves (*Canis simensis*). *Journal of Mammalogy*, **79**, 853–858.

Sillero-Zubiri C, Malcolm JR, Williams S, Marino J, Tefera Ashenafi Z, Laurenson MK, Gottelli D, Hood A, Macdonald DW, Wildt D and Ellis S (2000). *Ethiopian wolf conservation strategy workshop*. p. 61. IUCN/SSC Canid Specialist Group and Conservation Breeding Specialist Group, Dinsho, Ethiopia.

Sillero-Zubiri C, Hoffmann M and Macdonald DW eds (in press). *Canids: foxes, wolves, jackals and dogs. Status survey and conservation action plan*. IUCN/SSC Canid Specialist Group, Gland, Switzerland, and Cambridge, UK.

Silveira L (1999). *Ecologia e conservação dos mamíferos carnívoros do Parque Nacional das Emas, Goiás*. M.Sc. dissertation, Universidade Federal de Goiás, Brazil.

Silveira L, Jácomo ATA, Rodrigues FHG and Diniz-Filho JAF (1998). Bush dogs (*Speothos venaticus*) in Emas National Park, central Brazil. *Mammalia*, **62**, 446–449.

Silveira FT, Lainson R, Shaw JJ and Povoa MM (1982). Leishmaniasis in Brazil: XVIII. Further evidence incriminating the fox *Cerdocyon thous* L. as a reservoir of Amazonian visceral leishmaniasis. *Transactions of the Royal Society of Tropical Medicine and Hygiene*, **76**, 6.

Simberloff DS and Wilson EO (1969). Experimental zoogeography of islands: the colonization of empty islands. *Ecology*, **50**, 278–296.

Simonetti JA (1988). The carnivorous predatory guild of central Chile: a human-induced community trait? *Revista Chilena de Historia Natural*, **61**, 23–25.

Simpson GG (1945). The principles of classification and a classification of mammals. *Bulletin of the American Museum of Natural History*, **85**, 1–350.

Sinclair ARE (1975). The resource limitation of trophic levels in tropical grassland ecosystems. *Journal of Animal Ecology*, **44**, 497–520.

Sinclair ARE (1979). The Serengeti environment. In *Serengeti: dynamics of an ecosystem*.

Singer FJ and Mack JA (1999). Predicting the effects of wildfire and predators on ungulates. In TW Clark, A Peyton Curlee, SC Minta, and PM Kareiva (eds), *Carnivores in ecosystems. The Yellowstone experience*, pp. 189–237. Yale University Press, New Haven, CT, USA.

Singh BA (1998). *Tiger haven*. Oxford University Press, New Delhi, India.

Skead DM (1973). The incidence of calling in the black-backed jackal. *Journal of South African Wildlife Management Association*, **3**, 128–129.

Skinner JD and Smithers RHN (1990). *Mammals of the southern African sub-region*, 2nd Edition. University of Pretoria Press, Pretoria, South Africa.

Sklepkovych BO (1989). Cannibalism among Arctic foxes, *Alopex lagopus*, in the Swedish Lapplands. *Fauna Flora, Stockholm*, **84**, 145–150.

Slate J, Kruuk LEB, Marshall TC, Pemberton JM and Clutton-Brock TH (2000). Inbreeding depression influences lifetime breeding success in a wild population of red deer (Cervus elaphus). *Proceedings of the Royal Society of London Series B*, **267**, 1657–1662.

Slattery JP and Brien SJO (1995). Molecular phylogeny of the red panda (*Ailurus fulgens*). *Journal of Heredity*, **86**, 413–422.

Smith D, Meier T, Geffen E, Mech LD, Burch JW, Adams LG and Wayne RK (1997a). Is incest common in gray wolf packs? *Behavioral Ecology*, **8**, 384–391.

Smith DA, Ralls K, Davenport B, Adams B and Maldonado JE (2001). Canine assistants for conservationists. *Science*, **291**, 435.

Smith DW, Brewster WG and Bangs EE (1999). Wolves in the Greater Yellowstone ecosystem: restoration of a top carnivore in a complex management environment. In TW Clark, AP Curlee, SC Minta, and PM Karieva (eds), *Carnivores in ecosystems: the Yellowstone experience*, pp. 103–126. Yale University press, New Haven, CT, USA.

Smith DW, Peterson RO and Houston DB (2003a). Yellowstone after wolves. *Bioscience*, **53**, 330–340.

Smith GC and Harris S (1991). Rabies in urban foxes (*Vulpes vulpes*) in Britain: the use of spatial stochastic simulation model to examine the pattern of spread and evaluate the efficacy of different control regimes. *Philosophical Transactions of the Royal Society B*, **334**, 459–479.

Smith M, Budd KJ and Gross C (2003b). The distribution of Blanford's fox (*Vulpes cana* Blanford, 1877) in the United Arab Emirates. *Journal of Arid Environments*, **54**, 55–60.

Smith MJ (1990). *The role of bounties in pest management with specific reference to state dingo control programs*. Diploma dissertation. Charles Sturt University, Wagga Wagga, NSW, Australia.

Smith TB and Wayne RK (1996). *Molecular genetic approaches in conservation*. Oxford University Press, New York, USA.

Smith TB, Wayne RK, Girman DJ and Bruford MW (1997b). A role for ecotones in generating rainforest biodiversity. *Science*, **276**, 1855–1857.

Smithers RHN (1966). A southern bat-eared fox. *Animal Kingdom*, **69**, 163–167.

Smithers RHN (1971). The mammals of Botswana. *Museum Memoirs, National Museum of Rhodesia*, **4**, 1–340.

Smithers RHN (1983). *The mammals of the southern African subregion*, University of Pretoria, Pretoria, South Africa.

Smithers RHN and Wilson VJ (1979). *Checklist and atlas of the mammals of Zimbabwe-Rhodesia*. Salisbury: Trustees, National Museums and Monuments, Zimbabwe-Rhodesia.

Snaydon RW (1962). Micro-distribution of Trifolium repens L. and its relation to soil factors. *Journal of Ecology*, **50**, 133–143.

Snow C (1973). *Habitat management series for endangered species, Report No. 6, San Joaquin kit fox*. U.S. Bureau of Land Management Technical Note, Denver Service Center, Denver, CO, USA.

Sokal RR and Rohlf FJ (1981). *Biometry*. W. H. Freeman and Co., New York, USA.

Solomon N and French J (1997). *Cooperative breeding in mammals*. Cambridge University Press, Cambridge, UK.

Soulé M (1980). Thresholds for survival: maintaining fitness and evolutionary potential. In ME Soulé and BA Wilcox (eds), *Conservation biology. An Evolutionary-Ecological Approach*, pp. 151–169. Sinauer Associates, Sunderland, MA, USA.

Soulé ME and Terborgh J (1999). *Continental conservation: scientific foundations of regional reserve networks*. Island Press, Washington DC, USA.

Soulé ME, Bolger DT, Alberts AC, Wright J, Sorice M and Hill S (1988). Reconstructed dynamics of rapid extinctions of chaparral-requiring birds in urban habitat islands. *Conservation Biology*, **2**, 75–92.

Southern Rockies Ecosystem Project (2000). *Summary of base data and landscape variables for wolf habitat suitability on the Vermejo Park Ranch and surrounding areas*. Report to the Turner Endangered Species Fund, Bozeman, Montana, USA.

Southwood TRE (1977). Habitat, the templet for ecological strategies? *Journal of Animal Ecology*, **46**, 336–365.

Sovada MA and Scheick BK (1999). 1999 Annual Report, preliminary report to the swift fox conservation team: historic and recent distribution of swift foxes in North America. In CG Schmitt (ed), *Swift fox conservation team 1999 annual report*, pp. 80–118. New Mexico Department of Game and Fish, Albuquerque, NM, USA.

Sovada MA, Sargeant AB and Grier JW (1995). Differential effects of coyotes and red foxes on duck nest success. *Journal of Wildlife Management*, **59**, 1–9.

Sovada MA, Roy CC, Bright JB and Gillis JR (1998). Causes and rates of mortality of swift foxes in western Kansas. *Journal of Wildlife Management*, **62**, 1300–1306.

Sovada MA, Roy CC and Telesco DJ (2001). Seasonal food habits of swift foxes in cropland and rangeland habitats in western Kansas. *American Midland Naturalist*, **145**, 101–111.

Sovada MA, Slivinski CC and Woodward RO (2003). Home range, habitat use, pup dispersal and litter sizes of swift foxes in western Kansas. In M Sovada and L Carbyn (eds), *Ecology and conservation of swift foxes in a changing world*, pp. 149–160. Canadian Plains Research Center, University of Regina, Saskatchewan, Canada.

Spencer AW (1984). Food habits, grazing activities, and reproductive development of long-tailed voles, *Microtus longicaudus* (Merriam), in relation to snow cover in the mountains of Colorado. *Carnegie Museum of Natural History, Special Publication*, **10**, 67–90.

Spencer J and Burroughs R (1992). Antibody response to canine distemper vaccine in African wild dogs. *Journal of Wildlife Diseases*, **28**, 443–444.

Spencer KA and Egoscue HJ (1992). *Fleas of the San Joaquin kit fox* (Vulpes velox macrotis) *at Camp Roberts Army National Guard Training Site, California*. United States Department of Energy Topical Report No. EGG 10617-2161, National Technical Information Service, Springfield, VI, USA.

Spencer KA, Berry WH, Standley WG and O'Farrell TP (1992). *Reproduction of the San Joaquin kit fox* (Vulpes velox macrotis) *at Camp Roberts Army National Guard Training Site, California*. United States Department of Energy Topical Report No. EGG 10617-2154, National Technical Information Service, Springfield, VI, USA.

Spiegel LK (ed) (1996). *Studies of the San Joaquin kit fox in undeveloped and oil-developed areas*. California Energy Commission, Sacramento, CA, USA.

Spiegel LK and Tom J (1996). Reproduction of San Joaquin kit foxes in undeveloped and oil-developed habitats of Kern County, California. In LK Spiegel (ed), *Studies of the San Joaquin kit fox in undeveloped and oil-developed areas*, pp.53–69. California Energy Commission, Sacramento, CA, USA.

Stacey PB and Taper M (1992). Environmental variation and the persistence of small populations. *Ecological Applications*, **2**, 18–29.

Stamps JA and Buechner M (1985). The territorial defense hypothesis and the ecology of insular vertebrates. *The Quarterly Reviews of Biology*, **60**, 155–182.

Standley WG and McCue PM (1997). Prevalence of antibodies against selected diseases in San Joaquin kit foxes at Camp Roberts, California. *California Fish and Game*, **83**, 30–37.

Standley WG, Berry WH, O'Farrell TP and Kato TT (1992). *Mortality of the San Joaquin kit fox* (Vulpes velox macrotis) *at Camp Roberts Army National Guard Training Site, California*. United States Department of Energy Topical Report No. EGG 10617-2157, National Technical Information Service, Springfield, VI, USA.

Stearns SC (1992). *The evolution of life histories*. Oxford University Press, Oxford, UK.

Steck F, Wandeler A, Bichsel P, Capt S and Schneider L (1982). Oral immunization of foxes against rabies: a field study. *Zentralblatt für Veterinärische Medizin*, **29**, 372–396.

Steel RGD and Torrie JH (1980). *Principles and procedures of statistics: a biometrical approach*. McGraw-Hill Book Co., New York, USA.

Steinetz BG, Goldsmith LT and Leist G (1987). Plasma relaxin levels in pregnant and lactating dogs. *Biology of Reproduction*, **37**, 719–725.

Stenlund MH (1955). A field study of the timber wolf (*Canis lupus*) on the Superior National Forest, Minnesota. Minnesota Department of Conservation, *Technical Bulletin* No. 4.

Stenseth NC and Ims RA (1993). *The biology of lemmings*. Academic Press Ltd, London, UK.

Stephens MA (1982). Use of the von Mises distribution to analyze continuous proportions. *Biométrica*, **69**, 197–203.

Stephenson RO, Ballard WB, Smith CA and Richardson K (1995). Wolf biology and management in Alaska 1981–1991. In LN Carbyn, SH Fritts, and DR Seip (eds), *Ecology and conservation of wolves in a changing world*, pp. 43–54. Canadian Circumpolar Institute, Edmonton, Alberta, Canada.

Stewart PD, Anderson C and Macdonald DW (1997). A mechanism for passive range exclusion, evidence from the European badger (*Meles meles*). *Journal of Theoretical Biology*, **184**, 279–289.

Stewart PO and Macdonald DMW (2003). Badgers and badger fleas: Strategies and counter strategies. *Ethology*, **109**, 751–764.

Storm GL, Andrews RD, Phillips RL, Bishop RA, Siniff DB and Tester JR (1976). Morphology, reproduction, dispersal and mortality of mid-western red fox populations. *Wildlife Monographs*, **49**, 1–82.

Strachan R (in press). In DW Macdonald and C Hurst (eds), Conservation in the wider landscape: an approach to biodiversity restoration in the Upper Thames Tributaries Environmentally Sensitive Area. The second WildCRU Review, WildCRU, Oxford, UK.

Strahl SD, Silva JL and Goldstein IR (1992). The bush dog *Speothos venaticus* in Venezuela. *Mammalia*, **56**, 9–13.

Strand O, Linnell JDC, Krogstad S and Landa B (1999). Dietary and reproductive responses of Arctic foxes to changes in small rodent abundance. *Arctic*, **52**, 272–278.

Strand O, Landa A, Linnell JDC, Zimmermann B and Skogland T (2000). Social organization and parental behavior in the Arctic fox. *Journal of Mammalogy*, **81**, 223–233.

Stroganov SU (1969). *Carnivorous mammals of Siberia*, pp. 65–73. The Israel Program for Scientific Translations, Jerusalem, Israel.

Stromberg MR and Boyce MS (1986). Systematics and conservation of the swift fox, *Vulpes velox*, in North America. *Biological Conservation*, **35**, 97–110.

Strong DR (1992). Are trophic cascades all wet? Differentiation and donor-control in speciose ecosystems. *Ecology*, **73**, 747–754.

Stuart C (1981). Notes on the mammalian carnivores of the Cape Province, South Africa. *Bontebok*, **1**, 1–58.

Stuart C and Stuart T (1995). Canids in the southeastern Arabian Peninsula. *Canid News*, **3**, 30–32.

Sullivan TP (1977). Demography and dispersal in island and mainland populations of the deer mouse. *Peromyscus maniculatus*. *Ecology*, **58**, 964–978.

Sundqvist AK, Ellegren H, Olivier M and Vila C (2001). Y chromosome haplotyping in Scandinavian wolves (*Canis lupus*) based on microsatellite markers. *Molecular Ecology*, **10**, 1959–1966.

Sunquist ME, Sunquist F and Daneke DE (1989). Ecological separation in a Venezuelan llanos carnivore community. In KH Redford and JF Eisenberg (eds), *Advances in neotropical mammalogy*, pp. 197–232. The Sandhill Crane Press, Inc., Gainesville, FL, USA.

Suppo C, Naulin JM, Langlais M and Artois M (2000). A modelling approach to vaccination and contraception programmes for rabies control in fox populations. *Proceedings of the Royal Society, London, Series B*, **267**, 1575–1582.

Suresh HS, Dattaraja HS and Sukumar R (1996). Tree flora of Mudumalai sanctuary, Tamil Nadu, southern India. *Indian Forester*, **112**, 507–519.

Sutherland WJ (2000). *The conservation handbook: research, management and policy*. Blackwell Science, London, UK.

Swanepoel R, Barnard BJH, Meredith CD, Bishop GC, Brückner GK, Foggin CM and Hübschle OJ (1993). Rabies in southern Africa. In G Thomson and AA King (eds), *Rabies in southern and eastern Africa Onderstepoort Journal of Veterinary Research*, **60**, 325–346.

Taberlet P, Griffin S, Goossens B, Questiau S, Manceau V, Escaravage N, Waits LP and Bouvet J (1996a). Reliable genotyping of samples with very low DNA quantities using PCR. *Nucleic Acids Research*, **24**, 3189–3194.

Taberlet P, Gielly L and Bouvet J (1996b). *Etude génétique sur les loups du Mercantour*. Rapport pour la Direction de la Nature et des Paysages Ministére de l'Environment. Unpublished report.

Taberlet P, Waits LP, Luikart G (1999). Noninvasive genetic sampling: look before you leap. *Trends in Ecology and Evolution*, **14**, 323–327.

Taberlet P, Luikart G and Geffen E (2001). New methods for obtaining and analyzing genetic data from free-ranging carnivores. In JL Gittleman, SM Funk, DW Macdonald and RK Wayne (eds), *Carnivore conservation*, pp. 313–334. Cambridge University Press, Cambridge, UK.

Tamarin RH (1977). Dispersal in island and mainland voles. *Ecology*, **58**, 1044–1054.

Tamura K and Nei M (1993). Estimation of the number of nucleotide substitutions in the control region of mitochondrial DNA in humans and chimpanzees. *Molecular Biology and Evolution*, **10**, 512–526.

Tannerfeldt M and Angerbjörn A (1996). Life history strategies in a fluctuating environment: establishment and reproductive success in the Arctic fox. *Ecography*, **19**, 209–220.

Tannerfeldt M and Angerbjörn A (1998). Fluctuating resources and the evolution of litter size in the Arctic fox. *Oikos*, **83**, 545–559.

Tannerfeldt M, Angerbjörn A and Arvidson B (1994). The effect of summer feeding on juvenile Arctic fox survival—a field experiment. *Ecography*, **17**, 88–96.

Tannerfeldt M, Elmhagen B and Angerbjörn A (2002). Exclusion by interference competition? The relationship between red and Arctic foxes. *Oecologia*, **132**, 213–220.

Tannerfeldt M, Moehrenschlager A and Angerbjörn A (2003). Den Ecology of swift, kit and Arctic foxes: a review. In

M Sovada and L Carbyn (eds), *Ecology and conservation of swift foxes in a changing world*, pp.167–181. Canadian Plains Research Center, University of Regina, Saskatchewan, Canada.

Tapper SC (1992). *Game heritage; an ecological review of shooting and gamekeeping records*. The Game Conservancy Trust, Fordingbridge, UK.

Tapper SC, Potts GR and Brockless MH (1996). The effect of an experimental reduction in predation pressure on the breeding success and population density of grey partridges (*Perdix perdix*). *Journal of Applied Ecology*, **33**, 965–978.

Taylor BL, Chivers SJ, Sexton S and Dizon AE (2000). Evaluating dispersal estimates using mtDNA data: comparing analytical and simulation approaches. *Conservation Biology*, **14**, 1287–1297.

Taylor ME (1989). Locomotor adaptation by carnivores. In JL Gittleman (ed), *Carnivore behavior, ecology and evolution*, pp 382–409. Cornell University Press, Ithaca, New York, USA.

Tear TH, Scott JM, Hayward PH and Griffith B (1993). Status and prospects for success of the endangered species act: a look at recovery plan. *Science*, **262**, 976–977.

Tedford RH (1978). History of dogs and cats: a view from the fossil record. *Nutrition and management of dogs and cats*, chap. M23. Ralston Purina Co., St. Louis, Missouri, USA.

Tedford RH and Qiu Z-x (1996). A new canid genus from the Pliocene of Yushe, Shanxi Province. *Vertebrata PalAsiatica*, **34**, 27–40.

Tedford RH, Taylor BE and Wang X (1997). Phylogeny of the Caninae (*Carnivora: Canidea*): the living taxa. *American Museum Novitates*, **3146**, 1–37.

Tedford RH, Wang X and Taylor BE (in prep.). Phylogenetic systematics of the North American fossil Caninae (Carnivora: Canidae). *Bulletin of the American Museum of Natural History*.

Tembrock G (1962). Zur strukturanalyse des kampfverhaltens bei *Vulpes*. *Behaviour*, **19**, 261–282.

Terbough J, Lopez L, Nuñez N *et al.* (2001). Ecological meltdown in predator-free forest fragments. *Science*, **294**, 1923–1926.

Theberge JB and Theberge MT (1998). *Wolf country: eleven years tracking the Algonquin wolves*. McClelland and Stewart, Inc. Toronto, Ontario, Canada.

Theberge JB and Wedeles CHR (1989). Prey selection and habitat partitioning in sympatric coyote and red fox populations southwest Yukon. *Canadian Journal of Zoology*, **67**, 1285–1290.

Thenius E (1954). On the origins of the dholes. *Osterreich Zoologie Zietshcrift*, **5**, 377–388.

Thiel RP (1993). *The timber wolf in Wisconsin: the death and life of a magnificent predator*. University of Wisconsin Press, Madison, Wisconsin, USA.

Thompson BC (1978). Fence-crossing behavior exhibited by coyotes. *Wildlife Society Bulletin*, **6**, 14–17.

Thompson CM, Stackhouse EL, Roemer GW and Garcelon DK (1998). *Home range and density of the island fox in China Canyon, San Clemente Island, California*. Report to Department of the Navy, SW Division, Naval Facilities Engineering Command, San Diego, CA, USA.

Thompson DJ (1978b). Towards a realistic predator-prey model, the effect of temperature on the functional response and life history of larvae of the damselfly Ischnura elegans [Odon. Coenagrionidae]. *Journal of Animal*, **47**, 757–767.

Thomson PC (1992a). The behavioural ecology of dingoes in north-western Australia: II. Activity patterns, breeding season and pup rearing. *Wildlife Research*, **19**, 519–530.

Thomson PC (1992b). The behavioural ecology of dingoes in north-western Australia: III. Hunting and feeding behaviour, and diet. *Wildlife Research*, **19**, 531–541.

Thomson PC (1992c). The behavioural ecology of dingoes in north-western Australia: IV. Social and spatial organisation, and movements. *Wildlife Research*, **19**, 543–563.

Thomson PC and Marsack PR (1992). Aerial baiting of dingoes in arid pastoral areas with reference to rabies control. In P. O'Brien and G. Berry (eds), *Wildlife rabies contingency planning in Australia*, pp. 125–134. Bureau of Rural Resources Proceedings Number 11, Australian Government Publishing Service, Canberra, Australia.

Thomson R, Pritchard JK, Shen P, Oefner PJ and Feldman MW (2000). Recent common ancestry of human Y chromosomes: evidence from DNA sequence data. *Proceedings of the National Academy of Sciences USA*, **97**, 7360–7365.

Thurber JM and Peterson RO (1991). Changes in body size associated with range expansion in the coyote (*Canis latrans*). *Journal of Mammalogy*, **72**, 750–755.

Thurber JM and Peterson RO (1993). Effects of population density and pack size on the foraging ecology of gray wolves. *Journal of Mammalogy*, **74**, 879–889.

Till JA and Knowlton FF (1983). Efficacy of denning in alleviating coyote depredations upon domestic sheep. *Journal of Wildlife Management*, **47**, 1018–1025.

Tilson RL and Hamilton WJ, III. (1984). Social dominance and feeding patterns of spotted hyenas crocutacrocuta. *Animal Behaviour*, **32**, 715–724.

Timm SF (2001). *Summary report of the Santa Catalina Island fox captive breeding program*. Institute For Wildlife Studies, Arcata, CA, USA.

Timm SF, Stokely JM, Gehr TB, Peebles RL and Garcelon DK (2000). *Investigation into the decline of island foxes on Santa Catalina Island*, 29 pp. Institute for Wildlife Studies, Arcata, CA, USA.

Tirira D ed. (2001). *Libro rojo de los mamíferos del Ecuador*. SIMBIOE/Ecociencia/Ministerio del Ambiente/IUCN, Quito, Ecuador.

Todd AW and Keith LB (1983). Coyote demography during a snowshoe hare decline in Alberta. *Journal of Wildlife Management*, **47**, 394–404.

Todd AW, Keith LB and Fischer CA (1981). Population ecology of coyotes during a fluctuation of snowshoe hares. *Journal of Wildlife Management*, **45**, 629–640.

Tomback DF, Wachtel MA, Driscoll JW and Bekoff M (1989). Measuring dominance and constructing hierarchies; an example using mule deer. *Ethology*, **82**, 275–286.

Torres RV and Ferrusquía-Villafranca I (1981). *Cerdocyon* sp. nov. A (Mammalia, Carnivora) en Mexico y su significacion evolutiva y zoogeografica en relacion a los canidos sud-americanos. Anais II Congresso Latino-Americano de Paleontologia, pp., Porto Alegre, Brazil.

Trapp GR and Hallberg DL (1975). Ecology of the gray fox (*Urocyon cinereoargenteus*): a review. In MW Fox (ed), *The wild canids*, pp. 164–178. Van Nostrand Reinhold Company, New York, USA.

Trautman CG, Fredrickson LF and Carter AV (1974). Relationship of red foxes and other predators to populations of ring-necked pheasants and other prey in South Dakota. *Transaction of the North American Wildlife Conference*, **39**, 214–255.

Travaini A, Juste J, Novaro AJ and Capurro AF (2000a). Sexual dimorphism and sex identification in the South American culpeo fox, *Pseudalopex culpaeus* (Carnivora: Canidae). *Wildlife Research*, **27**, 669–674.

Travaini A, Zapata SC, Martínez-Peck R and Delibes M (2000b). Percepción y actitud humanas hacia la predación de ganado ovino por el zorro colorado (*Pseudalopex culpaeus*) en Santa Cruz, Patagonia argentina. *Mastozoologia Neotropical*, **7**, 117–129.

Treves A (2002). Wolf justice: managing human-carnivore conflict in the 21st century. *Wolf Print*, **13**, 6–9.

Treves A and Karanth KU (2003). Human-carnivore conflict and perspectives on carnivore management worldwide. *Conservation Biology*, **17**, 1491–1499.

Treves A and Woodroffe R (in press). Evaluation of lethal control and other removal techniques for the management of human-wildlife conflict. In R Woodroffe, S Thirgood, and A Rabinowitz (eds), *People and wildlife: conflict or coexistence?* Cambridge University Press, Cambridge, UK.

Treves A, Jurewicz RR, Naughton-Treves L, Rose RA, Willging RC and Wydeven AP (2002). Wolf depredation on domestic animals: control and compensation in Wisconsin, 1976–2000. *Wildlife Society Bulletin*, **30**, 231–241.

Treves A, Naughton-Treves L, Harper E, Mladenoff D, Rose R, Sickley T and Wydeven A (in press). Predicting human-carnivore conflict: a spatial model based on 25 years of wolf predation on livestock. *Conservation Biology*.

Trewhella WJ and Harris S (1988). A simulation model of the pattern of dispersal in urban fox (*Vulpes vulpes*) populations and its application for rabies control. *Journal of Applied Ecology*, **25**, 435–450.

Trewhella WJ, Harris S and McAllister FE (1988). Dispersal distance, home-range size and population density in the red fox (*Vulpes vulpes*): a quantitative analysis. *Journal of Applied Ecology*, **25**, 423–434.

Trivers RL (1972). Parental investment and sexual selection. In B Campbell (ed), *Sexual selection and the descent of man*, pp. 139–179. Aldine, Chicago, IL, USA.

Trivers RL (1974). Parent-offspring conflict. *American Zoologist*, **14**, 249–264.

Trivers RL and Willard DE (1973). Natural selection of parental ability to vary the sex ratio of offspring. *Science*, **179**, 90–92.

Trut LN (1999). Early canid domestication: the farm-fox experiment. *American Scientist*, **87**, 160–169.

Tsukada H (1997). A division between foraging range and territory related to food distribution in the red fox. *Journal of Ethology*, **15**, 27–37.

Tsukada M (1984). A vegetation map in the Japanese archipelago approximately 20,000 years B.P. *Japanese Journal of Ecology*, **34**, 203–208. (In Japanese.)

Tullar BF Jr and Berchielli LT Jr (1982). Comparison of red foxes and gray foxes in central New York with respect to certain features of behavior, movement, and mortality. *New York Fish and Game Journal*, **29**, 127–133.

Turnbull PCB, Bell RHV, Saigawa K, Munyenyembe FEC, Mulenga CK and Makala LHC (1991). Anthrax in wildlife in the Luangwa valley, Zambia. *The Veterinary Record*, **128**, 399–403.

Turner A and Antón M (1996). *The big cats and their fossil relatives*. Columbia University Press, New York, USA.

Tuyttens FAM and Macdonald DW (1998). Fertility control: an option for non-lethal control of wild carnivores? *Animal Welfare*, **7**, 339–364.

Tuyttens FAM and Macdonald DW (2000). Consequences of social perturbation for wildlife management and conservation Behaviour and conservation. In LM Gosling and WJ Sutherland (eds), *Behaviour and conservation*, pp. 315–329. Cambridge University Press, Cambridge, UK.

Tyndale-Biscoe CH (1994). Virus-vectored immunocontraception of feral mammals. *Reproduction, Fertility and Development*, **6**, 281–287.

Uresk DW and Sharps JC (1981). Denning habitat and diet of the swift fox in western South Dakota. *Great Basin Naturalist*, **46**, 249–253.

U.S. Department of Agriculture APHIS—WS (1998). *Final environmental assessement: predator damage management to protect the federally endangered San Clemente loggerhead shrike on San Clemente Island*. Department of the Navy, SW Division, Naval Facilities Engineering Command, San Diego, CA, USA.

U.S. Fish and Wildlife Service (1974). *United States list of endangered fauna, May 1974*, 22 pp. Washington DC, USA.

U.S. Fish and Wildlife Service (1976a). Determination that two species of butterflies are threatened species and two species of mammals are endangered species. *Federal Register*, **41**, 17736–17740.

U.S. Fish and Wildlife Service (1976b). Endangered status for 159 taxa of animals. *Federal Register*, **41**, 24062–24067.

U.S. Fish and Wildlife Service (1978a). *Recovery plan for the eastern timber wolf*, 79 pp. U. S. Fish and Wildlife Service, Washington DC, USA.

U.S. Fish and Wildlife Service (1978b). Predator damage in the west: a study of coyote management alternatives. U.S. Fish and Wildlife Service, Washington DC, USA.

U.S. Fish and Wildlife Service (1980). *Northern Rocky Mountain wolf recovery plan*, 67 pp. U.S. Fish and Wildlife Service, Denver, CO, USA.

U.S. Fish and Wildlife Service (1982). *Mexican wolf recovery plan* 115 pp. U.S. Fish and Wildlife Service, Albuquerque, NM, USA.

U.S. Fish and Wildlife Service (1984). *Red wolf recovery plan.* U.S. Fish and Wildlife Service, Atlanta, GE, USA.

U.S. Fish and Wildlife Service (1987a). *Restoring America's wildlife, 1937–1987: the first 50 years of the federal aid in wildlife restoration (Pittman-Robertson) Act.* U.S. Government Printing Office, Washington DC, USA.

U.S. Fish and Wildlife Service (1987b). *Northern Rocky Mountain wolf recovery plan,* 119 pp. U.S. Fish and Wildlife Service, Denver, CO, USA.

U.S. Fish and Wildlife Service (1989). *Red wolf recovery plan,* 110 pp. U.S. Fish and Wildlife Service, Atlanta, GE, USA.

U.S. Fish and Wildlife Service (1992a). Experimental release of red wolves into the Great Smoky Mountains National Park, 13 pp. Red Wolf Management Series Technical Report No. 8. U.S. Fish and Wildlife Service, Atlanta, GE, USA.

U.S. Fish and Wildlife Service (1992b). *Recovery plan for the eastern timber wolf,* 73 pp. U.S. Fish and Wildlife Service, Fort Snelling, MN, USA.

U.S. Fish and Wildlife Service (1994). *The reintroduction of grey wolves to Yellowstone National Park and central Idaho: final environmental impact statement.* U.S. Fish and Wildlife Service, Denver, CO, USA.

U.S. Fish and Wildlife Service (1996a). *Reintroduction of the Mexican wolf within its historic range in the southwestern United States: final environmental impact statement.* U.S. Fish and Wildlife Service, Albuquerque, NM, USA.

U.S. Fish and Wildlife Service (1996b). *Report to Congress on the recovery program for threatened and endangered species,* 27 pp. U.S. Government Printing Office, Washington DC, USA.

U.S. Fish and Wildlife Service (2001). Endangered and threatened wildlife and plants: annual notice of findings on recycled petitions. *Federal Register,* **66**, 1295–1300.

U.S. Fish and Wildlife Service (2002). *Summary of wolf population viability as determined by peer review.* U.S. Fish and Wildlife Service, Helena, MT, USA.

U.S. Fish and Wildlife Service, Nez Perce Tribe, National Park Service, and USDA Wildlife Services (2003). *Rocky Mountain wolf recovery 2002 annual report,* 61 pp. T Meier (ed), U.S. Fish and Wildlife Service, Ecological Services, Helena, MT, USA.

Valdespino C (2000). *The reproductive system of the fennec fox (Vulpes zerda).* Ph.D. dissertation, University of Missouri, St. Louis, Missouri, USA.

Valdespino C, Asa CS and Bauman JE (2002). Estrous cycles, copulation and pregnancy in the fennec fox (*Vulpes zerda*). *Journal of Mammalogy,* **83**, 99–109.

Valière N, Fumagalli L, Gielly L, Miquel C, Lequette B, Poulle M-L, Weber J-M, Arlettaz R and Taberlet P (2003). Long-distance wolf recolonization of France and Switzerland inferred from non-invasive genetic sampling over a period of 10 years. *Animal Conservation,* **6**, 83–92.

Vallopi L, Welsh D, Glaser D, Sharpe P, Garcelon D and Carter H (2000). *Final predictive ecological risk assessment for the potential reintroduction of bald eagles to the northern Channel Islands.* U.S. Fish and Wildlife Service, Region 1, Sacramento Fish and Wildlife Office. Prepared for the Southern California Damage Assessment Trustee Council.

Van Ballenberghe V (1972). Ecology, movements and population characteristics of timber wolves in northeastern Minnesota. Ph.D. dissertation, University of Minnesota, St. Paul, MN, USA.

Van Ballenberghe V and Erickson AW (1973). A wolf pack kills another wolf. *American Midland Naturalist,* **90**, 490–493.

van de Bildt MWG, Kuiken T, Visee AM, Lema S, Fitzjohn TR and Osterhaus ADME (2002). Distemper outbreak and its effect on African wild dog conservation. *Emerging Infectious Diseases,* **8**, 211–213.

Van der Merwe NJ (1953). The jackal. *Fauna and Flora, Transvaal,* **4**, 1–83.

van Heerden J and Kuhn F (1985). Reproduction in captive hunting dogs, *Lycaon pictus. South African Journal of Wildlife Research,* **15**, 80–84.

van Heerden J, Mills MGL, van Vuuren MJ, Kelly PJ and Dreyer MJ (1995). An investigation into the health status and diseases of wild dogs (*Lycaon pictus*) in the Kruger National Park. *Journal of the South African Veterinary Association,* **66**, 18–27.

van Heerden J, Bainbridge N, Burroughs REJ and Kriek NPJ (1989). Distemper-like disease and encephalitozoonosis in wild dogs (*Lycaon pictus* Temminck, 1820). *Onderstepoort Journal of Veterinary Research,* **48**, 19–21.

van Heerden J, Bingham J, Van Vuuren M, Burroughs REJ and Stylianides E (2002). Clinical and serological response of wild dogs (*Lycaon pictus*) to vaccination against canine distemper, canine parvovirus infection and rabies. *Journal of the South African Veterinary Association,* **73**, 8–12.

van Lawick H (1974). *Solo, the story of an African wild dog.* Houghton Mifflin, Boston, USA.

Van Lawick H and van Lawick-Goodall J (1970). Innocent killers. Houghton Mifflin, Boston, USA.

van Manen FT, Crawford BA and Clark JD (2000). Predicting red wolf release success in the southeastern United States. *Journal of Wildlife Management,* **64**, 895–902.

Van Valen L (1964). Nature of supernumeracy molars of *Otocyon. Journal of Mammology,* **45**, 284–286.

Van Valkenburgh B (1988). Trophic diversity in past and present guilds of large predatory mammals. *Paleobiology,* **14**, 155–173.

Van Valkenburgh B (1991). Iterative evolution of hypercarnivory in canids (Mammalia: Carnivora): evolutionary interactions among sympatric predators. *Paleobiology,* **17**, 340–362.

Van Valkenburgh B (1994). Extinction and replacement among predatory mammals in the North American Late Eocene—Oligocene: tracking a guild over twelve million years. *Historical Biology,* **8**, 1–22.

Van Valkenburgh B (1999). Major patterns in the history of carnivorous mammals. *Annual Review of Earth and Planetary Science,* **27**, 463–493.

Van Valkenburgh B (2001). The dog-eat-dog world of carnivores: a review of past and present carnivore community dynamics.

In C Stanford and HT Bunn (eds), *Meat-eating and human evolution*, pp. 101–121. Oxford University Press, Oxford, UK.

Van Valkenburgh B and Hertel F (1993). Tough times at La Brea: tooth breakage in large carnivores of the late Pleistocene. *Science*, **261**, 456–459.

Van Valkenburgh B and Hertel F (1998). The decline of North American predators during the Late Pleistocene. In JJ Saunders, BW Styles and GF Baryshnikov (eds), *Quaternary paleozoology in the northern hemisphere*, pp. 357–374. Illinois State Museum Scientific Papers 27, Springfield, IL, USA.

Van Valkenburgh B and Koepfli K-P (1993). Cranial and dental adaptations to predation in canids. In N Dunstone and ML Gorman (eds), *Mammals as predators*, pp. 15–37. Cranial and dental adaptations to predation in canids. *Symposium of the Zoological Society of London*, **65**, 15–37. Oxford University Press, Oxford, UK.

Van Valkenburgh B and Wayne R (1994). Shape divergence associated with size convergence in sympatric East African jackals. *Ecology*, **75**, 1567–1581.

Van Valkenburgh B, Sacco T and Wang X (2003). Pack hunting in Miocene borophagine dogs: evidence from craniodental morphology and body size. In LJ Flynn (ed), *Vertebrate fossils and their context: contributions in honor of Richard H. Tedford*, pp. 147–162, Bulletin of the American Museum of Natural History, 279, New York, USA.

Varman KS and Sukumar R (1993). Ecology of sambar in Mudumalai Sanctuary, southern India. In Ohtaishi N and Shenoy HI (eds), *Deer in China*, pp. 289–298, Elseiver Science Publishers, Amsterdam, Netherlands.

Varma KS and Sukumar R (1995). The line transect method for estimating densities of large mammals in a tropical deciduous forest: an evaluation of models and field methods. *Journal of Biosciences*, **20**, 273–287.

Vehrencamp SL (1983). A model for the evolution of despotic versus egalitarian societies. *Animal Behaviour*, **31**, 667–682.

Vellanoweth RL (1998). Earliest island fox remains on the southern Channel Islands: evidence from San Nicolas Island, California. *Journal of California Great Basin Anthropology*, **20**, 100–108.

Velloso AL, Wasser SK, Monfort SL and Dietz JM (1998). Longitudinal faecal steroid excretion in maned wolves (Chrysocyon brachyurus). *General and Comparative Endocrinology*, **112**, 96–107.

Venkataraman AB (1995). Do dholes (*Cuon alpinus*) live in packs in response to competition with or predation by large cats? *Current Science*, **69**, 934–936.

Venkataraman AB (1998). Male-biased sex ratios and their significance for cooperative breeding in dhole, *Cuon alpinus*, packs. *Ethology*, **104**, 671–684.

Venkataraman AB and Narendra Babu V (2001). Why are dholes abundant in the Nilgiri Foothills, southern India. Proceedings of the Canid Biology and Conservation Conference, 2001. Oxford University, Oxford, UK.

Venkataraman AB, Arumugam R and Sukumar R (1995). The foraging ecology of dhole (*Cuon alpinus*) in Mudumalai Sanctuary, southern India. *Journal of Zoology London*, **237**, 543–561.

Vickery PD, Hunter MLJ and Wells JV (1992). Evidence of incidental nest predation and its effects on nests of threatened grassland birds. *Oikos*, **63**, 1099–1104.

Vilà C. and Wayne RK (1999). Hybridization between wolves and dogs. *Conservation Biology*, **13**, 195–198.

Vilà C, Urios V and Castroviejo J (1995). Observations on the daily activitiy patterns in the Iberian wolf. In LN Carbyn, SH Fritts and DR Seip (eds), *Ecology and conservation of wolves in a changing world*, pp. 335–340. Canadian Circumpolar Institute, Edmonton, Alberta, Canada.

Vilà C, Savolainen P, Maldonado JE *et al.* (1997). The domestic dog has an ancient and genetically diverse origin. *Science*, **276**, 1687–1689.

Vilà C, Leonard JA, Iriarte A, O'Brien SJ, Johnson WE and Wayne RK (in press). Detecting the vanishing populations of the highly endangered Darwin's fox, *Pseudalopex fulvipes*, *Animal Conservation*.

Vilà C, Amorim IR, Leonard JA, Posada D, Castroviejo J, Petrucci-Fonseca F, Crandall KA, Ellegren H and Wayne RK (1999). Mitochondrial DNA phylogeography and population history of the grey wolf *Canis lupus*. *Molecular Ecology*, **8**, 2089–2103.

Vilà C, Walker C, Sundqvist A-K, Flagstad Ø, Andersone Z, Casulli A, Kojola I, Valdmann H, Halversone J and Ellegren H (2003a). Combined use of maternal, paternal and biparental genetic markers for the identification of wolf-dog hybrids. *Heredity*, **90**, 17–24.

Vilà C, Sundqvist A-K, Flagstad Ø, Seddon J, Björnerfeldt S, Kojola I, Casulli A, Sand H, Wabakken P and Ellegren H (2003b). Rescue of a severely bottlenecked wolf (*Canis lupus*) population by a single immigrant. *Proceedings of the Royal Society of London B*, **270**, 91–97.

Viranta S (1996). European Miocene Amphicyonidae—taxonomy, systematics and ecology. *Acta Zoologica Fennica*, **204**, 1–61.

Visee, AM (1996). *African wild dogs, Mkomazi game reserve, Tanzania—veterinary report.* Unpublished report, George Adamson Wildlife Preservation Trust.

Visee A, Fitzjohn T, Fitzjohn L and Lema S (2001). *African wild dog (Lycaon pictus) breeding program—Mkomazi game reserve, Tanzania. Report 1999–2001*. African Wild Dog Foundation, Schiedam.

Viswanathan GM, Afanasyev V, Buldyrev SV, Murphy EJ, Prince PA and Stanley HE (1996). Lévy flight search patterns of the wandering albatross. *Nature*, **381**, 413–415.

Voigt DR and Berg WE (1987). Coyote. In M Novak, JA Baker, ME Obbard and B Malloch (eds), *Wild fur bearer management and conservation in North America*, pp. 345–357. Ministry of Natural Resources, Ontario, Canada.

Voigt DR and Earle BD (1983). Avoidance of coyotes by red fox families. *Journal of Wildlife Management*, **47**, 852–857.

Voigt DR and Macdonald DW (1984). Variation in the spatial and social behaviour of the red fox, *Vulpes vulpes*. *Acta Zoologica Fennica*, **171**, 261–265.

Voigt DR, Tinline RL and Broekhoven LH (1985). A spatial simulation model for rabies control. In PJ Bacon (ed), *Population dynamics of rabies in wildlife*, pp. 311–349. Academic Press, London, UK.

von Schantz T (1981). Female cooperation, male competition, and dispersal in the red fox *Vulpes vulpes*. *Oikos*, **37**, 63–68.

von Schantz T (1984a). 'Non-breeders' in the red fox *Vulpes vulpes*: a case of resource surplus. *Oikos*, **42**, 59–65.

von Schantz T (1984b). Spacing strategies, kin selection, and population regulation in altricial vertebrates. *Oikos*, **42**, 48–58.

Vrana PB, Milinkovitich MC, Powell JR and Wheeler WC (1994). Higher level relationships of the arctoid Carnivora based on sequence data and "total evidence." *Molecular Phylogenetics and Evolution*, **3**, 47–58.

Vucetich JA and Creel S (1999). Ecological interactions social organization and extinction risk in African wild dogs. *Conservation Biology*, **13**, 1172–1182.

Vucetich JA and Peterson RO (2002). *The wolves and moose of Isle Royale*. www.isleroyalewolf.org

Vucetich JA and Peterson RO (2004). The influence of top-down, bottom-up, and abiotic factors on the moose (*Alces alces*) population of Isle Royale. *Proceedings of the Royal Society of London. Series B, Biological Sciences*, **271**, 183–189.

Vucetich JA and Peterson RO (in review). Rates of prey capture and population growth for wolves preying on moose. *Oikos*.

Vucetich JA and Waite TA (1999). Erosion of heterozygosity in fluctuating populations. *Conservation Biology*, **13**, 860–868.

Vucetich JA, Peterson RO and Waite TA (1997). Effects of social structure and prey dynamics on extinction risk in gray wolves. *Conservation Biology*, **11**, 957–965.

Vucetich JA, Waite AT, Qvarnemark L and Ibargueen S (2000). Population variability and extinction risk. *Conservation Biology*, **14**, 1704–1714.

Vucetich JA, Peterson RO and Schaefer CL (2002). The effect of prey and predator densities on wolf predation. *Ecology*, **83**, 3003–3013.

Vucetich JA, Peterson RO and Waite TA (2004). Raven scavenging favours group foraging in wolves. *Animal Behaviour*, 66.

Vucetich LM, Vucetich JA, Waite TA, Joshi CP and Peterson RO (2001). Genetic (RAPD) diversity in *Peromyscus maniculatus* in a naturally fragmentation landscape. *Molecular Ecology*, **10**, 35–40.

Wada MY, Suzuki T and Tsuchiya K (1998). Re-examination of the chromosome homology between two subspecies of Japanese raccoon dogs (*Nyctereutes procyonoides albus* and *N. p. viverrinus*). *Caryologia*, **51**, 13–18.

Wagner FH (1988). *Predator control and the sheep industry*. Regina Books, Claremont, California.

Waits LP, Luikart G and Taberlet P (2001). Estimating the probability of identity among genotypes in natural populations: cautions and guidelines. *Molecular Ecology*, **10**, 249–256.

Wallace AR (1869). *The Malay archipelago*. Harper and Brothers, New York, USA.

Walsh PB and Inglis JM (1989). Seasonal and diel rate of spontaneous vocalization in coyotes in south Texas. *Journal of Mammalogy*, **70**, 169–171.

Wandeler AI, Matter HC, Kappeler A and Budde A (1993). The ecology of dogs and canine rabies: a selective review. *Revue scientifique and technique OIE*, **12**, 51–71.

Wandeler P, Funk SM, Largiader CR, Gloor S and Breitenmoser U (2003). The city-fox phenomenon: genetic consequences of a recent colonization of urban habitat. *Molecular Ecology*, **12**, 647–656.

Wang J and Ryman N (2001). Genetic effects of multiple generations of supportive breeding. *Conservation Biology*, **15**, 1619–1631.

Wang X (1993). Transformation from plantigrady to digitigrady: functional morphology of locomotion in Hesperocyon (Canidae: Carnivora). *American Museum Novitates*, **3069**, 1–23.

Wang X (1994). Phylogenetic systematics of the Hesperocyoninae (Carnivora: Canidae). *Bulletin of the American Museum of Natural History*, **221**, 1–207.

Wang X (2003). New material of *Osbornodon* from the early Hemingfordian of Nebraska and Florida. In LJ Flynn (ed), *Vertebrate fossils and their context: Contributions in Honor of Richard H. Tedford. Bulletin of the American Museum of Natural History*, 279, pp. 163–176. American Museum of Natural History, New York, USA.

Wang X and Rothschild BM (1992). Multiple hereditary osteochondromata of Oligocene Hesperocyon (Carnivora: Canidae). *Journal of Vertebrate Paleontology*, **12**, 387–394.

Wang X and Tedford RH (1994). Basicranial anatomy and phylogeny of primitive canids and closely related miacids (Carnivora: Mammalia). *American Museum Novitates*, **3092**, 1–34.

Wang X and Tedford RH (1996). Canidae. In DR Prothero and RJ Emry (eds), *The terrestrial Eocene–Oligocene transition in North America, Pt. II: common vertebrates of the White river chronofauna*, pp. 433–452. Cambridge University Press, Cambridge, UK.

Wang X, Tedford RH and Taylor BE (1999). Phylogenetic systematics of the Borophaginae (Carnivora: Canidae). *Bulletin of the American Museum of Natural History*, **243**, 1–391.

Ward OG and Wurster-Hill DH (1989). Ecological studies of Japanese raccoon dogs, *Nyctereutes procionoides viverrinus*. *Journal of Mammalogy*, **70**, 330–334.

Ward OG and Wurster-Hill DH (1990). *Nyctereutes procyonoides*. *Mammalian Species*, **358**, 1–5.

Ward OG, Wurster-Hill DH, Ratty FJ and Song Y (1987). Comparative cytogenetics of Chinese and Japanese raccoon dogs, *Nyctereutes procyonoides*. *Cytogenetics and Cell Genetics*, **45**, 177–186.

Warrick GD and Cypher BL (1998). Factors affecting the spatial distribution of a kit fox population. *Journal of Wildlife Management*, **62**, 707–717.

Waser PM (1980). Small nocturnal carnivores: ecological studies in the Serengeti. *African Journal of Ecology*, **18**, 167–185.

Waser PM (1996). Patterns and consequences of dispersal in gregarious carnivores. In JL Gittleman (ed), *Carnivore behaviour, ecology and evolution*, Vol. 2, pp. 267–295. Cornell University Press, Ithaca, New York, USA.

Waser PM, Austad SN and Keane B (1986). When should animals tolerate inbreeding? *American Naturalist*, **128**, 529–537.

Waser PM, Creel SR and Lucas JR (1994). Death and disappearance: estimating mortality risks associated with philopatry and dispersal. *Behavioral Ecology*, **5**, 135–141.

Wasser SK, De-Lemos-Velloso A and Rodden MD (1995). Using fecal steroids to evaluate reproductive function in female maned wolves. *Journal of Wildlife Management*, **59**, 889–894.

Wayne RK (1995). Red wolves: to conserve or not to conserve. *Canid News*, **3**, 7–12.

Wayne RK (1993). Molecular evolution of the dog family. *Trends in Genetics*, **9**, 218–224.

Wayne RK (1996). Conservation genetics in the Canidae. In JC Avise and JL Hamrick (eds), *Conservation genetics: case histories from nature*, pp. 75–118. Chapman and Hall, New York, USA.

Wayne RK and Brown DM (2001). Hybridization and conservation of carnivores. In JL Gittleman, S Funk, DW Macdonald and RK Wayne (ed), *Carnivore conservation*, pp. 145–162. Cambridge University Press, Cambridge, UK.

Wayne RK and Gottelli D (1997). Systematics, population genetics and genetic management of the Ethiopian wolf. In C Sillero-Zubiri and DW Macdonald eds *The Ethiopian wolf: status survey and conservation action plan*, pp. 43–50. IUCN/SSC Canid Specialist Group, Gland, Switzerland and Cambridge, UK.

Wayne RK and Jenks SM (1991). Mitochondrial DNA analysis implying extensive hybridization of the endangered red wolf *Canis rufus*. *Nature*, **351**, 565–568.

Wayne RK and O'Brien SJ (1987). Allozyme divergence within the Canidae. *Systematic Zoology*, **36**, 339–355.

Wayne RK and Vilà C. (2003). Molecular genetic studies of wolves. In LD Mech and L Boitani (eds), *Wolves. Behavior, ecology, and conservation*. pp. 218–238. University of Chicago Press, Chicago, IL, USA.

Wayne RK, Allard MW and Honeycutt RL (1992). Mitochondrial DNA variability of the gray wolf: genetic consequences of population decline and habitat fragmentation. *Conservation Biology*, **6**, 559–569.

Wayne RK, Nash WG and O'Brien SJ (1987a). Chromosomal evolution of the Canidae. I. Species with high diploid numbers. *Cytogenetics and Cell Genetics*, **44**, 123–133.

Wayne RK, Nash WG and O'Brien SJ (1987b). Chromosomal evolution of the Canidae. II. Divergence from the primitive carnivore karyotype. *Cytogenetics and Cell Genetics*, **44**, 134–141.

Wayne RK, Kat PW, Fuller TK, Van Valkenburgh B and O'Brien SJ (1989a). Genetic and morphologic divergence among sympatric canids (Mammalia: Carnivora). *Journal of Heredity*, **80**, 447–454.

Wayne RK, Benveniste RE, Janczewski DN and O'Brien SJ (1989b). Molecular and biochemical evolution of the Carnivora. In JL Gittleman (ed), *Carnivore behavior, ecology, and evolution*, pp. 465–494. Cornell University Press, Ithaca, NY, USA.

Wayne RK, Meyer A, Lehman N, Van Valkenburgh B, Kat PW, Fuller TK, Girman D and O'Brien SJ (1990a). Large sequence divergence among mitochondrial DNA genotypes within populations of East African black-backed jackals. *Proceedings of the National Academy of Sciences USA*, **87**, 1772–1776.

Wayne RK, Van Valkenburgh B, Fuller TK and Kat PW (1990b). Allozyme and morphologic differences among highly divergent mtDNA haplotypes of black-backed jackals. In M Clegg and SJ O'Brien (eds), *Molecular evolution*, pp. 161–169. UCLA Symposia on Molecular and Cellular Biology. Liss, NY, USA.

Wayne RK, Gilbert DA, Eisenhawer A, Lehman N, Hansen K, Girman D, Peterson RO, Mech LD, Gogan PJP, Seal US and Krumenaker RJ (1991a). Conservation genetics of the endangered Isle Royale gray wolf. *Conservation Biology*, **5**, 41–51.

Wayne RK, George SB, Gilbert D, Collins PW, Kovach SD, Girman D and Lehman N (1991b). A morphologic and genetic study of the island fox, *Urocyon littoralis*. *Evolution*, **45**, 1849–1868.

Wayne RK, Geffen E, Girman DJ, Koeppfli KP, Lau LM and Marshall CR (1997). Molecular systematics of the Canidae. *Systematic Biology*, **46**, 622–653.

Wayne RK, Leonard JA and Cooper A (1999). Full of sound and fury: the recent history of ancient DNA. *Annual Review of Ecology and Systematics*, **30**, 457–477.

Weaver JL (1993). Refining the equation for interpreting prey occurrence in gray wolf scats. *Journal of Wildlife Management*, **57**, 534–538.

Webb SD (1985). Late Cenozoic mammal dispersals between the Americas. In FG Stehli and SD Webb (eds), *The great American biotic interchange*, Ch 14, Plenum Press, New York, USA.

Weise TF, Robinson WL, Hook RA and Mech LD (1979). An experimental translocation of the Eastern timber wolf. In E Klinghammer (ed), *The behaviour and ecology of wolves*, pp. 346–419. Garland STPM Press, New York, USA.

Wells MC and Bekoff M (1981). An observational study of scent-marking in coyotes, *Canis latrans*. *Animal Behaviour*, **29**, 332–350.

Wells MC and Bekoff M (1982). Predation by wild coyotes: behavioral and ecological analyses. *Journal of Mammalogy*, **63**, 118–127.

Werdelin L (1989). Constraint and adaptation in the bone-cracking canid Osteoborus (Mammalia: Canidae). *Paleobiology*, **15**, 387–401.

Werdelin L (2001). The earliest Canids in Africa. *Canid Biology and Conservation Conference, Oxford 17–21 September (2001). Programme and Abstracts*, p.110.

Werdelin L and Solounias N (1991). The Hyaenidae: taxonomy, systematics and evolution. *Fossils and Strata*, **30**, 1–104.

Werdelin L and Turner A (1996). Turnover in the guild of larger carnivores in Eurasia across the Miocene–Pliocene boundary. *Acta Zoologica Cracoviensia*, **39**, 585–592.

Whitby JE, Johnstone P and Sillero-Zubiri C (1997). Rabies virus in the decomposed brain of an Ethiopian wolf detected by nested reverse transcription-polymerase chain reaction. *Journal of Wildlife Diseases*, **33**, 912–915.

White GC (2000). Population viability analysis: data requirements and essential analyses. In L Boitani and TK Fuller (eds), *Research techniques in animal ecology*, pp. 288–331. University of Columbia Press, New York, USA.

White GC, Anderson DR, Burnham KP and Otis DL (1982). *Capture-recapture and removal methods for sampling closed populations*. Los Alamos National Laboratory, NM, USA.

White PA (1992). *Social organization and activity patterns of arctic foxes on St. Paul Island, Alaska*. M.Sc. dissertation. University of California, Berkeley, CA, USA.

White PCL and Harris S (1994). Encounters between red foxes (*Vulpes vulpes*): implications for territory maintenance, social cohesion and dispersal. *Journal of Animal Ecology*, **63**, 315–327.

White PCL, Saunders G and Harris S (1996). Spatio-temporal patterns of home range use by foxes (*Vulpes vulpes*) in urban environments. *Journal of Animal Ecology*, **65**, 121–125.

White PJ and Garrott RA (1997). Factors regulating kit fox populations. *Canadian Journal of Zoology*, **75**, 1982–1988.

White PJ and Garrott RA (1999). Population dynamics of kit foxes. *Canadian Journal of Zoology*, **77**, 486–493.

White PJ and Ralls K (1993). Reproduction and spacing patterns of kit foxes relative to changing pray availability. *Journal of Wildlife Management*, **57**, 861–867.

White PJ, Ralls K and Garrott RA (1994). Coyote-kit fox interactions as revealed by telemetry. *Canadian Journal of Zoology*, **72**, 1831–1836.

White PJ, Ralls K and Vanderbilt White CA (1995). Overlap in habitat and food use between coyotes and San Joaquin kit foxes. *The Southwestern Naturalist*, **40**, 342–349.

White PJ, Vanderbilt CA and Ralls K (1996). Functional and numerical responses of kit foxes to a short-term decline in mammalian prey. *Southwestern Naturalist*, **40**, 342–349.

White PJ, Berry WH, Eliason JJ and Hanson MT (2000). Catastrophic decrease in an isolated population of kit foxes. *The Southwestern Naturalist*, **45**, 204–211.

White PS, Tatum OL, Deaven LL and Longmire JL (1999). New, male-specific microsatellite markers from the human Y chromosome. *Genomics*, **57**, 433–437.

WHO (1988). World Health Organization. Report of WHO Consultation on Dog Ecology Studies Related to Rabies Control, 22–25 February 1988. World Health Organization, Geneva, Switzerland. WHO/Rab.Res./88.25.

WHO (1992). WHO expert committee on rabies. Technical Report Series 824. 8th Report. WHO, Geneva, Switzerland.

Wickens GE (1984). Flora. In JL Cloudsley-Thompson (ed), *Sahara desert*, pp. 67–75. Pergamon Press, Oxford, UK.

Wildt DE, Bush M, Goodrowe KL *et al.* (1987). Reproductive and genetic consequences of founding isolated lion populations. *Nature*, **329**, 328–331.

Wilkinson L, Hill M, Welna JP and Birkenbeuel GK (1992). SYSTAT for Windows: statistics, version 5 ed. SYSTAT Inc., Evanston, IL, USA.

Williams CL, Blejwas K, Johnston JJ and Jaeger MM (2003). Temporal genetic variation in a coyote (*Canis latrans*) population experiencing high turnover. *Journal of Mammalogy*, **84**, 177–184.

Williams JLH (1971). Notes on the Nilgiri Tahr. *Journal of the Bombay Natural History Society*, **68**, 824–827.

Wilson AC and Stanley Price MR (1994). Reintroduction as a reason for captive breeding. In PJS Olney, GM Mace and ATC Feistner (eds), *Creative conservation: interactive management of wild and captive animals*, pp. 243–264. Chapman and Hall, London, UK.

Wilson, Clarke AB, Coleman K and Dearstyne T (1994). Shyness and boldness in humans and other animals. *Trends in Ecology and Evolution*, **9**, 442–446.

Wilson DE and Reeder DM eds (1993). *Mammal species of the world: a taxonomic and geographic reference*, 2nd Edition Smithsonian Institution Press, Washington DC, USA.

Wilson DE, Cole FR, Nichols JD, Rudran R and Foster MS (1996). Measuring and monitoring biological diversity: standard methods for mammals. Smithsonian Institution Press, Washington DC, USA.

Wilson PJ, Grewal S, McFadden T, Chambers RC and White BN (2003). Mitochondrial DNA extracted from eastern North American wolves killed in the 1800s is not of gray wolf origin. *Canadian Journal of Zoology*, **81**, 936–940.

Wilson PJ, Grewal S, Lawford ID, Heal JNM, Granacki AG, Pennock D, Theberge JB, Theberge MT, Voigt DR, Waddell W, Chambers RE, Paquet PC, Goulet G, Cluff D and White BN (2000). DNA profiles of the eastern Canadian wolf and the red wolf provide evidence for a common evolutionary history independent of the gray wolf. *Canadian Journal of Zoology*, **78**, 2156–2166.

Wisconsin Department of Natural Resources (1999). *Wisconsin wolf management plan*, 74pp. Wisconsin Department of Natural Resources, Madison, WI, USA.

Wolfe ME and Allen DL (1973). Continued studies of the status, socializations, and relationships of Isle Royale wolves, 1967–1970. *Journal of Mammalogy*, **54**, 611–635.

Wolff JO (1992). Parents suppress reproduction and stimulate dispersal in opposite-sex juvenile white-footed mice. *Nature*, **359**, 409–410.

Wolff JO and Macdonald DW (2004). Promiscuous females protect their offspring. *Trends in Ecology and Evolution*, **19**, 127–134.

Wondolleck JM, Yaffe SL and Crowfoot JE (1994). A conflict management perspective. In TW Clark, RP Reading and AL Clarke (eds), *Endangered species recovery*, pp. 304–326. Island Press, Washington DC, USA.

Woodford MH and Rossiter PB (1994). Disease risks associated with wildlife translocation projects. In PJS Olney, GM Mace and ATC Feistner (eds), *Creative conservation: interactive management of wild and captive animals*, pp. 178–200. Chapman and Hall, London, UK.

Woodroffe R (1997). The conservation implications of immobilizing, radio-collaring and vaccinationg free-ranging wild dogs. In R Woodroffe, JR Ginsberg and DW Macdonald (eds), *The African wild dog: status survey and conservation action plan*, pp. 124–138. IUCN/SSC Canid Specialist Group, Gland, Switzerland and Cambridge, UK.

Woodroffe R (1999). Managing disease threats to wild mammals. *Animal Conservation*, **2**, 185–193.

Woodroffe R (2001a). Assessing the risks of intervention: immobilization, radio-collaring and vaccination of African wild dogs. *Oryx*, **35**, 234–244.

Woodroffe R (2001b). Strategies for carnivore conservation: lessons from temporary extinctions. In JL Gittleman, SM Funk, DW Macdonald and RK Wayne (eds), *Carnivore conservation*, pp. 61–92. Cambridge University Press, Cambridge, UK.

Woodroffe R (in prep). How risky is vaccination? Data from captive African wild dogs.

Woodroffe R and Ginsberg JR (1997). Past and future causes of wild dogs' population decline. In R Woodroffe, JR Ginsberg, and DW Macdonald (eds), *The African wild dog: status survey and conservation action plan*, pp.58–74. IUCN/SSC Canid Specialist Group, Gland, Switzerland and Cambridge, UK.

Woodroffe R and Ginsberg JR (1998). Edge effects and the extinction of populations inside protected areas. *Science*, **280**, 2126–2128.

Woodroffe R and Ginsberg JR (1999a). Conserving the African wild dog *Lycaon pictus*. I. Diagonising and treating causes of decline. *Oryx*, **33**, 132–142.

Woodroffe R and Ginsberg JR (1999b). Conserving the African wild dog *Lycaon pictus*. II. Is there a role for reintroduction? *Oryx*, **33**, 143–151.

Woodroffe R and Ginsberg JR (2000). Ranging behaviour and vulnerability to extinction in carnivores. In LM Gosling and WJ Sutherland (eds), *Behaviour and conservation*, pp. 125–140. Cambridge University Press, Cambridge, UK.

Woodroffe R and Macdonald DW (1993). Badger sociality—models of spatial grouping. In N Dunstone and ML Gorman, eds Mammals as predators. *Proceedings of the Symposia of the Zoological Society of London*, **65**, 145–169.

Woodroffe R, Ginsberg JR and Macdonald DW (1997). The African wild dog: status survey and conservation action plan. IUCN/SSC Canid Specialist Group, Gland, Switzerland and Cambridge, UK.

Woodroffe R, McNutt JW and Mills MGL (in press). African wild dog (*Lycaon pictus*). In C Sillero-Zubiri, M Hoffmann and DW Macdonald (eds), *Canids: foxes, wolves, jackals and dogs. Status survey and conservation action plan*. IUCN/SSC Canid Specialist Group, Gland, Switzerland, and Cambridge, UK.

Woollard T and Harris S (1990). A behavioural comparison of dispersing and non-dispersing foxes (*Vulpes vulpes*) and an evaluation of some dispersal hypotheses. *Journal of Animal Ecology*, **59**, 709–722.

World Commission on Environment and Development (1987). *Our common future*. Oxford University Press, Oxford, UK.

World Health Organisation (1995). *World Survey of rabies for the year 1995*. Available online at: http://www.who.int/emcdocuments/rabies/docs/whoemczoo971.pdf

Wozencraft WC (1989). Classification of the recent Carnivora. In JL Gittleman (ed), *Carnivore behavior, ecology and evolution*. Cornell University Press, Ithaca, NY, USA.

Wozencraft WC (1993). Order Carnivora. In DE Wilson and DM Reeder (eds), *Mammal species of the world: a taxonomic and geographic reference*. Smithsonian Institution Press, Washington DC, USA.

Wright H (2004). Monogamy in the bat-eared fox, *Otocyon megalotis*. Ph.D. dissertation, University of Warwick, UK.

Wydeven AP, Treves A, Brost B and Wiedenhoeft JE (2003). Characteristics of wolf packs depredating on domestic animals in Wisconsin, USA. *Conservation Biology*.

Wydeven AP, Schultz RN and Thiel RP (1995). Monitoring of a recovering gray wolf population in Wisconsin, 1979–1991. In LN Carbyn, SH Fritts and DR Seip (eds), *Ecology and conservation of wolves in a changing world*, pp. 147–156. Canadian Circumpolar Institute, Edmonton, Alberta, Canada.

Wydeven AP, Fuller TK, Weber W and MacDonald K (1998). The potential for wolf recovery in the northeastern United States via dispersal from southeastern Canada. *Wildlife Society Bulletin*, **26**, 776–784.

Wyman J (1967). *The jackals of the Serengeti. Animals*, **1967**, 79–83.

Wynne-Edwards KE and Lisk RD (1989). Differential effects of paternal presence on pup survival in two species of dwarf hamster (*Photopus sungorus* and *Phodopus campbelli*). *Physiology and Behaviour*, **45**, 49–53.

Yachimori S (1997). *Estimation of family relationship and behavioural changes among individuals constituting a family of the wild raccoon dogs*. Ph.D. dissertation, Nihon University, Japan. (In Japanese.)

Yahnke CJ (1994). Systematic relationships of Darwin's fox and the extinct Falkland Island wolf inferred from mtDNA sequence. M.A. dissertation, Northern Illinois University, IL, USA.

Yahnke CJ (1995). Metachromism and the insight of Wilfred Osgood: evidence of common ancestry for Darwin's fox and the Sechura fox. *Revista Chilena de Historia Natural*, **68**, 459–467.

Yahnke CJ, Johnson WE, Geffen E, Smith D, Hertel F, Roy MS, Bonacic CF, Fuller TK, Van VB and Wayne RK (1996). Darwin's fox: a distinct endangered species in a vanishing habitat. *Conservation biology*, **10**, 366–375.

Yalden DW (1983). The extent of high ground in Ethiopia compared to the rest of Africa. *Sinet: Ethiopian Journal of Science*, **6**, 35–38.

Yalden DW (1996). Historical dichotomies in the exploitation of mammals. In VJ Taylor and N Dunstone (eds), *The exploitation of mammal populations*, pp. 16–27. Chapman & Hall, London, UK.

Yalden DW and Largen MJ (1992). The endemic mammals of Ethiopia. *Mammal Review*, **22**, 115–150.

Yamamoto I (1987). Male parental care in the raccoon dog *Nyctereutes procyonoides* during the early rearing period. In Y Itô, JL Brown and J Kikkawa (eds), *Animal societies: theories and facts*, pp.189–195. Japan Societies Scientific Press, Tokyo, Japan.

Yamamoto Y (1991). *Diet and distribution of raccoon dog,* Nyctereutes procyonoides viverrinus, *in Kawasaki.* Kawasaki-city Natural Environment Survey Report II, 185–194. (In Japanese.)

Yamamoto Y (1993). Home range and diel activity pattern of the raccoon dog, *Nyctereutes procyonoides viverrinus* in Kawasaki. *Bulletin of the Kawasaki Municipal Science Museum for Youth*, **4**, 7–11. (In Japanese.)

Yamamoto Y (1994). Comparative analyses on food habits of Japanese marten, red fox, badger and raccoon dog in the Mt. Nyugasa, Nagano Prefecture, Japan. *Natural Environmental Scientific Research*, **7**, 45–52. (In Japanese with English summary.)

Yamamoto Y and Kinosita A (1994). Food composition of the raccoon dog *Nyctereutes procyonoides viverrinus* in Kawasaki. *Bulletin of the Kawasaki Municipal Science Museum for Youth*, **5**, 29–34. (In Japanese.)

Yamamoto Y, Terao K, Horiguchi T, Morita M and Yachimori S (1994). Home range and dispersal of the raccoon dog (*Nyctereutes procyonoides viverrinus*) in the Mt. Nyugasa, Nagano Prefecture, Japan. *Natural Environmental Scientific Research*, **7**, 53–61. (In Japanese.)

Yamamoto Y, Kinosita A and Higashimoto H (1995). Distribution and habitat selection of the raccoon dog *Nyctereutes procyonoides viverrinus* in Kawasaki city. *Bulletin of the Kawasaki Municipal Science Museum for Youth*, **6**, 83–88. (In Japanese.)

Yashiki H (1987). Ecological study on the raccoon dog, *Nyctereutes procyonoides viverrinus*, in Shiga Heights. *Bulletin of Institution of Nature Education, Shiga Heights, Shinshu University*, **24**, 43–53. (In Japanese.)

Yearsley EF and Samuel DE (1980). Use of reclaimed surface mines by foxes in West Virginia. *Journal of Wildlife Management*, **44**, 729–734.

Yodzis P (1994). Predator-prey theory and management of multispecies fisheries. *Ecological Applications*, **4**, 51–58.

Yoshioka M, Kishimoto M, Nigi H *et al.* (1990). Seasonal changes in serum levels of testosterone and progesterone on the Japanese raccoon dog, *Nyctereutes procyonoides viverrinus*. *Proceedings. Japanese Society For Comparative Endocrinology*, **5**, 17.

Young MK and Harig AL (2001). A critique of the recovery of greenback cutthroat trout. *Conservation Biology*, **15**, 1575–1584.

Young SP (1944). The wolves of North America. Part 1. Their history, life habits, economic status, and control. In SP Young and EA Goldman (eds), *The wolves of North America, Part 1*. Dover, New York, NY, American Wildlife Institute, Washington DC, USA.

Young SP (1970). *The last of the loners*. MacMillan, New York, USA.

Young SP and Goldman EA (1944). *The wolves of North America*. American Wildlife Institute, Washington DC, USA.

Young SP and Jackson HHT (1951). *The clever coyote*. Wildlife Management Institute, Washington DC, USA.

Young TP (1994). Natural die-offs of large mammals: implications for conservation. *Conservation Biology*, **8**, 410–418.

Zabel CJ (1986). *Reproductive behavior of the red fox (Vulpes vulpes): a longitudinal study of an island population*. Ph.D. dissertation, University of California, CA, USA.

Zabel CJ and Taggart SJ (1989). Shift in red fox, *Vulpes vulpes*, mating system associated with El Niño in the Bering Sea. *Animal Behaviour*, **38**, 830–838.

Zar JH (1996). *Biostatistical analysis*, 3rd Edition. Prentice Hall, Upper Saddle River, NJ, USA.

Zarnoch SJ, Anthony RG and Storm GS (1977). Computer simulated dynamics of a local red fox population. In RL Phillips and C Jonkel (eds), *Proceedings of the 1975 Predator Symposium*, pp. 253–268 Montana Forest Conservation Experimental Station, University of Montana, Missoula, MT, USA.

Zimen E (1976). On the regulation of pack size in wolves. *Zeitschrift für Tierpsychologie*, **40**, 300–341.

Zimen E (1981). *The wolf: his place in the natural world*. Souvenir Press, London, UK.

Zimen E (1982). A wolf pack sociogram. In FH Harrington and PC Paquet (eds), *Wolves of the world: perspectives of behavior, ecology, and conservation*, pp. 282–323. Noyes Publications, Park Ridge, NJ, USA.

Zimen E (1984). Long range movements of the red fox, *Vulpes vulpes* L. *Acta Zologica Tennica*, **171**, 267–270.

Zimen E and Boitani L (1975). Number and distribution of wolves in Italy. *Zeitschrift für Saugetierkunde*, **40**, 102–112.

Zimmerman AL (1998). *Reestablishment of swift fox* (Vulpes vulpes) *in northcentral Montana*. M.S. dissertation, Montana State University, Bozeman, MT, USA.

Zimmerman AL, Irby L and Giddings B (2003). The status and ecology of swift foxes in north central Montana. In M Sovada and L Carbyn (eds), *Ecology and conservation of swift foxes in a changing world*, pp. 49–59. Canadian Plains Research Center, University of Regina, Saskatchewan, Canada.

Zipf GK (1949). *Human behaviour and the principle of least effort*. Cambridge, MA, USA.

Zoellick BW, O'Farrell TP, McCue PM, Harris CE and Kato TT (1987). *Reproduction of the San Joaquin kit fox on Naval Petroleum Reserve #1, Elk Hills, California, 1980–1985*. U.S. Department of Energy Topical Report No. EGG 10282-2144.

Zuercher GL, Swarner M, Silveira L and Carrillo O (in press). Bush dog (*Spothos venaticus*). In C Sillero-Zubiri, M Hoffmann, and DW Macdonald (eds), *Canids: foxes, wolves, jackals and dogs. Status survey and conservation action plan*. IUCN/SSC Canid Specialist Group, Gland, Switzerland, and Cambridge, UK.

Zumbaugh DM, Choate JR and Fox LB (1985). Winter food habits of the swift fox on the central high plains. *Prairie Naturalist*, **17**, 41–47.

Zunino GE, Vaccaro OB, Canevari M and Gardner AL (1995). Taxonomy of the genus *Lycalopex* (Carnivora: Canidae) in Argentina. *Proceedings of the Biological Society of Washington*, **108**, 729–747.

Index